Facilities Maintenance & Repair Costs with RSMeans data

Brian Adams, Senior Editor

2020

27th annual edition

Chief Data Officer
Noam Reininger

Vice President, Data
Tim Duggan

Principal Engineer
Bob Mewis *(1, 4)*

Contributing Editors
Brian Adams *(21, 22)*
Paul Cowan
Christopher Babbitt
Sam Babbitt
Stephen Bell
Michelle Curran
Antonio D'Aulerio *(26, 27, 28, 48)*
Matthew Doheny *(8, 9, 10)*
John Gomes, CCT *(13, 41)*
Derrick Hale, PE *(2, 31, 32, 33, 34, 35, 44, 46)*
Barry Hutchinson
Joseph Kelble *(14, 23, 25)*
Scott Keller *(3, 5)*
Charles Kibbee
Gerard Lafond, PE
Thomas Lane *(6, 7)*
Thomas Lyons
Jake MacDonald *(11, 12)*
John Melin, P.E.
Elisa Mello
Matthew Sorrentino
Kevin Souza
David Yazbek

Production Manager
Debbie Panarelli

Production
Jonathan Forgit
Sharon Larsen
Sheryl Rose
Janice Thalin

Data Quality Manager
Joseph Ingargiola

Innovation
Ray Diwakar
Kedar Gaikwad
Srini Narla

Cover Design
Blaire Collins

Data Analytics
David Byars
Ellen D'amico
Thomas Hauger
Cameron Jagoe
Matthew Kelliher-Gibson
Renee Rudicil

**Numbers in italics are the divisional responsibilities for each editor. Please contact the designated editor directly with any questions.*

RSMeans data from Gordian
Construction Publishers & Consultants
1099 Hingham Street, Suite 201
Rockland, MA 02370
United States of America
1.800.448.8182
RSMeans.com

Printed in the United States of America
ISSN 1074-0953
ISBN 978-1-950656-07-3

0300 $949.99 per copy (in United States)
Price is subject to change without prior notice.

Related Data and Services

Our engineers recommend the following products and services to complement *Facilities Maintenance & Repair Costs with RSMeans data:*

Annual Cost Data Books

2020 Facilities Construction Costs with RSMeans data
2020 Electrical Costs with RSMeans data
2020 Mechanical Costs with RSMeans data
2020 Plumbing Costs with RSMeans data

Reference Books

Facilities Planning and Relocation
Life Cycle Costing for Facilities
Cost Planning & Estimating for Facilities Maintenance
Estimating Building Costs
RSMeans Estimating Handbook
Building Security: Strategies & Costs

Seminars and In-House Training

Facilities Maintenance & Repair Estimating
Unit Price Estimating
Training for our online estimating solution
Mechanical & Electrical Estimating
Scheduling with MSProject for Construction Professionals

RSMeans data

For access to the latest cost data, an intuitive search, and an easy-to-use estimate builder, take advantage of the time savings available from our online application.

To learn more visit: **RSMeans.com/online**

Enterprise Solutions

Building owners, facility managers, building product manufacturers, and attorneys across the public and private sectors engage with RSMeans data Enterprise to solve unique challenges where trusted construction cost data is critical.

To learn more visit: **RSMeans.com/Enterprise**

Custom Built Data Sets

Building and Space Models: Quickly plan construction costs across multiple locations based on geography, project size, building system component, product options, and other variables for precise budgeting and cost control.

Predictive Analytics: Accurately plan future builds with custom graphical interactive dashboards, negotiate future costs of tenant build-outs, and identify and compare national account pricing.

Consulting

Building Product Manufacturing Analytics: Validate your claims and assist with new product launches.

Third-Party Legal Resources: Used in cases of construction cost or estimate disputes, construction product failure vs. installation failure, eminent domain, class action construction product liability, and more.

API

For resellers or internal application integration, RSMeans data is offered via API. Deliver Unit, Assembly, and Square Foot Model data within your interface. To learn more about how you can provide your customers with the latest in localized construction cost data visit: **RSMeans.com/API**

Table of Contents

Foreword

The Value of RSMeans data from Gordian

Since 1942, RSMeans data has been the industry-standard materials, labor, and equipment cost information database for contractors, facility owners and managers, architects, engineers, and anyone else that requires the latest localized construction cost information. More than 75 years later, the objective remains the same: to provide facility and construction professionals with the most current and comprehensive construction cost database possible.

With the constant influx of new construction methods and materials, in addition to ever-changing labor and material costs, last year's cost data is not reliable for today's designs, estimates, or budgets. Gordian's cost engineers apply real-world construction experience to identify and quantify new building products and methodologies, adjust productivity rates, and adjust costs to local market conditions across the nation. This adds up to more than 22,000 hours in cost research annually. This unparalleled construction cost expertise is why so many facility and construction professionals rely on RSMeans data year over year.

About Gordian

Gordian originated in the spirit of innovation and a strong commitment to helping clients reach and exceed their construction goals. In 1982, Gordian's chairman and founder, Harry H. Mellon, created Job Order Contracting while serving as chief engineer at the Supreme Headquarters Allied Powers Europe. Job Order Contracting is a unique indefinite delivery/indefinite quantity (IDIQ) process, which enables facility owners to complete a substantial number of repair, maintenance, and construction projects with a single, competitively awarded contract. Realizing facility and infrastructure owners across various industries could greatly benefit from the time and cost saving advantages of this innovative construction procurement solution, he established Gordian in 1990.

Continuing the commitment to providing the most relevant and accurate facility and construction data, software, and expertise in the industry, Gordian enhanced the fortitude of its data with the acquisition of RSMeans in 2014. And in an effort to expand its facility management capabilities, Gordian acquired Sightlines, the leading provider of facilities benchmarking data and analysis, in 2015.

Our Offerings

Gordian is the leader in facility and construction cost data, software, and expertise for all phases of the building life cycle. From planning to design, procurement, construction, and operations, Gordian's solutions help clients maximize efficiency, optimize cost savings, and increase building quality with its highly specialized data engineers, software, and unique proprietary data sets.

Our Commitment

At Gordian, we do more than talk about the quality of our data and the usefulness of its application. We stand behind all of our RSMeans data—from historical cost indexes to construction materials and techniques—to craft current costs and predict future trends. If you have any questions about our products or services, please call us toll-free at 800.448.8182 or visit our website at gordian.com.

MasterFormat® 2016/ MasterFormat® 2018 Comparison Table

This table compares the 2016 edition of the Construction Specifications Institute's MasterFormat® to the expanded 2018 edition. For your convenience, all revised 2016 numbers and titles are listed along with the corresponding 2018 numbers and titles.

CSI 2016 MF ID	CSI 2016 MF Description	CSI 2018 MF ID	CSI 2018 MF Description
015632	Temporary Security	015733	Temporary Security
019308	Facility Maintenance Equipment	019308	Facilities Maintenance, Equipment
024200	Removal and Salvage of Construction Materials	024200	Removal and Diversion of Construction Materials
040130	Unit Masonry Cleaning	04012052	Cleaning Masonry
068010	Composite Decking	067300	Composite Decking
072127	Reflective Insulation	072153	Reflective Insulation
072610	Above-Grade Vapor Retarders	072613	Above-Grade Vapor Retarders
074473	Metal Faced Panels	074433	Metal Faced Panels
075430	Ketone Ethylene Ester Roofing	075416	Ketone Ethylene Ester Roofing
077280	Vents	077280	Vent Options
087125	Weatherstripping	087125	Door Weatherstripping
087530	Weatherstripping	087530	Window Weatherstripping
096223	Bamboo Flooring	096436	Bamboo Flooring
099103	Paint Restoration	090190	Maintenance of Painting and Coating
102833	Laundry Accessories	102823	Laundry Accessories
117610	Operating Room Equipment	117610	Equipment for Operating Rooms
117710	Radiology Equipment	117710	Equipment for Radiology
122310	Wood Interior Shutters	122313	Wood Interior Shutters
123580	Commercial Kitchen Casework	123539	Commercial Kitchen Casework
124636	Desk Accessories	124113	Desk Accessories
141210	Dumbwaiters	141000	Dumbwaiters
211113	Facility Water Distribution Piping	211113	Facility Fire Suppression Piping
233715	Louvers	233715	Air Outlets and Inlets, HVAC Louvers
260580	Wiring Connections	260583	Wiring Connections
270110	Operation and Maintenance of Communications Systems	270110	Operation and Maintenance of Communication Systems
272123	Data Communications Switches and Hubs	272129	Data Communications Switches and Hubs
283149	Carbon-Monoxide Detection Sensors	284611.21	Carbon-Monoxide Detection Sensors
284621	Fire Alarm	284620	Fire Alarm
316233	Drilled Micropiles	316333	Drilled Micropiles
323420	Fabricated Pedestrian Bridges	323413	Fabricated Pedestrian Bridges
337543	Shunt Reactors	337253	Shunt Reactors
350100	Operation and Maint. of Waterway & Marine Construction	350100	Operation and Maintenance of Waterway and Marine Construction

How the Cost Data Is Built: An Overview

A Complete Reference for the Facility Manager

This data set was designed as a complete reference and cost data source for facility managers, owners, engineers, contractors, and anyone who manages real estate. A complete facilities management program starts with a building audit. The audit describes the structures, their characteristics and major equipment, and apparent deficiencies. An audit is not a prerequisite to using any data contained in this data set, but it does provide an organizational framework. A list of equipment and deficiencies forms the basis for a repair budget and preventive maintenance program. An audit also provides facility size and usage criteria on which the general maintenance (cleaning) program will be based. Together, the five sections of this data set provide a framework and definitive data for developing a complete program for all facets of facilities maintenance.

Even if you have a regular building audit and preventive maintenance program in place, this information will be useful.

For the occasional user, the data set provides:

- a reference for time and material requirements for maintenance & repair tasks to fill in where your organization lacks its own historical data;
- preventive maintenance (PM) checklists with labor-hour standards that can be used to benchmark your program;
- a quick reference for general maintenance (cleaning) productivity rates to analyze proposals from outside contractors;
- an explanation of Life Cycle Cost analysis as a tool to assist in making the correct budget decisions; and
- data to validate line items in budgets.

For the frequent user, the data set supplies all of the above plus:

- guidance for getting the audit under way;
- a foundation for zero-based budgeting of maintenance & repair, preventive maintenance, and general maintenance;
- cost data that can be used to estimate preventive maintenance, deferred maintenance (repairs), and general maintenance (cleaning) programs; and
- detailed task descriptions for establishing a complete PM program.

How to Use the Cost Data: The Details

This section contains an in-depth explanation of how the data is arranged and how you can use it to determine a reliable maintenance & repair cost estimate. It includes information about how we develop our cost figures and how to completely prepare your estimate.

Section 1: Maintenance & Repair (M&R)

The Maintenance & Repair section is a listing of common maintenance tasks performed at facilities. The tasks include removal and replacement, repair, and refinishing. This section is organized according to the UNIFORMAT II system.

The purpose of Section 1 is to provide cost data and an approximate frequency of occurrence for each maintenance & repair task. This data is used by facility managers, owners, plant engineers, and contractors to prepare estimates for deferred maintenance projects. The data is also used to estimate labor-hours and material costs for in-house staff. The frequency listing facilitates preparation of a zero-based budget.

Under the System Description column, each task, including its components, is described. In some cases additional information (such as "2% of total roof area") appears in parentheses. The next column, Frequency, refers to how often one should expect, and therefore estimate, that this task will have to be performed. Labor-hours have been enhanced, where appropriate, by 30% to include setup and delay time. In some cases, the list of steps comprising the task includes most of the potential repairs that could be made on a system. For those tasks, individual items, as appropriate, can be selected from the all-encompassing list.

Maintenance & Repair

B30 ROOFING — B3013 Roof Covering

B3013 105 — Built-Up Roofing

	System Description	Freq. (Years)	Crew	Unit	Labor Hours	2020 Bare Costs				Total In-House	Total w/O&P
						Material	Labor	Equipment	Total		
0600	**Place new BUR membrane over existing**	20	G-5	Sq.							
	Set up, secure and take down ladder				.020		.92		.92	1.33	1.62
	Sweep / spud ballast clean				.500		23		23	33.50	40.50
	Vent existing membrane				.130		6		6	8.65	10.50
	Cut out buckled flashing				.024		1.10		1.10	1.60	1.94
	Install 2 ply membrane flashing				.027	.41	1.24		1.65	2.22	2.68
	Install 4 ply bituminous roofing				3.733	133	161	38	332	420	490
	Reinstall ballast				.381		17.60		17.60	25.50	31
	Clean up				.390		18.05		18.05	26	31.50
	Total				5.205	133.41	228.91	38	400.32	**518.80**	**609.74**
0700	**Total BUR roof replacement**	28	G-1	Sq.							
	Set up, secure and take down ladder				.020		.92		.92	1.33	1.62
	Sweep / spud ballast clean				.500		23		23	33.50	40.50
	Remove built-up roofing				2.500		106		106	137	170
	Remove insulation board				1.026		44		44	56	70
	Remove flashing				.024		1.10		1.10	1.60	1.94
	Install 2" perlite insulation				.879	109	41		150	178	207
	Install 2 ply membrane flashing				.027	.41	1.24		1.65	2.22	2.68
	Install 4 ply bituminous membrane				2.800	133	121	28.50	282.50	350	410
	Clean up				1.000		46		46	66.50	81
	Total										

B3013 120 — Modified Bituminous / Therm

	System Description	Freq. (Years)	Crew	U
0100	**Debris removal by hand & visual inspection**	1	[illegible]	[illegible]
	Set up, secure and take down ladder			
	Pick up trash / debris & clean up			
	Visual inspection			
	Total			

The Construction Specifications Institute (CSI) and Construction Specifications Canada (CSC) have produced the 2018 edition of MasterFormat®, a system of titles and numbers used extensively to organize construction information.

All unit prices in the RSMeans cost data are now arranged in the 50-division MasterFormat® 2018 system.

Section 2: Preventive Maintenance (PM)

The Preventive Maintenance section provides the framework for a complete PM program. The establishment of a program is frequently hindered by the lack of a comprehensive list of equipment, actual PM steps, and budget documentation. This section fulfills all these needs. The facility manager, plant engineer, or owner can use these schedules to establish labor-hours and a budget. The maintenance contractor can use these PM checklists as the basis for a comprehensive maintenance proposal or as an estimating aid when bidding PM contracts. Copies or print-outs of individual sheets can also be distributed to maintenance personnel to identify the procedures required. The hours listed are predicated on work being performed by experienced technicians familiar with the PM tasks and equipped with the proper tools and materials.

The PM section lists the tasks and their frequency, whether weekly, monthly, quarterly, semi-annually, or annually. The frequency of those procedures is based on non-critical usage (e.g., in "normal use" situations, not facilities such as surgical suites or computer rooms that demand absolute adherence to a limited range of environmental conditions). The labor-hours to perform each item on the checklist are listed in the next column. Beneath the table is cost data for the PM schedule. The data is shown both annually and annualized to provide the facility manager with an estimating range. If all tasks on the schedule are performed once a year, the *annually* line should be used in the PM estimate. The *annualized* data is used when all items on the schedule are performed at the frequency shown.

Preventive Maintenance

D30 HVAC | **D3025 130** | **Boiler, Hot Water, Oil/Gas/Comb.**

PM Components	Labor-hrs.	W	M	Q	S	A
PM System D3025 130 1950						
Boiler, hot water; oil, gas or combination fired, up to 120 MBH						
1 Check combustion chamber for air or gas leaks.	.077					√
2 Inspect and clean oil burner gun and ignition assembly, where applicable.	.658					√
3 Inspect fuel system for leaks and change fuel filter element, where applicable.	.098					√
4 Check fuel lines and connections for damage.	.023		√	√	√	√
5 Check for proper operational response of burner to thermostat controls.	.133			√	√	√
6 Check and lubricate burner and blower motors.	.079			√	√	√
7 Check main flame failure protection and main flame detection scanner on boiler equipped with spark ignition (oil burner).	.124		√	√	√	√
8 Check electrical wiring to burner controls and blower.	.079					√
9 Clean firebox (sweep and vacuum).	.577					√
10 Check operation of mercury control switches (i.e., steam pressure, hot water temperature limit, atomizing or combustion air proving, etc.).	.143		√	√	√	√
11 Check operation and condition of safety pressure relief valve.	.030		√	√	√	√
12 Check operation of boiler low water cut off devices.	.056		√	√	√	√
13 Check hot water pressure gauges.	.073		√	√	√	√
14 Inspect and clean water column sight glass (or replace).	.127		√	√	√	√
15 Clean fire side of water jacket boiler.	.433					√
16 Check condition of flue pipe, damper and exhaust stack.	.147			√	√	√
17 Check boiler operation through complete cycle, up to 30 minutes.	.650					√
18 Check fuel level with gauge pole, add as required.	.046		√	√	√	√
19 Clean area around boiler.	.066		√	√	√	√
20 Fill out maintenance checklist and report deficiencies.	.022		√	√	√	√
Total labor-hours/period			.710	1.069	1.069	3.641
Total labor-hours/year			5.678	2.138	1.069	3.641

		Cost Each					
		2020 Bare Costs				Total	Total
Description	Labor-hrs.	Material	Labor	Equip.	Total	In-House	w/O&P

Section 3: General Maintenance (GM)

This section provides labor-hour estimates and costs to perform day-to-day maintenance. The data is used to estimate cleaning times, compare and assess estimates submitted by maintenance companies, or to budget in-house staff. The information is divided into *Interior* and *Exterior* maintenance. A common maintenance laborer (*Clam*) is used to perform these tasks.

Section 4: Facilities Audits

The facilities audit is generally the basis from which the maintenance & repair, preventive maintenance, and general maintenance estimates are prepared. The audit provides the following:

1. A list of deficiencies for the maintenance & repair estimate
2. A list of equipment and other items from which to prepare a preventive maintenance estimate
3. A list of facilities' usage and size requirements to prepare a general maintenance estimate

The Introduction to Facilities Audits in this section explains the rationale for the audit and the steps required to complete it successfully.

This section also provides forms for listing all the organization's facilities and a separate detailed form for the specifics of each facility. Checklists are provided for major building components. Each checklist should be accompanied by a blank audit form. This form is used to record the audit findings, prioritize the deficiencies, and estimate the cost to remedy these deficiencies.

Reference Section

This section includes information on Equipment Rental Costs, Crew Listings, Travel Costs, Reference Tables, Life Cycle Costing, and a listing of abbreviations.

Equipment Rental Costs: This section contains the average costs to rent and operate hundreds of pieces of construction equipment.

Crew Listings: This section lists all the crews referenced. For the purposes of this data set, a crew is composed of more than one trade classification and/or the addition of power equipment to any trade classification. Power equipment is included in the cost of the crew. Costs are shown both with bare labor rates and with the installing contractor's overhead and profit added. For each, the total crew cost per eight-hour day and the composite cost per labor-hour are listed.

Travel Costs: This chart provides labor-hour costs for round-trip travel between a base of operation and the project site.

Reference Tables: In this section, you'll find reference tables, explanations, and estimating information that support how we develop the unit price data, technical data, and estimating procedures.

Life Cycle Costing: This section provides definitions and basic equations for performing life cycle analyses. A sample problem is solved to demonstrate the methodology.

Abbreviations: A listing of the abbreviations used throughout this data set, along with the terms they represent, is included in this section.

The Design Assumptions of this Data Set

This data set is designed to be as easy to use as possible. To that end, we have made certain assumptions and limited its scope.

1. Unlike any other data set, *Facilities Maintenance & Repair Costs with RSMeans data* was designed for estimating a wide range of maintenance tasks in diverse environments. Due to this diversity, the level of accuracy of the data is +/– 20%.
2. We have established material prices based on a national average.
3. We have computed labor costs based on a 30-city national average of union wage rates.
4. Except where major equipment or component replacement is described, the projects in this data set are small. Therefore, material prices and labor-hours have been enhanced to reflect the increased costs of small-scale work.
5. The PM frequencies are based on non-critical applications. Increased frequencies are required for critical environments.

General Maintenance

01 93 Facility Maintenance

01 93 04 – Landscaping Maintenance

01 93 04.15 Edging		Crew	Daily Output	Labor-Hours	Unit	Material	2020 Bare Costs Labor	Equipment	Total	Total In-House	Total Incl O&P
0010	**EDGING**										
0020	Hand edging, at walks	1 Clam	16	.500	C.L.F.		15.80		15.80	20.50	25.50
0030	At planting, mulch or stone beds		7	1.143			36		36	46.50	58
0040	Power edging, at walks		88	.091			2.87		2.87	3.71	4.60
0050	At planting, mulch or stone beds		24	.333			10.55		10.55	13.60	16.90
01 93 04.20 Flower, Shrub and Tree Care											
0010	**FLOWER, SHRUB & TREE CARE**										
0020	Flower or shrub beds, bark mulch, 3" deep hand spreader	1 Clam	100	.080	S.Y.	4.12	2.53		6.65	7.80	9.20
0030	Peat moss, 1" deep hand spreader		900	.009		5.15	.28		5.43	6.05	6.90
0040	Wood chips, 2" deep hand spreader		220	.036		1.68	1.15		2.83	3.34	3.94
0050	Cleaning		1	8	M.S.F.		253		253	325.00	405
0060	Fertilizing, dry granular, 3#/M.S.F.		85	.094		1.23	2.97		4.20	5.20	6.30
0070	Weeding, mulched bed		20	.400			12.65		12.65	16.35	20.50
0080	Unmulched bed		8	1			31.50		31.50	41.00	50.50
0090	Trees, pruning from ground, 1-1/2" caliper		84	.095	Ea.		3.01		3.01	3.89	4.82
0100	2" caliper		70	.114			3.61		3.61	4.67	5.80
0110	2-1/2" caliper		50	.160			5.05		5.05	6.55	8.10
0120	3" caliper		30	.267			8.45		8.45	10.90	13.50
0130	4" caliper	2 Clam	21	.762			24		24	31.00	38.50
0140	6" caliper		12	1.333			42		42	54.50	67.50
0150	9" caliper		7.50	2.133			67.50		67.50	87.00	108
0160	12" caliper		6.50	2.462			78		78	101.00	125
0170	Fertilize, slow release tablets	1 Clam	100	.080		2.30	2.53		4.83	5.80	6.95

How to Use the Cost Data: The Details

What's Behind the Numbers? The Development of Cost Data

RSMeans data engineers continually monitor developments in the construction industry in order to ensure reliable, thorough, and up-to-date cost information. While overall construction costs may vary relative to general economic conditions, price fluctuations within the industry are dependent upon many factors. Individual price variations may, in fact, be opposite to overall economic trends. Therefore, costs are constantly tracked and complete updates are performed yearly. Also, new items are frequently added in response to changes in materials and methods.

Costs in U.S. Dollars

All costs represent U.S. national averages and are given in U.S. dollars. The City Cost Index (CCI) with RSMeans data can be used to adjust costs to a particular location. The CCI for Canada can be used to adjust U.S. national averages to local costs in Canadian dollars. No exchange rate conversion is necessary because it has already been factored in.

G The processes or products identified by the green symbol in our publications have been determined to be environmentally responsible and/or resource-efficient solely by RSMeans data engineering staff. The inclusion of the green symbol does not represent compliance with any specific industry association or standard.

Material Costs

RSMeans data engineers contact manufacturers, dealers, distributors, and contractors all across the U.S. and Canada to determine national average material costs. If you have access to current material costs for your specific location, you may wish to make adjustments to reflect differences from the national average. Included within material costs are fasteners for a normal installation. RSMeans data engineers use manufacturers' recommendations, written specifications, and/or standard construction practices for the sizing and spacing of fasteners. Adjustments to material costs may be required for your specific application or location. The manufacturer's warranty is assumed. Extended warranties are not included in the material costs. **Material costs do not include sales tax.**

Labor Costs

Labor costs are based upon a mathematical average of trade-specific wages in 30 major U.S. cities. The type of wage (union, open shop, or residential) is identified on the inside back cover of printed publications or selected by the estimator when using the electronic products. Markups for the wages can also be found on the inside back cover of printed publications and/or under the labor references found in the electronic products.

- If wage rates in your area vary from those used, or if rate increases are expected within a given year, labor costs should be adjusted accordingly.

Labor costs reflect productivity based on actual working conditions. In addition to actual installation, these figures include time spent during a normal weekday on tasks, such as material receiving and handling, mobilization at the site, site movement, breaks, and cleanup.

Productivity data is developed over an extended period so as not to be influenced by abnormal variations and reflects a typical average.

Equipment Costs

Equipment costs include not only rental but also operating costs for equipment under normal use. The operating costs include parts and labor for routine servicing, such as the repair and replacement of pumps, filters, and worn lines. Normal operating expendables, such as fuel, lubricants, tires, and electricity (where applicable), are also included. Extraordinary operating expendables with highly variable wear patterns, such as diamond bits and blades, are excluded. These costs are included under materials. Equipment rental rates are obtained from industry sources throughout North America—contractors, suppliers, dealers, manufacturers, and distributors.

Rental rates can also be treated as reimbursement costs for contractor-owned equipment. Owned equipment costs include depreciation, loan payments, interest, taxes, insurance, storage, and major repairs.

Equipment costs do not include operators' wages.

Equipment Cost/Day—The cost of equipment required for each crew is included in the Crew Listings in the Reference Section (small tools that are considered essential everyday tools are not listed out separately). The Crew Listings itemize specialized tools and heavy equipment along with labor trades. The daily cost of itemized equipment included in a crew is based on dividing the weekly bare rental rate by 5 (number of working days per week), then adding the hourly operating cost times 8 (the number of hours per day). This Equipment Cost/Day is shown in the last column of the Equipment Rental Costs in the Reference Section.

Mobilization, Demobilization—The cost to move construction equipment from an equipment yard or rental company to the job site and back again is not included in equipment costs. Mobilization (to the site) and demobilization (from the site) costs can be found in the Unit Price Section. If a piece of equipment is already at the job site, it is not appropriate to utilize mobilization or demobilization costs again in an estimate.

Overhead and Profit

Total Cost including O&P for the installing contractor is shown in the last column of the Unit Price and/or Assemblies. This figure is the sum of the bare material cost plus 10% for profit, the bare labor cost plus total overhead and profit, and the bare equipment cost plus 10% for profit. Details for the calculation of overhead and profit on labor are shown on the inside back cover of the printed product and in the Reference Section of the electronic product.

General Conditions

Cost data in this data set are presented in two ways: Bare Costs and Total Cost including O&P (Overhead and Profit). General Conditions, or General Requirements, of the contract should also be added to the Total Cost including O&P when applicable. Costs for General Conditions are listed in Division 1 of the Unit Price Section and in the Reference Section.

General Conditions for the installing contractor may range from 0% to 10% of the Total Cost including O&P. For the general or prime contractor, costs for General Conditions may range from 5% to 15% of the Total Cost including O&P, with a figure of 10% as the most typical allowance. If applicable, the Assemblies and Models sections use costs that include the installing contractor's overhead and profit (O&P).

Factors Affecting Costs

Costs can vary depending upon a number of variables. Here's a listing of some factors that affect costs and points to consider.

Quality—The prices for materials and the workmanship upon which productivity is based represent sound construction work. They are also in line with industry standard and manufacturer specifications and are frequently used by federal, state, and local governments.

Overtime—We have made no allowance for overtime. If you anticipate premium time or work beyond normal working hours, be sure to make an appropriate adjustment to your labor costs.

Productivity—The productivity, daily output, and labor-hour figures for each line item are based on an eight-hour work day in daylight hours in moderate temperatures and up to a 14' working height unless otherwise indicated. For work that extends beyond normal work hours or is performed under adverse conditions, productivity may decrease.

Size of Project—The size, scope of work, and type of construction project will have a significant impact on cost. Economies of scale can reduce costs for large projects. Unit costs can often run higher for small projects.

Location—Material prices are for metropolitan areas. However, in dense urban areas, traffic and site storage limitations may increase costs. Beyond a 20-mile radius of metropolitan areas, extra trucking or transportation charges may also increase the material costs slightly. On the other hand, lower wage rates may be in effect. Be sure to consider both of these factors when preparing an estimate, particularly if the job site is located in a central city or remote rural location. In addition, highly specialized subcontract items may require travel and per-diem expenses for mechanics.

Other Factors—

- season of year
- contractor management
- weather conditions
- local union restrictions
- building code requirements
- availability of:
  - adequate energy
  - skilled labor
  - building materials
- owner's special requirements/restrictions
- safety requirements
- environmental considerations
- access

Unpredictable Factors—General business conditions influence "in-place" costs of all items. Substitute materials and construction methods may have to be employed. These may affect the installed cost and/or life cycle costs. Such factors may be difficult to evaluate and cannot necessarily be predicted on the basis of the job's location in a particular section of the country. Thus, where these factors apply, you may find significant but unavoidable cost variations for which you will have to apply a measure of judgment to your estimate.

Rounding of Costs

In printed publications only, all unit prices in excess of $5.00 have been rounded to make them easier to use and still maintain adequate precision of the results.

How Subcontracted Items Affect Costs

A considerable portion of all large construction jobs is usually subcontracted. In fact, the percentage done by subcontractors is constantly increasing and may run over 90%. Since the workers employed by these companies do nothing else but install their particular products, they soon become experts in that line. As a result, installation by these firms is accomplished so efficiently that the total in-place cost, even with the general contractor's overhead and profit, is no more, and often less, than if the principal contractor had handled the installation. Companies that deal with construction specialties are anxious to have their products perform well and, consequently, the installation will be the best possible.

Contingencies

The allowance for contingencies generally provides for unforeseen construction difficulties. On alterations or repair jobs, 20% is not too much. If drawings are final and only field contingencies are being considered, 2% or 3% is probably sufficient and often nothing needs to be added. Contractually, changes in plans will be covered by extras. The contractor should consider inflationary price trends and possible material shortages during the course of the job. These escalation factors are dependent upon both economic conditions and the anticipated time between the estimate and actual construction. If drawings are not complete or approved, or a budget cost is wanted, it is wise to add 5% to 10%. Contingencies, then, are a matter of judgment.

Important Estimating Considerations

The productivity, or daily output, of each craftsman or crew assumes a well-managed job where tradesmen with the proper tools and equipment, along with the appropriate construction materials, are present. Included are daily set-up and cleanup time, break time, and plan layout time. Unless otherwise indicated, time for material movement on site (for items that can be transported by hand) of up to 200' into the building and to the first or second floor is also included. If material has to be transported by other means, over greater distances, or to higher floors, an additional allowance should be considered by the estimator.

While horizontal movement is typically a sole function of distances, vertical transport introduces other variables that can significantly impact productivity. In an occupied building, the use of elevators (assuming access, size, and required protective measures are acceptable) must be understood at the time of the estimate. For new construction, hoist wait and cycle times can easily be 15 minutes and may result in scheduled access extending beyond the normal work day. Finally, all vertical transport will impose strict weight limits likely to preclude the use of any motorized material handling.

The productivity, or daily output, also assumes installation that meets manufacturer/designer/ standard specifications. A time allowance for quality control checks, minor adjustments, and any task required to ensure proper function or operation is also included. For items that require connections to services, time is included for positioning, leveling, securing the unit, and making all the necessary connections (and start up where applicable) to ensure a complete installation. Estimating of the services themselves (electrical, plumbing, water, steam, hydraulics, dust collection, etc.) is separate.

In some cases, the estimator must consider the use of a crane and an appropriate crew for the installation of large or heavy items. For those situations where a crane is not included in the assigned crew and as part of the line item cost, then equipment rental costs, mobilization and demobilization costs, and operator and support personnel costs must be considered.

Labor-Hours

The labor-hours expressed in this publication are derived by dividing the total daily labor-hours for the crew by the daily output. Based on average installation time and the assumptions listed above, the labor-hours include: direct labor, indirect labor, and nonproductive time. A typical day for a craftsman might include but is not limited to:

- Direct Work
  - Measuring and layout
  - Preparing materials
  - Actual installation
  - Quality assurance/quality control
- Indirect Work
  - Reading plans or specifications
  - Preparing space
  - Receiving materials
  - Material movement
  - Giving or receiving instruction
  - Miscellaneous
- Non-Work
  - Chatting
  - Personal issues
  - Breaks
  - Interruptions (i.e., sickness, weather, material or equipment shortages, etc.)

If any of the items for a typical day do not apply to the particular work or project situation, the estimator should make any necessary adjustments.

Final Checklist

Estimating can be a straightforward process provided you remember the basics. Here's a checklist of some of the steps you should remember to complete before finalizing your estimate.

Did you remember to:

- factor in the City Cost Index for your locale?
- take into consideration which items have been marked up and by how much?
- mark up the entire estimate sufficiently for your purposes?
- read the background information on techniques and technical matters that could impact your project time span and cost?
- include all components of your project in the final estimate?
- double check your figures for accuracy?
- call RSMeans data engineers if you have any questions about your estimate or the data you've used? Remember, Gordian stands behind all of our products, including our extensive RSMeans data solutions. If you have any questions about your estimate, about the costs you've used from our data, or even about the technical aspects of the job that may affect your estimate, feel free to call the Gordian RSMeans editors at 1.800.448.8182.

Maintenance & Repair

Table of Contents

Table of Contents (cont.)

C3023 Floor Finishes

C3033 Ceiling Finishes

D20 Plumbing

D2013 Plumbing Fixtures

D2023 Domestic Water Distribution

D2033 Sanitary Waste

D2043 Rain Water Drainage

D2093 Other Plumbing Systems

D30 HVAC

D3013 Energy Supply

D3023 Heat Generating Systems

D3033 Cooling Generating Systems

D3043 Distribution Systems

D3053 Terminal & Package Units

Table of Contents (cont.)

How to Use the Maintenance & Repair Assemblies Cost Tables

The following is a detailed explanation of a sample Maintenance & Repair Assemblies Cost Table. Next to each bold number that follows is the described item with the appropriate component of the sample entry following in parentheses.

Total system costs, as well as the individual component costs, are shown. In most cases, the intent is for the user to apply the total system costs. However, changes and adjustments to the components or partial use of selected components is also appropriate. In particular, selected equipment system tables in the mechanical section include complete listings of operations that are meant to be chosen from rather than used in total.

B30 ROOFING | **B3013** | **Roof Covering**

B3013 105 (1) **Built-Up Roofing**

	System Description	Freq. (Years)	Crew	Unit	Labor Hours	2020 Bare Costs (7) Material	Labor	Equipment	Total	Total In-House (8)	Total w/O&P (9)
0600	**Place new BUR membrane over existing**	20	G-5	Sq.							
	Set up, secure and take down ladder				.020		.92		.92	1.33	1.62
	Sweep / spud ballast clean				.500		23		23	33.50	40.50
	Vent existing membrane				.130		6		6	8.65	10.50
	Cut out buckled flashing				.024		1.10		1.10	1.60	1.94
	Install 2 ply membrane flashing				.027	.41	1.24		1.65	2.22	2.68
	Install 4 ply bituminous roofing				3.733	133	161	38	332	420	490
	Reinstall ballast				.381		17.60		17.60	25.50	31
	Clean up				.390		18.05		18.05	26	31.50
	Total				5.205	133.41	228.91	38	400.32	**518.80**	**609.74**
0700	**Total BUR roof replacement** (2)	28 (3)	G-1 (4)	Sq. (5)							
	Set up, secure and take down ladder				.020		.92		.92	1.33	1.62
	Sweep / spud ballast clean				.500		23		23	33.50	40.50
	Remove built-up roofing				2.500		106		106	137	170
	Remove insulation board				1.026		44		44	56	70
	Remove flashing				.024		1.10		1.10	1.60	1.94
	Install 2" perlite insulation				.879	109	41		150	178	207
	Install 2 ply membrane flashing				.027	.41	1.24		1.65	2.22	2.68
	Install 4 ply bituminous membrane				2.800	133	121	28.50	282.50	350	410
	Clean up				1.000		46		46	66.50	81
	Total				(6) 8.776	242.41	384.26	28.50	655.17	**826.15**	**984.74**

B3013 120 **Modified Bituminous / Thermoplastic**

	System Description	Freq. (Years)	Crew	Unit	Labor Hours	2020 Bare Costs Material	Labor	Equipment	Total	Total In-House	Total w/O&P
0100	**Debris removal by hand & visual inspection**	1	2 ROFC	M.S.F.							
	Set up, secure and take down ladder				.052		2.40		2.40	3.46	4.20
	Pick up trash / debris & clean up				.327		15.10		15.10	21.50	26.50
	Visual inspection				.327		15.10		15.10	21.50	26.50
	Total				.705		32.60		32.60	**46.46**	**57.20**
0200	**Non - destructive moisture inspection**	5	2 ROFC	M.S.F.							
	Set up, secure and take down ladder				.052		2.40		2.40	3.46	4.20
	Infrared inspection of roof membrane				2.133		98.50		98.50	142	172
	Total				2.185		100.90		100.90	**145.46**	**176.20**

1 System/Line Numbers (B3013 105 0700)

Each Maintenance & Repair Assembly has been assigned a unique identification number based on the UNIFORMAT II classification system.

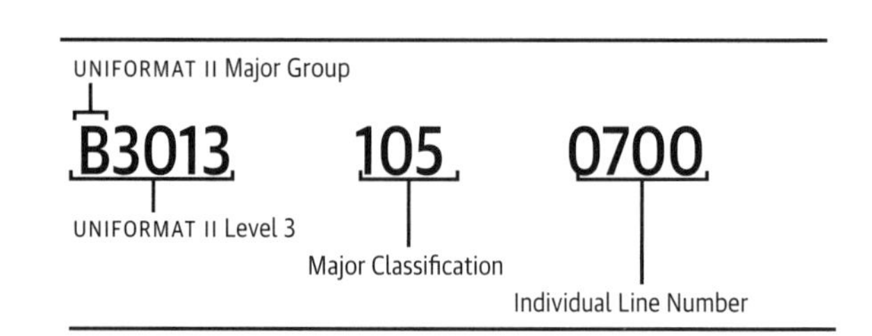

2 System Description

A one-line description of each system is given, followed by a description of each component that makes up the system. In selected items, a percentage factor for planning annual costs is also given.

3 Frequency (28)

The projected frequency of occurrence in years for each system is listed. For example, **28** indicates that this operation should be budgeted for once every 28 years. If the frequency is less than 1, then the operation should be planned for more than once a year. For example, a value of **.5** indicates a planned occurrence of 2 times a year.

4 Crew (G-1)

The "Crew" column designates the typical trade or crew used for the task, although other trades may be used for minor portions of the work. If the task can be accomplished by one trade and requires no power equipment, that trade and the number of workers are listed (for example, "4 Rofc"). If the task requires a composite crew, a crew code designation is listed (for example, "G-1"). You'll find full details on all composite crews in the Crew Listings.

- For a complete list of all trades utilized in this data set, as well as their abbreviations, see the inside back cover of the data set.

Crews - Maintenance

Crew No.	Bare Costs		In-House Costs		Incl. Subs O&P		Cost Per Labor-Hour		
Crew G-1	Hr.	Daily	Hr.	Daily	Hr.	Daily	Bare Costs	In House	Incl. O&P
1 Roofer Foreman (outside)	$48.20	$ 385.60	$69.40	$ 555.20	$84.35	$ 674.80	$43.17	$62.19	$75.55
4 Roofers Composition	46.20	1478.40	66.55	2129.60	80.85	2587.20			
2 Roofer Helpers	34.60	553.60	49.85	797.60	60.55	968.80			
1 Application Equipment		192.85		192.85		212.13			
1 Tar Kettle/Pot		207.85		207.85		228.63			
1 Crew Truck		166.25		166.25		182.88	10.12	10.12	11.14
56 L.H., Daily Totals		$2984.55		$4049.35		$4854.44	$53.30	$72.31	$86.69

5 Unit of Measure for Each System (Sq.)

The abbreviated designation indicates the unit of measure, as defined by industry standards, upon which the price of the component is based. In this example, "Cost per Square" is the unit of measure for this system or "assembly."

6 Labor-Hours (8.776)

The "Labor-Hours" figure represents the number of labor-hours required to perform one unit of work. To calculate the number of labor-hours required for a particular task, multiply the quantity of the item by the number of labor-hours shown. For example:

Quantity	X	Productivity Rate	=	Duration
30 Sq.	X	8.776 Labor-Hours/Sq.	=	263 Labor-Hours

7 Bare Costs:

Material (242.41)

This figure for the unit material cost for the line item is the "bare" material cost with no overhead and profit allowances included. *Costs shown reflect national average material prices for the current year and generally include delivery to the facility. Small purchases may require an additional delivery charge. No sales taxes are included.*

Labor (384.26)

The unit labor costs are derived by multiplying bare labor-hour costs for Crew G-1 and other trades by labor-hour units. In this case, the bare labor-hour cost for the G-1 crew is found in the Crew Section. (If a trade is listed or used for minor portions of the work, the hourly labor cost—the wage rate—is found on the inside back cover of this data set).

Equipment (28.50)

Equipment costs for each crew are listed in the description of each crew. The unit equipment cost is derived by multiplying the bare equipment hourly cost by the labor-hour units.

Total (655.17)

The total of the bare costs is the arithmetic total of the three previous columns: material, labor, and equipment.

Material	+	Labor	+	Equipment	=	Total
$242.41	+	$384.26	+	$28.50	=	$655.17

8 Total In-House Costs (826.15)

"Total In-House Costs" include suggested markups to bare costs for the direct overhead requirements of in-house maintenance staff. The figure in this column is the sum of three components: the bare material cost plus 10% (for purchasing and handling small quantities); the bare labor cost plus workers' compensation and average fixed overhead (per the labor rate table on the inside back cover of this data set or, if a crew is listed, from the Crew Listings); and the bare equipment cost.

9 Total Costs Including O&P (984.74)

"Total Costs Including O&P" include suggested markups to bare costs for an outside subcontractor's overhead and profit requirements. The figure in this column is the sum of three components: the bare material cost plus 25% for profit; the bare labor cost plus total overhead and profit (per the labor rate table on the inside back cover of the printed product or the Reference Section of the electronic product, or, if a crew is listed, from the Crew Listings); and the bare equipment cost plus 10% for profit.

A1033 110 Concrete, Unfinished

	System Description	Freq. (Years)	Crew	Unit	Labor Hours	2020 Bare Costs				Total In-House	Total w/O&P
						Material	Labor	Equipment	Total		
0010	**Minor repairs to concrete floor**	15	1 CEFI	S.F.							
	Repair spall				.612		26	5.45	31.45	39.50	47.50
	Repair concrete flooring				.061	3.22	3.06		6.28	7.40	8.85
	Total				.673	3.22	29.06	5.45	37.73	**46.90**	**56.35**
0020	**Replace unfinished concrete floor**	75	2 CEFI	C.S.F.							
	Break up concrete slab				1.664		76	14.80	90.80	114	137
	Load into truck				1.950		86.25	35.25	121.50	150	176
	Rebar #3 to #7				.139	11.75	7.80		19.55	23	27
	Edge form				.693	1.10	34.90		36	46.50	57.50
	Slab on grade 4" in place with finish				1.577	125.25	75	.75	201	234	276
	Expansion joint				.003	.05	.14		.19	.23	.28
	Cure concrete				.109	6.23	4.60		10.83	12.80	15.15
	Total				6.135	144.38	284.69	50.80	479.87	**580.53**	**688.93**

B1013 510 Exterior Concrete Stairs

	System Description	Freq. (Years)	Crew	Unit	Labor Hours	2020 Bare Costs				Total In-House	Total w/O&P
						Material	Labor	Equipment	Total		
1010	**Repair exterior concrete steps**	30	1 CEFI	S.F.							
	Repair spalls in steps				.104	14.85	5.20		20.05	23	26.50
	Total				.104	14.85	5.20		20.05	**23**	**26.50**
1020	**Replace exterior concrete steps**	75	1 CEFI	S.F.							
	Remove concrete				.520		22	4.64	26.64	33.50	40.50
	Install forms				.328	.97	16.95		17.92	23	28
	Pour concrete				.567	3.11	25	.64	28.75	36	44
	Total				1.416	4.08	63.95	5.28	73.31	**92.50**	**112.50**
1030	**Refinish exterior concrete steps**	3	1 PORD	S.F.							
	Prepare surface				.016		.71		.71	.91	1.13
	Refinish surface, brushwork, 1 coat				.013	.09	.56		.65	.81	.99
	Total				.029	.09	1.27		1.36	**1.72**	**2.12**
2030	**Refinish exterior concrete landing**	3	1 PORD	S.F.							
	Prepare surface				.016		.71		.71	.91	1.13
	Refinish surface, brushwork, 1 coat				.013	.09	.56		.65	.81	.99
	Total				.029	.09	1.27		1.36	**1.72**	**2.12**

B1013 520 Exterior Masonry Steps

	System Description	Freq. (Years)	Crew	Unit	Labor Hours	2020 Bare Costs				Total In-House	Total w/O&P
						Material	Labor	Equipment	Total		
1010	**Repair exterior masonry steps**	4	1 BRIC	S.F.							
	Remove damaged masonry				.074		3.13		3.13	4.04	5
	Clean area				.019	.05	.97		1.02	1.32	1.62
	Replace with new masonry				.236	4.21	11.35		15.56	19.45	23.50
	Total				.329	4.26	15.45		19.71	**24.81**	**30.12**

B1013 530 Exterior Wood Stairs

	System Description	Freq. (Years)	Crew	Unit	Labor Hours	2020 Bare Costs				Total In-House	Total w/O&P
						Material	Labor	Equipment	Total		
1030	**Refinish exterior wood steps**	3	1 PORD	S.F.							
	Prepare surface				.016		.71		.71	.91	1.13
	Wash / dry surface				.001		.05		.05	.07	.08
	Refinish surface, brushwork, 1 coat				.013	.09	.56		.65	.81	.99
	Total				.030	.09	1.32		1.41	**1.79**	**2.20**
1040	**Replace exterior wood steps**	30	2 CARP	S.F.							
	Remove stairs (to 6′ wide)				.100		4.21		4.21	5.45	6.75
	New stringers (to 5 risers)				.048	.86	2.55		3.41	4.24	5.15
	Risers				.083	.76	4.43		5.19	6.55	8.05
	Treads				.123	2.34	6.55		8.89	11.05	13.45
	Total				.354	3.96	17.74		21.70	**27.29**	**33.40**
2030	**Refinish exterior wood landing**	3	1 PORD	S.F.							
	Prepare surface				.016		.71		.71	.91	1.13
	Wash / dry surface				.001		.05		.05	.07	.08
	Refinish surface, brushwork, 1 coat				.013	.09	.56		.65	.81	.99
	Total				.030	.09	1.32		1.41	**1.79**	**2.20**

B1013 540 Exterior Wood Railing

	System Description	Freq. (Years)	Crew	Unit	Labor Hours	2020 Bare Costs				Total In-House	Total w/O&P
						Material	Labor	Equipment	Total		
1010	**Replace exterior wood balustrade**	20	1 CARP	L.F.							
	Remove damaged balusters and handrail				.087		3.65		3.65	4.71	5.85
	Install new balusters and handrail				.473	25	25		50	60	72
	Total				.559	25	28.65		53.65	**64.71**	**77.85**
1030	**Refinish exterior wood balustrade**	7	1 PORD	L.F.							
	Prepare balusters and handrails for painting				.052		2.31		2.31	2.95	3.67
	Wash / dry surface				.001		.05		.05	.07	.08
	Paint balusters and handrails, brushwork, primer + 1 coat				.089	.88	3.95		4.83	6	7.35
	Total				.142	.88	6.31		7.19	**9.02**	**11.10**

B1013 550 Exterior Metal Stairs

	System Description	Freq. (Years)	Crew	Unit	Labor Hours	2020 Bare Costs Material	Labor	Equipment	Total	Total In-House	Total w/O&P
1010	**Repair exterior metal steps**	15	1 SSWK	S.F.							
	Remove damaged stair tread				.099		5.72		5.72	7.60	9.30
	New stair tread				.104	52.34	6.01		58.35	65.50	75.50
	Total				.203	52.34	11.73		64.07	**73.10**	**84.80**
1040	**Replace exterior metal steps**	40	2 SSWK	S.F.							
	Remove stairs				.168		9.70		9.70	12.90	15.90
	New stairs				.239	167.31	13.73	7.29	188.33	210	239
	Total				.408	167.31	23.43	7.29	198.03	**222.90**	**254.90**
1050	**Refinish exterior metal steps**	3	1 PORD	S.F.							
	Prepare metal railing for painting				.019		.85		.85	1.09	1.35
	Wash / dry surface				.001		.05		.05	.07	.08
	Paint metal steps, brushwork, primer + 1 coat				.015	.13	.68		.81	1.01	1.24
	Total				.036	.13	1.58		1.71	**2.17**	**2.67**
2030	**Refinish exterior metal landing**	3	1 PORD	S.F.							
	Prepare metal railing for painting				.019		.85		.85	1.09	1.35
	Wash / dry surface				.001		.05		.05	.07	.08
	Paint metal landing, brushwork, primer + 1 coat				.015	.13	.68		.81	1.01	1.24
	Total				.036	.13	1.58		1.71	**2.17**	**2.67**

B1013 560 Exterior Metal Railing

	System Description	Freq. (Years)	Crew	Unit	Labor Hours	2020 Bare Costs Material	Labor	Equipment	Total	Total In-House	Total w/O&P
1010	**Replace exterior metal hand rail**	30	1 SSWK	L.F.							
	Remove damaged railings				.087		3.65		3.65	4.71	5.85
	Install new metal railings				.163	35	9.50	.75	45.25	52	60
	Total				.250	35	13.15	.75	48.90	**56.71**	**65.85**
1030	**Refinish exterior metal hand rail**	7	1 PORD	L.F.							
	Prepare metal railing for painting				.019		.85		.85	1.09	1.35
	Wash / dry surface				.001		.05		.05	.07	.08
	Paint metal railing, brushwork, primer + 1 coat				.015	.13	.68		.81	1.01	1.24
	Total				.036	.13	1.58		1.71	**2.17**	**2.67**

B1013 570 Exterior Wrought Iron Balustrade

	System Description	Freq. (Years)	Crew	Unit	Labor Hours	2020 Bare Costs Material	Labor	Equipment	Total	Total In-House	Total w/O&P
1010	**Repair exterior wrought iron balustrade**	20	1 SSWK	L.F.							
	Cut / remove damaged section				.325		13.70		13.70	17.70	22
	Reset new section				.867	96.50	50		146.50	173	203
	Total				1.192	96.50	63.70		160.20	**190.70**	**225**
1030	**Refinish exterior wrought iron balustrade**	7	1 PORD	L.F.							
	Prepare metal balustrade for painting				.019		.85		.85	1.09	1.35
	Wash / dry surface				.001		.05		.05	.07	.08
	Paint metal balustrade, brushwork, primer + 1 coat				.015	.13	.68		.81	1.01	1.24
	Total				.036	.13	1.58		1.71	**2.17**	**2.67**

B1013 580 Exterior Metal Fire Escapes

	System Description	Freq. (Years)	Crew	Unit	Labor Hours	2020 Bare Costs Material	Labor	Equipment	Total	Total In-House	Total w/O&P
1010	**Refinish exterior fire escape balcony, 2′ wide**	7	1 PORD	L.F.							
	Prepare metal fire escape balcony for painting				.100		4.44		4.44	5.70	7.05
	Brush on primer coat				.100	2.55	4.44		6.99	8.50	10.25
	Brush on finish coat				.100	2.25	4.44		6.69	8.15	9.85
	Total				.300	4.80	13.32		18.12	**22.35**	**27.15**
1020	**Replace exterior fire escape balcony, 2′ wide**	25	2 SSWK	L.F.							
	Remove metal fire escape balcony				.500		29		29	38.50	47
	Install new metal fire escape balcony				2.000	63.50	115		178.50	223	268
	Total				2.500	63.50	144		207.50	**261.50**	**315**
2010	**Refinish exterior fire escape stair and platform**	7	1 PORD	Flight							
	Prepare fire escape stair and platform for painting				2.400		107		107	136	169
	Brush on primer coat				2.400	61.20	106.56		167.76	204	246
	Brush on finish coat				2.400	54	106.56		160.56	196	237
	Total				7.200	115.20	320.12		435.32	**536**	**652**
2020	**Replace exterior fire escape stair and platform**	25	2 SSWK	Flight							
	Remove metal fire escape stair and platform				14.995		865		865	1,150	1,425
	Install new metal fire escape stair and platform				59.925	1,125	3,450		4,575	5,825	7,050
	Total				74.920	1,125	4,315		5,440	**6,975**	**8,475**

B1013 580 Exterior Metal Fire Escapes

	System Description	Freq. (Years)	Crew	Unit	Labor Hours	2020 Bare Costs				Total In-House	Total w/O&P
						Material	Labor	Equipment	Total		
3010	**Refinish exterior fire escape ladder**	7	1 PORD	V.L.F.							
	Prepare metal fire escape ladder for painting				.020		.89		.89	1.14	1.41
	Brush on primer coat				.020	.34	.89		1.23	1.51	1.84
	Brush on finish coat				.020	.30	.89		1.19	1.47	1.79
	Total				.060	.64	2.67		3.31	**4.12**	**5.04**
3020	**Replace exterior fire escape ladder**	25	2 SSWK	V.L.F.							
	Remove metal fire escape ladder				.250		14.40		14.40	19.15	23.50
	Install new metal fire escape ladder				.500	42	29		71	84.50	99.50
	Total				.750	42	43.40		85.40	**103.65**	**123**

B1013 910 Exterior Steel Decking

	System Description	Freq. (Years)	Crew	Unit	Labor Hours	2020 Bare Costs				Total In-House	Total w/O&P
						Material	Labor	Equipment	Total		
0020	**Replace exterior steel decking**	30	2 SSWK	S.F.							
	Remove decking				.011		.68	.21	.89	1.13	1.34
	Assemble and weld decking				.011	3.23	.68	.21	4.12	4.68	5.40
	Total				.023	3.23	1.36	.42	5.01	**5.81**	**6.74**

B1013 920 Exterior Metal Grating

	System Description	Freq. (Years)	Crew	Unit	Labor Hours	2020 Bare Costs				Total In-House	Total w/O&P
						Material	Labor	Equipment	Total		
0010	**Repair exterior metal floor grating - (2% of grating)**	10	1 SSWK	S.F.							
	Misc. welding of floor grating				.348	1.38	20.50	6.40	28.28	36	43
	Total				.348	1.38	20.50	6.40	28.28	**36**	**43**
0020	**Replace exterior metal floor grating**	30	2 SSWK	S.F.							
	Remove grating				.011		.68	.21	.89	1.13	1.34
	Assemble and weld grating				.059	19.70	3.46	.27	23.43	26.50	30.50
	Total				.071	19.70	4.14	.48	24.32	**27.63**	**31.84**

B2013 104 Copings

	System Description	Freq. (Years)	Crew	Unit	Labor Hours	2020 Bare Costs Material	Labor	Equipment	Total	Total In-House	Total w/O&P
2010	**Replace precast concrete coping, 12″ wide**	50	2 BRIC	L.F.							
	Set up and secure scaffold, 3 floors to roof				.429		22.74		22.74	29.50	36.50
	Remove coping				.050		2.11		2.11	2.72	3.37
	Install new precast concrete coping				.229	27.50	10.70		38.20	44.50	51.50
	Remove scaffold				.429		22.74		22.74	29.50	36.50
	Total				1.137	27.50	58.29		85.79	**106.22**	**127.87**
2020	**Replace limestone coping, 12″ wide**	50	2 BRIC	L.F.							
	Set up and secure scaffold, 3 floors to roof				.429		22.74		22.74	29.50	36.50
	Remove coping				.050		2.11		2.11	2.72	3.37
	Install new limestone coping				.178	16.80	8.35		25.15	29.50	34.50
	Remove scaffold				.429		22.74		22.74	29.50	36.50
	Total				1.086	16.80	55.94		72.74	**91.22**	**110.87**
2030	**Replace marble coping, 12″ wide**	50	2 BRIC	L.F.							
	Set up and secure scaffold, 3 floors to roof				.429		22.74		22.74	29.50	36.50
	Remove coping				.050		2.11		2.11	2.72	3.37
	Install new marble coping				.200	19.90	9.40		29.30	34	40
	Remove scaffold				.429		22.74		22.74	29.50	36.50
	Total				1.108	19.90	56.99		76.89	**95.72**	**116.37**
2040	**Replace terra cotta coping, 12″ wide**	50	2 BRIC	L.F.							
	Set up and secure scaffold, 3 floors to roof				.429		22.74		22.74	29.50	36.50
	Remove coping				.050		2.11		2.11	2.72	3.37
	Install new terra cotta coping				.200	8.60	9.40		18	21.50	26
	Remove scaffold				.429		22.74		22.74	29.50	36.50
	Total				1.108	8.60	56.99		35.59	**83.22**	**102.37**

B2013 109 Concrete Block

	System Description	Freq. (Years)	Crew	Unit	Labor Hours	2020 Bare Costs				Total In-House	Total w/O&P
						Material	Labor	Equipment	Total		
1010	**Repair 8″ concrete block wall - 1st floor**	25	1 BRIC	S.F.							
	Set up, secure and take down ladder				.130		6.75		6.75	8.85	10.95
	Remove damaged block				.183		7.68		7.68	9.95	12.30
	Wire truss reinforcing				.005	.30	.27		.57	.67	.81
	Install block				.111	2.93	5.35		8.28	10.20	12.25
	Total				.429	3.23	20.05		23.28	**29.67**	**36.31**
1020	**Replace 8″ concrete block wall - 1st floor**	60	2 BRIC	C.S.F.							
	Set up and secure scaffold				.750		39.75		39.75	51.50	64
	Remove damaged block				4.571		192		192	249	310
	Wire truss reinforcing				.260	14.75	13.50		28.25	34	40.50
	Install block				11.111	293	535		828	1,025	1,225
	Remove scaffold				.750		39.75		39.75	51.50	64
	Total				17.442	307.75	820		1,127.75	**1,411**	**1,703.50**
1030	**Waterproof concrete block wall - 1st floor**	10	1 ROFC	C.S.F.							
	Steam clean				2.971		126	27	153	192	231
	Spray surface, silicone				.260	40	12		52	61	71
	Total				3.231	40	138	27	205	**253**	**302**
2020	**Replace 8″ concrete block wall - 2nd floor**	60	2 BRIC	C.S.F.							
	Set up and secure scaffold				1.500		79.50		79.50	103	128
	Remove block				4.571		192		192	249	310
	Wire truss reinforcing				.260	14.75	13.50		28.25	34	40.50
	Install block				11.111	293	535		828	1,025	1,225
	Remove scaffold				1.500		79.50		79.50	103	128
	Total				18.942	307.75	899.50		1,207.25	**1,514**	**1,831.50**
2030	**Waterproof concrete block wall - 2nd floor**	10	G-8	C.S.F.							
	Position aerial lift truck, raise & lower platform				.250		11.55	12	23.55	30	33
	Steam clean				11.765		495	112	607	765	920
	Spray surface, silicone				.260	40	12	12	64	75	85
	Total				12.275	40	518.55	136	694.55	**870**	**1,038**

B2013 109 Concrete Block

	System Description	Freq. (Years)	Crew	Unit	Labor Hours	2020 Bare Costs				Total In-House	Total w/O&P
						Material	Labor	Equipment	Total		
3020	**Replace 8″ concrete block wall - 3rd floor**	60	2 BRIC	C.S.F.							
	Set up and secure scaffold				1.500		79.50		79.50	103	128
	Remove damaged block				4.571		192		192	249	310
	Wire truss reinforcing				.260	14.75	13.50		28.25	34	40.50
	Install block				11.111	293	535		828	1,025	1,225
	Remove scaffold				1.500		79.50		79.50	103	128
	Total				18.942	307.75	899.50		1,207.25	**1,514**	**1,831.50**
3030	**Waterproof concrete block wall - 3rd floor**	10	G-8	C.S.F.							
	Position aerial lift truck, raise & lower platform				.300		13.85	14.40	28.25	36	40.50
	Steam clean				11.765		495	112	607	765	920
	Spray surface, silicone				.260	40	12	12	64	75	85
	Total				12.325	40	520.85	138.40	699.25	**876**	**1,045.50**

B2013 114 Concrete Block Wall, Painted

	System Description	Freq. (Years)	Crew	Unit	Labor Hours	2020 Bare Costs				Total In-House	Total w/O&P
						Material	Labor	Equipment	Total		
1030	**Point & refinish block wall - 1st floor**	25	1 BRIC	C.S.F.							
	Set up and secure scaffold				.750		39.75		39.75	51.50	64
	Steam clean				2.600		109		109	141	175
	Clean & point block				3.924		204		204	267	330
	Paint surface, brushwork, 1 coat				1.040	22.10	46.80		68.90	83	100
	Remove scaffold				.750		39.75		39.75	51.50	64
	Total				9.064	22.10	439.30		461.40	**594**	**733**
2030	**Point & refinish block wall - 2nd floor**	25	1 BRIC	C.S.F.							
	Set up and secure scaffold				1.500		79.50		79.50	103	128
	Steam clean				2.600		109		109	141	175
	Clean & point block				3.924		204		204	267	330
	Paint surface, brushwork, 1 coat				1.040	22.10	46.80		68.90	83	100
	Remove scaffold				1.500		79.50		79.50	103	128
	Total				10.564	22.10	518.80		540.90	**697**	**861**

B2013 114 Concrete Block Wall, Painted

	System Description	Freq. (Years)	Crew	Unit	Labor Hours	2020 Bare Costs				Total In-House	Total w/O&P
						Material	Labor	Equipment	Total		
3020	**Point & refinish block wall - 3rd floor**	25	1 BRIC	C.S.F.							
	Set up and secure scaffold				2.250		119.25		119.25	155	192
	Steam clean				2.600		109		109	141	175
	Clean & point block				3.924		204		204	267	330
	Paint surface, brushwork, 1 coat				1.040	22.10	46.80		68.90	83	100
	Remove scaffold				2.250		119.25		119.25	155	192
	Total				12.064	22.10	598.30		620.40	**801**	**989**

B2013 118 Terra Cotta

	System Description	Freq. (Years)	Crew	Unit	Labor Hours	2020 Bare Costs				Total In-House	Total w/O&P
						Material	Labor	Equipment	Total		
1020	**Replace terra cotta wall - 1st floor**	50	2 BRIC	C.S.F.							
	Set up and secure scaffold				.750		39.75		39.75	51.50	64
	Remove terra cotta				1.040		44	9.30	53.30	67.50	81
	Replace terra cotta				10.400	630	498		1,128	1,350	1,600
	Remove scaffold				.750		39.75		39.75	51.50	64
	Total				12.940	630	621.50	9.30	1,260.80	**1,520.50**	**1,809**
2010	**Replace terra cotta wall - 2nd floor**	50	2 BRIC	C.S.F.							
	Set up and secure scaffold				1.500		79.50		79.50	103	128
	Remove terra cotta				1.040		44	9.30	53.30	67.50	81
	Replace terra cotta				10.400	630	498		1,128	1,350	1,600
	Remove scaffold				1.500		79.50		79.50	103	128
	Total				14.440	630	701	9.30	1,340.30	**1,623.50**	**1,937**
2020	**Replace terra cotta wall - 3rd floor**	50	2 BRIC	C.S.F.							
	Set up and secure scaffold				2.250		119.25		119.25	155	192
	Remove terra cotta				1.040		44	9.30	53.30	67.50	81
	Replace terra cotta				10.400	630	498		1,128	1,350	1,600
	Remove scaffold				2.250		119.25		119.25	155	192
	Total				15.940	630	780.50	9.30	1,419.80	**1,727.50**	**2,065**

B2013 119 Clay Brick

	System Description	Freq. (Years)	Crew	Unit	Labor Hours	2020 Bare Costs Material	Labor	Equipment	Total	Total In-House	Total w/O&P
1010	**Repair brick wall - 1st floor**	25	1 BRIC	S.F.							
	Set up, secure and take down ladder				.130		6.75		6.75	8.85	10.95
	Remove damage				.451		18.98		18.98	24.50	30.50
	Install brick				.236	4.21	11.35		15.56	19.45	23.50
	Total				.817	4.21	37.08		41.29	**52.80**	**64.95**
1020	**Point brick wall - 1st floor**	25	2 BRIC	C.S.F.							
	Set up and secure scaffold				1.333		59		59	75.50	94
	Steam clean				2.971		126	27	153	192	231
	Clean & point brick				10.000	4	520		524	685	845
	Remove scaffold				1.333		59		59	75.50	94
	Total				15.638	4	764	27	795	**1,028**	**1,264**
1030	**Waterproof brick wall - 1st floor**	10	1 ROFC	C.S.F.							
	Steam clean				2.971		126	27	153	192	231
	Spray surface silicone				.260	40	12		52	61	71
	Total				3.231	40	138	27	205	**253**	**302**
2010	**Replace brick wall - 2nd floor**	75	2 BRIC	C.S.F.							
	Set up and secure scaffold				1.500		79.50		79.50	103	128
	Remove damage				10.401		438		438	565	700
	Install brick				23.636	421	1,135		1,556	1,950	2,350
	Remove scaffold				1.500		79.50		79.50	103	128
	Total				37.037	421	1,732		2,153	**2,721**	**3,306**
2020	**Point brick wall - 2nd floor**	25	2 BRIC	C.S.F.							
	Set up and secure scaffold				1.500		79.50		79.50	103	128
	Steam clean				2.971		126	27	153	192	231
	Clean & point brick				10.000	4	520		524	685	845
	Remove scaffold				1.500		79.50		79.50	103	128
	Total				15.971	4	805	27	836	**1,083**	**1,332**
2030	**Waterproof brick wall - 2nd floor**	10	G-8	C.S.F.							
	Position aerial lift truck, raise & lower platform				.250		11.55	12	23.55	30	33
	Steam clean				11.765		495	112	607	765	920
	Spray surface, silicone				.260	40	12	12	64	75	85
	Total				12.275	40	518.55	136	694.55	**870**	**1,038**

B2013 119 Clay Brick

	System Description	Freq. (Years)	Crew	Unit	Labor Hours	2020 Bare Costs				Total In-House	Total w/O&P
						Material	Labor	Equipment	Total		
3010	**Replace brick wall - 3rd floor**	75	2 BRIC	C.S.F.							
	Set up and secure scaffold				1.500		79.50		79.50	103	128
	Remove damage				10.401		438		438	565	700
	Install brick				23.636	421	1,135		1,556	1,950	2,350
	Remove scaffold				1.500		79.50		79.50	103	128
	Total				37.037	421	1,732		2,153	**2,721**	**3,306**
3020	**Point brick wall - 3rd floor**	25	2 BRIC	C.S.F.							
	Set up and secure scaffold				1.500		79.50		79.50	103	128
	Steam clean				2.971		126	27	153	192	231
	Clean & point brick				10.000	4	520		524	685	845
	Remove scaffold				1.500		79.50		79.50	103	128
	Total				15.971	4	805	27	836	**1,083**	**1,332**
3030	**Waterproof brick wall - 3rd floor**	10	G-8	C.S.F.							
	Position aerial lift truck, raise & lcwer platform				.300		13.85	14.40	28.25	36	40.50
	Steam clean				11.765		495	112	607	765	920
	Spray surface, silicone				.260	40	12	12	64	75	85
	Total				12.325	40	520.85	138.40	699.25	**876**	**1,045.50**

B2013 121 Clay Brick, Painted

	System Description	Freq. (Years)	Crew	Unit	Labor Hours	2020 Bare Costs				Total In-House	Total w/O&P
						Material	Labor	Equipment	Total		
1010	**Repair painted brick wall - 1st floor**	25	1 BRIC	S.F.							
	Set up, secure and take down ladder				.130		6.75		6.75	8.85	10.95
	Remove damage				.451		18.98		18.98	24.50	30.50
	Install brick				.236	4.21	11.35		15.56	19.45	23.50
	Paint surface, brushwork, 1 coat				.010	.22	.47		.69	.84	1.01
	Total				.827	4.43	37.55		41.98	**53.64**	**65.96**

B2013 121 Clay Brick, Painted

	System Description	Freq. (Years)	Crew	Unit	Labor Hours	2020 Bare Costs				Total In-House	Total w/O&P
						Material	Labor	Equipment	Total		
1030	**Point brick wall - 1st floor**	25	1 BRIC	C.S.F.							
	Set up and secure scaffold				.500		26.50		26.50	34.50	42.50
	Steam clean				2.971		126	27	153	192	231
	Clean & point brick				10.000	4	520		524	685	845
	Paint surface, brushwork, 1 coat				1.040	22.10	46.80		68.90	83	100
	Remove scaffold				.500		26.50		26.50	34.50	42.50
	Total				15.011	26.10	745.80	27	798.90	**1,029**	**1,261**
2010	**Replace brick wall - 2nd floor**	75	1 BRIC	C.S.F.							
	Set up and secure scaffold				1.500		79.50		79.50	103	128
	Remove damage				10.401		438		438	565	700
	Install brick				23.636	421	1,135		1,556	1,950	2,350
	Paint surface, brushwork, 1 coat				1.040	22.10	46.80		68.90	83	100
	Remove scaffold				1.500		79.50		79.50	103	128
	Total				38.077	443.10	1,778.80		2,221.90	**2,804**	**3,406**
2020	**Point and refinish brick wall - 2nd floor**	25	1 BRIC	C.S.F.							
	Set up and secure scaffold				1.500		79.50		79.50	103	128
	Steam clean				2.971		126	27	153	192	231
	Clean & point brick				10.000	4	520		524	685	845
	Paint surface brushwork, 1 coat				1.040	22.10	46.80		68.90	83	100
	Remove scaffold				1.500		79.50		79.50	103	128
	Total				17.011	26.10	851.80	27	904.90	**1,166**	**1,432**
3010	**Replace brick wall - 3rd floor**	75	1 BRIC	C.S.F.							
	Set up and secure scaffold				1.500		79.50		79.50	103	128
	Remove damage				10.401		438		438	565	700
	Install brick				23.636	421	1,135		1,556	1,950	2,350
	Paint surface brushwork, 1 coat				.010	.22	.47		.69	.84	1.01
	Remove scaffold				1.500		79.50		79.50	103	128
	Total				37.047	421.22	1,732.47		2,153.69	**2,721.84**	**3,307.01**

B20 EXTERIOR CLOSURE | B2013 | Exterior Walls

B2013 121 — Clay Brick, Painted

	System Description	Freq. (Years)	Crew	Unit	Labor Hours	2020 Bare Costs Material	Labor	Equipment	Total	Total In-House	Total w/O&P
3020	**Point and refinish brick wall - 3rd floor**	25	1 BRIC	C.S.F.							
	Set up and secure scaffold				1.500		79.50		79.50	103	128
	Steam clean				2.971		126	27	153	192	231
	Clean & point brick				10.000	4	520		524	685	845
	Paint surface, brushwork, 1 coat				1.040	22.10	46.80		68.90	83	100
	Remove scaffold				1.500		79.50		79.50	103	128
	Total				17.011	26.10	851.80	27	904.90	**1,166**	**1,432**

B2013 123 — Field Stone

	System Description	Freq. (Years)	Crew	Unit	Labor Hours	2020 Bare Costs Material	Labor	Equipment	Total	Total In-House	Total w/O&P
1010	**Repair field stone wall - 1st floor**	25	1 BRIC	S.F.							
	Set up, secure and take down ladder				.390		16.70		16.70	21.50	26.50
	Remove damaged stones				.173		7.35	1.55	8.90	11.20	13.50
	Reset stones				.416	22	21.50	5.55	49.05	58	68
	Total				.979	22	45.55	7.10	74.65	**90.70**	**108**
1020	**Point stone wall, 8″ x 8″ stones - 1st floor**	25	2 BRIC	C.S.F.							
	Set up and secure scaffold				1.333		59		59	75.50	94
	Steam clean				2.971		126	27	153	192	231
	Clean & point stonework joints				5.714	76	297		373	475	575
	Remove scaffold				1.333		59		59	75.50	94
	Total				11.352	76	541	27	644	**818**	**994**
2020	**Point stone wall, 8″ x 8″ stones - 2nd floor**	25	2 BRIC	C.S.F.							
	Set up and secure scaffold				1.500		79.50		79.50	103	128
	Steam clean				2.971		126	27	153	192	231
	Clean & point stonework joints				5.714	76	297		373	475	575
	Remove scaffold				1.500		79.50		79.50	103	128
	Total				11.685	76	582	27	685	**873**	**1,062**

B2013 123 Field Stone

	System Description	Freq. (Years)	Crew	Unit	Labor Hours	2020 Bare Costs				Total In-House	Total w/O&P
						Material	Labor	Equipment	Total		
3020	**Point stone wall, 8″ x 8″ stones - 3rd floor**	25	2 BRIC	C.S.F.							
	Set up and secure scaffold				1.500		79.50		79.50	103	128
	Steam clean				2.971		126	27	153	192	231
	Clean & point stonework joints				5.714	76	297		373	475	575
	Remove scaffold				1.500		79.50		79.50	103	128
	Total				11.685	76	582	27	685	**873**	**1,062**

B2013 140 Glass Block

	System Description	Freq. (Years)	Crew	Unit	Labor Hours	2020 Bare Costs				Total In-House	Total w/O&P
						Material	Labor	Equipment	Total		
1020	**Replace glass block wall - 1st floor**	75	2 BRIC	C.S.F.							
	Set up and secure scaffold				.750		39.75		39.75	51.50	64
	Remove blocks				5.200		221	46	267	335	405
	Replace blocks				45.217	2,300	2,150		4,450	5,350	6,375
	Remove scaffold				.750		39.75		39.75	51.50	64
	Total				51.917	2,300	2,450.50	46	4,796.50	**5,788**	**6,908**
2020	**Replace glass block wall - 2nd floor**	75	2 BRIC	C.S.F.							
	Set up and secure scaffold				1.500		79.50		79.50	103	128
	Remove blocks				5.200		221	46	267	335	405
	Replace blocks				45.217	2,300	2,150		4,450	5,350	6,375
	Remove scaffold				1.500		79.50		79.50	103	128
	Total				53.417	2,300	2,530	46	4,876	**5,891**	**7,036**
3020	**Replace glass block wall - 3rd floor**	75	2 BRIC	C.S.F.							
	Set up and secure scaffold				2.250		119.25		119.25	155	192
	Remove blocks				5.200		221	46	267	335	405
	Replace blocks				45.217	2,300	2,150		4,450	5,350	6,375
	Remove scaffold				2.250		119.25		119.25	155	192
	Total				54.917	2,300	2,609.50	46	4,955.50	**5,995**	**7,164**

B2013 141 Aluminum Siding

	System Description	Freq. (Years)	Crew	Unit	Labor Hours	2020 Bare Costs				Total In-House	Total w/O&P
						Material	Labor	Equipment	Total		
1010	**Replace aluminum siding - 1st floor**	35	1 CARP	C.S.F.							
	Set up, secure and take down ladder				.520		27.60		27.60	35.50	44
	Remove damaged siding				1.434		60		60	78	97
	Install siding				5.368	186	280		466	560	675
	Total				7.322	186	367.60		553.60	**673.50**	**816**
1020	**Refinish aluminum siding - 1st floor**	20	1 PORD	C.S.F.							
	Set up, secure and take down ladder				1.300		57.50		57.50	74	91.50
	Prepare surface				1.159		51		51	66	82
	Refinish surface, brushwork, primer + 1 coat				1.231	14	55		69	85	105
	Total				3.690	14	163.50		177.50	**225**	**278.50**
2010	**Replace aluminum siding - 2nd floor**	35	1 CARP	C.S.F.							
	Set up and secure scaffold				1.500		79.50		79.50	103	128
	Remove damaged siding				1.434		60		60	78	97
	Install siding				5.368	186	280		466	560	675
	Remove scaffold				1.500		79.50		79.50	103	128
	Total				9.802	186	499		685	**844**	**1,028**
2020	**Refinish aluminum siding - 2nd floor**	20	1 PORD	C.S.F.							
	Set up and secure scaffold				1.500		79.50		79.50	103	128
	Prepare surface				1.159		51		51	66	82
	Refinish surface, brushwork, primer + 1 coat				1.231	14	55		69	85	105
	Remove scaffold				1.500		79.50		79.50	103	128
	Total				5.390	14	265		279	**357**	**443**
3010	**Replace aluminum siding - 3rd floor**	35	1 CARP	C.S.F.							
	Set up and secure scaffold				2.250		119.25		119.25	155	192
	Remove damaged siding				1.434		60		60	78	97
	Install siding				5.368	186	280		466	560	675
	Remove scaffold				2.250		119.25		119.25	155	192
	Total				11.302	186	578.50		764.50	**948**	**1,156**

B2013 141 Aluminum Siding

	System Description	Freq. (Years)	Crew	Unit	Labor Hours	2020 Bare Costs				Total In-House	Total w/O&P
						Material	Labor	Equipment	Total		
3020	**Refinish aluminum siding - 3rd floor**	20	1 PORD	C.S.F.							
	Set up and secure scaffold				2.250		119.25		119.25	155	192
	Prepare surface				1.159		51		51	66	82
	Refinish surface, brushwork, primer + 1 coat				1.231	14	55		69	85	105
	Remove scaffold				2.250		119.25		119.25	155	192
	Total				6.890	14	344.50		358.50	**461**	**571**
4120	**Spray refinish aluminum siding - 1st floor**	20	1 PORD	C.S.F.							
	Set up, secure and take down ladder				1.300		57.50		57.50	74	91.50
	Prepare surface				1.159		51		51	66	82
	Protect/mask unpainted surfaces				.031	.30	1.40		1.70	2	2.58
	Spray full primer coat				.444	9	20		29	39	46.50
	Spray finish coat				.500	9	23		32	42	51.50
	Total				3.434	18.30	152.90		171.20	**223**	**274.08**
4220	**Spray refinish aluminum siding - 2nd floor**	20	2 PORD	C.S.F.							
	Set up and secure scaffold				1.500		79.50		79.50	103	123
	Prepare surface				1.159		51		51	66	82
	Protect/mask unpainted surfaces				.031	.30	1.40		1.70	2	2.58
	Spray full primer coat				.444	9	20		29	39	46.50
	Spray finish coat				.500	9	23		32	42	51.50
	Remove scaffold				1.500		79.50		79.50	103	128
	Total				5.134	18.30	254.40		272.70	**355**	**438.58**
4320	**Spray refinish aluminum siding - 3rd floor**	20	2 PORD	C.S.F.							
	Set up and secure scaffold				2.250		119.25		119.25	155	192
	Prepare surface				1.159		51		51	66	82
	Protect/mask unpainted surfaces				.031	.30	1.40		1.70	2	2.58
	Spray full primer coat				.444	9	20		29	39	46.50
	Spray finish coat				.500	9	23		32	42	51.50
	Remove scaffold				2.250		119.25		119.25	155	192
	Total				6.634	18.30	333.90		352.20	**459**	**566.58**

B2013 142 Steel Siding

	System Description	Freq. (Years)	Crew	Unit	Labor Hours	2020 Bare Costs Material	Labor	Equipment	Total	Total In-House	Total w/O&P
1010	**Replace steel siding - 1st floor**	35	1 CARP	C.S.F.							
	Set up, secure and take down ladder				.520		27.60		27.60	35.50	44
	Remove damaged siding				1.434		60		60	78	97
	Install siding				5.266	169	275		444	535	645
	Total				7.220	169	362.60		531.60	**648.50**	**786**
1020	**Refinish steel siding - 1st floor**	20	1 PORD	C.S.F.							
	Set up, secure and take down ladder				1.300		57.50		57.50	74	91.50
	Prepare surface				1.159		51		51	66	82
	Refinish surface, brushwork, primer + 1 coat				1.231	14	55		69	85	105
	Total				3.690	14	163.50		177.50	**225**	**278.50**
2010	**Replace steel siding - 2nd floor**	35	1 CARP	C.S.F.							
	Set up and secure scaffold				1.500		79.50		79.50	103	128
	Remove damaged siding				1.434		60		60	78	97
	Install siding				5.266	169	275		444	535	645
	Remove scaffold				1.500		79.50		79.50	103	128
	Total				9.700	169	494		663	**819**	**998**
2020	**Refinish steel siding - 2nd floor**	20	1 PORD	C.S.F.							
	Set up and secure scaffold				1.500		79.50		79.50	103	128
	Prepare surface				1.159		51		51	66	82
	Refinish surface, brushwork, primer + 1 coat				1.231	14	55		69	85	105
	Remove scaffold				1.500		79.50		79.50	103	128
	Total				5.390	14	265		279	**357**	**443**
3010	**Replace steel siding - 3rd floor**	35	1 CARP	C.S.F.	11.302						
	Set up and secure scaffold				2.250		119.25		119.25	155	192
	Remove damaged siding				1.434		60		60	78	97
	Install siding				5.266	169	275		444	535	645
	Remove scaffold				2.250		119.25		119.25	155	192
	Total				11.200	169	573.50		742.50	**923**	**1,126**

B2013 142 Steel Siding

	System Description	Freq. (Years)	Crew	Unit	Labor Hours	2020 Bare Costs				Total In-House	Total w/O&P
						Material	Labor	Equipment	Total		
3020	**Refinish steel siding - 3rd floor**	20	1 PORD	C.S.F.							
	Set up and secure scaffold				2.250		119.25		119.25	155	192
	Prepare surface				1.159		51		51	66	82
	Refinish surface, brushwork, primer + 1 coat				1.231	14	55		69	85	105
	Remove scaffold				2.250		119.25		119.25	155	192
	Total				6.890	14	344.50		358.50	**461**	**571**
4120	**Spray refinish steel siding - 1st floor**	20	1 PORD	C.S.F.							
	Set up, secure and take down ladder				1.300		57.50		57.50	74	91.50
	Prepare surface				1.159		51		51	66	82
	Protect/mask unpainted surfaces				.031	.30	1.40		1.70	2	2.58
	Spray full primer coat				.444	9	20		29	39	46.50
	Spray finish coat				.500	9	23		32	42	51.50
	Total				3.434	18.30	152.90		171.20	**223**	**274.08**
4220	**Spray refinish steel siding - 2nd floor**	20	2 PORD	C.S.F.							
	Set up and secure scaffold				1.500		79.50		79.50	103	128
	Prepare surface				1.159		51		51	66	82
	Protect/mask unpainted surfaces				.031	.30	1.40		1.70	2	2.58
	Spray full primer coat				.444	9	20		29	39	46.50
	Spray finish coat				.500	9	23		32	42	51.50
	Remove scaffold				1.500		79.50		79.50	103	128
	Total				5.134	18.30	254.40		272.70	**355**	**438.58**
4320	**Spray refinish steel siding - 3rd floor**	20	2 PORD	C.S.F.							
	Set up and secure scaffold				2.250		119.25		119.25	155	192
	Prepare surface				1.159		51		51	66	82
	Protect/mask unpainted surfaces				.031	.30	1.40		1.70	2	2.58
	Spray full primer coat				.444	9	20		29	39	46.50
	Spray finish coat				.500	9	23		32	42	51.50
	Remove scaffold				2.250		119.25		119.25	155	192
	Total				6.634	18.30	333.90		352.20	**459**	**566.58**

B2013 143 Masonite Panel, Sealed

	System Description	Freq. (Years)	Crew	Unit	Labor Hours	2020 Bare Costs				Total In-House	Total w/O&P
						Material	Labor	Equipment	Total		
1010	**Replace hardboard panels - 1st floor**	12	1 CARP	C.S.F.							
	Set up, secure and take down ladder				.520		27.60		27.60	35.50	44
	Remove damaged panel				1.434		60		60	78	97
	Cut new hardboard panel to fit				2.773	271	147		418	490	575
	Total				4.727	271	234.60		505.60	**603.50**	**716**
2010	**Replace hardboard panels - 2nd floor**	12	1 CARP	C.S.F.							
	Set up and secure scaffold				1.500		79.50		79.50	103	128
	Remove damaged panel				1.434		60		60	78	97
	Cut new hardboard panel to fit				2.773	271	147		418	490	575
	Remove scaffold				1.500		79.50		79.50	103	128
	Total				7.207	271	366		637	**774**	**928**
3010	**Replace hardboard panels - 3rd floor**	12	1 CARP	C.S.F.							
	Set up and secure scaffold				2.250		119.25		119.25	155	192
	Remove damaged panel				1.434		60		60	78	97
	Cut new hardboard panel to fit				2.773	271	147		418	490	575
	Remove scaffold				2.250		119.25		119.25	155	192
	Total				8.707	271	445.50		716.50	**878**	**1,056**

B2013 145 Wood, Clapboard Finished, 1 Coat

	System Description	Freq. (Years)	Crew	Unit	Labor Hours	2020 Bare Costs				Total In-House	Total w/O&P
						Material	Labor	Equipment	Total		
1010	**Replace & finish wood clapboards - 1st floor**	25	1 CARP	C.S.F.							
	Set up, secure and take down ladder				.520		27.60		27.60	35.50	44
	Remove damage				2.600		138		138	179	221
	Replace wood clapboards				2.711	343.08	143.85		486.93	565	660
	Refinish surface, brushwork, primer + 1 coat				2.000	21	89		110	137	167
	Total				7.831	364.08	398.45		762.53	**916.50**	**1,092**

B2013 145 — Wood, Clapboard Finished, 1 Coat

	System Description	Freq. (Years)	Crew	Unit	Labor Hours	2020 Bare Costs				Total In-House	Total w/O&P
						Material	Labor	Equipment	Total		
2010	**Replace & finish wood clapboards - 2nd floor**	25	1 CARP	C.S.F.							
	Set up and secure scaffold				1.500		79.50		79.50	103	128
	Remove damage				2.600		138		138	179	221
	Replace wood clapboards				3.524	446	187		633	735	860
	Refinish surface, brushwork, primer + 1 coat				2.000	21	89		110	137	167
	Remove scaffold				1.500		79.50		79.50	103	128
	Total				11.124	467	573		1,040	**1,257**	**1,504**
3010	**Replace & finish wood clapboards - 3rd floor**	25	1 CARP	C.S.F.							
	Set up and secure scaffold				2.250		119.25		119.25	155	192
	Remove damage				2.600		138		138	179	221
	Replace wood clapboards				3.489	441.58	185.15		626.73	725	850
	Refinish surface, brushwork, primer + 1 coat				2.000	21	89		110	137	167
	Remove scaffold				2.250		119.25		119.25	155	192
	Total				12.589	462.58	650.65		1,113.23	**1,351**	**1,622**

B2013 147 — Wood Shingles, Unfinished

	System Description	Freq. (Years)	Crew	Unit	Labor Hours	2020 Bare Costs				Total In-House	Total w/O&P
						Material	Labor	Equipment	Total		
1010	**Repair wood shingles - 1st floor**	12	1 CARP	S.F.							
	Set up, secure and take down ladder				.052		2.76		2.76	3.57	4.42
	Remove damaged shingles				.039		1.64		1.64	2.12	2.63
	Replace shingles				.056	2.52	2.99		5.51	6.65	7.95
	Total				.147	2.52	7.39		9.91	**12.34**	**15**
1020	**Replace wood shingles - 1st floor**	40	2 CARP	C.S.F.							
	Set up and secure scaffold				.750		39.75		39.75	51.50	64
	Remove damaged shingles				2.600		109		109	141	175
	Replace shingles				4.324	194	230		424	510	615
	Remove scaffold				.750		39.75		39.75	51.50	64
	Total				8.424	194	418.50		612.50	**754**	**918**

B2013 147 Wood Shingles, Unfinished

	System Description	Freq. (Years)	Crew	Unit	Labor Hours	2020 Bare Costs				Total In-House	Total w/O&P
						Material	Labor	Equipment	Total		
2010	**Replace wood shingles - 2nd floor**	40	1 CARP	C.S.F.							
	Set up and secure scaffold				1.000		53		53	68.50	85
	Remove damaged shingles				2.600		109		109	141	175
	Replace shingles				4.324	194	230		424	510	615
	Remove scaffold				1.000		53		53	68.50	85
	Total				8.924	194	445		639	**788**	**960**
3010	**Replace wood shingles - 3rd floor**	40	1 CARP	C.S.F.							
	Set up and secure scaffold				1.500		79.50		79.50	103	128
	Remove damaged shingles				2.600		109		109	141	175
	Replace shingles				4.324	194	230		424	510	615
	Remove scaffold				1.500		79.50		79.50	103	128
	Total				9.924	194	498		692	**857**	**1,046**

B2013 148 Wood Shingles, Finished

	System Description	Freq. (Years)	Crew	Unit	Labor Hours	2020 Bare Costs				Total In-House	Total w/O&P
						Material	Labor	Equipment	Total		
1020	**Replace & finish wood shingles - 1st floor**	45	1 CARP	C.S.F.							
	Set up, secure and take down ladder				1.500		79.50		79.50	103	128
	Remove damage				2.600		109		109	141	175
	Replace shingles				4.324	194	230		424	510	615
	Refinish surface, brushwork, primer + 1 coat				2.000	21	89		110	137	167
	Total				10.424	215	507.50		722.50	**891**	**1,085**
1030	**Refinish wood shingles - 1st floor**	5	1 PORD	C.S.F.							
	Set up and secure scaffold				.750		39.75		39.75	51.50	64
	Prepare surface				1.667		74		74	95	118
	Refinish surface, brushwork, primer + 1 coat				2.000	21	89		110	137	167
	Remove scaffold				.750		39.75		39.75	51.50	64
	Total				5.167	21	242.50		263.50	**335**	**413**

B2013 148 Wood Shingles, Finished

	System Description	Freq. (Years)	Crew	Unit	Labor Hours	2020 Bare Costs				Total In-House	Total w/O&P
						Material	Labor	Equipment	Total		
2020	**Refinish wood shingles - 2nd floor**	5	1 PORD	C.S.F.							
	Set up and secure scaffold				1.500		79.50		79.50	103	128
	Prepare surface				1.667		74		74	95	118
	Refinish surface, brushwork, primer + 1 coat				2.000	21	89		110	137	167
	Remove scaffold				1.500		79.50		79.50	103	128
	Total				6.667	21	322		343	**438**	**541**
3020	**Refinish wood shingles - 3rd floor**	5	1 PORD	C.S.F.							
	Set up and secure scaffold				2.250		119.25		119.25	155	192
	Prepare surface				1.667		74		74	95	118
	Refinish surface, brushwork, primer + 1 coat				2.000	21	89		110	137	167
	Remove scaffold				2.250		119.25		119.25	155	192
	Total				8.167	21	401.50		422.50	**542**	**669**
4120	**Spray refinish wood siding - 1st floor**	5	1 PORD	C.S.F.							
	Set up, secure and take down ladder				1.300		57.50		57.50	74	91.50
	Prepare surface				1.159		51		51	66	82
	Protect/mask unpainted surfaces				.031	.30	1.40		1.70	2	2.58
	Spray full primer coat				.410	12	18		30	36	44
	Spray finish coat				.410	17	18		35	42	50.50
	Total				3.310	29.30	145.90		175.20	**220**	**270.58**
4220	**Spray refinish wood siding - 2nd floor**	5	2 PORD	C.S.F.							
	Set up and secure scaffold				1.500		79.50		79.50	103	128
	Prepare surface				1.159		51		51	66	82
	Protect/mask unpainted surfaces				.031	.30	1.40		1.70	2	2.53
	Spray full primer coat				.410	12	18		30	36	44
	Spray finish coat				.410	17	18		35	42	50.50
	Remove scaffold				1.500		79.50		79.50	103	128
	Total				5.010	29.30	247.40		276.70	**352**	**435.03**

B2013 148 Wood Shingles, Finished

	System Description	Freq. (Years)	Crew	Unit	Labor Hours	2020 Bare Costs				Total In-House	Total w/O&P
						Material	Labor	Equipment	Total		
4320	**Spray refinish wood siding - 3rd floor**	5	2 PORD	C.S.F.							
	Set up and secure scaffold				2.250		119.25		119.25	155	192
	Prepare surface				1.159		51		51	66	82
	Protect/mask unpainted surfaces				.031	.30	1.40		1.70	2	2.58
	Spray full primer coat				.410	12	18		30	36	44
	Spray finish coat				.410	17	18		35	42	50.50
	Remove scaffold				2.250		119.25		119.25	155	192
	Total				6.510	29.30	326.90		356.20	**456**	**563.08**

B2013 150 Fiberglass Panel, Rigid

	System Description	Freq. (Years)	Crew	Unit	Labor Hours	2020 Bare Costs				Total In-House	Total w/O&P
						Material	Labor	Equipment	Total		
1010	**Replace fiberglass panels - 1st floor**	9	1 CARP	C.S.F.							
	Set up, secure and take down ladder				.520		27.60		27.60	35.50	44
	Remove damaged fiberglass panel				1.434		60		60	78	97
	Install new fiberglass panel				4.727	378	247		625	730	865
	Total				6.681	378	334.60		712.60	**843.50**	**1,006**
2010	**Replace fiberglass panels - 2nd floor**	9	1 CARP	C.S.F.							
	Set up and secure scaffold				1.500		79.50		79.50	103	128
	Remove damaged fiberglass panel				1.434		60		60	78	97
	Install new fiberglass panel				4.727	378	247		625	730	865
	Remove scaffold				1.500		79.50		79.50	103	128
	Total				9.161	378	466		844	**1,014**	**1,218**
3010	**Replace fiberglass panels - 3rd floor**	9	1 CARP	C.S.F.							
	Set up and secure scaffold				2.250		119.25		119.25	155	192
	Remove damaged fiberglass panel				1.434		60		60	78	97
	Install new fiberglass panel				4.727	378	247		625	730	865
	Remove scaffold				2.250		119.25		119.25	155	192
	Total				10.661	378	545.50		923.50	**1,118**	**1,346**

B2013 152 Synthetic Veneer Plaster

	System Description	Freq. (Years)	Crew	Unit	Labor Hours	2020 Bare Costs				Total In-House	Total w/O&P
						Material	Labor	Equipment	Total		
1030	**Refinish veneer plaster - 1st floor**	10	1 BRIC	C.S.F.							
	Set up and secure scaffold				.750		39.75		39.75	51.50	64
	Pressure wash surface				.260		11	2	13	17	21
	Refinish surface				1.270	158	56		214	246	288
	Remove scaffold				.750		39.75		39.75	51.50	64
	Total				3.030	158	146.50	2	306.50	**366**	**437**
2020	**Refinish veneer plaster - 2nd floor**	10	1 BRIC	C.S.F.							
	Set up and secure scaffold				1.500		79.50		79.50	103	128
	Pressure wash surface				.260		11	2	13	17	21
	Refinish surface				1.270	158	56		214	246	288
	Remove scaffold				1.500		79.50		79.50	103	128
	Total				4.530	158	226	2	386	**469**	**565**
3020	**Refinish veneer plaster - 3rd floor**	10	1 BRIC	C.S.F.							
	Set up and secure scaffold				2.250		119.25		119.25	155	192
	Pressure wash surface				.260		11	2	13	17	21
	Refinish surface				1.270	158	56		214	246	288
	Remove scaffold				2.250		119.25		119.25	155	192
	Total				6.030	158	305.50	2	465.50	**573**	**693**

B2013 156 Exterior Soffit Gypsum Board, Painted

	System Description	Freq. (Years)	Crew	Unit	Labor Hours	2020 Bare Costs				Total In-House	Total w/O&P
						Material	Labor	Equipment	Total		
1030	**Refinish exterior painted gypsum board - 1st floor soffit**	5	1 PORD	C.S.F.							
	Set up and secure scaffold				.750		39.75		39.75	51.50	64
	Prepare surface				1.159		51		51	66	82
	Paint surface, roller, primer + 1 coat				1.231	11	55		66	82	101
	Remove scaffold				.750		39.75		39.75	51.50	64
	Total				3.890	11	185.50		196.50	**251**	**311**

B2013 156 Exterior Soffit Gypsum Board, Painted

	System Description	Freq. (Years)	Crew	Unit	Labor Hours	2020 Bare Costs				Total In-House	Total w/O&P
						Material	Labor	Equipment	Total		
1040	**Replace exterior painted gypsum board - 1st floor soffit**	25	2 CARP	C.S.F.							
	Set up and secure scaffold				.750		39.75		39.75	51.50	64
	Remove old exterior gypsum board				2.737		115		115	149	185
	Install new exterior gypsum board				3.467	71	184		255	315	385
	Paint surface, roller, primer + 1 coat				1.231	11	55		66	82	101
	Remove scaffold				.750		39.75		39.75	51.50	64
	Total				8.935	82	433.50		515.50	**649**	**799**
2010	**Replace exterior painted gypsum board - 2nd floor soffit**	25	1 CARP	C.S.F.							
	Set up and secure scaffold				1.500		79.50		79.50	103	128
	Remove old exterior gypsum board				2.737		115		115	149	185
	Install new exterior gypsum board				3.467	71	184		255	315	385
	Paint surface, roller, primer + 1 coat				2.462	22	110		132	164	202
	Remove scaffold				1.500		79.50		79.50	103	128
	Total				11.666	93	568		661	**834**	**1,028**
2020	**Refinish exterior painted gypsum board - 2nd floor soffit**	5	1 PORD	C.S.F.							
	Set up and secure scaffold				1.500		79.50		79.50	103	128
	Prepare surface				1.159		51		51	66	82
	Paint surface, roller, primer + 1 coat				1.231	11	55		66	82	101
	Remove scaffold				1.500		79.50		79.50	103	128
	Total				5.390	11	265		276	**354**	**439**
3010	**Replace exterior painted gypsum board - 3rd floor soffit**	25	1 CARP	C.S.F.							
	Set up and secure scaffold				2.250		119.25		119.25	155	192
	Remove old exterior gypsum board				2.737		115		115	149	185
	Install new exterior gypsum board				3.467	71	184		255	315	385
	Paint surface, roller, primer + 1 coat				3.693	33	165		198	246	300
	Remove scaffold				2.250		119.25		119.25	155	192
	Total				14.397	104	702.50		806.50	**1,020**	**1,254**
3020	**Refinish exterior painted gypsum board - 3rd floor soffit**	5	1 PORD	C.S.F.							
	Set up and secure scaffold				2.250		119.25		119.25	155	192
	Prepare surface				1.159		51		51	66	82
	Paint surface, roller, primer + 1 coat				1.231	11	55		66	82	101
	Remove scaffold				2.250		119.25		119.25	155	192
	Total				6.890	11	344.50		355.50	**458**	**567**

B2013 157 Overhang, Exterior Entry

	System Description	Freq. (Years)	Crew	Unit	Labor Hours	2020 Bare Costs Material	Labor	Equipment	Total	Total In-House	Total w/O&P
1030	**Refinish wood overhang**	5	1 PORD	S.F.							
	Prepare surface				.016		.71		.71	.91	1.13
	Wash / dry surface				.001		.05		.05	.07	.08
	Refinish surface, brushwork, primer + 1 coat				.020	.21	.89		1.10	1.37	1.67
	Total				.037	.21	1.65		1.86	**2.35**	**2.88**

B2013 158 Recaulking

	System Description	Freq. (Years)	Crew	Unit	Labor Hours	2020 Bare Costs Material	Labor	Equipment	Total	Total In-House	Total w/O&P
1010	**Recaulk expansion & control joints**	20	1 BRIC	L.F.							
	Set up, secure and take down ladder				.130		6.75		6.75	8.85	10.95
	Cut out and recaulk, silicone				.067	.69	3.47		4.16	5.30	6.45
	Total				.197	.69	10.22		10.91	**14.15**	**17.40**
2010	**Recaulk door**	20	1 BRIC	L.F.							
	Set up, secure and take down ladder				.130		6.75		6.75	8.85	10.95
	Cut out and recaulk, silicone				.067	.69	3.47		4.16	5.30	6.45
	Total				.197	.69	10.22		10.91	**14.15**	**17.40**
3010	**Recaulk window, 4′ x 6′ - 1st floor**	20	2 BRIC	Ea.							
	Set up and secure scaffold				1.333		69.50		69.50	90.50	112
	Cut out and recaulk, silicone				1.600	16.56	83.28		99.84	127	155
	Remove scaffold				1.333		69.50		69.50	90.50	112
	Total				4.267	16.56	222.28		238.84	**308**	**379**
3020	**Recaulk window, 4′ x 6′ - 2nd floor**	20	2 BRIC	Ea.							
	Set up and secure scaffold				1.500		79.50		79.50	103	128
	Cut out and recaulk, silicone				1.600	16.56	83.28		99.84	127	155
	Remove scaffold				1.500		79.50		79.50	103	128
	Total				4.600	16.56	242.28		258.84	**333**	**411**
3030	**Recaulk window, 4′ x 6′ - 3rd floor**	20	2 BRIC	Ea.							
	Set up and secure scaffold				1.500		79.50		79.50	103	128
	Cut out and recaulk, silicone				1.600	16.56	83.28		99.84	127	155
	Remove scaffold				1.500		79.50		79.50	103	128
	Total				4.600	16.56	242.28		258.84	**333**	**411**

B2013 310 Wood Louvers and Shutters

	System Description	Freq. (Years)	Crew	Unit	Labor Hours	2020 Bare Costs				Total In-House	Total w/O&P
						Material	Labor	Equipment	Total		
1010	**Repair wood louver in frame 1st floor**	6	1 CARP	Ea.							
	Set up, secure and take down ladder				.001		.06		.06	.08	.10
	Remove louver				.473		19.90		19.90	25.50	32
	Remove damaged stile				.078		4.17		4.17	5.40	6.70
	Remove damaged slat				.039		2.06		2.06	2.66	3.30
	Install new slat				.077	.80	4.10		4.90	6.20	7.60
	Install new stile				.157	2.73	8.34		11.07	13.75	16.75
	Reinstall louver				.075		3.97		3.97	5.15	6.35
	Total				.900	3.53	42.60		46.13	**58.74**	**72.80**
1030	**Refinish wood louver - 1st floor**	5	1 PORD	Ea.							
	Set up and secure scaffold				.444		19.75		19.75	25	31.50
	Prepare louver for painting				.434		19.20		19.20	24.50	30.50
	Paint louvered surface (to 16 S.F.), brushwork, primer + 1 coat				.800	2.43	35.50		37.93	48	59.50
	Remove scaffold				.444		19.75		19.75	25	31.50
	Total				2.123	2.43	94.20		96.63	**122.50**	**153**
1040	**Replace wood louver - 1st floor**	45	1 CARP	Ea.							
	Set up and secure scaffold				.444		23.50		23.50	30.50	38
	Remove louver				.473		19.90		19.90	25.50	32
	Install new wood louver				.520	152.50	27.75		180.25	203	235
	Remove scaffold				.444		23.50		23.50	30.50	38
	Total				1.882	152.50	94.65		247.15	**289.50**	**343**
2010	**Repair wood louver - 2nd floor**	6	1 CARP	Ea.							
	Set up and secure scaffold				.889		47		47	61	75.50
	Remove louver				.473		19.90		19.90	25.50	32
	Remove damaged stile				.078		4.17		4.17	5.40	6.70
	Remove damaged slat				.039		2.06		2.06	2.66	3.30
	Install new slat				.077	.80	4.10		4.90	6.20	7.60
	Install new stile				.157	2.73	8.34		11.07	13.75	16.75
	Reinstall louver				.075		3.97		3.97	5.15	6.35
	Remove scaffold				.889		47		47	61	75.50
	Total				2.677	3.53	136.54		140.07	**180.66**	**223.70**

B2013 310 Wood Louvers and Shutters

	System Description	Freq. (Years)	Crew	Unit	Labor Hours	2020 Bare Costs				Total In-House	Total w/O&P
						Material	Labor	Equipment	Total		
2030	**Refinish wood louver - 2nd floor**	4	1 PORD	Ea.							
	Set up and secure scaffold				.889		39.50		39.50	50.50	62.50
	Prepare louver for painting				.434		19.20		19.20	24.50	30.50
	Paint louvered surface (to 16 S.F.), brushwork, primer + 1 coat				.800	2.43	35.50		37.93	48	59.50
	Remove scaffold				.889		39.50		39.50	50.50	62.50
	Total				3.012	2.43	133.70		136.13	**173.50**	**215**
2040	**Replace wood louver - 2nd floor**	45	1 CARP	Ea.							
	Set up and secure scaffold				.889		47		47	61	75.50
	Remove louver				.473		19.90		19.90	25.50	32
	Install new wood louver				.520	152.50	27.75		180.25	203	235
	Remove scaffold				.889		47		47	61	75.50
	Total				2.771	152.50	141.65		294.15	**350.50**	**418**
3010	**Repair wood louver - 3rd floor**	6	1 CARP	Ea.							
	Set up and secure scaffold				1.333		71		71	91.50	114
	Remove louver				.473		19.90		19.90	25.50	32
	Remove damaged stile				.078		4.17		4.17	5.40	6.70
	Remove damaged slat				.039		2.06		2.06	2.66	3.30
	Install new slat				.077	.80	4.10		4.90	6.20	7.60
	Install new stile				.157	2.73	8.34		11.07	13.75	16.75
	Reinstall louver				.075		3.97		3.97	5.15	6.35
	Remove scaffold				1.333		71		71	91.50	114
	Total				3.565	3.53	184.54		188.07	**241.66**	**300.70**
3030	**Refinish wood louver - 3rd floor**	4	1 PORD	Ea.							
	Set up and secure scaffold				1.333		59		59	75.50	94
	Prepare louver for painting				.434		19.20		19.20	24.50	30.50
	Paint louvered surface (to 16 S.F.), brushwork, primer + 1 coat				.800	2.43	35.50		37.93	48	59.50
	Remove scaffold				1.333		59		59	75.50	94
	Total				3.901	2.43	172.70		175.13	**223.50**	**278**
3040	**Replace wood louver - 3rd floor**	45	1 CARP	Ea.							
	Set up and secure scaffold				1.333		71		71	91.50	114
	Remove louver				.473		19.90		19.90	25.50	32
	Install new wood louver				.520	152.50	27.75		180.25	203	235
	Remove scaffold				1.333		71		71	91.50	114
	Total				3.660	152.50	189.65		342.15	**411.50**	**495**

B2013 320 Aluminum Louver

	System Description	Freq. (Years)	Crew	Unit	Labor Hours	2020 Bare Costs				Total In-House	Total w/O&P
						Material	Labor	Equipment	Total		
1030	**Refinish aluminum louver - 1st floor**	5	1 PORD	Ea.							
	Set up and secure scaffold				.444		19.75		19.75	25	31.50
	Prepare louver for painting				.434		19.20		19.20	24.50	30.50
	Paint louvered surface (to 16 S.F.), brushwork, primer + 1 coat				.800	2.43	35.50		37.93	48	59.50
	Remove scaffold				.444		19.75		19.75	25	31.50
	Total				2.123	2.43	94.20		96.63	**122.50**	**153**
1040	**Replace aluminum louver - 1st floor**	60	1 CARP	Ea.							
	Set up and secure scaffold				.444		23.50		23.50	30.50	38
	Remove louver				.473		19.90		19.90	25.50	32
	Install new louver				1.783	264	111		375	430	505
	Remove scaffold				.444		23.50		23.50	30.50	38
	Total				3.145	264	177.90		441.90	**516.50**	**613**
2020	**Refinish aluminum louver - 2nd floor**	12	1 PORD	Ea.							
	Set up and secure scaffold				.889		47		47	61	75.50
	Prepare louver for painting				.434		19.20		19.20	24.50	30.50
	Paint louvered surface (to 16 S.F.), brushwork, primer + 1 coat				.800	2.43	35.50		37.93	48	59.50
	Remove scaffold				.889		47		47	61	75.50
	Total				3.012	2.43	148.70		151.13	**194.50**	**241**
2040	**Replace aluminum louver - 2nd floor**	60	1 CARP	Ea.							
	Set up and secure scaffold				.889		47		47	61	75.50
	Remove louver				.473		19.90		19.90	25.50	32
	Install new louver				1.783	264	111		375	430	505
	Remove scaffold				.889		47		47	61	75.50
	Total				4.034	264	224.90		488.90	**577.50**	**688**
3030	**Refinish aluminum louver - 3rd floor**	5	1 PORD	Ea.							
	Set up and secure scaffold				1.333		59		59	75.50	94
	Prepare louver for painting				.434		19.20		19.20	24.50	30.50
	Paint louvered surface (to 16 S.F.), brushwork, primer + 1 coat				.800	2.43	35.50		37.93	48	59.50
	Remove scaffold				1.333		59		59	75.50	94
	Total				3.901	2.43	172.70		175.13	**223.50**	**278**

B2013 320 Aluminum Louver

	System Description	Freq. (Years)	Crew	Unit	Labor Hours	2020 Bare Costs Material	Labor	Equipment	Total	Total In-House	Total w/O&P
3040	**Replace aluminum louver - 3rd floor**	60	1 CARP	Ea.							
	Set up and secure scaffold				1.333		71		71	91.50	114
	Remove louver				.473		19.90		19.90	25.50	32
	Install new louver				1.783	264	111		375	430	505
	Remove scaffold				1.333		71		71	91.50	114
	Total				4.922	264	272.90		536.90	**638.50**	**765**

B2013 330 Steel Louver

	System Description	Freq. (Years)	Crew	Unit	Labor Hours	2020 Bare Costs Material	Labor	Equipment	Total	Total In-House	Total w/O&P
1030	**Refinish steel louver - 1st floor**	5	1 PORD	Ea.							
	Set up and secure scaffold				.444		19.75		19.75	25	31.50
	Prepare louver for painting				.434		19.20		19.20	24.50	30.50
	Paint louvered surface (to 16 S.F.), brushwork, primer + 1 coat				.800	2.43	35.50		37.93	48	59.50
	Remove scaffold				.444		19.75		19.75	25	31.50
	Total				2.123	2.43	94.20		96.63	**122.50**	**153**
1040	**Replace steel louver - 1st floor**	40	1 CARP	Ea.							
	Set up and secure scaffold				.444		23.50		23.50	30.50	38
	Remove louver				.473		19.90		19.90	25.50	32
	Install new louver				1.783	264	111		375	430	505
	Remove scaffold				.444		23.50		23.50	30.50	38
	Total				3.145	264	177.90		441.90	**516.50**	**613**
2030	**Refinish steel louver - 2nd floor**	5	1 PORD	Ea.							
	Set up and secure scaffold				.889		39.50		39.50	50.50	62.50
	Prepare louver for painting				.434		19.20		19.20	24.50	30.50
	Paint louvered surface (to 16 S.F.), brushwork, primer + 1 coat				.800	2.43	35.50		37.93	48	59.50
	Remove scaffold				.889		39.50		39.50	50.50	62.50
	Total				3.012	2.43	133.70		136.13	**173.50**	**215**

B2013 330 Steel Louver

	System Description	Freq. (Years)	Crew	Unit	Labor Hours	2020 Bare Costs				Total In-House	Total w/O&P
						Material	Labor	Equipment	Total		
2040	**Replace steel louver - 2nd floor**	40	1 CARP	Ea.							
	Set up and secure scaffold				.889		47		47	61	75.50
	Remove louver				.473		19.90		19.90	25.50	32
	Install new louver				1.783	264	111		375	430	505
	Remove scaffold				.889		47		47	61	75.50
	Total				4.034	264	224.90		488.90	**577.50**	**688**
3030	**Refinish steel louver - 3rd floor**	5	1 PORD	Ea.							
	Set up and secure scaffold				1.333		59		59	75.50	94
	Prepare louver for painting				.434		19.20		19.20	24.50	30.50
	Paint louvered surface (to 16 S.F.), brushwork, primer + 1 coat				.800	2.43	35.50		37.93	48	59.50
	Remove scaffold				1.333		59		59	75.50	94
	Total				3.901	2.43	172.70		175.13	**223.50**	**278**
3040	**Replace steel louver - 3rd floor**	40	1 CARP	Ea.							
	Set up and secure scaffold				1.333		71		71	91.50	114
	Remove louver				.473		19.90		19.90	25.50	32
	Install new louver				1.783	264	111		375	430	505
	Remove scaffold				1.333		71		71	91.50	114
	Total				4.922	264	272.90		536.90	**638.50**	**765**

B2023 102 Steel Frame, Operating

	System Description	Freq. (Years)	Crew	Unit	Labor Hours	2020 Bare Costs				Total In-House	Total w/O&P
						Material	Labor	Equipment	Total		
1010	**Replace glass - 1st floor (1% of glass)**	1	1 CARP	S.F.							
	Remove damaged glass				.052		2.19		2.19	2.83	3.50
	Install new glass				.043	6.65	2.21		8.86	10.15	11.85
	Total				.095	6.65	4.40		11.05	**12.98**	**15.35**
1020	**Repair 3'-9" x 5'-5" steel frame window - 1st floor**	20	1 CARP	Ea.							
	Set up and secure scaffold				.444		23.50		23.50	30.50	38
	Remove window gaskets				.250		13.30		13.30	17.20	21.50
	Install new gaskets				.564	117.38	30.08		147.46	168	195
	Misc. hardware replacement				.333	78	17.70		95.70	109	126
	Remove scaffold				.444		23.50		23.50	30.50	38
	Total				2.037	195.38	108.08		303.46	**355.20**	**418.50**
1030	**Refinish 3'-9" x 5'-5" steel frame window - 1st floor**	5	1 PORD	Ea.							
	Set up, secure and take down ladder				.267		11.85		11.85	15.15	18.90
	Prepare window for painting				.473		21		21	27	33.50
	Paint window, brushwork, primer + 1 coat				1.000	2.20	44.50		46.70	59	73.50
	Total				1.739	2.20	77.35		79.55	**101.15**	**125.90**
1040	**Replace 3'-9" x 5'-5" steel frame window - 1st floor**	45	1 CARP	Ea.							
	Set up and secure scaffold				.444		23.50		23.50	30.50	38
	Remove window				.650		27.50		27.50	35.50	44
	Install new window with frame & glazing (to 21 S.F.)				2.078	1,275	120		1,395	1,550	1,800
	Install sealant				.764	8.62	39.80		48.42	61.50	75
	Remove scaffold				.444		23.50		23.50	30.50	38
	Total				4.381	1,283.62	234.30		1,517.92	**1,708**	**1,995**
1050	**Finish new 3'-9" x 5'-5" steel frame window - 1st floor**	45	1 PORD	Ea.							
	Set up, secure and take down ladder				.267		11.85		11.85	15.15	18.90
	Prepare window for painting				.236		10.50		10.50	13.40	16.75
	Paint window, brushwork, primer + 1 coat				1.000	2.20	44.50		46.70	59	73.50
	Total				1.503	2.20	66.85		69.05	**87.55**	**109.15**

B2023 102 Steel Frame, Operating

	System Description	Freq. (Years)	Crew	Unit	Labor Hours	2020 Bare Costs				Total In-House	Total w/O&P
						Material	Labor	Equipment	Total		
2010	**Replace glass - 2nd floor (1% of glass)**	1	1 CARP	S.F.							
	Set up and secure scaffold				.889		47		47	61	75.50
	Remove damaged glass				.052		2.19		2.19	2.83	3.50
	Install new glass				.043	6.65	2.21		8.86	10.15	11.85
	Remove scaffold				.889		47		47	61	75.50
	Total				1.873	6.65	98.40		105.05	**134.98**	**166.35**
2020	**Repair 3′-9″ x 5′-5″ steel frame window - 2nd floor**	20	1 CARP	Ea.							
	Set up and secure scaffold				.889		47		47	61	75.50
	Remove window gaskets				.250		13.30		13.30	17.20	21.50
	Install new gaskets				.564	117.38	30.08		147.46	168	195
	Misc. hardware replacement				.333	78	17.70		95.70	109	126
	Remove scaffold				.889		47		47	61	75.50
	Total				2.925	195.38	155.08		350.46	**416.20**	**493.50**
2030	**Refinish 3′-9″ x 5′-5″ steel frame window - 2nd floor**	5	1 PORD	Ea.							
	Set up and secure scaffold				.889		39.50		39.50	50.50	62.50
	Prepare window for painting				.473		21		21	27	33.50
	Paint window, brushwork, primer + 1 coat				1.000	2.20	44.50		46.70	59	73.50
	Remove scaffold				.889		39.50		39.50	50.50	62.50
	Total				3.251	2.20	144.50		146.70	**187**	**232**
2040	**Replace 3′-9″ x 5′-5″ steel frame window - 2nd floor**	45	1 CARP	Ea.							
	Set up and secure scaffold				.889		47		47	61	75.50
	Remove window				.650		27.50		27.50	35.50	44
	Install new window with frame & glazing (to 21 S.F.)				2.078	1,275	120		1,395	1,550	1,800
	Install sealant				.764	8.62	39.80		48.42	61.50	75
	Remove scaffold				.889		47		47	61	75.50
	Total				5.270	1,283.62	281.30		1,564.92	**1,769**	**2,070**
2050	**Finish new 3′-9″ x 5′-5″ steel frame window - 2nd floor**	45	1 PORD	Ea.							
	Set up and secure scaffold				.889		39.50		39.50	50.50	62.50
	Prepare window for painting				.236		10.50		10.50	13.40	16.75
	Paint window, brushwork, primer + 1 coat				1.000	2.20	44.50		46.70	59	73.50
	Remove scaffold				.889		39.50		39.50	50.50	62.50
	Total				3.014	2.20	134		136.20	**173.40**	**215.25**

B2023 102 Steel Frame, Operating

	System Description	Freq. (Years)	Crew	Unit	Labor Hours	2020 Bare Costs Material	Labor	Equipment	Total	Total In-House	Total w/O&P
3010	**Replace glass - 3rd floor (1% of glass)**	1	1 CARP	S.F.							
	Set up and secure scaffold				.123		6.55		6.55	8.45	10.50
	Remove damaged glass				.052		2.19		2.19	2.83	3.50
	Install new glass				.043	6.65	2.21		8.86	10.15	11.85
	Remove scaffold				.123		6.55		6.55	8.45	10.50
	Total				.341	6.65	17.50		24.15	**29.88**	**36.35**
3020	**Repair 3'-9" x 5'-5" steel frame window - 3rd floor**	20	1 CARP	Ea.							
	Set up and secure scaffold				1.333		71		71	91.50	114
	Remove window gaskets				.250		13.30		13.30	17.20	21.50
	Install new gaskets				.564	117.38	30.08		147.46	168	195
	Misc. hardware replacement				.333	78	17.70		95.70	109	126
	Remove scaffold				1.333		71		71	91.50	114
	Total				3.814	195.38	203.08		398.46	**477.20**	**570.50**
3030	**Refinish 3'-9" x 5'-5" steel frame window - 3rd floor**	5	1 PORD	Ea.							
	Set up and secure scaffold				1.333		59		59	75.50	94
	Prepare window for painting				.473		21		21	27	33.50
	Paint window, brushwork, primer + 1 coat				1.000	2.20	44.50		46.70	59	73.50
	Remove scaffold				1.333		59		59	75.50	94
	Total				4.139	2.20	183.50		185.70	**237**	**295**
3040	**Replace 3'-9" x 5'-5" steel frame window - 3rd floor**	45	1 CARP	Ea.							
	Set up and secure scaffold				1.333		71		71	91.50	114
	Remove window				.650		27.50		27.50	35.50	44
	Install new window with frame & glazing (to 21 S.F.)				2.078	1,275	120		1,395	1,550	1,800
	Install sealant				.764	8.62	39.80		48.42	61.50	75
	Remove scaffold				1.333		71		71	91.50	114
	Total				6.159	1,283.62	329.30		1,612.92	**1,830**	**2,147**
3050	**Finish new 3'-9" x 5'-5" steel frame window - 3rd floor**	45	1 PORD	Ea.							
	Set up and secure scaffold				1.333		59		59	75.50	94
	Prepare window for painting				.236		10.50		10.50	13.40	16.75
	Paint window, brushwork, primer + 1 coat				1.000	2.20	44.50		46.70	59	73.50
	Remove scaffold				1.333		59		59	75.50	94
	Total				3.903	2.20	173		175.20	**223.40**	**278.25**

B2023 103 Steel Frame, Fixed

	System Description	Freq. (Years)	Crew	Unit	Labor Hours	2020 Bare Costs				Total In-House	Total w/O&P
						Material	Labor	Equipment	Total		
1010	**Replace glass - 1st floor (1% of glass)**	1	1 CARP	S.F.							
	Remove damaged glass				.052		2.19		2.19	2.83	3.50
	Install new glass				.043	6.65	2.21		8.86	10.15	11.85
	Total				.095	6.65	4.40		11.05	**12.98**	**15.35**
1020	**Repair 2′-0″ x 3′-0″ steel frame window - 1st flr.**	20	1 CARP	Ea.							
	Set up and secure scaffold				.444		23.50		23.50	30.50	38
	Remove window gaskets				.250		13.30		13.30	17.20	21.50
	Install new gaskets				.308	64	16.40		80.40	91.50	106
	Remove scaffold				.444		23.50		23.50	30.50	38
	Total				1.447	64	76.70		140.70	**169.70**	**203.50**
1030	**Refinish 2′-0″ x 3′-0″ st. frame window - 1st flr.**	5	1 PORD	Ea.							
	Set up, secure and take down ladder				.267		11.85		11.85	15.15	18.90
	Prepare window for painting				.473		21		21	27	33.50
	Paint window, brushwork, primer + 1 coat				1.000	2.20	44.50		46.70	59	73.50
	Total				1.739	2.20	77.35		79.55	**101.15**	**125.90**
1040	**Replace 2′-0″ x 3′-0″ steel frame window - 1st flr.**	45	1 CARP	Ea.							
	Set up and secure scaffold				.444		23.50		23.50	30.50	38
	Remove window				.650		27.50		27.50	35.50	44
	Install new window with frame & glazing (2′-0″ x 3′-0″)				2.078	219	120		339	400	470
	Install sealant				.764	8.62	39.80		48.42	61.50	75
	Remove scaffold				.444		23.50		23.50	30.50	38
	Total				4.381	227.62	234.30		461.92	**558**	**665**
2010	**Replace glass - 2nd floor (1% of glass)**	1	1 CARP	S.F.							
	Set up and secure scaffold				.044		2.35		2.35	3.05	3.78
	Remove damaged glass				.052		2.19		2.19	2.83	3.50
	Install new glass				.043	6.65	2.21		8.86	10.15	11.85
	Remove scaffold				.044		2.35		2.35	3.05	3.78
	Total				.184	6.65	9.10		15.75	**19.08**	**22.91**

B2023 103 Steel Frame, Fixed

	System Description	Freq. (Years)	Crew	Unit	Labor Hours	2020 Bare Costs Material	Labor	Equipment	Total	Total In-House	Total w/O&P
2020	**Repair 2'-0" x 3'-0" steel frame window - 2nd flr.**	20	1 CARP	Ea.							
	Set up and secure scaffold				.889		47		47	61	75.50
	Remove window gaskets				.250		13.30		13.30	17.20	21.50
	Install new gaskets				.308	64	16.40		80.40	91.50	106
	Remove scaffold				.889		47		47	61	75.50
	Total				2.335	64	123.70		187.70	**230.70**	**278.50**
2030	**Refinish 2'-0" x 3'-0" steel frame window - 2nd flr.**	5	1 PORD	Ea.							
	Set up and secure scaffold				.889		39.50		39.50	50.50	62.50
	Prepare window for painting				.473		21		21	27	33.50
	Paint window, brushwork, primer + 1 coat				1.000	2.20	44.50		46.70	59	73.50
	Remove scaffold				.889		39.50		39.50	50.50	62.50
	Total				3.251	2.20	144.50		146.70	**187**	**232**
2040	**Replace 2'-0" x 3'-0" steel frame window - 2nd flr.**	45	1 CARP	Ea.							
	Set up and secure scaffold				.889		47		47	61	75.50
	Remove window				.650		27.50		27.50	35.50	44
	Install new window with frame & glazing (2'-0" x 3'-0")				2.078	219	120		339	400	470
	Install sealant				.764	8.62	39.80		48.42	61.50	75
	Remove scaffold				.889		47		47	61	75.50
	Total				5.270	227.62	281.30		508.92	**619**	**740**
3010	**Replace glass - 3rd floor (1% of glass)**	1	1 CARP	S.F.							
	Set up and secure scaffold				.123		6.55		6.55	8.45	10.50
	Remove damaged glass				.052		2.19		2.19	2.83	3.50
	Install new glass				.043	6.65	2.21		8.86	10.15	11.85
	Remove scaffold				.123		6.55		6.55	8.45	10.50
	Total				.341	6.65	17.50		24.15	**29.88**	**36.35**
3020	**Repair 2'-0" x 3'-0" steel frame window - 3rd flr.**	20	1 CARP	Ea.							
	Set up and secure scaffold				1.333		71		71	91.50	114
	Remove window gaskets				.250		13.30		13.30	17.20	21.50
	Install new gaskets				.308	64	16.40		80.40	91.50	106
	Remove scaffold				1.333		71		71	91.50	114
	Total				3.224	64	171.70		235.70	**291.70**	**355.50**

B2023 103 Steel Frame, Fixed

	System Description	Freq. (Years)	Crew	Unit	Labor Hours	2020 Bare Costs				Total In-House	Total w/O&P
						Material	Labor	Equipment	Total		
3030	**Refinish 2′-0″ x 3′-0″ steel frame window - 3rd floor**	5	1 PORD	Ea.							
	Set up and secure scaffold				1.333		59		59	75.50	94
	Prepare window for painting				.473		21		21	27	33.50
	Paint window, brushwork, primer + 1 coat				1.000	2.20	44.50		46.70	59	73.50
	Remove scaffold				1.333		59		59	75.50	94
	Total				4.139	2.20	183.50		185.70	**237**	**295**
3040	**Replace 2′-0″ x 3′-0″ steel frame window - 3rd floor**	45	1 CARP	Ea.							
	Set up and secure scaffold				1.333		71		71	91.50	114
	Remove window				.650		27.50		27.50	35.50	44
	Install new window with frame & glazing (2′-0″ x 3′-0″)				2.078	219	120		339	400	470
	Install sealant				.764	8.62	39.80		48.42	61.50	75
	Remove scaffold				1.333		71		71	91.50	114
	Total				6.159	227.62	329.30		556.92	**680**	**817**

B2023 110 Aluminum Window, Operating

	System Description	Freq. (Years)	Crew	Unit	Labor Hours	2020 Bare Costs				Total In-House	Total w/O&P
						Material	Labor	Equipment	Total		
1010	**Replace glass - 1st flr. (1% of glass)**	1	1 CARP	S.F.							
	Remove damaged glass				.052		2.19		2.19	2.83	3.50
	Install new glass				.043	6.65	2.21		8.86	10.15	11.85
	Total				.095	6.65	4.40		11.05	**12.98**	**15.35**
1020	**Repair 3′ x 4′ aluminum window - 1st floor**	20	1 CARP	Ea.							
	Set up and secure scaffold				.444		23.50		23.50	30.50	38
	Remove window gaskets				.250		13.30		13.30	17.20	21.50
	Install new gaskets				.431	89.60	22.96		112.56	128	149
	Misc. hardware replacement				.333	78	17.70		95.70	109	126
	Remove scaffold				.444		23.50		23.50	30.50	38
	Total				1.903	167.60	100.96		268.56	**315.20**	**372.50**

B2023 110 Aluminum Window, Operating

	System Description	Freq. (Years)	Crew	Unit	Labor Hours	2020 Bare Costs				Total In-House	Total w/O&P
						Material	Labor	Equipment	Total		
1030	**Replace 3' x 4' aluminum window - 1st floor**	50	1 CARP	Ea.							
	Set up and secure scaffold				.444		23.50		23.50	30.50	38
	Remove window				.650		27.50		27.50	35.50	44
	Install new window with frame & glazing (to 12 S.F.)				2.078	510	120		630	720	835
	Install sealant				.555	6.27	28.93		35.20	44.50	54.50
	Remove scaffold				.444		23.50		23.50	30.50	38
	Total				4.173	516.27	223.43		739.70	**861**	**1,009.50**
2010	**Replace glass - 2nd flr. (1% of glass)**	1	1 CARP	S.F.							
	Set up and secure scaffold				.044		2.35		2.35	3.05	3.78
	Remove damaged glass				.052		2.19		2.19	2.83	3.50
	Install new glass				.043	6.65	2.21		8.86	10.15	11.85
	Remove scaffold				.044		2.35		2.35	3.05	3.78
	Total				.184	6.65	9.10		15.75	**19.08**	**22.91**
2020	**Repair 3' x 4' aluminum window - 2nd floor**	20	1 CARP	Ea.							
	Set up and secure scaffold				.889		47		47	61	75.50
	Remove window gaskets				.250		13.30		13.30	17.20	21.50
	Install new gaskets				.431	89.60	22.96		112.56	128	149
	Misc. hardware replacement				.333	78	17.70		95.70	109	126
	Remove scaffold				.889		47		47	61	75.50
	Total				2.792	167.60	147.96		315.56	**376.20**	**447.50**
2030	**Replace 3' x 4' aluminum window - 2nd floor**	50	1 CARP	Ea.							
	Set up and secure scaffold				.889		47		47	61	75.50
	Remove window				.650		27.50		27.50	35.50	44
	Install new window with frame & glazing (to 12 S.F.)				2.078	510	120		630	720	835
	Install sealant				.555	6.27	28.93		35.20	44.50	54.50
	Remove scaffold				.889		47		47	61	75.50
	Total				5.062	516.27	270.43		786.70	**922**	**1,084.50**
3010	**Replace glass - 3rd floor (1% of glass)**	1	1 CARP	S.F.							
	Set up and secure scaffold				.123		6.55		6.55	8.45	10.50
	Remove damaged glass				.052		2.19		2.19	2.83	3.50
	Install new glass				.043	6.65	2.21		8.86	10.15	11.85
	Remove scaffold				.123		6.55		6.55	8.45	10.50
	Total				.341	6.65	17.50		24.15	**29.88**	**36.35**

B2023 110 Aluminum Window, Operating

	System Description	Freq. (Years)	Crew	Unit	Labor Hours	2020 Bare Costs				Total In-House	Total w/O&P
						Material	Labor	Equipment	Total		
3020	**Repair 3′ x 4′ aluminum window - 3rd floor**	20	1 CARP	Ea.							
	Set up and secure scaffold				1.333		71		71	91.50	114
	Remove window gaskets				.250		13.30		13.30	17.20	21.50
	Install new gaskets				.431	89.60	22.96		112.56	128	149
	Misc. hardware replacement				.333	78	17.70		95.70	109	126
	Remove scaffold				1.333		71		71	91.50	114
	Total				3.681	167.60	195.96		363.56	**437.20**	**524.50**
3030	**Replace 3′ x 4′ aluminum window - 3rd floor**	50	1 CARP	Ea.							
	Set up and secure scaffold				1.333		71		71	91.50	114
	Remove window				.650		27.50		27.50	35.50	44
	Install new window with frame & glazing (to 12 S.F.)				2.078	510	120		630	720	835
	Install sealant				.555	6.27	28.93		35.20	44.50	54.50
	Remove scaffold				1.333		71		71	91.50	114
	Total				5.950	516.27	318.43		834.70	**983**	**1,161.50**

B2023 111 Aluminum Window, Fixed

	System Description	Freq. (Years)	Crew	Unit	Labor Hours	2020 Bare Costs				Total In-House	Total w/O&P
						Material	Labor	Equipment	Total		
1010	**Replace glass - 1st floor (1% of glass)**	1	1 CARP	S.F.							
	Remove damaged glass				.052		2.19		2.19	2.83	3.50
	Install new glass				.043	6.65	2.21		8.86	10.15	11.85
	Total				.095	6.65	4.40		11.05	**12.98**	**15.35**
1015	**Replace glass - 1st floor, 1″ insulating panel with heat reflective glass**	30	2 GLAZ	S.F.							
	Remove damaged glass				.052		2.19		2.19	2.83	3.50
	Tinted insulated glass				.213	24	10.90		34.90	40.50	47.50
	Total				.265	24	13.09		37.09	**43.33**	**51**
1020	**Repair 2′-0″ x 3′-0″ aluminum window - 1st floor**	20	1 CARP	Ea.							
	Set up and secure scaffold				.444		23.50		23.50	30.50	38
	Remove window gaskets				.250		13.30		13.30	17.20	21.50
	Install new gaskets				.308	64	16.40		80.40	91.50	106
	Remove scaffold				.444		23.50		23.50	30.50	38
	Total				1.447	64	76.70		140.70	**169.70**	**203.50**

B2023 111 Aluminum Window, Fixed

	System Description	Freq. (Years)	Crew	Unit	Labor Hours	2020 Bare Costs				Total In-House	Total w/O&P
						Material	Labor	Equipment	Total		
1030	**Replace 2′-0″ x 3′-0″ aluminum window - 1st floor**	50	1 CARP	Ea.							
	Set up and secure scaffold				.444		23.50		23.50	30.50	38
	Remove window				.650		27.50		27.50	35.50	44
	Install new window with frame & glazing (2′-0″ x 3′-0″)				2.078	219	120		339	400	470
	Install sealant				.555	6.27	28.93		35.20	44.50	54.50
	Remove scaffold				.444		23.50		23.50	30.50	38
	Total				4.173	225.27	223.43		448.70	**541**	**644.50**
2010	**Replace glass - 2nd floor (1% of glass)**	1	1 CARP	S.F.							
	Set up and secure scaffold				.889		47		47	61	75.50
	Remove damaged glass				.052		2.19		2.19	2.83	3.50
	Install new glass				.043	6.65	2.21		8.86	10.15	11.85
	Remove scaffold				.889		47		47	61	75.50
	Total				1.873	6.65	98.40		105.05	**134.98**	**166.35**
2020	**Repair 2′-0″ x 3′-0″ aluminum window - 2nd floor**	20	1 CARP	Ea.							
	Set up and secure scaffold				.889		47		47	61	75.50
	Remove window gaskets				.250		13.30		13.30	17.20	21.50
	Install new gaskets				.308	64	16.40		80.40	91.50	106
	Remove scaffold				.889		47		47	61	75.50
	Total				2.335	64	123.70		187.70	**230.70**	**278.50**
2030	**Replace 2′-0″ x 3′-0″ aluminum window - 2nd floor**	50	1 CARP	Ea.							
	Set up and secure scaffold				.889		47		47	61	75.50
	Remove window				.650		27.50		27.50	35.50	44
	Install new window with frame & glazing (2′-0″ x 3′-0″)				2.078	219	120		339	400	470
	Install sealant				.555	6.27	28.93		35.20	44.50	54.50
	Remove scaffold				.889		47		47	61	75.50
	Total				5.062	225.27	270.43		495.70	**602**	**719.50**
3010	**Replace glass - 3rd floor (1% of glass)**	1	1 CARP	S.F.							
	Set up and secure scaffold				.123		6.55		6.55	8.45	10.50
	Remove damaged glass				.052		2.19		2.19	2.83	3.50
	Install new glass				.043	6.65	2.21		8.86	10.15	11.85
	Remove scaffold				.123		6.55		6.55	8.45	10.50
	Total				.341	6.65	17.50		24.15	**29.88**	**36.35**

B2023 111 Aluminum Window, Fixed

	System Description	Freq. (Years)	Crew	Unit	Labor Hours	2020 Bare Costs				Total In-House	Total w/O&P
						Material	Labor	Equipment	Total		
3020	**Repair 2′-0″ x 3′-0″ aluminum window - 3rd floor**	20	1 CARP	Ea.							
	Set up and secure scaffold				1.333		71		71	91.50	114
	Remove window gaskets				.250		13.30		13.30	17.20	21.50
	Install new gaskets				.308	64	16.40		80.40	91.50	106
	Remove scaffold				1.333		71		71	91.50	114
	Total				3.224	64	171.70		235.70	**291.70**	**355.50**
3030	**Replace 2′-0″ x 3′-0″ aluminum window - 3rd floor**	50	1 CARP	Ea.							
	Set up and secure scaffold				1.333		71		71	91.50	114
	Remove window				.650		27.50		27.50	35.50	44
	Install new window with frame & glazing (2′-0″ x 3′-0″)				2.078	219	120		339	400	470
	Install sealant				.555	6.27	28.93		35.20	44.50	54.50
	Remove scaffold				1.333		71		71	91.50	114
	Total				5.950	225.27	318.43		543.70	**663**	**796.50**

B2023 112 Wood Frame, Operating

	System Description	Freq. (Years)	Crew	Unit	Labor Hours	2020 Bare Costs				Total In-House	Total w/O&P
						Material	Labor	Equipment	Total		
1010	**Replace glass - 1st floor (1% of glass)**	1	1 CARP	S.F.							
	Remove damaged glass				.052		2.19		2.19	2.83	3.50
	Install new glass				.043	6.65	2.21		8.86	10.15	11.85
	Total				.095	6.65	4.40		11.05	**12.98**	**15.35**
1020	**Repair 2′-3″ x 6′-0″ wood frame window - 1st flr.**	15	1 CARP	Ea.							
	Set up and secure scaffold				.444		23.50		23.50	30.50	38
	Remove putty				.667		35.50		35.50	46	57
	Install new putty				.333	6.55	17.70		24.25	30	36.50
	Misc. hardware replacement				.333	78	17.70		95.70	109	126
	Remove scaffold				.444		23.50		23.50	30.50	38
	Total				2.222	84.55	117.90		202.45	**246**	**295.50**
1030	**Refinish 2′-3″ x 6′-0″ wood frame window - 1st flr.**	4	1 PORD	Ea.							
	Set up, secure and take down ladder				.267		11.85		11.85	15.15	18.80
	Prepare window for painting				.473		21		21	27	33.50
	Paint window, brushwork, primer + 1 coat				1.000	2.20	44.50		46.70	59	73.50
	Total				1.739	2.20	77.35		79.55	**101.15**	**125.80**

B2023 112 Wood Frame, Operating

	System Description	Freq. (Years)	Crew	Unit	Labor Hours	2020 Bare Costs				Total In-House	Total w/O&P
						Material	Labor	Equipment	Total		
1040	**Replace 2'-3" x 6'-0" wood frame window - 1st flr.**	40	1 CARP	Ea.							
	Set up and secure scaffold				.444		23.50		23.50	30.50	38
	Remove window				.473		19.90		19.90	25.50	32
	Install new window with frame & glazing (to 14 S.F.)				1.143	360	60.50		420.50	475	550
	Install sealant				.625	7.05	32.55		39.60	50.50	61.50
	Remove scaffold				.444		23.50		23.50	30.50	38
	Total				3.130	367.05	159.95		527	**612**	**719.50**
2010	**Replace glass - 2nd floor (1% of glass)**	1	1 CARP	S.F.							
	Set up and secure scaffold				.044		2.35		2.35	3.05	3.78
	Remove damaged glass				.052		2.19		2.19	2.83	3.50
	Install new glass				.043	6.65	2.21		8.86	10.15	11.85
	Remove scaffold				.044		2.35		2.35	3.05	3.78
	Total				.184	6.65	9.10		15.75	**19.08**	**22.91**
2020	**Repair 2'-3" x 6'-0" wood frame window - 2nd flr.**	15	1 CARP	Ea.							
	Set up and secure scaffold				.889		47		47	61	75.50
	Remove putty				.667		35.50		35.50	46	57
	New putty				.333	6.55	17.70		24.25	30	36.50
	Misc. hardware replacement				.333	78	17.70		95.70	109	126
	Remove scaffold				.889		47		47	61	75.50
	Total				3.111	84.55	164.90		249.45	**307**	**370.50**
2030	**Refinish 2'-3" x 6'-0" wood frame window - 2nd flr.**	5	1 PORD	Ea.							
	Set up and secure scaffold				.889		39.50		39.50	50.50	62.50
	Prepare window for painting				.473		21		21	27	33.50
	Paint window, brushwork, primer + 1 coat				1.000	2.20	44.50		46.70	59	73.50
	Remove scaffold				.889		39.50		39.50	50.50	62.50
	Total				3.251	2.20	144.50		146.70	**187**	**232**
2040	**Replace 2'-3" x 6'-0" wood frame window - 2nd flr.**	40	1 CARP	Ea.							
	Set up and secure scaffold				.889		47		47	61	75.50
	Remove window				.473		19.90		19.90	25.50	32
	Install new window with frame & glazing (to 14 S.F.)				1.143	360	60.50		420.50	475	550
	Install sealant				.625	7.05	32.55		39.60	50.50	61.50
	Remove scaffold				.889		47		47	61	75.50
	Total				4.018	367.05	206.95		574	**673**	**794.50**

B2023 112 Wood Frame, Operating

	System Description	Freq. (Years)	Crew	Unit	Labor Hours	2020 Bare Costs				Total In-House	Total w/O&P
						Material	Labor	Equipment	Total		
3010	**Replace glass - 3rd floor (1% of glass)**	1	1 CARP	S.F.							
	Set up and secure scaffold				.123		6.55		6.55	8.45	10.50
	Remove damaged glass				.052		2.19		2.19	2.83	3.50
	Install new glass				.043	6.65	2.21		8.86	10.15	11.85
	Remove scaffold				.123		6.55		6.55	8.45	10.50
	Total				.341	6.65	17.50		24.15	**29.88**	**36.35**
3020	**Repair 2′-3″ x 6′-0″ wood frame window - 3rd flr.**	15	1 CARP	Ea.							
	Set up and secure scaffold				1.333		71		71	91.50	114
	Remove putty				.667		35.50		35.50	46	57
	Install new putty				.333	6.55	17.70		24.25	30	36.50
	Misc. hardware replacement				.333	78	17.70		95.70	109	126
	Remove scaffold				1.333		71		71	91.50	114
	Total				4.000	84.55	212.90		297.45	**368**	**447.50**
3030	**Refinish 2′-3″ x 6′-0″ wood frame window - 3rd flr.**	5	1 PORD	Ea.							
	Set up and secure scaffold				1.333		59		59	75.50	94
	Prepare window for painting				.473		21		21	27	33.50
	Paint window, brushwork, primer + 1 coat				1.000	2.20	44.50		46.70	59	73.50
	Remove scaffold				1.333		59		59	75.50	94
	Total				4.139	2.20	183.50		185.70	**237**	**295**
3040	**Replace 2′-3″ x 6′-0″ wood frame window - 3rd flr.**	40	1 CARP	Ea.							
	Set up and secure scaffold				1.333		71		71	91.50	114
	Remove window				.473		19.90		19.90	25.50	32
	Install new window with frame & glazing (to 14 S.F.)				1.143	360	60.50		420.50	475	550
	Install sealant				.625	7.05	32.55		39.60	50.50	61.50
	Remove scaffold				1.333		71		71	91.50	114
	Total				4.907	367.05	254.95		622	**734**	**871.50**

B2023 113 — Wood Frame, Fixed

	System Description	Freq. (Years)	Crew	Unit	Labor Hours	2020 Bare Costs Material	Labor	Equipment	Total	Total In-House	Total w/O&P
1010	**Replace glass - 1st floor**	1	1 CARP	S.F.							
	Remove damaged glass				.052		2.19		2.19	2.83	3.50
	Install new glass				.043	6.65	2.21		8.86	10.15	11.85
	Total				.095	6.65	4.40		11.05	**12.98**	**15.35**
1020	**Repair 3'-6" x 4'-0" wood frame window - 1st flr.**	20	1 CARP	Ea.							
	Set up and secure scaffold				.444		23.50		23.50	30.50	38
	Remove putty				.667		35.50		35.50	46	57
	New putty				.333	6.55	17.70		24.25	30	36.50
	Remove scaffold				.444		23.50		23.50	30.50	38
	Total				1.889	6.55	100.20		106.75	**137**	**169.50**
1030	**Refinish 3'-6" x 4'-0" wood frame window - 1st flr.**	5	1 PORD	Ea.							
	Set up, secure and take down ladder				.444		19.75		19.75	25	31.50
	Prepare window for painting				.473		21		21	27	33.50
	Paint window, brushwork, primer + 1 coat				1.000	2.20	44.50		46.70	59	73.50
	Total				1.917	2.20	85.25		87.45	**111**	**138.50**
1040	**Replace 3'-6" x 4'-0" wood frame window - 1st flr.**	40	1 CARP	Ea.							
	Set up and secure scaffold				.444		23.50		23.50	30.50	38
	Remove window				.473		19.90		19.90	25.50	32
	Install new window with frame & glazing (to 14 S.F.)				1.733	440	92		532	605	700
	Install sealant				.625	7.05	32.55		39.60	50.50	61.50
	Remove scaffold				.444		23.50		23.50	30.50	38
	Total				3.720	447.05	191.45		638.50	**742**	**869.50**
2010	**Replace glass - 2nd floor (1% of glass)**	1	1 CARP	S.F.							
	Set up and secure scaffold				.044		2.35		2.35	3.05	3.78
	Remove damaged glass				.052		2.19		2.19	2.83	3.50
	Install new glass				.043	6.65	2.21		8.86	10.15	11.85
	Remove scaffold				.044		2.35		2.35	3.05	3.78
	Total				.184	6.65	9.10		15.75	**19.08**	**22.91**

B2023 113 Wood Frame, Fixed

	System Description	Freq. (Years)	Crew	Unit	Labor Hours	2020 Bare Costs				Total In-House	Total w/O&P
						Material	Labor	Equipment	Total		
2020	**Repair 3′-6″ x 4′-0″ wood frame window - 2nd flr.**	20	1 CARP	Ea.							
	Set up and secure scaffold				.889		47		47	61	75.50
	Remove putty				.667		35.50		35.50	46	57
	Install new putty				.333	6.55	17.70		24.25	30	36.50
	Remove scaffold				.889		47		47	61	75.50
	Total				2.778	6.55	147.20		153.75	**198**	**244.50**
2030	**Refinish 3′-6″ x 4′-0″ wood frame window - 2nd flr.**	5	1 PORD	Ea.							
	Set up and secure scaffold				.889		39.50		39.50	50.50	62.50
	Prepare window for painting				.473		21		21	27	33.50
	Paint window, brushwork, primer + 1 coat				1.000	2.20	44.50		46.70	59	73.50
	Remove scaffold				.889		39.50		39.50	50.50	62.50
	Total				3.251	2.20	144.50		146.70	**187**	**232**
2040	**Replace 3′-6″ x 4′-0″ wood frame window - 2nd flr.**	40	1 CARP	Ea.							
	Set up and secure scaffold				.889		47		47	61	75.50
	Remove window				.473		19.90		19.90	25.50	32
	Install new window with frame & glazing (to 14 S.F.)				1.733	440	92		532	605	700
	Install sealant				.625	7.05	32.55		39.60	50.50	61.50
	Remove scaffold				.889		47		47	61	75.50
	Total				4.609	447.05	238.45		685.50	**803**	**944.50**
3010	**Replace glass - 3rd floor (1% of glass)**	1	1 CARP	S.F.							
	Set up and secure scaffold				.123		6.55		6.55	8.45	10.50
	Remove damaged glass				.052		2.19		2.19	2.83	3.50
	Install new glass				.043	6.65	2.21		8.86	10.15	11.85
	Remove scaffold				.123		6.55		6.55	8.45	10.50
	Total				.341	6.65	17.50		24.15	**29.88**	**36.35**
3020	**Repair 3′-6″ x 4′-0″ wood frame window - 3rd flr.**	20	1 CARP	Ea.							
	Set up and secure scaffold				1.333		71		71	91.50	114
	Remove putty				.667		35.50		35.50	46	57
	Install new putty				.333	6.55	17.70		24.25	30	36.50
	Remove scaffold				1.333		71		71	91.50	114
	Total				3.667	6.55	195.20		201.75	**259**	**321.50**

B2023 113 Wood Frame, Fixed

	System Description	Freq. (Years)	Crew	Unit	Labor Hours	2020 Bare Costs				Total In-House	Total w/O&P
						Material	Labor	Equipment	Total		
3030	**Refinish 3'-6" x 4'-0" wood frame window - 3rd flr.**	5	1 PORD	Ea.							
	Set up and secure scaffold				1.333		59		59	75.50	94
	Prepare window for painting				.473		21		21	27	33.50
	Paint window, brushwork, primer + 1 coat				1.000	2.20	44.50		46.70	59	73.50
	Remove scaffold				1.333		59		59	75.50	94
	Total				4.139	2.20	183.50		185.70	**237**	**295**
3040	**Replace 3'-6" x 4'-0" wood frame window - 3rd flr.**	40	1 CARP	Ea.							
	Set up and secure scaffold				1.333		71		71	91.50	114
	Remove window				.473		19.90		19.90	25.50	32
	Install new window with frame & glazing (to 14 S.F.)				1.733	440	92		532	605	700
	Install sealant				.625	7.05	32.55		39.60	50.50	61.50
	Remove scaffold				1.333		71		71	91.50	114
	Total				5.498	447.05	286.45		733.50	**864**	**1,021.50**

B2023 116 Aluminum Frame Storm Window

	System Description	Freq. (Years)	Crew	Unit	Labor Hours	2020 Bare Costs				Total In-House	Total w/O&P
						Material	Labor	Equipment	Total		
1010	**Replace glass - (1% of glass)**	1	1 CARP	S.F.							
	Remove glass				.052		2.19		2.19	2.83	3.50
	Install glass / storm window				.173	8.40	8.85		17.25	20.50	24.50
	Total				.225	8.40	11.04		19.44	**23.33**	**28**
1030	**Refinish 3'-0" x 4'-0" aluminum frame storm window - 1st floor**	10	1 PORD	Ea.							
	Set up and secure scaffold				.444		19.75		19.75	25	31.50
	Prepare window frame surface				.473		21		21	27	33.50
	Refinish window frame surface, brushwork, primer + 1 coat				.615	1.22	27.50		28.72	36.50	45
	Remove scaffold				.444		19.75		19.75	25	31.50
	Total				1.977	1.22	88		89.22	**113.50**	**141.50**
1040	**Replace 3'-0" x 4'-0" aluminum frame storm window - 1st floor**	50	1 CARP	Ea.							
	Set up and secure scaffold				.444		23.50		23.50	30.50	38
	Remove old window				.385		16.20		16.20	21	26
	Install new window (to 12 S.F.)				.743	126	39.50		165.50	190	221
	Remove scaffold				.444		23.50		23.50	30.50	38
	Total				2.017	126	102.70		228.70	**272**	**323**

B2023 116 Aluminum Frame Storm Window

	System Description	Freq. (Years)	Crew	Unit	Labor Hours	2020 Bare Costs Material	Labor	Equipment	Total	Total In-House	Total w/O&P
2030	**Refinish 3′-0″ x 4′-0″ aluminum frame storm window - 2nd floor**	10	1 PORD	Ea.							
	Set up and secure scaffold				.889		39.50		39.50	50.50	62.50
	Prepare window frame surface				.473		21		21	27	33.50
	Refinish window frame surface, brushwork, primer + 1 coat				.615	1.22	27.50		28.72	36.50	45
	Remove scaffold				.889		39.50		39.50	50.50	62.50
	Total				2.866	1.22	127.50		128.72	**164.50**	**203.50**
2040	**Replace 3′-0″ x 4′-0″ aluminum frame storm window - 2nd floor**	50	1 CARP	Ea.							
	Set up and secure scaffold				.889		47		47	61	75.50
	Remove old window				.385		16.20		16.20	21	26
	Install new window (to 12 S.F.)				.743	126	39.50		165.50	190	221
	Remove scaffold				.889		47		47	61	75.50
	Total				2.906	126	149.70		275.70	**333**	**398**
3030	**Refinish 3′-0″ x 4′-0″ aluminum frame storm window - 3rd floor**	10	1 PORD	Ea.							
	Set up and secure scaffold				1.333		59		59	75.50	94
	Prepare window frame surface				.473		21		21	27	33.50
	Refinish window frame surface, brushwork, primer + 1 coat				.615	1.22	27.50		28.72	36.50	45
	Remove scaffold				1.333		59		59	75.50	94
	Total				3.755	1.22	166.50		167.72	**214.50**	**266.50**
3040	**Replace 3′-0″ x 4′-0″ aluminum frame storm window - 3rd floor**	50	1 CARP	Ea.							
	Set up and secure scaffold				1.333		71		71	91.50	114
	Remove old window				.385		16.20		16.20	21	26
	Install new window (to 12 S.F.)				.743	126	39.50		165.50	190	221
	Remove scaffold				1.333		71		71	91.50	114
	Total				3.795	126	197.70		323.70	**394**	**475**

B2023 117 Steel Frame Storm Window

	System Description	Freq. (Years)	Crew	Unit	Labor Hours	2020 Bare Costs Material	Labor	Equipment	Total	Total In-House	Total w/O&P
1010	**Replace glass - (1% of glass)**	1	1 CARP	S.F.							
	Remove glass				.052		2.19		2.19	2.83	3.50
	Install glass / storm window				.173	8.40	8.85		17.25	20.50	24.50
	Total				.225	8.40	11.04		19.44	**23.33**	**28**

B2023 117 Steel Frame Storm Window

	System Description	Freq. (Years)	Crew	Unit	Labor Hours	2020 Bare Costs				Total In-House	Total w/O&P
						Material	Labor	Equipment	Total		
1030	**Refinish 3'-0" x 4'-0" steel frame storm window - 1st floor**	5	1 PORD	Ea.							
	Set up and secure scaffold				.444		19.75		19.75	25	31.50
	Prepare window frame surface				.473		21		21	27	33.50
	Refinish window frame surface, brushwork, primer + 1 coat				.615	1.22	27.50		28.72	36.50	45
	Remove scaffold				.444		19.75		19.75	25	31.50
	Total				1.977	1.22	88		89.22	**113.50**	**141.50**
1040	**Replace 3'-0" x 4'-0" steel frame storm window - 1st floor**	25	1 CARP	Ea.							
	Set up and secure scaffold				.444		23.50		23.50	30.50	38
	Remove old window				.385		16.20		16.20	21	26
	Install new window (to 12 S.F.)				.743	126	39.50		165.50	190	221
	Remove scaffold				.444		23.50		23.50	30.50	38
	Total				2.017	126	102.70		228.70	**272**	**323**
2030	**Refinish 3'-0" x 4'-0" steel frame storm window - 2nd floor**	5	1 PORD	Ea.							
	Set up and secure scaffold				.889		39.50		39.50	50.50	62.50
	Prepare window frame surface				.473		21		21	27	33.50
	Refinish window frame surface, brushwork, primer + 1 coat				.615	1.22	27.50		28.72	36.50	45
	Remove scaffold				.889		39.50		39.50	50.50	62.50
	Total				2.866	1.22	127.50		128.72	**164.50**	**203.50**
2040	**Replace 3'-0" x 4'-0" steel frame storm window - 2nd floor**	25	1 CARP	Ea.							
	Set up and secure scaffold				.889		47		47	61	75.50
	Remove old window				.385		16.20		16.20	21	26
	Install new window (to 12 S.F.)				.743	126	39.50		165.50	190	221
	Remove scaffold				.889		47		47	61	75.50
	Total				2.906	126	149.70		275.70	**333**	**398**
3030	**Refinish 3'-0" x 4'-0" steel frame storm window - 3rd floor**	5	1 PORD	Ea.							
	Set up and secure scaffold				1.333		59		59	75.50	94
	Prepare window frame surface				.473		21		21	27	33.50
	Refinish window frame surface, brushwork, primer + 1 coat				.615	1.22	27.50		28.72	36.50	45
	Remove scaffold				1.333		59		59	75.50	94
	Total				3.755	1.22	166.50		167.72	**214.50**	**266.50**

B2023 117 Steel Frame Storm Window

	System Description	Freq. (Years)	Crew	Unit	Labor Hours	2020 Bare Costs Material	Labor	Equipment	Total	Total In-House	Total w/O&P
3040	**Replace 3′-0″ x 4′-0″ steel frame storm window - 3rd floor**	25	1 CARP	Ea.							
	Set up and secure scaffold				1.333		71		71	91.50	114
	Remove old window				.385		16.20		16.20	21	26
	Install new window (to 12 S.F.)				.743	126	39.50		165.50	190	221
	Remove scaffold				1.333		71		71	91.50	114
	Total				3.795	126	197.70		323.70	**394**	**475**

B2023 119 Wood Frame Storm Window

	System Description	Freq. (Years)	Crew	Unit	Labor Hours	2020 Bare Costs Material	Labor	Equipment	Total	Total In-House	Total w/O&P
1010	**Replace glass - (1% of glass)**	1	1 CARP	S.F.							
	Remove glass				.052		2.19		2.19	2.83	3.50
	Install new float glass				.043	6.65	2.21		8.86	10.15	11.85
	Total				.095	6.65	4.40		11.05	**12.98**	**15.35**
1020	**Repair 3′-0″ x 5′-0″ wood frame storm window**	10	1 CARP	Ea.							
	Remove putty				.667		35.50		35.50	46	57
	Install new putty				.333	6.55	17.70		24.25	30	36.50
	Total				1.000	6.55	53.20		59.75	**76**	**93.50**
1030	**Refinish 3′-0″ x 5′-0″ wood frame storm window**	5	1 PORD	Ea.							
	Prepare window frame surface				.473		21		21	27	33.50
	Refinish window frame surface, brushwork, primer + 1 coat				.615	1.22	27.50		28.72	36.50	45
	Total				1.088	1.22	48.50		49.72	**63.50**	**78.50**
1040	**Replace 3′-0″ x 5′-0″ wood frame storm window - 1st floor**	40	1 CARP	Ea.							
	Set up and secure scaffold				.444		23.50		23.50	30.50	38
	Remove old window				.473		19.90		19.90	25.50	32
	Install new window (to 15 S.F.)				.693	176	36.80		212.80	241	279
	Remove scaffold				.444		23.50		23.50	30.50	38
	Total				2.055	176	103.70		279.70	**327.50**	**387**

B2023 119 Wood Frame Storm Window

	System Description	Freq. (Years)	Crew	Unit	Labor Hours	2020 Bare Costs				Total In-House	Total w/O&P
						Material	Labor	Equipment	Total		
2040	**Replace 3'-0" x 5'-0" wood frame storm window - 2nd floor**	40	1 CARP	Ea.							
	Set up and secure scaffold				.889		47		47	61	75.50
	Remove old window				.473		19.90		19.90	25.50	32
	Install new window (to 15 S.F.)				.693	176	36.80		212.80	241	279
	Remove scaffold				.889		47		47	61	75.50
	Total				2.944	176	150.70		326.70	**388.50**	**462**
3040	**Replace 3'-0" x 5'-0" wood frame storm window - 3rd floor**	40	1 CARP	Ea.							
	Set up and secure scaffold				1.333		71		71	91.50	114
	Remove old window				.473		19.90		19.90	25.50	32
	Install new window (to 15 S.F.)				.693	176	36.80		212.80	241	279
	Remove scaffold				1.333		71		71	91.50	114
	Total				3.833	176	198.70		374.70	**449.50**	**539**

B2023 122 Vinyl Clad Wood Window, Operating

	System Description	Freq. (Years)	Crew	Unit	Labor Hours	2020 Bare Costs				Total In-House	Total w/O&P
						Material	Labor	Equipment	Total		
1030	**Repl. 3'-0" x 4'-0" vinyl clad window - 1st flr.**	60	1 CARP	Ea.							
	Set up and secure scaffold				.444		23.50		23.50	30.50	38
	Remove window				.473		19.90		19.90	25.50	32
	Install new window (3'-0" x 4'-0")				1.156	405	61.50		466.50	525	605
	Remove scaffold				.444		23.50		23.50	30.50	38
	Total				2.518	405	128.40		533.40	**611.50**	**713**
2030	**Repl. 3'-0" x 4'-0" vinyl clad window - 2nd flr.**	60	1 CARP	Ea.							
	Set up and secure scaffold				.889		47		47	61	75.50
	Remove window				.473		19.90		19.90	25.50	32
	Install new window (3'-0" x 4'-0")				1.156	405	61.50		466.50	525	605
	Remove scaffold				.889		47		47	61	75.50
	Total				3.407	405	175.40		580.40	**672.50**	**788**

B2023 122 Vinyl Clad Wood Window, Operating

	System Description	Freq. (Years)	Crew	Unit	Labor Hours	2020 Bare Costs Material	Labor	Equipment	Total	Total In-House	Total w/O&P
3030	**Repl. 3′-0″ x 4′-0″ vinyl clad window - 3rd flr.**	60	1 CARP	Ea.							
	Set up and secure scaffold				1.333		71		71	91.50	114
	Remove window				.473		19.90		19.90	25.50	32
	Install new window (3′0″ x 4′0″)				1.156	405	61.50		466.50	525	605
	Remove scaffold				1.333		71		71	91.50	114
	Total				4.295	405	223.40		628.40	**733.50**	**865**

B2023 123 Vinyl Clad Wood Window, Fixed

	System Description	Freq. (Years)	Crew	Unit	Labor Hours	2020 Bare Costs Material	Labor	Equipment	Total	Total In-House	Total w/O&P
1030	**Repl. 4′-0″ x 4′-0″ vinyl clad window - 1st flr.**	60	1 CARP	Ea.							
	Set up and secure scaffold				.444		23.50		23.50	30.50	38
	Remove window				.473		19.90		19.90	25.50	32
	Install new window (4′-0″ x 4′-0″)				1.733	535	92		627	710	815
	Remove scaffold				.444		23.50		23.50	30.50	38
	Total				3.095	535	158.90		693.90	**796.50**	**923**
2030	**Repl. 4′-0″ x 4′-0″ vinyl clad window - 2nd flr.**	60	1 CARP	Ea.							
	Set up and secure scaffold				.889		47		47	61	75.50
	Remove window				.473		19.90		19.90	25.50	32
	Install new window (4′-0″ x 4′-0″)				1.733	535	92		627	710	815
	Remove scaffold				.889		47		47	61	75.50
	Total				3.984	535	205.90		740.90	**857.50**	**998**
3030	**Repl. 4′-0″ x 4′-0″ vinyl clad window - 3rd flr.**	60	1 CARP	Ea.							
	Set up and secure scaffold				1.333		71		71	91.50	114
	Remove window				.473		19.90		19.90	25.50	32
	Install new window (4′-0″ x 4′-0″)				1.733	535	92		627	710	815
	Remove scaffold				1.333		71		71	91.50	114
	Total				4.873	535	253.90		788.90	**918.50**	**1,075**

B2023 150 Glass Block, Fixed

	System Description	Freq. (Years)	Crew	Unit	Labor Hours	2020 Bare Costs				Total In-House	Total w/O&P
						Material	Labor	Equipment	Total		
1020	**Repair window - 1st floor (2% of glass)**	8	1 BRIC	S.F.							
	Cut out damaged block				.433		18.38	3.88	22.26	28	34
	Replace glass block				.452	23	21.50		44.50	53.50	64
	Total				.886	23	39.88	3.88	66.76	**81.50**	**98**
1030	**Replace glass block window - 1st floor**	75	1 BRIC	S.F.							
	Set up, secure and take down ladder				.267		13.90		13.90	18.15	22.50
	Demolish glass block				.074		3.13		3.13	4.04	5
	Replace glass block				.452	23	21.50		44.50	53.50	64
	Total				.793	23	38.53		61.53	**75.69**	**91.50**
2020	**Repair window - 2nd floor (2% of glass)**	8	1 BRIC	S.F.							
	Set up and secure scaffold				.889		46.50		46.50	60.50	75
	Cut out damaged block				.433		18.38	3.88	22.26	28	34
	Replace glass block				.452	23	21.50		44.50	53.50	64
	Remove scaffold				.889		46.50		46.50	60.50	75
	Total				2.663	23	132.88	3.88	159.76	**202.50**	**248**
2030	**Replace glass block window - 2nd floor**	75	1 BRIC	S.F.							
	Set up and secure scaffold				.889		46.50		46.50	60.50	75
	Demolish glass block				.074		3.13		3.13	4.04	5
	Replace glass block				.452	23	21.50		44.50	53.50	64
	Remove scaffold				.889		46.50		46.50	60.50	75
	Total				2.304	23	117.63		140.63	**178.54**	**219**
3020	**Repair window - 3rd floor (2% of glass)**	8	1 BRIC	S.F.							
	Set up and secure scaffold				1.333		69.50		69.50	90.50	112
	Cut out damaged block				.433		18.38	3.88	22.26	28	34
	Replace glass block				.452	23	21.50		44.50	53.50	64
	Remove scaffold				1.333		69.50		69.50	90.50	112
	Total				3.552	23	178.88	3.88	205.76	**262.50**	**322**
3030	**Replace glass block window - 3rd floor**	75	1 BRIC	S.F.							
	Set up and secure scaffold				1.333		69.50		69.50	90.50	112
	Demolish glass block				.074		3.13		3.13	4.04	5
	Replace glass block				.452	23	21.50		44.50	53.50	64
	Remove scaffold				1.333		69.50		69.50	90.50	112
	Total				3.193	23	163.63		186.63	**238.54**	**293**

B2023 160 Aluminum Shutter

	System Description	Freq. (Years)	Crew	Unit	Labor Hours	2020 Bare Costs				Total In-House	Total w/O&P
						Material	Labor	Equipment	Total		
1030	**Refinish aluminum shutter - 1st floor**	5	1 PORD	Ea.							
	Set up and secure scaffold				.444		19.75		19.75	25	31.50
	Prepare shutter for painting				.434		19.20		19.20	24.50	30.50
	Paint shutter surface (to 16 S.F.), brushwork, primer + 1 coat				.800	2.43	35.50		37.93	48	59.50
	Remove scaffold				.444		19.75		19.75	25	31.50
	Total				2.123	2.43	94.20		96.63	**122.50**	**153**
1040	**Replace aluminum shutter - 1st floor**	60	1 CARP	Ea.							
	Set up and secure scaffold				.444		23.50		23.50	30.50	38
	Remove shutter				.260		13.75		13.75	17.85	22.50
	Install new shutter				.520	109.50	27.75		137.25	156	181
	Remove scaffold				.444		23.50		23.50	30.50	38
	Total				1.669	109.50	88.50		198	**234.85**	**279.50**
2030	**Refinish aluminum shutter - 2nd floor**	5	1 PORD	Ea.							
	Set up and secure scaffold				.889		39.50		39.50	50.50	62.50
	Prepare shutter for painting				.434		19.20		19.20	24.50	30.50
	Paint shutter surface (to 16 S.F.), brushwork, primer + 1 coat				.800	2.43	35.50		37.93	48	59.50
	Remove scaffold				.889		39.50		39.50	50.50	62.50
	Total				3.012	2.43	133.70		136.13	**173.50**	**215**
2040	**Replace aluminum shutter - 2nd floor**	60	1 CARP	Ea.							
	Set up and secure scaffold				.889		47		47	61	75.50
	Remove shutter				.260		13.75		13.75	17.85	22.50
	Install new shutter				.520	109.50	27.75		137.25	156	181
	Remove scaffold				.889		47		47	61	75.50
	Total				2.558	109.50	135.50		245	**295.85**	**354.50**
3030	**Refinish aluminum shutter - 3rd floor**	5	1 PORD	Ea.							
	Set up and secure scaffold				1.333		59		59	75.50	94
	Prepare shutter for painting				.434		19.20		19.20	24.50	30.50
	Paint shutter surface (to 16 S.F.), brushwork, primer + 1 coat				.800	2.43	35.50		37.93	48	59.50
	Remove scaffold				1.333		59		59	75.50	94
	Total				3.901	2.43	172.70		175.13	**223.50**	**278**

B2023 160 Aluminum Shutter

	System Description	Freq. (Years)	Crew	Unit	Labor Hours	2020 Bare Costs				Total In-House	Total w/O&P
						Material	Labor	Equipment	Total		
3040	**Replace aluminum shutter - 3rd floor**	60	1 CARP	Ea.							
	Set up and secure scaffold				1.333		71		71	91.50	114
	Remove shutter				.260		13.75		13.75	17.85	22.50
	Install new shutter				.520	109.50	27.75		137.25	156	181
	Remove scaffold				1.333		71		71	91.50	114
	Total				3.447	109.50	183.50		293	**356.85**	**431.50**

B2023 162 Steel Shutter

	System Description	Freq. (Years)	Crew	Unit	Labor Hours	2020 Bare Costs				Total In-House	Total w/O&P
						Material	Labor	Equipment	Total		
1030	**Refinish steel shutter - 1st floor**	5	1 PORD	Ea.							
	Set up and secure scaffold				.444		19.75		19.75	25	31.50
	Prepare shutter for painting				.434		19.20		19.20	24.50	30.50
	Paint shutter surface (to 16 S.F.), brushwork, primer + 1 coat				.800	2.43	35.50		37.93	48	59.50
	Remove scaffold				.444		19.75		19.75	25	31.50
	Total				2.123	2.43	94.20		96.63	**122.50**	**153**
1040	**Replace steel shutter - 1st floor**	40	1 CARP	Ea.							
	Set up and secure scaffold				.444		23.50		23.50	30.50	38
	Remove old shutter				.308		17.94	1.44	19.38	25.50	31
	Install new steel shutter				.617	40.80	36	2.82	79.62	95.50	113
	Remove scaffold				.444		23.50		23.50	30.50	38
	Total				1.814	40.80	100.94	4.26	146	**182**	**220**
2030	**Refinish steel shutter - 2nd floor**	5	1 PORD	Ea.							
	Set up and secure scaffold				.889		39.50		39.50	50.50	62.50
	Prepare shutter for painting				.434		19.20		19.20	24.50	30.50
	Paint shutter surface (to 16 S.F.), brushwork, primer + 1 coat				.800	2.43	35.50		37.93	48	59.50
	Remove scaffold				.889		39.50		39.50	50.50	62.50
	Total				3.012	2.43	133.70		136.13	**173.50**	**215**

B2023 162 Steel Shutter

	System Description	Freq. (Years)	Crew	Unit	Labor Hours	2020 Bare Costs				Total In-House	Total w/O&P
						Material	Labor	Equipment	Total		
2040	**Replace steel shutter - 2nd floor**	40	1 CARP	Ea.							
	Set up and secure scaffold				.889		47		47	61	75.50
	Remove old shutter				.308		17.94	1.44	19.38	25.50	31
	Install new steel shutter				.617	40.80	36	2.82	79.62	95.50	113
	Remove scaffold				.889		47		47	61	75.50
	Total				2.703	40.80	147.94	4.26	193	**243**	**295**
3030	**Refinish steel shutter - 3rd floor**	5	1 PORD	Ea.							
	Set up and secure scaffold				1.333		59		59	75.50	94
	Prepare shutter for painting				.434		19.20		19.20	24.50	30.50
	Paint shutter surface (to 16 S.F.), brushwork, primer + 1 coat				.800	2.43	35.50		37.93	48	59.50
	Remove scaffold				1.333		59		59	75.50	94
	Total				3.901	2.43	172.70		175.13	**223.50**	**278**
3040	**Replace steel shutter - 3rd floor**	40	1 CARP	Ea.							
	Set up and secure scaffold				1.333		71		71	91.50	114
	Remove old shutter				.308		17.94	1.44	19.38	25.50	31
	Install new steel shutter				.617	40.80	36	2.82	79.62	95.50	113
	Remove scaffold				1.333		71		71	91.50	114
	Total				3.592	40.80	195.94	4.26	241	**304**	**372**

B2023 164 Wood Shutter

	System Description	Freq. (Years)	Crew	Unit	Labor Hours	2020 Bare Costs				Total In-House	Total w/O&P
						Material	Labor	Equipment	Total		
1010	**Repair wood shutter - 1st floor**	6	1 CARP	Ea.							
	Remove shutter				.260		13.75		13.75	17.85	22.50
	Remove damaged panel				.157		8.34		8.34	10.80	13.40
	Install new panel				.314	5.46	16.68		22.14	27.50	33.50
	Install shutter				.520		27.75		27.75	35.50	44.50
	Total				1.251	5.46	66.52		71.98	**91.65**	**113.90**

B2023 164 Wood Shutter

	System Description	Freq. (Years)	Crew	Unit	Labor Hours	2020 Bare Costs				Total In-House	Total w/O&P
						Material	Labor	Equipment	Total		
1030	**Refinish wood shutter - 1st floor**	5	1 PORD	Ea.							
	Set up and secure scaffold				.444		19.75		19.75	25	31.50
	Prepare shutter for painting				.434		19.20		19.20	24.50	30.50
	Paint shutter surface (to 16 S.F.), brushwork, primer + 1 coat				.800	2.43	35.50		37.93	48	59.50
	Remove scaffold				.444		19.75		19.75	25	31.50
	Total				2.123	2.43	94.20		96.63	**122.50**	**153**
1040	**Replace wood shutter - 1st floor**	40	1 CARP	Ea.							
	Set up and secure scaffold				.444		23.50		23.50	30.50	38
	Remove shutter				.260		13.75		13.75	17.85	22.50
	Install new wood shutter				.520	152.50	27.75		180.25	203	235
	Remove scaffold				.444		23.50		23.50	30.50	38
	Total				1.669	152.50	88.50		241	**281.85**	**333.50**
2010	**Repair wood shutter - 2nd floor**	6	1 CARP	Ea.							
	Set up and secure scaffold				.889		47		47	61	75.50
	Remove shutter				.260		13.75		13.75	17.85	22.50
	Remove damaged panel				.157		8.34		8.34	10.80	13.40
	Install new panel				.314	5.46	16.68		22.14	27.50	33.50
	Install shutter				.520		27.75		27.75	35.50	44.50
	Remove scaffold				.889		47		47	61	75.50
	Total				3.029	5.46	130.52		165.98	**213.65**	**264.90**
2030	**Refinish wood shutter - 2nd floor**	5	1 PORD	Ea.							
	Set up and secure scaffold				.889		39.50		39.50	50.50	62.50
	Prepare shutter for painting				.434		19.20		19.20	24.50	30.50
	Paint shutter surface (to 16 S.F.), brushwork, primer + 1 coat				.800	2.43	35.50		37.93	48	59.50
	Remove scaffold				.889		39.50		39.50	50.50	62.50
	Total				3.012	2.43	133.70		136.13	**173.50**	**215**
2040	**Replace wood shutter - 2nd floor**	40	1 CARP	Ea.							
	Set up and secure scaffold				.889		47		47	61	75.50
	Remove shutter				.260		13.75		13.75	17.85	22.50
	Install new wood shutter				.520	152.50	27.75		180.25	203	235
	Remove scaffold				.889		47		47	61	75.50
	Total				2.558	152.50	135.50		288	**342.85**	**408.50**

B2023 164 Wood Shutter

	System Description	Freq. (Years)	Crew	Unit	Labor Hours	2020 Bare Costs				Total In-House	Total w/O&P
						Material	Labor	Equipment	Total		
3010	**Repair wood shutter - 3rd floor**	6	1 CARP	Ea.							
	Set up and secure scaffold				1.333		71		71	91.50	114
	Remove shutter				.260		13.75		13.75	17.85	22.50
	Remove damaged panel				.157		8.34		8.34	10.80	13.40
	Install new panel				.314	5.46	16.68		22.14	27.50	33.50
	Install shutter				.520		27.75		27.75	35.50	44.50
	Remove scaffold				1.333		71		71	91.50	114
	Total				3.917	5.46	208.52		213.98	**274.65**	**341.90**
3030	**Refinish wood shutter - 3rd floor**	5	1 PORD	Ea.							
	Set up and secure scaffold				1.333		59		59	75.50	94
	Prepare shutter for painting				.434		19.20		19.20	24.50	30.50
	Paint shutter surface (to 16 S.F.), brushwork, primer + 1 coat				.800	2.43	35.50		37.93	48	59.50
	Remove scaffold				1.333		59		59	75.50	94
	Total				3.901	2.43	172.70		175.13	**223.50**	**278**
3040	**Replace wood shutter - 3rd floor**	40	1 CARP	Ea.							
	Set up and secure scaffold				1.333		71		71	91.50	114
	Remove shutter				.260		13.75		13.75	17.85	22.50
	Install new wood shutter				.520	152.50	27.75		180.25	203	235
	Remove scaffold				1.333		71		71	91.50	114
	Total				3.447	152.50	183.50		336	**403.85**	**485.50**

B2023 166 Metal Window Grating

	System Description	Freq. (Years)	Crew	Unit	Labor Hours	2020 Bare Costs				Total In-House	Total w/O&P
						Material	Labor	Equipment	Total		
1020	**Refinish 3′-0″ x 4′-0″ metal window grating**	5	1 PORD	Ea.							
	Prepare window grating surface				.308		13.65		13.65	17.45	21.50
	Paint grated surface, brushwork, primer + 1 coat				.167	1.92	7.44		9.36	11.50	14.15
	Total				.474	1.92	21.09		23.01	**28.95**	**35.65**
1030	**Replace 3′-0″ x 4′-0″ metal window grating - 1st floor**	30	1 SSWK	Ea.							
	Remove old grating				.800		33.50		33.50	43.50	54
	Fabricate and install new window grating				6.275	258	362		620	765	915
	Total				7.075	258	395.50		653.50	**808.50**	**969**

B2023 166 Metal Window Grating

	System Description	Freq. (Years)	Crew	Unit	Labor Hours	2020 Bare Costs Material	Labor	Equipment	Total	Total In-House	Total w/O&P
2030	**Replace 3'-0" x 4'-0" metal window grating - 2nd floor**	30	1 SSWK	Ea.							
	Set up and secure scaffold				.889		51		51	68	84
	Remove old grating				.800		33.50		33.50	43.50	54
	Fabricate and install new window grating				6.275	258	362		620	765	915
	Remove scaffold				.889		51		51	68	84
	Total				8.852	258	497.50		755.50	**944.50**	**1,137**
3030	**Replace 3'-0" x 4'-0" metal window grating - 3rd floor**	30	1 SSWK	Ea.							
	Set up and secure scaffold				1.333		77		77	102	126
	Remove old window grating				.800		33.50		33.50	43.50	54
	Fabricate and install new window grating				6.275	258	362		620	765	915
	Remove scaffold				1.333		77		77	102	126
	Total				9.741	258	549.50		807.50	**1,012.50**	**1,221**

B2023 168 Metal Wire Mesh Cover

	System Description	Freq. (Years)	Crew	Unit	Labor Hours	2020 Bare Costs Material	Labor	Equipment	Total	Total In-House	Total w/O&P
1020	**Refinish 3'-0" x 4'-0" metal wire mesh window cover**	5	1 PORD	Ea.							
	Prepare metal mesh surface				.103		4.55		4.55	5.80	7.25
	Paint metal mesh surface, brushwork, primer + 1 coat				.167	1.92	7.44		9.36	11.50	14.15
	Total				.269	1.92	11.99		13.91	**17.30**	**21.40**
1030	**Repl. 3'-0" x 4'-0" metal wire mesh window cover - 1st floor**	30	1 SSWK	Ea.							
	Set up and secure scaffold				.444		25.50		25.50	34	42
	Remove old mesh				.800		33.50		33.50	43.50	54
	Install new wire mesh & frame (to 12 S.F.)				.452	62.50	24		86.50	100	117
	Remove scaffold				.444		25.50		25.50	34	42
	Total				2.141	62.50	108.50		171	**211.50**	**255**
2030	**Repl. 3'-0" x 4'-0" metal wire mesh window cover - 2nd floor**	30	1 SSWK	Ea.							
	Set up and secure scaffold				.889		51		51	68	84
	Remove old mesh				.800		33.50		33.50	43.50	54
	Install new wire mesh & frame (to 12 S.F.)				.452	62.50	24		86.50	100	117
	Remove scaffold				.889		51		51	68	84
	Total				3.030	62.50	159.50		222	**279.50**	**339**

B2023 168 Metal Wire Mesh Cover

	System Description	Freq. (Years)	Crew	Unit	Labor Hours	2020 Bare Costs				Total In-House	Total w/O&P
						Material	Labor	Equipment	Total		
3030	**Repl. 3′-0″ x 4′-0″ metal wire mesh window cover - 3rd floor**	30	1 SSWK	Ea.							
	Set up and secure scaffold				1.333		77		77	102	126
	Remove old mesh				.800		33.50		33.50	43.50	54
	Install new wire mesh & frame (:o 12 S.F.)				.452	62.50	24		86.50	100	117
	Remove scaffold				1.333		77		77	102	126
	Total				3.919	62.50	211.50		274	**347.50**	**423**

B2023 170 Aluminum Frame Window Screen

	System Description	Freq. (Years)	Crew	Unit	Labor Hours	2020 Bare Costs				Total In-House	Total w/O&P
						Material	Labor	Equipment	Total		
1010	**Repair aluminum frame window screen**	20	1 CARP	S.F.							
	Remove screen from mountings				.052		3		3	3.98	4.91
	Repair damaged screen				.017	3.53	1		4.53	5.20	6.05
	Remount screen				.104		6		6	7.95	9.80
	Total				.173	3.53	10		13.53	**17.13**	**20.76**
1020	**Refinish 3′-0″ x 4′-0″ aluminum frame window screen**	5	1 PORD	Ea.							
	Prepare screen frame surface				.473		21		21	27	33.50
	Refinish screen frame surface, brushwork, primer + 1 coat				.615	1.22	27.50		28.72	36.50	45
	Total				1.088	1.22	48.50		49.72	**63.50**	**78.50**
1030	**Replace 3′-0″ x 4′-0″ aluminum frame window screen - 1st floor**	50	1 CARP	Ea.							
	Set up and secure scaffold				.444		23.50		23.50	30.50	38
	Remove screen from mountings				.624		36		36	48	59
	Install new screen				1.248	282	72		354	405	470
	Remove scaffold				.444		23.50		23.50	30.50	38
	Total				2.761	282	155		437	**514**	**605**
2030	**Replace 3′-0″ x 4′-0″ aluminum frame window screen - 2nd floor**	50	1 CARP	Ea.							
	Set up and secure scaffold				.889		47		47	61	75.50
	Remove screen from mountings				.624		36		36	48	59
	Install new screen				1.248	282	72		354	405	470
	Remove scaffold				.889		47		47	61	75.50
	Total				3.650	282	202		484	**575**	**680**

B2023 170 Aluminum Frame Window Screen

	System Description	Freq. (Years)	Crew	Unit	Labor Hours	2020 Bare Costs				Total In-House	Total w/O&P
						Material	Labor	Equipment	Total		
3030	**Replace 3′-0″ x 4′-0″ aluminum frame window screen - 3rd floor**	50	1 CARP	Ea.							
	Set up and secure scaffold				1.333		71		71	91.50	114
	Remove screen from mountings				.624		36		36	48	59
	Install new screen				1.248	282	72		354	405	470
	Remove scaffold				1.333		71		71	91.50	114
	Total				4.539	282	250		532	**636**	**757**

B2023 172 Steel Frame Window Screen

	System Description	Freq. (Years)	Crew	Unit	Labor Hours	2020 Bare Costs				Total In-House	Total w/O&P
						Material	Labor	Equipment	Total		
1010	**Repair steel frame window screen**	10	1 CARP	S.F.							
	Remove screen from mountings				.052		3		3	3.98	4.91
	Repair damaged screen				.017	3.53	1		4.53	5.20	6.05
	Remount screen				.104		6		6	7.95	9.80
	Total				.173	3.53	10		13.53	**17.13**	**20.76**
1020	**Refinish 3′-0″ x 4′-0″ steel frame window screen**	5	1 PORD	Ea.							
	Prepare screen frame surface				.473		21		21	27	33.50
	Refinish screen frame surface, brushwork, primer + 1 coat				.615	1.22	27.50		28.72	36.50	45
	Total				1.088	1.22	48.50		49.72	**63.50**	**78.50**
1030	**Replace 3′-0″ x 4′-0″ steel frame window screen - 1st floor**	40	1 CARP	Ea.							
	Set up and secure scaffold				.444		23.50		23.50	30.50	38
	Remove screen from mountings				.624		36		36	48	59
	Install new screen				1.248	55.56	72		127.56	157	187
	Remove scaffold				.444		23.50		23.50	30.50	38
	Total				2.761	55.56	155		210.56	**266**	**322**
2030	**Replace 3′-0″ x 4′-0″ steel frame window screen - 2nd floor**	40	1 CARP	Ea.							
	Set up and secure scaffold				.889		47		47	61	75.50
	Remove screen from mountings				.624		36		36	48	59
	Install new screen				1.248	55.56	72		127.56	157	187
	Remove scaffold				.889		47		47	61	75.50
	Total				3.650	55.56	202		257.56	**327**	**397**

B2023 172 Steel Frame Window Screen

	System Description	Freq. (Years)	Crew	Unit	Labor Hours	2020 Bare Costs Material	Labor	Equipment	Total	Total In-House	Total w/O&P
3030	**Replace 3′-0″ x 4′-0″ steel frame window screen - 3rd floor**	40	1 CARP	Ea.							
	Set up and secure scaffold				1.333		71		71	91.50	114
	Remove screen from mountings				.624		36		36	48	59
	Install new screen				1.248	55.56	72		127.56	157	187
	Remove scaffold				1.333		71		71	91.50	114
	Total				4.539	55.56	250		305.56	**388**	**474**

B2023 174 Wood Frame Window Screen

	System Description	Freq. (Years)	Crew	Unit	Labor Hours	2020 Bare Costs Material	Labor	Equipment	Total	Total In-House	Total w/O&P
1010	**Repair wood frame window screen**	10	1 CARP	S.F.							
	Remove screen from mountings				.052		3		3	3.98	4.91
	Repair damaged screen				.017	3.53	1		4.53	5.20	6.05
	Remount screen				.104		6		6	7.95	9.80
	Total				.173	3.53	10		13.53	**17.13**	**20.76**
1020	**Refinish 3′-0″ x 4′-0″ wood frame window screen**	5	1 PORD	Ea.							
	Prepare screen frame surface				.473		21		21	27	33.50
	Refinish screen frame surface, brushwork, primer + 1 coat				.615	1.22	27.50		28.72	36.50	45
	Total				1.088	1.22	48.50		49.72	**63.50**	**78.50**
1030	**Replace 3′-0″ x 4′-0″ wood frame window screen - 1st floor**	40	1 CARP	Ea.							
	Set up and secure scaffold				.444		23.50		23.50	30.50	38
	Remove screen from mountings				.624		36		36	48	59
	Install new screen				.666	101.40	35.40		136.80	157	183
	Remove scaffold				.444		23.50		23.50	30.50	38
	Total				2.179	101.40	118.40		219.80	**266**	**318**
2030	**Replace 3′-0″ x 4′-0″ wood frame window screen - 2nd floor**	40	1 CARP	Ea.							
	Set up and secure scaffold				.889		47		47	61	75.50
	Remove screen from mountings				.624		36		36	48	59
	Install new screen				.666	101.40	35.40		136.80	157	183
	Remove scaffold				.889		47		47	61	75.50
	Total				3.067	101.40	165.40		266.80	**327**	**393**

B2023 174 Wood Frame Window Screen

	System Description	Freq. (Years)	Crew	Unit	Labor Hours	2020 Bare Costs				Total In-House	Total w/O&P
						Material	Labor	Equipment	Total		
3030	**Replace 3′-0″ x 4′-0″ wood frame window screen - 3rd floor**	40	1 CARP	Ea.							
	Set up and secure scaffold				1.333		71		71	91.50	114
	Remove screen from mountings				.624		36		36	48	59
	Install new screen				.666	101.40	35.40		136.80	157	183
	Remove scaffold				1.333		71		71	91.50	114
	Total				3.956	101.40	213.40		314.80	**388**	**470**

B2023 324 Plate Glass

	System Description	Freq. (Years)	Crew	Unit	Labor Hours	2020 Bare Costs				Total In-House	Total w/O&P
						Material	Labor	Equipment	Total		
1020	**Replace plate glass storefront - 1st floor**	50	2 CARP	C.S.F.							
	Remove plate glass and frame				3.467		184		184	238	295
	Replace plate glass and frame				16.000	3,200	815		4,015	4,575	5,300
	Total				19.467	3,200	999		4,199	**4,813**	**5,595**
2020	**Replace plate glass storefront - 2nd floor**	50	2 CARP	C.S.F.							
	Set up and secure scaffold				1.500		79.50		79.50	103	128
	Remove plate glass and frame				3.467		184		184	238	295
	Replace plate glass and frame				16.000	3,200	815		4,015	4,575	5,300
	Remove scaffold				1.500		79.50		79.50	103	128
	Total				22.467	3,200	1,158		4,358	**5,019**	**5,851**
3020	**Replace plate glass storefront - 3rd floor**	50	2 CARP	C.S.F.							
	Set up and secure scaffold				2.250		119.25		119.25	155	192
	Remove plate glass and frame				3.467		184		184	238	295
	Replace plate glass and frame				16.000	3,200	815		4,015	4,575	5,300
	Remove scaffold				2.250		119.25		119.25	155	192
	Total				23.967	3,200	1,237.50		4,437.50	**5,123**	**5,979**

B2033 111 Glazed Aluminum

	System Description	Freq. (Years)	Crew	Unit	Labor Hours	2020 Bare Costs				Total In-House	Total w/O&P
						Material	Labor	Equipment	Total		
1020	**Repair aluminum storefront door**	12	1 CARP	Ea.							
	Remove weatherstripping				.015		.78		.78	1.01	1.25
	Remove door closer				.167		8.85		8.85	11.45	14.20
	Remove door hinge				.167		8.85		8.85	11.45	14.20
	Install new weatherstrip				3.478	52	185		237	296	360
	Install new door closer				.868	105.50	46		151.50	176	206
	Install new hinge (50% of total hinges / 12 years)				.167		8.85		8.85	11.45	14.20
	Hinge, 5 x 5, brass base					69.50			69.50	76.50	87
	Oil / lubricate door closer				.065		3.43		3.43	4.43	5.50
	Oil / lubricate hinges				.065		3.43		3.43	4.43	5.50
	Total				4.990	227	265.19		492.19	**592.72**	**707.85**
1030	**Replace 3′-0″ x 7′-0″ aluminum storefront doors**	50	1 CARP	Ea.							
	Remove door				.650		27.50		27.50	35.50	44
	Remove door frame				1.300		69		69	89.50	111
	Install new door, frame, and hardware				10.458	940	605		1,545	1,825	2,175
	Install new glass				3.640	166.95	185.85		352.80	425	505
	Total				16.048	1,106.95	887.35		1,994.30	**2,375**	**2,835**
2010	**Replace insulating glass - (3% of glass)**	1	1 CARP	S.F.							
	Remove damaged glass				.052		2.19		2.19	2.83	3.50
	Install new insulating glass				.277	34	14.15		48.15	55.50	65
	Total				.329	34	16.34		50.34	**58.33**	**68.50**
2020	**Repair aluminum storefront sliding door**	12	1 CARP	Ea.							
	Remove weatherstripping				.015		.78		.78	1.01	1.25
	Oil / lubricate door closer				.065		3.43		3.43	4.43	5.50
	Oil / inspect wheel & track				.065		3.43		3.43	4.43	5.50
	Install new door track / glide wheels (25% of total / 12 years)				.440	15.25	23.38		38.63	47	56.50
	Install new weatherstrip				4.167	53.50	221		274.50	345	420
	Total				4.750	68.75	252.02		320.77	**401.87**	**488.75**

B2033 130 Glazed Wood

	System Description	Freq. (Years)	Crew	Unit	Labor Hours	2020 Bare Costs				Total In-House	Total w/O&P
						Material	Labor	Equipment	Total		
1020	**Repair glazed wood doors**	12	1 CARP	Ea.							
	Remove weatherstripping				.015		.78		.78	1.01	1.25
	Remove door closer				.167		8.85		8.85	11.45	14.20
	Remove door hinge				.167		8.85		8.85	11.45	14.20
	Install new weatherstrip				3.478	52	185		237	296	360
	Install new door closer				.868	105.50	46		151.50	176	206
	Install new hinge (50% of total hinges / 12 years)				.167		3.85		8.85	11.45	14.20
	Hinge, 5 x 5, brass base					69.50			69.50	76.50	87
	Oil / lubricate door closer				.065		3.43		3.43	4.43	5.50
	Oil / lubricate hinges				.065		3.43		3.43	4.43	5.50
	Total				4.990	227	265.19		492.19	**592.72**	**707.85**
1040	**Replace 3′-0″ x 7′-0″ glazed wood door**	40	1 CARP	Ea.							
	Remove door				.650		27.50		27.50	35.50	44
	Remove door frame				.743		39.50		39.50	51	63.50
	Install new door frame				.943	136.85	50.15		187	215	251
	Install new oak door sill				.832	86	44.20		130.20	152	178
	Install exterior door				2.105	880	112		992	1,125	1,275
	Total				5.274	1,102.85	273.35		1,376.20	**1,578.50**	**1,811.50**
1050	**Refinish glazed wood doors**	4	1 PORD	Ea.							
	Paint door, door frame & trim, brushwork, primer + 2 coats				1.760	17.30	78		95.30	119	146
	Total				1.760	17.30	78		95.30	**119**	**146**
2010	**Replace insulating glass - (8% of glass)**	1	1 CARP	S.F.							
	Remove damaged glass				.052		2.19		2.19	2.83	3.50
	Install new insulated glass				.277	34	14.15		48.15	55.50	65
	Total				.329	34	16.34		50.34	**58.33**	**68.50**
2020	**Repair glazed wood sliding door**	14	1 CARP	Ea.							
	Remove weatherstripping				.015		.78		.78	1.01	1.25
	Oil / lubricate door closer				.065		3.43		3.43	4.43	5.50
	Oil / inspect wheel & track				.065		3.43		3.43	4.43	5.50
	Install new door track / glide wheels (25% of total / 14 years)				.440	15.25	23.38		38.63	47	56.50
	Install new weatherstrip				4.167	53.50	221		274.50	345	420
	Total				4.750	68.75	252.02		320.77	**401.87**	**488.75**

B2033 130 Glazed Wood

	System Description	Freq. (Years)	Crew	Unit	Labor Hours	2020 Bare Costs				Total In-House	Total w/O&P
						Material	Labor	Equipment	Total		
2030	**Prepare and refinish glazed wood sliding door**	4	1 PORD	Ea.							
	Prepare surface				.310		13.75		13.75	17.60	22
	Paint door surface, brushwork, primer + 2 coats				.310	7.54	13.78		21.32	26	31.50
	Total				.621	7.54	27.53		35.07	**43.60**	**53.50**
2040	**Replace 6′-0″ x 7′-0″ glazed wood sliding door**	40	2 CARP	Ea.							
	Remove sliding door				1.733		92		92	119	148
	Install new sliding door				5.200	1,525	276		1,801	2,025	2,350
	Total				6.933	1,525	368		1,893	**2,144**	**2,498**
2050	**Refinish glazed wood sliding door & frame**	4	1 PORD	Ea.							
	Prepare surface				.465		20.63		20.63	26.50	33
	Paint door, door frame & trim				1.760	17.30	78		95.30	119	146
	Total				2.225	17.30	98.63		115.93	**145.50**	**179**

B2033 206 Solid Core, Painted

	System Description	Freq. (Years)	Crew	Unit	Labor Hours	2020 Bare Costs				Total In-House	Total w/O&P
						Material	Labor	Equipment	Total		
1010	**Repair solid core door, painted**	12	1 CARP	Ea.							
	Remove weatherstripping				.015		.78		.78	1.01	1.25
	Remove door hinge				.250		13.28		13.28	17.20	21.50
	Install new weatherstrip				3.478	52	185		237	296	360
	Install new hinge (50% of total hinges / 12 years)				.250		13.28		13.28	17.20	21.50
	Hinge, 5 x 5, brass base					104.25			104.25	115	130
	Oil / lubricate hinges				.065		3.43		3.43	4.43	5.50
	Total				4.058	156.25	215.77		372.02	**450.84**	**539.75**
1020	**Prepare and refinish solid core door, painted**	4	1 PORD	Ea.							
	Prepare door surface				.333		14.80		14.80	18.90	23.50
	Paint door surface, brushwork, primer + 2 coats				.549	13.34	24.38		37.72	46	55.50
	Total				.883	13.34	39.18		52.52	**64.90**	**79**

	System Description	Freq. (Years)	Crew	Unit	Labor Hours	2020 Bare Costs				Total In-House	Total w/O&P
						Material	Labor	Equipment	Total		
1030	**Replace 3'-0" x 7'-0" solid core door, painted**	40	1 CARP	Ea.							
	Remove door				.650		27.50		27.50	35.50	44
	Remove door frame				.743		39.50		39.50	51	63.50
	Install new door frame				.943	136.85	50.15		187	215	251
	Install new oak door sill				.832	86	44.20		130.20	152	178
	Install new solid core door				1.143	182	60.50		242.50	279	325
	Install new weatherstrip				4.167	53.50	221		274.50	345	420
	Install new lockset				.867	81	46		127	149	175
	Total				9.345	539.35	488.85		1,028.20	**1,226.50**	**1,456.50**
2010	**Repair solid core sliding wood door**	14	1 CARP	Ea.							
	Remove weatherstripping				.015		.78		.78	1.01	1.25
	Oil / lubricate door closer				.065		3.43		3.43	4.43	5.50
	Oil / inspect wheel & track				.065		3.43		3.43	4.43	5.50
	Install new door track / glide wheels (25% of total / 14 years)				.440	15.25	23.38		38.63	47	56.50
	Install new weatherstrip				4.167	53.50	221		274.50	345	420
	Total				4.750	68.75	252.02		320.77	**401.87**	**488.75**
2030	**Replace 3'-0" x 7'-0" solid core sliding wood door**	40	1 CARP	Ea.							
	Remove sliding door				1.733		92		92	119	148
	Install new sliding door frame				.998	124.20	53.10		177.30	205	240
	Install new coor track / glide wheels				1.760	61	93.50		154.50	188	226
	Install new solid core door				1.143	182	60.50		242.50	279	325
	Install new weatherstrip				3.478	52	185		237	296	360
	Install new lockset				.867	81	46		127	149	175
	Total				9.980	500.20	530.10		1,030.30	**1,236**	**1,474**
3010	**Replace sliding door glass - (3% of glass)**	1	1 CARP	S.F.							
	Remove damaged glass				.052		2.19		2.19	2.83	3.50
	Install new glass				.173	7.95	8.85		16.80	20	24
	Total				.225	7.95	11.04		18.99	**22.83**	**27.50**

B2033 206 Solid Core, Painted

	System Description	Freq. (Years)	Crew	Unit	Labor Hours	2020 Bare Costs				Total In-House	Total w/O&P
						Material	Labor	Equipment	Total		
3040	**Replace 3′-0″ x 7′-0″ solid core door, w/ safety glass, painted**	40	1 CARP	Ea.							
	Remove door				.650		27.50		27.50	35.50	44
	Remove door frame				.743		39.50		39.50	51	63.50
	Install new door frame				.943	136.85	50.15		187	215	251
	Install new oak door sill				.832	86	44.20		130.20	152	178
	Install new solid core door				1.143	182	60.50		242.50	279	325
	Install new glass (3% of total glass/year)				.201	9.22	10.27		19.49	23.50	28
	Install new weatherstrip				4.167	53.50	221		274.50	345	420
	Install new lockset				.867	81	46		127	149	175
	Total				9.546	548.57	499.12		1,047.69	**1,250**	**1,484.50**

B2033 210 Aluminum Doors

	System Description	Freq. (Years)	Crew	Unit	Labor Hours	2020 Bare Costs				Total In-House	Total w/O&P
						Material	Labor	Equipment	Total		
1010	**Repair aluminum door**	12	1 CARP	Ea.							
	Remove door closer				.167		8.85		8.85	11.45	14.20
	Remove door hinge				.250		13.28		13.28	17.20	21.50
	Install new door closer				1.735	211	92		303	350	410
	Install new hinge (50% of total hinges / 12 years)				.250		13.28		13.28	17.20	21.50
	Hinge, 5 x 5, brass base					104.25			104.25	115	130
	Oil / lubricate door closer				.065		3.43		3.43	4.43	5.50
	Oil / lubricate hinges				.065		3.43		3.43	4.43	5.50
	Total				2.531	315.25	134.27		449.52	**519.71**	**608.20**
1020	**Replace 3′-0″ x 7′-0″ flush aluminum door**	50	1 CARP	Ea.							
	Remove door				.650		27.50		27.50	35.50	44
	Remove cylinder lock				.400		21.50		21.50	27.50	34
	Reinstall cylinder lock				1.156		61.50		61.50	79.50	98.50
	Install new door				1.224	1,625	65		1,690	1,875	2,125
	Total				3.431	1,625	175.50		1,800.50	**2,017.50**	**2,301.50**
2010	**Replace wire glass - (3% of glass)**	1	1 CARP	S.F.							
	Remove damaged glass				.052		2.19		2.19	2.83	3.50
	Install new wire glass				.179	31.90	9.11		41.01	47	54.50
	Total				.231	31.90	11.30		43.20	**49.83**	**58**

B2033 210 Aluminum Doors

	System Description	Freq. (Years)	Crew	Unit	Labor Hours	2020 Bare Costs Material	Labor	Equipment	Total	Total In-House	Total w/O&P
2020	**Repair aluminum frame and door**	12	1 CARP	Ea.							
	Remove weatherstripping				.015		.78		.78	1.01	1.25
	Remove door closer				.167		8.85		8.85	11.45	14.20
	Remove door hinge				.250		13.28		13.28	17.20	21.50
	Install new weatherstrip				3.478	52	185		237	296	360
	Install new door closer				1.735	211	92		303	350	410
	Install new hinge (50% of total hinges / 12 years)				.250		13.28		13.28	17.20	21.50
	Hinge, 5 x 5, brass base					104.25			104.25	115	130
	Oil / lubricate door closer				.065		3.43		3.43	4.43	5.50
	Oil / lubricate hinges				.065		3.43		3.43	4.43	5.50
	Total				6.024	367.25	320.05		687.30	**816.72**	**969.45**
2030	**Replace 3′-0″ x 7′-0″ aluminum door, incl. vision lite**	50	1 CARP	Ea.							
	Remove door closer				.167		8.85		8.85	11.45	14.20
	Remove door				.650		27.50		27.50	35.50	44
	Install new door				1.224	1,625	65		1,690	1,875	2,125
	Reinstall door closer				1.735		92		92	119	148
	Install new glass				.217	9.94	11.06		21	25	30
	Total				3.993	1,634.94	204.41		1,839.35	**2,065.95**	**2,361.20**
3010	**Replace float glass - (3% of glass)**	1	1 CARP	S.F.							
	Remove damaged glass				.052		2.19		2.19	2.83	3.50
	Install new glass				.201	9.22	10.27		19.49	23.50	28
	Total				.253	9.22	12.46		21.68	**26.33**	**31.50**
3020	**Repair aluminum sliding door**	12	1 CARP	Ea.							
	Remove weatherstripping				.015		.78		.78	1.01	1.25
	Oil / lubricate door closer				.065		3.43		3.43	4.43	5.50
	Install new door closer (25% of total / 12 years)				.434	52.75	23		75.75	88	103
	Oil / inspect wheel & track				.065		3.43		3.43	4.43	5.50
	Install new door track / glide wheels (25% of total / 12 years)				.440	15.25	23.38		38.63	47	56.50
	Install new weatherstrip				4.167	53.50	221		274.50	345	420
	Total				5.184	121.50	275.02		396.52	**489.87**	**591.75**

B2033 210 Aluminum Doors

	System Description	Freq. (Years)	Crew	Unit	Labor Hours	2020 Bare Costs				Total In-House	Total w/O&P
						Material	Labor	Equipment	Total		
3030	**Replace 3′-0″ x 7′-0″ aluminum sliding door**	50	1 CARP	Ea.							
	Remove door				.650		27.50		27.50	35.50	44
	Install new door track / glide wheels				1.760	61	93.50		154.50	188	226
	Install new aluminum door				1.224	1,625	65		1,690	1,875	2,125
	Total				3.634	1,686	186		1,872	**2,098.50**	**2,395**
4020	**Repair flush aluminum door, wood core**	12	1 CARP	Ea.							
	Remove weatherstripping				.015		.78		.78	1.01	1.25
	Remove door closer				.167		8.85		8.85	11.45	14.20
	Remove door hinge				.250		13.28		13.28	17.20	21.50
	Install new weatherstrip				3.478	52	185		237	296	360
	Install new door closer				1.735	211	92		303	350	410
	Install new hinge (50% of total hinges / 12 years)				.250		13.28		13.28	17.20	21.50
	Hinge, 5 x 5, brass base					104.25			104.25	115	130
	Oil / lubricate door closer				.065		3.43		3.43	4.43	5.50
	Oil / lubricate hinges				.065		3.43		3.43	4.43	5.50
	Total				6.024	367.25	320.05		687.30	**816.72**	**969.45**
4030	**Replace 3′-0″ x 7′-0″ aluminum door, wood core**	50	1 CARP	Ea.							
	Remove door				.650		27.50		27.50	35.50	44
	Remove door closer				.167		8.85		8.85	11.45	14.20
	Install new aluminum door				1.224	1,625	65		1,690	1,875	2,125
	Install door closer				1.735		92		92	119	148
	Total				3.777	1,625	193.35		1,818.35	**2,040.95**	**2,331.20**
5010	**Replace tempered glass - (3% of glass)**	1	1 CARP	S.F.							
	Remove damaged glass				.052		2.19		2.19	2.83	3.50
	Install new glass				.201	9.22	10.27		19.49	23.50	28
	Total				.253	9.22	12.46		21.68	**26.33**	**31.50**

B2033 210 Aluminum Doors

	System Description	Freq. (Years)	Crew	Unit	Labor Hours	2020 Bare Costs				Total In-House	Total w/O&P
						Material	Labor	Equipment	Total		
5020	**Repair aluminum door, insulated**	12	1 CARP	Ea.							
	Remove weatherstripping				.015		.78		.78	1.01	1.25
	Remove door closer				.167		8.85		8.85	11.45	14.20
	Remove door hinge				.250		13.28		13.28	17.20	21.50
	Install new weatherstrip				3.478	52	185		237	296	360
	Install new door closer				1.735	211	92		303	350	410
	Install new hinge (50% of total hinges / 12 years)				.250		13.28		13.28	17.20	21.50
	Hinge, 5 x 5, brass base					104.25			104.25	115	130
	Oil / lubricate door closer				.065		3.43		3.43	4.43	5.50
	Oil / lubricate hinges				.065		3.43		3.43	4.43	5.50
	Total				6.024	367.25	320.05		687.30	**816.72**	**969.45**
5030	**Replace 3′-0″ x 7′-0″ aluminum door, insulated**	50	1 CARP	Ea.							
	Remove door				.650		27.50		27.50	35.50	44
	Remove door closer				.167		8.85		8.85	11.45	14.20
	Install new aluminum door				1.224	1,625	65		1,690	1,875	2,125
	Install door closer				1.735		92		92	119	148
	Total				3.777	1,625	193.35		1,818.35	**2,040.95**	**2,331.20**

B2033 220 Steel, Painted

	System Description	Freq. (Years)	Crew	Unit	Labor Hours	2020 Bare Costs				Total In-House	Total w/O&P
						Material	Labor	Equipment	Total		
1010	**Repair steel door, painted**	14	1 CARP	Ea.							
	Remove weatherstripping				.015		.78		.78	1.01	1.25
	Remove door closer				.167		8.85		8.85	11.45	14.20
	Remove door hinge				.250		13.28		13.28	17.20	21.50
	Install new weatherstrip				3.478	52	185		237	296	360
	Install new door closer				1.735	211	92		303	350	410
	Install new hinge (50% of total hinges / 14 years)				.250		13.28		13.28	17.20	21.50
	Hinge, 5 x 5, brass base					104.25			104.25	115	130
	Oil / lubricate door closer				.065		3.43		3.43	4.43	5.50
	Oil / lubricate hinges				.065		3.43		3.43	4.43	5.50
	Total				6.024	367.25	320.05		687.30	**816.72**	**969.45**

B2033 220 Steel, Painted

	System Description	Freq. (Years)	Crew	Unit	Labor Hours	2020 Bare Costs				Total In-House	Total w/O&P
						Material	Labor	Equipment	Total		
1020	**Refinish 3′-0″ x 7′-0″ steel door, painted**	4	1 PORD	Ea.							
	Paint on finished metal surface, brushwork, primer + 1 coat				.800	5.10	35.50		40.60	51	63
	Total				.800	5.10	35.50		40.60	**51**	**63**
1030	**Replace 3′-0″ x 7′-0″ steel door, painted**	45	1 CARP	Ea.							
	Remove door				.650		27.50		27.50	35.50	44
	Remove cylinder lock				.400		21.50		21.50	27.50	34
	Install new door				1.224	515	65		580	650	750
	Reinstall cylinder lock				1.156		61.50		61.50	79.50	98.50
	Total				3.431	515	175.50		690.50	**792.50**	**926.50**
2010	**Replace tempered glass - (3% of glass)**	1	1 CARP	S.F.							
	Remove damaged glass				.052		2.19		2.19	2.83	3.50
	Install new glass				.201	9.22	10.27		19.49	23.50	28
	Total				.253	9.22	12.46		21.68	**26.33**	**31.50**
2030	**Prepare and refinish 3′-0″ x 7′-0″ steel door, painted**	4	1 PORD	Ea.							
	Prepare surface				.333		14.80		14.80	18.90	23.50
	Refinish door surface, brushwork, primer + 1 coat				1.333	4.82	59		63.82	81	100
	Total				1.667	4.82	73.80		78.62	**99.90**	**123.50**
2040	**Replace 3′-0″ x 7′-0″ steel door w/ wire glass, painted**	45	1 CARP	Ea.							
	Remove door				.650		27.50		27.50	35.50	44
	Remove door closer				.167		8.85		8.85	11.45	14.20
	Remove door frame & trim				1.300		69		69	89.50	111
	Install new steel door				1.224	475	65		540	605	700
	Install new steel frame				1.301	214	69		283	325	380
	Install new wire glass				.179	38.28	9.11		47.39	54	62.50
	Reinstall door closer				1.735		92		92	119	148
	Total				6.556	727.28	340.46		1,067.74	**1,239.45**	**1,459.70**

B2033 220 Steel, Painted

	System Description	Freq. (Years)	Crew	Unit	Labor Hours	2020 Bare Costs				Total In-House	Total w/O&P
						Material	Labor	Equipment	Total		
3010	**Repair steel sliding door, painted**	14	1 CARP	Ea.							
	Remove weatherstripping				.015		.78		.78	1.01	1.25
	Oil / lubricate door closer				.065		3.43		3.43	4.43	5.50
	Install new door closer (25% of total / 14 years)				.434	52.75	23		75.75	88	103
	Oil / inspect wheel & track				.065		3.43		3.43	4.43	5.50
	Install new door track / glide wheels (25% of total / 14 years)				.440	15.25	23.38		38.63	47	56.50
	Install new weatherstrip				4.167	53.50	221		274.50	345	420
	Total				5.184	121.50	275.02		396.52	**489.87**	**591.75**
3020	**Refinish 3′-0″ x 7′-0″ steel sliding door, painted**	4	1 PORD	Ea.							
	Prepare surface				.333		14.80		14.80	18.90	23.50
	Refinish door surface, brushwork, primer + 1 coat				1.333	4.82	59		63.82	81	100
	Total				1.667	4.82	73.80		78.62	**99.90**	**123.50**
3030	**Replace 3′-0″ x 7′-0″ steel sliding door, painted**	45	1 CARP	Ea.							
	Remove door				.650		27.50		27.50	35.50	44
	Remove door closer				.167		8.85		8.85	11.45	14.20
	Install new steel door				1.224	475	65		540	605	700
	Install door closer				1.735		92		92	119	148
	Total				3.777	475	193.35		668.35	**770.95**	**906.20**
4010	**Replace steel sliding door glass - (3% of glass)**	1	1 CARP	S.F.							
	Remove damaged glass				.052		2.19		2.19	2.83	3.50
	Install new glass				.201	9.22	10.27		19.49	23.50	28
	Total				.253	9.22	12.46		21.68	**26.33**	**31.50**
4020	**Repair steel door, insulated core, painted**	14	1 CARP	Ea.							
	Remove weatherstripping				.015		.78		.78	1.01	1.25
	Remove door closer				.167		8.85		8.85	11.45	14.20
	Remove door hinge				.167		8.85		8.85	11.45	14.20
	Install new weatherstrip				3.478	52	185		237	296	360
	Install new door closer				1.735	211	92		303	350	410
	Install new hinge (50% of total hinges / 14 years)				.167		8.85		8.85	11.45	14.20
	Hinge, 5 x 5, brass base					69.50			69.50	76.50	87
	Oil / lubricate door closer				.065		3.43		3.43	4.43	5.50
	Oil / lubricate hinges				.065		3.43		3.43	4.43	5.50
	Total				5.857	332.50	311.19		643.69	**766.72**	**911.85**

B2033 220 Steel, Painted

	System Description	Freq. (Years)	Crew	Unit	Labor Hours	2020 Bare Costs				Total In-House	Total w/O&P
						Material	Labor	Equipment	Total		
4030	**Refinish 3'-0" x 7'-0" steel door, insulated core, painted**	4	1 PORD	Ea.							
	Prepare surface				.333		14.80		14.80	18.90	23.50
	Refinish door surface, brushwork, primer + 1 coat				1.333	4.82	59		63.82	81	100
	Total				1.667	4.82	73.80		78.62	**99.90**	**123.50**
4040	**Replace 3'-0" x 7'-0" steel door, insulated core, painted**	45	1 CARP	Ea.							
	Remove door				.650		27.50		27.50	35.50	44
	Remove door closer				.167		8.85		8.85	11.45	14.20
	Remove door frame & trim				1.300		69		69	89.50	111
	Install new steel door				1.224	475	65		540	605	700
	Install new steel frame				1.301	214	69		283	325	380
	Install new insulated glass				.322	39.44	16.41		55.85	64.50	75.50
	Install door closer				1.735		92		92	119	148
	Total				6.699	728.44	347.76		1,076.20	**1,249.95**	**1,472.70**

B2033 221 Steel, Unpainted

	System Description	Freq. (Years)	Crew	Unit	Labor Hours	2020 Bare Costs				Total In-House	Total w/O&P
						Material	Labor	Equipment	Total		
1010	**Repair steel door, unpainted**	14	1 CARP	Ea.							
	Remove weatherstripping				.015		.78		.78	1.01	1.25
	Remove door closer				.167		8.85		8.85	11.45	14.20
	Remove door hinge				.167		8.85		8.85	11.45	14.20
	Install new weatherstrip				3.478	52	185		237	296	360
	Install new door closer				1.735	211	92		303	350	410
	Install new hinge (50% of total hinges / 14 years)				.167		8.85		8.85	11.45	14.20
	Hinge, 5 x 5, brass base					69.50			69.50	76.50	87
	Oil / lubricate door closer				.065		3.43		3.43	4.43	5.50
	Oil / lubricate hinges				.065		3.43		3.43	4.43	5.50
	Total				5.857	332.50	311.19		643.69	**766.72**	**911.85**
1020	**Replace 3'-0" x 7'-0" steel door, unpainted**	45	1 CARP	Ea.							
	Remove door				.650		27.50		27.50	35.50	44
	Remove door frame & trim				1.300		69		69	89.50	111
	Install new steel door				1.224	615	65		680	760	875
	Install new steel frame				1.301	214	69		283	325	380
	Total				4.475	829	230.50		1,059.50	**1,210**	**1,410**

B2033 221 Steel, Unpainted

	System Description	Freq. (Years)	Crew	Unit	Labor Hours	2020 Bare Costs				Total In-House	Total w/O&P
						Material	Labor	Equipment	Total		
2010	**Replace wire glass - (3% of glass)**	1	1 CARP	S.F.							
	Remove damaged glass				.052		2.19		2.19	2.83	3.50
	Install new wire glass				.179	31.90	9.11		41.01	47	54.50
	Total				.231	31.90	11.30		43.20	**49.83**	**58**
2020	**Repair steel door with safety glass, unpainted**	14	1 CARP	Ea.							
	Remove weatherstripping				.015		.78		.78	1.01	1.25
	Remove door closer				.167		8.85		8.85	11.45	14.20
	Remove door hinge				.167		8.85		8.85	11.45	14.20
	Install new weatherstrip				3.478	52	185		237	296	360
	Install new door closer				1.735	211	92		303	350	410
	Install new hinge (50% of total hinges / 14 years)				.167		8.85		8.85	11.45	14.20
	Hinge, 5 x 5, brass base					69.50			69.50	76.50	87
	Oil / lubricate door closer				.065		3.43		3.43	4.43	5.50
	Oil / lubricate hinges				.065		3.43		3.43	4.43	5.50
	Total				5.857	332.50	311.19		643.69	**766.72**	**911.85**
2030	**Repl. 3′-0″ x 7′-0″ steel door with safety glass, unpainted**	45	1 CARP	Ea.							
	Remove door				.650		27.50		27.50	35.50	44
	Remove door closer				.167		8.85		8.85	11.45	14.20
	Remove door frame & trim				1.300		69		69	89.50	111
	Install new steel door				1.301	605	69		674	755	865
	Install new steel frame				1.301	214	69		283	325	380
	Install new wire glass				.308	66	15.70		81.70	93	108
	Install door closer				1.735		92		92	119	148
	Total				6.762	885	351.05		1,236.05	**1,428.45**	**1,670.20**
3010	**Repair steel sliding door, unpainted**	14	1 CARP	Ea.							
	Remove weatherstripping				.015		.78		.78	1.01	1.25
	Oil / lubricate door closer				.065		3.43		3.43	4.43	5.50
	Install new door closer (25% of total / 14 years)				.434	52.75	23		75.75	88	103
	Oil / inspect wheel & track				.065		3.43		3.43	4.43	5.50
	New door track / glide wheels (25% of total / 14 years)				.440	15.25	23.38		38.63	47	56.50
	Total				1.018	68	54.02		122.02	**144.87**	**171.75**

B2033 221 Steel, Unpainted

	System Description	Freq. (Years)	Crew	Unit	Labor Hours	2020 Bare Costs				Total In-House	Total w/O&P
						Material	Labor	Equipment	Total		
3020	**Replace 3′-0″ x 7′-0″ steel sliding door, unpainted**	45	1 CARP	Ea.							
	Remove door				.650		27.50		27.50	35.50	44
	Remove door closer				.167		8.85		8.85	11.45	14.20
	Install new steel door				1.224	615	65		680	760	875
	Reinstall door closer				1.735		92		92	119	148
	Total				3.776	615	193.35		808.35	**925.95**	**1,081.20**
4010	**Replace tempered glass - (3% of glass)**	1	1 CARP	S.F.							
	Remove damaged glass				.052		2.19		2.19	2.83	3.50
	Install new glass				.201	9.22	10.27		19.49	23.50	28
	Total				.253	9.22	12.46		21.68	**26.33**	**31.50**
4020	**Repair steel door, insulated core, unpainted**	14	1 CARP	Ea.							
	Remove weatherstripping				.015		.78		.78	1.01	1.25
	Remove door closer				.167		8.85		8.85	11.45	14.20
	Remove door hinge				.167		8.85		8.85	11.45	14.20
	Install new weatherstrip				3.478	52	185		237	296	360
	Install new door closer (50% of total / 14 years)				.868	105.50	46		151.50	176	206
	Install new hinge (50% of total hinges / 14 years)				.167		8.85		8.85	11.45	14.20
	Hinge, 5 x 5, brass base					69.50			69.50	76.50	87
	Oil / lubricate door closer				.065		3.43		3.43	4.43	5.50
	Oil / lubricate hinges				.065		3.43		3.43	4.43	5.50
	Total				4.990	227	265.19		492.19	**592.72**	**707.85**
4030	**Repl. 3′-0″ x 7′-0″ steel door, insulated core, unpainted**	45	1 CARP	Ea.							
	Remove door				.650		27.50		27.50	35.50	44
	Remove door closer				.167		8.85		8.85	11.45	14.20
	Remove door frame & trim				1.300		69		69	89.50	111
	Install new insulated steel door				1.301	690	69		759	850	975
	Install new steel frame				1.301	214	69		283	325	380
	Install new insulated glass				.322	39.44	16.41		55.85	64.50	75.50
	Reinstall door closer				1.735		92		92	119	148
	Total				6.776	943.44	351.76		1,295.20	**1,494.95**	**1,747.70**

B2033 410 Steel Roll-Up, Single

	System Description	Freq. (Years)	Crew	Unit	Labor Hours	2020 Bare Costs				Total In-House	Total w/O&P
						Material	Labor	Equipment	Total		
1010	**Repair 12′ x 12′ steel single roll-up door**	10	1 CARP	Ea.							
	Remove overhead type door section				1.303		69.25		69.25	89.50	111
	Install new roll-up door section				4.348	312.50	250		562.50	675	805
	Total				5.651	312.50	319.25		631.75	**764.50**	**916**
1020	**Refinish 12′ x 12′ steel single roll-up door**	5	1 PORD	Ea.							
	Set up, secure and take down ladder				.258		11.45		11.45	14.65	18.20
	Prepare door surface				2.000		88.80		88.80	114	141
	Paint overhead door, brushwork, primer + 1 coat, 1 side				1.493	34.02	66.78		100.80	122	148
	Total				3.751	34.02	167.03		201.05	**250.65**	**307.20**
1030	**Replace 12′ x 12′ steel single roll-up door**	35	2 CARP	Ea.							
	Remove overhead type door				5.212		277		277	360	445
	Install new roll-up door				17.391	1,250	1,000		2,250	2,700	3,225
	Total				22.603	1,250	1,277		2,527	**3,060**	**3,670**

B2033 412 Steel Roll-Up, Double

	System Description	Freq. (Years)	Crew	Unit	Labor Hours	2020 Bare Costs				Total In-House	Total w/O&P
						Material	Labor	Equipment	Total		
1020	**Refinish 12′ x 24′ steel double roll-up door**	5	1 PORD	Ea.							
	Set up, secure and take down ladder				.533		23.50		23.50	30.50	37.50
	Prepare door surface				4.000		177.60		177.60	227	282
	Paint overhead door, brushwork, primer + 1 coat, 1 side				2.986	68.04	133.56		201.60	244	297
	Total				7.519	68.04	334.66		402.70	**501.50**	**616.50**
1040	**Replace 12′ x 24′ steel double roll-up door**	35	2 CARP	Ea.							
	Remove overhead type door				10.423		554		554	715	890
	Install new door				34.783	2,500	2,000		4,500	5,425	6,425
	Total				45.206	2,500	2,554		5,054	**6,140**	**7,315**

B2033 413 Steel Garage Door, Single Leaf, Spring

	System Description	Freq. (Years)	Crew	Unit	Labor Hours	2020 Bare Costs				Total In-House	Total w/O&P
						Material	Labor	Equipment	Total		
1010	**Repair 9′ x 7′ steel, single leaf, garage door**	10	1 CARP	Ea.							
	Repair damaged track				.068		3.63		3.63	4.70	5.80
	Remove cremone lockset				.400		21.50		21.50	27.50	34
	Install new cremone lockset				1.040	163	55.50		218.50	251	292
	Retension door coil spring				1.143		60.50		60.50	78.50	97.50
	Total				2.651	163	141.13		304.13	**361.70**	**429.30**
1020	**Refinish 9′ x 7′ steel, single leaf, garage door**	5	1 PORD	Ea.							
	Set up, secure and take down ladder				.258		11.45		11.45	14.65	18.20
	Prepare surface				2.000		88.80		88.80	114	141
	Paint garage door, brushwork, primer + 1 coat, 1 side				1.493	34.02	66.78		100.80	122	148
	Total				3.751	34.02	167.03		201.05	**250.65**	**307.20**
1030	**Replace 9′ x 7′ steel, single leaf, garage door**	35	2 CARP	Ea.							
	Remove door				2.602		138		138	179	222
	Install new door				2.602	1,100	138		1,238	1,400	1,600
	Total				5.203	1,100	276		1,376	**1,579**	**1,822**

B2033 420 Aluminum Roll-Up, Single

	System Description	Freq. (Years)	Crew	Unit	Labor Hours	2020 Bare Costs				Total In-House	Total w/O&P
						Material	Labor	Equipment	Total		
1010	**Repair 12′ x 12′ aluminum single roll-up door**	10	1 CARP	Ea.							
	Remove door section				1.303		69.25		69.25	89.50	111
	Install new door section				3.478	856.25	185		1,041.25	1,175	1,375
	Total				4.781	856.25	254.25		1,110.50	**1,264.50**	**1,486**
1020	**Refinish 12′ x 12′ aluminum single roll-up door**	5	1 PORD	Ea.							
	Set up, secure and take down ladder				.258		11.45		11.45	14.65	18.20
	Prepare surface				2.000		88.80		88.80	114	141
	Paint overhead type door				1.493	34.02	66.78		100.80	122	148
	Total				3.751	34.02	167.03		201.05	**250.65**	**307.20**
1030	**Replace 12′ x 12′ aluminum single roll-up door**	35	2 CARP	Ea.							
	Paint overhead door, brushwork, primer + 1 coat, 1 side				1.706	38.88	76.32		115.20	140	170
	Install new door				13.913	3,425	740		4,165	4,725	5,450
	Total				15.619	3,463.88	816.32		4,280.20	**4,865**	**5,620**

B2033 422 Aluminum Roll-Up, Double

	System Description	Freq. (Years)	Crew	Unit	Labor Hours	2020 Bare Costs				Total In-House	Total w/O&P
						Material	Labor	Equipment	Total		
1010	**Repair 12′ x 24′ aluminum double roll-up door**	10	1 CARP	Ea.							
	Remove door section				2.606		138.50		138.50	179	223
	Install new door section				6.957	1,712.50	370		2,082.50	2,350	2,725
	Total				9.562	1,712.50	508.50		2,221	**2,529**	**2,948**
1020	**Refinish 12′ x 24′ aluminum double roll-up door**	5	1 PORD	Ea.							
	Set up, secure and take down ladder				.533		23.50		23.50	30.50	37.50
	Prepare surface				4.000		177.60		177.60	227	282
	Paint overhead door, brushwork, primer + 1 coat, 1 side				2.986	68.04	133.56		201.60	244	297
	Total				7.519	68.04	334.66		402.70	**501.50**	**616.50**
1030	**Replace 12′ x 24′ aluminum double roll-up door**	35	2 CARP	Ea.							
	Remove overhead type door				10.423		554		554	715	890
	Install new door				27.826	6,850	1,480		8,330	9,450	10,900
	Total				38.250	6,850	2,034		8,884	**10,165**	**11,790**

B2033 423 Aluminum Garage Door, Single Leaf, Spring

	System Description	Freq. (Years)	Crew	Unit	Labor Hours	2020 Bare Costs				Total In-House	Total w/O&P
						Material	Labor	Equipment	Total		
1010	**Repair 9′ x 7′ aluminum, single leaf, garage door**	10	1 CARP	Ea.							
	Repair damaged track				.684		36.30		36.30	47	58
	Remove cremone lockset				.400		21.50		21.50	27.50	34
	Install new cremone lockset				1.040	163	55.50		218.50	251	292
	Retension door coil spring				1.143		60.50		60.50	78.50	97.50
	Total				3.267	163	173.80		336.80	**404**	**481.50**
1020	**Refinish 9′ x 7′ aluminum, single leaf, garage door**	5	1 PORD	Ea.							
	Set up, secure and take down ladder				.258		11.45		11.45	14.65	18.20
	Prepare surface				2.000		88.80		88.80	114	141
	Paint garage door, brushwork, primer + 1 coat, 1 side				.747	17.01	33.39		50.40	61	74
	Total				3.005	17.01	133.64		150.65	**189.65**	**233.20**
1030	**Replace 9′ x 7′ aluminum, single leaf, garage door**	35	2 CARP	Ea.							
	Remove door				2.602		138		138	179	222
	Install new door				2.602	1,100	138		1,238	1,400	1,600
	Total				5.203	1,100	276		1,376	**1,579**	**1,822**

B2033 430 Wood Roll-Up, Single

	System Description	Freq. (Years)	Crew	Unit	Labor Hours	2020 Bare Costs				Total In-House	Total w/O&P
						Material	Labor	Equipment	Total		
1010	**Repair 12′ x 12′ wood single roll-up door**	8	1 CARP	Ea.							
	Remove door section				1.303		69.25		69.25	89.50	111
	Install new door section				3.478	693.75	185		878.75	1,000	1,150
	Total				4.781	693.75	254.25		948	**1,089.50**	**1,261**
1020	**Refinish 12′ x 12′ wood single roll-up door**	5	1 PORD	Ea.							
	Set up, secure and take down ladder				.258		11.45		11.45	14.65	18.20
	Prepare door surface				2.000		88.80		88.80	114	141
	Paint overhead door, brushwork, primer + 1 coat, 1 side				3.150	16.38	139.86		156.24	197	242
	Total				5.408	16.38	240.11		256.49	**325.65**	**401.20**
1030	**Replace 12′ x 12′ wood single roll-up door**	16	2 CARP	Ea.							
	Remove overhead type door				5.212		277		277	360	445
	Install new door				13.913	2,775	740		3,515	4,000	4,650
	Total				19.125	2,775	1,017		3,792	**4,360**	**5,095**

B2033 432 Wood Roll-Up, Double

	System Description	Freq. (Years)	Crew	Unit	Labor Hours	2020 Bare Costs				Total In-House	Total w/O&P
						Material	Labor	Equipment	Total		
1010	**Repair 12′ x 24′ wood double roll-up door**	8	1 CARP	Ea.							
	Remove door section				2.606		138.50		138.50	179	223
	Install new door section				6.957	1,387.50	370		1,757.50	2,000	2,325
	Total				9.562	1,387.50	508.50		1,896	**2,179**	**2,548**
1020	**Refinish 12′ x 24′ wood double roll-up door**	5	1 PORD	Ea.							
	Set up, secure and take down ladder				.533		23.50		23.50	30.50	37.50
	Prepare surface				4.000		177.60		177.60	227	282
	Paint overhead door, brushwork, primer + 1 coat, 1 side				6.300	32.76	279.72		312.48	395	485
	Total				10.833	32.76	480.82		513.58	**652.50**	**804.50**
1030	**Replace 12′ x 24′ wood double roll-up door**	16	2 CARP	Ea.							
	Remove overhead type door				10.423		554		554	715	890
	Install new door				27.826	5,550	1,480		7,030	8,025	9,300
	Total				38.250	5,550	2,034		7,584	**8,740**	**10,190**

B2033 433 Wood Garage Door, Single Leaf, Spring

	System Description	Freq. (Years)	Crew	Unit	Labor Hours	2020 Bare Costs				Total In-House	Total w/O&P
						Material	Labor	Equipment	Total		
1010	**Repair 9′ x 7′ wood, single leaf, garage door**	8	1 CARP	Ea.							
	Repair damaged track				.068		3.63		3.63	4.70	5.80
	Remove cremone lockset				.400		21.50		21.50	27.50	34
	Install new cremone lockset				1.040	163	55.50		218.50	251	292
	Retension door coil spring				1.143		60.50		60.50	78.50	97.50
	Total				2.651	163	141.13		304.13	**361.70**	**429.30**
1020	**Refinish 9′ x 7′ wood, single leaf, garage door**	5	1 PORD	Ea.							
	Set up, secure and take down ladder				.258		11.45		11.45	14.65	18.20
	Prepare surface				2.000		88.80		88.80	114	141
	Paint garage door, brushwork, primer + 1 coat, 1 side				3.150	16.38	139.86		156.24	197	242
	Total				5.408	16.38	240.11		256.49	**325.65**	**401.20**
1030	**Replace 9′ x 7′ wood, single leaf, garage door**	16	2 CARP	Ea.							
	Remove door				2.602		138		138	179	222
	Install new door				2.602	785	138		923	1,050	1,200
	Total				5.203	785	276		1,061	**1,229**	**1,422**

B2033 513 Electric Bifolding Hangar Door

	System Description	Freq. (Years)	Crew	Unit	Labor Hours	2020 Bare Costs				Total In-House	Total w/O&P
						Material	Labor	Equipment	Total		
0120	**Remove and replace electric bi-folding hangar door motor**	15	2 SKWK	Ea.							
	Hangar door,electric bi-folding door, replace motor				3.200	495	176		671	770	900
	Total				3.200	495	176		671	**770**	**900**
0140	**Remove and replace electric bi-folding hangar door cables**	15	2 SKWK	Ea.							
	Hangar door, elec bi-folding door, replace cables				8.000	149	440		589	725	885
	Total				8.000	149	440		589	**725**	**885**
0160	**Remove and replace electric bi-folding hangar door**	20	2 SKWK	Ea.							
	Hangar door demo				77.568		4,464		4,464	5,925	7,325
	Hangar door, bi-fldg dr, remove and replace 20′ x 80′				116.368	36,000	6,704		42,704	48,500	56,000
	Total				193.936	36,000	11,168		47,168	**54,425**	**63,325**

B2033 810 Hinges, Brass

	System Description	Freq. (Years)	Crew	Unit	Labor Hours	2020 Bare Costs				Total In-House	Total w/O&P
						Material	Labor	Equipment	Total		
1010	**Replace brass door hinge**	60	1 CARP	Ea.							
	Remove door				.650		27.50		27.50	35.50	44
	Remove door hinge				.167		8.85		8.85	11.45	14.20
	Install new hinge				.167		8.85		8.85	11.45	14.20
	Full mortise hinge, 5 x 5, brass base					69.50			69.50	76.50	87
	Reinstall exterior door				1.388		74		74	95.50	118
	Total				2.371	69.50	119.20		188.70	**230.40**	**277.40**

B2033 820 Lockset, Brass

	System Description	Freq. (Years)	Crew	Unit	Labor Hours	2020 Bare Costs				Total In-House	Total w/O&P
						Material	Labor	Equipment	Total		
1010	**Replace brass door lockset**	30	1 CARP	Ea.							
	Remove mortise lock set				.421		22.50		22.50	29	36
	Mortise lock				1.301	345	69		414	470	540
	Total				1.722	345	91.50		436.50	**499**	**576**

B2033 830 Door Closer, Brass

	System Description	Freq. (Years)	Crew	Unit	Labor Hours	2020 Bare Costs				Total In-House	Total w/O&P
						Material	Labor	Equipment	Total		
1010	**Replace brass door closer**	15	1 CARP	Ea.							
	Remove door closer				.167		8.85		8.85	11.45	14.20
	Install new door closer				1.735	211	92		303	350	410
	Total				1.902	211	100.85		311.85	**361.45**	**424.20**

B2033 840 Door Opener

	System Description	Freq. (Years)	Crew	Unit	Labor Hours	2020 Bare Costs				Total In-House	Total w/O&P
						Material	Labor	Equipment	Total		
1010	**Automatic door opener on existing door**	7	1 ELEC	Ea.							
	Cutout demolition of exterior wall for actuator button				1.739		74	15.55	89.55	113	135
	Install 1/2" E.M.T. conduit				2.462	30.40	151.20		181.60	219	270
	Install conductor, type THWN copper, #14				1.040	10.01	63.70		73.71	89.50	111
	Install junction box				.533	11.75	32.50		44.25	53	65
	Install outlet				.381	10.65	23.50		34.15	40.50	49.50
	Install plate				.100	2.72	6.15		8.87	10.55	12.85
	Automatic opener, button operation				20.000	3,925	1,100		5,025	5,725	6,650
	Total				26.255	3,990.53	1,451.05	15.55	5,457.13	**6,250.55**	**7,293.35**

B2033 850 Deadbolt, Brass

	System Description	Freq. (Years)	Crew	Unit	Labor Hours	2020 Bare Costs				Total In-House	Total w/O&P
						Material	Labor	Equipment	Total		
1010	**Replace brass deadbolt**	20	1 CARP	Ea.							
	Remove deadbolt				.154		8.20		8.20	10.55	13.10
	Install new deadbolt				1.156	193	61.50		254.50	292	340
	Total				1.309	193	69.70		262.70	**302.55**	**353.10**

B2033 860 Weatherstripping, Brass

	System Description	Freq. (Years)	Crew	Unit	Labor Hours	2020 Bare Costs				Total In-House	Total w/O&P
						Material	Labor	Equipment	Total		
1010	**Replace brass door weatherstripping**	20	1 CARP	Ea.							
	Remove weatherstripping				.015		.78		.78	1.01	1.25
	Install new weatherstrip				3.478	52	185		237	296	360
	Total				3.493	52	185.78		237.78	**297.01**	**361.25**

B2033 870 Panic Device

	System Description	Freq. (Years)	Crew	Unit	Labor Hours	2020 Bare Costs Material	Labor	Equipment	Total	Total In-House	Total w/O&P
1010	**Replace door panic device**	25	1 CARP	Ea.							
	Remove panic bar				.421		22.50		22.50	29	36
	Install new panic bar				2.080	1,125	111		1,236	1,375	1,575
	Total				2.501	1,125	133.50		1,258.50	**1,404**	**1,611**

B2033 905 Hollow Core, Painted

	System Description	Freq. (Years)	Crew	Unit	Labor Hours	2020 Bare Costs Material	Labor	Equipment	Total	Total In-House	Total w/O&P
1010	**Repair hollow core door**	12	1 CARP	Ea.							
	Remove weatherstripping				.015		.78		.78	1.01	1.25
	Remove door hinge				.167		8.85		8.85	11.45	14.20
	Install new weatherstrip				3.478	52	185		237	296	360
	Install new hinge (50% of total hinges / 12 years)				.167		8.85		8.85	11.45	14.20
	Hinge, 5 x 5, brass base					69.50			69.50	76.50	87
	Oil / lubricate hinges				.065		3.43		3.43	4.43	5.50
	Total				3.891	121.50	206.91		328.41	**400.84**	**482.15**
1020	**Refinish 3′-0″ x 7′-0″ hollow core door**	4	1 PORD	Ea.							
	Prepare door surface				.333		14.80		14.80	18.90	23.50
	Paint door surface, brushwork, primer + 1 coat				.549	13.34	24.38		37.72	46	55.50
	Total				.883	13.34	39.18		52.52	**64.90**	**79**
1030	**Replace 3′-0″ x 7′-0″ hollow core door**	30	1 CARP	Ea.							
	Remove door				.650		27.50		27.50	35.50	44
	Remove door frame				.743		39.50		39.50	51	63.50
	Install new door frame				.943	136.85	50.15		187	215	251
	Install new oak door sill				.832	86	44.20		130.20	152	178
	Install new hollow core door				1.000	174	53		227	260	305
	Install new weatherstrip				3.478	52	185		237	296	360
	Install new lockset				.867	81	46		127	149	175
	Total				8.514	529.85	445.35		975.20	**1,158.50**	**1,376.50**

B2033 905 Hollow Core, Painted

	System Description	Freq. (Years)	Crew	Unit	Labor Hours	2020 Bare Costs Material	Labor	Equipment	Total	Total In-House	Total w/O&P
2010	**Repair hollow core sliding wood door**	14	1 CARP	Ea.							
	Remove weatherstripping				.015		.78		.78	1.01	1.25
	Oil / lubricate door closer				.065		3.43		3.43	4.43	5.50
	Oil / inspect wheel & track				.065		3.43		3.43	4.43	5.50
	Install new door track / glide wheels (25% of total / 14 years)				.440	15.25	23.38		38.63	47	56.50
	Install new weatherstrip				4.167	53.50	221		274.50	345	420
	Total				4.750	68.75	252.02		320.77	**401.87**	**488.75**
2030	**Replace 3′-0″ x 7′-0″ hollow core sliding wood door**	30	1 CARP	Ea.							
	Remove sliding door				1.733		92		92	119	148
	Install new sliding door frame				4.622	475	269	21	765	905	1,050
	Install new hollow core door				1.000	174	53		227	260	305
	Install new weatherstrip				3.478	52	185		237	296	360
	Lockset				.867	81	46		127	149	175
	Total				11.701	782	645	21	1,448	**1,729**	**2,038**

B2033 925 Metal Grated

	System Description	Freq. (Years)	Crew	Unit	Labor Hours	2020 Bare Costs Material	Labor	Equipment	Total	Total In-House	Total w/O&P
1010	**Repair metal grated door**	15	1 SSWK	Ea.							
	Remove damaged metal bar				.836		48.30		48.30	64	79
	Oil / lubricate door closer				.065		3.72		3.72	4.94	6.10
	Oil / lubricate hinges				.065		3.43		3.43	4.43	5.50
	Metal grating repair (3% of door area / 15 years)				.320	1.66	19	5.8[illegible]	26.54	33.50	39.50
	Replace damaged metal bar				1.600	14.40	93.50	14.7[illegible]	122.60	156	187
	Total				2.885	16.06	167.95	20.5[illegible]	204.59	**262.87**	**317.10**
1020	**Prepare and refinish metal grated painted door**	4	1 PORD	Ea.							
	Prepare grated door surface				.308		13.65		13.65	17.45	21.50
	Paint grated door surface, brushwork, 1 coat				.319	3.68	14.26		17.94	22	27
	Total				.627	3.68	27.91		31.59	**39.45**	**48.50**

B2033 925 Metal Grated

	System Description	Freq. (Years)	Crew	Unit	Labor Hours	2020 Bare Costs Material	Labor	Equipment	Total	Total In-House	Total w/O&P
1030	**Replace 3'-0" x 7'-0" metal grated painted door**	30	2 SSWK	Ea.							
	Remove door				.650		27.50		27.50	35.50	44
	Remove door closer				.167		8.85		8.85	11.45	14.20
	Remove door hinge				.500		26.55		26.55	34.50	42.50
	Install new grated door				15.686	645	905		1,550	1,900	2,275
	Install new hinge (25% of total hinges / 30 years)				.500		26.55		26.55	34.50	42.50
	Hinge, 5 x 5					104.25			104.25	115	130
	Total				17.503	749.25	994.45		1,743.70	**2,130.95**	**2,548.20**

B2033 926 Metal Wire Mesh Door

	System Description	Freq. (Years)	Crew	Unit	Labor Hours	2020 Bare Costs Material	Labor	Equipment	Total	Total In-House	Total w/O&P
1010	**Repair metal wire mesh door**	15	1 SSWK	Ea.							
	Remove wire mesh				1.750		100.80		100.80	134	165
	Wire mesh					10.50			10.50	11.55	13.15
	Install wire mesh				7.000		403.20		403.20	535	660
	Oil / lubricate door closer				.065		3.72		3.72	4.94	6.10
	Oil / lubricate hinges				.065		3.43		3.43	4.43	5.50
	Total				8.879	10.50	511.15		521.65	**689.92**	**849.75**
1020	**Prepare and refinish metal wire mesh door**	4	1 PORD	Ea.							
	Prepare metal mesh surface				.312		13.85		13.85	17.70	22
	Paint metal mesh door surface, brushwork, 1 coat				.319	3.68	14.26		17.94	22	27
	Total				.631	3.68	28.11		31.79	**39.70**	**49**
1030	**Replace metal wire mesh door**	30	2 SSWK	Ea.							
	Remove wire mesh door				.640		37		37	49	60.50
	Remove wire mesh door frame				2.602		150		150	199	246
	Install new woven wire swinging door				3.471	294	184		478	560	665
	Install new metal door frame				3.200	259	186	14.70	459.70	550	645
	Install new hinge (25% of total hinges / 30 years)				.500		26.55		26.55	34.50	42.50
	Hinge, 5 x 5					104.25			104.25	115	130
	Install new deadlock night latch				1.039	58	55		113	135	161
	Total				11.451	715.25	638.55	14.70	1,368.50	**1,642.50**	**1,950**

B2033 927 Louvered Door

	System Description	Freq. (Years)	Crew	Unit	Labor Hours	2020 Bare Costs Material	Labor	Equipment	Total	Total In-House	Total w/O&P
1010	**Repair aluminum louvered door**	12	1 CARP	Ea.							
	Remove door				.650		27.50		27.50	35.50	44
	Remove weatherstripping				.015		.78		.78	1.01	1.25
	Remove damaged louver				.137		7.25		7.25	9.40	11.65
	Install new louver				.274	17.85	14.55		32.40	38.50	46
	Oil / lubricate door closer				.065		3.43		3.43	4.43	5.50
	Install new weatherstrip				3.478	52	185		237	296	360
	Reinstall exterior door				1.388		74		74	95.50	118
	Total				6.006	69.85	312.51		382.36	**480.34**	**586.40**
1020	**Prepare and refinish aluminum louvered door**	4	1 PORD	Ea.							
	Prepare door surface				.333		14.80		14.80	18.90	23.50
	Paint door surface, brushwork, primer + 1 coat				.549	13.34	24.38		37.72	46	55.50
	Total				.883	13.34	39.18		52.52	**64.90**	**79**
1040	**Replace 3′-0″ x 7′-0″ aluminum louvered door**	50	1 CARP	Ea.							
	Remove door				.650		27.50		27.50	35.50	44
	Remove door closer				.167		8.85		8.85	11.45	14.20
	Install new aluminum door				1.224	1,625	65		1,690	1,875	2,125
	Install new door closer				1.735	211	92		303	350	410
	Total				3.777	1,836	193.35		2,029.35	**2,271.95**	**2,593.20**
2010	**Repair steel louvered door**	14	1 CARP	Ea.							
	Remove door				.650		27.50		27.50	35.50	44
	Remove weatherstripping				.015		.78		.78	1.01	1.25
	Remove damaged louver				.137		7.25		7.25	9.40	11.65
	Install new louver				.274	17.85	14.55		32.40	38.50	46
	Oil / lubricate door closer				.065		3.43		3.43	4.43	5.50
	Install new weatherstrip				3.478	52	185		237	296	360
	Reinstall exterior door				1.388		74		74	95.50	118
	Total				6.006	69.85	312.51		382.36	**480.34**	**586.40**
2020	**Prepare and refinish steel louvered door**	4	1 PORD	Ea.							
	Prepare door surfaces				.333		14.80		14.80	18.90	23.50
	Paint door surface, brushwork, primer + 1 coat				.549	13.34	24.38		37.72	46	55.50
	Total				.883	13.34	39.18		52.52	**64.90**	**79**

B2033 927 Louvered Door

	System Description	Freq. (Years)	Crew	Unit	Labor Hours	2020 Bare Costs				Total In-House	Total w/O&P
						Material	Labor	Equipment	Total		
2040	**Replace 3′-0″ x 7′-0″ steel louvered door**	45	1 CARP	Ea.							
	Remove door				.650		27.50		27.50	35.50	44
	Remove door closer				.167		8.85		8.85	11.45	14.20
	Install new steel door, incl. louver				1.224	615	65		680	760	875
	Install new door closer				1.735	211	92		303	350	410
	Total				3.776	826	193.35		1,019.35	**1,156.95**	**1,343.20**
3010	**Repair wood louvered door**	7	1 CARP	Ea.							
	Remove door				.650		27.50		27.50	35.50	44
	Remove weatherstripping				.015		.78		.78	1.01	1.25
	Remove damaged louver				.137		7.25		7.25	9.40	11.65
	Install new louver				.274	17.85	14.55		32.40	38.50	46
	Oil / lubricate door closer				.065		3.43		3.43	4.43	5.50
	Install new weatherstrip				3.478	52	185		237	296	360
	Reinstall exterior door				1.388		74		74	95.50	118
	Total				6.006	69.85	312.51		382.36	**480.34**	**586.40**
3020	**Prepare and refinish wood louvered door**	4	1 PORD	Ea.							
	Prepare door surfaces				.333		14.80		14.80	18.90	23.50
	Paint door surface, brushwork, primer + 1 coat				.549	13.34	24.38		37.72	46	55.50
	Total				.883	13.34	39.18		52.52	**64.90**	**79**
3040	**Replace 3′-0″ x 7′-0″ wood louvered door**	40	1 CARP	Ea.							
	Remove door				.650		27.50		27.50	35.50	44
	Remove door closer				.167		8.85		8.85	11.45	14.20
	Install new solid core door				1.143	182	60.50		242.50	279	325
	Add for installed wood louver					243			243	267	305
	Install new door closer				1.735	211	92		303	350	410
	Total				3.695	636	188.85		824.85	**942.95**	**1,098.20**

B20 EXTERIOR CLOSURE | B2033 | Exterior Doors

B2033 935 | Aluminum Storm/Screen Door

	System Description	Freq. (Years)	Crew	Unit	Labor Hours	2020 Bare Costs				Total In-House	Total w/O&P
						Material	Labor	Equipment	Total		
1010	**Repair aluminum storm/screen door**	5	1 CARP	Ea.							
	Remove weatherstripping				.015		.78		.78	1.01	1.25
	Remove screen / storm door threshold				.519		27.60		27.60	35.50	44
	Remove door closer				.167		8.85		8.85	11.45	14.20
	Install new screen / storm door closer				1.301	29.50	69		98.50	122	148
	Install new weatherstrip				.870	13	46.25		59.25	74	90.50
	Install new plain aluminum threshold				.520	11.10	27.50		38.60	48	58.50
	Total				3.391	53.60	179.98		233.58	**291.96**	**356.45**
1020	**Replace aluminum storm/screen door**	20	1 CARP	Ea.							
	Remove door				.650		27.50		27.50	35.50	44
	Install new anodized aluminum screen / storm door				1.487	192	79		271	315	365
	Total				2.137	192	106.50		298.50	**350.50**	**409**

B2033 960 | Aluminized Steel Pedestrian Gate

	System Description	Freq. (Years)	Crew	Unit	Labor Hours	2020 Bare Costs				Total In-House	Total w/O&P
						Material	Labor	Equipment	Total		
1010	**Repair aluminized steel pedestrian gate**	10	1 CARP	Ea.							
	Remove gate from hinges				.286		15.20		15.20	19.65	24.50
	Oil / lubricate hinges				.065		3.43		3.43	4.43	5.50
	Install new 6′ aluminized steel fencing				.360	85.50	15.75	3.75	105	119	136
	Reinstall gate				.615		32.50		32.50	42.50	52.50
	Total				1.326	85.50	66.88	3.75	156.13	**185.58**	**218.50**
1030	**Prepare and refinish pedestrian gate**	5	1 PORD	Ea.							
	Lay drop cloth				.090		4		4	5.10	6.30
	Prepare / brush surface				.042		1.86		1.86	2.34	2.94
	Paint gate, brushwork, 1 coat				.150	1.44	6.66		8.10	10.10	12.40
	Remove drop cloth				.090		4		4	5.10	6.30
	Total				.371	1.44	16.52		17.96	**22.64**	**27.94**
1040	**Replace pedestrian gate**	40	1 CARP	Ea.							
	Remove old gate from hinges				.286		15.20		15.20	19.65	24.50
	Install new 6′ x 3′ aluminized steel gate				3.117	217	137	32.50	386.50	450	525
	Total				3.403	217	152.20	32.50	401.70	**469.65**	**549.50**

B2033 961 Pedestrian Gate, Galvanized Steel

	System Description	Freq. (Years)	Crew	Unit	Labor Hours	2020 Bare Costs				Total In-House	Total w/O&P
						Material	Labor	Equipment	Total		
1010	**Repair steel gate**	10	1 CARP	Ea.							
	Remove gate from hinges				.286		15.20		15.20	19.65	24.50
	Oil / lubricate hinges				.065		3.43		3.43	4.43	5.50
	Install new 6′ galvanized steel fencing				.360	64.50	15.75	3.75	84	95.50	110
	Reinstall gate				.615		32.50		32.50	42.50	52.50
	Total				1.326	64.50	66.88	3.75	135.13	**162.08**	**192.50**
1030	**Prepare and refinish steel gate**	5	1 PORD	Ea.							
	Lay drop cloth				.090		4		4	5.10	6.30
	Prepare / brush surface				.042		1.86		1.86	2.34	2.94
	Paint gate, brushwork, 1 coat				.150	1.44	6.66		8.10	10.10	12.40
	Remove drop cloth				.090		4		4	5.10	6.30
	Total				.371	1.44	16.52		17.96	**22.64**	**27.94**
1040	**Replace steel gate**	40	1 CARP	Ea.							
	Remove old gate from hinges				.286		15.20		15.20	19.65	24.50
	Install new 6′ x 3′ galvanized steel gate				3.117	207	137	32.50	376.50	440	515
	Total				3.403	207	152.20	32.50	391.70	**459.65**	**539.50**

B2033 962 Pedestrian Gate, Wood

	System Description	Freq. (Years)	Crew	Unit	Labor Hours	2020 Bare Costs				Total In-House	Total w/O&P
						Material	Labor	Equipment	Total		
1010	**Repair wood gate**	7	1 CARP	Ea.							
	Remove gate from hinges				.286		15.20		15.20	19.65	24.50
	Oil / lubricate hinges				.065		3.43		3.43	4.43	5.50
	Replace damaged fencing				1.171	135	51.30	12.18	198.48	228	264
	Reinstall gate				.615		32.50		32.50	42.50	52.50
	Total				2.136	135	102.43	12.18	249.61	**294.58**	**346.50**
1030	**Prepare and refinish wood gate**	4	1 PORD	Ea.							
	Lay drop cloth				.090		4		4	5.10	6.30
	Prepare / brush surface				.042		1.86		1.86	2.34	2.94
	Paint gate, brushwork, primer + 1 coat				.138	1.62	6.12		7.74	9.70	11.75
	Remove drop cloth				.090		4		4	5.10	6.30
	Total				.359	1.62	15.98		17.60	**22.24**	**27.29**

B2033 962 Pedestrian Gate, Wood

	System Description	Freq. (Years)	Crew	Unit	Labor Hours	2020 Bare Costs				Total In-House	Total w/O&P
						Material	Labor	Equipment	Total		
1040	**Replace wood gate**	25	1 CARP	Ea.							
	Remove old gate from hinges				.286		15.20		15.20	19.65	24.50
	New 6′-0″ x 3′-0″ wood gate				3.468	250	152	36	438	510	595
	Total				3.754	250	167.20	36	453.20	**529.65**	**619.50**

B2033 963 Pedestrian Gate, Wrought Iron

	System Description	Freq. (Years)	Crew	Unit	Labor Hours	2020 Bare Costs				Total In-House	Total w/O&P
						Material	Labor	Equipment	Total		
1010	**Repair wrought iron gate**	11	1 CARP	Ea.							
	Remove gate from hinges				.286		15.20		15.20	19.65	24.50
	Oil / lubricate hinges				.065		3.43		3.43	4.43	5.50
	Repair hand forged wrought iron				2.600	289.50	150		439.50	520	610
	Reinstall gate				.615		32.50		32.50	42.50	52.50
	Total				3.566	289.50	201.13		490.63	**586.58**	**692.50**
1030	**Prepare and refinish wrought iron gate**	5	1 PORD	Ea.							
	Lay drop cloth				.090		4		4	5.10	6.30
	Prepare / brush surface				.125		5.58		5.58	7	8.80
	Paint gate, brushwork, primer + 1 coat				.167	1.62	7.38		9	11.15	13.75
	Remove drop cloth				.090		4		4	5.10	6.30
	Total				.471	1.62	20.96		22.58	**28.35**	**35.15**
1040	**Replace wrought iron gate**	45	1 CARP	Ea.							
	Remove old gate from hinges				.286		15.20		15.20	19.65	24.50
	Install new hand forged wrought iron gate				3.467	386	200		586	690	810
	Total				3.753	386	215.20		601.20	**709.65**	**834.50**

B3013 105 Built-Up Roofing

	System Description	Freq. (Years)	Crew	Unit	Labor Hours	2020 Bare Costs Material	Labor	Equipment	Total	Total In-House	Total w/O&P
0100	**Debris removal and visual inspection**	0.50	2 ROFC	M.S.F.							
	Set up, secure and take down ladder				.052		2.40		2.40	3.46	4.20
	Pick up trash / debris & clean-up				.327		15.10		15.10	21.50	26.50
	Visual inspection				.327		15.10		15.10	21.50	26.50
	Total				.705		32.60		32.60	**46.46**	**57.20**
0200	**Non-destructive moisture inspection**	5	2 ROFC	M.S.F.							
	Set up, secure and take down ladder				.052		2.40		2.40	3.46	4.20
	Infrared inspection of roof membrane				2.133		98.50		98.50	142	172
	Total				2.185		100.90		100.90	**145.46**	**176.20**
0300	**Minor BUR membrane repairs - (2% of roof area)**	1	G-5	Sq.							
	Set up, secure and take down ladder				1.000		46		46	67	81
	Sweep / spud ballast clean				.640		29.50		29.50	42.50	51.50
	Cross cut incision through bitumen				.258		11.90		11.90	17.15	21
	Install base sheet and 2 plies of glass mopped				3.294	108	142	33.50	283.50	360	420
	Reinstall ballast in bitumen				.390		18.05		18.05	26	31.50
	Clean up				.390		18.05		18.05	26	31.50
	Total				5.973	108	265.50	33.50	407	**538.65**	**636.50**
0400	**BUR flashing repairs - (2 S.F. per sq. repaired)**	1	2 ROFC	S.F.							
	Set up, secure and take down ladder				.010		.46		.46	.67	.81
	Sweep / spud ballast clean				.009		.42		.42	.60	.73
	Cut out buckled flashing				.020		.92		.92	1.33	1.62
	Install 2 ply membrane flashing				.013	.21	.62		.83	1.12	1.34
	Reinstall ballast in bitumen				.005		.24		.24	.35	.42
	Clean up				.004		.18		.18	.27	.32
	Total				.062	.21	2.84		3.05	**4.34**	**5.24**
0500	**Minor BUR membrane replacement - (25% of roof area)**	15	G-5	Sq.							
	Set up, secure and take down ladder				.078		3.61		3.61	5.20	6.30
	Remove 5 ply built-up roof				3.252		138		138	179	221
	Remove roof insulation board				1.333		57		57	73	91
	Install 2″ polystyrene insulation				.641	68	30		98	118	137
	Install 4 ply bituminous roofing				3.733	133	161	38	332	420	490
	Clean up				1.000		46		46	66.50	81
	Total				10.037	201	435.61	38	674.61	**861.70**	**1,026.30**

B3013 105 Built-Up Roofing

	System Description	Freq. (Years)	Crew	Unit	Labor Hours	2020 Bare Costs Material	Labor	Equipment	Total	Total In-House	Total w/O&P
0600	**Place new BUR membrane over existing**	20	G-5	Sq.							
	Set up, secure and take down ladder				.020		.92		.92	1.33	1.62
	Sweep / spud ballast clean				.500		23		23	33.50	40.50
	Vent existing membrane				.130		6		6	8.65	10.50
	Cut out buckled flashing				.024		1.10		1.10	1.60	1.94
	Install 2 ply membrane flashing				.027	.41	1.24		1.65	2.22	2.68
	Install 4 ply bituminous roofing				3.733	133	161	38	332	420	490
	Reinstall ballast				.381		17.60		17.60	25.50	31
	Clean up				.390		18.05		18.05	26	31.50
	Total				5.205	133.41	228.91	38	400.32	**518.80**	**609.74**
0700	**Total BUR roof replacement**	28	G-1	Sq.							
	Set up, secure and take down ladder				.020		.92		.92	1.33	1.62
	Sweep / spud ballast clean				.500		23		23	33.50	40.50
	Remove built-up roofing				2.500		106		106	137	170
	Remove insulation board				1.026		44		44	56	70
	Remove flashing				.024		1.10		1.10	1.60	1.94
	Install 2″ perlite insulation				.879	109	41		150	178	207
	Install 2 ply membrane flashing				.027	.41	1.24		1.65	2.22	2.68
	Install 4 ply bituminous membrane				2.800	133	121	28.50	282.50	350	410
	Clean up				1.000		46		46	66.50	81
	Total				8.776	242.41	384.26	28.50	655.17	**826.15**	**984.74**

B3013 120 Modified Bituminous / Thermoplastic

	System Description	Freq. (Years)	Crew	Unit	Labor Hours	2020 Bare Costs Material	Labor	Equipment	Total	Total In-House	Total w/O&P
0100	**Debris removal by hand & visual inspection**	1	2 ROFC	M.S.F.							
	Set up, secure and take down ladder				.052		2.40		2.40	3.46	4.20
	Pick up trash / debris & clean up				.327		15.10		15.10	21.50	26.50
	Visual inspection				.327		15.10		15.10	21.50	26.50
	Total				.705		32.60		32.60	**46.46**	**57.20**
0200	**Non - destructive moisture inspection**	5	2 ROFC	M.S.F.							
	Set up, secure and take down ladder				.052		2.40		2.40	3.46	4.20
	Infrared inspection of roof membrane				2.133		98.50		98.50	142	172
	Total				2.185		100.90		100.90	**145.46**	**176.20**

B3013 120 Modified Bituminous / Thermoplastic

	System Description	Freq. (Years)	Crew	Unit	Labor Hours	2020 Bare Costs				Total In-House	Total w/O&P
						Material	Labor	Equipment	Total		
0300	**Minor thermoplastic membrane repairs - (2% of roof area)**	1	G-5	Sq.							
	Set up, secure and take down ladder				1.000		46		46	67	81
	Clean away loose surfacing				.258		11.90		11.90	17.15	21
	Install 150 mil mod. bit., fully adhered				2.597	84	109	13	206	263	310
	Clean up				.390		18.05		18.05	26	31.50
	Total				4.245	84	184.95	13	281.95	**373.15**	**443.50**
0500	**Thermoplastic flashing repairs - (2 S.F. per sq. repaired)**	1	2 ROFC	S.F.							
	Set up, secure and take down ladder				.010		.46		.46	.67	.81
	Clean away loose surfacing				.003		.12		.12	.17	.21
	Remove flashing				.020		.92		.92	1.33	1.62
	Install 120 mil mod. bit. flashing				.019	.80	.78	.09	1.67	2.10	2.47
	Clean up				.004		.18		.18	.27	.32
	Total				.055	.80	2.46	.09	3.35	**4.54**	**5.43**
0600	**Minor thermoplastic membrane replacement - (25% of roof area)**	20	G-5	Sq.							
	Set up, secure and take down ladder				.078		3.61		3.61	5.20	6.30
	Remove existing membrane / insulation				4.571		211		211	305	370
	Install 2″ perlite insulation				1.140	109	53		162	196	228
	Clean adjacent membrane / patch				1.143		53		53	76	92.50
	Install fully adhered 150 mil membrane				2.597	84	109	13	206	263	310
	Clean up				1.000		46		46	66.50	81
	Total				10.529	193	475.61	13	681.61	**911.70**	**1,087.80**
0700	**Total thermoplastic roof replacement**	25	G-1	Sq.							
	Set up, secure and take down ladder				.020		.92		.92	1.33	1.62
	Remove existing membrane / insulation				3.501		162		162	233	283
	Remove flashing				.026		1.20		1.20	1.74	2.10
	Install 2″ perlite insulation				.879	109	41		150	178	207
	Install flashing				.037	1.60	1.56	.18	3.34	4.20	4.94
	Install fully adhered 180 mil membrane				2.000	96	84	10	190	238	278
	Clean up				1.000		46		46	66.50	81
	Total				7.463	206.60	336.68	10.18	553.46	**722.77**	**857.66**

B3013 125 Thermosetting

	System Description	Freq. (Years)	Crew	Unit	Labor Hours	2020 Bare Costs Material	Labor	Equipment	Total	Total In-House	Total w/O&P
0100	**Debris removal by hand & visual inspection**	1	2 ROFC	M.S.F.							
	Set up, secure and take down ladder				.052		2.40		2.40	3.46	4.20
	Pick up trash / debris & clean up				.327		15.10		15.10	21.50	26.50
	Visual inspection				.327		15.10		15.10	21.50	26.50
	Total				.705		32.60		32.60	**46.46**	**57.20**
0200	**Non - destructive moisture inspection**	5	2 ROFC	M.S.F.							
	Set up, secure and take down ladder				.052		2.40		2.40	3.46	4.20
	Infrared inspection of roof membrane				2.133		98.50		98.50	142	172
	Total				2.185		100.90		100.90	**145.46**	**176.20**
0300	**Minor thermoset membrane repairs - (2% of roof area)**	1	G-5	Sq.							
	Set up, secure and take down ladder				1.000		46		46	67	81
	Broom roof surface clean				.258		11.90		11.90	17.15	21
	Remove / vent damaged membrane				.258		11.90		11.90	17.15	21
	Install thermosetting patch				2.597	84	109	13	206	263	310
	Clean up				.390		18.05		18.05	26	31.50
	Total				4.503	84	196.85	13	293.85	**390.30**	**464.50**
0400	**Thermoset flashing repairs - (2 S.F. per sq. repaired)**	1	2 ROFC	S.F.							
	Set up, secure and take down ladder				.010		.46		.46	.67	.81
	Remove flashing				.020		.92		.92	1.33	1.62
	Prime surfaces				.003		.14		.14	.20	.24
	Install 2 ply membrane flashing				.007	.10	.31		.41	.55	.67
	Clean up				.004		.18		.18	.27	.32
	Total				.044	.10	2.01		2.11	**3.02**	**3.66**
0500	**Minor thermoset membrane replacement - (25% of roof area)**	10	G-5	Sq.							
	Set up, secure and take down ladder				.078		3.61		3.61	5.20	6.30
	Broom roof surface clean				.258		11.90		11.90	17.15	21
	Remove, strip-in & seal joint				.593		27.50		27.50	39.50	48
	Install base sheet and 2 plies of glass mopped				3.294	108	142	33.50	283.50	360	420
	Clean up				.390		18.05		18.05	26	31.50
	Total				4.613	108	203.06	33.50	344.56	**447.85**	**526.80**

B3013 125 Thermosetting

	System Description	Freq. (Years)	Crew	Unit	Labor Hours	2020 Bare Costs				Total In-House	Total w/O&P
						Material	Labor	Equipment	Total		
0600	**Total thermoset roof replacement**	20	G-1	Sq.							
	Set up, secure and take down ladder				.020		.92		.92	1.33	1.62
	Remove existing membrane / insulation				3.501		162		162	233	283
	Remove flashing				.026		1.20		1.20	1.74	2.10
	Install 2" perlite insulation				.879	109	41		150	178	207
	Install new flashing				.020	.41	.92		1.33	1.78	2.14
	Install thermosetting membrane				2.800	180	121	28	329	405	470
	Clean up				1.000		46		46	66.50	81
	Total				8.246	289.41	373.04	28	690.45	**887.35**	**1,046.86**

B3013 127 EPDM (Ethylene Propylene Diene Monomer)

	System Description	Freq. (Years)	Crew	Unit	Labor Hours	2020 Bare Costs				Total In-House	Total w/O&P
						Material	Labor	Equipment	Total		
0600	**Total EPDM roof replacement**	25	G-5	Sq.							
	Set up, secure and take down ladder				.020		.92		.92	1.33	1.62
	Remove existing membrane / insulation				3.501		162		162	233	283
	Remove flashing				.026		1.20		1.20	1.74	2.10
	Install 2" perlite insulation				.879	109	41		150	178	207
	Install new flashing				.020	.41	.92		1.33	1.78	2.14
	Fully adhered with adhesive				1.538	120	64.50	7.40	191.90	233	271
	Clean up				1.000		46		46	66.50	81
	Total				6.985	229.41	316.54	7.40	553.35	**715.35**	**847.86**

B3013 130 Metal Panel Roofing

	System Description	Freq. (Years)	Crew	Unit	Labor Hours	2020 Bare Costs				Total In-House	Total w/O&P
						Material	Labor	Equipment	Total		
0100	**Debris removal by hand & visual inspection**	1	2 ROFC	M.S.F.							
	Set up, secure and take down ladder				.100		4.62		4.62	6.65	8.10
	Visual inspection				.327		15.10		15.10	21.50	26.50
	Total				.427		19.72		19.72	**28.15**	**34.60**

B3013 130 Metal Panel Roofing

	System Description	Freq. (Years)	Crew	Unit	Labor Hours	2020 Bare Costs				Total In-House	Total w/O&P
						Material	Labor	Equipment	Total		
0300	**Minor metal roof finish repairs - (2% of roof area)**	5	2 ROFC	S.F.							
	Set up, secure and take down ladder				.010		.46		.46	.67	.81
	Wire brush surface to remove oxide				.020		.92		.92	1.33	1.62
	Prime and paint prepared surface				.016	.73	.74		1.47	1.86	2.20
	Clean up				.004		.18		.18	.27	.32
	Total				.050	.73	2.30		3.03	**4.13**	**4.95**
0400	**Metal roof flashing replacement - (2 S.F. per sq. repaired)**	1	2 ROFC	S.F.							
	Set up, secure and take down ladder				.010		.46		.46	.67	.81
	Dismantle seam				.016		.74		.74	1.06	1.29
	Remove metal roof panels				.060		2.76		2.76	3.99	4.86
	Remove damaged metal flashing				.019		.88		.88	1.26	1.54
	Install new aluminum flashing				.072	1.38	3.33		4.71	6.30	7.60
	Replace metal panels				.129		5.97		5.97	8.60	10.45
	Clean up				.004		.18		.18	.27	.32
	Total				.310	1.38	14.32		15.70	**22.15**	**26.87**
0500	**Minor metal roof panel replacement - (2.5% of roof area)**	20	2 ROFC	S.F.							
	Set up, secure and take down ladder				.010		.46		.46	.67	.81
	Dismantle seam				.016		.74		.74	1.06	1.29
	Remove metal roof panels				.020		.92		.92	1.33	1.62
	Install new metal panel				.044	3.06	2.29		5.35	6.30	7.45
	Re-assemble seam				.016		.74		.74	1.06	1.29
	Prime and paint surface				.016	.73	.74		1.47	1.86	2.20
	Clean up				.004		.18		.18	.27	.32
	Total				.126	3.79	6.07		9.86	**12.55**	**14.98**
0600	**Total metal roof panel replacement**	30	2 ROFC	Sq.							
	Set up, secure and take down ladder				.020		.92		.92	1.33	1.62
	Remove existing roofing panel				3.600		166		166	240	291
	Remove flashing metal				.038		1.76		1.76	2.52	3.08
	Install new aluminum flashing				.072	1.38	3.33		4.71	6.30	7.60
	Install new metal panel				4.384	306	229		535	630	745
	Clean up				.390		18.05		18.05	26	31.50
	Total				8.504	307.38	419.06		726.44	**906.15**	**1,079.80**

B3013 140 Slate Tile Roofing

	System Description	Freq. (Years)	Crew	Unit	Labor Hours	2020 Bare Costs Material	Labor	Equipment	Total	Total In-House	Total w/O&P
0100	**Visual inspection**	3	2 ROFC	M.S.F.							
	Set up, secure and take down ladder				.100		4.62		4.62	6.65	8.10
	Visual inspection				.327		15.10		15.10	21.50	26.50
	Total				.427		19.72		19.72	**28.15**	**34.60**
0200	**Minor slate tile replacement - (2.5% of roof area)**	20	2 ROTS	S.F.							
	Set up, secure and take down ladder				.010		.46		.46	.67	.81
	Remove and replace damaged slates				.800	7.15	37		44.15	61	74
	Clean up				.013		.60		.60	.87	1.05
	Total				.823	7.15	38.06		45.21	**62.54**	**75.86**
0300	**Slate tile flashing repairs - (2 S.F. per sq. repaired)**	20	2 ROTS	S.F.							
	Set up, secure and take down ladder				.010		.46		.46	.67	.81
	Removal of adjacent slate tiles				.100		4.62		4.62	6.65	8.10
	Remove damaged metal flashing				.052		2.40		2.40	3.46	4.20
	Install 16 oz. copper flashing				.089	8.15	4.11		12.26	14.90	17.40
	Reinstall tiles in position				.134		6.18		6.18	8.90	10.80
	Clean up				.013		.60		.60	.87	1.05
	Total				.398	8.15	18.37		26.52	**35.45**	**42.36**
0400	**Total slate tile roof replacement**	70	2 ROTS	Sq.							
	Set up, secure and take down ladder				.020		.92		.92	1.33	1.62
	Remove slate shingles				1.600		68		68	88	109
	Install 16 oz. copper flashing				.178	16.30	8.22		24.52	30	35
	Install 15# felt				.138	10.20	6.35		16.55	20.50	24
	Install slate roofing				4.571	505	211		716	860	1,000
	Clean up				1.333		61.50		61.50	88.50	108
	Total				7.840	531.50	355.99		887.49	**1,088.33**	**1,277.62**

B3013 142 Mineral Fiber Steep Roofing

	System Description	Freq. (Years)	Crew	Unit	Labor Hours	2020 Bare Costs Material	Labor	Equipment	Total	Total In-House	Total w/O&P
0300	**Visual inspection**	3	2 ROFC	M.S.F.							
	Set up, secure and take down ladder				.100		4.62		4.62	6.65	8.10
	Visual inspection				.327		15.10		15.10	21.50	26.50
	Total				.427		19.72		19.72	**28.15**	**34.60**

B3013 144 Clay Tile Roofing

	System Description	Freq. (Years)	Crew	Unit	Labor Hours	2020 Bare Costs Material	Labor	Equipment	Total	Total In-House	Total w/O&P
0300	**Visual inspection**	3	2 ROFC	M.S.F.							
	Set up, secure and take down ladder				.100		4.62		4.62	6.65	8.10
	Visual inspection				.327		15.10		15.10	21.50	26.50
	Total				.427		19.72		19.72	**28.15**	**34.60**
0400	**Minor clay tile repairs - (2% of roof area)**	20	2 ROFC	S.F.							
	Set up, secure and take down ladder				.010		.46		.46	.67	.81
	Remove damaged tiles				.020		.93		.93	1.33	1.62
	Install new clay tiles				.073	5.35	3.35		8.70	10.75	12.60
	Clean up				.013		.60		.60	.87	1.05
	Total				.116	5.35	5.34		10.69	**13.62**	**16.08**
0500	**Clay tile flashing repairs - (2 S.F. per sq. repaired)**	1	2 ROFC	S.F.							
	Set up, secure and take down ladder				.010		.46		.46	.67	.81
	Remove adjacent tiles				.208		9.60		9.60	13.85	16.80
	Remove damaged metal flashing				.052		2.40		2.40	3.46	4.20
	Install 16 oz copper flashing				.178	16.30	8.22		24.52	30	35
	Reinstall clay tiles				.073		3.35		3.35	4.84	5.90
	Clean up				.013		.60		.60	.87	1.05
	Total				.533	16.30	24.63		40.93	**53.69**	**63.76**
0600	**Total clay tile roof replacement**	70	2 ROFC	Sq.							
	Set up, secure and take down ladder				.020		.92		.92	1.33	1.62
	Remove clay tile roofing				3.200		148		148	213	259
	Remove damaged metal flashing				.040		1.85		1.85	2.66	3.23
	Install 16 oz copper flashing				.178	16.30	8.22		24.52	30	35
	Install felt				.138	10.20	6.35		16.55	20.50	24
	Install new clay tiles				7.273	535	335		870	1,075	1,250
	Clean up				1.333		61.50		61.50	88.50	108
	Total				12.182	561.50	561.84		1,123.34	**1,430.99**	**1,680.85**

B3013 146 Asphalt Shingle Roofing

	System Description	Freq. (Years)	Crew	Unit	Labor Hours	2020 Bare Costs				Total In-House	Total w/O&P
						Material	Labor	Equipment	Total		
0100	**Debris removal by hand & visual inspection**	1	2 ROFC	M.S.F.							
	Set up, secure and take down ladder				.100		4.62		4.62	6.65	8.10
	Visual inspection				.327		15.10		15.10	21.50	26.50
	Total				.427		19.72		19.72	**28.15**	**34.60**
0300	**Minor asphalt shingle repair - (2% of roof area)**	1	2 ROFC	S.F.							
	Set up, secure and take down ladder				.010		.46		.46	.67	.81
	Remove damaged shingles				.024		1		1	1.29	1.60
	Install asphalt shingles				.036	1.03	1.64		2.67	3.50	4.16
	Clean up				.003		.12		.12	.17	.21
	Total				.072	1.03	3.22		4.25	**5.63**	**6.78**
0500	**Asphalt shingle flashing repairs - (2 S.F. per sq. repaired)**	1	2 ROFC	S.F.							
	Set up, secure and take down ladder				.010		.46		.46	.67	.81
	Remove adjoining shingles				.046		1.94		1.94	2.52	3.12
	Remove damaged metal flashing				.019		.88		.88	1.26	1.54
	Install new aluminum flashing				.072	1.38	3.33		4.71	6.30	7.60
	Install asphalt shingles				.071	2.06	3.28		5.34	7	8.30
	Clean up				.003		.12		.12	.17	.21
	Total				.221	3.44	10.01		13.45	**17.92**	**21.58**
0600	**Install new asphalt shingles over existing**	20	2 ROFC	Sq.							
	Set up, secure and take down ladder				.020		.92		.92	1.33	1.62
	Remove shingles at eaves and ridge				.460		21.20		21.20	30.50	37
	Install new aluminum flashing				.144	2.76	6.66		9.42	12.65	15.15
	Install asphalt shingles				1.778	103	82		185	232	273
	Clean up				.258		11.90		11.90	17.15	21
	Total				2.660	105.76	122.68		228.44	**293.63**	**347.77**
0700	**Total asphalt shingle roof replacement**	40	2 ROFC	Sq.							
	Set up, secure and take down ladder				.020		.92		.92	1.33	1.62
	Remove existing shingles				2.353		100		100	129	160
	Remove damaged metal flashing				.038		1.76		1.76	2.52	3.08
	Install 15# felt				.138	10.20	6.35		16.55	20.50	24
	Install new aluminum flashing				.144	2.76	6.66		9.42	12.65	15.15
	Install asphalt shingles				1.778	103	82		185	232	273
	Clean up				.258		11.90		11.90	17.15	21
	Total				4.729	115.96	209.59		325.55	**415.15**	**497.85**

B3013 160 Roll Roofing

	System Description	Freq. (Years)	Crew	Unit	Labor Hours	2020 Bare Costs Material	Labor	Equipment	Total	Total In-House	Total w/O&P
0100	**Debris removal by hand & visual inspection**	1	2 ROFC	M.S.F.							
	Set up, secure and take down ladder				.100		4.62		4.62	6.65	8.10
	Pick up trash / debris & clean up				.327		15.10		15.10	21.50	26.50
	Visual inspection				.327		15.10		15.10	21.50	26.50
	Total				.753		34.82		34.82	**49.65**	**61.10**
0300	**Minor roofing repairs - (2% of roof area)**	1	2 ROFC	S.F.							
	Set up, secure and take down ladder				.010		.46		.46	.67	.81
	Cut out damaged area				.015		.69		.69	1	1.21
	Install roll roofing with mastic				.027	.75	1.17	.28	2.20	2.80	3.27
	Clean up				.003		.12		.12	.17	.21
	Total				.055	.75	2.44	.28	3.47	**4.64**	**5.50**
0400	**Flashing repairs - (2 S.F. per sq. repaired)**	1	2 ROFC	S.F.							
	Set up, secure and take down ladder				.010		.46		.46	.67	.81
	Remove adjacent roofing				.015		.69		.69	1	1.21
	Remove damaged metal flashing				.019		.88		.88	1.26	1.54
	Install aluminum flashing				.072	1.38	3.33		4.71	6.30	7.60
	Install roll roofing with mastic				.054	1.49	2.34	.55	4.38	5.60	6.55
	Clean up				.003		.12		.12	.17	.21
	Total				.173	2.87	7.82	.55	11.24	**15**	**17.92**
0500	**Minor replacement - (25% of roof area)**	15	2 ROFC	Sq.							
	Set up, secure and take down ladder				.078		3.61		3.61	5.20	6.30
	Remove strip-in				1.067		49.50		49.50	71	86
	Renew striping with 2 ply felts				2.917	121	126	29.50	276.50	345	405
	Clean up				.258		11.90		11.90	17.15	21
	Total				4.319	121	191.01	29.50	341.51	**438.35**	**518.30**
0600	**Total roof replacement**	20	2 ROFC	Sq.							
	Set up, secure and take down ladder				.020		.92		.92	1.33	1.62
	Remove existing shingles				1.333		61.50		61.50	88.50	108
	Remove flashing				.019		.88		.88	1.26	1.54
	Install aluminum flashing				.144	2.76	6.66		9.42	12.65	15.15
	Install 4 ply roll roofing				2.917	121	126	29.50	276.50	345	405
	Clean up				.258		11.90		11.90	17.15	21
	Total				4.691	123.76	207.86	29.50	361.12	**465.89**	**552.31**

B3013 165 Corrugated Fiberglass Panel Roofing

	System Description	Freq. (Years)	Crew	Unit	Labor Hours	2020 Bare Costs				Total In-House	Total w/O&P
						Material	Labor	Equipment	Total		
0100	**Debris removal & visual inspection**	5	2 ROFC	M.S.F.							
	Set up, secure and take down ladder				.100		4.62		4.62	6.65	8.10
	Visual inspection				.327		15.10		15.10	21.50	26.50
	Total				.427		19.72		19.72	**28.15**	**34.60**
0200	**Fiberglass roof flashing repairs - (2 S.F. per sq. repaired)**	15	2 ROFC	S.F.							
	Set up, secure and take down ladder				.010		.46		.46	.67	.81
	Cut flashing				.010		.46		.46	.67	.81
	Install new flashing				.036	.33	1.68		2.01	2.78	3.35
	Clean up				.004		.18		.18	.27	.32
	Total				.060	.33	2.78		3.11	**4.39**	**5.29**
0400	**Minor fiberglass roof panel replacement - (25% of roof area)**	15	2 ROFC	S.F.							
	Set up, secure and take down ladder				.010		.46		.46	.67	.81
	Remove damaged panel				.030		1.39		1.39	2	2.43
	Install new fiberglass panel				.064	2.55	3.34		5.89	7.05	8.50
	Install strip-in with joint seal				.029	.02	1.50		1.52	1.98	2.45
	Clean up				.004		.18		.18	.27	.32
	Total				.137	2.57	6.87		9.44	**11.97**	**14.51**
0600	**Total fiberglass roof replacement**	20	2 ROFC	Sq.							
	Set up, secure and take down ladder				.020		.92		.92	1.33	1.62
	Remove existing fiberglass panels				2.500		116		116	166	202
	Install new flashing				.073	.66	3.36		4.02	5.55	6.70
	Install new fiberglass panel				6.400	255	334		589	705	850
	Apply joint sealer to seams				.029	.02	1.50		1.52	1.98	2.45
	Clean up				.390		18.05		18.05	26	31.50
	Total				9.412	255.68	473.83		729.51	**905.86**	**1,094.27**

B3013 170 Roof Edges

	System Description	Freq. (Years)	Crew	Unit	Labor Hours	2020 Bare Costs Material	Labor	Equipment	Total	Total In-House	Total w/O&P
1010	**Replace Roof edges, aluminum, duranodic, .050″ thick, 6″ face**	25	2 ROFC	L.F.							
	Set up, secure, and take down ladder				.010		.46		.46	.67	.81
	Remove adjacent roofing				.015		.69		.69	1	1.21
	Remove roof edges				.080		3.37		3.37	4.35	5.40
	2″ x 8″ Edge blocking				.025	1.23	1.30		2.53	3.04	3.64
	Cants, 4″ x 4″, treated timber, cut diagonally				.025	1.89	1.14		3.03	3.72	4.35
	Aluminum gravel stop with Duranodic finish				.089	12.38	5.54		17.92	20.50	24
	Roof covering, splice sheet				.011	2.49	.48	.06	3.03	3.49	4.01
	Total				.254	17.99	12.98	.06	31.03	**36.77**	**43.42**
1020	**Replace Roof edges, copper, 20 oz., 8″ face**	35	2 ROFC	L.F.							
	Set up, secure, and take down ladder				.010		.46		.46	.67	.81
	Remove adjacent roofing				.015		.69		.69	1	1.21
	Remove roof edge				.080		3.37		3.37	4.35	5.40
	2″ x 8″ Edge blocking				.025	1.23	1.30		2.53	3.04	3.64
	Cants, 4″ x 4″, treated timber, cut diagonally				.025	1.89	1.14		3.03	3.72	4.35
	Copper gravel stop				.139	16.30	6.42		22.72	27	31.50
	Roof covering, splice sheet				.011	2.49	.48	.06	3.03	3.49	4.01
	Total				.305	21.91	13.86	.06	35.83	**43.27**	**50.92**
1030	**Replace Roof edges, galvanized, 20 ga., 6″ face**	25	2 ROFC	L.F.							
	Set up, secure, and take down ladder				.010		.46		.46	.67	.81
	Remove adjacent roofing				.015		.69		.69	1	1.21
	Remove roof edge				.080		3.37		3.37	4.35	5.40
	2″ x 8″ edge blocking				.025	1.23	1.30		2.53	3.04	3.64
	Cants, 4″ x 4″, treated timber, cut diagonally				.025	1.89	1.14		3.03	3.72	4.35
	Galvanized steel gravel stop				.092	1.83	4.26		6.09	8.15	9.75
	Roof covering, splice sheet				.011	2.49	.48	.06	3.03	3.49	4.01
	Total				.258	7.44	11.70	.06	19.20	**24.42**	**29.17**

B3013 620 Gutters and Downspouts

	System Description	Freq. (Years)	Crew	Unit	Labor Hours	2020 Bare Costs				Total In-House	Total w/O&P
						Material	Labor	Equipment	Total		
1010	**Replace aluminum gutter, enameled, 5″ K type, .027 ″ thick**	40	2 SHEE	L.F.							
	Set up, secure and take down ladder				.010		.46		.46	.67	.81
	Remove gutter				.033		1.40		1.40	1.81	2.25
	Install new gutter				.064	2.94	3.99		6.93	8.25	9.95
	Clean up				.003		.12		.12	.17	.21
	Total				.110	2.94	5.97		8.91	**10.90**	**13.22**
1020	**Replace aluminum gutter, enameled, 5″ K type, .032 ″ thick**	40	2 SHEE	L.F.							
	Set up, secure and take down ladder				.010		.46		.46	.67	.81
	Remove gutter				.033		1.40		1.40	1.81	2.25
	Install new gutter				.064	3.79	3.99		7.78	9.20	11
	Clean up				.003		.12		.12	.17	.21
	Total				.110	3.79	5.97		9.76	**11.85**	**14.27**
1060	**Replace aluminum downspout, 2″ x 3″, .024″ thick**	25	2 SHEE	L.F.							
	Set up, secure and take down ladder				.010		.46		.46	.67	.81
	Remove downspout				.023		.96		.96	1.24	1.54
	Install new downspout				.044	2.17	2.77		4.94	5.85	7.05
	Clean up				.003		.12		.12	.17	.21
	Total				.080	2.17	4.31		6.48	**7.93**	**9.61**
1070	**Replace aluminum downspout, 3″ x 4″, .024″ thick**	25	2 SHEE	L.F.							
	Set up, secure and take down ladder				.010		.46		.46	.67	.81
	Remove downspout				.023		.96		.96	1.24	1.54
	Install new downspout				.057	2.29	3.56		5.85	7	8.45
	Clean up				.003		.12		.12	.17	.21
	Total				.093	2.29	5.10		7.39	**9.08**	**11.01**
2010	**Replace galvanized steel gutter, 5″ box type, 28 ga.**	40	2 SHEE	L.F.							
	Set up, secure and take down ladder				.010		.46		.46	.67	.81
	Remove gutter				.033		1.40		1.40	1.81	2.25
	Install new gutter				.064	2.40	3.99		6.39	7.65	9.25
	Clean up				.003		.12		.12	.17	.21
	Total				.110	2.40	5.97		8.37	**10.30**	**12.52**

B3013 620 Gutters and Downspouts

	System Description	Freq. (Years)	Crew	Unit	Labor Hours	2020 Bare Costs				Total In-House	Total w/O&P
						Material	Labor	Equipment	Total		
2020	**Replace galvanized steel gutter, 6″ box type, 26 ga.**	40	2 SHEE	L.F.							
	Set up, secure and take down ladder				.010		.46		.46	.67	.81
	Remove gutter				.033		1.40		1.40	1.81	2.25
	Install new gutter				.064	2.64	3.99		6.63	7.90	9.55
	Clean up				.003		.12		.12	.17	.21
	Total				.110	2.64	5.97		8.61	**10.55**	**12.82**
2060	**Replace round corrugated galvanized downspout, 3″ diameter**	25	2 SHEE	L.F.							
	Set up, secure and take down ladder				.010		.46		.46	.67	.81
	Remove downspout				.023		.96		.96	1.24	1.54
	Install new downspout				.042	2.30	2.62		4.92	5.85	7
	Clean up				.003		.12		.12	.17	.21
	Total				.078	2.30	4.16		6.46	**7.93**	**9.56**
2070	**Replace round corrugated galvanized downspout, 4″ diameter**	25	2 SHEE	L.F.							
	Set up, secure and take down ladder				.010		.46		.46	.67	.81
	Remove downspout				.023		.96		.96	1.24	1.54
	Install new downspout				.055	2.19	3.44		5.63	6.75	8.15
	Clean up				.003		.12		.12	.17	.21
	Total				.091	2.19	4.98		7.17	**8.83**	**10.71**
3020	**Replace vinyl box gutter, 5″ wide**	20	1 CARP	L.F.	.119						
	Set up, secure and take down ladder				.010		.46		.46	.67	.81
	Remove gutter				.033		1.40		1.40	1.81	2.25
	Install new gutter				.070	1.51	3.70		5.21	6.45	7.80
	Clean up				.003		.12		.12	.17	.21
	Total				.116	1.51	5.68		7.19	**9.10**	**11.07**
3060	**Replace vinyl downspout, 2″ x 3″**	15	1 CARP	L.F.	.074						
	Set up, secure and take down ladder				.010		.46		.46	.67	.81
	Remove downspout				.023		.96		.96	1.24	1.54
	Install new downspout				.038	2.17	2.37		4.54	5.35	6.45
	Clean up				.003		.12		.12	.17	.21
	Total				.074	2.17	3.91		6.08	**7.43**	**9.01**

B3013 620 Gutters and Downspouts

	System Description	Freq. (Years)	Crew	Unit	Labor Hours	2020 Bare Costs				Total In-House	Total w/O&P
						Material	Labor	Equipment	Total		
4020	**Replace 6″ copper box gutters, 16 oz.**	50	1 SHEE	L.F.							
	Set up, secure and take down ladder				.010		.46		.46	.67	.81
	Remove gutter				.033		1.40		1.40	1.81	2.25
	Install new gutter				.064	8.45	3.99		12.44	14.30	16.80
	Clean up				.003		.12		.12	.17	.21
	Total				.110	8.45	5.97		14.42	**16.95**	**20.07**
4060	**Replace copper downspouts, 2″ x 3″, 16 oz.**	40	1 SHEE	L.F.	.119						
	Set up, secure and take down ladder				.010		.46		.46	.67	.81
	Remove gutter				.023		.96		.96	1.24	1.54
	Install new gutter				.042	9.20	2.62		11.82	13.40	15.60
	Clean up				.003		.12		.12	.17	.21
	Total				.078	9.20	4.16		13.36	**15.48**	**18.16**

B3023 110 Single Unit Glass Skylight

	System Description	Freq. (Years)	Crew	Unit	Labor Hours	2020 Bare Costs				Total In-House	Total w/O&P
						Material	Labor	Equipment	Total		
1010	**Repair glass skylight glazing**	6	1 CARP	S.F.							
	Set up, secure and take down scaffold				.130		6.90		6.90	8.95	11.05
	Remove glazing				.069		2.92		2.92	3.77	4.68
	Install new glazing				.291	27.50	14.85		42.35	49.50	58
	Total				.490	27.50	24.67		52.17	**62.22**	**73.73**
1020	**Replace continuous skylight and structure**	40	2 CARP	C.S.F.							
	Set up and secure scaffold				.750		39.75		39.75	51.50	64
	Remove unit skylight				5.202		276		276	355	445
	Inspect / remove flashing				3.152		146		146	210	255
	Replace flashing				3.152	205	146		351	435	510
	Install new unit				12.960	2,300	675		2,975	3,400	3,950
	Remove scaffold				.750		39.75		39.75	51.50	64
	Total				25.966	2,505	1,322.50		3,827.50	**4,503**	**5,288**
1040	**Replace skylight and structure, double glazed, 10 to 20 S.F.**	40	G-3	S.F.							
	Set up, secure and take down ladder				.010		.46		.46	.67	.81
	Remove skylight, plstc domes,flush/curb mtd				.041		2.15		2.15	2.78	3.45
	10 S.F. to 20 S.F., double				.102	28.50	5.30		33.80	38	44
	Total				.152	28.50	7.91		36.41	**41.45**	**48.26**
1080	**Replace skylight and structure, double glazed, 30 to 65 S.F.**	40	G-3	S.F.							
	Set up, secure and take down ladder				.010		.46		.46	.67	.81
	Remove skylight, plstc domes,flush/curb mtd				.041		2.15		2.15	2.78	3.45
	30 S.F. to 65 S.F., double				.069	29	3.59		32.59	36.50	42
	Total				.119	29	6.20		35.20	**39.95**	**46.26**

B3023 120 Hatches

	System Description	Freq. (Years)	Crew	Unit	Labor Hours	2020 Bare Costs				Total In-House	Total w/O&P
						Material	Labor	Equipment	Total		
1020	**Replace aluminum roof hatch & structure**	40	2 CARP	Ea.							
	Set up, secure and take down scaffold				.130		6.90		6.90	8.95	11.05
	Remove roof hatch				1.000		42		42	54.50	67.50
	Install 2′-6″ x 3′-0″ aluminum hatch				3.200	900	167		1,067	1,200	1,400
	Total				4.330	900	215.90		1,115.90	**1,263.45**	**1,478.55**

B3023 120 Hatches

	System Description	Freq. (Years)	Crew	Unit	Labor Hours	2020 Bare Costs				Total In-House	Total w/O&P
						Material	Labor	Equipment	Total		
1120	**Replace galvanized roof hatch & structure**	40	2 CARP	Ea.							
	Set up, secure and take down scaffold				.130		6.90		6.90	8.95	11.05
	Selective demo, therm & moist protect, roof acc, roof hatch, to 10 S.F.				1.000		42		42	54.50	67.50
	Install 2′-6″ x 3′-0″ galvanized roof hatch				3.200	850	167		1,017	1,150	1,325
	Total				4.330	850	215.90		1,065.90	**1,213.45**	**1,403.55**

B3023 130 Smoke Hatches

	System Description	Freq. (Years)	Crew	Unit	Labor Hours	2020 Bare Costs				Total In-House	Total w/O&P
						Material	Labor	Equipment	Total		
1010	**Replace galvanized smoke hatch single unit 4′ x 4′**	40	2 CARP	Ea.							
	Set up and secure scaffold				.750		39.75		39.75	51.50	64
	Inspect / remove flashing				.032		1.46		1.46	2.10	2.55
	Replace flashing				.032	2.05	1.46		3.51	4.36	5.10
	Galvanized steel cover and frame				2.462	1,725	128		1,853	2,050	2,350
	Remove scaffold				.750		39.75		39.75	51.50	64
	Total				4.025	1,727.05	210.42		1,937.47	**2,159.46**	**2,485.65**
1020	**Replace aluminum smoke hatch single unit, 4′ x 8′**	40	2 CARP	Ea.							
	Set up and secure scaffold				.750		39.75		39.75	51.50	64
	Smoke vent 4′ x 8′				4.000		168		168	218	270
	Inspect / remove flashing				.756		35.04		35.04	50.50	61
	Replace flashing				.756	49.20	35.04		84.24	105	123
	4′ x 8′ aluminum cover and frame				5.246	3,275	274		3,549	3,950	4,525
	Remove scaffold				.750		39.75		39.75	51.50	64
	Total				12.259	3,324.20	591.58		3,915.78	**4,426.50**	**5,107**

C1013 110 Concrete Block, Painted

	System Description	Freq. (Years)	Crew	Unit	Labor Hours	2020 Bare Costs				Total In-House	Total w/O&P
						Material	Labor	Equipment	Total		
0010	**Repair 8″ concrete block wall - (2% of walls)**	25	1 BRIC	C.S.F.							
	Remove block				6.420		273	57	330	415	500
	Replace block				11.111	293	535		828	1,025	1,225
	Total				17.531	293	808	57	1,158	**1,440**	**1,725**
0020	**Refinish concrete block wall**	4	1 PORD	C.S.F.							
	Prepare surface				.250		11		11	14	18
	Fill & paint block, brushwork				1.376	22	61		83	102	125
	Place and remove mask and drops				.370		16		16	21	26
	Total				1.996	22	88		110	**137**	**169**
0040	**Replace 8″ concrete block wall**	75	2 BRIC	C.S.F.							
	Set up and secure scaffold				.750		39.75		39.75	51.50	64
	Remove block				6.420		273	57	330	415	500
	Replace block				11.111	293	535		828	1,025	1,225
	Remove scaffold				.750		39.75		39.75	51.50	64
	Total				19.031	293	887.50	57	1,237.50	**1,543**	**1,853**

C1013 115 Glazed CMU Interior Wall Finish

	System Description	Freq. (Years)	Crew	Unit	Labor Hours	2020 Bare Costs				Total In-House	Total w/O&P
						Material	Labor	Equipment	Total		
0010	**Repair 4″ glazed C.M.U. wall - (2% of walls)**	25	1 BRIC	C.S.F.							
	Remove damaged blocks				5.200		221	46	267	335	405
	Install new blocks				15.072	1,085	720		1,805	2,150	2,525
	Total				20.272	1,085	941	46	2,072	**2,485**	**2,930**
0020	**Replace 4″ glazed C.M.U. wall**	75	2 BRIC	C.S.F.							
	Set up and secure scaffold				.750		39.75		39.75	51.50	64
	Remove blocks				5.200		221	46	267	335	405
	Install new blocks				15.072	1,085	720		1,805	2,150	2,525
	Remove scaffold				.750		39.75		39.75	51.50	64
	Total				21.772	1,085	1,020.50	46	2,151.50	**2,588**	**3,058**

C1013 120 Glass Block

	System Description	Freq. (Years)	Crew	Unit	Labor Hours	2020 Bare Costs				Total In-House	Total w/O&P
						Material	Labor	Equipment	Total		
0010	**Repair glass block wall - (2% of walls)**	25	1 BRIC	C.S.F.							
	Remove damaged glass block				17.829		765		765	985	1,225
	Repair / replace blocks				45.217	2,300	2,150		4,450	5,350	6,375
	Total				63.046	2,300	2,915		5,215	**6,335**	**7,600**
0020	**Replace glass block wall**	75	2 BRIC	C.S.F.							
	Set up and secure scaffold				.750		39.75		39.75	51.50	64
	Remove damaged glass block				17.829		765		765	985	1,225
	Repair / replace blocks				45.217	2,300	2,150		4,450	5,350	6,375
	Remove scaffold				.750		39.75		39.75	51.50	64
	Total				64.546	2,300	2,994.50		5,294.50	**6,438**	**7,728**

C1013 125 Stone Veneer

	System Description	Freq. (Years)	Crew	Unit	Labor Hours	2020 Bare Costs				Total In-House	Total w/O&P
						Material	Labor	Equipment	Total		
0010	**Repair 4″ stone veneer wall - (2% of walls)**	25	1 BRIC	C.S.F.							
	Remove damaged stone				7.429	57	387		444	570	695
	Replace stone				41.600	1,820	1,995		3,815	4,600	5,475
	Total				49.029	1,877	2,382		4,259	**5,170**	**6,170**
0020	**Replace 4″ stone veneer wall**	75	2 BRIC	C.S.F.							
	Set up and secure scaffold				.750		39.75		39.75	51.50	64
	Remove damaged stone				7.429	57	387		444	570	695
	Replace stone				41.600	1,820	1,995		3,815	4,600	5,475
	Remove scaffold				.750		39.75		39.75	51.50	64
	Total				50.529	1,877	2,461.50		4,338.50	**5,273**	**6,298**

C1013 230 Demountable Partitions

	System Description	Freq. (Years)	Crew	Unit	Labor Hours	2020 Bare Costs Material	Labor	Equipment	Total	Total In-House	Total w/O&P
1010	**Remove and reinstall demountable partitions**	5	2 CARP	C.L.F.							
	Remove demountable partitions for reuse				16.000		850		850	1,100	1,350
	Vinyl clad gypsum wall system, material only to replace damaged					1,220			1,220	1,350	1,525
	Install new and reused vinyl clad gypsum in new configuration				34.286		1,820		1,820	2,350	2,900
	18 ga. prefinished 3′-0″ x 6′-8″ to 4-7/8″ deep frame				3.000	636	159		795	905	1,050
	Reinstall doors				3.200		169.50		169.50	220	273
	Cleanup				1.000		53		53	69	85
	Total				57.486	1,856	3,051.50		4,907.50	**5,994**	**7,183**

C1013 544 Channel Frame Wire Mesh Wall

	System Description	Freq. (Years)	Crew	Unit	Labor Hours	2020 Bare Costs Material	Labor	Equipment	Total	Total In-House	Total w/O&P
0010	**Repair channel frame wire mesh wall - (2% of walls)**	20	1 SSWK	C.S.F.							
	Remove damaged 1-1/2″ diamond mesh panel				12.000		515		515	665	820
	Replace with new 1-1/2″ diamond mesh panel				2.857	497.50	151.25		648.75	745	865
	Total				14.857	497.50	666.25		1,163.75	**1,410**	**1,685**
0020	**Refinish channel frame wire mesh wall**	5	1 PORD	C.S.F.							
	Prepare surface				.370		16		16	21	26
	Scrape surface				1.250		56		56	71	88
	Refinish surface, brushwork, primer + 1 coat				1.538	51	68		119	143	172
	Total				3.158	51	140		191	**235**	**286**

C1013 730 Plate Glass Wall - Interior

	System Description	Freq. (Years)	Crew	Unit	Labor Hours	2020 Bare Costs Material	Labor	Equipment	Total	Total In-House	Total w/O&P
0010	**Repair plate glass interior wall - (2% of total)**	25	1 CARP	C.S.F.							
	Remove plate glass & frame				6.933		292		292	375	470
	Repair damaged glass				23.111	8,100	1,255		9,355	10,600	12,200
	Total				30.044	8,100	1,547		9,647	**10,975**	**12,670**

C1013 730 Plate Glass Wall - Interior

	System Description	Freq. (Years)	Crew	Unit	Labor Hours	2020 Bare Costs				Total In-House	Total w/O&P
						Material	Labor	Equipment	Total		
0020	**Replace plate glass interior wall**	50	2 CARP	C.S.F.							
	Set up and secure scaffold				.750		39.75		39.75	51.50	64
	Remove plate glass & frame				6.933		292		292	375	470
	Repair damaged glass				23.111	8,100	1,255		9,355	10,600	12,200
	Remove scaffold				.750		39.75		39.75	51.50	64
	Total				31.544	8,100	1,626.50		9,726.50	**11,078**	**12,798**

C1023 108 Fully Glazed Wooden Doors

	System Description	Freq. (Years)	Crew	Unit	Labor Hours	2020 Bare Costs				Total In-House	Total w/O&P
						Material	Labor	Equipment	Total		
1010	**Replace insulating glass (3% of glass)**	1	1 CARP	S.F.							
	Remove damaged glass				.052		2.19		2.19	2.83	3.50
	Install new insulating glass				.277	35	14.15		49.15	56.50	66.50
	Total				.329	35	16.34		51.34	**59.33**	**70**
1020	**Repair fully glazed wood door**	10	1 CARP	Ea.							
	Remove lockset				.520		27.50		27.50	35.50	44.50
	Replace lockset				1.040	163	55.50		218.50	251	292
	Oil hinges				.065		3.43		3.43	4.43	5.50
	Oil door closer				.065		3.43		3.43	4.43	5.50
	Total				1.689	163	89.86		252.86	**295.36**	**347.50**
1030	**Refinish 3′-0″ x 7′-0″ fully glazed wood door**	4	1 PORD	Ea.							
	Prepare interior door for painting				.433		19.25		19.25	24.50	30.50
	Paint door, roller + brush, 1 coat				1.333	2.65	59		61.65	78.50	97.50
	Total				1.767	2.65	78.25		80.90	**103**	**128**
1040	**Replace 3′-0″ x 7′-0″ fully glazed wood door**	40	1 CARP	Ea.							
	Remove interior door				.520		22		22	28.50	35
	Remove interior door frame				1.300		69		69	89.50	111
	Remove door closer				.167		8.85		8.85	11.45	14.20
	Install new wood door and frame				2.600	762.50	138		900.50	1,025	1,175
	Reinstall door closer				1.735		92		92	119	148
	Total				6.322	762.50	329.85		1,092.35	**1,273.45**	**1,483.20**

C1023 110 Steel Interior Door, Painted

	System Description	Freq. (Years)	Crew	Unit	Labor Hours	2020 Bare Costs				Total In-House	Total w/O&P
						Material	Labor	Equipment	Total		
1010	**Repair steel painted door**	14	1 CARP	Ea.							
	Remove lockset				.520		27.50		27.50	35.50	44.50
	Replace lockset				1.040	163	55.50		218.50	251	292
	Oil hinges				.065		3.43		3.43	4.43	5.50
	Oil door closer				.065		3.43		3.43	4.43	5.50
	Total				1.689	163	89.86		252.86	**295.36**	**347.50**

C1023 110 Steel Interior Door, Painted

	System Description	Freq. (Years)	Crew	Unit	Labor Hours	2020 Bare Costs				Total In-House	Total w/O&P
						Material	Labor	Equipment	Total		
1020	**Refinish 3′-0″ x 7′-0″ steel door, painted**	4	1 PORD	Ea.							
	Prepare interior door for painting				.433		19.25		19.25	24.50	30.50
	Paint door, roller + brush, 1 coat				.800	5.10	35.50		40.60	51	63
	Total				1.233	5.10	54.75		59.85	**75.50**	**93.50**
1030	**Replace 3′-0″ x 7′-0″ steel door, painted**	60	1 CARP	Ea.							
	Remove interior door				.520		22		22	28.50	35
	Remove interior door frame				1.300		69		69	89.50	111
	Remove door closer				.167		8.85		8.85	11.45	14.20
	Install new steel door				1.224	615	65		680	760	875
	Install new steel door frame				1.000	214	53		267	305	355
	Reinstall door closer				1.735		92		92	119	148
	Total				5.946	829	309.85		1,138.85	**1,313.45**	**1,538.20**
2010	**Safety glass replace, (3% of glass)**	1	1 CARP	S.F.							
	Remove damaged glass				.052		2.19		2.19	2.83	3.50
	Install new safety glass				.173	7.95	8.85		16.80	20	24
	Total				.225	7.95	11.04		18.99	**22.83**	**27.50**
2030	**Refinish, 3′-0″ x 7′-0″ steel door w/ safety glass**	4	1 PORD	Ea.							
	Prepare interior door for painting				.433		19.25		19.25	24.50	30.50
	Paint door, roller + brush, 1 coat				.800	5.10	35.50		40.60	51	63
	Total				1.233	5.10	54.75		59.85	**75.50**	**93.50**
2040	**Replace 3′-0″ x 7′-0″ steel door & frame w/ vision lite**	60	1 CARP	Ea.							
	Remove interior door				.520		22		22	28.50	35
	Remove interior door frame				1.300		69		69	89.50	111
	Remove door closer				.167		8.85		8.85	11.45	14.20
	Install new steel door				1.224	615	65		680	760	875
	Install new steel door frame				1.301	214	69		283	325	380
	Reinstall door closer				1.735		92		92	119	148
	Vision lite					111			111	122	139
	Total				6.246	940	325.85		1,265.85	**1,455.45**	**1,702.20**

C1023 110 Steel Interior Door, Painted

	System Description	Freq. (Years)	Crew	Unit	Labor Hours	2020 Bare Costs				Total In-House	Total w/O&P
						Material	Labor	Equipment	Total		
3010	**Repair 3′-0″ x 7′-0″ steel sliding door**	14	1 CARP	Ea.							
	Oil door closer				.065		3.43		3.43	4.43	5.50
	Oil / inspect track / glide wheels				.065		3.43		3.43	4.43	5.50
	Remove door track / glide wheels (25% of total / 14 years)				.220		11.63		11.63	15.10	18.75
	Install new door track / glide wheels (25% of total / 14 years)				.440	15.25	23.38		38.63	47	56.50
	Total				.789	15.25	41.87		57.12	**70.96**	**86.25**
3020	**Refinish 3′-0″ x 7′-0″ steel sliding door**	4	1 PORD	Ea.							
	Prepare interior door for painting				.433		19.25		19.25	24.50	30.50
	Paint door, roller + brush, 1 coat				.800	5.10	35.50		40.60	51	63
	Total				1.233	5.10	54.75		59.85	**75.50**	**93.50**
3030	**Replace 3′-0″ x 7′-0″ steel sliding door & frame**	60	1 CARP	Ea.							
	Remove interior door				.520		22		22	28.50	35
	Remove interior door frame				1.300		69		69	89.50	111
	Remove door closer				.167		8.85		8.85	11.45	14.20
	Install new steel door				1.224	615	65		680	760	875
	Install new steel sliding door frame				3.200	259	186	14.70	459.70	550	645
	Install new door track / glide wheels				1.760	61	93.50		154.50	188	226
	Install door closer				1.735		92		92	119	148
	Total				9.905	935	536.35	14.70	1,486.05	**1,746.45**	**2,054.20**
4010	**Replace wire glass (3% of glass)**	1	1 CARP	S.F.							
	Remove damaged glass				.052		2.19		2.19	2.83	3.50
	Install new wire glass				.179	31.90	9.11		41.01	47	54.50
	Total				.231	31.90	11.30		43.20	**49.83**	**58**
4030	**Refinish 3′-0″ x 7′-0″ steel half glass door**	4	1 PORD	Ea.							
	Prepare interior door for painting				.433		19.25		19.25	24.50	30.50
	Paint door, roller + brush, 1 coat				.800	5.10	35.50		40.60	51	63
	Total				1.233	5.10	54.75		59.85	**75.50**	**93.50**

C1023 110 Steel Interior Door, Painted

	System Description	Freq. (Years)	Crew	Unit	Labor Hours	2020 Bare Costs				Total In-House	Total w/O&P
						Material	Labor	Equipment	Total		
4040	**Replace 3′-0″ x 7′-0″ steel half glass door & frame**	60	1 CARP	Ea.							
	Remove interior door				.520		22		22	28.50	35
	Remove door closer				.167		8.85		8.85	11.45	14.20
	Remove interior door frame				1.300		69		69	89.50	111
	Install new steel door frame				1.301	214	69		283	325	380
	Install new half glass interior door				1.224	475	65		540	605	700
	Install new safety glass				1.040	47.70	53.10		100.80	121	145
	Reinstall door closer				1.735		92		92	119	148
	Total				7.287	736.70	378.95		1,115.65	**1,299.45**	**1,533.20**

C1023 111 Steel Interior Door, Unpainted

	System Description	Freq. (Years)	Crew	Unit	Labor Hours	2020 Bare Costs				Total In-House	Total w/O&P
						Material	Labor	Equipment	Total		
1010	**Repair steel door, unpainted**	14	1 CARP	Ea.							
	Remove lockset				.520		27.50		27.50	35.50	44.50
	Replace lockset				1.040	163	55.50		218.50	251	292
	Oil hinges				.065		3.43		3.43	4.43	5.50
	Oil door closer				.065		3.43		3.43	4.43	5.50
	Total				1.689	163	89.86		252.86	**295.36**	**347.50**
1020	**Replace 3′-0″ x 7′-0″ steel door & frame, unpainted**	60	1 CARP	Ea.							
	Remove interior door				.520		22		22	28.50	35
	Remove interior door frame				1.300		69		69	89.50	111
	Remove door closer				.167		8.85		8.85	11.45	14.20
	Install new steel door				1.224	615	65		680	760	875
	Install new steel door frame				1.000	214	53		267	305	355
	Reinstall door closer				1.735		92		92	119	148
	Total				5.946	829	309.85		1,138.85	**1,313.45**	**1,538.20**
2010	**Replace safety glass (3% of glass)**	1	1 CARP	S.F.							
	Remove damaged glass				.052		2.19		2.19	2.83	3.50
	Install new safety glass				.173	7.95	8.85		16.80	20	24
	Total				.225	7.95	11.04		18.99	**22.83**	**27.50**

C1023 111 Steel Interior Door, Unpainted

	System Description	Freq. (Years)	Crew	Unit	Labor Hours	2020 Bare Costs				Total In-House	Total w/O&P
						Material	Labor	Equipment	Total		
2030	**Replace 3′-0″ x 7′-0″ steel half glass door & frame, unpainted**	60	1 CARP	Ea.							
	Remove interior door				.520		22		22	28.50	35
	Remove interior door frame				1.300		69		69	89.50	111
	Remove door closer				.167		8.85		8.85	11.45	14.20
	Install new half glass interior door				1.224	475	65		540	605	700
	Install new safety glass				1.560	71.55	79.65		151.20	181	217
	Install new steel door frame				1.301	214	69		283	325	380
	Reinstall door closer				1.735		92		92	119	148
	Total				7.807	760.55	405.50		1,166.05	**1,359.45**	**1,605.20**
3010	**Repair 3′-0″ x 7′-0″ steel sliding door, unpainted**	14	1 CARP	Ea.							
	Oil door closer				.065		3.43		3.43	4.43	5.50
	Oil / inspect track / glide wheels				.065		3.43		3.43	4.43	5.50
	Remove door track / glide wheels (25% of total / 14 years)				.220		11.63		11.63	15.10	18.75
	Install new door track / glide wheels (25% of total / 14 years)				.440	15.25	23.38		38.63	47	56.50
	Total				.789	15.25	41.87		57.12	**70.96**	**86.25**
3020	**Replace 3′-0″ x 7′-0″ steel sliding door, unpainted**	60	1 CARP	Ea.							
	Remove interior door				.520		22		22	28.50	35
	Remove interior door frame				1.300		69		69	89.50	111
	Remove door closer				.167		8.85		8.85	11.45	14.20
	Install new steel door				1.224	615	65		680	760	875
	Install new steel sliding door frame				3.200	259	186	14.70	459.70	550	645
	Install new door track / glide wheels				1.760	61	93.50		154.50	188	226
	Reinstall door closer				1.735		92		92	119	148
	Total				9.905	935	536.35	14.70	1,486.05	**1,746.45**	**2,054.20**
4010	**Replace wire glass (3% of glass)**	1	1 CARP	S.F.							
	Remove damaged glass				.052		2.19		2.19	2.83	3.50
	Install new wire glass				.179	31.90	9.11		41.01	47	54.50
	Total				.231	31.90	11.30		43.20	**49.83**	**58**

C1023 112 Bi-Fold Interior Doors

	System Description	Freq. (Years)	Crew	Unit	Labor Hours	2020 Bare Costs				Total In-House	Total w/O&P
						Material	Labor	Equipment	Total		
2010	**Repair 2′-6″ x 6′-8″ bi-fold louvered door, wood**	15	1 CARP	Ea.							
	Remove interior door				.520		22		22	28.50	35
	Remove damaged stile				.078		4.17		4.17	5.40	6.70
	Remove damaged slat				.039		2.06		2.06	2.66	3.30
	Install new slat				.077	.80	4.10		4.90	6.20	7.60
	Install new stile				.157	2.73	8.34		11.07	13.75	16.75
	Reinstall door				.667		35.50		35.50	46	57
	Oil / lubricate door				.065		3.43		3.43	4.43	5.50
	Total				1.602	3.53	79.60		83.13	**106.94**	**131.85**
2020	**Refinish 2′-6″ x 6′-8″ bi-fold louvered door, wood**	8	1 PORD	Ea.							
	Prepare door for finish				.967		43		43	55	68
	Paint door, roller + brush, 1 coat				1.143	5.95	50.50		56.45	71.50	88
	Total				2.110	5.95	93.50		99.45	**126.50**	**156**
2040	**Replace 2′-6″ x 6′-8″ wood louver bi-fold door & frame**	24	1 CARP	Ea.							
	Remove doors				.520		22		22	28.50	35
	Install new door				1.600	305	85		390	445	515
	Install new interior door frame				.998	97.20	53.10		150.30	176	206
	Total				3.118	402.20	160.10		562.30	**649.50**	**756**

C1023 114 Aluminum Interior Doors

	System Description	Freq. (Years)	Crew	Unit	Labor Hours	2020 Bare Costs				Total In-House	Total w/O&P
						Material	Labor	Equipment	Total		
1010	**Repair aluminum door**	12	1 CARP	Ea.							
	Remove lockset				.520		27.50		27.50	35.50	44.50
	Replace lockset				1.040	163	55.50		218.50	251	292
	Oil hinges				.065		3.43		3.43	4.43	5.50
	Oil door closer				.065		3.43		3.43	4.43	5.50
	Total				1.689	163	89.86		252.86	**295.36**	**347.50**

C1023 114 Aluminum Interior Doors

	System Description	Freq. (Years)	Crew	Unit	Labor Hours	2020 Bare Costs				Total In-House	Total w/O&P
						Material	Labor	Equipment	Total		
1020	**Replace 3′-0″ x 7′-0″ aluminum door & frame**	50	1 CARP	Ea.							
	Remove interior door				.520		22		22	28.50	35
	Remove interior door frame				1.300		69		69	89.50	111
	Remove door closer				.167		8.85		8.85	11.45	14.20
	Install new aluminum door				1.224	1,625	65		1,690	1,875	2,125
	Install new aluminum door frame				2.971	705	171		876	1,000	1,150
	Reinstall door closer				1.735		92		92	119	148
	Total				7.917	2,330	427.85		2,757.85	**3,123.45**	**3,583.20**
2010	**Replace safety glass (3% of glass)**	1	1 CARP	S.F.							
	Remove damaged glass				.052		2.19		2.19	2.83	3.50
	Install new safety glass				.173	7.95	8.85		16.80	20	24
	Total				.225	7.95	11.04		18.99	**22.83**	**27.50**
2030	**Replace 3′-0″ x 7′-0″ aluminum door & frame w/ vision lite**	50	1 CARP	Ea.							
	Remove interior door				.520		22		22	28.50	35
	Remove interior door frame				1.300		69		69	89.50	111
	Remove door closer				.167		8.85		8.85	11.45	14.20
	Install new aluminum door				1.224	1,625	65		1,690	1,875	2,125
	Install new aluminum door frame				2.971	705	171		876	1,000	1,150
	Reinstall door closer				1.735		92		92	119	148
	Vision lite					111			111	122	139
	Total				7.917	2,441	427.85		2,868.85	**3,245.45**	**3,722.20**
3010	**Replace insulating glass (3% of glass)**	1	1 CARP	S.F.							
	Remove damaged glass				.052		2.19		2.19	2.83	3.50
	Install new insulating glass				.277	35	14.15		49.15	56.50	66.50
	Total				.329	35	16.34		51.34	**59.33**	**70**
3020	**Repair 3′-0″ x 7′-0″ aluminum sliding door**	14	1 CARP	Ea.							
	Oil door closer				.065		3.43		3.43	4.43	5.50
	Oil / inspect track / glide wheels				.065		3.43		3.43	4.43	5.50
	Remove door track / glide wheels (25% of total / 14 years)				.220		11.63		11.63	15.10	18.75
	Install new door track / glide wheels (25% of total / 14 years)				.440	15.25	23.38		38.63	47	56.50
	Total				.789	15.25	41.87		57.12	**70.96**	**86.25**

C1023 114 Aluminum Interior Doors

	System Description	Freq. (Years)	Crew	Unit	Labor Hours	2020 Bare Costs				Total In-House	Total w/O&P
						Material	Labor	Equipment	Total		
3030	**Replace 3′-0″ x 7′-0″ aluminum sliding door**	50	1 CARP	Ea.							
	Remove interior door				.520		22		22	28.50	35
	Remove interior door frame				1.300		69		69	89.50	111
	Remove door closer				.167		8.85		8.85	11.45	14.20
	Install new aluminum door				1.224	1,625	65		1,690	1,875	2,125
	Install new aluminum door frame				2.971	705	171		876	1,000	1,150
	Install new door track / glide wheels				1.760	61	93.50		154.50	188	226
	Install door closer				1.735		92		92	119	148
	Total				9.677	2,391	521.35		2,912.35	**3,311.45**	**3,809.20**

C1023 116 Fully Glazed Aluminum Doors

	System Description	Freq. (Years)	Crew	Unit	Labor Hours	2020 Bare Costs				Total In-House	Total w/O&P
						Material	Labor	Equipment	Total		
1010	**Replace insulating glass (3% of glass)**	1	1 CARP	S.F.							
	Remove damaged glass				.052		2.19		2.19	2.83	3.50
	Install new insulating glass				.277	35	14.15		49.15	56.50	66.50
	Total				.329	35	16.34		51.34	**59.33**	**70**
1020	**Repair 3′-0″ x 7′-0″ fully glazed aluminum door**	12	1 CARP	Ea.							
	Remove lockset				.520		27.50		27.50	35.50	44.50
	Replace lockset				1.040	163	55.50		218.50	251	292
	Oil hinges				.065		3.43		3.43	4.43	5.50
	Oil door closer				.065		3.43		3.43	4.43	5.50
	Total				1.689	163	89.86		252.86	**295.36**	**347.50**
1030	**Replace 3′-0″ x 7′-0″ fully glazed aluminum door**	50	1 CARP	Ea.							
	Remove interior door				.520		22		22	28.50	35
	Remove interior door frame				1.300		69		69	89.50	111
	Remove door closer				.167		8.85		8.85	11.45	14.20
	Install new aluminum door and frame				10.458	940	605		1,545	1,825	2,175
	Install new safety glass				3.640	166.95	185.85		352.80	425	505
	Reinstall door closer				1.735		92		92	119	148
	Total				17.819	1,106.95	982.70		2,089.65	**2,498.45**	**2,988.20**

C1023 120 Hollow Core Interior Doors

	System Description	Freq. (Years)	Crew	Unit	Labor Hours	2020 Bare Costs				Total In-House	Total w/O&P
						Material	Labor	Equipment	Total		
1010	**Repair hollow core wood door**	7	1 CARP	Ea.							
	Remove lockset				.520		27.50		27.50	35.50	44.50
	Replace lockset				1.040	163	55.50		218.50	251	292
	Oil hinges				.065		3.43		3.43	4.43	5.50
	Oil door closer				.065		3.43		3.43	4.43	5.50
	Total				1.689	163	89.86		252.86	**295.36**	**347.50**
1020	**Refinish 3′-0″ x 7′-0″ hollow core wood door**	4	1 PORD	Ea.							
	Prepare interior door for painting				.433		19.25		19.25	24.50	30.50
	Paint door, roller + brush, 1 coat				.533	5.95	23.50		29.45	37	45
	Total				.967	5.95	42.75		48.70	**61.50**	**75.50**
1030	**Replace 3′-0″ x 7′-0″ hollow core wood door**	30	1 CARP	Ea.							
	Remove interior door				.520		22		22	28.50	35
	Remove interior door frame				1.300		69		69	89.50	111
	Install new interior wood door and frame				.941	165	50		215	246	286
	Total				2.761	165	141		306	**364**	**432**

C1023 121 Solid Core Interior Doors

	System Description	Freq. (Years)	Crew	Unit	Labor Hours	2020 Bare Costs				Total In-House	Total w/O&P
						Material	Labor	Equipment	Total		
1010	**Repair solid core wood door**	11	1 CARP	Ea.							
	Remove lockset				.520		27.50		27.50	35.50	44.50
	Replace lockset				1.040	163	55.50		218.50	251	292
	Oil hinges				.065		3.43		3.43	4.43	5.50
	Oil door closer				.065		3.43		3.43	4.43	5.50
	Total				1.689	163	89.86		252.86	**295.36**	**347.50**
1020	**Refinish 3′-0″ x 7′-0″ solid core wood door**	4	1 PORD	Ea.							
	Prepare interior door for painting				.433		19.25		19.25	24.50	30.50
	Paint door, roller + brush, 1 coat				.533	5.95	23.50		29.45	37	45
	Total				.967	5.95	42.75		48.70	**61.50**	**75.50**

C1023 121 Solid Core Interior Doors

	System Description	Freq. (Years)	Crew	Unit	Labor Hours	2020 Bare Costs				Total In-House	Total w/O&P
						Material	Labor	Equipment	Total		
1030	**Replace 3'-0" x 7'-0" solid core wood door**	40	1 CARP	Ea.							
	Remove interior door				.520		22		22	28.50	35
	Remove interior door frame				.650		34.45		34.45	44.50	55.50
	Install new interior wood door & frame				.941	370	50		420	470	545
	Total				2.111	370	106.45		476.45	**543**	**635.50**
2010	**Repair solid core sliding wood door**	14	1 CARP	Ea.							
	Oil door closer				.065		3.43		3.43	4.43	5.50
	Oil / inspect track / glide wheels				.065		3.43		3.43	4.43	5.50
	Remove door track / glide wheels (25% of total / 14 years)				.220		11.63		11.63	15.10	18.75
	Install new door track / glide wheels (25% of total / 14 years)				.440	15.25	23.38		38.63	47	56.50
	Total				.789	15.25	41.87		57.12	**70.96**	**86.25**
2030	**Replace 3'-0" x 7'-0" solid core sliding wood door**	40	1 CARP	Ea.							
	Remove interior door				.520		22		22	28.50	35
	Remove interior door frame				.650		34.45		34.45	44.50	55.50
	Remove door closer				.167		8.85		8.85	11.45	14.20
	Install new interior wood door and frame				.941	370	50		420	470	545
	Install new door track / glide wheels				1.760	61	93.50		154.50	188	226
	Reinstall door closer				1.735		92		92	119	148
	Total				5.773	431	300.80		731.80	**861.45**	**1,023.70**
3040	**Replace 3'-0" x 7'-0" solid core wood door, w/ safety glass**	40	1 CARP	Ea.							
	Remove interior door				.520		22		22	28.50	35
	Remove interior door frame				.650		34.45		34.45	44.50	55.50
	Install new interior wood door and frame				.941	370	50		420	470	545
	Install new safety glass lite					92.50			92.50	102	116
	Total				2.111	462.50	106.45		568.95	**645**	**751.50**

C1023 129 Interior Gates

	System Description	Freq. (Years)	Crew	Unit	Labor Hours	2020 Bare Costs				Total In-House	Total w/O&P
						Material	Labor	Equipment	Total		
1030	**Prepare and refinish interior metal gate**	5	1 PORD	Ea.							
	Lay drop cloth				.090		4		4	5.10	6.30
	Prepare / brush surface				.125		5.58		5.58	7	8.80
	Paint gate, brushwork, primer + 1 coat				.167	1.62	7.38		9	11.15	13.75
	Remove drop cloth				.090		4		4	5.10	6.30
	Total				.471	1.62	20.96		22.58	**28.35**	**35.15**

C1023 324 Hinges, Brass

	System Description	Freq. (Years)	Crew	Unit	Labor Hours	2020 Bare Costs				Total In-House	Total w/O&P
						Material	Labor	Equipment	Total		
0010	**Replace 1 1/2 pair brass hinges**	60	1 CARP	Ea.							
	Remove door				.650		27.50		27.50	35.50	44
	Remove hinges				.500		26.55		26.55	34.50	42.50
	Replace hinges				.500		26.55		26.55	34.50	42.50
	Hinges, 1 1/2 pair					208.50			208.50	229	261
	Install door				1.224		65		65	84	104
	Total				2.875	208.50	145.60		354.10	**417.50**	**494**

C1023 325 Lockset, Brass

	System Description	Freq. (Years)	Crew	Unit	Labor Hours	2020 Bare Costs				Total In-House	Total w/O&P
						Material	Labor	Equipment	Total		
0010	**Replace brass lockset**	30	1 CARP	Ea.							
	Remove lockset				.400		21.50		21.50	27.50	34
	Install lockset				1.156	239	61.50		300.50	340	395
	Total				1.556	239	83		322	**367.50**	**429**

C1023 326 Door Closer, Brass

	System Description	Freq. (Years)	Crew	Unit	Labor Hours	2020 Bare Costs				Total In-House	Total w/O&P
						Material	Labor	Equipment	Total		
0010	**Replace brass closer**	15	1 CARP	Ea.							
	Remove door closer				.167		8.85		8.85	11.45	14.20
	Install new door closer				1.735	211	92		303	350	410
	Total				1.902	211	100.85		311.85	**361.45**	**424.20**

C1023 327 Deadbolt, Brass

	System Description	Freq. (Years)	Crew	Unit	Labor Hours	2020 Bare Costs				Total In-House	Total w/O&P
						Material	Labor	Equipment	Total		
0010	**Replace brass deadbolt**	20	1 CARP	Ea.							
	Remove deadbolt				.578		30.50		30.50	39.50	49
	Replace deadbolt				1.156	193	61.50		254.50	292	340
	Total				1.733	193	92		285	**331.50**	**389**

C1023 328 Weatherstripping, Brass

	System Description	Freq. (Years)	Crew	Unit	Labor Hours	2020 Bare Costs				Total In-House	Total w/O&P
						Material	Labor	Equipment	Total		
0010	**Replace brass weatherstripping**	20	1 CARP	Ea.							
	Remove weatherstripping				.015		.78		.78	1.01	1.25
	Install new weatherstripping				3.478	52	185		237	296	360
	Total				3.493	52	185.78		237.78	**297.01**	**361.25**

C1023 329 Panic Bar

	System Description	Freq. (Years)	Crew	Unit	Labor Hours	2020 Bare Costs				Total In-House	Total w/O&P
						Material	Labor	Equipment	Total		
0010	**Replace panic bar**	25	1 CARP	Ea.							
	Remove panic bar				1.040		55.50		55.50	71.50	88.50
	Install panic bar				2.080	1,125	111		1,236	1,375	1,575
	Total				3.120	1,125	166.50		1,291.50	**1,446.50**	**1,663.50**

C1033 110 Toilet Partitions

	System Description	Freq. (Years)	Crew	Unit	Labor Hours	2020 Bare Costs				Total In-House	Total w/O&P
						Material	Labor	Equipment	Total		
1010	**Replace toilet partitions, laminate clad-overhead braced, per stall**	20	2 CARP	Ea.							
	Remove overhead braced laminate clad partitions				1.333		71		71	91.50	114
	Install overhead braced laminate clad partitions				3.478	810	185		995	1,125	1,300
	Cleanup				.500		26.50		26.50	34.50	42.50
	Total				5.312	810	282.50		1,092.50	**1,251**	**1,456.50**
1020	**Replace toilet partitions, painted metal-overhead braced, per stall**	20	2 CARP	Ea.							
	Remove overhead braced painted metal partitions				1.333		71		71	91.50	114
	Install overhead braced painted metal partitions				3.478	385	185		570	660	775
	Cleanup				.500		26.50		26.50	34.50	42.50
	Total				5.312	385	282.50		667.50	**786**	**931.50**
1030	**Replace toilet partitions, porcelain enamel-overhead braced, per stall**	20	2 CARP	Ea.							
	Remove overhead braced partitions				1.333		71		71	91.50	114
	Install overhead braced porcelain enamel partitions				3.478	605	185		790	905	1,050
	Cleanup				.500		26.50		26.50	34.50	42.50
	Total				5.312	605	282.50		887.50	**1,031**	**1,206.50**
1040	**Replace toilet partitions, phenolic-overhead braced, per stall**	20	2 CARP	Ea.							
	Remove overhead braced partitions				1.333		71		71	91.50	114
	Install phenolic overhead braced partition				3.478	740	185		925	1,050	1,225
	Cleanup				.500		26.50		26.50	34.50	42.50
	Total				5.312	740	282.50		1,022.50	**1,176**	**1,381.50**
1050	**Replace toilet partitions, stainless steel-overhead braced, per stall**	30	2 CARP	Ea.							
	Remove overhead braced partitions				1.333		71		71	91.50	114
	Install stainless steel overhead braced partition				2.667	1,050	142		1,192	1,350	1,550
	Cleanup				.500		26.50		26.50	34.50	42.50
	Total				4.500	1,050	239.50		1,289.50	**1,476**	**1,706.50**
1060	**Replace toilet partitions, stainless steel, ceiling hung, per stall**	30	2 CARP	Ea.							
	Remove ceiling hung partitions				1.333		71		71	91.50	114
	Install stainless steel ceiling hung partition				4.000	1,100	213		1,313	1,475	1,725
	Cleanup				.500		26.50		26.50	34.50	42.50
	Total				5.833	1,100	310.50		1,410.50	**1,601**	**1,881.50**

C1033 110 Toilet Partitions

	System Description	Freq. (Years)	Crew	Unit	Labor Hours	2020 Bare Costs				Total In-House	Total w/O&P
						Material	Labor	Equipment	Total		
2020	**Replace urinal screen, stainless steel**	30	2 CARP	Ea.							
	Remove stainless steel urinal screen				1.000		42		42	54.50	67.50
	Install stainless steel urinal screen				2.000	535	106		641	725	840
	Cleanup				.500		26.50		26.50	34.50	42.50
	Total				3.500	535	174.50		709.50	**814**	**950**
3010	**Replace shower stall, fiberglass, per stall**	30	2 SHEE	Ea.							
	Remove shower base, partition and door				2.000		84		84	109	135
	Install shower base, partition and door				3.556	755	222		977	1,100	1,300
	Cleanup				.500		26.50		26.50	34.50	42.50
	Total				6.056	755	332.50		1,087.50	**1,243.50**	**1,477.50**

C1033 210 Metal Lockers

	System Description	Freq. (Years)	Crew	Unit	Labor Hours	2020 Bare Costs				Total In-House	Total w/O&P
						Material	Labor	Equipment	Total		
1010	**Replace metal lockers, single tier**	20	1 SHEE	Ea.							
	Remove lockers				.533		22.50		22.50	29	36
	Install 12″ x 18″ x 72″ single enameled locker				.400	230	25		255	284	325
	Cleanup				.500		26.50		26.50	34.50	42.50
	Total				1.433	230	74		304	**347.50**	**403.50**

C20 STAIRS | C2013 Stair Construction

C2013 110 Interior Concrete Steps

	System Description	Freq. (Years)	Crew	Unit	Labor Hours	2020 Bare Costs				Total In-House	Total w/O&P
						Material	Labor	Equipment	Total		
0010	**Repair interior concrete steps**	15	1 CEFI	S.F.							
	Repair spalls in steps				.104	14.85	5.20		20.05	23	26.50
	Total				.104	14.85	5.20		20.05	**23**	**26.50**
0020	**Replace interior concrete steps**	100	2 CEFI	S.F.							
	Remove concrete				.520		22	4.6[illegible]	26.64	33.50	40.50
	Install new concrete steps				.520	5.40	27	.2[illegible]	32.69	40.50	50
	Total				1.040	5.40	49	4.9[illegible]	59.33	**74**	**90.50**

C2013 115 Interior Masonry Steps, Painted

	System Description	Freq. (Years)	Crew	Unit	Labor Hours	2020 Bare Costs				Total In-House	Total w/O&P
						Material	Labor	Equipment	Total		
0010	**Repair interior masonry steps, painted**	20	1 BRIC	S.F.							
	Remove damaged masonry				.083		3.50		3.50	4.53	5.60
	Replace with new masonry				.232	4.46	10.90		15.36	19.10	23
	Refinish surface				.014	.22	.61		.83	1.02	1.25
	Total				.329	4.68	15.01		19.69	**24.65**	**29.85**
0040	**Replace interior masonry steps, painted**	50	2 BRIC	S.F.							
	Remove masonry steps				4.000		168		168	218	270
	Clean work area				.015		.65	.1[illegible]	.80	1	1.20
	Install new masonry steps				.891	93.50	43		136.50	158	185
	Total				4.907	93.50	211.65	.1[illegible]	305.30	**377**	**456.20**

C2013 130 Interior Metal Steps

	System Description	Freq. (Years)	Crew	Unit	Labor Hours	2020 Bare Costs				Total In-House	Total w/O&P
						Material	Labor	Equipment	Total		
0010	**Repair interior metal steps**	15	1 SSWK	S.F.							
	Remove damaged tread				.347		20		20	26.50	32.50
	Replace with cast iron tread, 1/2" thick				.120	60.39	6.93		67.32	75.50	87
	Total				.467	60.39	26.93		87.32	**102**	**119.50**

C2013 130 Interior Metal Steps

	System Description	Freq. (Years)	Crew	Unit	Labor Hours	2020 Bare Costs Material	Labor	Equipment	Total	Total In-House	Total w/O&P
0030	**Refinish interior metal steps**	9	1 PORD	S.F.							
	Prepare steps for painting				.026		1.15		1.15	1.48	1.83
	Wipe surface				.001		.05		.05	.07	.08
	Refinish surface, roller + brush, 1 coat				.025	.38	1.11		1.49	1.84	2.24
	Total				.052	.38	2.31		2.69	**3.39**	**4.15**
0040	**Replace interior metal steps**	50	2 SSWK	S.F.							
	Remove steps				.264		11.06		11.06	14.35	17.80
	Install pre-erected steel steps				.163	138	9.50	.75	148.25	165	189
	Prepare steps for painting				.026		1.15		1.15	1.48	1.83
	Wipe surface				.001		.05		.05	.07	.08
	Refinish surface				.025	.38	1.11		1.49	1.84	2.24
	Total				.479	138.38	22.87	.75	162	**182.74**	**210.95**

C2013 435 Interior Metal Stair Railing

	System Description	Freq. (Years)	Crew	Unit	Labor Hours	2020 Bare Costs Material	Labor	Equipment	Total	Total In-House	Total w/O&P
0030	**Refinish interior metal stair railing**	7	1 PORD	S.F.							
	Prepare surface				.019		.85		.85	1.09	1.35
	Refinish surface, roller + brush, 1 coat				.015	.13	.68		.81	1.01	1.24
	Total				.035	.13	1.53		1.66	**2.10**	**2.59**
0040	**Replace interior metal stair railing**	45	2 SSWK	L.F.							
	Remove railing				.098		5.70	.45	6.15	8.05	9.80
	Install new railing				.195	25	11.35	.90	37.25	43.50	51
	Total				.293	25	17.05	1.35	43.40	**51.55**	**60.80**

C2013 437 Interior Iron Stair Railing

	System Description	Freq. (Years)	Crew	Unit	Labor Hours	2020 Bare Costs Material	Labor	Equipment	Total	Total In-House	Total w/O&P
0030	**Refinish interior wrought iron stair railing**	7	1 PORD	L.F.							
	Prepare surface				.019		.85		.85	1.09	1.35
	Refinish surface, roller + brush, 1 coat				.015	.13	.68		.81	1.01	1.24
	Total				.035	.13	1.53		1.66	**2.10**	**2.59**

C2013 437 Interior Iron Stair Railing

	System Description	Freq. (Years)	Crew	Unit	Labor Hours	2020 Bare Costs				Total In-House	Total w/O&P
						Material	Labor	Equipment	Total		
0040	**Replace interior wrought iron stair railing**	45	2 SSWK	L.F.							
	Remove railing				.087		3.65		3.65	4.71	5.85
	Install new railing				1.300	33	75		108	136	164
	Total				1.387	33	78.65		111.65	**140.71**	**169.85**

C2013 440 Interior Wood Steps

	System Description	Freq. (Years)	Crew	Unit	Labor Hours	2020 Bare Costs				Total In-House	Total w/O&P
						Material	Labor	Equipment	Total		
0030	**Refinish interior wood steps**	3	1 PORD	S.F.							
	Prepare surface				.026		1.15		1.15	1.48	1.83
	Refinish surface, roller + brush, 1 coat				.026	.07	1.15		1.22	1.56	1.92
	Total				.052	.07	2.30		2.37	**3.04**	**3.75**
0040	**Replace interior wood steps**	40	1 CARP	S.F.							
	Remove steps				.172		7.26		7.26	9.35	11.55
	Replace steps				.325	92.50	17.25		109.75	124	143
	Total				.497	92.50	24.51		117.01	**133.35**	**154.55**

C2013 445 Interior Wood Stair Railing

	System Description	Freq. (Years)	Crew	Unit	Labor Hours	2020 Bare Costs				Total In-House	Total w/O&P
						Material	Labor	Equipment	Total		
0030	**Refinish interior wood stair railing**	7	1 PORD	L.F.							
	Prepare railings for refinishing				.026		1.15		1.15	1.48	1.83
	Refinish surface, roller + brush, 1 coat				.026	.07	1.15		1.22	1.56	1.92
	Total				.052	.07	2.30		2.37	**3.04**	**3.75**
0040	**Replace interior wood stair railing**	40	1 CARP	L.F.							
	Remove railing				.087		3.65		3.65	4.71	5.85
	Replace railing				.130	2.24	6.90		9.14	11.40	13.85
	Total				.217	2.24	10.55		12.79	**16.11**	**19.70**

C2023 110 Carpeted Steps

	System Description	Freq. (Years)	Crew	Unit	Labor Hours	2020 Bare Costs				Total In-House	Total w/O&P
						Material	Labor	Equipment	Total		
0020	**Replace carpeted steps**	8	2 TILF	S.F.							
	Remove old carpet				.010		.44		.44	.57	.70
	Install new carpet				.069	15.35	3.40		18.75	21	24.50
	Additional labor				.229	30.69	11.32		42.01	48	56
	Total				.308	46.04	15.16		61.20	**69.57**	**81.20**

C2023 120 Rubber Steps

	System Description	Freq. (Years)	Crew	Unit	Labor Hours	2020 Bare Costs				Total In-House	Total w/O&P
						Material	Labor	Equipment	Total		
0020	**Replace rubber steps**	18	2 TILF	L.F.							
	Remove rubber covering				.015		.63		.63	.81	1
	Prepare floor for installation				.063		3.14		3.14	3.95	4.92
	Install new rubber tread				.090	14.35	4.48		18.83	21.50	25
	Install new rubber riser				.042	8.55	2.06		10.61	12	13.90
	Total				.210	22.90	10.31		33.21	**38.26**	**44.82**

C2023 130 Terrazzo Steps

	System Description	Freq. (Years)	Crew	Unit	Labor Hours	2020 Bare Costs				Total In-House	Total w/O&P
						Material	Labor	Equipment	Total		
0020	**Replace terrazzo steps**	50	2 MSTZ	S.F.							
	Remove old terrazzo tile				.052		2.19		2.19	2.83	3.51
	Clean area				.020		1.06		1.06	1.37	1.70
	Replace with new tread				.743	3.13	33.50	12.25	48.88	59	70
	Total				.815	3.13	36.75	12.25	52.13	**63.20**	**75.21**

C3013 106 Vinyl Wall Covering

	System Description	Freq. (Years)	Crew	Unit	Labor Hours	2020 Bare Costs				Total In-House	Total w/O&P
						Material	Labor	Equipment	Total		
0010	**Repair med. wt. vinyl wall covering - (2% of walls)**	1	1 PORD	C.S.F.							
	Remove damaged vinyl wall covering				1.156		52		52	66	82
	Install vinyl wall covering				2.391	106	107		213	253	300
	Total				3.547	106	159		265	**319**	**382**
0020	**Replace medium weight vinyl wall covering**	15	1 PORD	C.S.F.							
	Set up and secure scaffold				.750		39.75		39.75	51.50	64
	Remove vinyl wall covering				1.156		52		52	66	82
	Install vinyl wall covering				2.391	106	107		213	253	300
	Remove scaffold				.750		39.75		39.75	51.50	64
	Total				5.047	106	238.50		344.50	**422**	**510**

C3013 202 Wallpaper

	System Description	Freq. (Years)	Crew	Unit	Labor Hours	2020 Bare Costs				Total In-House	Total w/O&P
						Material	Labor	Equipment	Total		
0010	**Repair wallpaper - (2% of walls)**	8	1 PORD	S.Y.							
	Remove damaged wallpaper				.104		4.68		4.68	5.95	7.40
	Repair / replace wallpaper				.175	11.52	7.83		19.35	22.50	27
	Total				.279	11.52	12.51		24.03	**28.45**	**34.40**
0020	**Replace wallpaper**	20	1 PORD	S.Y.							
	Set up, secure and take down scaffold				.130		6.90		6.90	8.95	11.05
	Remove damaged wallpaper				.104		4.68		4.68	5.95	7.40
	Replace wallpaper				.175	11.52	7.83		19.35	22.50	27
	Set up, secure and take down scaffold				.130		6.90		6.90	8.95	11.05
	Total				.539	11.52	26.31		37.83	**46.35**	**56.50**

C3013 206 Fabric Interior Wall Finish

	System Description	Freq. (Years)	Crew	Unit	Labor Hours	2020 Bare Costs				Total In-House	Total w/O&P
						Material	Labor	Equipment	Total		
0010	**Repair fabric wall finish**	9	1 PORD	S.Y.							
	Set up, secure and take down ladder				.130		5.80		5.80	7.40	9.20
	Remove fabric wall covering				.104		4.68		4.68	5.95	7.40
	Repair / replace fabric				.234	11.70	10.44		22.14	26	31
	Total				.468	11.70	20.92		32.62	**39.35**	**47.60**
0020	**Replace fabric wall finish**	50	1 PORD	S.Y.							
	Set up and secure scaffold				1.350		71.55		71.55	93	115
	Remove fabric wall covering				.104		4.68		4.68	5.95	7.40
	Replace fabric				.234	11.70	10.44		22.14	26	31
	Remove scaffold				1.350		71.55		71.55	93	115
	Total				3.038	11.70	158.22		169.92	**217.95**	**268.40**

C3013 208 Cork Tile

	System Description	Freq. (Years)	Crew	Unit	Labor Hours	2020 Bare Costs				Total In-House	Total w/O&P
						Material	Labor	Equipment	Total		
0010	**Repair 12″ x 12″ x 3/16″ cork tile wall - (2% of walls)**	8	1 CARP	C.S.F.							
	Remove old tiles				.620		26		26	34	42
	Replace with new tile				4.426	426	197		623	720	845
	Total				5.046	426	223		649	**754**	**887**
0020	**Replace 12″ x 12″ x 3/16″ cork tile wall**	16	2 CARP	C.S.F.							
	Set up and secure scaffold				.750		39.75		39.75	51.50	64
	Remove old tiles				.620		26		26	34	42
	Replace with new tile				4.426	426	197		623	720	845
	Remove scaffold				.750		39.75		39.75	51.50	64
	Total				6.546	426	302.50		728.50	**857**	**1,015**

C3013 210 Acoustical Tile

	System Description	Freq. (Years)	Crew	Unit	Labor Hours	2020 Bare Costs				Total In-House	Total w/O&P
						Material	Labor	Equipment	Total		
0010	**Repair acoustical tile - (2% of walls)**	25	1 CARP	C.S.F.							
	Remove damaged tile				1.067		45		45	58	72
	Install new tile				2.737	402	145		547	630	735
	Total				3.804	402	190		592	**688**	**807**
0030	**Refinish acoustical tile**	10	1 PORD	C.S.F.							
	Wipe surface				.894	7	40		47	59	72
	Prepare surface				.894	7	40		47	59	72
	Refinish surface				1.529	27	68		95	117	142
	Total				3.317	41	148		189	**235**	**286**
0040	**Replace acoustical tile**	60	2 CARP	C.S.F.							
	Set up and secure scaffold				.750		39.75		39.75	51.50	64
	Remove old tiles				1.067		45		45	58	72
	Install new tile				2.737	402	145		547	630	735
	Remove scaffold				.750		39.75		39.75	51.50	64
	Total				5.304	402	269.50		671.50	**791**	**935**

C3013 212 Stucco

	System Description	Freq. (Years)	Crew	Unit	Labor Hours	2020 Bare Costs				Total In-House	Total w/O&P
						Material	Labor	Equipment	Total		
0010	**Repair stucco wall - (2% of walls)**	20	1 BRIC	S.Y.							
	Prepare surface				.023		.99		.99	1.26	1.62
	Repair stucco				.529	3.87	24.39	1.53	29.79	37.50	45.50
	Place and remove mask and drops				.067		2.88		2.88	3.78	4.68
	Total				.618	3.87	28.26	1.53	33.66	**42.54**	**51.80**
0030	**Refinish stucco wall**	4	1 PORD	S.Y.							
	Wash surface				.023		.99		.99	1.26	1.62
	Prepare surface				.033		1.44		1.44	1.89	2.34
	Refinish surface, brushwork, 2 coats				.138	2.43	6.12		8.55	10.55	12.75
	Place and remove mask and drops				.033		1.44		1.44	1.89	2.34
	Total				.227	2.43	9.99		12.42	**15.59**	**19.05**

C3013 212 Stucco

	System Description	Freq. (Years)	Crew	Unit	Labor Hours	2020 Bare Costs Material	Labor	Equipment	Total	Total In-House	Total w/O&P
0040	**Replace stucco wall**	75	2 BRIC	S.Y.							
	Set up, secure and take down scaffold				.130		6.90		6.90	8.95	11.05
	Remove stucco				.279		14.85		14.85	19.15	24
	Replace stucco				1.200	6.80	56.50	2.93	66.23	83	101
	Set up, secure and take down scaffold				.130		6.90		6.90	8.95	11.05
	Total				1.739	6.80	85.15	2.93	94.88	**120.05**	**147.10**
4120	**Spray refinish stucco wall**	5	1 PORD	S.Y.							
	Wash surface				.023		.99		.99	1.26	1.62
	Prepare surface				.033		1.44		1.44	1.89	2.34
	Refinish surface, spray, 2 coats				.049	2.16	2.16		4.32	5.15	6.20
	Place and remove mask and drops				.033		1.44		1.44	1.89	2.34
	Total				.138	2.16	6.03		8.19	**10.19**	**12.50**

C3013 213 Plaster

	System Description	Freq. (Years)	Crew	Unit	Labor Hours	2020 Bare Costs Material	Labor	Equipment	Total	Total In-House	Total w/O&P
0010	**Repair plaster wall - (2% of walls)**	13	1 PLAS	S.Y.							
	Remove damage				.468		19.71		19.71	25.50	31.50
	Replace two coat plaster finish, incl. lath				.752	7.40	35.50	1.83	44.73	55.50	67.50
	Total				1.220	7.40	55.21	1.83	64.44	**81**	**99**
0030	**Refinish plaster wall**	4	1 PORD	S.Y.							
	Prepare surface				.033		1.44		1.44	1.89	2.34
	Paint / seal surface, brushwork, 1 coat				.063	.63	2.79		3.42	4.23	5.20
	Place and remove mask and drops				.033		1.44		1.44	1.89	2.34
	Total				.129	.63	5.67		6.30	**8.01**	**9.88**
0040	**Replace plaster wall**	75	2 PLAS	S.Y.							
	Set up, secure and take down scaffold				.130		6.90		6.90	8.95	11.05
	Remove material				.468		19.71		19.71	25.50	31.50
	Replace two coat plaster wall, incl. lath				.752	7.40	35.50	1.83	44.73	55.50	67.50
	Total				1.350	7.40	62.11	1.83	71.34	**89.95**	**110.05**

C3013 214 Drywall

	System Description	Freq. (Years)	Crew	Unit	Labor Hours	2020 Bare Costs Material	Labor	Equipment	Total	Total In-House	Total w/O&P
0010	**Repair 5/8″ drywall - (2% of walls)**	20	1 CARP	S.F.							
	Remove damage				.008		.34		.34	.44	.54
	Replace 5/8″ drywall, taped and finished				.022	.33	1.15		1.48	1.84	2.25
	Total				.030	.33	1.49		1.82	**2.28**	**2.79**
0030	**Refinish drywall**	4	1 PORD	S.F.							
	Place and remove mask and drops				.004		.16		.16	.21	.26
	Prepare surface				.004		.16		.16	.21	.26
	Paint surface, brushwork, 1 coat				.007	.07	.31		.38	.47	.58
	Total				.014	.07	.63		.70	**.89**	**1.10**
0040	**Replace 5/8″ drywall**	75	2 CARP	S.F.							
	Set up, secure and take down ladder				.014		.77		.77	.99	1.23
	Remove drywall				.008		.34		.34	.44	.54
	Replace 5/8″ drywall, taped and finished				.022	.33	1.15		1.48	1.84	2.25
	Total				.044	.33	2.26		2.59	**3.27**	**4.02**
0050	**Office painting, 10′ x 12′, 10′ high walls**	5	1 PORD	Ea.							
	Spread drop cloths				.007		.33		.33	.42	.52
	Prepare drywall partitions				1.628		70.40		70.40	92.50	114
	Clean drywall partitions				.220	.89	9.24		10.13	12.95	15.85
	Paint drywall partitions, roller + brush, 1 coat				3.062	30.80	136.40		167.20	207	254
	Remove drop cloths				.007		.33		.33	.42	.52
	Total				4.925	31.69	216.70		248.39	**313.29**	**384.89**
0060	**Office painting, 10′ x 15′, 10′ high walls**	5	1 PORD	Ea.							
	Spread drop cloths				.009		.41		.41	.53	.65
	Prepare drywall partitions				1.850		80		80	105	130
	Clean drywall partitions				.250	1.01	10.50		11.51	14.70	18
	Paint drywall partitions, roller + brush, 1 coat				3.480	35	155		190	235	289
	Remove drop cloths				.009		.41		.41	.53	.65
	Total				5.599	36.01	246.32		282.33	**355.76**	**438.30**

C3013 214 Drywall

	System Description	Freq. (Years)	Crew	Unit	Labor Hours	2020 Bare Costs				Total In-House	Total w/O&P
						Material	Labor	Equipment	Total		
2030	**Refinish drywall, 12′ to 24′ high**	5	2 PORD	S.F.							
	Set up and secure scaffold				.015		.80		.80	1.03	1.28
	Place and remove mask and drops				.004		.16		.16	.21	.26
	Prepare surface				.004		.16		.16	.21	.26
	Paint surface, roller + brushwork, 1 coat				.007	.07	.31		.38	.47	.58
	Remove scaffold				.015		.80		.80	1.03	1.28
	Total				.044	.07	2.23		2.30	**2.95**	**3.66**
3030	**Refinish drywall, over 24′ high**	5	2 PORD	S.F.							
	Set up and secure scaffold				.023		1.19		1.19	1.55	1.92
	Place and remove mask and drops				.004		.16		.16	.21	.26
	Prepare surface				.004		.16		.16	.21	.26
	Paint surface, roller + brushwork, 1 coat				.007	.07	.31		.38	.47	.58
	Remove scaffold				.023		1.19		1.19	1.55	1.92
	Total				.059	.07	3.01		3.08	**3.99**	**4.94**

C3013 215 Fiberglass Panels, Rigid

	System Description	Freq. (Years)	Crew	Unit	Labor Hours	2020 Bare Costs				Total In-House	Total w/O&P
						Material	Labor	Equipment	Total		
0010	**Repair glass cloth fiberglass panels - (2% of walls)**	9	1 CARP	C.S.F.							
	Remove damaged fiberglass panels				2.600		109		109	141	175
	Install new fiberglass panel				6.710	915	357		1,272	1,475	1,725
	Total				9.310	915	466		1,381	**1,616**	**1,900**
0040	**Replace glass cloth fiberglass panels**	35	2 CARP	C.S.F.							
	Set up and secure scaffold				.750		39.75		39.75	51.50	64
	Remove old panels				2.600		109		109	141	175
	Remove old furring				.516		22		22	28	35
	Install new furring				2.101	47	112		159	196	238
	Install new glass cloth fiberglass panels				6.710	915	357		1,272	1,475	1,725
	Remove scaffold				.750		39.75		39.75	51.50	64
	Total				13.427	962	679.50		1,641.50	**1,943**	**2,301**

C3013 220 Tile

	System Description	Freq. (Years)	Crew	Unit	Labor Hours	2020 Bare Costs				Total In-House	Total w/O&P
						Material	Labor	Equipment	Total		
0010	**Repair 4″ x 4″ thin set ceramic tile - (2% of walls)**	10	1 TILF	C.S.F.							
	Remove damaged tiles				2.600		109		109	141	175
	Replace tile / grout				10.947	289	483		772	925	1,125
	Total				13.547	289	592		881	**1,066**	**1,300**
0020	**Replace 4″ x 4″ thin set ceramic tile**	75	1 TILF	C.S.F.							
	Set up and secure scaffold				.750		39.75		39.75	51.50	64
	Remove tiles				2.600		109		109	141	175
	Replace tile / grout				10.947	289	483		772	925	1,125
	Remove scaffold				.750		39.75		39.75	51.50	64
	Total				15.047	289	671.50		960.50	**1,169**	**1,428**

C3013 230 Plywood Paneling

	System Description	Freq. (Years)	Crew	Unit	Labor Hours	2020 Bare Costs				Total In-House	Total w/O&P
						Material	Labor	Equipment	Total		
0010	**Repair plywood paneling - (2% of walls)**	10	1 CARP	C.S.F.							
	Remove damaged paneling				1.156		49		49	63	78
	Replace paneling				4.952	124	263		387	475	575
	Total				6.108	124	312		436	**538**	**653**
0020	**Refinish plywood paneling**	10	1 PORD	C.S.F.							
	Place mask and drops				.741		33		33	42	52
	Sand & paint, brushwork, 1 coat				.990	7	44		51	64	79
	Remove mask and drops				.741		33		33	42	52
	Total				2.472	7	110		117	**148**	**183**
0040	**Replace plywood paneling**	30	2 CARP	C.S.F.							
	Set up and secure scaffold				.750		39.75		39.75	51.50	64
	Remove damaged paneling				1.156		49		49	63	78
	Replace paneling				4.952	124	263		387	475	575
	Remove scaffold				.750		39.75		39.75	51.50	64
	Total				7.608	124	391.50		515.50	**641**	**781**

C3013 234 Wainscot

	System Description	Freq. (Years)	Crew	Unit	Labor Hours	2020 Bare Costs				Total In-House	Total w/O&P
						Material	Labor	Equipment	Total		
0010	**Repair wainscot - (2% of walls)**	10	1 CARP	C.S.F.							
	Remove damaged wainscot				1.733		73		73	94	117
	Replace wainscot				16.000	1,680	850		2,530	2,950	3,450
	Total				17.733	1,680	923		2,603	**3,044**	**3,567**
0020	**Refinish wainscot**	4	1 PORD	C.S.F.							
	Place mask and drops				.741		33		33	42	52
	Sand & paint, brushwork, 1 coat				.990	7	44		51	64	79
	Remove mask and drops				.741		33		33	42	52
	Total				2.472	7	110		117	**148**	**183**
0040	**Replace wainscot**	40	2 CARP	C.S.F.							
	Remove wainscot				1.733		73		73	94	117
	Replace wainscot				16.000	1,680	850		2,530	2,950	3,450
	Total				17.733	1,680	923		2,603	**3,044**	**3,567**

C3013 240 Stainless Steel Interior Finish

	System Description	Freq. (Years)	Crew	Unit	Labor Hours	2020 Bare Costs				Total In-House	Total w/O&P
						Material	Labor	Equipment	Total		
0020	**Replace stainless steel wall**	75	2 SSWK	C.S.F.							
	Set up, secure and take down ladder				1.500		79.50		79.50	103	128
	Remove existing stainless steel sheets				2.342		99		99	127	158
	Replace stainless steel sheets				13.419	510	620		1,130	1,450	1,725
	Total				17.261	510	798.50		1,308.50	**1,680**	**2,011**

C3013 410 Fireplace Mantles, Wood

	System Description	Freq. (Years)	Crew	Unit	Labor Hours	2020 Bare Costs				Total In-House	Total w/O&P
						Material	Labor	Equipment	Total		
0030	**Refinish wood surface - fireplaces**	7	1 PORD	S.F.							
	Prepare surface				.003		.15		.15	.19	.24
	Varnish, 3 coats, brushwork, sanding included				.025	.29	1.09		1.38	1.72	2.10
	Total				.028	.29	1.24		1.53	**1.91**	**2.34**

C3023 112 Concrete, Finished

	System Description	Freq. (Years)	Crew	Unit	Labor Hours	2020 Bare Costs				Total In-House	Total w/O&P
						Material	Labor	Equipment	Total		
0020	**Refinish concrete floor**	25	2 CEFI	C.S.F.							
	Add topping to existing floor 1 ″				6.933	74	314	[illegible]	442	540	650
	Total				6.933	74	314	[illegible]	442	**540**	**650**

C3023 405 Epoxy Flooring

	System Description	Freq. (Years)	Crew	Unit	Labor Hours	2020 Bare Costs				Total In-House	Total w/O&P
						Material	Labor	Equipment	Total		
0020	**Replace epoxy flooring**	15	2 CEFI	C.S.F.							
	Strip existing flooring				3.200		170		170	220	272
	Repair and seal floor				12.735	452	555	[illegible]	1,021	1,225	1,475
	Total				15.935	452	725	[illegible]	1,191	**1,445**	**1,747**

C3023 410 Vinyl Tile

	System Description	Freq. (Years)	Crew	Unit	Labor Hours	2020 Bare Costs				Total In-House	Total w/O&P
						Material	Labor	Equipment	Total		
0020	**Replace vinyl tile flooring**	18	1 TILF	S.Y.							
	Remove damaged floor tile				.234		9.81		9.81	12.70	15.75
	Prepare surface				.645		28.53		28.53	36	44.50
	Install new tiles				.187	11.07	9.27		20.34	24	28.50
	Total				1.067	11.07	47.61		58.68	**72.70**	**88.75**

C3023 412 Vinyl Sheet

	System Description	Freq. (Years)	Crew	Unit	Labor Hours	2020 Bare Costs				Total In-House	Total w/O&P
						Material	Labor	Equipment	Total		
0020	**Replace vinyl sheet flooring**	18	1 TILF	S.Y.							
	Remove damaged floor tile				.234		9.81		9.81	12.70	15.75
	Prepare surface				.645		28.53		28.53	36	44.50
	Install new vinyl sheet				.407	37.35	20.16		57.51	66.50	78.50
	Total				1.286	37.35	58.50		95.85	**115.20**	**138.75**

C3023 414 Rubber Tile

	System Description	Freq. (Years)	Crew	Unit	Labor Hours	2020 Bare Costs Material	Labor	Equipment	Total	Total In-House	Total w/O&P
0020	**Replace rubber tile floor**	18	1 TILF	S.Y.							
	Remove damaged floor tile				.234		9.81		9.81	12.70	15.75
	Prepare surface				.645		28.53		28.53	36	44.50
	Install new tiles				1.040	105.75	51.30		157.05	181	213
	Total				1.920	105.75	89.64		195.39	**229.70**	**273.25**

C3023 418 Rubber / Vinyl Trim

	System Description	Freq. (Years)	Crew	Unit	Labor Hours	2020 Bare Costs Material	Labor	Equipment	Total	Total In-House	Total w/O&P
0010	**Replace rubber cove base**	9	1 TILF	L.F.							
	Remove damaged base				.017		.73		.73	.94	1.17
	Clean up debris				.001		.05		.05	.06	.08
	Install new base				.025	1.43	1.26		2.69	3.15	3.76
	Total				.044	1.43	2.04		3.47	**4.15**	**5.01**

C3023 420 Ceramic Tile

	System Description	Freq. (Years)	Crew	Unit	Labor Hours	2020 Bare Costs Material	Labor	Equipment	Total	Total In-House	Total w/O&P
0010	**Ceramic tile floor repairs - (2% of floors)**	15	1 TILF	C.S.F.							
	Regrout ceramic tile floors				16.640	5	735		740	930	1,150
	Total				16.640	5	735		740	**930**	**1,150**
0020	**Replace 2″ x 2″ thin set ceramic tile floor**	50	1 TILF	C.S.F.							
	Remove damaged floor tile				2.600		109		109	141	175
	Prepare surface				7.172		317		317	400	495
	Install new tiles				10.947	635	483		1,118	1,300	1,550
	Total				20.719	635	909		1,544	**1,841**	**2,220**

C3023 428 Ceramic Trim

	System Description	Freq. (Years)	Crew	Unit	Labor Hours	2020 Bare Costs				Total In-House	Total w/O&P
						Material	Labor	Equipment	Total		
0020	**Replace ceramic trim**	50	2 TILF	L.F.							
	Remove ceramic tile trim				.015		.63		.63	.81	1
	Clean up debris				.010		.52		.52	.68	.84
	Install new ceramic tile trim				.229	4.22	10.10		14.32	17.35	21
	Total				.253	4.22	11.25		15.47	**18.84**	**22.84**

C3023 430 Terrazzo

	System Description	Freq. (Years)	Crew	Unit	Labor Hours	2020 Bare Costs				Total In-House	Total w/O&P
						Material	Labor	Equipment	Total		
0100	**Terrazzo floor repairs - (2% of floors)**	15	1 MSTZ	S.F.							
	Chip existing terrazzo				.089		3.75		3.75	4.85	6
	Place terrazzo				.166	3.45	7.50	[illegible]	13.71	16.25	19.05
	Remove debris				.004		.18		.18	.24	.29
	Total				.260	3.45	11.43	[illegible]	17.64	**21.34**	**25.34**
0200	**Replace terrazzo floor**	75	2 MSTZ	C.S.F.							
	Break-up existing terrazzo				5.943		250		250	325	400
	Remove debris				.433		18.25		18.25	23.50	29.50
	Load into truck				.650		28.75	11.75	40.50	50	59
	Place terrazzo				16.640	345	750	276	1,371	1,625	1,900
	Total				23.666	345	1,047	237.75	1,679.75	**2,023.50**	**2,388.50**

C3023 438 Terrazzo Trim

	System Description	Freq. (Years)	Crew	Unit	Labor Hours	2020 Bare Costs				Total In-House	Total w/O&P
						Material	Labor	Equipment	Total		
0020	**Replace precast terrazzo trim**	75	2 MSTZ	L.F.							
	Remove old terrazzo trim				.023		.98		.98	1.26	1.56
	Clean area				.010		.52		.52	.68	.84
	Install new precast terrazzo trim				.416	24.50	20.50		45	53	62.50
	Total				.449	24.50	22		46.50	**54.94**	**64.90**

C3023 440 Quarry Tile

	System Description	Freq. (Years)	Crew	Unit	Labor Hours	2020 Bare Costs				Total In-House	Total w/O&P
						Material	Labor	Equipment	Total		
0010	**Quarry tile floor repairs - (2% of floors)**	15	1 TILF	S.F.							
	Regrout quarry tile floor				.166	.05	7.35		7.40	9.30	11.55
	Total				.166	.05	7.35		7.40	**9.30**	**11.55**
0020	**Replace quarry tile floor**	50	2 TILF	S.F.							
	Remove damaged floor tile				.026		1.09		1.09	1.41	1.75
	Prepare surface				.072		3.17		3.17	3.98	4.96
	Install new tiles				.109	6.35	4.83		11.18	13.05	15.50
	Total				.207	6.35	9.09		15.44	**18.44**	**22.21**

C3023 450 Brick

	System Description	Freq. (Years)	Crew	Unit	Labor Hours	2020 Bare Costs				Total In-House	Total w/O&P
						Material	Labor	Equipment	Total		
0020	**Replace thickset brick floor**	60	2 BRIC	S.F.							
	Remove brick				.035		1.46		1.46	1.89	2.34
	Prepare 2" thick mortar bed				.260	1.18	12.20		13.38	17.25	21
	Install new brick				.219	10	9.65		19.65	23	27.50
	Total				.514	11.18	23.31		34.49	**42.14**	**50.84**

C3023 460 Marble

	System Description	Freq. (Years)	Crew	Unit	Labor Hours	2020 Bare Costs				Total In-House	Total w/O&P
						Material	Labor	Equipment	Total		
0020	**Replace thinset marble floor**	50	2 TILF	S.F.							
	Remove marble				.035		1.46		1.46	1.89	2.34
	Prepare floor				.260	1.18	12.20		13.38	17.25	21
	Replace marble flooring				.347	11.65	15.30		26.95	32	38.50
	Total				.641	12.83	28.96		41.79	**51.14**	**61.84**

C3023 470 Plywood

	System Description	Freq. (Years)	Crew	Unit	Labor Hours	2020 Bare Costs Material	Labor	Equipment	Total	Total In-House	Total w/O&P
0020	**Replace plywood floor**	40	2 CARP	C.S.F.							
	Remove existing flooring				1.733		92		92	119	148
	Install new plywood flooring 3/4"				1.600	149	85		234	274	320
	Total				3.333	149	177		326	**393**	**468**

C3023 472 Wood Parquet

	System Description	Freq. (Years)	Crew	Unit	Labor Hours	2020 Bare Costs Material	Labor	Equipment	Total	Total In-House	Total w/O&P
0020	**Sand and refinish parquet floor**	10	1 CARP	S.F.							
	Sand floor				.035	.23	1.48		1.71	2.17	2.67
	Refinish existing floor, brushwork, 2 coats				.026	.23	1.09		1.32	1.66	2.04
	Total				.061	.46	2.57		3.03	**3.83**	**4.71**
0030	**Replace 5/16" oak parquet floor**	40	2 CARP	S.F.							
	Remove flooring				.023		1.23		1.23	1.59	1.97
	Repair underlayment				.014	1.25	.74		1.99	2.33	2.74
	Install new parquet flooring				.065	5.50	3.45		8.95	10.50	12.45
	Total				.102	6.75	5.42		12.17	**14.42**	**17.16**

C3023 474 Maple Strip

	System Description	Freq. (Years)	Crew	Unit	Labor Hours	2020 Bare Costs Material	Labor	Equipment	Total	Total In-House	Total w/O&P
0020	**Sand and refinish maple strip floor**	10	1 CARP	S.F.							
	Sand floor				.026	.23	1.09		1.32	1.66	2.04
	Refinish maple strip floor, brushwork, 2 coats				.035	.23	1.48		1.71	2.17	2.67
	Total				.061	.46	2.57		3.03	**3.83**	**4.71**
0030	**Replace maple floor**	40	2 CARP	S.F.							
	Remove flooring				.023		1.23		1.23	1.59	1.97
	Repair underlayment				.014	1.25	.74		1.99	2.33	2.74
	Install new maple flooring				.061	5.10	3.25		8.35	9.80	11.60
	Total				.098	6.35	5.22		11.57	**13.72**	**16.31**

C3023 478 Wood Trim

	System Description	Freq. (Years)	Crew	Unit	Labor Hours	2020 Bare Costs				Total In-House	Total w/O&P
						Material	Labor	Equipment	Total		
0010	**Repair 1″ x 3″ wood trim**	13	1 CARP	L.F.							
	Remove wood				.021		.88		.88	1.13	1.40
	Install new trim				.173	5.04	9.20		14.24	17.50	21
	Total				.194	5.04	10.08		15.12	**18.63**	**22.40**
0030	**Refinish 1″ x 3″ wood trim**	7	1 PORD	L.F.							
	Prepare surface				.003		.15		.15	.19	.24
	Varnish, 3 coats, brushwork, sanding included				.025	.29	1.09		1.38	1.72	2.10
	Total				.028	.29	1.24		1.53	**1.91**	**2.34**
0040	**Replace 1″ x 3″ wood trim**	75	2 CARP	L.F.							
	Remove wood				.021		.88		.88	1.13	1.40
	Install new trim				.173	5.04	9.20		14.24	17.50	21
	Prepare surface				.003		.15		.15	.19	.24
	Varnish, 3 coats, brushwork, sanding included				.025	.29	1.09		1.38	1.72	2.10
	Total				.222	5.33	11.32		16.65	**20.54**	**24.74**

C3023 510 Carpet

	System Description	Freq. (Years)	Crew	Unit	Labor Hours	2020 Bare Costs				Total In-House	Total w/O&P
						Material	Labor	Equipment	Total		
0020	**Replace carpet**	8	2 TILF	S.Y.							
	Remove damaged carpet				.094		3.96		3.96	5.15	6.30
	Install new carpet				.138	41	6.85		47.85	53.50	62
	Total				.232	41	10.81		51.81	**58.65**	**68.30**

C3033 105 Plaster

	System Description	Freq. (Years)	Crew	Unit	Labor Hours	2020 Bare Costs Material	Labor	Equipment	Total	Total In-House	Total w/O&P
0010	**Repair plaster ceiling - (2% of ceilings)**	12	1 PLAS	S.Y.							
	Set up, secure and take down scaffold				.130		6.90		6.90	8.95	11.05
	Remove damaged ceiling				.267		11.25		11.25	14.60	18
	Replace plaster, 2 coats on lath				.752	6.80	35.50	1.83	44.13	55	66.50
	Total				1.149	6.80	53.65	1.83	62.28	**78.55**	**95.55**
0030	**Refinish plaster ceiling**	10	1 PORD	S.Y.							
	Set up, secure and take down scaffold				.130		6.90		6.90	8.95	11.05
	Place mask and drops				.033		1.44		1.44	1.89	2.34
	Sand & paint				.089	.63	3.96		4.59	5.75	7.10
	Total				.252	.63	12.30		12.93	**16.59**	**20.49**
0040	**Replace plaster ceiling**	75	2 PLAS	S.Y.							
	Set up, secure and take down scaffold				.130		6.90		6.90	8.95	11.05
	Remove ceiling				.267		11.25		11.25	14.60	18
	Replace plaster, 2 coats on lath				.752	6.80	35.50	1.83	44.13	55	66.50
	Set up, secure and take down scaffold				.130		6.90		6.90	8.95	11.05
	Total				1.279	6.80	60.55	1.83	69.18	**87.50**	**106.60**

C3033 107 Gypsum Wall Board

	System Description	Freq. (Years)	Crew	Unit	Labor Hours	2020 Bare Costs Material	Labor	Equipment	Total	Total In-House	Total w/O&P
0010	**Repair gypsum board ceiling - (2% of ceilings)**	20	1 CARP	C.S.F.							
	Set up, secure and take down scaffold				1.300		69		69	89.50	111
	Remove damaged gypsum board				2.737		115		115	149	185
	Replace 5/8″ gypsum board, taped and finished				3.382	45	180		225	282	345
	Total				7.419	45	364		409	**520.50**	**641**
0020	**Refinish gypsum board ceiling, up to 12′ high**	20	1 PORD	C.S.F.							
	Set up, secure and take down scaffold				1.300		69		69	89.50	111
	Wash surface				.250		11		11	14	18
	Sand & paint				.990	7	44		51	64	79
	Place and remove mask and drops				.370		16		16	21	26
	Total				2.910	7	140		147	**188.50**	**234**

C3033 107 Gypsum Wall Board

	System Description	Freq. (Years)	Crew	Unit	Labor Hours	2020 Bare Costs				Total In-House	Total w/O&P
						Material	Labor	Equipment	Total		
0040	**Replace gypsum board ceiling, up to 12′ high**	40	2 CARP	C.S.F.							
	Set up and secure scaffold				.750		39.75		39.75	51.50	64
	Remove damaged gypsum board				2.737		115		115	149	185
	Replace 5/8″ gypsum board ceiling, taped and finished				3.382	45	180		225	282	345
	Remove scaffold				.750		39.75		39.75	51.50	64
	Total				7.619	45	374.50		419.50	**534**	**658**
2020	**Refinish gypsum board ceiling, 12′ to 24′ high**	5	2 PORD	C.S.F.							
	Set up and secure scaffold				1.500		79.50		79.50	103	128
	Place and remove mask and drops				.370		16		16	21	26
	Prepare surface				.370		16		16	21	26
	Paint surface, roller + brushwork, 1 coat				.696	7	31		38	47	58
	Remove scaffold				1.500		79.50		79.50	103	128
	Total				4.436	7	222		229	**295**	**366**
3020	**Refinish gypsum board ceiling, over 24′ high**	5	2 PORD	C.S.F.							
	Set up and secure scaffold				2.250		119.25		119.25	155	192
	Place and remove mask and drops				.370		16		16	21	26
	Prepare surface				.370		16		16	21	26
	Paint surface, roller + brushwork, 1 coat				.696	7	31		38	47	58
	Remove scaffold				2.250		119.25		119.25	155	192
	Total				5.936	7	301.50		308.50	**399**	**494**

C3033 108 Acoustic Tile

	System Description	Freq. (Years)	Crew	Unit	Labor Hours	2020 Bare Costs				Total In-House	Total w/O&P
						Material	Labor	Equipment	Total		
0010	**Acoustic tile repairs - (2% of ceilings)**	9	1 CARP	C.S.F.							
	Set up, secure and take down scaffold				1.300		69		69	89.50	111
	Remove damaged tile				1.067		45		45	58	72
	Install new tile				2.737	402	145		547	630	735
	Total				5.104	402	259		661	**777.50**	**918**

C3033 108 Acoustic Tile

	System Description	Freq. (Years)	Crew	Unit	Labor Hours	2020 Bare Costs Material	Labor	Equipment	Total	Total In-House	Total w/O&P
0020	**Replace acoustic tile ceiling, non fire-rated**	20	1 CARP	C.S.F.							
	Set up, secure and take down scaffold				.750		39.75		39.75	51.50	64
	Remove old ceiling tiles				.500		21		21	27	34
	Remove old ceiling grid				.625		26		26	34	42
	Install new ceiling grid				1.000	93	53		146	171	201
	Install new ceiling tiles				1.280	85	68		153	182	215
	Sweep and clean debris				.520		30.50		30.50	39	48.50
	Total				4.675	178	238.25		416.25	**504.50**	**604.50**
0030	**Refinish acoustic tile ceiling/grid (unoccupied area)**	5	2 PORD	C.S.F.							
	Protect exposed surfaces and mask light fixtures				.080		3.55		3.55	4.54	5.65
	Refinish acoustic tiles, roller + brush, 1 coat				.267	.01	11.80		11.81	15.15	18.80
	Total				.347	.01	15.35		15.36	**19.69**	**24.45**
0040	**Refinish acoustic tile ceiling/grid (occupied area)**	5	2 PORD	C.S.F.							
	Protect exposed surfaces and mask light fixtures				.320		14.20		14.20	18.15	22.50
	Refinish acoustic tiles, roller + brush, 1 coat				.320	.01	14.20		14.21	18.15	22.50
	Total				.640	.01	28.40		28.41	**36.30**	**45**
2030	**Refinish acoustic tile ceiling, 12′ to 24′ high**	5	2 PORD	C.S.F.							
	Set up and secure scaffold				1.500		79.50		79.50	103	128
	Place and remove mask and drops				.100		2		2	3	4
	Paint surface, roller + brushwork, 1 coat				.267	.01	11.80		11.81	15.15	18.80
	Remove scaffold				1.500		79.50		79.50	103	128
	Total				3.367	.01	172.80		172.81	**224.15**	**278.80**
3030	**Refinish acoustic tile ceiling, over 24′ high**	5	2 PORD	C.S.F.							
	Set up and secure scaffold				2.250		119.25		119.25	155	192
	Place and remove mask and drops				.100		2		2	3	4
	Paint surface, roller + brushwork, 1 coat				.267	.01	11.80		11.81	15.15	18.80
	Remove scaffold				2.250		119.25		119.25	155	192
	Total				4.867	.01	252.30		252.31	**328.15**	**406.80**

C3033 109 Acoustic Tile, Fire-Rated

	System Description	Freq. (Years)	Crew	Unit	Labor Hours	2020 Bare Costs Material	Labor	Equipment	Total	Total In-House	Total w/O&P
0010	**Replace acoustic tile ceiling, fire-rated**	20	1 CARP	C.S.F.							
	Set up, secure and take down scaffold				.750		39.75		39.75	51.50	64
	Remove old ceiling tiles				.500		21		21	27	34
	Remove old grid system				.625		26		26	34	42
	Install new grid system				1.200	111.60	63.60		175.20	205	242
	Install new ceiling tiles				1.185	128	63		191	222	261
	Sweep & clean debris				.520		30.50		30.50	39	48.50
	Total				4.780	239.60	243.85		483.45	**578.50**	**691.50**

C3033 120 Wood

	System Description	Freq. (Years)	Crew	Unit	Labor Hours	2020 Bare Costs Material	Labor	Equipment	Total	Total In-House	Total w/O&P
0010	**Repair wood ceiling - (2% of ceilings)**	10	1 CARP	C.S.F.							
	Set up, secure and take down scaffold				1.300		69		69	89.50	111
	Remove damaged ceiling				1.455		61		61	79	98
	Replace wood				5.200	193	276		469	570	685
	Total				7.955	193	406		599	**738.50**	**894**
0030	**Refinish wood ceiling**	6	1 PORD	C.S.F.							
	Set up and secure scaffold				.375		19.88		19.88	26	32
	Wash surface				.286		13		13	16	20
	Sand & paint, brushwork, 1 coat				.990	7	44		51	64	79
	Place and remove mask and drops				.370		16		16	21	26
	Remove scaffold				.375		19.88		19.88	26	32
	Total				2.396	7	112.76		119.76	**153**	**189**
0040	**Replace wood ceiling**	50	2 CARP	C.S.F.							
	Set up and secure scaffold				.750		39.75		39.75	51.50	64
	Remove damaged ceiling				1.455		61		61	79	98
	Replace wood				5.200	193	276		469	570	685
	Remove scaffold				.750		39.75		39.75	51.50	64
	Total				8.155	193	416.50		609.50	**752**	**911**

D2013 110 Tankless Water Closet

	System Description	Freq. (Years)	Crew	Unit	Labor Hours	2020 Bare Costs Material	Labor	Equipment	Total	Total In-House	Total w/O&P
0010	**Replace flush valve diaphragm**	10	1 PLUM	Ea.							
	Turn valve off and on				.010		.67		.67	.83	1.04
	Remove / replace diaphragm				.300	1.93	19.35		21.28	26	32.50
	Check for leaks				.013		.84		.84	1.04	1.30
	Total				.323	1.93	20.86		22.79	**27.87**	**34.84**
0015	**Rebuild flush valve**	20	1 PLUM	Ea.							
	Turn valve off and on				.010		.67		.67	.83	1.04
	Remove flush valve				.100		6.45		6.45	7.95	9.95
	Rebuild valve				.350	65	22.50		87.50	99.50	116
	Reinstall valve				1.000		64.50		64.50	79.50	99.50
	Check operation				.024		1.55		1.55	1.91	2.39
	Total				1.484	65	95.67		160.67	**189.69**	**228.88**
0020	**Unplug clogged line**	5	1 PLUM	Ea.							
	Turn valve off and on				.010		.67		.67	.83	1.04
	Remove water closet				2.078		134		134	166	207
	Unplug line				.200		12.90		12.90	15.95	19.95
	Reinstall water closet				.510		33		33	40.50	51
	Check operation				.167		10.75		10.75	13.30	16.65
	Total				2.965		191.32		191.32	**236.58**	**295.64**
0030	**Replace tankless water closet**	35	2 PLUM	Ea.							
	Turn valve off and on				.010		.67		.67	.83	1.04
	Remove flush valve				.100		6.45		6.45	7.95	9.95
	Remove water closet				.460		29.50		29.50	36.50	46
	Install wall-hung water closet				2.759	1,100	160		1,260	1,400	1,625
	Check operation				.150		9.65		9.65	11.95	14.95
	Total				3.479	1,100	206.27		1,306.27	**1,457.23**	**1,696.94**
0040	**Replace tankless flush valve**	25	1 PLUM	Ea.							
	Turn valve off and on				.010		.67		.67	.83	1.04
	Remove flush valve				.100		6.45		6.45	7.95	9.95
	Install water closet flush valve				1.000	146	64.50		210.50	240	282
	Check operation				.024		1.55		1.55	1.91	2.39
	Total				1.134	146	73.17		219.17	**250.69**	**295.38**

D2013 110 Tankless Water Closet

	System Description	Freq. (Years)	Crew	Unit	Labor Hours	2020 Bare Costs				Total In-House	Total w/O&P
						Material	Labor	Equipment	Total		
0050	**Replace wax ring gasket**	5	1 PLUM	Ea.							
	Turn valve off and on				.010		.67		.67	.83	1.04
	Remove floor mounted tankless bowl				.667		43		43	53	66.50
	Install new wax gasket				.083	1.67	5.35		7.02	8.50	10.40
	Install floor mounted tankless bowl				1.000		64.50		64.50	79.50	99.50
	Check operation				.150		9.65		9.65	11.95	14.95
	Total				1.910	1.67	123.17		124.84	**153.78**	**192.39**
0080	**Replace stainless steel detention water closet flush valve actuator**	20	1 SKWK	Ea.							
	Turn valve off and on				.010		.67		.67	.83	1.04
	Remove flush valve actuator				.100		6.45		6.45	7.95	9.95
	Remove flush valve				.100		6.45		6.45	7.95	9.95
	Detention equipment, replace toilet actuator				1.000	274	55		329	370	430
	Check operation				.024		1.55		1.55	1.91	2.39
	Total				1.234	274	70.12		344.12	**388.64**	**453.33**
0090	**Replace stainless steel detention water closet flush valve**	20	1 SKWK	Ea.							
	Turn valve off and on				.010		.67		.67	.83	1.04
	Remove flush valve				.100		6.45		6.45	7.95	9.95
	Detention equipment, replace water closet valve				1.000	320	55		375	420	490
	Check operation				.024		1.55		1.55	1.91	2.39
	Total				1.134	320	63.67		383.67	**430.69**	**503.38**

D2013 130 Flush-Tank Water Closet

	System Description	Freq. (Years)	Crew	Unit	Labor Hours	2020 Bare Costs				Total In-House	Total w/O&P
						Material	Labor	Equipment	Total		
0010	**Unplug clogged line**	5	1 PLUM	Ea.							
	Turn valve off and on				.010		.67		.67	.83	1.04
	Remove water closet				2.078		134		134	166	207
	Unplug line				.200		12.90		12.90	15.95	19.95
	Reinstall water closet				.510		33		33	40.50	51
	Check operation				.167		10.75		10.75	13.30	16.65
	Total				2.965		191.32		191.32	**236.58**	**295.64**

D2013 130 Flush-Tank Water Closet

	System Description	Freq. (Years)	Crew	Unit	Labor Hours	2020 Bare Costs Material	Labor	Equipment	Total	Total In-House	Total w/O&P
0020	**Replace washer / diaphragm in ball cock**	5	1 PLUM	Ea.							
	Turn valve on and off				.010		.67		.67	.83	1.04
	Remove and replace washer / diaphragm				.180	2.56	11.60		14.16	17.15	21
	Check operation				.024		1.55		1.55	1.91	2.39
	Total				.214	2.56	13.82		16.38	**19.89**	**24.43**
0030	**Replace valve and ball cock assembly**	15	1 PLUM	Ea.							
	Turn valve off and on				.010		.67		.67	.83	1.04
	Remove flush valve and ball cock assembly				.300		19.35		19.35	24	30
	Install flush valve and ball cock assembly				.610	12	39.50		51.50	62	76
	Check operation				.167		10.75		10.75	13.30	16.65
	Total				1.087	12	70.27		82.27	**100.13**	**123.69**
0040	**Install gasket between tank and bowl**	20	1 PLUM	Ea.							
	Turn valve off and on				.010		.67		.67	.83	1.04
	Remove tank and gasket				.200		12.90		12.90	15.95	19.95
	Install new gasket and reinstall tank				.300	3.15	19.35		22.50	27.50	34
	Inspect connection				.009		.59		.59	.73	.91
	Total				.520	3.15	33.51		36.66	**45.01**	**55.90**
0050	**Replace two piece water closet**	35	2 PLUM	Ea.							
	Turn water off and on				.010		.67		.67	.83	1.04
	Remove floor-mounted water closet				1.000		64.50		64.50	79.50	99.50
	Install new bowl, tank and seat				3.019	219	175		394	455	545
	Check for leaks				.013		.84		.84	1.04	1.30
	Total				4.042	219	241.01		460.01	**536.37**	**646.84**
0060	**Replace one piece water closet**	35	2 PLUM	Ea.							
	Turn water off and on				.010		.67		.67	.83	1.04
	Remove floor-mounted water closet				1.000		64.50		64.50	79.50	99.50
	Install floor-mounted water closet with seat				3.019	900	175		1,075	1,200	1,400
	Check for leaks				.013		.84		.84	1.04	1.30
	Total				4.042	900	241.01		1,141.01	**1,281.37**	**1,501.84**

D2013 210 Urinal

	System Description	Freq. (Years)	Crew	Unit	Labor Hours	2020 Bare Costs				Total In-House	Total w/O&P
						Material	Labor	Equipment	Total		
0010	**Replace flush valve diaphragm**	7	1 PLUM	Ea.							
	Turn valve off and on				.010		.67		.67	.83	1.04
	Remove / replace diaphragm				.300	1.93	19.35		21.28	26	32.50
	Check for leaks				.013		.84		.84	1.04	1.30
	Total				.323	1.93	20.86		22.79	**27.87**	**34.84**
0015	**Rebuild flush valve**	20	1 PLUM	Ea.							
	Turn valve off and on				.010		.67		.67	.83	1.04
	Remove flush valve				.100		6.45		6.45	7.95	9.95
	Rebuild valve				.350	65	22.50		87.50	99.50	116
	Reinstall valve				1.000		64.50		64.50	79.50	99.50
	Check operation				.024		1.55		1.55	1.91	2.39
	Total				1.484	65	95.67		160.67	**189.69**	**228.88**
0020	**Unplug line**	5	1 PLUM	Ea.							
	Turn valve off and on				.010		.67		.67	.83	1.04
	Remove and reinstall bowl				1.600		103		103	128	159
	Unplug line				.200		12.90		12.90	15.95	19.95
	Check operation				.167		10.75		10.75	13.30	16.65
	Total				1.977		127.32		127.32	**158.08**	**196.64**
0030	**Replace wall-hung urinal**	35	2 PLUM	Ea.							
	Turn valve off and on				.010		.67		.67	.83	1.04
	Remove urinal				2.667		172		172	213	266
	Fit up urinal to wall				.196		12.65		12.65	15.60	19.55
	Install flushometer pipe				.096		6.20		6.20	7.65	9.55
	Install urinal				5.333	315	310		625	730	875
	Check operation				.334		21.50		21.50	26.50	33.50
	Total				8.636	315	523.02		838.02	**993.58**	**1,204.64**

D2013 310 Lavatory, Iron, Enamel

	System Description	Freq. (Years)	Crew	Unit	Labor Hours	2020 Bare Costs				Total In-House	Total w/O&P
						Material	Labor	Equipment	Total		
0010	**Replace washer in spud connection**	7	1 PLUM	Ea.							
	Turn valve off and on				.022		1.42		1.42	1.76	2.20
	Remove gasket or washer				.017		1.09		1.09	1.35	1.69
	Install new gasket / washer set				.026	4.48	1.68		6.16	7	8.20
	Install spud connection				.026		1.67		1.67	2.07	2.59
	Inspect connection				.009		.59		.59	.73	.91
	Total				.100	4.48	6.45		10.93	**12.91**	**15.59**
0020	**Replace washer in faucet**	2	1 PLUM	Ea.							
	Turn valves on and off				.022		1.42		1.42	1.76	2.20
	Remove valve stem and washer				.091		5.85		5.85	7.25	9.05
	Install washer				.026	.49	1.68		2.17	2.62	3.21
	Install stem				.017		1.09		1.09	1.35	1.69
	Test faucet				.010		.67		.67	.83	1.04
	Total				.166	.49	10.71		11.20	**13.81**	**17.19**
0040	**Replace faucets**	10	1 PLUM	Ea.							
	Turn valves on and off				.022		1.42		1.42	1.76	2.20
	Remove faucets and tubing				.381		24.50		24.50	30.50	38
	Install detention sink faucets and tubing				1.040	67.50	67		134.50	157	188
	Test faucets				.010		.67		.67	.83	1.04
	Total				1.453	67.50	93.59		161.09	**190.09**	**229.24**
0050	**Clean out strainer and P trap**	2	1 PLUM	Ea.							
	Remove strainer and P trap				.157		10.10		10.10	12.50	15.65
	Clean and reinstall strainer and trap				.320		20.50		20.50	25.50	32
	Total				.477		30.60		30.60	**38**	**47.65**
0060	**Replace lavatory**	40	2 PLUM	Ea.							
	Shut off water (hot and cold)				.022		1.42		1.42	1.76	2.20
	Disconnect and remove wall-hung lavatory				.800		51.50		51.50	64	79.50
	Install new wall-hung lavatory with trim, (18″ x 15″)				2.807	415	181		596	680	800
	Total				3.629	415	233.92		648.92	**745.76**	**881.70**

D2013 330 Lavatory, Vitreous China

	System Description	Freq. (Years)	Crew	Unit	Labor Hours	2020 Bare Costs				Total In-House	Total w/O&P
						Material	Labor	Equipment	Total		
0010	**Replace washer in spud connection**	7	1 PLUM	Ea.							
	Turn valve off and on				.022		1.42		1.42	1.76	2.20
	Loosen locknuts				.011		.71		.71	.88	1.10
	Remove spud connection				.017		1.09		1.09	1.35	1.69
	Remove gasket or washer				.017		1.09		1.09	1.35	1.69
	Clean spud seat				.021		1.34		1.34	1.66	2.07
	Install new gasket / washer set				.026	4.48	1.68		6.16	7	8.20
	Install spud connection				.026		1.67		1.67	2.07	2.59
	Tighten locknut				.014		.92		.92	1.14	1.43
	Inspect connection				.009		.59		.59	.73	.91
	Total				.163	4.48	10.51		14.99	**17.94**	**21.88**
0020	**Replace washer in faucet**	2	1 PLUM	Ea.							
	Turn valves on and off				.022		1.42		1.42	1.76	2.20
	Remove packing nut and install				.023		1.51		1.51	1.86	2.33
	Remove valve stem and cap				.017		1.09		1.09	1.35	1.69
	Remove seat washer and install				.023		1.51		1.51	1.86	2.33
	Pry out washer				.017		1.09		1.09	1.35	1.69
	Install washer				.026	.49	1.68		2.17	2.62	3.21
	Install stem				.017		1.09		1.09	1.35	1.69
	Tighten main gland valves				.012		.75		.75	.93	1.17
	Test faucet				.010		.67		.67	.83	1.04
	Total				.168	.49	10.81		11.30	**13.91**	**17.35**
0040	**Replace faucets**	10	1 PLUM	Ea.							
	Turn valves on and off				.022		1.42		1.42	1.76	2.20
	Remove faucets and tubing				.381		24.50		24.50	30.50	38
	Install faucets and tubing				1.040	67.50	67		134.50	157	188
	Test faucets				.010		.67		.67	.83	1.04
	Total				1.453	67.50	93.59		161.09	**190.09**	**229.24**
0050	**Clean out strainer and P trap**	2	1 PLUM	Ea.							
	Remove strainer and P trap				.157		10.10		10.10	12.50	15.65
	Clean and reinstall strainer and trap				.320		20.50		20.50	25.50	32
	Total				.477		30.60		30.60	**38**	**47.65**

D2013 330 Lavatory, Vitreous China

	System Description	Freq. (Years)	Crew	Unit	Labor Hours	2020 Bare Costs				Total In-House	Total w/O&P
						Material	Labor	Equipment	Total		
0060	**Replace lavatory**	35	2 PLUM	Ea.							
	Shut off water (hot and cold)				.022		1.42		1.42	1.76	2.20
	Disconnect and remove wall-hung lavatory				1.485		95.50		95.50	118	148
	Install new lavatory (to 20″ x 17″)				2.971	170	172		342	400	480
	Install faucets and tubing				1.040	67.50	67		134.50	157	188
	Total				5.518	237.50	335.92		573.42	**676.76**	**818.20**

D2013 350 Lavatory, Enameled Steel

	System Description	Freq. (Years)	Crew	Unit	Labor Hours	2020 Bare Costs				Total In-House	Total w/O&P
						Material	Labor	Equipment	Total		
0010	**Replace washer in spud connection**	7	1 PLUM	Ea.							
	Turn valve off and on				.022		1.42		1.42	1.76	2.20
	Loosen locknut				.014		.92		.92	1.14	1.43
	Remove spud connection				.017		1.09		1.09	1.35	1.69
	Remove gasket or washer				.017		1.09		1.09	1.35	1.69
	Clean spud seat				.021		1.34		1.34	1.66	2.07
	Install new gasket / washer set				.026	4.48	1.68		6.16	7	8.20
	Install spud connection				.026		1.67		1.67	2.07	2.59
	Tighten locknut				.014		.92		.92	1.14	1.43
	Inspect connection				.009		.59		.59	.73	.91
	Total				.166	4.48	10.72		15.20	**18.20**	**22.21**
0020	**Replace washer in faucet**	2	1 PLUM	Ea.							
	Turn valves on and off				.022		1.42		1.42	1.76	2.20
	Remove valve stem and washer				.091		5.85		5.85	7.25	9.05
	Install washer				.026	.49	1.68		2.17	2.62	3.21
	Install stem				.017		1.09		1.09	1.35	1.69
	Test faucet				.010		.67		.67	.83	1.04
	Total				.166	.49	10.71		11.20	**13.81**	**17.19**
0040	**Replace faucets**	10	1 PLUM	Ea.							
	Turn valves on and off				.022		1.42		1.42	1.76	2.20
	Remove faucets and tubing				.381		24.50		24.50	30.50	38
	Install faucets and tubing				1.040	67.50	67		134.50	157	188
	Test faucets				.010		.67		.67	.83	1.04
	Total				1.453	67.50	93.59		161.09	**190.09**	**229.24**

D2013 350 Lavatory, Enameled Steel

	System Description	Freq. (Years)	Crew	Unit	Labor Hours	2020 Bare Costs				Total In-House	Total w/O&P
						Material	Labor	Equipment	Total		
0050	**Clean out strainer and P trap**	2	1 PLUM	Ea.							
	Remove strainer and P trap				.157		10.10		10.10	12.50	15.65
	Clean and reinstall strainer and trap				.320		20.50		20.50	25.50	32
	Total				.477		30.60		30.60	**38**	**47.65**
0060	**Replace lavatory**	35	2 PLUM	Ea.							
	Shut off water (hot and cold)				.022		1.42		1.42	1.76	2.20
	Disconnect and remove wall-hung lavatory				1.793		116		116	143	179
	Install new lavatory (to 20″ x 17″)				3.586	126	208		334	395	480
	Install faucets and tubing				1.040	67.50	67		134.50	157	188
	Total				6.441	193.50	392.42		585.92	**696.76**	**849.20**

D2013 410 Sink, Iron Enamel

	System Description	Freq. (Years)	Crew	Unit	Labor Hours	2020 Bare Costs				Total In-House	Total w/O&P
						Material	Labor	Equipment	Total		
0010	**Replace faucet washer**	2	1 PLUM	Ea.							
	Turn valves on and off				.022		1.42		1.42	1.76	2.20
	Remove valve stem and washer				.091		5.85		5.85	7.25	9.05
	Install washer				.026	.49	1.68		2.17	2.62	3.21
	Install stem				.017		1.09		1.09	1.35	1.69
	Test faucet				.010		.67		.67	.83	1.04
	Total				.166	.49	10.71		11.20	**13.81**	**17.19**
0020	**Clean trap**	3	1 PLUM	Ea.							
	Drain trap				.021		1.36		1.36	1.68	2.10
	Remove / install trap				.043		2.79		2.79	3.45	4.31
	Inspect trap				.031		2.01		2.01	2.49	3.11
	Clean trap				.021		1.34		1.34	1.66	2.07
	Total				.116		7.50		7.50	**9.28**	**11.59**
0030	**Replace faucets**	10	1 PLUM	Ea.							
	Turn valves on and off				.022		1.42		1.42	1.76	2.20
	Remove faucets and tubing				.381		24.50		24.50	30.50	38
	Install faucets and tubing				1.040	67.50	67		134.50	157	188
	Test faucets				.010		.67		.67	.83	1.04
	Total				1.453	67.50	93.59		161.09	**190.09**	**229.24**

D2013 410 Sink, Iron Enamel

	System Description	Freq. (Years)	Crew	Unit	Labor Hours	2020 Bare Costs Material	Labor	Equipment	Total	Total In-House	Total w/O&P
0040	**Unstop sink**	2	1 PLUM	Ea.							
	Unstop plugged drain				.571		37		37	45.50	57
	Total				.571		37		37	**45.50**	**57**
0060	**Replace kitchen sink, P.E.C.I.**	10	1 PLUM	Ea.							
	Remove PE sink				1.625		94.50		94.50	117	146
	Install countertop porcelain enameled kitchen sink, 24" x 21"				2.857	310	166		476	545	645
	Check for leaks				.013		.84		.84	1.04	1.30
	Total				4.495	310	261.34		571.34	**663.04**	**792.30**
0070	**Replace laundry sink, P.E.C.I.**	35	2 PLUM	Ea.							
	Remove PE sink				1.625		94.50		94.50	117	146
	Install free-standing porcelain enameled laundry sink, 24" x 21"				2.667	615	155		770	870	1,000
	Total				4.292	615	249.50		864.50	**987**	**1,146**

D2013 420 Sink, Enameled Steel

	System Description	Freq. (Years)	Crew	Unit	Labor Hours	2020 Bare Costs Material	Labor	Equipment	Total	Total In-House	Total w/O&P
0010	**Replace faucet washer**	2	1 PLUM	Ea.							
	Turn valves on and off				.022		1.42		1.42	1.76	2.20
	Remove valve stem and washer				.091		5.85		5.85	7.25	9.05
	Install washer				.026	.49	1.68		2.17	2.62	3.21
	Install stem				.017		1.09		1.09	1.35	1.69
	Test faucet				.010		.67		.67	.83	1.04
	Total				.166	.49	10.71		11.20	**13.81**	**17.19**
0020	**Clean trap**	3	1 PLUM	Ea.							
	Drain trap				.021		1.36		1.36	1.68	2.10
	Remove / install trap				.043		2.79		2.79	3.45	4.31
	Inspect trap				.031		2.01		2.01	2.49	3.11
	Clean trap				.021		1.34		1.34	1.66	2.07
	Total				.116		7.50		7.50	**9.28**	**11.59**

D2013 420 Sink, Enameled Steel

	System Description	Freq. (Years)	Crew	Unit	Labor Hours	2020 Bare Costs				Total In-House	Total w/O&P
						Material	Labor	Equipment	Total		
0030	**Replace faucets**	10	1 PLUM	Ea.							
	Turn valves on and off				.022		1.42		1.42	1.76	2.20
	Remove faucets and tubing				.381		24.50		24.50	30.50	38
	Install faucets and tubing				1.040	67.50	67		134.50	157	188
	Test faucets				.010		.67		.67	.83	1.04
	Total				1.453	67.50	93.59		161.09	**190.09**	**229.24**
0040	**Unstop sink**	2	1 PLUM	Ea.							
	Unstop plugged drain				.571		37		37	45.50	57
	Total				.571		37		37	**45.50**	**57**
0060	**Replace sink, enameled steel**	35	2 PLUM	Ea.							
	Remove sink				1.793		104		104	129	161
	Install countertop enameled steel sink, 24″ x 21″				3.714	535	215		750	855	1,000
	Total				5.507	535	319		854	**984**	**1,161**

D2013 430 Sink, Stainless Steel

	System Description	Freq. (Years)	Crew	Unit	Labor Hours	2020 Bare Costs				Total In-House	Total w/O&P
						Material	Labor	Equipment	Total		
0010	**Replace faucet washer**	2	1 PLUM	Ea.							
	Turn valves on and off				.022		1.42		1.42	1.76	2.20
	Remove valve stem and washer				.091		5.85		5.85	7.25	9.05
	Install washer				.026	.49	1.68		2.17	2.62	3.21
	Install stem				.017		1.09		1.09	1.35	1.69
	Test faucet				.010		.67		.67	.83	1.04
	Total				.166	.49	10.71		11.20	**13.81**	**17.19**
0020	**Clean trap**	3	1 PLUM	Ea.							
	Drain trap				.021		1.36		1.36	1.68	2.10
	Remove / install trap				.043		2.79		2.79	3.45	4.31
	Inspect trap				.031		2.01		2.01	2.49	3.11
	Clean trap				.021		1.34		1.34	1.66	2.07
	Total				.116		7.50		7.50	**9.28**	**11.59**

D2013 430 Sink, Stainless Steel

	System Description	Freq. (Years)	Crew	Unit	Labor Hours	2020 Bare Costs Material	Labor	Equipment	Total	Total In-House	Total w/O&P
0030	**Replace faucets**	10	1 PLUM	Ea.							
	Turn valves on and off				.022		1.42		1.42	1.76	2.20
	Remove faucets and tubing				.381		24.50		24.50	30.50	38
	Install faucets and tubing				1.040	67.50	67		134.50	157	188
	Test faucets				.010		.67		.67	.83	1.04
	Total				1.453	67.50	93.59		161.09	**190.09**	**229.24**
0040	**Unstop sink**	2	1 PLUM	Ea.							
	Unstop plugged drain				.571		37		37	45.50	57
	Total				.571		37		37	**45.50**	**57**
0060	**Replace sink, stainless steel**	40	2 PLUM	Ea.							
	Remove sink				1.625		94.50		94.50	117	146
	Install countertop stainless steel sink, 25″ x 22″				3.714	690	215		905	1,025	1,200
	Total				5.339	690	309.50		999.50	**1,142**	**1,346**
0070	**Replace stainless steel detention sink faucets**	15	1 SKWK	Ea.							
	Turn valves on and off				.022		1.42		1.42	1.76	2.20
	Remove faucets and tubing				.333		21.50		21.50	26.50	33
	Detention equipment, replace faucet valve and tubing				2.000	450	129		579	655	760
	Check for leaks				.013		.84		.84	1.04	1.30
	Total				2.368	450	152.76		602.76	**684.30**	**796.50**

D2013 440 Sink, Plastic

	System Description	Freq. (Years)	Crew	Unit	Labor Hours	2020 Bare Costs Material	Labor	Equipment	Total	Total In-House	Total w/O&P
0010	**Replace faucet washer**	2	1 PLUM	Ea.							
	Turn valves on and off				.022		1.42		1.42	1.76	2.20
	Remove valve stem and washer				.091		5.85		5.85	7.25	9.05
	Install washer				.026	.49	1.68		2.17	2.62	3.21
	Install stem				.017		1.09		1.09	1.35	1.69
	Test faucet				.010		.67		.67	.83	1.04
	Total				.166	.49	10.71		11.20	**13.81**	**17.19**

D2013 440 Sink, Plastic

	System Description	Freq. (Years)	Crew	Unit	Labor Hours	2020 Bare Costs				Total In-House	Total w/O&P
						Material	Labor	Equipment	Total		
0020	**Clean trap**	3	1 PLUM	Ea.							
	Drain trap				.021		1.36		1.36	1.68	2.10
	Remove / install trap				.043		2.79		2.79	3.45	4.31
	Inspect trap				.031		2.01		2.01	2.49	3.11
	Clean trap				.021		1.34		1.34	1.66	2.07
	Total				.116		7.50		7.50	**9.28**	**11.59**
0030	**Replace faucets**	10	1 PLUM	Ea.							
	Turn valves on and off				.022		1.42		1.42	1.76	2.20
	Remove faucets and tubing				.381		24.50		24.50	30.50	38
	Install faucets and tubing				1.040	67.50	67		134.50	157	188
	Test faucets				.010		.67		.67	.83	1.04
	Total				1.453	67.50	93.59		161.09	**190.09**	**229.24**
0040	**Unstop, sink**	2	1 PLUM	Ea.							
	Unstop plugged drain				.571		37		37	45.50	57
	Total				.571		37		37	**45.50**	**57**
0060	**Replace sink and fittings, polyethylene**	15	2 PLUM	Ea.							
	Remove sink				3.467		201		201	249	310
	Install bench mounted polyethylene sink, 18-1/2" x 18-1/2"				4.000	375	232		607	700	830
	Total				7.467	375	433		808	**949**	**1,140**
0080	**Replace lab sink and fittings, polyethylene**	15	2 PLUM	Ea.	8.800						
	Remove sink				3.467		201		201	249	310
	Install bench mounted PE lab sink, 23-1/2" x 20-1/2"				5.333	1,250	310		1,560	1,750	2,050
	Total				8.800	1,250	511		1,761	**1,999**	**2,360**

D2013 444 Laboratory Sink

	System Description	Freq. (Years)	Crew	Unit	Labor Hours	2020 Bare Costs				Total In-House	Total w/O&P
						Material	Labor	Equipment	Total		
0010	**Replace faucet washer**	2	1 PLUM	Ea.							
	Turn valves on and off				.022		1.42		1.42	1.76	2.20
	Remove valve stem and washer				.091		5.85		5.85	7.25	9.05
	Install washer				.026	.49	1.68		2.17	2.62	3.21
	Install stem				.017		1.09		1.09	1.35	1.69
	Test faucet				.010		.67		.67	.83	1.04
	Total				.166	.49	10.71		11.20	**13.81**	**17.19**
0030	**Replace faucets**	10	1 PLUM	Ea.							
	Turn valves on and off				.022		1.42		1.42	1.76	2.20
	Remove faucets and tubing				.381		24.50		24.50	30.50	38
	Install faucets and tubing				1.040	67.50	67		134.50	157	188
	Test faucets				.010		.67		.67	.83	1.04
	Total				1.453	67.50	93.59		161.09	**190.09**	**229.24**
0060	**Replace sink and fittings, stainless steel, lab.**	15	2 PLUM	Ea.							
	Remove lab. sink				3.467		201		201	249	310
	Install bench mounted stainless steel lab. sink, 47″ x 24″				5.333	1,250	310		1,560	1,750	2,050
	Total				8.800	1,250	511		1,761	**1,999**	**2,360**

D2013 450 Laundry Sink, Plastic

	System Description	Freq. (Years)	Crew	Unit	Labor Hours	2020 Bare Costs				Total In-House	Total w/O&P
						Material	Labor	Equipment	Total		
0020	**Replace washer in faucet**	2	1 PLUM	Ea.							
	Turn valves on and off				.022		1.42		1.42	1.76	2.20
	Remove valve stem and washer				.091		5.85		5.85	7.25	9.05
	Install washer				.026	.49	1.68		2.17	2.62	3.21
	Install stem				.017		1.09		1.09	1.35	1.69
	Test faucet				.010		.67		.67	.83	1.04
	Total				.166	.49	10.71		11.20	**13.81**	**17.19**

D2013 450 Laundry Sink, Plastic

	System Description	Freq. (Years)	Crew	Unit	Labor Hours	2020 Bare Costs				Total In-House	Total w/O&P
						Material	Labor	Equipment	Total		
0040	**Replace faucets**	10	1 PLUM	Ea.							
	Turn valves on and off				.022		1.42		1.42	1.76	2.20
	Remove faucets and tubing				.381		24.50		24.50	30.50	38
	Install faucets and tubing				1.040	67.50	67		134.50	157	188
	Test faucets				.010		.67		.67	.83	1.04
	Total				1.453	67.50	93.59		161.09	**190.09**	**229.24**
0050	**Clean out strainer and P trap**	2	1 PLUM	Ea.							
	Remove strainer and P trap				.157		10.10		10.10	12.50	15.65
	Clean and reinstall strainer and trap				.320		20.50		20.50	25.50	32
	Total				.477		30.60		30.60	**38**	**47.65**
0060	**Replace laundry sink**	20	2 PLUM	Ea.							
	Shut off water (hot and cold)				.022		1.42		1.42	1.76	2.20
	Disconnect and remove laundry sink				1.600		93		93	115	143
	Install sink (23″ x 18″, single)				3.200	145	186		331	390	470
	Install faucets and tubing				1.040	67.50	67		134.50	157	188
	Total				5.862	212.50	347.42		559.92	**663.76**	**803.20**

D2013 455 Laundry Sink, Stone

	System Description	Freq. (Years)	Crew	Unit	Labor Hours	2020 Bare Costs				Total In-House	Total w/O&P
						Material	Labor	Equipment	Total		
0060	**Replace laundry sink, molded stone**	40	2 PLUM	Ea.							
	Remove molded stone sink				1.625		94.50		94.50	117	146
	Install laundry sink, w/trim, molded stone, 22″x23″,sgl compt				2.667	176	155		331	385	460
	Total				4.292	176	249.50		425.50	**502**	**606**

D2013 460 Service/Utility Sink

	System Description	Freq. (Years)	Crew	Unit	Labor Hours	2020 Bare Costs				Total In-House	Total w/O&P
						Material	Labor	Equipment	Total		
0010	**Replace faucet washer**	2	1 PLUM	Ea.							
	Turn valves on and off				.022		1.42		1.42	1.76	2.20
	Remove valve stem and washer				.091		5.85		5.85	7.25	9.05
	Install washer				.026	.49	1.68		2.17	2.62	3.21
	Install stem				.017		1.09		1.09	1.35	1.69
	Test faucet				.010		.67		.67	.83	1.04
	Total				.166	.49	10.71		11.20	**13.81**	**17.19**
0020	**Clean trap**	3	1 PLUM	Ea.							
	Drain trap				.021		1.36		1.36	1.68	2.10
	Remove / install trap				.043		2.79		2.79	3.45	4.31
	Inspect trap				.031		2.01		2.01	2.49	3.11
	Clean trap				.021		1.34		1.34	1.66	2.07
	Total				.116		7.50		7.50	**9.28**	**11.59**
0030	**Replace faucets**	10	1 PLUM	Ea.							
	Turn valves on and off				.022		1.42		1.42	1.76	2.20
	Remove faucets and tubing				.381		24.50		24.50	30.50	38
	Install faucets and tubing				1.040	67.50	67		134.50	157	188
	Test faucets				.010		.67		.67	.83	1.04
	Total				1.453	67.50	93.59		161.09	**190.09**	**229.24**
0040	**Unstop sink**	2	1 PLUM	Ea.							
	Unstop plugged drain				.571		37		37	45.50	57
	Total				.571		37		37	**45.50**	**57**
0060	**Replace service sink, P.E.C.I.**	35	1 PLUM	Ea.							
	Remove sink				1.625		94.50		94.50	117	146
	Install service/utility sink, 22″ x 18″				5.212	930	300		1,230	1,400	1,625
	Total				6.837	930	394.50		1,324.50	**1,517**	**1,771**

D2013 470 Group Wash Fountain

	System Description	Freq. (Years)	Crew	Unit	Labor Hours	2020 Bare Costs				Total In-House	Total w/O&P
						Material	Labor	Equipment	Total		
0060	**Replace group wash fountain, 54″ diameter**	20	Q-2	Ea.							
	Shut off water (hot and cold)				.022		1.42		1.42	1.76	2.20
	Disconnect and remove group wash fountain, 54″ dia.				4.800		289		289	355	445
	Install group wash fountain, terrazzo, 54″ diameter				9.600	10,600	575		11,175	12,400	14,100
	Install faucets and tubing				1.040	67.50	67		134.50	157	188
	Total				15.462	10,667.50	932.42		11,599.92	**12,913.76**	**14,735.20**

D2013 510 Bathtub, Cast Iron Enamel

	System Description	Freq. (Years)	Crew	Unit	Labor Hours	2020 Bare Costs				Total In-House	Total w/O&P
						Material	Labor	Equipment	Total		
0010	**Inspect / clean shower head**	3	1 PLUM	Ea.							
	Turn water off and on				.010		.67		.67	.83	1.04
	Remove shower head				.217		13.95		13.95	17.25	21.50
	Check shower head				.009		.59		.59	.73	.91
	Install shower head				.433		28		28	34.50	43
	Total				.669		43.21		43.21	**53.31**	**66.45**
0020	**Replace mixing valve barrel**	2	1 PLUM	Ea.							
	Turn valves on and off				.022		1.42		1.42	1.76	2.20
	Remove mixing valve cover / barrel				.650		42		42	52	65
	Install new mixing valve barrel / replace cover				.650	162	42		204	230	268
	Test valve				.010		.67		.67	.83	1.04
	Total				1.332	162	86.09		248.09	**284.59**	**336.24**
0030	**Replace mixing valve**	10	1 PLUM	Ea.							
	Turn valves off and on				.022		1.42		1.42	1.76	2.20
	Remove mixing valve				.867		56		56	69	86.50
	Install new mixing valve				1.733	144	112		256	297	355
	Test valve				.010		.67		.67	.83	1.04
	Total				2.633	144	170.09		314.09	**368.59**	**444.74**
0070	**Replace tub**	40	2 PLUM	Ea.							
	Remove cast iron tub				1.891		110		110	136	170
	Install cast iron enameled bathtub				3.782	1,275	219		1,494	1,675	1,925
	Total				5.673	1,275	329		1,604	**1,811**	**2,095**

D2013 530 Bathtub, Enameled Steel

	System Description	Freq. (Years)	Crew	Unit	Labor Hours	2020 Bare Costs				Total In-House	Total w/O&P
						Material	Labor	Equipment	Total		
0010	**Inspect / clean shower head**	3	1 PLUM	Ea.							
	Turn water off and on				.010		.67		.67	.83	1.04
	Remove shower head				.217		13.95		13.95	17.25	21.50
	Check shower head				.009		.59		.59	.73	.91
	Install shower head				.433		28		28	34.50	43
	Total				.669		43.21		43.21	**53.31**	**66.45**
0020	**Replace mixing valve barrel**	2	1 PLUM	Ea.							
	Turn valves on and off				.022		1.42		1.42	1.76	2.20
	Remove mixing valve cover / barrel				.650		42		42	52	65
	Install new mixing valve barrel / replace cover				.650	162	42		204	230	268
	Test valve				.010		.67		.67	.83	1.04
	Total				1.332	162	86.09		248.09	**284.59**	**336.24**
0030	**Replace mixing valve**	10	1 PLUM	Ea.							
	Turn valves off and on				.022		1.42		1.42	1.76	2.20
	Remove mixing valve				.867		56		56	69	86.50
	Install new mixing valve				1.733	144	112		256	297	355
	Test valve				.010		.67		.67	.83	1.04
	Total				2.633	144	170.09		314.09	**368.59**	**444.74**
0070	**Replace tub**	35	2 PLUM	Ea.							
	Remove tub				1.891		110		110	136	170
	Install enameled steel bathtub				3.782	515	219		734	840	985
	Total				5.673	515	329		844	**976**	**1,155**

D2013 550 Bathtub, Fiberglass

	System Description	Freq. (Years)	Crew	Unit	Labor Hours	2020 Bare Costs				Total In-House	Total w/O&P
						Material	Labor	Equipment	Total		
0010	**Inspect / clean shower head**	3	1 PLUM	Ea.							
	Turn water off and on				.010		.67		.67	.83	1.04
	Remove shower head				.217		13.95		13.95	17.25	21.50
	Check shower head				.009		.59		.59	.73	.91
	Install shower head				.433		28		28	34.50	43
	Total				.669		43.21		43.21	**53.31**	**66.45**

D2013 550 Bathtub, Fiberglass

	System Description	Freq. (Years)	Crew	Unit	Labor Hours	2020 Bare Costs				Total In-House	Total w/O&P
						Material	Labor	Equipment	Total		
0020	**Replace mixing valve barrel**	2	1 PLUM	Ea.							
	Turn valves on and off				.022		1.42		1.42	1.76	2.20
	Remove mixing valve cover / barrel				.650		42		42	52	65
	Install new mixing valve barrel / replace cover				.650	162	42		204	230	268
	Test valve				.010		.67		.67	.83	1.04
	Total				1.332	162	86.09		248.09	**284.59**	**336.24**
0030	**Replace mixing valve**	10	1 PLUM	Ea.							
	Turn valves off and on				.022		1.42		1.42	1.76	2.20
	Remove mixing valve				.867		56		56	69	86.50
	Install new mixing valve				1.733	144	112		256	297	355
	Test valve				.010		.67		.67	.83	1.04
	Total				2.633	144	170.09		314.09	**368.59**	**444.74**
0070	**Replace tub**	20	2 PLUM	Ea.							
	Remove bathtub				2.600		151		151	186	233
	Install fiberglass bathtub with shower wall surround				5.200	790	300		1,090	1,250	1,450
	Total				7.800	790	451		1,241	**1,436**	**1,683**

D2013 710 Shower, Terrazzo

	System Description	Freq. (Years)	Crew	Unit	Labor Hours	2020 Bare Costs				Total In-House	Total w/O&P
						Material	Labor	Equipment	Total		
0010	**Inspect / clean shower head**	3	1 PLUM	Ea.							
	Turn water off and on				.010		.67		.67	.83	1.04
	Remove shower head				.217		13.95		13.95	17.25	21.50
	Check shower head				.009		.59		.59	.73	.91
	Install shower head				.433		28		28	34.50	43
	Total				.669		43.21		43.21	**53.31**	**66.45**
0020	**Replace mixing valve barrel**	2	1 PLUM	Ea.							
	Turn valves on and off				.022		1.42		1.42	1.76	2.20
	Remove mixing valve cover / barrel				.650		42		42	52	65
	Install new mixing valve barrel / replace cover				.650	162	42		204	230	268
	Test valve				.010		.67		.67	.83	1.04
	Total				1.332	162	86.09		248.09	**284.59**	**336.24**

D2013 710 Shower, Terrazzo

	System Description	Freq. (Years)	Crew	Unit	Labor Hours	2020 Bare Costs Material	Labor	Equipment	Total	Total In-House	Total w/O&P
0030	**Replace mixing valve**	10	1 PLUM	Ea.							
	Turn valves off and on				.022		1.42		1.42	1.76	2.20
	Remove mixing valve				.867		56		56	69	86.50
	Install new mixing valve				1.733	144	112		256	297	355
	Test valve				.010		.67		.67	.83	1.04
	Total				2.633	144	170.09		314.09	**368.59**	**444.74**
0060	**Replace terrazzo shower surface**	30	2 PLUM	Ea.							
	Remove old terrazzo surface				3.004		126.75		126.75	163	203
	Install precast terrazzo tiles				9.014	334.75	445.25		780	930	1,125
	Install curb				2.080	141	103.50		244.50	284	335
	Total				14.098	475.75	675.50		1,151.25	**1,377**	**1,663**

D2013 730 Shower, Enameled Steel

	System Description	Freq. (Years)	Crew	Unit	Labor Hours	2020 Bare Costs Material	Labor	Equipment	Total	Total In-House	Total w/O&P
0010	**Inspect / clean shower head**	3	1 PLUM	Ea.							
	Turn water off and on				.010		.67		.67	.83	1.04
	Remove shower head				.217		13.95		13.95	17.25	21.50
	Check shower head				.009		.59		.59	.73	.91
	Install shower head				.433		28		28	34.50	43
	Total				.669		43.21		43.21	**53.31**	**66.45**
0020	**Replace mixing valve barrel**	2	1 PLUM	Ea.							
	Turn valves on and off				.022		1.42		1.42	1.76	2.20
	Remove mixing valve cover / barrel				.650		42		42	52	65
	Install new mixing valve barrel / replace cover				.650	162	42		204	230	268
	Test valve				.010		.67		.67	.83	1.04
	Total				1.332	162	86.09		248.09	**284.59**	**336.24**
0030	**Replace mixing valve**	10	1 PLUM	Ea.							
	Turn valves off and on				.022		1.42		1.42	1.76	2.20
	Remove mixing valve				.867		56		56	69	86.50
	Install new mixing valve				1.733	144	112		256	297	355
	Test valve				.010		.67		.67	.83	1.04
	Total				2.633	144	170.09		314.09	**368.59**	**444.74**

D2013 730 Shower, Enameled Steel

	System Description	Freq. (Years)	Crew	Unit	Labor Hours	2020 Bare Costs				Total In-House	Total w/O&P
						Material	Labor	Equipment	Total		
0060	**Replace shower**	35	2 PLUM	Ea.							
	Remove shower				5.200		300		300	375	465
	Install shower				4.522	1,300	262		1,562	1,750	2,025
	Total				9.722	1,300	562		1,862	**2,125**	**2,490**

D2013 750 Shower, Fiberglass

	System Description	Freq. (Years)	Crew	Unit	Labor Hours	2020 Bare Costs				Total In-House	Total w/O&P
						Material	Labor	Equipment	Total		
0010	**Inspect / clean shower head**	3	1 PLUM	Ea.							
	Turn water off and on				.010		.67		.67	.83	1.04
	Remove shower head				.217		13.95		13.95	17.25	21.50
	Check shower head				.009		.59		.59	.73	.91
	Install shower head				.433		28		28	34.50	43
	Total				.669		43.21		43.21	**53.31**	**66.45**
0020	**Replace mixing valve barrel**	2	1 PLUM	Ea.							
	Turn valves on and off				.022		1.42		1.42	1.76	2.20
	Remove mixing valve cover / barrel				.650		42		42	52	65
	Install new mixing valve barrel / replace cover				.650	162	42		204	230	268
	Test valve				.010		.67		.67	.83	1.04
	Total				1.332	162	86.09		248.09	**284.59**	**336.24**
0030	**Replace mixing valve**	10	1 PLUM	Ea.							
	Turn valves off and on				.022		1.42		1.42	1.76	2.20
	Remove mixing valve				.867		56		56	69	86.50
	Install new mixing valve				1.733	144	112		256	297	355
	Test valve				.010		.67		.67	.83	1.04
	Total				2.633	144	170.09		314.09	**368.59**	**444.74**
0060	**Replace shower and fittings**	20	2 PLUM	Ea.							
	Remove shower				4.334		251		251	310	390
	Install shower, fiberglass				3.302	465	192		657	750	875
	Total				7.635	465	443		908	**1,060**	**1,265**

D2013 770 Shower, Misc.

	System Description	Freq. (Years)	Crew	Unit	Labor Hours	2020 Bare Costs				Total In-House	Total w/O&P
						Material	Labor	Equipment	Total		
0010	**Inspect / clean shower head**	3	1 PLUM	Ea.							
	Turn water off and on				.010		.67		.67	.83	1.04
	Remove shower head				.217		13.95		13.95	17.25	21.50
	Check shower head				.009		.59		.59	.73	.91
	Install shower head				.433		28		28	34.50	43
	Total				.669		43.21		43.21	**53.31**	**66.45**
0020	**Replace mixing valve barrel**	2	1 PLUM	Ea.							
	Turn valves on and off				.022		1.42		1.42	1.76	2.20
	Remove mixing valve cover / barrel				.650		42		42	52	65
	Install new mixing valve barrel / replace cover				.650	162	42		204	230	268
	Test valve				.010		.67		.67	.83	1.04
	Total				1.332	162	86.09		248.09	**284.59**	**336.24**
0030	**Replace mixing valve**	10	1 PLUM	Ea.							
	Turn valves off and on				.022		1.42		1.42	1.76	2.20
	Remove mixing valve				.867		56		56	69	86.50
	Install new mixing valve				1.733	144	112		256	297	355
	Test valve				.010		.67		.67	.83	1.04
	Total				2.633	144	170.09		314.09	**368.59**	**444.74**
0060	**Replace shower and fittings, aluminum**	25	2 PLUM	Ea.							
	Remove shower				4.334		251		251	310	390
	Install shower				3.302	465	192		657	750	875
	Total				7.635	465	443		908	**1,060**	**1,265**
0070	**Replace shower head with water conserving head**	10	1 PLUM	Ea.							
	Remove shower head				.217		13.95		13.95	17.25	21.50
	Install new water conserving head				.400	91.50	26		117.50	133	154
	Turn water on and off				.010		.67		.67	.83	1.04
	Total				.627	91.50	40.62		132.12	**151.08**	**176.54**
0120	**Replace shower, C.M.U.**	20	D-8	Ea.							
	Remove C.M.U. shower enclosure				3.595		152.88	[illegible]	184.80	233	280
	Install C.M.U. shower enclosure				6.772	183.12	324.80		507.92	625	755
	Install built-in head, arm and valve				2.000	104	129		233	274	330
	Total				12.367	287.12	606.68	[illegible]	925.72	**1,132**	**1,365**

D2013 770 Shower, Misc.

	System Description	Freq. (Years)	Crew	Unit	Labor Hours	2020 Bare Costs				Total In-House	Total w/O&P
						Material	Labor	Equipment	Total		
0200	**Replace shower, glazed C.M.U.**	25	D-8	Ea.							
	Remove C.M.U. shower enclosure				4.697		200	42	242	305	365
	Install glazed C.M.U. shower enclosure				9.393	733.60	450.80		1,184.40	1,400	1,650
	Install built-in head, arm and valve				2.000	104	129		233	274	330
	Total				16.090	837.60	779.80	42	1,659.40	**1,979**	**2,345**
0280	**Replace shower surface, ceramic tile**	30	D-7	Ea.							
	Remove ceramic tile shower enclosure surface				3.066		129		129	167	207
	Install ceramic tile in shower enclosure				6.130	161.84	270.48		432.32	520	625
	Install built-in head, arm and valve				2.000	104	129		233	274	330
	Total				11.197	265.84	528.48		794.32	**961**	**1,162**

D2013 810 Drinking Fountain

	System Description	Freq. (Years)	Crew	Unit	Labor Hours	2020 Bare Costs				Total In-House	Total w/O&P
						Material	Labor	Equipment	Total		
0010	**Check / minor repairs**	1	1 PLUM	Ea.							
	Check unit, make minor adjustments				.667		43		43	53	66.50
	Total				.667		43		43	**53**	**66.50**
0020	**Repair internal leaks**	4	1 PLUM	Ea.							
	Open unit, repair leaks, close				.615		39.50		39.50	49	61.50
	Total				.615		39.50		39.50	**49**	**61.50**
0030	**Correct water pressure**	2	1 PLUM	Ea.							
	Adjust water pressure, close				.571		37		37	45.50	57
	Total				.571		37		37	**45.50**	**57**
0040	**Replace refrigerant**	2	1 PLUM	Ea.							
	Add freon to unit				.133	24	8.75		32.75	37	43.50
	Total				.133	24	8.75		32.75	**37**	**43.50**
0050	**Repair drain leak**	4	1 PLUM	Ea.							
	Remove old pipe				.052		3.35		3.35	4.14	5.20
	Install new pipe				.186	12.70	11.95		24.65	29	34.50
	Total				.238	12.70	15.30		28	**33.14**	**39.70**

D2013 810 Drinking Fountain

	System Description	Freq. (Years)	Crew	Unit	Labor Hours	2020 Bare Costs				Total In-House	Total w/O&P
						Material	Labor	Equipment	Total		
0070	**Replace fountain**	10	2 PLUM	Ea.							
	Remove old water cooler				2.600		151		151	186	233
	Install new wall mounted water cooler				5.200	700	300		1,000	1,150	1,350
	Total				7.800	700	451		1,151	**1,336**	**1,583**

D2013 910 Emergency Shower Station

	System Description	Freq. (Years)	Crew	Unit	Labor Hours	2020 Bare Costs				Total In-House	Total w/O&P
						Material	Labor	Equipment	Total		
0020	**Inspect and clean shower head**	3	1 PLUM	Ea.							
	Turn water off and on				.010		.67		.67	.83	1.04
	Remove shower head				.217		13.95		13.95	17.25	21.50
	Check shower head				.009		.59		.59	.73	.91
	Install shower head				.433		28		28	34.50	43
	Total				.669		43.21		43.21	**53.31**	**66.45**
0030	**Replace shower**	25	2 PLUM	Ea.							
	Remove emergency shower				2.600		151		151	186	233
	Install emergency shower				5.200	390	300		690	800	955
	Total				7.800	390	451		841	**986**	**1,188**

D2013 920 Emergency Eye Wash

	System Description	Freq. (Years)	Crew	Unit	Labor Hours	2020 Bare Costs				Total In-House	Total w/O&P
						Material	Labor	Equipment	Total		
0020	**Inspect and clean spray heads**	3	1 PLUM	Ea.							
	Turn water off and on				.010		.67		.67	.83	1.04
	Remove spray head				.217		13.95		13.95	17.25	21.50
	Check shower head				.009		.59		.59	.73	.91
	Install spray head				.433		28		28	34.50	43
	Total				.669		43.21		43.21	**53.31**	**66.45**
0030	**Replace eye wash station**	25	2 PLUM	Ea.							
	Remove old eye wash station				2.600		151		151	186	233
	Install eye wash station				5.200	490	300		790	910	1,075
	Total				7.800	490	451		941	**1,096**	**1,308**

D2023 110 Pipe & Fittings, Copper

	System Description	Freq. (Years)	Crew	Unit	Labor Hours	2020 Bare Costs				Total In-House	Total w/O&P
						Material	Labor	Equipment	Total		
0010	**Resolder joint**	10	1 PLUM	Ea.							
	Measure, cut & ream both ends				.053		3.44		3.44	4.25	5.30
	Solder fitting (3/4″ coupling)				.495	2.71	32		34.71	42.50	53
	Total				.549	2.71	35.44		38.15	**46.75**	**58.30**
0020	**Replace 3/4″ copper pipe and fittings**	20	2 PLUM	L.F.							
	Remove old pipe				.071		4.58		4.58	5.65	7.10
	Install 3/4″ copper tube with couplings and hangers				.141	8.65	9.05		17.70	20.50	25
	Total				.212	8.65	13.63		22.28	**26.15**	**32.10**
0030	**Replace 1″ copper pipe and fittings**	25	2 PLUM	L.F.							
	Remove old pipe				.073		4.68		4.68	5.80	7.25
	Install 1″ copper tube with couplings and hangers				.158	12.80	10.15		22.95	26.50	31.50
	Total				.230	12.80	14.83		27.63	**32.30**	**38.75**
0050	**Replace 1-1/2″ copper pipe and fittings**	25	2 PLUM	L.F.							
	Remove old pipe				.104		6.70		6.70	8.30	10.35
	Install 1 1/2″ copper tubing, couplings, hangers				.208	18.60	13.40		32	37	44
	Total				.312	18.60	20.10		38.70	**45.30**	**54.35**
0060	**Replace 2″ copper pipe and fittings**	25	2 PLUM	L.F.							
	Remove old pipe				.130		8.40		8.40	10.35	12.95
	Install 2″ copper tube with couplings and hangers				.260	27	16.75		43.75	50.50	60
	Total				.390	27	25.15		52.15	**60.85**	**72.95**
0070	**Replace 4″ copper pipe and fittings**	25	2 PLUM	L.F.							
	Remove old pipe				.274		15.90		15.90	19.65	24.50
	Install 4″ copper tube with couplings and hangers				.547	97.50	32		129.50	147	171
	Total				.821	97.50	47.90		145.40	**166.65**	**195.50**
0080	**Replace 8″ copper pipe and fittings**	25	2 PLUM	L.F.							
	Remove old pipe				.694		41.50		41.50	51.50	64.50
	Install 8″ copper tube with couplings and hangers				.918	455	55		510	570	655
	Total				1.612	455	96.50		551.50	**621.50**	**719.50**

D2023 120 Pipe & Fittings, Steel/Iron, Threaded

	System Description	Freq. (Years)	Crew	Unit	Labor Hours	2020 Bare Costs				Total In-House	Total w/O&P
						Material	Labor	Equipment	Total		
0030	**Replace 3/4″ threaded steel pipe and fittings**	75	2 PLUM	L.F.							
	Remove old pipe				.086		5.55		5.55	6.85	8.55
	Install 3/4″ diameter steel pipe, couplings, hangers				.131	4.40	8.45		12.85	15.30	18.55
	Total				.217	4.40	14		18.40	**22.15**	**27.10**
0040	**Replace 1-1/2″ threaded steel pipe and fittings**	75	2 PLUM	L.F.							
	Remove old pipe				.130		7.55		7.55	9.30	11.65
	Install 1-1/2″ diameter steel pipe, couplings, hangers				.260	8.75	15.10		23.85	28.50	34.50
	Total				.390	8.75	22.65		31.40	**37.80**	**46.15**
0050	**Replace 2″ threaded steel pipe and fittings**	75	2 PLUM	L.F.							
	Remove old pipe				.163		9.45		9.45	11.70	14.60
	Install 2″ diameter steel pipe, couplings, hangers				.325	6.90	18.85		25.75	31	37.50
	Total				.488	6.90	28.30		35.20	**42.70**	**52.10**
0060	**Replace 4″ threaded steel pipe and fittings**	75	2 PLUM	L.F.							
	Remove old pipe				.289		16.75		16.75	20.50	26
	Install 4″ diameter steel pipe, couplings, hangers				.578	23	33.50		56.50	66.50	81
	Total				.867	23	50.25		73.25	**87**	**107**

D2023 130 Solar Piping: Pipe & Fittings, PVC

	System Description	Freq. (Years)	Crew	Unit	Labor Hours	2020 Bare Costs				Total In-House	Total w/O&P
						Material	Labor	Equipment	Total		
0010	**Reglue joint, install 1″ tee**	10	1 PLUM	Ea.							
	Cut exist pipe, install tee, 1″				.867	1.39	56		57.39	70.50	88
	Inspect joints				.126		8.10		8.10	10.05	12.55
	Total				.993	1.39	64.10		65.49	**80.55**	**100.55**
0110	**Reglue joint, install 1-1/4″ tee**	10	1 PLUM	Ea.							
	Cut exist pipe, install tee, 1-1/4″				.945	2.16	61		63.16	77.50	96.50
	Inspect joints				.126		8.10		8.10	10.05	12.55
	Total				1.071	2.16	69.10		71.26	**87.55**	**109.05**

D2023 130 Solar Piping: Pipe & Fittings, PVC

	System Description	Freq. (Years)	Crew	Unit	Labor Hours	2020 Bare Costs				Total In-House	Total w/O&P
						Material	Labor	Equipment	Total		
0210	**Reglue joint, install 1-1/2″ tee**	10	1 PLUM	Ea.							
	Cut exist pipe, install tee, 1-1/2″				1.040	2.64	67		69.64	86	107
	Inspect joints				.126		8.10		8.10	10.05	12.55
	Total				1.166	2.64	75.10		77.74	**96.05**	**119.55**
0310	**Reglue joint, install 2″ tee**	10	Q-1	Ea.							
	Cut exist pipe, install tee, 2″				1.224	3.84	71		74.84	92	115
	Inspect joints				.126		8.10		8.10	10.05	12.55
	Total				1.350	3.84	79.10		82.94	**102.05**	**127.55**
1020	**Install 10′ section PVC 1″ diameter**	20	1 PLUM	Ea.							
	Remove broken pipe				1.131		73		73	90	113
	Install 10′ new PVC pipe 1″ diameter				2.939	122.85	189.15		312	370	445
	Inspect joints				.084		5.40		5.40	6.70	8.35
	Total				4.154	122.85	267.55		390.40	**466.70**	**566.35**
1120	**Install 10′ PVC 1-1/4″ diameter**	20	1 PLUM	Ea.							
	Remove broken pipe				1.239		80		80	98.50	124
	Install 10′ new PVC pipe 1-1/4″ diameter				3.219	129.35	207.35		336.70	400	480
	Inspect joints				.084		5.40		5.40	6.70	8.35
	Total				4.542	129.35	292.75		422.10	**505.20**	**612.35**
1220	**Install 10′ PVC 1-1/2″ diameter**	20	1 PLUM	Ea.							
	Remove broken pipe				1.445		93		93	115	144
	Install 10′ new PVC pipe 1-1/2″ diameter				3.756	133.25	241.80		375.05	445	545
	Inspect joints				.084		5.40		5.40	6.70	8.35
	Total				5.284	133.25	340.20		473.45	**566.70**	**697.35**
1320	**Install 10′ section PVC 2″ diameter**	20	Q-1	Ea.							
	Remove broken pipe				1.763		102.50		102.50	127	158
	Install 10′ new PVC pipe 2″ diameter				4.583	157.95	266.50		424.45	505	605
	Inspect joints				.084		5.40		5.40	6.70	8.35
	Total				6.430	157.95	374.40		532.35	**638.70**	**771.35**

D2023 130 Solar Piping: Pipe & Fittings, PVC

	System Description	Freq. (Years)	Crew	Unit	Labor Hours	2020 Bare Costs				Total In-House	Total w/O&P
						Material	Labor	Equipment	Total		
2030	**Replace 1000′ PVC pipe 1″ diameter**	30	1 PLUM	M.L.F.							
	Remove broken pipe				113.080		7,300		7,300	9,000	11,300
	Install 1000′ new PVC pipe 1″ diameter				226.080	9,450	14,550		24,000	28,400	34,300
	Inspect joints				8.399		540		540	670	835
	Total				347.559	9,450	22,390		31,840	**38,070**	**46,435**
2130	**Replace 1000′ PVC pipe 1-1/4″ diameter**	30	1 PLUM	M.L.F.							
	Remove broken pipe				123.850		8,000		8,000	9,875	12,400
	Install 1000′ new PVC pipe 1-1/4″ diameter				247.620	9,950	15,950		25,900	30,700	36,900
	Inspect joints				8.399		540		540	670	835
	Total				379.869	9,950	24,490		34,440	**41,245**	**50,135**
2230	**Replace 1000′ PVC pipe 1-1/2″ diameter**	30	1 PLUM	M.L.F.							
	Remove broken pipe				144.490		9,300		9,300	11,500	14,400
	Install 1000′ new PVC pipe 1-1/2″ diameter				288.890	10,250	18,600		28,850	34,300	41,800
	Inspect joints				8.399		540		540	670	835
	Total				441.779	10,250	28,440		38,690	**46,470**	**57,035**
2330	**Replace 1000′ PVC pipe 2″ diameter**	30	Q-1	M.L.F.							
	Remove broken pipe				176.320		10,250		10,250	12,700	15,800
	Install 1000′ new PVC pipe 2″ diameter				352.540	12,150	20,500		32,650	38,700	46,700
	Inspect joints				8.399		540		540	670	835
	Total				537.259	12,150	31,290		43,440	**52,070**	**63,335**

D2023 150 Valve, Non-Drain, Less Than 1-1/2″

	System Description	Freq. (Years)	Crew	Unit	Labor Hours	2020 Bare Costs				Total In-House	Total w/O&P
						Material	Labor	Equipment	Total		
0020	**Replace old valve**	10	1 PLUM	Ea.							
	Remove old valve				.400		26		26	32	40
	Install new valve				.800	385	51.50		436.50	485	560
	Total				1.200	385	77.50		462.50	**517**	**600**

D2023 152 — Valve, Non-Drain, 2″

	System Description	Freq. (Years)	Crew	Unit	Labor Hours	2020 Bare Costs Material	Labor	Equipment	Total	Total In-House	Total w/O&P
0020	**Replace old valve**	10	1 PLUM	Ea.							
	Remove old valve (2″)				.800		46.50		46.50	57.50	71.50
	Install new valve (to 2″)				.727	615	47		662	735	840
	Total				1.527	615	93.50		708.50	**792.50**	**911.50**

D2023 154 — Valve, Non-Drain, 3″

	System Description	Freq. (Years)	Crew	Unit	Labor Hours	2020 Bare Costs Material	Labor	Equipment	Total	Total In-House	Total w/O&P
0020	**Replace old valve**	10	1 PLUM	Ea.							
	Remove old valve				.800		46.50		46.50	57.50	71.50
	Install new valve				1.600	2,025	93		2,118	2,350	2,675
	Total				2.400	2,025	139.50		2,164.50	**2,407.50**	**2,746.50**

D2023 156 — Valve, Non-Drain, 4″ and Larger

	System Description	Freq. (Years)	Crew	Unit	Labor Hours	2020 Bare Costs Material	Labor	Equipment	Total	Total In-House	Total w/O&P
0020	**Replace old valve, 4″**	10	2 PLUM	Ea.							
	Remove old flanged gate valve, 4″				5.200		315		315	385	485
	Install new flanged gate valve, 4″				7.038	1,000	425		1,425	1,625	1,900
	Total				12.239	1,000	740		1,740	**2,010**	**2,385**
0030	**Replace old valve, 6″**	10	2 PLUM	Ea.							
	Remove old flanged gate valve, 6″				4.615		278		278	345	430
	Install new flanged gate valve, 6″				10.399	1,725	625		2,350	2,675	3,125
	Total				15.014	1,725	903		2,628	**3,020**	**3,555**
0040	**Replace old valve, 8″**	10	2 PLUM	Ea.							
	Remove old flanged gate valve, 8″				3.871		233		233	288	360
	Install new flanged gate valve, 8″				12.371	2,950	745		3,695	4,175	4,850
	Total				16.242	2,950	978		3,928	**4,463**	**5,210**

D2023 156 Valve, Non-Drain, 4″ and Larger

	System Description	Freq. (Years)	Crew	Unit	Labor Hours	2020 Bare Costs				Total In-House	Total w/O&P
						Material	Labor	Equipment	Total		
0050	**Replace old valve, 10″**	10	3 PLUM	Ea.							
	Remove old flanged gate valve, 10″				3.380		203		203	251	315
	Install new flanged gate valve, 10″				14.201	5,175	855		6,030	6,750	7,800
	Total				17.581	5,175	1,058		6,233	**7,001**	**8,115**

D2023 160 Insulation, Pipe

	System Description	Freq. (Years)	Crew	Unit	Labor Hours	2020 Bare Costs				Total In-House	Total w/O&P
						Material	Labor	Equipment	Total		
0020	**Remove old insulation & replace with new (1/2″ pipe size, 1″ wall)**	15	1 PLUM	L.F.							
	Remove pipe insulation				.025		1.60		1.60	1.97	2.47
	Install fiberglass pipe insulation (1″ wall, 1/2″ pipe size)				.087	.92	4.57		5.49	6.85	8.40
	Total				.111	.92	6.17		7.09	**8.82**	**10.87**
0030	**Remove old insulation & replace with new (3/4″ pipe size, 1″ wall)**	15	1 PLUM	L.F.							
	Remove pipe insulation				.025		1.60		1.60	1.97	2.47
	Install fiberglass pipe insulation (1″ wall, 3/4″ pipe size)				.090	.98	4.77		5.75	7.15	8.80
	Total				.115	.98	6.37		7.35	**9.12**	**11.27**
0040	**Remove old insulation & replace with new (1-1/2″ pipe size, 1″ wall)**	15	1 PLUM	L.F.							
	Remove pipe insulation				.025		1.60		1.60	1.97	2.47
	Install fiberglass pipe insulation (1″ wall, 1-1/2″ pipe size)				.099	1.25	5.25		6.50	8.05	9.85
	Total				.124	1.25	6.85		8.10	**10.02**	**12.32**
0050	**Remove old insulation & replace with new (1/2″ pipe size, 3/4″ wall)**	15	1 PLUM	L.F.							
	Remove pipe insulation				.025		1.60		1.60	1.97	2.47
	Install flexible rubber tubing (3/4″ wall, 1/2″ pipe size)				.117	1.13	6.85		7.98	10	12.25
	Total				.142	1.13	8.45		9.58	**11.97**	**14.72**
0060	**Remove old insulation & replace with new (3/4″ pipe size, 3/4″ wall)**	15	1 PLUM	L.F.							
	Remove pipe insulation				.025		1.60		1.60	1.97	2.47
	Install flexible rubber tubing (3/4″ wall, 3/4″ pipe size)				.117	2.13	6.85		8.98	11.10	13.50
	Total				.142	2.13	8.45		10.58	**13.07**	**15.97**

D2023 160 Insulation, Pipe

	System Description	Freq. (Years)	Crew	Unit	Labor Hours	2020 Bare Costs				Total In-House	Total w/O&P
						Material	Labor	Equipment	Total		
0070	**Remove old insulation & replace w/new (1-1/2″ pipe size, 3/4″ wall)**	15	1 PLUM	L.F.							
	Remove pipe insulation				.025		1.60		1.60	1.97	2.47
	Install flexible rubber tubing (3/4″ wall, 1-1/2″ pipe size)				.120	3.07	7		10.07	12.30	14.95
	Total				.144	3.07	8.60		11.67	**14.27**	**17.42**

D2023 210 Water Heater, Gas / Oil, 30 Gallon

	System Description	Freq. (Years)	Crew	Unit	Labor Hours	2020 Bare Costs				Total In-House	Total w/O&P
						Material	Labor	Equipment	Total		
0010	**Overhaul heater**	5	1 PLUM	Ea.							
	Turn valves off and on				.022		1.42		1.42	1.76	2.20
	Remove thermostat, check, and insulate				.182		11.70		11.70	14.50	18.10
	Remove burner, clean and install				.229		14.75		14.75	18.20	23
	Clean pilot and thermocouple				.296		19.10		19.10	23.50	29.50
	Remove burner door and install				.033		2.10		2.10	2.59	3.24
	Check after maintenance				.216		13.95		13.95	17.25	21.50
	Drain and fill tank				.615		39.50		39.50	49	61.50
	Total				1.593		102.52		102.52	**126.80**	**159.04**
0020	**Clean and service**	1	1 PLUM	Ea.							
	Check / adjust fuel burner				2.667		172		172	213	266
	Total				2.667		172		172	**213**	**266**
0030	**Replace water heater**	10	2 PLUM	Ea.							
	Remove old heater				2.600		168		168	207	259
	Install 30 gallon gas water heater				5.202	2,100	335		2,435	2,725	3,150
	Check after installation				.216		13.95		13.95	17.25	21.50
	Total				8.018	2,100	516.95		2,616.95	**2,949.25**	**3,430.50**

D2023 212 Water Heater, Gas / Oil, 70 Gallon

	System Description	Freq. (Years)	Crew	Unit	Labor Hours	2020 Bare Costs				Total In-House	Total w/O&P
						Material	Labor	Equipment	Total		
0010	**Overhaul heater**	5	1 PLUM	Ea.							
	Remove hot water heater and insulate				.667		43		43	53	66.50
	Remove thermostat, check and install				.182		11.70		11.70	14.50	18.10
	Remove burner, clean and install				.229		14.75		14.75	18.20	23
	Clean pilot and thermocouple				.296		19.10		19.10	23.50	29.50
	Remove burner door and install				.033		2.10		2.10	2.59	3.24
	Drain and refill tank				.615		39.50		39.50	49	61.50
	Turn valve off and on				.211		13.55		13.55	16.80	21
	Total				2.232		143.70		143.70	**177.59**	**222.84**
0020	**Clean & service**	1	1 PLUM	Ea.							
	Check / adjust fuel burner				2.667		172		172	213	266
	Total				2.667		172		172	**213**	**266**
0030	**Replace hot water heater**	12	2 PLUM	Ea.							
	Remove old heater				2.600		168		168	207	259
	Install water heater, oil fired, 70 gallon				6.932	2,375	445		2,820	3,175	3,650
	Total				9.532	2,375	613		2,988	**3,382**	**3,909**

D2023 214 Water Heater, Gas / Oil, 1150 GPH

	System Description	Freq. (Years)	Crew	Unit	Labor Hours	2020 Bare Costs				Total In-House	Total w/O&P
						Material	Labor	Equipment	Total		
0010	**Minor repairs, adjustments**	2	2 PLUM	Ea.							
	Reset, adjust controls				1.333		86		86	106	133
	Total				1.333		86		86	**106**	**133**
0020	**Clean & service**	2	2 PLUM	Ea.							
	Drain tank, clean, and service				8.000		515		515	640	795
	Total				8.000		515		515	**640**	**795**
0030	**Replace hot water heater**	20	2 PLUM	Ea.							
	Remove water heater				26.016		1,500		1,500	1,875	2,325
	Install water heater, gas fired, 1150 GPH				51.948	32,500	3,025		35,525	39,500	45,300
	Total				77.964	32,500	4,525		37,025	**41,375**	**47,625**

D2023 220 Water Heater, Electric, 120 Gallon

	System Description	Freq. (Years)	Crew	Unit	Labor Hours	2020 Bare Costs				Total In-House	Total w/O&P
						Material	Labor	Equipment	Total		
0010	**Drain and flush**	7	1 PLUM	Ea.							
	Drain, clean and readjust				4.000		258		258	320	400
	Total				4.000		258		258	**320**	**400**
0020	**Check operation**	3	1 PLUM	Ea.							
	Check for relief valve				.010		.67		.67	.83	1.04
	Check for proper water temperature				.003		.17		.17	.21	.26
	Fill out maintenance report				.022		1.42		1.42	1.76	2.20
	Total				.035		2.26		2.26	**2.80**	**3.50**
0030	**Replace water heater**	15	2 PLUM	Ea.							
	Remove old heater				4.334		279		279	345	430
	Install water heater				8.667	13,700	560		14,260	15,800	18,000
	Total				13.001	13,700	839		14,539	**16,145**	**18,430**

D2023 222 Water Heater, Electric, 300 Gallon

	System Description	Freq. (Years)	Crew	Unit	Labor Hours	2020 Bare Costs				Total In-House	Total w/O&P
						Material	Labor	Equipment	Total		
0010	**Drain and flush**	7	1 PLUM	Ea.							
	Drain, clean and adjust				4.000		258		258	320	400
	Total				4.000		258		258	**320**	**400**
0020	**Check operation**	3	1 PLUM	Ea.							
	Check relief valve				.010		.67		.67	.83	1.04
	Check for proper water temperature				.003		.17		.17	.21	.26
	Fill out maintenance report				.022		1.42		1.42	1.76	2.20
	Total				.035		2.26		2.26	**2.80**	**3.50**
0030	**Replace water heater**	15	2 PLUM	Ea.							
	Remove old heater				8.000		465		465	575	715
	Install water heater				16.000	67,500	930		68,430	75,500	86,000
	Total				24.000	67,500	1,395		68,895	**76,075**	**86,715**

D2023 224 Water Heater, Electric, 1000 Gallon

	System Description	Freq. (Years)	Crew	Unit	Labor Hours	2020 Bare Costs				Total In-House	Total w/O&P
						Material	Labor	Equipment	Total		
0010	**Drain and flush**	7	1 PLUM	Ea.							
	Drain, clean and adjust				4.000		258		258	320	400
	Total				4.000		258		258	**320**	**400**
0020	**Check operation**	3	1 PLUM	Ea.							
	Check relief valve				.010		.67		.67	.83	1.04
	Check for proper water temperature				.003		.17		.17	.21	.26
	Fill out maintenance report				.022		1.42		1.42	1.76	2.20
	Total				.035		2.26		2.26	**2.80**	**3.50**
0030	**Replace water heater**	15	2 PLUM	Ea.							
	Remove heater				22.284		1,350		1,350	1,650	2,075
	Install water heater				44.610	124,500	2,675		127,175	140,500	160,000
	Total				66.894	124,500	4,025		128,525	**142,150**	**162,075**

D2023 226 Water Heater, Electric, 2000 Gallon

	System Description	Freq. (Years)	Crew	Unit	Labor Hours	2020 Bare Costs				Total In-House	Total w/O&P
						Material	Labor	Equipment	Total		
0010	**Drain and flush**	7	1 PLUM	Ea.							
	Drain, clean and adjust				4.000		258		258	320	400
	Total				4.000		258		258	**320**	**400**
0020	**Check operation**	3	1 PLUM	Ea.							
	Check relief valve				.010		.67		.67	.83	1.04
	Check for proper water temperature				.003		.17		.17	.21	.26
	Fill out maintenance report				.022		1.42		1.42	1.76	2.20
	Total				.035		2.26		2.26	**2.80**	**3.50**
0030	**Replace water heater**	15	2 PLUM	Ea.							
	Remove heater				31.209		1,875		1,875	2,325	2,900
	Install water heater				62.338	154,500	3,750		158,250	174,500	199,000
	Total				93.547	154,500	5,625		160,125	**176,825**	**201,900**

D2023 230 Steam Converter, Domestic Hot Water

	System Description	Freq. (Years)	Crew	Unit	Labor Hours	2020 Bare Costs				Total In-House	Total w/O&P
						Material	Labor	Equipment	Total		
0020	**Inspect for leaks**	1	1 PLUM	Ea.							
	Check relief valve operation				.038		2.43		2.43	3.01	3.76
	Check for proper water temperature				.054		3.51		3.51	4.34	5.40
	Total				.092		5.94		5.94	**7.35**	**9.16**
0030	**Replace steam converter**	20	2 PLUM	Ea.							
	Remove heat exchanger				1.733		112		112	138	173
	Install new heat exchanger				3.467	2,750	205		2,955	3,275	3,750
	Total				5.200	2,750	317		3,067	**3,413**	**3,923**

D2023 240 Storage Tank, Domestic Hot Water

	System Description	Freq. (Years)	Crew	Unit	Labor Hours	2020 Bare Costs				Total In-House	Total w/O&P
						Material	Labor	Equipment	Total		
0020	**Replace storage tank, glass lined, P.E., 80 gal.**	50	2 PLUM	Ea.							
	Remove hot water storage tank				.578		37		37	46	57.50
	Install new glass lined, P.E., 80 Gal. hot water storage tank				1.156	4,050	74.50		4,124.50	4,550	5,175
	Total				1.733	4,050	111.50		4,161.50	**4,596**	**5,232.50**

D2023 245 Solar Storage Tank, 1000 Gallon

	System Description	Freq. (Years)	Crew	Unit	Labor Hours	2020 Bare Costs				Total In-House	Total w/O&P
						Material	Labor	Equipment	Total		
0010	**Replace 1000 gallon solar storage tank**	20	Q-9	Ea.							
	Remove tank				8.000		450		450	565	705
	Install 1000 gallon solar storage tank				16.000	7,425	895		8,320	9,300	10,700
	Total				24.000	7,425	1,345		8,770	**9,865**	**11,405**

D2023 250 Expansion Chamber

	System Description	Freq. (Years)	Crew	Unit	Labor Hours	2020 Bare Costs Material	2020 Bare Costs Labor	2020 Bare Costs Equipment	2020 Bare Costs Total	Total In-House	Total w/O&P
0010	**Refill**	5	1 PLUM	Ea.							
	Refill air chamber				.039		2.51		2.51	3.11	3.89
	Total				.039		2.51		2.51	**3.11**	**3.89**
0020	**Remove old chamber, install new**	10	1 PLUM	Ea.							
	Remove / install expansion device, 15 gal.				.941	855	55.50		910.50	1,000	1,150
	Total				.941	855	55.50		910.50	**1,000**	**1,150**

D2023 260 Circulation Pump, 1/12 HP

	System Description	Freq. (Years)	Crew	Unit	Labor Hours	2020 Bare Costs Material	2020 Bare Costs Labor	2020 Bare Costs Equipment	2020 Bare Costs Total	Total In-House	Total w/O&P
0020	**Pump maintenance, 1/12 H.P.**	0.50	1 PLUM	Ea.							
	Check for leaks				.010		.67		.67	.83	1.04
	Check pump operation				.012		.75		.75	.93	1.17
	Lubricate pump and motor				.047		3.02		3.02	3.73	4.66
	Check suction or discharge pressure				.004		.25		.25	.31	.39
	Check packing glands, tighten				.014		.92		.92	1.14	1.43
	Fill out inspection report				.022		1.42		1.42	1.76	2.20
	Total				.109		7.03		7.03	**8.70**	**10.89**
0030	**Replace pump / motor assembly**	10	2 PLUM	Ea.							
	Remove flanged connection pump				1.733		101		101	124	155
	Install new flanged connection pump (1/12 H.P.)				3.467	710	201		911	1,025	1,200
	Total				5.200	710	302		1,012	**1,149**	**1,355**

D2023 261 Circulation Pump, 1/8 HP

	System Description	Freq. (Years)	Crew	Unit	Labor Hours	2020 Bare Costs				Total In-House	Total w/O&P
						Material	Labor	Equipment	Total		
0020	**Pump maintenance, 1/8 H.P.**	1	1 PLUM	Ea.							
	Check for leaks				.010		.67		.67	.83	1.04
	Check pump operation				.012		.75		.75	.93	1.17
	Lubricate pump and motor				.047		3.02		3.02	3.73	4.66
	Check suction or discharge pressure				.004		.25		.25	.31	.39
	Check packing glands, tighten				.014		.92		.92	1.14	1.43
	Fill out inspection report				.022		1.42		1.42	1.76	2.20
	Total				.109		7.03		7.03	**8.70**	**10.89**
0030	**Replace pump / motor assembly**	20	2 PLUM	Ea.							
	Remove flanged connection pump				1.733		101		101	124	155
	Install new flanged connection pump (1/8 H.P.)				3.467	1,250	201		1,451	1,625	1,875
	Total				5.200	1,250	302		1,552	**1,749**	**2,030**

D2023 262 Circulation Pump, 1/6 HP

	System Description	Freq. (Years)	Crew	Unit	Labor Hours	2020 Bare Costs				Total In-House	Total w/O&P
						Material	Labor	Equipment	Total		
0020	**Pump maintenance, 1/6 H.P.**	1	1 PLUM	Ea.							
	Check for leaks				.010		.67		.67	.83	1.04
	Check pump operation				.012		.75		.75	.93	1.17
	Lubricate pump and motor				.047		3.02		3.02	3.73	4.66
	Check suction or discharge pressure				.004		.25		.25	.31	.39
	Check packing glands, tighten				.014		.92		.92	1.14	1.43
	Fill out inspection report				.022		1.42		1.42	1.76	2.20
	Total				.109		7.03		7.03	**8.70**	**10.89**
0030	**Replace pump / motor assembly**	20	2 PLUM	Ea.							
	Remove flanged connection pump				2.080		134		134	166	207
	Install new flanged connection pump (1/6 H.P.)				4.160	1,750	241		1,991	2,225	2,575
	Total				6.240	1,750	375		2,125	**2,391**	**2,782**

D2023 264 Circulation Pump, 1/2 HP

	System Description	Freq. (Years)	Crew	Unit	Labor Hours	2020 Bare Costs				Total In-House	Total w/O&P
						Material	Labor	Equipment	Total		
0020	**Pump maintenance, 1/2 H.P.**	1	1 PLUM	Ea.							
	Check for leaks				.010		.67		.67	.83	1.04
	Check pump operation				.012		.75		.75	.93	1.17
	Check alignment				.053		3.44		3.44	4.25	5.30
	Lubricate pump and motor				.047		3.02		3.02	3.73	4.66
	Check suction or discharge pressure				.004		.25		.25	.31	.39
	Check packing glands, tighten				.014		.92		.92	1.14	1.43
	Fill out inspection report				.022		1.42		1.42	1.76	2.20
	Total				.163		10.47		10.47	**12.95**	**16.19**
0030	**Replace pump / motor assembly, partial**	20	2 PLUM	Ea.							
	Remove flanged connection pump				2.600		151		151	186	233
	Install new flanged connection pump (1/2 H.P.)				5.200	2,950	300		3,250	3,625	4,150
	Total				7.800	2,950	451		3,401	**3,811**	**4,383**

D2023 266 Circulation Pump, Bronze 1 HP

	System Description	Freq. (Years)	Crew	Unit	Labor Hours	2020 Bare Costs				Total In-House	Total w/O&P
						Material	Labor	Equipment	Total		
0020	**Pump maintenance, 1 H.P.**	0.50	1 PLUM	Ea.							
	Check for leaks				.010		.67		.67	.83	1.04
	Check pump operation				.012		.75		.75	.93	1.17
	Lubricate pump and motor				.047		3.02		3.02	3.73	4.66
	Check suction or discharge pressure				.004		.25		.25	.31	.39
	Check packing glands, tighten				.014		.92		.92	1.14	1.43
	Fill out inspection report				.022		1.42		1.42	1.76	2.20
	Total				.109		7.03		7.03	**8.70**	**10.89**
0040	**Replace pump / motor assembly**	20	2 PLUM	Ea.							
	Remove pump / motor				2.600		151		151	186	233
	Install new flanged connection pump, bronze (1 H.P.)				5.200	4,775	300		5,075	5,625	6,425
	Total				7.800	4,775	451		5,226	**5,811**	**6,658**

D2023 267 Circulation Pump, C.I. 1-1/2 HP

	System Description	Freq. (Years)	Crew	Unit	Labor Hours	2020 Bare Costs				Total In-House	Total w/O&P
						Material	Labor	Equipment	Total		
0020	**Pump maintenance, 1-1/2 H.P.**	0.50	1 PLUM	Ea.							
	Check for leaks				.010		.67		.67	.83	1.04
	Check pump operation				.012		.75		.75	.93	1.17
	Check alignment				.053		3.44		3.44	4.25	5.30
	Lubricate pump and motor				.047		3.02		3.02	3.73	4.66
	Check suction / discharge pressure				.004		.25		.25	.31	.39
	Check packing glands, tighten				.014		.92		.92	1.14	1.43
	Fill out inspection report				.022		1.42		1.42	1.76	2.20
	Total				.163		10.47		10.47	**12.95**	**16.19**
0040	**Replace pump / motor assembly**	20	2 PLUM	Ea.							
	Remove flanged connection pump				2.600		151		151	186	233
	Install new flanged connection pump, C.I. (1-1/2 H.P.)				5.200	2,400	300		2,700	3,025	3,475
	Total				7.800	2,400	451		2,851	**3,211**	**3,708**

D2023 268 Circulation Pump, 3 HP

	System Description	Freq. (Years)	Crew	Unit	Labor Hours	2020 Bare Costs				Total In-House	Total w/O&P
						Material	Labor	Equipment	Total		
0020	**Pump maintenance, 3 H.P.**	0.50	Q-1	Ea.							
	Check for leaks				.012		.74		.74	.92	1.16
	Check pump operation				.012		.75		.75	.93	1.17
	Check alignment				.059		3.82		3.82	4.72	5.90
	Lubricate pump and motor				.052		3.36		3.36	4.14	5.20
	Check suction / discharge pressure				.004		.25		.25	.31	.39
	Check packing glands, tighten				.016		1.02		1.02	1.27	1.59
	Fill out inspection report				.022		1.42		1.42	1.76	2.20
	Total				.176		11.36		11.36	**14.05**	**17.61**
0040	**Replace pump / motor assembly**	20	Q-1	Ea.							
	Remove flanged connection pump				4.400		255.20		255.20	315	395
	Install new flanged connection pump, 3 HP				8.889	10,400	515		10,915	12,100	13,800
	Total				13.289	10,400	770.20		11,170.20	**12,415**	**14,195**

D2023 310 Hose Bibb

	System Description	Freq. (Years)	Crew	Unit	Labor Hours	2020 Bare Costs Material	Labor	Equipment	Total	Total In-House	Total w/O&P
0020	**Replace old valve with new**	10	1 PLUM	Ea.							
	Remove hose bibb				.217		13.95		13.95	17.25	21.50
	Install new hose bibb				.433	12.05	28		40.05	48	58
	Total				.650	12.05	41.95		54	**65.25**	**79.50**

D2023 320 Water Meter

	System Description	Freq. (Years)	Crew	Unit	Labor Hours	2020 Bare Costs Material	Labor	Equipment	Total	Total In-House	Total w/O&P
0010	**Overhaul**	13	1 PLUM	Ea.							
	Open meter, replace worn parts				.300	1.93	19.35		21.28	26	32.50
	Total				.300	1.93	19.35		21.28	**26**	**32.50**
0020	**Remove old 5/8″ meter, install new**	25	1 PLUM	Ea.							
	Remove 5/8″ water meter				.325		21		21	26	32.50
	Install 5/8″ threaded water meter				.650	54	42		96	111	133
	Total				.975	54	63		117	**137**	**165.50**
0025	**Remove old 3/4″ meter, install new**	25	1 PLUM	Ea.							
	Remove 3/4″ water meter				.371		24		24	29.50	37
	Install 3/4″ threaded water meter				.743	98.50	48		146.50	168	197
	Total				1.114	98.50	72		170.50	**197.50**	**234**
0030	**Remove old 1″ meter, install new**	25	1 PLUM	Ea.							
	Remove 1″ water meter				.433		28		28	34.50	43
	Install 1″ threaded water meter				.867	149	56		205	233	273
	Total				1.300	149	84		233	**267.50**	**316**
0035	**Remove old 1-1/2″ meter, install new**	25	1 PLUM	Ea.							
	Remove 1-1/2″ water meter				.650		42		42	52	65
	Install 1-1/2″ threaded water meter				1.300	365	84		449	505	585
	Total				1.950	365	126		491	**557**	**650**
0040	**Remove old 2″ meter, install new**	25	1 PLUM	Ea.							
	Remove 2″ water meter				.867		56		56	69	86.50
	Install 2″ threaded water meter				1.733	495	112		607	685	790
	Total				2.600	495	168		663	**754**	**876.50**

D2023 320 Water Meter

	System Description	Freq. (Years)	Crew	Unit	Labor Hours	2020 Bare Costs				Total In-House	Total w/O&P
						Material	Labor	Equipment	Total		
0045	**Remove old 3″ meter, install new**	25	Q-1	Ea.							
	Remove 3″ water meter				3.467		201		201	249	310
	Install 3″ flanged water meter				6.932	2,650	400		3,050	3,400	3,925
	Total				10.399	2,650	601		3,251	**3,649**	**4,235**
0050	**Remove old 4″ meter, install new**	25	Q-1	Ea.							
	Remove 4″ water meter				6.932		400		400	495	620
	Install 4″ flanged water meter				13.865	4,250	805		5,055	5,675	6,575
	Total				20.797	4,250	1,205		5,455	**6,170**	**7,195**
0055	**Remove old 6″ meter, install new**	25	Q-1	Ea.							
	Remove 6″ water meter				10.403		605		605	745	935
	Install 6″ flanged water meter				20.806	6,850	1,200		8,050	9,025	10,400
	Total				31.209	6,850	1,805		8,655	**9,770**	**11,335**
0060	**Remove old 8″ meter, install new**	25	Q-1	Ea.							
	Remove 8″ water meter				12.998		755		755	930	1,175
	Install 8″ flanged water meter				26.016	10,700	1,500		12,200	13,600	15,700
	Total				39.014	10,700	2,255		12,955	**14,530**	**16,875**

D2023 370 Water Softener

	System Description	Freq. (Years)	Crew	Unit	Labor Hours	2020 Bare Costs				Total In-House	Total w/O&P
						Material	Labor	Equipment	Total		
0030	**Replace softener**	15	2 PLUM	Ea.							
	Remove old unit				2.600		168		168	207	259
	Install new water softener				5.200	940	335		1,275	1,450	1,700
	Total				7.800	940	503		1,443	**1,657**	**1,959**

D2033 110 Pipe & Fittings, Cast Iron

	System Description	Freq. (Years)	Crew	Unit	Labor Hours	2020 Bare Costs				Total In-House	Total w/O&P
						Material	Labor	Equipment	Total		
0020	**Unclog main drain**	10	1 PLUM	Ea.							
	Unclog main drain using auger				.630		40.50		40.50	50	63
	Total				.630		40.50		40.50	**50**	**63**
0030	**Replace cast iron pipe, 4″**	40	2 PLUM	L.F.							
	Remove old pipe				.189		10.95		10.95	13.55	16.95
	Install new 4″ cast iron pipe				.378	25	22		47	54.50	65.50
	Total				.567	25	32.95		57.95	**68.05**	**82.45**

D2033 130 Pipe & Fittings, PVC

	System Description	Freq. (Years)	Crew	Unit	Labor Hours	2020 Bare Costs				Total In-House	Total w/O&P
						Material	Labor	Equipment	Total		
0010	**Unclog floor drain**	20	1 PLUM	Ea.							
	Unclog plugged floor drain				.650		42		42	52	65
	Total				.650		42		42	**52**	**65**
0020	**Unclog 4″ - 12″ diameter PVC main drain per L.F.**	10	1 PLUM	L.F.							
	Unclog main drain using auger				.052		3.33		3.33	4.11	5.15
	Total				.052		3.33		3.33	**4.11**	**5.15**
0040	**Repair joint**	10	1 PLUM	Ea.							
	Remove plastic fitting (2 connections)				.650		37.50		37.50	46.50	58.50
	Connect fitting (2 connections)				1.300	9.60	75.50		85.10	104	129
	Inspect joints				.163		10.50		10.50	13	16.25
	Total				2.113	9.60	123.50		133.10	**163.50**	**203.75**
0060	**Replace pipe, 1-1/2″**	30	2 PLUM	L.F.							
	Remove broken pipe				.145		8.70		8.70	10.80	13.50
	Install new pipe, couplings, hangers				.289	10.25	18.60		28.85	34.50	42
	Inspect joints				.327		21		21	26	32.50
	Total				.760	10.25	48.30		58.55	**71.30**	**88**

D2033 130 Pipe & Fittings, PVC

	System Description	Freq. (Years)	Crew	Unit	Labor Hours	2020 Bare Costs				Total In-House	Total w/O&P
						Material	Labor	Equipment	Total		
0080	**Replace pipe, 2″**	30	2 PLUM	L.F.							
	Remove broken pipe				.176		10.20		10.20	12.65	15.80
	Install new pipe, couplings, hangers				.353	12.15	20.50		32.65	38.50	46.50
	Inspect joints				.327		21		21	26	32.50
	Total				.855	12.15	51.70		63.85	**77.15**	**94.80**
0100	**Replace pipe, 4″**	30	2 PLUM	L.F.							
	Remove broken pipe				.217		12.55		12.55	15.55	19.45
	Install new pipe, couplings, hangers				.433	14.80	25		39.80	47.50	57.50
	Inspect joints				.327		21		21	26	32.50
	Total				.977	14.80	58.55		73.35	**89.05**	**109.45**
0120	**Replace pipe, 6″**	30	2 PLUM	L.F.							
	Remove broken pipe				.267		15.45		15.45	19.15	24
	New pipe, couplings, hangers				.533	26.50	34.50		61	71.50	86
	Inspect joints				.327		21		21	26	32.50
	Total				1.126	26.50	70.95		97.45	**116.65**	**142.50**
0140	**Replace pipe, 8″**	30	2 PLUM	L.F.							
	Remove broken pipe				.325		19.55		19.55	24	30
	Install new pipe, couplings, hangers				.650	37	39		76	89	107
	Inspect joints				.327		21		21	26	32.50
	Total				1.302	37	79.55		116.55	**139**	**169.50**

D2033 305 Pipe & Fittings

	System Description	Freq. (Years)	Crew	Unit	Labor Hours	2020 Bare Costs				Total In-House	Total w/O&P
						Material	Labor	Equipment	Total		
3010	**Unclog floor drain**	10	1 PLUM	Ea.							
	Unclog plugged floor drain				4.000		262		262	325	405
	Solvent					30.50			30.50	33.50	38
	Total				4.000	30.50	262		292.50	**358.50**	**443**
3020	**Unclog 4″ - 12″ diameter main drain per L.F.**	10	2 PLUM	L.F.							
	Unclog main drain using auger				.052		3.33		3.33	4.11	5.15
	Total				.052		3.33		3.33	**4.11**	**5.15**

D2033 310 Floor Drain W/O Bucket

	System Description	Freq. (Years)	Crew	Unit	Labor Hours	2020 Bare Costs				Total In-House	Total w/O&P
						Material	Labor	Equipment	Total		
0010	**Clean drain**	4	1 PLUM	Ea.							
	Remove debris from drain				1.590		103		103	127	158
	Total				1.590		103		103	**127**	**158**
0030	**Replace floor drain**	40	1 PLUM	Ea.							
	Remove floor drain				1.040		67		67	83	104
	Install floor drain				2.080	980	121		1,101	1,225	1,400
	Total				3.120	980	188		1,168	**1,308**	**1,504**

D2033 330 Floor Drain With Bucket

	System Description	Freq. (Years)	Crew	Unit	Labor Hours	2020 Bare Costs				Total In-House	Total w/O&P
						Material	Labor	Equipment	Total		
0010	**Clean out bucket**	5	1 PLUM	Ea.							
	Remcve debris from bucket				4.000		258		258	320	400
	Total				4.000		258		258	**320**	**400**
0030	**Replace floor drain**	40	1 PLUM	Ea.							
	Remove floor drain				1.040		67		67	83	104
	Install floor drain				2.080	980	121		1,101	1,225	1,400
	Total				3.120	980	188		1,168	**1,308**	**1,504**

D20 PLUMBING — D2043 Rain Water Drainage

D2043 110 — Distribution: Gutters, Pipe

	System Description	Freq. (Years)	Crew	Unit	Labor Hours	2020 Bare Costs Material	2020 Bare Costs Labor	2020 Bare Costs Equipment	2020 Bare Costs Total	Total In-House	Total w/O&P
1010	**General maintenance & repair**	1	1 PLUM	M.L.F.							
	Inspect rainwater pipes				.104		6.70		6.70	8.30	10.35
	Clean out pipes				4.000		258		258	320	400
	Total				4.104		264.70		264.70	**328.30**	**410.35**
1020	**Replace pipe or gutter**	20	1 PLUM	L.F.	.650						
	Remove 4″ PVC roof drain system				.217		12.55		12.55	15.55	19.45
	Install 4″ roof drain system				.333	19.20	19.35		38.55	45	54
	Total				.550	19.20	31.90		51.10	**60.55**	**73.45**

D2043 210 — Drain: Roof, Scupper, Area

	System Description	Freq. (Years)	Crew	Unit	Labor Hours	2020 Bare Costs Material	2020 Bare Costs Labor	2020 Bare Costs Equipment	2020 Bare Costs Total	Total In-House	Total w/O&P
1010	**General maintenance & repair**	1	1 PLUM	Ea.							
	Inspect drain				.104		6.70		6.70	8.30	10.35
	Clean drain				.400		26		26	32	40
	Total				.504		32.70		32.70	**40.30**	**50.35**
1020	**Replace drain**	40	1 PLUM	Ea.							
	Remove drain				.800		46.50		46.50	57.50	71.50
	Install roof drain				1.600	510	93		603	675	780
	Total				2.400	510	139.50		649.50	**732.50**	**851.50**

D2043 310 Rainwater Sump Pump

	System Description	Freq. (Years)	Crew	Unit	Labor Hours	2020 Bare Costs				Total In-House	Total w/O&P
						Material	Labor	Equipment	Total		
1020	**Pump maintenance, 87 GPM**	1	1 PLUM	Ea.							
	Check for leaks				.010		.67		.67	.83	1.04
	Check pump operation				.012		.75		.75	.93	1.17
	Check alignment				.053		3.44		3.44	4.25	5.30
	Lubricate pump and motor				.047		3.02		3.02	3.73	4.66
	Check suction or discharge pressure				.004		.25		.25	.31	.39
	Check packing glands, tighten				.014		.92		.92	1.14	1.43
	Fill out inspection report				.022		1.42		1.42	1.76	2.20
	Total				.163		10.47		10.47	**12.95**	**16.19**
1030	**Replace sump pump / motor assembly**	20	2 PLUM	Ea.							
	Remove sump pump assembly				1.300		84		84	104	130
	Install new cast iron sump pump (87 GPM)				2.078	320	134		454	520	605
	Total				3.378	320	218		538	**624**	**735**

D2093 910 — Pipe & Fittings, Industrial Gas

	System Description	Freq. (Years)	Crew	Unit	Labor Hours	2020 Bare Costs				Total In-House	Total w/O&P
						Material	Labor	Equipment	Total		
1010	**General maintenance**	2	1 PLUM	M.L.F.							
	General maintenance				.500		32		32	40	50
	Total				.500		32		32	**40**	**50**
1030	**Replace pipe and fittings**	75	2 PLUM	L.F.							
	Shut off valve				1.067		69		69	85	106
	Remove old pipe				.162		9.40		9.40	11.60	14.55
	Install 2″ black steel pipe, couplings, and hangers				.325	6.90	18.85		25.75	31	37.50
	Check for leaks				.889		57.50		57.50	71	88.50
	Total				2.443	6.90	154.75		161.65	**198.60**	**246.55**

D2093 920 — Pipe & Fittings, Anesthesia

	System Description	Freq. (Years)	Crew	Unit	Labor Hours	2020 Bare Costs				Total In-House	Total w/O&P
						Material	Labor	Equipment	Total		
1010	**Resolder joint**	12	1 PLUM	Ea.							
	Measure, cut & clean both ends				.053		3.44		3.44	4.25	5.30
	Attach fitting				.578	5.30	37		42.30	52	64
	Total				.631	5.30	40.44		45.74	**56.25**	**69.30**
1030	**Replace pipe and fittings**	25	2 PLUM	L.F.							
	Remove old pipe				.069		4.47		4.47	5.55	6.90
	Install new 3/4″ copper tubing, couplings, hangers				.141	8.65	9.05		17.70	20.50	25
	Total				.210	8.65	13.52		22.17	**26.05**	**31.90**

D2093 930 — Pipe & Fittings, Oxygen

	System Description	Freq. (Years)	Crew	Unit	Labor Hours	2020 Bare Costs				Total In-House	Total w/O&P
						Material	Labor	Equipment	Total		
1010	**Resolder joint**	12	1 PLUM	Ea.							
	Measure, cut & clean both ends				.289		18.65		18.65	23	29
	Attach fitting				.578	5.30	37		42.30	52	64
	Total				.867	5.30	55.65		60.95	**75**	**93**

D2093 930 Pipe & Fittings, Oxygen

	System Description	Freq. (Years)	Crew	Unit	Labor Hours	2020 Bare Costs				Total In-House	Total w/O&P
						Material	Labor	Equipment	Total		
1030	**Replace pipe and fittings**	25	2 PLUM	L.F.							
	Remove old pipe				.069		4.47		4.47	5.55	6.90
	Install new 3/4" copper tubing, couplings, hangers				.141	8.65	9.05		17.70	20.50	25
	Total				.210	8.65	13.52		22.17	**26.05**	**31.90**

D2093 940 Pipe & Fittings, Compressed Air

	System Description	Freq. (Years)	Crew	Unit	Labor Hours	2020 Bare Costs				Total In-House	Total w/O&P
						Material	Labor	Equipment	Total		
1010	**General maintenance**	2	1 PLUM	M.L.F.							
	General maintenance				.500		32		32	40	50
	Total				.500		32		32	**40**	**50**
1030	**Replace pipe and fittings**	75	2 PLUM	L.F.							
	Shut off valve				1.067		69		69	85	106
	Remove old pipe				.052		3.35		3.35	4.14	5.20
	Install 2" black steel pipe, couplings, and hangers				.325	6.90	18.85		25.75	31	37.50
	Check for leaks				.889		57.50		57.50	71	88.50
	Total				2.333	6.90	148.70		155.60	**191.14**	**237.20**

D2093 946 Compressed Air Systems, Compressors

	System Description	Freq. (Years)	Crew	Unit	Labor Hours	2020 Bare Costs				Total In-House	Total w/O&P
						Material	Labor	Equipment	Total		
1004	**Check and adjust 3/4 H.P. compressor**	1	1 SPRI	Ea.							
	Check compressor oil level				.022		1.42		1.42	1.76	2.20
	Check V-belt tension				.038		2.43		2.43	3.01	3.76
	Drain moisture from air tank				.033		2.10		2.10	2.59	3.24
	Clean air intake filter on compressor				.178		11.45		11.45	14.15	17.70
	Clean reusable oil filter				.182		11.70		11.70	14.50	18.10
	Check pressure relief valve				.108		6.95		6.95	8.60	10.75
	Clean cooling fins				.105		6.80		6.80	8.40	10.50
	Check motor				.060		3.85		3.85	4.76	5.95
	Check control circuits				.182		11.70		11.70	14.50	18.10
	Operation checks				.222		14.30		14.30	17.70	22
	Fill out maintenance report				.022		1.42		1.42	1.76	2.20
	Total				1.151		74.12		74.12	**91.73**	**114.50**
1006	**Replace 3/4 H.P. compressor**	25	1 SPRI	Ea.							
	Remove air compressor				4.457		273		273	335	420
	Install new 3/4 H.P. compressor				6.250	1,200	395		1,595	1,800	2,100
	Total				10.707	1,200	668		1,868	**2,135**	**2,520**
1010	**Check and adjust 2 H.P. compressor**	1	1 PLUM	Ea.	1.151						
	Check compressor oil level				.022		1.42		1.42	1.76	2.20
	Check V-belt tension				.038		2.43		2.43	3.01	3.76
	Drain moisture from air tank				.033		2.10		2.10	2.59	3.24
	Clean air intake filter on compressor				.178		11.45		11.45	14.15	17.70
	Clean reusable oil filter				.182		11.70		11.70	14.50	18.10
	Check pressure relief valve				.108		6.95		6.95	8.60	10.75
	Clean cooling fins				.105		6.80		6.80	8.40	10.50
	Check motor				.060		3.85		3.85	4.76	5.95
	Check control circuits				.182		11.70		11.70	14.50	18.10
	Operation checks				.222		14.30		14.30	17.70	22
	Fill out maintenance report				.022		1.42		1.42	1.76	2.20
	Total				1.151		74.12		74.12	**91.73**	**114.50**
1030	**Replace 2 H.P. compressor**	25	2 PLUM	Ea.							
	Remove air compressor				4.457		273		273	335	420
	Install new 2 H.P. compressor				8.915	3,175	545		3,720	4,175	4,825
	Total				13.372	3,175	818		3,993	**4,510**	**5,245**

D2093 946 Compressed Air Systems, Compressors

	System Description	Freq. (Years)	Crew	Unit	Labor Hours	2020 Bare Costs				Total In-House	Total w/O&P
						Material	Labor	Equipment	Total		
3010	**Check and adjust 10 H.P. compressor**	1	1 PLUM	Ea.							
	Check compressor oil level				.022		1.42		1.42	1.76	2.20
	Check tension on V-belt				.038		2.43		2.43	3.01	3.76
	Drain moisture from air tank				.033		2.10		2.10	2.59	3.24
	Clean air intake filter on compressor				.178		11.45		11.45	14.15	17.70
	Clean reusable oil filter				.182		11.70		11.70	14.50	18.10
	Check operation of pressure valve				.108		6.95		6.95	8.60	10.75
	Clean cooling fins				.105		6.80		6.80	8.40	10.50
	Check motor				.060		3.85		3.85	4.76	5.95
	Check compressor circuits				.182		11.70		11.70	14.50	18.10
	Perform operation				.222		14.30		14.30	17.70	22
	Fill out maintenance report				.022		1.42		1.42	1.76	2.20
	Total				1.151		74.12		74.12	**91.73**	**114.50**
3030	**Replace 10 H.P. compressor**	25	2 PLUM	Ea.							
	Remove compressor				17.335		1,025		1,025	1,275	1,575
	Install 2-stage compressor 10 H.P.				34.632	6,275	2,050		8,325	9,425	11,000
	Total				51.967	6,275	3,075		9,350	**10,700**	**12,575**
4010	**Check and adjust 25 H.P. compressor**	1	1 PLUM	Ea.							
	Check compressor oil level				.022		1.42		1.42	1.76	2.20
	Check tension on V-belt				.038		2.43		2.43	3.01	3.76
	Drain moisture from air tank				.033		2.10		2.10	2.59	3.24
	Clean air intake filter on compressor				.178		11.45		11.45	14.15	17.70
	Clean reusable oil filter				.182		11.70		11.70	14.50	18.10
	Check operation of pressure valve				.108		6.95		6.95	8.60	10.75
	Clean cooling fins				.105		6.80		6.80	8.40	10.50
	Check motor				.060		3.85		3.85	4.76	5.95
	Check compressor circuits				.182		11.70		11.70	14.50	18.10
	Perform operation				.222		14.30		14.30	17.70	22
	Fill out maintenance report				.022		1.42		1.42	1.76	2.20
	Total				1.151		74.12		74.12	**91.73**	**114.50**
4030	**Replace 25 H.P. compressor**	25	2 PLUM	Ea.							
	Remove compressor				26.002		1,600		1,600	1,975	2,450
	Install 2-stage compressor 25 H.P.				51.948	16,300	3,175		19,475	21,900	25,300
	Total				77.950	16,300	4,775		21,075	**23,875**	**27,750**

D2093 946 Compressed Air Systems, Compressors

	System Description	Freq. (Years)	Crew	Unit	Labor Hours	2020 Bare Costs				Total In-House	Total w/O&P
						Material	Labor	Equipment	Total		
5030	**Check operation**										
	Check operation of unit	1	1 STPI	Ea.	.320		21		21	26	32.50
	Total				.320		21		21	**26**	**32.50**

D3013 110 Fuel Oil Storage Tank, 275 Gallon

	System Description	Freq. (Years)	Crew	Unit	Labor Hours	2020 Bare Costs Material	Labor	Equipment	Total	Total In-House	Total w/O&P
0010	**Replace 275 gallon fuel storage tank**	30	Q-5	Ea.							
	Remove 275 gallon tank				2.081		123		123	152	190
	Install 275 gallon tank				4.156	510	245		755	865	1,025
	Total				6.236	510	368		878	**1,017**	**1,215**

D3013 150 Fuel Level Meter

	System Description	Freq. (Years)	Crew	Unit	Labor Hours	2020 Bare Costs Material	Labor	Equipment	Total	Total In-House	Total w/O&P
0010	**Preventive maintenance**	5	1 STPI	Ea.							
	Check fuel level meter calibration				.500		33		33	40.50	50.50
	Total				.500		33		33	**40.50**	**50.50**
0020	**Replace remote tank fuel gauge**	20	1 STPI	Ea.							
	Remove remote read gauge				2.081		136		136	169	211
	Replace remote read gauge 5″ pointer travel				4.167	4,225	273		4,498	4,975	5,700
	Total				6.247	4,225	409		4,634	**5,144**	**5,911**

D3013 160 Oil Filter

	System Description	Freq. (Years)	Crew	Unit	Labor Hours	2020 Bare Costs Material	Labor	Equipment	Total	Total In-House	Total w/O&P
0010	**Preventive maintenance**	1	1 STPI	Ea.							
	Replace filter element				.052	4.30	3.40		7.70	8.95	10.65
	Total				.052	4.30	3.40		7.70	**8.95**	**10.65**
0020	**Replace filter housing**	30	1 STPI	Ea.							
	Replace filter housing				.520	43	34		77	89.50	106
	Total				.520	43	34		77	**89.50**	**106**

D3013 170 Fuel Oil Storage: Pipe & Fittings, Copper

	System Description	Freq. (Years)	Crew	Unit	Labor Hours	2020 Bare Costs				Total In-House	Total w/O&P
						Material	Labor	Equipment	Total		
0010	**Remake flare type joint**	10	1 STPI	M.L.F.							
	Remake flare type joint				.286		18.75		18.75	23	29
	Total				.286		18.75		18.75	**23**	**29**
0020	**Install 10′ sect. 3/8″ type L copper per M.L.F.**	20	1 PLUM	Ea.							
	Remove section 3/8″ diameter copper				.619		39.90		39.90	49.50	61.50
	Install 10′ section new 3/8″ diameter type L				1.238	33.40	80		113.40	135	165
	Total				1.857	33.40	119.90		153.30	**184.50**	**226.50**
0030	**Install 10′ sect. 1/2″ type L copper per M.L.F.**	20	1 PLUM	Ea.							
	Remove section 1/2″ diameter copper				.642		41.40		41.40	51	64
	Install 10′ section new 1/2″ diameter type L				1.284	36.80	82.50		119.30	143	174
	Total				1.926	36.80	123.90		160.70	**194**	**238**
0040	**Install 10′ sect. 5/8″ type L copper per M.L.F.**	20	1 PLUM	Ea.							
	Remove section 5/8″ diameter copper				.658		42.40		42.40	52.50	65.50
	Install 10′ section new 5/8″ diameter type L				1.316	57.50	85		142.50	168	203
	Total				1.975	57.50	127.40		184.90	**220.50**	**268.50**
0050	**Install 10′ sect. 3/4″ type L copper per M.L.F.**	20	1 PLUM	Ea.							
	Remove section 3/4″ diameter copper				.684		44.10		44.10	54.50	68
	Install 10′ section new 3/4″ diameter type L				1.369	47.10	88		135.10	161	195
	Total				2.053	47.10	132.10		179.20	**215.50**	**263**
0060	**Install 10′ section 1″ type L copper per M.L.F.**	20	1 PLUM	Ea.							
	Remove section 1″ diameter copper				.765		49.30		49.30	61	76
	Install 10′ section new 1″ diameter type L				1.529	76	98.50		174.50	206	248
	Total				2.294	76	147.80		223.80	**267**	**324**
0130	**Replace 1000′ type L 3/8″ copper**	25	1 PLUM	M.L.F.							
	Remove section 3/8″ diameter copper				61.920		3,990		3,990	4,950	6,150
	Install 1000′ section new 3/8″ diameter type L				123.800	3,340	8,000		11,340	13,500	16,500
	Total				185.720	3,340	11,990		15,330	**18,450**	**22,650**
0140	**Replace 1000′ type L 1/2″ copper**	25	1 PLUM	M.L.F.							
	Remove section 1/2″ diameter copper				64.220		4,140		4,140	5,125	6,400
	Install 1000′ section new 1/2″ diameter type L				128.390	3,680	8,250		11,930	14,300	17,400
	Total				192.610	3,680	12,390		16,070	**19,425**	**23,800**

D3013 170 Fuel Oil Storage: Pipe & Fittings, Copper

	System Description	Freq. (Years)	Crew	Unit	Labor Hours	2020 Bare Costs				Total In-House	Total w/O&P
						Material	Labor	Equipment	Total		
0150	**Replace 1000′ type L 5/8″ copper**	25	1 PLUM	M.L.F.							
	Remove section 5/8″ diameter copper				65.840		4,240		4,240	5,250	6,550
	Install 1000′ section new 5/8″ diameter type L				131.640	5,750	8,500		14,250	16,800	20,300
	Total				197.480	5,750	12,740		18,490	**22,050**	**26,850**
0160	**Replace 1000′ type L 3/4″ copper**	25	1 PLUM	M.L.F.							
	Remove section 3/4″ diameter copper				68.440		4,410		4,410	5,450	6,800
	Install 1000′ section new 3/4″ diameter type L				136.850	4,710	8,800		13,510	16,100	19,500
	Total				205.290	4,710	13,210		17,920	**21,550**	**26,300**
0170	**Replace 1000′ type L 1″ copper**	25	1 PLUM	M.L.F.							
	Remove section 1″ diameter copper				76.480		4,930		4,930	6,100	7,600
	Install 1000′ section new 1″ diameter type L				152.930	7,600	9,850		17,450	20,600	24,800
	Total				229.410	7,600	14,780		22,380	**26,700**	**32,400**

D3013 210 Natural Gas: Pipe & Fittings, Steel/Iron

	System Description	Freq. (Years)	Crew	Unit	Labor Hours	2020 Bare Costs				Total In-House	Total w/O&P
						Material	Labor	Equipment	Total		
0010	**Install new 2″ gasket, 1 per M.L.F.**	30	1 STPI	Ea.							
	Disconnect joint, remove gasket & clean faces				.803		52.70		52.70	65	81.50
	Install gasket and make up bolts				.800	8.90	52.50		61.40	74.50	92
	Total				1.603	8.90	105.20		114.10	**139.50**	**173.50**
0020	**Install new 3″ gasket, 1 per M.L.F.**	30	1 STPI	Ea.							
	Disconnect joint, remove gasket & clean faces				.945		62		62	76.50	96
	Install gasket and make up bolts				.945	9.10	62		71.10	86.50	107
	Total				1.890	9.10	124		133.10	**163**	**203**
0030	**Install new 4″ gasket, 1 per M.L.F.**	30	1 STPI	Ea.							
	Disconnect joint, remove gasket & clean faces				1.300		85		85	105	132
	Install gasket and make up bolts				1.300	17.90	84		101.90	123	152
	Total				2.600	17.90	169		186.90	**228**	**284**

D3013 210 Natural Gas: Pipe & Fittings, Steel/Iron

	System Description	Freq. (Years)	Crew	Unit	Labor Hours	2020 Bare Costs				Total In-House	Total w/O&P
						Material	Labor	Equipment	Total		
0040	**Install new 6″ gasket, 1 per M.L.F.**	30	1 STPI	Ea.							
	Disconnect joint, remove gasket & clean faces				1.733		114		114	140	176
	Install gasket and make up bolts				1.733	28.50	114		142.50	172	212
	Total				3.466	28.50	228		256.50	**312**	**388**
0110	**Replace 10′ of buried 2″ diam. st. pipe/M.L.F.**	12	Q-4	Ea.							
	Check break or leak				.016		1.05		1.05	1.30	1.62
	Turn valve on and off				.010		.66		.66	.81	1.01
	Excavate by hand				5.771		242.20		242.20	315	390
	Remove broken 2″ diameter section				.743		45.60	2.50	48.10	59	73
	Install 10′ of pipe coated & wrapped				1.143	80	70	3.80	153.80	179	213
	Check for leaks, additional section				.900		59		59	73	91
	Backfill by hand				2.101		88.88		88.88	114	142
	Total				10.684	80	507.39	6.30	593.69	**742.11**	**911.63**
0120	**Replace 10′ of buried 3″ diam. st. pipe/M.L.F.**	12	Q-4	Ea.							
	Check break or leak				.016		1.05		1.05	1.30	1.62
	Turn valve on and off				.010		.66		.66	.81	1.01
	Excavate by hand				5.771		242.20		242.20	315	390
	Remove broken 3″ diameter section				.800		49.10	2.70	51.80	63.50	79
	Install 10′ of pipe coated & wrapped				1.231	132	75.50	4.10	211.60	243	287
	Check for leaks, additional section				.900		59		59	73	91
	Backfill by hand				2.101		88.88		88.88	114	142
	Total				10.829	132	516.39	6.80	655.19	**810.61**	**991.63**
0130	**Replace 10′ of buried 4″ diam. st. pipe/M.L.F.**	12	B-35	Ea.							
	Check break or leak				.016		1.05		1.05	1.30	1.62
	Turn valve on and off				.010		.66		.66	.81	1.01
	Excavate by hand				5.771		242.20		242.20	315	390
	Remove broken 4″ diameter section				1.224		64.50	20.40	84.90	104	124
	Install 10′ of pipe coated & wrapped				1.882	171.50	99	31.40	301.90	350	405
	Check for leaks, additional section				.900		59		59	73	91
	Backfill by hand				2.101		88.88		88.88	114	142
	Total				11.905	171.50	555.29	51.80	778.59	**958.11**	**1,154.63**

D3013 210 Natural Gas: Pipe & Fittings, Steel/Iron

	System Description	Freq. (Years)	Crew	Unit	Labor Hours	2020 Bare Costs				Total In-House	Total w/O&P
						Material	Labor	Equipment	Total		
0140	**Replace 10′ of buried 6″ diam. st. pipe/M.L.F.**	12	B-35	Ea.							
	Check break or leak				.016		1.05		1.05	1.30	1.62
	Turn valve on and off				.010		.66		.66	.81	1.01
	Excavate by hand				5.771		242.20		242.20	315	390
	Remove broken 6″ diameter section				1.734		91	28[illegible]	119.90	147	175
	Install 10′ of pipe coated & wrapped				2.667	305	140	44[illegible]	489.50	560	650
	Check for leaks, additional section				.900		59		59	73	91
	Backfill by hand				2.101		88.88		88.88	114	142
	Total				13.199	305	622.79	73[illegible]	1,001.19	**1,211.11**	**1,450.63**
0210	**Replace 1000 L.F. of buried 2″ diam. st. pipe**	75	Q-4	M.L.F.							
	Check break or leak				1.600		105		105	130	162
	Turn valve on and off				1.000		65.50		65.50	81	101
	Excavate with 1/2 C.Y. backhoe				23.111		1,144.44	335.56	1,480	1,825	2,175
	Remove broken 2″ diameter section				74.250		4,560	250	4,810	5,900	7,325
	Install 1000′ of pipe coated & wrapped				114.290	8,000	7,000	380	15,380	17,900	21,300
	Check for leaks, additional section				9.000		590		590	730	910
	Backfill with 1 C.Y. bucket				8.667		448.89	217.[illegible]	666.67	810	950
	Total				231.918	8,000	13,913.83	1,183.[illegible]	23,097.17	**27,376**	**32,923**
0220	**Replace 1000 L.F. of buried 3″ diam. st. pipe**	75	Q-4	M.L.F.							
	Check break or leak				1.600		105		105	130	162
	Turn valve on and off				1.000		65.50		65.50	81	101
	Excavate with 1/2 C.Y. backhoe				23.111		1,144.44	335.[illegible]	1,480	1,825	2,175
	Remove broken 3″ diameter section				80.000		4,910	27[illegible]	5,180	6,350	7,900
	Install 1000′ of pipe coated & wrapped				123.080	13,200	7,550	41[illegible]	21,160	24,300	28,700
	Check for leaks, additional section				9.000		590		590	730	910
	Backfill with 1 C.Y. bucket				8.667		448.89	217.[illegible]	666.67	810	950
	Total				246.458	13,200	14,813.83	1,23[illegible]	29,247.17	**34,226**	**40,898**
0230	**Replace 1000 L.F. of buried 4″ diam. st. pipe**	75	B-35	M.L.F.							
	Check break or leak				1.600		105		105	130	162
	Turn valve on and off				1.000		65.50		65.50	81	101
	Excavate with 1/2 C.Y. backhoe				23.111		1,144.44	335.[illegible]	1,480	1,825	2,175
	Remove broken 4″ diameter section				122.390		6,450	2,04[illegible]	8,490	10,400	12,400
	Install 1000′ of pipe coated & wrapped				188.240	17,150	9,900	3,14[illegible]	30,190	34,800	40,500
	Check for leaks, additional section				9.000		590		590	730	910
	Backfill with 1 C.Y. bucket				8.667		448.89	217.[illegible]	666.67	810	950
	Total				354.008	17,150	18,703.83	5,73[illegible]	41,587.17	**48,776**	**57,198**

D3013 210 Natural Gas: Pipe & Fittings, Steel/Iron

	System Description	Freq. (Years)	Crew	Unit	Labor Hours	2020 Bare Costs				Total In-House	Total w/O&P
						Material	Labor	Equipment	Total		
0240	**Replace 1000 L.F. of buried 6″ diam. st. pipe**	75	B-35	M.L.F.							
	Check break or leak				1.600		105		105	130	162
	Turn valve on and off				1.000		65.50		65.50	81	101
	Excavate with 1/2 C.Y. backhoe				23.111		1,144.44	335.56	1,480	1,825	2,175
	Remove broken 6″ diameter section				173.390		9,100	2,890	11,990	14,700	17,500
	Install 1000′ of pipe coated & wrapped				266.670	30,500	14,000	4,450	48,950	56,000	65,000
	Check for leaks, additional section				9.000		590		590	730	910
	Backfill with 1 C.Y. bucket				8.667		448.89	217.78	666.67	810	950
	Total				483.438	30,500	25,453.83	7,893.34	63,847.17	**74,276**	**86,798**
0410	**Replace 10′ of hung 2″ diam. st. pipe/M.L.F.**	12	Q-1	Ea.							
	Check break or leak				.016		1.05		1.05	1.30	1.62
	Turn valve on and off				1.000		66		66	81	101
	Remove broken 2″ diameter section				1.626		94.50		94.50	117	146
	Install 10′ of black 2″ diameter steel pipe				3.250	69	188.50		257.50	310	375
	Check for leaks, additional section				.900		59		59	73	91
	Total				6.792	69	409.05		478.05	**582.30**	**714.62**
0420	**Replace 10′ of hung 3″ diam. st. pipe/M.L.F.**	12	Q-15	Ea.							
	Check break or leak				.016		1.05		1.05	1.30	1.62
	Turn valve on and off				.010		.66		.66	.81	1.01
	Remove broken 3″ diameter section				2.420		140.50	16.10	156.60	191	233
	Install 10′ of black 3″ diameter steel pipe				4.837	154.50	280	32.20	466.70	550	665
	Check for leaks, additional section				.900		59		59	73	91
	Total				8.182	154.50	481.21	48.30	684.01	**816.11**	**991.63**
0430	**Replace 10′ of hung 4″ diam. st. pipe/M.L.F.**	12	Q-15	Ea.							
	Check break or leak				.016		1.05		1.05	1.30	1.62
	Turn valve on and off				.010		.66		.66	.81	1.01
	Remove broken 4″ diameter section				2.812		163	18.70	181.70	222	271
	Install 10′ of black 4″ diameter steel pipe				5.622	240	325	37.40	602.40	710	845
	Check for leaks, additional section				.900		59		59	73	91
	Total				9.359	240	548.71	56.10	844.81	**1,007.11**	**1,209.63**

D3013 210 Natural Gas: Pipe & Fittings, Steel/Iron

	System Description	Freq. (Years)	Crew	Unit	Labor Hours	2020 Bare Costs				Total In-House	Total w/O&P
						Material	Labor	Equipment	Total		
0440	**Replace 10′ of hung 6″ diam. st. pipe/M.L.F.**	12	Q-16	Ea.							
	Check break or leak				.016		1.05		1.05	1.30	1.62
	Turn valve on and off				.010		.66		.66	.81	1.01
	Remove broken 6″ diameter section				4.335		260	19.20	279.20	345	425
	Install 10′ of black 6″ diameter steel pipe				8.667	470	520	38.40	1,028.40	1,200	1,425
	Check for leaks, additional section				.900		59		59	73	91
	Total				13.928	470	840.71	57.60	1,368.31	**1,620.11**	**1,943.63**
0510	**Replace 1000 L.F. of hung 2″ diam. steel pipe**	75	Q-1	M.L.F.							
	Check break or leak				1.600		105		105	130	162
	Turn valve on and off				100.000		6,550		6,550	8,100	10,100
	Remove broken 2″ diameter section				162.550		9,450		9,450	11,700	14,600
	Install 1000′ of black 2″ diameter steel pipe				325.000	6,900	18,850		25,750	30,900	37,600
	Check for leaks, additional section				9.000		590		590	730	910
	Total				598.150	6,900	35,545		42,445	**51,560**	**63,372**
0520	**Replace 1000 L.F. of hung 3″ diam. steel pipe**	75	Q-15	M.L.F.							
	Check break or leak				1.600		105		105	130	162
	Turn valve on and off				1.000		65.50		65.50	81	101
	Remove broken 3″ diameter section				241.950		14,050	1,610	15,660	19,100	23,300
	Install 1000′ of black 3″ diameter steel pipe				483.680	15,450	28,000	3,220	46,670	55,000	66,500
	Check for leaks, additional section				9.000		590		590	730	910
	Total				737.230	15,450	42,810.50	4,830	63,090.50	**75,041**	**90,973**
0530	**Replace 1000 L.F. of hung 4″ diam. steel pipe**	75	Q-15	M.L.F.							
	Check break or leak				1.600		105		105	130	162
	Turn valve on and off				1.000		65.50		65.50	81	101
	Remove broken 4″ diameter section				281.150		16,300	1,870	18,170	22,200	27,100
	Install 1000′ of black 4″ diameter steel pipe				562.190	24,000	32,500	3,740	60,240	71,000	84,500
	Check for leaks, additional section				9.000		590		590	730	910
	Total				854.940	24,000	49,560.50	5,610	79,170.50	**94,141**	**112,773**

D3013 210 Natural Gas: Pipe & Fittings, Steel/Iron

	System Description	Freq. (Years)	Crew	Unit	Labor Hours	2020 Bare Costs				Total In-House	Total w/O&P
						Material	Labor	Equipment	Total		
0540	**Replace 1000 L.F. of hung 6″ diam. steel pipe**	75	Q-16	M.L.F.							
	Check break or leak				1.600		105		105	130	162
	Turn valve on and off				1.000		65.50		65.50	81	101
	Remove broken 6″ diameter section				433.450		26,000	1,920	27,920	34,400	42,600
	Install 1000′ of black 6″ diameter steel pipe				866.740	47,000	52,000	3,840	102,840	120,500	143,500
	Check for leaks, additional section				9.000		590		590	730	910
	Total				1311.790	47,000	78,760.50	5,760	131,520.50	**155,841**	**187,273**

D3013 240 Natural Gas: Pressure Reducing Valve

	System Description	Freq. (Years)	Crew	Unit	Labor Hours	2020 Bare Costs				Total In-House	Total w/O&P
						Material	Labor	Equipment	Total		
0010	**Check gas pressure**	5	1 STPI	Ea.							
	Check pressure				.130		8.50		8.50	10.55	13.20
	Total				.130		8.50		8.50	**10.55**	**13.20**
0110	**Replace pressure regulator 1/2″ diam. pipe**	14	1 STPI	Ea.							
	Turn valve on and off				.010		.66		.66	.81	1.01
	Remove 1/2″ diameter regulator				.217		14.20		14.20	17.55	22
	Replace 1/2″ diameter pipe size regulator				.433	59.50	28.50		88	101	118
	Total				.660	59.50	43.36		102.86	**119.36**	**141.01**
0120	**Replace pressure regulator 1″ diam. pipe**	14	1 STPI	Ea.							
	Turn valve on and off				.010		.66		.66	.81	1.01
	Remove 1″ diameter regulator				.274		17.95		17.95	22	28
	Replace 1″ diameter pipe size regulator				.547	110	36		146	165	193
	Total				.831	110	54.61		164.61	**187.81**	**222.01**
0130	**Replace pressure regulator 1-1/2″ diam. pipe**	14	1 STPI	Ea.							
	Turn valve on and off				.010		.66		.66	.81	1.01
	Remove 1-1/2″ diameter regulator				.400		26		26	32.50	40.50
	Replace 1-1/2″ diameter pipe size regulator				.800	725	52.50		777.50	860	985
	Total				1.210	725	79.16		804.16	**893.31**	**1,026.51**

D3013 240 Natural Gas: Pressure Reducing Valve

	System Description	Freq. (Years)	Crew	Unit	Labor Hours	2020 Bare Costs				Total In-House	Total w/O&P
						Material	Labor	Equipment	Total		
0140	**Replace pressure regulator 2″ diam. pipe**	14	1 STPI	Ea.							
	Turn valve on and off				.010		.66		.66	.81	1.01
	Remove 2″ diameter regulator				.473		31		31	38.50	48
	Replace 2″ diameter pipe size regulator				.946	725	62		787	875	1,000
	Total				1.428	725	93.66		818.66	**914.31**	**1,049.01**

D3013 260 LPG Distribution: Pipe & Fittings, Steel/Iron

	System Description	Freq. (Years)	Crew	Unit	Labor Hours	2020 Bare Costs				Total In-House	Total w/O&P
						Material	Labor	Equipment	Total		
0120	**Replace 10′ st. pipe 1/2″ diam. per M.L.F.**	12	1 PLUM	Ea.							
	Check break or leak				.016		1.05		1.05	1.30	1.62
	Turn valve on and off				.010		.66		.66	.81	1.01
	Remove broken 1/2″ diameter section				.826		53		53	66	82.50
	Install section of pipe 1/2″ diameter				1.651	40.30	106.50		146.80	176	215
	Check for leaks, additional section				.900		59		59	73	91
	Total				3.402	40.30	220.21		260.51	**317.11**	**391.13**
0220	**Replace 10′ st. pipe 3/4″ diam. per M.L.F.**	12	1 PLUM	Ea.							
	Check break or leak				.016		1.05		1.05	1.30	1.62
	Turn valve on and off				.010		.66		.66	.81	1.01
	Remove broken 3/4″ diameter section				.853		55		55	68	85
	Install section of pipe 3/4″ diameter				1.312	44	84.50		128.50	153	186
	Check for leaks, additional section				.900		59		59	73	91
	Total				3.090	44	200.21		244.21	**296.11**	**364.63**
0320	**Replace 10′ st. pipe 1″ diam. per M.L.F.**	12	1 PLUM	Ea.							
	Check break or leak				.016		1.05		1.05	1.30	1.62
	Turn valve on and off				.010		.66		.66	.81	1.01
	Remove broken 1″ diameter section				.982		63.50		63.50	78	98
	Install section of pipe 1″ diameter				1.962	78	126.50		204.50	242	293
	Check for leaks, additional section				.900		59		59	73	91
	Total				3.870	78	250.71		328.71	**395.11**	**484.63**

D3013 260 LPG Distribution: Pipe & Fittings, Steel/Iron

	System Description	Freq. (Years)	Crew	Unit	Labor Hours	2020 Bare Costs				Total In-House	Total w/O&P
						Material	Labor	Equipment	Total		
0420	**Replace 10′ st. pipe 1-1/4″ diam. M.L.F.**	12	Q-1	Ea.							
	Check break or leak				.016		1.05		1.05	1.30	1.62
	Turn valve on and off				.010		.66		.66	.81	1.01
	Remove broken 1-1/4″ diameter section				1.169		68		68	84	105
	Install section of pipe 1-1/4″ diameter				2.337	82	135.50		217.50	258	315
	Check for leaks, additional section				.900		59		59	73	91
	Total				4.432	82	264.21		346.21	**417.11**	**513.63**
0520	**Replace 10′ st. pipe 1-1/2″ diam. M.L.F.**	12	Q-1	Ea.							
	Check break or leak				.016		1.05		1.05	1.30	1.62
	Turn valve on and off				.010		.66		.66	.81	1.01
	Remove broken 1-1/2″ diameter section				1.300		75.50		75.50	93.50	117
	Install section of pipe 1-1/2″ diameter				2.600	87.50	151		238.50	283	345
	Check for leaks, additional section				.900		59		59	73	91
	Total				4.826	87.50	287.21		374.71	**451.61**	**555.63**
0620	**Replace 10′ section st. pipe 2″ diam. M.L.F.**	12	Q-1	Ea.							
	Check break or leak				.016		1.05		1.05	1.30	1.62
	Turn valve on and off				.010		.66		.66	.81	1.01
	Remove broken 2″ diameter section				1.626		94.50		94.50	117	146
	Install section of pipe 2″ diameter				3.250	69	188.50		257.50	310	375
	Check for leaks, additional section				.900		59		59	73	91
	Total				5.802	69	343.71		412.71	**502.11**	**614.63**
1130	**Replace 1000′ of 1/2″ diameter steel pipe**	75	Q-1	M.L.F.							
	Check break or leak				1.600		105		105	130	162
	Turn valve on and off				1.000		65.50		65.50	81	101
	Remove broken 1/2″ diameter section				82.560		5,300		5,300	6,575	8,250
	Install section of pipe 1/2″ diameter				165.080	4,030	10,650		14,680	17,600	21,500
	Check for leaks, additional section				9.000		590		590	730	910
	Total				259.240	4,030	16,710.50		20,740.50	**25,116**	**30,923**

D3013 260 LPG Distribution: Pipe & Fittings, Steel/Iron

	System Description	Freq. (Years)	Crew	Unit	Labor Hours	2020 Bare Costs				Total In-House	Total w/O&P
						Material	Labor	Equipment	Total		
1230	**Replace 1000′ of 3/4″ diameter steel pipe**	75	Q-1	M.L.F.							
	Check break or leak				1.600		105		105	130	162
	Turn valve on and off				1.000		65.50		65.50	81	101
	Remove broken 3/4″ diameter section				85.270		5,500		5,500	6,800	8,500
	Install section of pipe 3/4″ diameter				131.150	4,400	8,450		12,850	15,300	18,600
	Check for leaks, additional section				9.000		590		590	730	910
	Total				228.020	4,400	14,710.50		19,110.50	**23,041**	**28,273**
1330	**Replace 1000′ of 1″ diameter steel pipe**	75	Q-1	M.L.F.							
	Check break or leak				1.600		105		105	130	162
	Turn valve on and off				1.000		65.50		65.50	81	101
	Remove broken 1″ diameter section				98.150		6,350		6,350	7,825	9,800
	Install section of pipe 1″ diameter				196.220	7,800	12,650		20,450	24,200	29,300
	Check for leaks, additional section				9.000		590		590	730	910
	Total				305.970	7,800	19,760.50		27,560.50	**32,966**	**40,273**
1430	**Replace 1000′ of 1-1/4″ diameter steel pipe**	75	Q-1	M.L.F.							
	Check break or leak				1.600		105		105	130	162
	Turn valve on and off				1.000		65.50		65.50	81	101
	Remove broken 1-1/4″ diameter section				116.890		6,800		6,800	8,375	10,500
	Install section of pipe 1-1/4″ diameter				233.710	8,200	13,550		21,750	25,800	31,300
	Check for leaks, additional section				9.000		590		590	730	910
	Total				362.200	8,200	21,110.50		29,310.50	**35,116**	**42,973**
1530	**Replace 1000′ of 1-1/2″ diameter steel pipe**	75	Q-1	M.L.F.							
	Check break or leak				1.600		105		105	130	162
	Turn valve on and off				1.000		65.50		65.50	81	101
	Remove broken 1-1/2″ diameter section				130.040		7,550		7,550	9,325	11,700
	Install section of pipe 1-1/2″ diameter				260.000	8,750	15,100		23,850	28,300	34,400
	Check for leaks, additional section				9.000		590		590	730	910
	Total				401.640	8,750	23,410.50		32,160.50	**38,566**	**47,273**

D3013 260 LPG Distribution: Pipe & Fittings, Steel/Iron

	System Description	Freq. (Years)	Crew	Unit	Labor Hours	2020 Bare Costs				Total In-House	Total w/O&P
						Material	Labor	Equipment	Total		
1630	**Replace 1000′ of 2″ diameter steel pipe**	75	Q-1	M.L.F.							
	Check break or leak				1.600		105		105	130	162
	Turn valve on and off				1.000		65.50		65.50	81	101
	Remove broken 2″ diameter section				162.550		9,450		9,450	11,700	14,600
	Install section of pipe 2″ diameter				325.000	6,900	13,850		25,750	30,900	37,600
	Check for leaks, additional section				9.000		590		590	730	910
	Total				499.150	6,900	29,060.50		35,960.50	**43,541**	**53,373**

D3013 601 Solar Panel, 3′ x 8′

	System Description	Freq. (Years)	Crew	Unit	Labor Hours	2020 Bare Costs				Total In-House	Total w/O&P
						Material	Labor	Equipment	Total		
0010	**Replace solar panel 3′ x 8′**	15	Q-1	Ea.							
	Remove panel				1.891		110		110	136	170
	Install 3′ x 8′ solar panel aluminum frame				3.783	1,025	219		1,244	1,400	1,625
	Total				5.674	1,025	329		1,354	**1,536**	**1,795**

D3023 180 Boiler, Gas

	System Description	Freq. (Years)	Crew	Unit	Labor Hours	2020 Bare Costs				Total In-House	Total w/O&P
						Material	Labor	Equipment	Total		
1010	**Repair boiler, gas, 250 MBH**	7	1 STPI	Ea.							
	Remove / replace burner blower				.600		39.50		39.50	48.50	61
	Remove burner blower bearing				1.000		65.50		65.50	81	101
	Replace bearings				2.000	70	131		201	239	291
	Remove burner blower motor				.976		64		64	79	99
	Replace burner blower motor				1.951	320	120		440	500	585
	Remove burner fireye				.195		12.80		12.80	15.80	19.80
	Replace burner fireye				.300	260	19.65		279.65	310	355
	Remove burner gas regulator				.274		17.95		17.95	22	28
	Replace burner gas regulator				.547	110	36		146	165	193
	Remove burner auto gas valve				.274		17.95		17.95	22	27.50
	Replace burner auto gas valve				.421	183	27.50		210.50	235	271
	Remove burner solenoid valve				.279		17.65		17.65	22	27.50
	Replace burner solenoid valve				.557	620	35.50		655.50	725	830
	Repair controls				.500		33		33	40.50	50.50
	Total				9.874	1,563	638		2,201	**2,504.80**	**2,939.30**
1060	**Replace boiler, gas, 250 MBH**	30	Q-7	Ea.							
	Remove boiler				21.888		1,375		1,375	1,700	2,100
	Replace boiler, 250 MBH				43.776	3,825	2,725		6,550	7,575	9,000
	Total				65.663	3,825	4,100		7,925	**9,275**	**11,100**
2010	**Repair boiler, gas, 2000 MBH**	7	Q-5	Ea.							
	Remove / replace burner blower				.600		39.50		39.50	48.50	61
	Remove burner blower bearing				1.000		65.50		65.50	81	101
	Replace bearings				2.000	70	131		201	239	291
	Remove burner blower motor				.976		64		64	79	99
	Replace burner blower motor				1.951	247	120		367	420	495
	Remove burner fireye				.195		12.80		12.80	15.80	19.80
	Replace burner fireye				.300	260	19.65		279.65	310	355
	Remove burner gas regulator				.727		43		43	53	66.50
	Replace burner gas regulator				1.455	1,400	86		1,486	1,650	1,875
	Remove burner auto gas valve				.800		46.50		46.50	57.50	71.50
	Replace burner auto gas valve				1.231	395	71.50		466.50	525	605
	Remove burner solenoid valve				2.600		165		165	204	255
	Replace burner solenoid valve				5.199	1,200	330		1,530	1,725	2,000
	Repair controls				.500		33		33	40.50	50.50
	Total				19.534	3,572	1,227.45		4,799.45	**5,448.30**	**6,345.30**

D3023 180 Boiler, Gas

	System Description	Freq. (Years)	Crew	Unit	Labor Hours	2020 Bare Costs				Total In-House	Total w/O&P
						Material	Labor	Equipment	Total		
2070	**Replace boiler, gas, 2000 MBH**	30	Q-7	Ea.							
	Remove boiler				57.762		3,600		3,600	4,450	5,575
	Replace boiler, 2000 MBH				110.000	23,300	6,850		30,150	34,100	39,700
	Total				167.762	23,300	10,450		33,750	**38,550**	**45,275**
3010	**Repair boiler, gas, 10,000 MBH**	7	Q-5	Ea.							
	Remove / replace burner blower				1.000		65.50		65.50	81	101
	Remove burner blower bearing				1.000		65.50		65.50	81	101
	Replace bearings				2.000	70	131		201	239	291
	Remove burner motor				1.000		65.50		65.50	81	101
	Replace burner motor				1.951	207	120		327	375	445
	Remove burner fireye				.195		12.80		12.80	15.80	19.80
	Replace burner fireye				.300	260	19.65		279.65	310	355
	Remove burner gas regulator				1.300		76.50		76.50	95	119
	Replace burner gas regulator				2.600	2,500	153		2,653	2,950	3,350
	Remove burner auto gas valve				5.195		310		310	385	485
	Replace burner auto gas valve				8.000	6,250	480		6,730	7,475	8,550
	Remove burner solenoid valve				2.600		165		165	204	255
	Replace burner solenoid valve				10.399	3,750	660		4,410	4,950	5,725
	Repair controls				.500		33		33	40.50	50.50
	Total				38.040	13,037	2,357.45		15,394.45	**17,282.30**	**19,948.30**
3070	**Replace boiler, gas, 10,000 MBH**	30	Q-7	Ea.							
	Remove boiler				364.000		22,680		22,680	28,100	35,100
	Replace boiler, gas, 10,000 MBH				722.400	179,200	45,080		224,280	253,000	293,500
	Total				1086.400	179,200	67,760		246,960	**281,100**	**328,600**

D3023 182 Boiler, Coal

	System Description	Freq. (Years)	Crew	Unit	Labor Hours	2020 Bare Costs				Total In-House	Total w/O&P
						Material	Labor	Equipment	Total		
1010	**Repair boiler, coal, 4600 MBH**	20	4 STPI	Ea.							
	Remove coal feeder				2.703		176.80		176.80	219	273
	Replace coal feeder				4.054	1,736.80	265.20		2,002	2,250	2,575
	Remove vibrating grates				33.592		2,204.80		2,204.80	2,725	3,400
	Replace vibrating grates				62.816	22,152	4,118.40		26,270.40	29,500	34,000
	Remove motor				.832		54.60		54.60	67.50	84
	Replace motor				.450	283.40	27.56		310.96	345	395
	Repair controls				5.200		340.60		340.60	420	530
	Total				109.647	24,172.20	7,187.96		31,360.16	**35,526.50**	**41,257**
1050	**Replace boiler, coal, 4600 MBH**	30	5 STPI	Ea.							
	Remove boiler				122.304		8,008		8,008	9,925	12,400
	Replace boiler, coal, 4600 MBH				244.712	110,552	16,016		126,568	141,500	163,000
	Total				367.016	110,552	24,024		134,576	**151,425**	**175,400**

D3023 184 Boiler, Oil

	System Description	Freq. (Years)	Crew	Unit	Labor Hours	2020 Bare Costs				Total In-House	Total w/O&P
						Material	Labor	Equipment	Total		
1010	**Repair boiler, oil, 250 MBH**	7	Q-5	Ea.							
	Remove / replace burner blower				.600		39.50		39.50	48.50	61
	Remove burner blower bearing				1.000		65.50		65.50	81	101
	Replace bearings				2.000	70	131		201	239	291
	Remove burner blower motor				.800		52.50		52.50	65	81
	Replace burner blower motor				1.951	320	120		440	500	585
	Remove burner fireye				.195		12.80		12.80	15.80	19.80
	Replace burner fireye				.300	260	19.65		279.65	310	355
	Remove burner ignition transformer				.250		16.40		16.40	20.50	25.50
	Replace burner ignition transformer				.500	100	33		133	151	176
	Remove burner ignition electrode				.200		13.10		13.10	16.20	20.50
	Replace burner ignition electrode				.400	14.30	26		40.30	48	58.50
	Remove burner oil pump				.350		23		23	28.50	35.50
	Replace burner oil pump				.700	128	46		174	198	231
	Remove burner nozzle				.167		10.95		10.95	13.50	16.90
	Replace burner nozzle				.300	6.50	19.65		26.15	31.50	38.50
	Repair controls				.600		39.50		39.50	48.50	61
	Total				10.313	898.80	668.55		1,567.35	**1,815**	**2,157.20**

D3023 184 Boiler, Oil

	System Description	Freq. (Years)	Crew	Unit	Labor Hours	2020 Bare Costs				Total In-House	Total w/O&P
						Material	Labor	Equipment	Total		
1060	**Replace boiler, oil, 250 MBH**	30	Q-7	Ea.							
	Remove boiler				24.465		1,525		1,525	1,875	2,350
	Replace boiler, oil, 250 MBH				48.930	3,600	3,050		6,650	7,725	9,225
	Total				73.394	3,600	4,575		8,175	**9,600**	**11,575**
1066	**Replace boiler, oil, 300 MBH**	30	Q-7	Ea.							
	Remove boiler				29.091		1,825		1,825	2,250	2,800
	Replace boiler, oil, 300 MBH				59.480	4,600	3,700		8,300	9,650	11,500
	Total				88.570	4,600	5,525		10,125	**11,900**	**14,300**
1070	**Replace boiler, oil, 420 MBH**	30	Q-7	Ea.							
	Remove boiler				35.955		2,250		2,250	2,775	3,475
	Replace boiler, oil, 420 MBH				71.749	5,700	4,475		10,175	11,800	14,100
	Total				107.704	5,700	6,725		12,425	**14,575**	**17,575**
1080	**Replace boiler, oil, 530 MBH**	30	Q-7	Ea.							
	Remove boiler				42.440		2,650		2,650	3,275	4,100
	Replace boiler, oil, 530 MBH				84.881	7,200	5,300		12,500	14,500	17,200
	Total				127.321	7,200	7,950		15,150	**17,775**	**21,300**
2010	**Repair boiler, oil, 2000 MBH**	7	1 STPI	Ea.							
	Remove / replace burner blower				.600		39.50		39.50	48.50	61
	Remove burner blower bearing				1.000		65.50		65.50	81	101
	Replace bearings				2.000	70	131		201	239	291
	Remove burner blower motor				.976		64		64	79	99
	Replace burner blower motor				1.951	247	120		367	420	495
	Remove burner fireye				.195		12.80		12.80	15.80	19.80
	Replace burner fireye				.300	260	19.65		279.65	310	355
	Remove burner ignition transformer				.286		18.75		18.75	23	29
	Replace burner ignition transformer				.571	99.50	37.50		137	156	182
	Remove burner ignition electrode				.200		13.10		13.10	16.20	20.50
	Replace burner ignition electrode				.400	14.30	26		40.30	48	58.50
	Remove burner oil pump				.400		26		26	32.50	40.50
	Replace burner oil pump				.800	150	52.50		202.50	230	269
	Remove burner nozzle				.167		10.95		10.95	13.50	16.90
	Replace burner nozzle				.333	10.25	22		32.25	38.50	47
	Repair controls				.600		39.50		39.50	48.50	61
	Total				10.779	851.05	698.75		1,549.80	**1,799.50**	**2,146.20**

D3023 184 Boiler, Oil

	System Description	Freq. (Years)	Crew	Unit	Labor Hours	2020 Bare Costs				Total In-House	Total w/O&P
						Material	Labor	Equipment	Total		
2060	**Replace boiler, oil, 2000 MBH**	30	Q-7	Ea.							
	Remove boiler				57.762		3,600		3,600	4,450	5,575
	Replace boiler, oil, 2000 MBH				149.000	17,400	9,275		26,675	30,600	36,200
	Total				206.762	17,400	12,875		30,275	**35,050**	**41,775**
2070	**Replace boiler, oil, 3000 MBH**	30	Q-7	Ea.							
	Remove boiler				110.000		6,875		6,875	8,525	10,600
	Replace boiler, oil, 3000 MBH				221.000	23,400	13,800		37,200	42,800	50,500
	Total				331.000	23,400	20,675		44,075	**51,325**	**61,100**
3010	**Repair boiler, oil, 10,000 MBH**	7	1 STPI	Ea.							
	Remove / replace burner blower				.600		39.50		39.50	48.50	61
	Remove burner blower bearing				1.000		65.50		65.50	81	101
	Replace bearings				2.000	70	131		201	239	291
	Remove burner blower motor				.976		64		64	79	99
	Replace burner blower motor				2.312	805	142		947	1,050	1,225
	Remove burner fireye				.195		12.80		12.80	15.80	19.80
	Replace burner fireye				.300	260	19.65		279.65	310	355
	Remove burner ignition transformer				.333		22		22	27	34
	Replace burner ignition transformer				.667	99.50	43.50		143	163	192
	Remove burner ignition electrode				.267		17.50		17.50	21.50	27
	Replace burner ignition electrode				.444	14.40	29		43.40	52	63
	Remove burner oil pump				.500		33		33	40.50	50.50
	Replace burner oil pump				1.000	128	65.50		193.50	222	261
	Remove burner nozzle				.167		10.95		10.95	13.50	16.90
	Replace burner nozzle				.400	10.55	26		36.55	44	53.50
	Repair controls				.600		39.50		39.50	48.50	61
	Total				11.761	1,387.45	761.40		2,148.85	**2,455.30**	**2,910.70**
3060	**Replace boiler, oil, 10,000 MBH**	30	Q-7	Ea.							
	Remove boiler				373.100		23,247		23,247	28,800	36,000
	Replace boiler, oil, 10,000 MBH				522.340	147,087.50	32,574.50		179,662	202,000	234,000
	Total				895.440	147,087.50	55,821.50		202,909	**230,800**	**270,000**

D3023 186 Boiler, Gas/Oil

	System Description	Freq. (Years)	Crew	Unit	Labor Hours	2020 Bare Costs Material	Labor	Equipment	Total	Total In-House	Total w/O&P
1010	**Repair boiler, gas/oil, 2000 MBH**	7	Q-5	Ea.							
	Remove/replace burner blower				.600		39.50		39.50	48.50	61
	Remove burner blower bearing				1.000		65.50		65.50	81	101
	Replace bearings				2.000	70	131		201	239	291
	Remove burner blower motor				.976		64		64	79	99
	Replace burner blower motor				1.951	247	120		367	420	495
	Remove burner fireye				.195		12.80		12.80	15.80	19.80
	Replace burner fireye				.300	260	19.65		279.65	310	355
	Remove burner ignition transformer				.333		22		22	27	34
	Replace burner ignition transformer				.571	99.50	37.50		137	156	182
	Remove burner ignition electrode				.200		13.10		13.10	16.20	20.50
	Replace burner ignition electrode				.400	14.30	26		40.30	48	58.50
	Remove burner oil pump				.500		33		33	40.50	50.50
	Replace burner oil pump				.800	150	52.50		202.50	230	269
	Remove burner nozzle				.167		10.95		10.95	13.50	16.90
	Replace burner nozzle				.333	10.25	22		32.25	38.50	47
	Remove burner gas regulator				.727		43		43	53	66.50
	Replace burner gas regulator				1.455	1,400	86		1,486	1,650	1,875
	Remove burner auto gas valve				.800		46.50		46.50	57.50	71.50
	Replace burner auto gas valve				1.231	395	71.50		466.50	525	605
	Remove burner solenoid valve				2.600		165		165	204	255
	Replace burner solenoid valve				5.199	1,200	330		1,530	1,725	2,000
	Repair controls				.600		39.50		39.50	48.50	61
	Total				22.939	3,846.05	1,451		5,297.05	**6,026**	**7,034.20**
1050	**Replace boiler, gas/oil, 2000 MBH**	30	Q-7	Ea.							
	Remove boiler				57.762		3,600		3,600	4,450	5,575
	Replace boiler, gas/oil, 2000 MBH				116.000	39,900	7,200		47,100	53,000	61,000
	Total				173.762	39,900	10,800		50,700	**57,450**	**66,575**

D3023 186 Boiler, Gas/Oil

	System Description	Freq. (Years)	Crew	Unit	Labor Hours	2020 Bare Costs				Total In-House	Total w/O&P
						Material	Labor	Equipment	Total		
2010	**Repair boiler, gas/oil, 20,000 MBH**	7	Q-5	Ea.							
	Remove/replace burner blower				.600		39.50		39.50	48.50	61
	Remove burner blower bearing				1.000		65.50		65.50	81	101
	Replace bearings				2.000	70	131		201	239	291
	Remove burner blower motor				.976		60		60	73.50	92
	Replace burner blower motor				1.951	320	120		440	500	585
	Remove burner fireye				.195		12.80		12.80	15.80	19.80
	Replace burner fireye				.300	260	19.65		279.65	310	355
	Remove burner ignition transformer				.333		22		22	27	34
	Replace burner ignition transformer				.667	99.50	43.50		143	163	192
	Remove burner ignition electrode				.267		17.50		17.50	21.50	27
	Replace burner ignition electrode				.444	14.40	29		43.40	52	63
	Remove burner oil pump				.500		33		33	40.50	50.50
	Replace burner oil pump				2.602	207	154		361	415	495
	Remove burner nozzle				.167		10.95		10.95	13.50	16.90
	Replace burner nozzle				.571	10.55	37.50		48.05	58	71
	Remove burner gas regulator				1.300		76.50		76.50	95	119
	Replace burner gas regulator				2.600	2,500	153		2,653	2,950	3,350
	Remove burner auto gas valve				5.195		310		310	385	485
	Replace burner auto gas valve				8.000	6,250	480		6,730	7,475	8,550
	Remove burner solenoid valve				2.600		165		165	204	255
	Replace burner solenoid valve				10.399	3,750	660		4,410	4,950	5,725
	Repair controls				.600		39.50		39.50	48.50	61
	Total				43.266	13,481.45	2,679.90		16,161.35	**18,165.80**	**20,999.20**
2050	**Replace boiler, gas/oil, 20,000 MBH**	30	Q-7	Ea.							
	Remove boiler				1527.360		95,460		95,460	118,000	147,500
	Replace boiler, gas/oil, 20,000 MBH				3156.840	304,880	196,840		501,720	579,000	686,000
	Total				4684.200	304,880	292,300		597 180	**697,000**	**833,500**

D3023 198 Blowoff System

	System Description	Freq. (Years)	Crew	Unit	Labor Hours	2020 Bare Costs				Total In-House	Total w/O&P
						Material	Labor	Equipment	Total		
1010	**Repair boiler blowoff system**	10	1 STPI	Ea.							
	Repair leak				1.000		65.50		65.50	81	101
	Total				1.000		65.50		65.50	**81**	**101**

D3023 198 Blowoff System

	System Description	Freq. (Years)	Crew	Unit	Labor Hours	2020 Bare Costs				Total In-House	Total w/O&P
						Material	Labor	Equipment	Total		
1020	**Replace boiler blowoff system**	15	Q-5	Ea.							
	Remove boiler blowoff				2.773		164		164	202	253
	Replace boiler blowoff				5.556	6,975	330		7,305	8,075	9,225
	Total				8.329	6,975	494		7,469	**8,277**	**9,478**

D3023 292 Chemical Feed System

	System Description	Freq. (Years)	Crew	Unit	Labor Hours	2020 Bare Costs				Total In-House	Total w/O&P
						Material	Labor	Equipment	Total		
1010	**Repair chemical feed**	15	1 STPI	Ea.							
	Repair controls				.040		2.62		2.62	3.24	4.05
	Remove / replace agitator motor				1.951	280	128		408	465	550
	Remove / replace pump seals / bearings				2.078	35	134		169	204	251
	Remove / replace pump motor				1.951	247	120		367	420	495
	Total				6.020	562	384.62		946.62	**1,092.24**	**1,300.05**
1030	**Replace chemical feed**	15	2 STPI	Ea.							
	Remove / replace feeder				2.500	635	161		796	900	1,050
	Total				2.500	635	161		796	**900**	**1,050**

D3023 294 Feed Water Supply

	System Description	Freq. (Years)	Crew	Unit	Labor Hours	2020 Bare Costs				Total In-House	Total w/O&P
						Material	Labor	Equipment	Total		
1010	**Repair feed water supply pump**	15	1 STPI	Ea.							
	Repair controls				.600		39.50		39.50	48.50	61
	Remove / replace pump seals / bearings				1.600		105		105	130	162
	Remove pump motor				2.078		136		136	168	211
	Replace pump motor				4.167	3,875	273		4,148	4,600	5,275
	Remove / replace pump coupling				1.000		65.50		65.50	81	101
	Total				9.445	3,875	619		4,494	**5,027.50**	**5,810**
1030	**Replace feed water pump**	15	Q-2	Ea.							
	Remove pump				11.111		670		670	825	1,025
	Replace pump				22.222	26,600	1,325		27,925	30,900	35,300
	Total				33.333	26,600	1,995		28,595	**31,725**	**36,325**

D3023 296 Deaerator

	System Description	Freq. (Years)	Crew	Unit	Labor Hours	2020 Bare Costs Material	Labor	Equipment	Total	Total In-House	Total w/O&P
1010	**Repair deaerator**	10	1 STPI	Ea.							
	Repair controls				1.000		65.50		65.50	81	101
	Total				1.000		65.50		65.50	**81**	**101**
1030	**Replace deaerator**	20	4 STPI	Ea.							
	Remove unit				7.500		460		460	565	710
	Replace unit				180.000	33,400	11,800		45,200	51,500	60,000
	Total				187.500	33,400	12,260		45,660	**52,065**	**60,710**

D3023 298 Separators for 9000 Ton Chilled Water System

	System Description	Freq. (Years)	Crew	Unit	Labor Hours	2020 Bare Costs Material	Labor	Equipment	Total	Total In-House	Total w/O&P
1010	**Clean separator strainer**	10	Q-6	Ea.							
	Clean separator strainer				10.000		655		655	810	1,000
	Total				10.000		655		655	**810**	**1,000**
1030	**Replace separator**	20	Q-6	Ea.							
	Remove unit				22.500		1,380		1,380	1,700	2,125
	Replace separator				42.857	68,400	2,625		71,025	78,500	89,500
	Total				65.357	68,400	4,005		72,405	**80,200**	**91,625**

D3023 310 Metal Flue / Chimney

	System Description	Freq. (Years)	Crew	Unit	Labor Hours	2020 Bare Costs Material	Labor	Equipment	Total	Total In-House	Total w/O&P
0010	**Replace metal flue, all fuel SS, 6″ diameter**	15	Q-9	L.F.							
	Replace flue, all fuel stainless steel, 6″ diameter				.387	87.73	21.68		109.41	124	144
	Total				.387	87.73	21.68		109.41	**124**	**144**
0020	**Replace metal flue, all fuel SS, 10″ diameter**	15	Q-9	L.F.							
	Replace flue, all fuel stainless steel, 10″ diameter				.483	105.13	27.12		132.25	150	174
	Total				.483	105.13	27.12		132.25	**150**	**174**

D3023 310 Metal Flue / Chimney

	System Description	Freq. (Years)	Crew	Unit	Labor Hours	2020 Bare Costs Material	Labor	Equipment	Total	Total In-House	Total w/O&P
0030	**Replace metal flue, all fuel SS, 20″ diameter**	15	Q-10	L.F.							
	Replace flue, all fuel stainless steel, 20″ diameter				.967	197.20	56.55		253.75	288	335
	Total				.967	197.20	56.55		253.75	**288**	**335**
0040	**Replace metal flue, all fuel SS, 32″ diameter**	15	Q-10	L.F.							
	Replace flue, all fuel stainless steel, 32″ diameter				1.289	329.15	74.68		403.83	455	530
	Total				1.289	329.15	74.68		403.83	**455**	**530**
0050	**Replace metal flue, all fuel SS, 48″ diameter**	15	Q-10	L.F.							
	Replace flue, all fuel stainless steel, 48″ diameter				1.832	485.75	106.58		592.33	670	775
	Total				1.832	485.75	106.58		592.33	**670**	**775**
2010	**Replace metal flue, gas vent, galvanized, 48″ diameter**	10	Q-10	L.F.							
	Replace flue, gas vent, galvanized, 48″ diameter				1.895	1,725	110.25		1,835.25	2,025	2,325
	Total				1.895	1,725	110.25		1,835.25	**2,025**	**2,325**

D3023 388 Pneumatic Coal Spreader

	System Description	Freq. (Years)	Crew	Unit	Labor Hours	2020 Bare Costs Material	Labor	Equipment	Total	Total In-House	Total w/O&P
1010	**Repair spreader, pneumatic coal**	10	2 STPI	Ea.							
	Repair controls				1.000		65.50		65.50	81	101
	Remove conveyor bearing				1.111		73		73	90	113
	Replace bearings				2.222	78.50	146		224.50	266	325
	Remove conveyor motor				2.597		170		170	211	263
	Replace conveyor motor				5.195	3,625	320		3,945	4,375	5,025
	Remove / replace blower				1.200		78.50		78.50	97.50	122
	Remove blower bearings				1.000		65.50		65.50	81	101
	Replace bearings				2.000	70	131		201	239	291
	Remove blower motor				2.000		131		131	162	203
	Replace blower motor				4.000	2,125	245		2,370	2,650	3,025
	Total				22.326	5,898.50	1,425.50		7,324	**8,252.50**	**9,569**
1060	**Replace coal spreader**	12	4 STPI	Ea.							
	Remove spreader				65.041		4,275		4,275	5,275	6,600
	Replace spreader				130.000	8,125	8,525		16,650	19,500	23,400
	Total				195.041	8,125	12,800		20,925	**24,775**	**30,000**

D3023 390 Fuel Oil Equipment

	System Description	Freq. (Years)	Crew	Unit	Labor Hours	2020 Bare Costs Material	2020 Bare Costs Labor	2020 Bare Costs Equipment	2020 Bare Costs Total	Total In-House	Total w/O&P
1010	**Repair fuel oil equipment, pump**	10	1 STPI	Ea.							
	Remove pump seals / bearings				.600		39.50		39.50	48.50	61
	Replace pump seals / bearings				1.500	33.50	98.50		132	158	194
	Remove / replace pump coupling				.900		59		59	73	91
	Remove / replace impeller / shaft				.900		59		59	73	91
	Remove / replace motor				.650		42.50		42.50	52.50	66
	Remove strainers				.289		18.95		18.95	23.50	29.50
	Replace strainers				.578	23.50	38		61.50	72.50	88
	Total				5.417	57	355.45		412.45	**501**	**620.50**
1030	**Replace fuel oil 25 GPH pump / motor set**	15	Q-5	Ea.							
	Remove equipment				1.733		102		102	126	158
	Replace equipment				3.468	1,600	205		1,805	2,025	2,325
	Total				5.201	1,600	307		1,907	**2,151**	**2,483**
1040	**Replace fuel oil 45 GPH pump / motor set**	15	Q-5	Ea.							
	Remove equipment				1.733		102		102	126	158
	Replace equipment				3.468	1,600	205		1,805	2,025	2,325
	Total				5.201	1,600	307		1,907	**2,151**	**2,483**
1050	**Replace fuel oil 90 GPH pump / motor set**	15	Q-5	Ea.							
	Remove equipment				2.081		123		123	152	190
	Replace equipment				4.161	1,600	246		1,846	2,075	2,375
	Total				6.242	1,600	369		1,969	**2,227**	**2,565**
1060	**Replace fuel oil 160 GPH pump / motor set**	15	Q-5	Ea.							
	Remove equipment				2.597		153		153	189	237
	Replace equipment				5.195	1,700	305		2,005	2,250	2,600
	Total				7.792	1,700	458		2,158	**2,439**	**2,837**

D3033 115 Cooling Tower

	System Description	Freq. (Years)	Crew	Unit	Labor Hours	2020 Bare Costs				Total In-House	Total w/O&P
						Material	Labor	Equipment	Total		
1010	**Repair cooling tower, 50 ton**	10	2 STPI	Ea.							
	Repair controls				1.000		65.50		65.50	81	101
	Remove bearings				1.200		78.50		78.50	97.50	122
	Replace bearings				2.000	70	131		201	239	291
	Remove fan motor				1.155		75.50		75.50	93.50	117
	Replace fan motor				2.312	805	142		947	1,050	1,225
	Remove / replace float valve				.500	16.30	29.50		45.80	54.50	66
	Total				8.167	891.30	522		1,413.30	**1,615.50**	**1,922**
1030	**Replace cooling tower, 50 ton**	15	Q-6	Ea.							
	Remove cooling tower				10.397		636		636	785	985
	Replace cooling tower, 50 ton				20.779	12,180	1,260		13,440	15,000	17,200
	Total				31.177	12,180	1,896		14,076	**15,785**	**18,185**
2010	**Repair cooling tower, 100 ton**	10	2 STPI	Ea.							
	Repair controls				1.000		65.50		65.50	81	101
	Remove bearings				4.800		314		314	390	490
	Replace bearings				8.000	280	524		804	955	1,150
	Remove fan motors				4.621		302		302	375	470
	Replace fan motors				9.249	3,220	568		3,788	4,250	4,900
	Remove / replace float valve				.575	18.75	33.93		52.68	62.50	76
	Total				28.245	3,518.75	1,807.43		5,326.18	**6,113.50**	**7,187**
2030	**Replace cooling tower, 100 ton**	15	Q-6	Ea.							
	Remove cooling tower				14.286		875		875	1,075	1,350
	Replace cooling tower, 100 ton				28.571	17,000	1,750		18,750	20,900	24,000
	Total				42.857	17,000	2,625		19,625	**21,975**	**25,350**
3010	**Repair cooling tower, 300 ton**	10	2 STPI	Ea.							
	Repair controls				1.000		65.50		65.50	81	101
	Remove bearings				7.200		471		471	585	730
	Replace bearings				13.333	471	876		1,347	1,600	1,950
	Remove fan motors				6.931		453		453	560	700
	Replace fan motors				13.865	3,570	852		4,422	4,975	5,775
	Remove / replace float valve				1.250	40.75	73.75		114.50	136	165
	Total				43.580	4,081.75	2,791.25		6,873	**7,937**	**9,421**

D3033 115 Cooling Tower

	System Description	Freq. (Years)	Crew	Unit	Labor Hours	2020 Bare Costs				Total In-House	Total w/O&P
						Material	Labor	Equipment	Total		
3030	**Replace cooling tower, 300 ton**	15	Q-6	Ea.							
	Remove cooling tower				36.273		2,220		2,220	2,750	3,425
	Replace cooling tower, 300 ton				72.582	29,550	4,440		33,990	38,000	43,800
	Total				108.855	29,550	6,660		36,210	**40,750**	**47,225**
4010	**Repair cooling tower, 1000 ton**	10	2 STPI	Ea.							
	Repair controls				1.000		65.50		65.50	81	101
	Remove bearings				12.000		785		735	975	1,225
	Replace bearings				23.529	1,860	1,540		3,400	3,950	4,700
	Remove fan motors				11.552		755		755	935	1,175
	Replace fan motors				25.974	11,750	1,590		13,340	14,900	17,100
	Remove / replace float valve				2.350	76.61	138.65		215.26	256	310
	Total				76.406	13,686.61	4,874.15		18,560.76	**21,097**	**24,611**
4030	**Replace cooling tower, 1000 ton**	15	Q-6	Ea.							
	Remove cooling tower				104.350		6,400		6,400	7,900	9,850
	Replace cooling tower, 1000 ton				208.700	74,500	12,750		87,250	97,500	113,000
	Total				313.050	74,500	19,150		93,650	**105,400**	**122,850**

D3033 130 Chiller, Water Cooled, Reciprocating

	System Description	Freq. (Years)	Crew	Unit	Labor Hours	2020 Bare Costs				Total In-House	Total w/O&P
						Material	Labor	Equipment	Total		
1010	**Repair water cooled chiller, 20 ton**	10	Q-6	Ea.							
	Repair controls				1.000		65.50		65.50	81	101
	Remove compressor				21.463		1,320		1,320	1,625	2,025
	Replace compressor				42.857	23,700	2,625		26,325	29,300	33,700
	Remove / replace evaporator tube				1.200	305	73.50		378.50	425	495
	Remove / replace condenser tube				1.200	197	73.50		270.50	305	360
	Replace refrigerant				2.667	480	175		655	745	870
	Total				70.387	24,682	4,332.50		29,014.50	**32,481**	**37,551**
1030	**Replace chiller, water cooled, 20 ton, scroll**	20	Q-7	Ea.							
	Remove chiller				50.633		3,150		3,150	3,900	4,875
	Replace chiller, water cooled, 20 ton				88.889	21,200	5,550		26,750	30,200	35,100
	Total				139.522	21,200	8,700		29,900	**34,100**	**39,975**

D3033 130 Chiller, Water Cooled, Reciprocating

	System Description	Freq. (Years)	Crew	Unit	Labor Hours	2020 Bare Costs				Total In-House	Total w/O&P
						Material	Labor	Equipment	Total		
2010	**Repair water cooled chiller, 50 ton**	10	Q-6	Ea.							
	Repair controls				1.000		65.50		65.50	81	101
	Remove compressor				35.610		2,171.75		2,171.75	2,700	3,375
	Replace compressor				71.006	33,000	4,350		37,350	41,700	48,000
	Remove / replace evaporator tube				1.200	305	73.50		378.50	425	495
	Remove / replace condenser tube				1.200	197	73.50		270.50	305	360
	Replace refrigerant				4.000	720	262.50		982.50	1,125	1,300
	Total				114.016	34,222	6,996.75		41,218.75	**46,336**	**53,631**
2030	**Replace chiller, water cooled 50 ton, scroll**	20	Q-7	Ea.							
	Remove chiller				74.074		4,625		4,625	5,725	7,150
	Replace chiller, water cooled, 50 ton				114.000	31,300	7,125		38,425	43,200	50,000
	Total				188.074	31,300	11,750		43,050	**48,925**	**57,150**
3010	**Repair water cooled chiller, 100 ton**	10	Q-6	Ea.							
	Repair controls				1.000		65.50		65.50	81	101
	Remove compressor				77.922		4,775		4,775	5,900	7,375
	Replace compressor				154.000	38,000	9,600		47,600	53,500	62,500
	Remove / replace evaporator tube				4.798	915	299		1,214	1,375	1,600
	Remove / replace condenser tube				4.798	915	299		1,214	1,375	1,600
	Replace refrigerant				8.000	1,440	525		1,965	2,225	2,600
	Total				250.517	41,270	15,563.50		56,833.50	**64,456**	**75,776**
3030	**Replace chiller, water cooled, 100 ton, scroll**	20	Q-7	Ea.							
	Remove chiller				117.000		7,300		7,300	9,000	11,300
	Replace chiller, water cooled, 100 ton				178.000	69,500	11,100		80,600	90,000	104,000
	Total				295.000	69,500	18,400		87,900	**99,000**	**115,300**
5010	**Repair water cooled chiller, 200 ton**	10	Q-7	Ea.							
	Repair controls				1.000		65.50		65.50	81	101
	Remove compressor				152.099		9,509.50		9,509.50	11,700	14,700
	Replace compressor				304.920	66,836	18,942		85,778	97,000	113,000
	Remove / replace evaporator tube				9.610	1,700	600		2,300	2,600	3,050
	Remove / replace condenser tube				9.610	1,700	600		2,300	2,600	3,050
	Replace refrigerant				13.333	2,400	875		3,275	3,725	4,350
	Total				490.571	72,636	30,592		103,228	**117,706**	**138,251**

D3033 130 Chiller, Water Cooled, Reciprocating

	System Description	Freq. (Years)	Crew	Unit	Labor Hours	2020 Bare Costs				Total In-House	Total w/O&P
						Material	Labor	Equipment	Total		
5030	**Replace chiller, water cooled, 200 ton**	20	Q-7	Ea.							
	Remove chiller				163.000		10,200		10,200	12,600	15,800
	Replace chiller, water cooled, 200 ton				327.000	94,500	20,400		114,900	129,000	149,500
	Total				490.000	94,500	30,600		125,100	**141,600**	**165,300**

D3033 135 Chiller, Air Cooled, Reciprocating

	System Description	Freq. (Years)	Crew	Unit	Labor Hours	2020 Bare Costs				Total In-House	Total w/O&P
						Material	Labor	Equipment	Total		
1010	**Repair recip. chiller, air cooled, 20 ton**	10	2 STPI	Ea.							
	Repair controls				.800		52.50		52.50	65	81
	Remove fan bearing				3.600		235.50		235.50	292	365
	Replace bearings				6.000	210	393		603	715	870
	Remove fan motor				2.927		192		192	237	297
	Replace fan motor				5.854	960	360		1,320	1,500	1,750
	Remove compressor				24.490		1,500		1,500	1,850	2,325
	Replace compressor				48.780	28,700	2,975		31,675	35,300	40,500
	Replace refrigerant				2.667	480	175		655	745	870
	Total				95.117	30,350	5,883		36,233	**40,704**	**47,058**
1030	**Replace chiller, 20 ton**	20	Q-7	Ea.							
	Remove chiller				42.857		2,625		2,625	3,250	4,050
	Replace chiller, air cooled, reciprocating, 20 ton				119.000	23,300	7,425		30,725	34,800	40,600
	Total				161.857	23,300	10,050		33,350	**38,050**	**44,650**
2010	**Repair recip. chiller, air cooled, 50 ton**	10	Q-6	Ea.							
	Repair controls				1.000		65.50		65.50	81	101
	Remove fan bearing				3.600		235.50		235.50	292	365
	Replace bearings				6.667	235.50	438		673.50	800	970
	Remove fan motor				3.096		190.50		190.50	233	293
	Replace fan motor				6.936	2,415	426		2,841	3,175	3,675
	Remove compressor				60.976		3,750		3,750	4,625	5,750
	Replace compressor				156.000	34,300	9,525		43,825	49,500	57,500
	Remove / replace compressor motor				7.201		442		442	545	680
	Replace refrigerant				4.000	720	262.50		982.50	1,125	1,300
	Total				249.475	37,670.50	15,335		53,005.50	**60,376**	**70,634**

D3033 135 Chiller, Air Cooled, Reciprocating

	System Description	Freq. (Years)	Crew	Unit	Labor Hours	2020 Bare Costs Material	Labor	Equipment	Total	Total In-House	Total w/O&P
2030	**Replace chiller, 50 ton**	20	Q-7	Ea.							
	Remove chiller				73.395		4,575		4,575	5,650	7,075
	Replace chiller, air cooled, reciprocating, 50 ton				147.000	40,700	9,150		49,850	56,000	65,000
	Total				220.395	40,700	13,725		54,425	**61,650**	**72,075**
3010	**Repair chiller, air cooled, 100 ton**	10	Q-6	Ea.							
	Repair controls				1.000		65.50		65.50	81	101
	Remove fan bearing				4.200		276		276	340	425
	Replace bearings				6.667	235.50	438		673.50	800	970
	Remove fan motor				3.900		240		240	294	370
	Replace fan motor				7.792	3,525	477		4,002	4,475	5,150
	Remove compressor				97.403		5,968.75		5,968.75	7,375	9,225
	Replace compressor				192.500	47,500	12,000		59,500	67,000	78,000
	Remove / replace compressor motor				9.359		574		574	705	880
	Replace refrigerant				6.667	1,200	437.50		1,637.50	1,850	2,175
	Total				329.487	52,460.50	20,476.75		72,937.25	**82,920**	**97,296**
3030	**Replace chiller, 100 ton**	20	Q-7	Ea.							
	Remove chiller				84.211		5,250		5,250	6,500	8,125
	Replace chiller, air cooled, reciprocating, 100 ton				128.000	72,500	7,975		80,475	89,500	103,000
	Total				212.211	72,500	13,225		85,725	**96,000**	**111,125**

D3033 137 Chiller, Water Cooled, Scroll

	System Description	Freq. (Years)	Crew	Unit	Labor Hours	2020 Bare Costs Material	Labor	Equipment	Total	Total In-House	Total w/O&P
4010	**Repair water cooled chiller, 5 ton**	10	1 STPI	Ea.							
	Repair controls				.500		33		33	40.50	50.50
	Remove fan bearing				1.200		78.50		78.50	97.50	122
	Replace bearings				2.000	70	131		201	239	291
	Remove compressor				2.477		162		162	201	251
	Replace compressor				4.954	520	325		845	975	1,150
	Replace refrigerant				.667	120	43.75		163.75	186	218
	Total				11.797	710	773.25		1,483.25	**1,739**	**2,082.50**

D3033 137 Chiller, Water Cooled, Scroll

	System Description	Freq. (Years)	Crew	Unit	Labor Hours	2020 Bare Costs				Total In-House	Total w/O&P
						Material	Labor	Equipment	Total		
4030	**Replace chiller, water cooled, 5 ton**	20	Q-5	Ea.							
	Remove chiller				18.182		1,075		1,075	1,325	1,650
	Replace chiller, water cooled, 5 ton				36.364	3,475	2,150		5,625	6,475	7,675
	Total				54.545	3,475	3,225		6,700	**7,800**	**9,325**
5010	**Repair water cooled chiller, 10 ton**	10	2 STPI	Ea.							
	Repair controls				1.000		65.50		65.50	81	101
	Remove fan bearing				3.600		235.50		235.50	292	365
	Replace bearings				6.000	210	393		603	715	870
	Remove compressor				4.954		324		324	400	500
	Replace compressor				9.907	1,040	650		1,690	1,950	2,300
	Replace refrigerant				1.333	240	87.50		327.50	370	435
	Total				26.794	1,490	1,755.50		3,245.50	**3,808**	**4,571**
5030	**Replace chiller, water cooled, 10 ton**	20	Q-6	Ea.							
	Remove chiller				43.478		2,650		2,650	3,300	4,125
	Replace chiller, water cooled, 10 ton				86.957	6,575	5,325		11,900	13,800	16,400
	Total				130.435	6,575	7,975		14,550	**17,100**	**20,525**
6010	**Repair water cooled chiller, 15 ton**	10	Q-5	Ea.							
	Repair controls				1.000		65.50		65.50	81	101
	Remove fan bearing				24.000		1,320		1,320	1,700	2,100
	Replace bearings				6.000	210	393		603	715	870
	Remove compressor				6.932		410		410	505	630
	Replace compressor				13.865	2,350	820		3,170	3,600	4,200
	Replace refrigerant				2.000	360	131.25		491.25	560	655
	Total				53.797	2,920	3,139.75		6,059.75	**7,161**	**8,556**
6030	**Replace chiller, water cooled, 15 ton, scroll**	20	Q-6	Ea.							
	Remove chiller				42.857		2,625		2,625	3,250	4,050
	Replace chiller, water cooled, 15 ton				66.667	20,500	4,075		24,575	27,600	31,900
	Total				109.524	20,500	6,700		27,200	**30,850**	**35,950**

D3033 140 Chiller, Hermetic Centrifugal

	System Description	Freq. (Years)	Crew	Unit	Labor Hours	2020 Bare Costs				Total In-House	Total w/O&P
						Material	Labor	Equipment	Total		
1010	**Repair centrifugal chiller, 100 ton**	10	Q-7	Ea.							
	Repair controls				1.000		65.50		65.50	81	101
	Remove compressor				48.780		3,000		3,000	3,700	4,600
	Replace compressor				109.600	27,200	6,860		34,060	38,400	44,600
	Remove / replace condenser tube				4.800	788	294		1,082	1,225	1,450
	Remove / replace evaporator tube				4.800	1,220	294		1,514	1,700	1,975
	Remove / replace refrigerant filter				1.600	202	100		302	345	405
	Remove refrigerant				.440		28.80		28.80	35.50	44.50
	Replace refrigerant				8.000	1,440	525		1,965	2,225	2,600
	Total				179.020	30,850	11,167.30		42,017.30	**47,711.50**	**55,775.50**
1030	**Replace centrifugal chiller, 100 ton**	20	Q-7	Ea.							
	Remove chiller				112.000		6,975		6,975	8,625	10,800
	Replace chiller, centrifugal, 100 ton				224.000	92,000	14,000		106,000	118,500	136,500
	Total				336.000	92,000	20,975		112,975	**127,125**	**147,300**
2010	**Repair centrifugal chiller, 300 ton**	10	Q-7	Ea.							
	Repair controls				1.000		65.50		65.50	81	101
	Remove compressor				94.390		5,805		5,805	7,150	8,900
	Replace compressor				187.000	36,300	11,700		48,000	54,500	63,500
	Remove / replace condenser tube				12.000	1,970	735		2,705	3,075	3,600
	Remove / replace evaporator tube				12.000	3,050	735		3,785	4,275	4,950
	Remove / replace refrigerant filter				1.200	915	75	4	994	1,100	1,275
	Remove refrigerant				1.319		87		87	107	134
	Replace refrigerant				21.333	3,840	1,400		5,240	5,950	6,950
	Total				330.242	46,075	20,602.50	4	66,681.50	**76,238**	**89,410**
2030	**Replace centrifugal chiller, 300 ton**	20	Q-7	Ea.							
	Remove chiller				150.000		9,375		9,375	11,600	14,500
	Replace chiller, 300 ton				302.000	152,500	18,800		171,300	191,000	219,500
	Total				452.000	152,500	28 175		180,675	**202,600**	**234,000**

D3033 140 Chiller, Hermetic Centrifugal

	System Description	Freq. (Years)	Crew	Unit	Labor Hours	2020 Bare Costs				Total In-House	Total w/O&P
						Material	Labor	Equipment	Total		
3010	**Repair centrifugal chiller, 1000 ton**	10	Q-7	Ea.							
	Repair controls				1.000		65.50		65.50	81	101
	Remove compressor				153.659		9,450		9,450	11,600	14,500
	Replace compressor				296.000	78,500	18,500		97,000	109,000	126,500
	Remove / replace condenser tube				16.800	2,758	1,099		3,857	4,400	5,150
	Remove / replace evaporator tube				12.600	4,270	826		5,096	5,725	6,600
	Remove / replace refrigerant filter				1.600	915	105		1,020	1,125	1,300
	Remove refrigerant				3.516		232		232	285	355
	Replace refrigerant				29.333	5,280	1,925		7,205	8,175	9,575
	Total				514.508	91,723	32,202.50		123,925.50	**140,391**	**164,081**
3030	**Replace centrifugal chiller, 1000 ton**	20	Q-7	Ea.							
	Remove chiller				242.000		15,100		15,100	18,700	23,400
	Replace chiller, 1000 ton				485.000	404,500	30,300		434,800	482,500	552,500
	Total				727.000	404,500	45,400		449,900	**501,200**	**575,900**
6010	**Repair centrifugal chiller, 9000 ton**	10	Q-7	Ea.							
	Repair controls				1.000		65.50		65.50	81	101
	Remove compressor				1083.600		66,360		66,360	82,000	102,000
	Replace compressor				2112.000	540,000	131,400		671,400	757,000	878,500
	Remove / replace condenser tube				151.204	24,822	9,891		34,713	39,600	46,400
	Remove / replace evaporator tube				113.399	38,430	7,434		45,864	51,500	59,500
	Remove / replace refrigerant filter				14.400	8,235	945		9,180	10,200	11,800
	Remove refrigerant				31.648		2,088		2,088	2,575	3,200
	Replace refrigerant				263.993	47,520	17,325		64,845	73,500	86,000
	Total				3771.244	659,007	235,508.50		894,515.50	**1,016,456**	**1,187,501**
6030	**Replace centrifugal chiller, 9000 ton**	20	Q-7	Ea.							
	Remove chiller				1452.000		90,600		90,600	112,000	140,500
	Replace chiller, 9000 ton				3312.000	3,267,000	206,400		3,473,400	3,849,000	4,402,000
	Total				4764.000	3,267,000	297,000		3,564,000	**3,961,000**	**4,542,500**

D3033 142 Chiller, Open Centrifugal

	System Description	Freq. (Years)	Crew	Unit	Labor Hours	2020 Bare Costs				Total In-House	Total w/O&P
						Material	Labor	Equipment	Total		
1010	**Repair open centrifugal chiller, 300 ton**	10	Q-7	Ea.							
	Repair controls				1.000		65.50		65.50	81	101
	Remove compressor				104.878		6,450		6,450	7,925	9,900
	Replace compressor				205.700	39,930	12,870		52,800	60,000	70,000
	Remove / replace condenser tube				12.000	1,970	735		2,705	3,075	3,600
	Remove / replace evaporator tube				12.000	3,050	735		3,785	4,275	4,950
	Remove / replace refrigerant filter				1.200	915	75	4	994	1,100	1,275
	Remove refrigerant				20.000		1,100		1,100	1,400	1,750
	Replace refrigerant				21.333	3,840	1,400		5,240	5,950	6,950
	Total				378.111	49,705	23,430.50	4	73,139.50	**83,806**	**98,526**
1030	**Replace open centrifugal chiller, 300 ton**	20	Q-7	Ea.							
	Remove chiller				128.000		7,980		7,980	9,875	12,400
	Replace chiller, 300 ton				257.400	167,750	16,060		183,810	204,500	234,500
	Total				385.400	167,750	24,040		191,790	**214,375**	**246,900**
2010	**Repair open centrifugal chiller, 1000 ton**	10	Q-7	Ea.							
	Repair controls				1.000		65.50		65.50	81	101
	Remove compressor				165.854		10,200		10,200	12,500	15,600
	Replace compressor				325.600	86,350	20,350		106,700	120,000	139,500
	Remove / replace condenser tube				16.800	2,758	1,029		3,787	4,300	5,050
	Remove / replace evaporator tube				16.800	4,270	1,029		5,299	5,975	6,925
	Remove / replace refrigerant filter				1.600	1,375	100	5.35	1,480.35	1,650	1,875
	Remove refrigerant				35.204		2,304		2,304	2,850	3,550
	Replace refrigerant				34.666	6,240	2,275		8,515	9,675	11,300
	Total				597.523	100,993	37,352.50	5.35	138,350.85	**157,031**	**183,901**
2030	**Replace open centrifugal chiller, 1000 ton**	20	Q-7	Ea.							
	Remove chiller				219.800		13,720		13,720	16,900	21,200
	Replace chiller, 1000 ton				409.200	444,950	25,520		470,470	521,000	595,500
	Total				629.000	444,950	39,240		484,190	**537,900**	**616,700**

D3033 145 Chiller, Absorption

	System Description	Freq. (Years)	Crew	Unit	Labor Hours	2020 Bare Costs Material	Labor	Equipment	Total	Total In-House	Total w/O&P
1010	**Repair chiller, absorption, 100 ton**	10	2 STPI	Ea.							
	Repair controls				1.000		65.50		65.50	81	101
	Remove / replace concentrator tube				.900	305	53		358	400	465
	Remove / replace condenser tube				6.000	1,525	367.50		1,892.50	2,125	2,475
	Remove / replace evaporator tube				6.000	985	367.50		1,352.50	1,525	1,800
	Remove / replace absorber tube				.900	305	53		358	400	465
	Remove / replace gang pump				1.600	7,975	94.50		8,069.50	8,900	10,100
	Remove gang pump motor				3.704		227		227	279	350
	Replace gang pump motor				7.407	6,150	455		6,605	7,325	8,400
	Add refrigerant				8.000	440	525		965	1,125	1,350
	Total				35.511	17,685	2,208		19,893	**22,160**	**25,506**
1030	**Replace chiller, absorption, 100 ton**	20	Q-7	Ea.							
	Remove chiller				155.000		9,700		9,700	12,000	15,000
	Replace chiller				311.000	132,500	19,400		151,900	169,500	195,500
	Total				466.000	132,500	29,100		161,600	**181,500**	**210,500**
2010	**Repair chiller, absorption, 350 ton**	10	2 STPI	Ea.							
	Repair controls				1.000		65.50		65.50	81	101
	Remove / replace concentrator tube				1.800	610	106		716	800	925
	Remove / replace condenser tube				7.200	1,830	441		2,271	2,550	2,975
	Remove / replace evaporator tube				7.200	1,182	441		1,323	1,850	2,150
	Remove / replace absorber tube				1.800	610	106		716	800	925
	Remove / replace gang pump				1.600	7,975	94.50		8,069.50	8,900	10,100
	Remove gang pump motor				3.704		227		227	279	350
	Replace gang pump motor				7.407	6,150	455		6,605	7,325	8,400
	Add refrigerant				8.000	440	525		965	1,125	1,350
	Total				39.711	18,797	2,461		21,258	**23,710**	**27,276**
2030	**Replace chiller, absorption, 350 ton**	20	Q-7	Ea.							
	Remove chiller				204.000		12,700		12,700	15,700	19,700
	Replace chiller				400.000	375,500	25,000		400,500	444,000	508,000
	Total				604.000	375,500	37,700		413,200	**459,700**	**527,700**

D3033 145 Chiller, Absorption

	System Description	Freq. (Years)	Crew	Unit	Labor Hours	2020 Bare Costs				Total In-House	Total w/O&P
						Material	Labor	Equipment	Total		
3010	**Repair chiller, absorption, 950 ton**	10	4 STPI	Ea.							
	Repair controls				1.000		65.50		65.50	81	101
	Remove / replace concentrator tube				3.600	1,220	212		1,432	1,600	1,850
	Remove / replace condenser tube				12.000	3,050	735		3,785	4,275	4,950
	Remove / replace evaporator tube				12.000	1,970	735		2,705	3,075	3,600
	Remove / replace absorber tube				3.600	1,220	212		1,432	1,600	1,850
	Remove / replace gang pump				1.600	7,975	94.50		8,069.50	8,900	10,100
	Remove gang pump motor				3.704		227		227	279	350
	Replace gang pump motor				8.696	7,850	535		8,385	9,300	10,600
	Add refrigerant				8.000	440	525		965	1,125	1,350
	Total				54.198	23,725	3,341		27,066	**30,235**	**34,751**
3030	**Replace chiller, absorption, 950 ton**	20	Q-7	Ea.							
	Remove chiller				267.000		16,600		16,600	20,600	25,700
	Replace chiller				533.000	768,000	33,300		801,300	886,000	1,011,500
	Total				800.000	768,000	49,900		817,900	**906,600**	**1,037,200**

D3033 210 Air Cooled Condenser

	System Description	Freq. (Years)	Crew	Unit	Labor Hours	2020 Bare Costs				Total In-House	Total w/O&P
						Material	Labor	Equipment	Total		
1010	**Repair condenser, air cooled, 5 ton**	10	1 STPI	Ea.							
	Repair controls				.500		33		33	40.50	50.50
	Remove fan motor				1.800		118		118	146	182
	Replace fan motor				2.311	345	142		487	555	650
	Total				4.611	345	293		638	**741.50**	**882.50**
1030	**Replace condenser, air cooled, 5 ton**	15	Q-5	Ea.							
	Remove condenser				5.195		305		305	380	475
	Replace condenser, air cooled, 5 ton				10.390	5,100	615		5,715	6,375	7,325
	Total				15.584	5,100	920		6,020	**6,755**	**7,800**
2010	**Repair condenser, air cooled, 20 ton**	10	2 STPI	Ea.							
	Repair controls				1.000		65.50		65.50	81	101
	Remove fan motor				2.310		151		151	187	234
	Replace fan motor				4.622	740	284		1,024	1,150	1,350
	Total				7.932	740	500.50		1,240.50	**1,418**	**1,685**

D3033 210 Air Cooled Condenser

	System Description	Freq. (Years)	Crew	Unit	Labor Hours	2020 Bare Costs Material	Labor	Equipment	Total	Total In-House	Total w/O&P
2030	**Replace condenser, air cooled, 20 ton**	15	Q-5	Ea.							
	Remove condenser				10.390		615		615	760	950
	Replace condenser, air cooled, 20 ton				20.779	12,200	1,225		13,425	14,900	17,200
	Total				31.169	12,200	1,840		14,040	**15,660**	**18,150**
3010	**Repair condenser, air cooled, 50 ton**	10	2 STPI	Ea.							
	Repair controls				1.000		65.50		65.50	81	101
	Remove fan motor				3.466		226.50		226.50	281	350
	Replace fan motor				6.936	2,415	426		2,841	3,175	3,675
	Total				11.402	2,415	718		3,133	**3,537**	**4,126**
3030	**Replace condenser, air cooled, 50 ton**	15	Q-6	Ea.							
	Remove condenser				38.961		2,375		2,375	2,950	3,675
	Replace condenser, air cooled, 50 ton				77.922	26,800	4,775		31,575	35,400	40,900
	Total				116.883	26,800	7,150		33,950	**38,350**	**44,575**
4010	**Repair condenser, air cooled, 100 ton**	10	2 STPI	Ea.							
	Repair controls				1.000		65.50		65.50	81	101
	Remove fan motor				6.931		453		453	560	700
	Replace fan motor				13.873	4,830	852		5,682	6,350	7,350
	Total				21.804	4,830	1,370.50		6,200.50	**6,991**	**8,151**
4030	**Replace condenser, air cooled, 100 ton**	15	Q-7	Ea.							
	Remove condenser				69.264		4,325		4,325	5,350	6,675
	Replace condenser, air cooled, 100 ton				139.000	54,500	8,650		63,150	70,500	81,500
	Total				208.264	54,500	12,975		67,475	**75,850**	**88,175**

D3033 260 Evaporative Condenser

	System Description	Freq. (Years)	Crew	Unit	Labor Hours	2020 Bare Costs				Total In-House	Total w/O&P
						Material	Labor	Equipment	Total		
1010	**Repair evaporative condenser, 20 ton**	10	2 STPI	Ea.							
	Repair controls				1.000		65.50		65.50	81	101
	Remove fan motors				2.310		151		151	187	234
	Replace fan motors				4.622	740	284		1,024	1,150	1,350
	Remove pump and motor				20.779		1,200		1,200	1,500	1,875
	Replace pump and motor				41.558	4,725	2,400		7,125	8,175	9,625
	Remove / replace pump nozzles				5.000	51.50	295		346.50	420	520
	Remove / replace float valve				.500	16.30	29.50		45.80	54.50	66
	Total				75.770	5,532.80	4,425		9,957.80	**11,567.50**	**13,771**
1030	**Replace evaporative condenser, 20 ton**	15	Q-5	Ea.							
	Remove condenser				22.130		1,300		1,300	1,625	2,025
	Replace evaporative condenser, 20 ton				44.199	13,000	2,600		15,600	17,500	20,300
	Total				66.329	13,000	3,900		16,900	**19,125**	**22,325**
2010	**Repair evaporative condenser, 100 ton**	10	2 STPI	Ea.							
	Repair controls				1.000		65.50		65.50	81	101
	Remove fan motors				6.931		453		453	560	700
	Replace fan motors				13.873	4,830	852		5,682	6,350	7,350
	Remove pump and motor				20.779		1,200		1,200	1,500	1,875
	Replace pump and motor				41.558	4,725	2,400		7,125	8,175	9,625
	Remove / replace nozzles				25.000	257.50	1,475		1,732.50	2,100	2,600
	Remove / replace float valve				.500	16.30	29.50		45.80	54.50	66
	Total				109.642	9,828.80	6,475		16,303.80	**18,820.50**	**22,317**
2030	**Replace evaporative condenser, 100 ton**	15	Q-7	Ea.							
	Remove condenser				57.762		3,600		3,600	4,450	5,575
	Replace evaporative condenser, 100 ton				116.000	19,600	7,200		26,800	30,500	35,600
	Total				173.762	19,600	10,800		30,400	**34,950**	**41,175**

D3033 260 Evaporative Condenser

	System Description	Freq. (Years)	Crew	Unit	Labor Hours	2020 Bare Costs Material	Labor	Equipment	Total	Total In-House	Total w/O&P
3010	**Repair evaporative condenser, 300 ton**	10	2 STPI	Ea.							
	Repair controls				1.000		65.50		65.50	81	101
	Remove fan motors				11.552		755		755	935	1,175
	Replace fan motors				23.121	8,050	1,420		9,470	10,600	12,200
	Remove pump and motor				20.779		1,200		1,200	1,500	1,875
	Replace pump and motor				24.000	5,500	1,450		6,950	7,825	9,100
	Remove / replace nozzles				35.000	360.50	2,065		2,425.50	2,950	3,625
	Remove / replace float valve				1.000	32.60	59		91.60	109	132
	Total				116.453	13,943.10	7,014.50		20,957.60	**24,000**	**28,208**
3030	**Replace evaporative condenser, 300 ton**	15	Q-7	Ea.							
	Remove condenser				152.471		9,517.50		9,517.50	11,800	14,700
	Replace evaporative condenser, 300 ton				304.560	61,398	18,954		80,352	91,000	106,000
	Total				457.031	61,398	28,471.50		89,869.50	**102,800**	**120,700**

D3043 120 Fan Coil

	System Description	Freq. (Years)	Crew	Unit	Labor Hours	2020 Bare Costs				Total In-House	Total w/O&P
						Material	Labor	Equipment	Total		
1010	**Repair fan coil unit, 1 ton**	10	1 STPI	Ea.							
	Remove fan motor				1.200		78.50		78.50	97.50	122
	Replace fan motor				1.951	247	120		367	420	495
	Total				3.151	247	198.50		445.50	**517.50**	**617**
1030	**Replace fan coil unit, 1 ton**	15	Q-5	Ea.							
	Remove fan coil unit				1.733		102		102	126	158
	Replace fan coil unit, 1 ton				3.467	820	205		1,025	1,150	1,350
	Total				5.200	820	307		1,127	**1,276**	**1,508**
2010	**Repair fan coil unit, 3 ton**	10	1 STPI	Ea.							
	Remove fan motor				1.200		78.50		78.50	97.50	122
	Replace fan motor				1.951	320	120		440	500	585
	Total				3.151	320	198.50		518.50	**597.50**	**707**
2030	**Replace fan coil unit, 3 ton**	15	Q-5	Ea.							
	Remove fan coil unit				1.905		112		112	139	174
	Replace fan coil unit, 3 ton				5.195	1,875	305		2,180	2,450	2,825
	Total				7.100	1,875	417		2,292	**2,589**	**2,999**
3010	**Repair fan coil unit, 5 ton**	10	1 STPI	Ea.							
	Remove fan motor				1.200		78.50		78.50	97.50	122
	Replace fan motor				2.311	345	142		487	555	650
	Total				3.511	345	220.50		565.50	**652.50**	**772**
3030	**Replace fan coil unit, 5 ton**	15	Q-5	Ea.							
	Remove fan coil unit				2.363		139		139	172	216
	Replace fan coil unit, 5 ton				5.714	1,925	335		2,260	2,525	2,925
	Total				8.078	1,925	474		2,399	**2,697**	**3,141**
4010	**Repair fan coil unit, 10 ton**	10	1 STPI	Ea.							
	Remove fan motor				1.400		92		92	113	142
	Replace fan motor				2.312	805	142		947	1,050	1,225
	Total				3.712	805	234		1,039	**1,163**	**1,367**

D3043 120 Fan Coil

	System Description	Freq. (Years)	Crew	Unit	Labor Hours	2020 Bare Costs Material	Labor	Equipment	Total	Total In-House	Total w/O&P
4030	**Replace fan coil unit, 10 ton**	15	Q-6	Ea.							
	Remove fan coil unit				2.192		129		129	160	200
	Replace fan coil unit, 10 ton				23.077	2,750	1,400		4,150	4,775	5,625
	Total				25.269	2,750	1,529		4,279	**4,935**	**5,825**
5010	**Repair fan coil unit, 20 ton**	10	1 STPI	Ea.							
	Remove fan motor				1.600		105		105	130	162
	Replace fan motor				2.476	985	152		1,137	1,275	1,475
	Total				4.076	985	257		1,242	**1,405**	**1,637**
5030	**Replace fan coil unit, 20 ton**	15	Q-6	Ea.							
	Remove fan coil unit				2.614		154		154	191	239
	Replace fan coil unit, 20 ton				39.024	4,175	2,400		6,575	7,550	8,925
	Total				41.639	4,175	2,554		6,729	**7,741**	**9,164**
6010	**Repair fan coil unit, 30 ton**	10	1 STPI	Ea.							
	Remove fan motor				1.800		118		118	146	182
	Replace fan motor				2.597	1,175	159		1,334	1,500	1,725
	Total				4.398	1,175	277		1,452	**1,646**	**1,907**
6030	**Replace fan coil unit, 30 ton**	15	Q-6	Ea.							
	Remove fan coil unit				2.737		161		161	200	250
	Replace fan coil unit, 30 ton				51.948	7,550	3,175		10,725	12,200	14,400
	Total				54.685	7,550	3,336		10,886	**12,400**	**14,650**

D3043 122 Fan Coil, DX Air Conditioner, Cooling Only

	System Description	Freq. (Years)	Crew	Unit	Labor Hours	2020 Bare Costs				Total In-House	Total w/O&P
						Material	Labor	Equipment	Total		
1010	**Repair fan coil, DX 1-1/2 ton, cooling only**	10	1 STPI	Ea.							
	Repair controls				.300		19.65		19.65	24.50	30.50
	Remove / replace supply fan				.650		42.50		42.50	52.50	66
	Remove supply fan bearing				1.000		65.50		65.50	81	101
	Replace bearings				2.000	70	131		201	239	291
	Remove supply fan motor				1.156		71		71	87	109
	Replace supply fan motor				1.951	280	128		408	465	550
	Remove compressor				2.477		162		162	201	251
	Replace compressor				3.714	276	243		519	605	720
	Replace refrigerant				.267	48	17.50		65.50	74.50	87
	Total				13.515	674	880.15		1,554.15	**1,829.50**	**2,205.50**
1040	**Replace fan coil, DX 1-1/2 ton, no heat**	15	Q-5	Ea.							
	Remove fan coil unit				2.312		136		136	169	211
	Replace fan coil unit, 1-1/2 ton, no heat				4.160	685	245		930	1,050	1,225
	Total				6.472	685	381		1,066	**1,219**	**1,436**
2010	**Repair fan coil, DX 2 ton, cooling only**	10	1 STPI	Ea.							
	Repair controls				.300		19.65		19.65	24.50	30.50
	Remove / replace supply fan				.650		42.50		42.50	52.50	66
	Remove supply fan bearing				1.000		65.50		65.50	81	101
	Replace bearings				2.000	70	131		201	239	291
	Remove supply fan motor				1.156		71		71	87	109
	Replace supply fan motor				1.951	247	120		367	420	495
	Remove compressor				2.477		162		162	201	251
	Replace compressor				4.000	284	262		546	635	760
	Replace refrigerant				.400	72	26.25		98.25	112	131
	Total				13.934	673	899.90		1,572.90	**1,852**	**2,234.50**
2040	**Replace fan coil, DX 2 ton, no heat**	15	Q-5	Ea.							
	Remove fan coil unit				2.407		142		142	176	220
	Replace fan coil unit, DX 2 ton, no heat				4.334	670	256		926	1,050	1,225
	Total				6.741	670	398		1,068	**1,226**	**1,445**

D3043 122 Fan Coil, DX Air Conditioner, Cooling Only

	System Description	Freq. (Years)	Crew	Unit	Labor Hours	2020 Bare Costs				Total In-House	Total w/O&P
						Material	Labor	Equipment	Total		
3010	**Repair fan coil, DX 2-1/2 ton**	10	1 STPI	Ea.							
	Repair controls				.300		19.65		19.65	24.50	30.50
	Remove / replace supply fan				.650		42.50		42.50	52.50	66
	Remove supply fan bearing				1.000		65.50		65.50	81	101
	Replace bearings				2.000	70	131		201	239	291
	Remove supply fan motor				1.156		71		71	87	109
	Replace supply fan motor				1.951	207	120		327	375	445
	Remove compressor				2.477		162		162	201	251
	Replace compressor				4.334	335	284		619	720	860
	Replace refrigerant				.533	96	35		131	149	174
	Total				14.401	708	930.65		1,638.65	**1,929**	**2,327.50**
3040	**Replace fan coil, DX 2-1/2 ton, no heat**	15	Q-5	Ea.							
	Remove fan coil unit				2.626		155		155	192	240
	Replace fan coil, DX 2-1/2 ton, no heat				4.727	785	279		1,064	1,200	1,400
	Total				7.352	785	434		1,219	**1,392**	**1,640**
4010	**Repair fan coil, DX 3 ton, cooling only**	10	1 STPI	Ea.							
	Repair controls				.300		19.65		19.65	24.50	30.50
	Remove / replace supply fan				.650		42.50		42.50	52.50	66
	Remove supply fan bearing				1.000		65.50		65.50	81	101
	Replace bearings				2.000	70	131		201	239	291
	Remove supply fan motor				1.156		71		71	87	109
	Replace supply fan motor				1.951	207	120		327	375	445
	Remove compressor				2.477		162		162	201	251
	Replace compressor				4.522	370	296		666	775	925
	Replace refrigerant				.533	96	35		131	149	174
	Total				14.590	743	942.65		1,685.65	**1,984**	**2,392.50**
4040	**Replace fan coil, DX 3 ton, no heat**	15	Q-5	Ea.							
	Remove fan coil unit				3.042		179		179	222	278
	Replace fan coil, DX 3 ton, no heat				5.474	960	325		1,285	1,450	1,700
	Total				8.516	960	504		1,464	**1,672**	**1,978**

D3043 122 Fan Coil, DX Air Conditioner, Cooling Only

	System Description	Freq. (Years)	Crew	Unit	Labor Hours	2020 Bare Costs				Total In-House	Total w/O&P
						Material	Labor	Equipment	Total		
5050	**Repair fan coil, DX 5 ton, cooling only**	10	1 STPI	Ea.							
	Repair controls				.300		19.65		19.65	24.50	30.50
	Remove / replace supply fan				.650		42.50		42.50	52.50	66
	Remove supply fan bearing				1.000		65.50		65.50	81	101
	Replace bearings				2.000	70	131		201	239	291
	Remove supply fan motor				1.156		71		71	87	109
	Replace supply fan motor				1.951	207	120		327	375	445
	Remove compressor				2.477		162		162	201	251
	Replace compressor				4.954	520	325		845	975	1,150
	Replace refrigerant				.533	96	35		131	149	174
	Total				15.021	893	971.65		1,864.65	**2,184**	**2,617.50**
6060	**Replace fan coil, DX 5 ton, no heat**	15	Q-5	Ea.							
	Remove fan coil unit				3.852		227		227	281	350
	Replace fan coil, DX 5 ton, no heat				6.932	1,250	410		1,660	1,875	2,200
	Total				10.784	1,250	637		1,887	**2,156**	**2,550**
7070	**Repair fan coil, DX 10 ton, cooling only**	10	Q-6	Ea.							
	Repair controls				.300		19.65		19.65	24.50	30.50
	Remove / replace supply fan				.650		42.50		42.50	52.50	66
	Remove supply fan bearing				1.000		65.50		65.50	81	101
	Replace bearings				2.000	70	131		201	239	291
	Remove supply fan motor				1.156		71		71	87	109
	Replace supply fan motor				2.311	370	142		512	580	680
	Remove compressor				21.661		1,325		1,325	1,650	2,050
	Replace compressor				6.932	1,175	410		1,585	1,800	2,100
	Replace refrigerant				.800	144	52.50		196.50	223	261
	Total				36.810	1,759	2,259.15		4,018.15	**4,737**	**5,688.50**
8080	**Replace fan coil, DX 10 ton, no heat**	15	Q-6	Ea.							
	Remove fan coil unit				6.667		410		410	505	630
	Replace fan coil, DX 10 ton, no heat				12.000	2,250	735		2,985	3,375	3,950
	Total				18.667	2,250	1,145		3,395	**3,880**	**4,580**

D3043 122 Fan Coil, DX Air Conditioner, Cooling Only

	System Description	Freq. (Years)	Crew	Unit	Labor Hours	2020 Bare Costs Material	Labor	Equipment	Total	Total In-House	Total w/O&P
9090	**Repair fan coil, DX 20 ton, cooling only**	10	Q-6	Ea.							
	Repair controls				.300		19.65		19.65	24.50	30.50
	Remove / replace supply fan				.650		42.50		42.50	52.50	66
	Remove supply fan bearing				1.000		65.50		65.50	81	101
	Replace bearings				2.000	70	131		201	239	291
	Remove supply fan motor				1.156		71		71	87	109
	Replace supply fan motor				2.476	985	152		1,137	1,275	1,475
	Remove compressor				24.365		1,500		1,500	1,850	2,300
	Replace compressor				8.000	1,375	470		1,845	2,100	2,450
	Replace refrigerant				1.333	240	87.50		327.50	370	435
	Total				41.281	2,670	2,539.15		5,209.15	**6,079**	**7,257.50**
9590	**Replace fan coil, DX 20 ton, no heat**	15	Q-6	Ea.							
	Remove fan coil unit				24.793		1,525		1,525	1,875	2,350
	Replace fan coil, DX 20 ton, no heat				44.610	4,325	2,725		7,050	8,125	9,625
	Total				69.403	4,325	4,250		8,575	**10,000**	**11,975**
9600	**Repair fan coil, DX 30 ton, cooling only**	10	Q-6	Ea.							
	Repair controls				.300		19.65		19.65	24.50	30.50
	Remove / replace supply fan				.650		42.50		42.50	52.50	66
	Remove supply fan bearing				1.000		65.50		65.50	81	101
	Replace bearings				2.000	70	131		201	239	291
	Remove supply fan motor				1.156		71		71	87	109
	Replace supply fan motor				2.476	985	152		1,137	1,275	1,475
	Remove compressor				24.490		1,500		1,500	1,850	2,325
	Replace compressor				16.000	2,750	940		3,690	4,200	4,900
	Replace refrigerant				1.733	312	113.75		425.75	485	565
	Total				49.805	4,117	3,035.40		7,152.40	**8,294**	**9,862.50**
9610	**Replace fan coil, DX 30 ton, no heat**	15	Q-6	Ea.							
	Remove fan coil unit				25.974		1,600		1,600	1,975	2,450
	Replace fan coil, DX 20 ton, no heat				48.000	5,675	2,925		8,600	9,875	11,600
	Total				73.974	5,675	4,525		10,200	**11,850**	**14,050**

D3043 124 Fan Coil, DX Air Conditioner W/ Heat

	System Description	Freq. (Years)	Crew	Unit	Labor Hours	2020 Bare Costs				Total In-House	Total w/O&P
						Material	Labor	Equipment	Total		
1010	**Replace fan coil, DX 1-1/2 ton, with heat**	15	Q-5	Ea.							
	Remove fan coil unit				2.312		136		136	169	211
	Replace fan coil unit, 1-1/2 ton, with heat				6.474	959	382.20		1,341.20	1,525	1,775
	Total				8.786	959	518.20		1,477.20	**1,694**	**1,986**
2010	**Replace fan coil, DX 2 ton, with heat**	15	Q-5	Ea.							
	Remove fan coil unit				2.407		142		142	176	220
	Replace fan coil unit, DX 2 ton, with heat				6.741	938	397.60		1,335.60	1,525	1,800
	Total				9.148	938	539.60		1,477.60	**1,701**	**2,020**
3010	**Replace fan coil, DX 2-1/2 ton, with heat**	15	Q-5	Ea.							
	Remove fan coil unit				2.626		155		155	192	240
	Replace fan coil, DX 2-1/2 ton, with heat				7.354	1,099	434		1,533	1,750	2,050
	Total				9.979	1,099	589		1,688	**1,942**	**2,290**
4010	**Replace fan coil, DX 3 ton, with heat**	15	Q-5	Ea.							
	Remove fan coil unit				3.042		179		179	222	278
	Replace fan coil, DX 3 ton, with heat				8.514	1,344	504		1,848	2,100	2,450
	Total				11.556	1,344	683		2,027	**2,322**	**2,728**
5010	**Replace fan coil, DX 5 ton, with heat**	10	Q-5	Ea.							
	Remove fan coil unit				3.852		227		227	281	350
	Replace fan coil, DX 5 ton, with heat				10.785	1,750	637		2,387	2,700	3,175
	Total				14.637	1,750	864		2,614	**2,981**	**3,525**
6010	**Replace fan coil, DX 10 ton, with heat**	10	Q-6	Ea.							
	Remove fan coil unit				6.667		410		410	505	630
	Replace fan coil, DX 10 ton, with heat				18.667	3,150	1,141		4,291	4,875	5,700
	Total				25.333	3,150	1,551		4,701	**5,380**	**6,330**
7010	**Replace fan coil, DX 20 ton, with heat**	10	Q-6	Ea.							
	Remove fan coil unit				24.793		1,525		1,525	1,875	2,350
	Replace fan coil, DX 20 ton, with heat				69.421	6,055	4,235		10,290	11,900	14,100
	Total				94.215	6,055	5,760		11,815	**13,775**	**16,450**

D3043 128 Unit Ventilator

	System Description	Freq. (Years)	Crew	Unit	Labor Hours	2020 Bare Costs				Total In-House	Total w/O&P
						Material	Labor	Equipment	Total		
1010	**Repair unit ventilator, 750 CFM, 2 ton**	10	1 STPI	Ea.							
	Repair controls				.200		13.10		13.10	16.20	20.50
	Remove / replace fan				.650		42.50		42.50	52.50	66
	Remove fan motor				1.156		71		71	87	109
	Replace fan motor				1.951	247	120		367	420	495
	Total				3.957	247	246.60		493.60	**575.70**	**690.50**
1030	**Replace unit vent., 750 CFM, heat/cool coils**	15	Q-6	Ea.							
	Remove unit ventilator				7.792		475		475	590	735
	Replace unit, 750 CFM, with heat / cool coils				15.605	4,750	955		5,705	6,400	7,425
	Total				23.397	4,750	1,430		6,180	**6,990**	**8,160**
2010	**Repair unit ventilator, 1250 CFM, 3 ton**	10	1 STPI	Ea.							
	Repair controls				.200		13.10		13.10	16.20	20.50
	Remove / replace fan				.650		42.50		42.50	52.50	66
	Remove fan motor				1.156		71		71	87	109
	Replace fan motor				1.951	320	120		440	500	585
	Total				3.957	320	246.60		566.60	**655.70**	**780.50**
2030	**Replace unit vent., 1250 CFM, heat/cool coils**	15	Q-6	Ea.							
	Remove unit ventilator				11.142		680		680	845	1,050
	Replace unit , 1250 CFM, with heat / cool coils				22.284	5,725	1,375		7,100	7,975	9,250
	Total				33.426	5,725	2,055		7,780	**8,820**	**10,300**
2040	**Repair unit ventilator, 2000 CFM, 5 ton**	10	1 STPI	Ea.							
	Repair controls				.300		19.65		19.65	24.50	30.50
	Remove / replace fan				.650		42.50		42.50	52.50	66
	Remove fan motor				1.156		71		71	87	109
	Replace fan motor				2.311	345	142		487	555	650
	Total				4.417	345	275.15		620.15	**719**	**855.50**
2050	**Replace unit vent., 2000 CFM, heat/cool coils**	15	Q-6	Ea.							
	Remove unit ventilator				31.209		1,900		1,900	2,350	2,950
	Replace unit, 2000 CFM, with heat / cool coils				62.338	7,200	3,825		11,025	12,600	14,900
	Total				93.547	7,200	5,725		12,925	**14,950**	**17,850**

D3043 140 Duct Heater

	System Description	Freq. (Years)	Crew	Unit	Labor Hours	2020 Bare Costs				Total In-House	Total w/O&P
						Material	Labor	Equipment	Total		
0020	**Maintenance and inspection**	0.50	1 ELEC	Ea.							
	Inspect and clean duct heater				1.143		70		70	86	108
	Total				1.143		70		70	**86**	**108**
0030	**Replace duct heater**	15	1 ELEC	Ea.							
	Remove duct heater				.667		41		41	50.50	63
	Duct heater 12 KW				2.000	1,925	123		2,048	2,275	2,600
	Total				2.667	1,925	164		2,089	**2,325.50**	**2,663**

D3043 210 Draft Fan

	System Description	Freq. (Years)	Crew	Unit	Labor Hours	2020 Bare Costs				Total In-House	Total w/O&P
						Material	Labor	Equipment	Total		
1010	**Repair fan, induced draft, 2000 CFM**	10	1 STPI	Ea.							
	Remove bearings				1.000		65.50		65.50	81	101
	Replace bearings				2.000	70	131		201	239	291
	Total				3.000	70	196.50		266.50	**320**	**392**
1030	**Replace fan, induced draft, 2000 CFM**	20	Q-9	Ea.							
	Remove fan				3.151		177		177	222	277
	Replace fan, induced draft, 2000 CFM				6.304	2,625	355		2,980	3,325	3,825
	Total				9.456	2,625	532		3,157	**3,547**	**4,102**
2010	**Repair fan, induced draft, 6700 CFM**	10	1 STPI	Ea.							
	Remove bearings				1.000		65.50		65.50	81	101
	Replace bearings				2.000	70	131		201	239	291
	Total				3.000	70	196.50		266.50	**320**	**392**
2030	**Replace fan, induced draft, 6700 CFM**	20	Q-9	Ea.							
	Remove fan				4.522		254		254	320	395
	Replace fan, induced draft, 6700 CFM				9.045	3,200	505		3,705	4,150	4,800
	Total				13.567	3,200	759		3,959	**4,470**	**5,195**
3010	**Repair fan, induced draft, 17,700 CFM**	10	1 STPI	Ea.							
	Remove bearings				1.000		65.50		65.50	81	101
	Replace bearings				2.222	78.50	146		224.50	266	325
	Total				3.222	78.50	211.50		290	**347**	**426**

D3043 210 Draft Fan

	System Description	Freq. (Years)	Crew	Unit	Labor Hours	2020 Bare Costs				Total In-House	Total w/O&P
						Material	Labor	Equipment	Total		
3030	**Replace fan, induced draft, 17,700 CFM**	20	Q-9	Ea.							
	Remove fan				12.998		730		730	915	1,150
	Replace fan, induced draft, 17,700 CFM				26.016	9,625	1,450		11,075	12,400	14,300
	Total				39.014	9,625	2,180		11,805	**13,315**	**15,450**

D3043 220 Exhaust Fan

	System Description	Freq. (Years)	Crew	Unit	Labor Hours	2020 Bare Costs				Total In-House	Total w/O&P
						Material	Labor	Equipment	Total		
1010	**Replace fan & motor, propeller exh., 375 CFM**	15	Q-20	Ea.							
	Remove fan and motor				1.300		74.50		74.50	93	116
	Replace fan & motor, propeller exhaust 375 CFM				2.600	236	149		385	445	525
	Total				3.900	236	223.50		459.50	**538**	**641**
1030	**Replace fan & motor, propeller exh., 1000 CFM**	15	Q-20	Ea.							
	Remove fan and motor				1.625		93		93	116	145
	Replace fan & motor, propeller exhaust 1,323 CFM				3.250	390	186		576	660	780
	Total				4.875	390	279		669	**776**	**925**
1040	**Replace fan & motor, propeller exh., 4700 CFM**	15	Q-20	Ea.							
	Remove fan and motor				2.600		149		149	186	232
	Replace fan & motor, propeller exhaust 4,700 CFM				5.200	1,375	297		1,672	1,875	2,175
	Total				7.800	1,375	446		1,321	**2,061**	**2,407**
2030	**Replace roof mounted exhaust fan, 800 CFM**	20	Q-20	Ea.							
	Remove fan & motor				2.600		149		149	186	232
	Replace fan & motor, roof mounted exhaust, 800 CFM				5.200	1,050	297		1,347	1,525	1,775
	Total				7.800	1,050	446		1,496	**1,711**	**2,007**
2040	**Replace roof mounted exhaust fan, 2000 CFM**	20	Q-20	Ea.							
	Remove fan & motor				3.250		186		186	232	290
	Replace fan & motor, roof mounted exhaust, 2000 CFM				6.500	1,900	370		2,270	2,550	2,950
	Total				9.750	1,900	556		2,456	**2,782**	**3,240**

D3043 220 Exhaust Fan

	System Description	Freq. (Years)	Crew	Unit	Labor Hours	2020 Bare Costs				Total In-House	Total w/O&P
						Material	Labor	Equipment	Total		
2050	**Replace roof mounted exhaust fan, 8500 CFM**	20	Q-20	Ea.							
	Remove fan & motor				4.333		248		248	310	385
	Replace fan & motor, roof mounted exhaust, 8500 CFM				8.666	3,250	495		3,745	4,200	4,850
	Total				12.998	3,250	743		3,993	**4,510**	**5,235**
2060	**Replace roof mounted exhaust fan, 20,300 CFM**	20	Q-20	Ea.							
	Remove fan & motor				13.004		745		745	930	1,150
	Replace fan & motor, roof mounted exhaust, 20,300 CFM				26.008	8,950	1,475		10,425	11,700	13,500
	Total				39.012	8,950	2,220		11,170	**12,630**	**14,650**
3010	**Replace utility set, belt drive, 800 CFM**	10	Q-20	Ea.							
	Remove fan and motor				2.167		124		124	155	193
	Replace utility set, belt drive, 800 CFM				4.334	1,050	248		1,298	1,475	1,700
	Total				6.500	1,050	372		1,422	**1,630**	**1,893**
3020	**Replace utility set, belt drive, 3600 CFM**	10	Q-20	Ea.							
	Remove fan and motor				3.250		186		186	232	290
	Replace utility set, belt drive, 3600 CFM				6.500	2,200	370		2,570	2,875	3,325
	Total				9.750	2,200	556		2,756	**3,107**	**3,615**
3030	**Replace utility set, belt drive, 11,000 CFM**	10	Q-20	Ea.							
	Remove fan and motor				6.500		370		370	465	580
	Replace utility set, belt drive, 11,000 CFM				13.004	5,775	745		6,520	7,275	8,375
	Total				19.504	5,775	1,115		6,890	**7,740**	**8,955**
3040	**Replace utility set, belt drive, 20,000 CFM**	10	Q-20	Ea.							
	Remove fan and motor				16.247		930		930	1,150	1,450
	Replace utility set, belt drive, 20,000 CFM				32.520	7,750	1,850		9,600	10,800	12,600
	Total				48.767	7,750	2,780		10,530	**11,950**	**14,050**
4010	**Replace axial flow fan, 3900 CFM**	10	Q-20	Ea.							
	Remove fan and motor				3.823		218		218	273	340
	Replace axial flow fan, 3900 CFM				7.648	1,475	435		1,910	2,175	2,525
	Total				11.472	1,475	653		2,128	**2,448**	**2,865**

D3043 220 Exhaust Fan

	System Description	Freq. (Years)	Crew	Unit	Labor Hours	2020 Bare Costs				Total In-House	Total w/O&P
						Material	Labor	Equipment	Total		
4020	**Replace axial flow fan, 6400 CFM**	10	Q-20	Ea.							
	Remove fan and motor				4.643		265		265	330	415
	Replace axial flow fan, 6400 CFM				9.285	2,150	530		2,680	3,025	3,525
	Total				13.928	2,150	795		2,945	**3,355**	**3,940**
4030	**Replace axial flow fan, 16,900 CFM**	10	Q-20	Ea.							
	Remove fan and motor				6.502		370		370	465	580
	Replace axial flow fan, 16,900 CFM				13.004	2,875	745		3,620	4,100	4,750
	Total				19.506	2,875	1,115		3,990	**4,565**	**5,330**
4040	**Replace axial flow fan, 29,000 CFM**	10	Q-20	Ea.							
	Remove fan and motor				9.001		515		515	645	805
	Replace axial flow fan, 29,000 CFM				17.986	4,325	1,025		5,350	6,050	7,000
	Total				26.987	4,325	1,540		5,865	**6,695**	**7,805**

D3043 250 Fireplaces, Clay Flue

	System Description	Freq. (Years)	Crew	Unit	Labor Hours	2020 Bare Costs				Total In-House	Total w/O&P
						Material	Labor	Equipment	Total		
0010	**Replace baked clay flue, architectural**	75	2 BRIC	L.F.							
	Set up and secure scaffold				.022		1.15		1.15	1.50	1.86
	Masonry demolition				.083		3.50		3.50	4.53	5.60
	Remove damaged flue				.002		.12		.12	.15	.19
	Install new flue lining				.224	11.70	10.50		22.20	26.50	31.50
	Install new chimney				1.518	49.50	71		120.50	147	177
	Remove scaffold				.022		1.15		1.15	1.50	1.86
	Total				1.872	61.20	87.42		148.62	**181.18**	**218.01**

D3043 252 Fireplaces, Metal Flue

	System Description	Freq. (Years)	Crew	Unit	Labor Hours	2020 Bare Costs				Total In-House	Total w/O&P
						Material	Labor	Equipment	Total		
0020	**Replace metal pipe flue, architectural**	50	2 SSWK	L.F.							
	Remove damaged pipe				.008		.48		.48	.64	.79
	Install new vent chimney, prefabricated metal				.306	9.35	17.15		26.50	32	38.50
	Total				.314	9.35	17.63		26.98	**32.64**	**39.29**

D3043 310 Steam Converter, Commercial

	System Description	Freq. (Years)	Crew	Unit	Labor Hours	2020 Bare Costs Material	Labor	Equipment	Total	Total In-House	Total w/O&P
0010	**Repair steam converter**	5	1 STPI	Ea.							
	Repair heat exchanger				5.944		390		390	480	600
	Total				5.944		390		390	**480**	**600**
0020	**Inspect for leaks**	2	1 STPI	Ea.							
	Check relief valve operation				.038		2.47		2.47	3.06	3.82
	Check for proper water temperature				.055		3.58		3.58	4.43	5.55
	Total				.092		6.05		6.05	**7.49**	**9.37**
0030	**Replace steam converter**	30	Q-5	Ea.							
	Remove exchanger				2.081		123		123	152	190
	Install new exchanger 10 GPM of 40F to 180F				4.160	4,125	245		4,370	4,850	5,525
	Total				6.241	4,125	368		4,493	**5,002**	**5,715**

D3043 320 Flash Tank, 24 Gallon

	System Description	Freq. (Years)	Crew	Unit	Labor Hours	2020 Bare Costs Material	Labor	Equipment	Total	Total In-House	Total w/O&P
0010	**Repair flash tank**	5	1 STPI	Ea.							
	Repair flash tank				5.944		390		390	480	600
	Total				5.944		390		390	**480**	**600**
0030	**Replace flash tank 24 gal**	15	Q-5	Ea.							
	Remove old tank				.743		44		44	54	68
	Install new tank 24 gallon				1.486	760	87.50		847.50	945	1,075
	Total				2.229	760	131.50		891.50	**999**	**1,143**

D3043 330 Steam Regulator Valve

	System Description	Freq. (Years)	Crew	Unit	Labor Hours	2020 Bare Costs Material	Labor	Equipment	Total	Total In-House	Total w/O&P
0010	**Replace steam regulator valve 1-1/2″ diameter**	6	1 STPI	Ea.							
	Remove steam valve				.400		26		26	32.50	40.50
	Install steam regulator valve 1-1/2″ diameter				.800	4,325	52.50		4,377.50	4,825	5,475
	Total				1.200	4,325	78.50		4,403.50	**4,857.50**	**5,515.50**

D3043 330 Steam Regulator Valve

	System Description	Freq. (Years)	Crew	Unit	Labor Hours	2020 Bare Costs				Total In-House	Total w/O&P
						Material	Labor	Equipment	Total		
0110	**Replace steam regulator valve 2″ diameter**	6	1 STPI	Ea.							
	Remove steam valve				.473		31		31	38.50	48
	Install steam regulator valve 2″ diameter				.945	5,275	62		5,337	5,875	6,700
	Total				1.418	5,275	93		5,368	**5,913.50**	**6,748**
0210	**Replace steam regulator valve 2-1/2″ diameter**	6	Q-5	Ea.							
	Remove steam valve				.867		51		51	63	79
	Install steam regulator valve 2-1/2″ diameter				1.733	6,575	102		6,677	7,350	8,375
	Total				2.600	6,575	153		6,728	**7,413**	**8,454**
0310	**Replace steam regulator valve 3″ diameter**	6	Q-5	Ea.							
	Remove steam valve				.946		56		56	69	86.50
	Install steam regulator valve 3″ diameter flanged				1.891	8,275	112		8,387	9,250	10,500
	Total				2.836	8,275	168		8,443	**9,319**	**10,586.50**

D3043 340 Condensate Meter

	System Description	Freq. (Years)	Crew	Unit	Labor Hours	2020 Bare Costs				Total In-House	Total w/O&P
						Material	Labor	Equipment	Total		
0010	**Repair meter**	15	1 STPI	Ea.							
	Replace drum				1.538	850	99		949	1,050	1,225
	Replace bearings				2.597	139	167		306	360	435
	Total				4.136	989	266		1,255	**1,410**	**1,660**
0030	**Replace condensate meter 500 lb./hr.**	30	1 STPI	Ea.							
	Remove meter				.372		24.50		24.50	30	37.50
	Replace meter 500#/hr				.743	4,300	48.50		4,348.50	4,800	5,450
	Total				1.114	4,300	73		4,373	**4,830**	**5,487.50**
0130	**Replace condensate meter 1500 lb./hr.**	30	1 STPI	Ea.							
	Remove meter				.743		48.50		48.50	60	75.50
	Replace meter 1500#/hr				1.486	4,825	97.50		4,922.50	5,425	6,175
	Total				2.229	4,825	146		4,971	**5,485**	**6,250.50**

D3043 350 Steam Traps

	System Description	Freq. (Years)	Crew	Unit	Labor Hours	2020 Bare Costs				Total In-House	Total w/O&P
						Material	Labor	Equipment	Total		
1030	**Replace steam trap, 15 PSIG, 3/4″ threaded**	7	1 STPI	Ea.							
	Remove old CI body float & thermostatic trap				.325		21.50		21.50	26.50	33
	Install new CI body float & thermostatic trap				.650	168	42.50		210.50	237	276
	Total				.975	168	64		232	**263.50**	**309**
1040	**Replace steam trap, 15 PSIG, 1″ threaded**	7	1 STPI	Ea.							
	Remove old CI body float & thermostatic trap				.347		22.50		22.50	28	35
	Install new CI body float & thermostatic trap				.693	193	45.50		238.50	269	310
	Total				1.040	193	68		261	**297**	**345**
1050	**Replace steam trap, 15 PSIG, 1-1/4″ threaded**	7	1 STPI	Ea.							
	Remove old CI body float & thermostatic trap				.400		26		26	32.50	40.50
	Install new CI body float & thermostatic trap				.800	239	52.50		291.50	330	380
	Total				1.200	239	78.50		317.50	**362.50**	**420.50**
1060	**Replace steam trap, 15 PSIG, 1-1/2″ threaded**	7	1 STPI	Ea.							
	Remove old CI body float & thermostatic trap				.578		38		38	47	58.50
	Install new CI body float & thermostatic trap				1.156	360	76		436	490	565
	Total				1.733	360	114		474	**537**	**623.50**
1070	**Replace steam trap, 15 PSIG, 2″ threaded**	7	1 STPI	Ea.							
	Remove old CI body float & thermostatic trap				.867		57		57	70	88
	Install new CI body float & thermostatic trap				1.733	1,000	114		1,114	1,250	1,425
	Total				2.600	1,000	171		1,171	**1,320**	**1,513**

D3043 410 Radiator Valve

	System Description	Freq. (Years)	Crew	Unit	Labor Hours	2020 Bare Costs				Total In-House	Total w/O&P
						Material	Labor	Equipment	Total		
0010	**Replace radiator valve 1/2″ angle union**	50	1 STPI	Ea.							
	Remove radiator valve				.217		14.20		14.20	17.55	22
	Install radiator valve 1/2″ angle union				.433	67	28.50		95.50	109	128
	Total				.650	67	42.70		109.70	**126.55**	**150**

D3043 410 Radiator Valve

	System Description	Freq. (Years)	Crew	Unit	Labor Hours	2020 Bare Costs				Total In-House	Total w/O&P
						Material	Labor	Equipment	Total		
0020	**Replace radiator valve 3/4″ angle union**	50	1 STPI	Ea.							
	Remove radiator valve				.260		17.05		17.05	21	26.50
	Install radiator valve 3/4″ angle union				.520	73.50	34		107.50	123	144
	Total				.780	73.50	51.05		124.55	**144**	**170.50**
0030	**Replace radiator valve 1″ angle union**	50	1 STPI	Ea.							
	Remove radiator valve				.274		17.95		17.95	22	28
	Install radiator valve 1″ angle union				.547	84	36		120	137	161
	Total				.821	84	53.95		137.95	**159**	**189**
0040	**Replace radiator valve 1-1/4″ angle union**	50	1 STPI	Ea.							
	Remove radiator valve				.347		22.50		22.50	28	35
	Install radiator valve 1-1/4″ angle union				.693	109	45.50		154.50	176	207
	Total				1.040	109	68		177	**204**	**242**

D3043 420 Cast Iron Radiator, 10′ Section

	System Description	Freq. (Years)	Crew	Unit	Labor Hours	2020 Bare Costs				Total In-House	Total w/O&P
						Material	Labor	Equipment	Total		
0010	**Replace C.I. radiator 4 tube 25″H 10′ section**	50	Q-5	Section							
	Remove radiator				1.818		107.50		107.50	133	166
	Install C.I. radiator 4 tube 25″ high 10′ section				2.167	475	128		603	680	790
	Total				3.985	475	235.50		710.50	**813**	**956**

D3043 430 Baseboard Radiation, 10′ Section

	System Description	Freq. (Years)	Crew	Unit	Labor Hours	2020 Bare Costs				Total In-House	Total w/O&P
						Material	Labor	Equipment	Total		
0010	**Replace radiator, baseboard 10′**	20	Q-5	Ea.							
	Remove radiator				2.262		133.50		133.50	165	205
	Install 10′ baseboard radiator				4.522	490	265		755	870	1,025
	Total				6.783	490	398.50		888.50	**1,035**	**1,230**

D3043 440 Finned Radiator, Wall, 10′ Section

	System Description	Freq. (Years)	Crew	Unit	Labor Hours	2020 Bare Costs				Total In-House	Total w/O&P
						Material	Labor	Equipment	Total		
0010	**Replace finned radiator**	20	Q-5	Ea.							
	Replace radiator				3.468		205		205	253	315
	Install 10′ finned tube, 2 tier, 1-1/4″ cop tube				6.933	620	410		1,030	1,200	1,400
	Total				10.401	620	615		1,235	**1,453**	**1,715**

D3043 450 Duct Coil, 1-Row, Hot Water

	System Description	Freq. (Years)	Crew	Unit	Labor Hours	2020 Bare Costs				Total In-House	Total w/O&P
						Material	Labor	Equipment	Total		
1020	**Replace coil, hot water boost, 12″ x 24″**	25	Q-5	Ea.							
	Remove coil				.858		50.50		50.50	62.50	78.50
	Replace coil, hot water duct, 12″ x 24″				1.111	800	65.50		865.50	960	1,100
	Total				1.969	800	116		916	**1,022.50**	**1,178.50**
1030	**Replace coil, hot water boost, 24″ x 24″**	25	Q-5	Ea.							
	Remove coil				1.717		101		101	125	157
	Replace coil, hot water duct, 24″ x 24″				3.433	1,175	203		1,378	1,550	1,775
	Total				5.150	1,175	304		1,479	**1,675**	**1,932**
1040	**Replace coil, hot water boost, 24″ x 36″**	25	Q-5	Ea.							
	Remove coil				2.581		152		152	188	235
	Replace coil, hot water duct, 24″ x 36″				5.161	1,225	305		1,530	1,725	2,000
	Total				7.742	1,225	457		1,682	**1,913**	**2,235**
1050	**Replace coil, hot water boost, 36″ x 36″**	25	Q-5	Ea.							
	Remove coil				3.867		228		228	282	355
	Replace coil, hot water duct, 36″ x 36″				7.729	1,325	455		1,780	2,025	2,350
	Total				11.596	1,325	683		2,008	**2,307**	**2,705**

D3043 510 — Pipe & Fittings, Steel/Iron, Flanged

	System Description	Freq. (Years)	Crew	Unit	Labor Hours	2020 Bare Costs				Total In-House	Total w/O&P
						Material	Labor	Equipment	Total		
0010	**Install new gasket, 4″ pipe size**	25	1 PLUM	Ea.							
	Disconnect joint / remove flange gasket				1.301		84		84	104	130
	Install new gasket				1.300	17.90	84		101.90	123	152
	Total				2.601	17.90	168		185.90	**227**	**282**
0030	**Replace 2″ flanged steel pipe and fittings**	75	2 PLUM	L.F.							
	Remove old pipe				.231		13.40		13.40	16.55	20.50
	Install flanged 2″ steel pipe with hangers				.356	15.25	20.50	[illegible]	38.11	45	53.50
	Total				.587	15.25	33.90	[illegible]	51.51	**61.55**	**74**
0040	**Replace 4″ flanged steel pipe and fittings**	75	2 PLUM	L.F.							
	Remove old pipe				.400		23		23	28.50	36
	Install flanged 4″ steel pipe with hangers				.800	33	46.50	[illegible]	84.80	99.50	119
	Total				1.200	33	69.50	[illegible]	107.80	**128**	**155**
0050	**Replace 6″ flanged steel pipe and fittings**	75	2 PLUM	L.F.							
	Remove old pipe				.624		36		36	45	56
	Install flanged 6″ steel pipe with hangers				1.248	62.50	75	[illegible]	143.05	168	200
	Total				1.872	62.50	111	[illegible]	179.05	**213**	**256**
0060	**Replace 8″ flanged steel pipe and fittings**	75	2 PLUM	L.F.							
	Remove old pipe				.821		49.50		49.50	61	76.50
	Install flanged 8″ steel pipe with hangers				1.642	101	99	[illegible]	207.25	241	287
	Total				2.463	101	148.50	[illegible]	256.75	**302**	**363.50**

D3043 520 — Valves

	System Description	Freq. (Years)	Crew	Unit	Labor Hours	2020 Bare Costs				Total In-House	Total w/O&P
						Material	Labor	Equipment	Total		
1010	**Repack gate valve gland, 3/8″ - 1-1/2″**	10	1 STPI	Ea.							
	Remove / replace gate valve packing				.219	11.65	12.90		24.55	29	34.50
	Total				.219	11.65	12.90		24.55	**29**	**34.50**

D3043 520 Valves

	System Description	Freq. (Years)	Crew	Unit	Labor Hours	2020 Bare Costs				Total In-House	Total w/O&P
						Material	Labor	Equipment	Total		
1020	**Replace gate valve, partial, 3/8″ - 1-1/2″**	20	1 STPI	Ea.							
	Turn valve off and on				.017		1.11		1.11	1.38	1.72
	Remove old				.530		34.50		34.50	43	53.50
	Install new, 3/8″ - 1-1/2″ size				.800	269	52.50		321.50	360	415
	Total				1.347	269	88.11		357.11	**404.38**	**470.22**
2010	**Repack gate valve gland, 2″ - 3″**	10	1 STPI	Ea.							
	Remove / replace gate valve packing				.291	15.49	17.16		32.65	38.50	46
	Total				.291	15.49	17.16		32.65	**38.50**	**46**
2020	**Replace gate valve, partial, 2″ - 3″**	20	Q-1	Ea.							
	Turn valve off and on				.017		1.11		1.11	1.38	1.72
	Remove valve parts				.440		29		29	35.50	44.50
	Replace valve parts, 2″ - 3″				1.600	865	93		958	1,075	1,225
	Total				2.057	865	123.11		988.11	**1,111.88**	**1,271.22**
2060	**Repack gate valve gland, 8″ - 12″**	5	1 PLUM	Ea.							
	Replace valve packing, 8″ to 12″ diameter				1.600	46.50	103		149.50	179	217
	Total				1.600	46.50	103		149.50	**179**	**217**
3010	**Repack drain valve gland, 3/4″**	3	1 STPI	Ea.							
	Remove / replace drain valve packing				.219	11.65	12.90		24.55	29	34.50
	Total				.219	11.65	12.90		24.55	**29**	**34.50**
3020	**Replace drain valve stem assembly, 3/4″**	8	1 STPI	Ea.							
	Turn valve off and on				.017		1.11		1.11	1.38	1.72
	Remove valve stem assembly				.267		17.50		17.50	21.50	27
	Replace valve stem assembly				.267	16.50	17.50		34	40	47.50
	Total				.550	16.50	36.11		52.61	**62.88**	**76.22**
3030	**Replace drain valve, 3/4″**	20	1 STPI	Ea.							
	Turn valve off and on				.017		1.11		1.11	1.38	1.72
	Remove old valve				.308		20		20	25	31
	Install new valve, 3/4″				.308	16.50	20		36.50	43	51.50
	Total				.632	16.50	41.11		57.61	**69.38**	**84.22**

D3043 530 Circulator Pump

	System Description	Freq. (Years)	Crew	Unit	Labor Hours	2020 Bare Costs Material	2020 Bare Costs Labor	2020 Bare Costs Equipment	2020 Bare Costs Total	Total In-House	Total w/O&P
1010	**Repair circulator pump, 1/12 - 3/4 H.P.**	5	1 STPI	Ea.							
	Loosen coupling from pump shaft				.009		.59		.59	.73	.91
	Remove 4 stud bolts in motor				.035		2.29		2.29	2.84	3.55
	Remove motor				.052		3.41		3.41	4.21	5.25
	Remove impeller nut from shaft				.033		2.16		2.16	2.67	3.34
	Remove seal assembly				.019		1.25		1.25	1.54	1.93
	Inspect bearing brackets face				.005		.33		.33	.41	.51
	Clean bearing				.016		1.05		1.05	1.30	1.62
	Inspect shaft				.040	43.50	2.36		45.86	51	58
	Clean shaft				.017		1.11		1.11	1.38	1.72
	Install new seal and rubber				.013		.85		.85	1.05	1.32
	Install seal spring and washers				.006		.39		.39	.49	.61
	Install impeller nut on shaft				.040	10.85	2.36		13.21	14.85	17.20
	Clean pump body gasket area				.037		2.43		2.43	3	3.75
	Install new gasket (ready made)				.010		.66		.66	.81	1.01
	Install impeller				.009		.59		.59	.73	.91
	Install 4 stud bolts				.035		2.29		2.29	2.84	3.55
	Assemble motor to bearing				.024		1.57		1.57	1.95	2.43
	Install 4 stud bolts				.027		1.77		1.77	2.19	2.74
	Tighten coupling				.009		.59		.59	.73	.91
	Final tightening of bolts				.027		1.77		1.77	2.19	2.74
	Test run				.042		2.75		2.75	3.40	4.26
	Total				.505	54.35	32.57		86.92	**100.31**	**118.26**
1030	**Replace circulator pump, 1/12 - 3/4 H.P.**	15	Q-1	Ea.							
	Remove pump and motor				2.597		151		151	186	233
	Install new pump and motor, 1/12-3/4 H.P.				5.195	3,000	300		3,300	3,675	4,225
	Total				7.792	3,000	451		3,451	**3,861**	**4,458**

D3043 530 Circulator Pump

	System Description	Freq. (Years)	Crew	Unit	Labor Hours	2020 Bare Costs Material	Labor	Equipment	Total	Total In-House	Total w/O&P
2010	**Repair circulator pump, 1 H.P.**	5	1 STPI	Ea.							
	Loosen coupling from pump shaft				.009		.59		.59	.73	.91
	Remove 4 stud bolts in motor				.035		2.29		2.29	2.84	3.55
	Remove motor				.052		3.41		3.41	4.21	5.25
	Remove impeller nut from shaft				.033		2.16		2.16	2.67	3.34
	Remove seal assembly				.019		1.25		1.25	1.54	1.93
	Inspect bearing brackets face				.005		.33		.33	.41	.51
	Clean bearing				.016		1.05		1.05	1.30	1.62
	Inspect shaft				.040	43.50	2.36		45.86	51	58
	Clean shaft				.017		1.11		1.11	1.38	1.72
	Install new seal and rubber				.013		.85		.85	1.05	1.32
	Install seal spring and washers				.006		.39		.39	.49	.61
	Install impeller nut on shaft				.040	10.85	2.36		13.21	14.85	17.20
	Clean pump body gasket area				.037		2.43		2.43	3	3.75
	Install new gasket (ready made)				.013		.85		.85	1.05	1.32
	Install impeller				.009		.59		.59	.73	.91
	Install 4 stud bolts				.035		2.29		2.29	2.84	3.55
	Assemble motor to bearing				.024		1.57		1.57	1.95	2.43
	Install 4 stud bolts				.027		1.77		1.77	2.19	2.74
	Tighten coupling				.009		.59		.59	.73	.91
	Final tightening of bolts				.027		1.77		1.77	2.19	2.74
	Test run				.042		2.75		2.75	3.40	4.26
	Total				.508	54.35	32.76		87.11	**100.55**	**118.57**
2030	**Replace circulator. pump, 1 H.P.**	15	Q-1	Ea.							
	Remove pump				2.597		151		151	186	233
	Install new pump, 1 H.P.				5.200	4,775	300		5,075	5,625	6,425
	Total				7.797	4,775	451		5,226	**5,811**	**6,658**

D3043 540 Expansion Tank

	System Description	Freq. (Years)	Crew	Unit	Labor Hours	2020 Bare Costs Material	Labor	Equipment	Total	Total In-House	Total w/O&P
0010	**Refill expansion tank**	5	1 STPI	Ea.							
	Refill air chamber				.200		13.10		13.10	16.20	20.50
	Total				.200		13.10		13.10	**16.20**	**20.50**

D3043 540 Expansion Tank

	System Description	Freq. (Years)	Crew	Unit	Labor Hours	2020 Bare Costs				Total In-House	Total w/O&P
						Material	Labor	Equipment	Total		
0020	**Replace expansion tank, 24 gal capacity**	50	Q-5	Ea.							
	Remove expansion tank				.743		44		44	54	68
	Install expansion tank, 24 gallon capacity				1.486	1,550	87.50		1,637.50	1,825	2,075
	Total				2.228	1,550	131.50		1,681.50	**1,879**	**2,143**
0120	**Replace expansion tank, 60 gal capacity**	50	Q-5	Ea.							
	Remove expansion tank				1.300		76.50		76.50	95	119
	Install expansion tank, 60 gallon capacity				2.602	2,175	154		2,329	2,575	2,950
	Total				3.901	2,175	230.50		2,405.50	**2,670**	**3,069**
0220	**Replace expansion tank, 175 gal capacity**	50	Q-5	Ea.							
	Remove expansion tank				2.597		153		153	189	237
	Install expansion tank, 175 gallon capacity				5.195	4,900	305		5,205	5,775	6,600
	Total				7.792	4,900	458		5,358	**5,964**	**6,837**
0320	**Replace expansion tank, 400 gal capacity**	50	Q-5	Ea.							
	Remove expansion tank				3.721		220		220	271	340
	Install expansion tank, 400 gallon capacity				7.442	12,600	440		13,040	14,400	16,400
	Total				11.163	12,600	660		13,260	**14,671**	**16,740**

D3043 550 Pipe Insulation

	System Description	Freq. (Years)	Crew	Unit	Labor Hours	2020 Bare Costs				Total In-House	Total w/O&P
						Material	Labor	Equipment	Total		
1010	**Repair damaged pipe insulation, fiberglass 1/2″**	5	Q-14	Ea.							
	Remove 2′ length old insulation				.086		4.56		4.56	5.80	7.25
	Install insulation, fiberglass, 1″ wall - 1/2″ diam.				.173	1.84	9.14		10.98	13.70	16.80
	Total				.260	1.84	13.70		15.54	**19.50**	**24.05**
1110	**Repair damaged pipe insulation, fiberglass 3/4″**	5	Q-14	Ea.							
	Remove 2′ length old insulation				.090		4.78		4.78	6.10	7.55
	Install insulation, fiberglass, 1″ wall - 3/4″ diam.				.181	1.96	9.54		11.50	14.35	17.55
	Total				.271	1.96	14.32		16.28	**20.45**	**25.10**

D3043 550 Pipe Insulation

	System Description	Freq. (Years)	Crew	Unit	Labor Hours	2020 Bare Costs				Total In-House	Total w/O&P
						Material	Labor	Equipment	Total		
1120	**Repair damaged pipe insulation, fiberglass 1″**	5	Q-14	Ea.							
	Remove 2′ length old insulation				.095		5		5	6.40	7.90
	Install insulation, fiberglass, 1″ wall - 1″ diam.				.189	2.16	10		12.16	15.10	18.50
	Total				.284	2.16	15		17.16	**21.50**	**26.40**
1130	**Repair damaged pipe insulation, fbgs 1-1/4″**	5	Q-14	Ea.							
	Remove 2′ length old insulation				.099		5.22		5.22	6.65	8.25
	Install insulation, fiberglass, 1″ wall - 1-1/4″ diam.				.198	2.32	10.40		12.72	15.85	19.40
	Total				.296	2.32	15.62		17.94	**22.50**	**27.65**
1140	**Repair damaged pipe insulation, fbgs 1-1/2″**	5	Q-14	Ea.							
	Remove 2′ length old insulation				.099		5.22		5.22	6.65	8.25
	Install insulation, fiberglass, 1″ wall - 1-1/2″ diam.				.198	2.50	10.50		13	16.10	19.75
	Total				.297	2.50	15.72		18.22	**22.75**	**28**
1150	**Repair damaged pipe insulation, fiberglass 2″**	5	Q-14	Ea.							
	Remove 2′ length old insulation				.104		5.48		5.48	7	8.70
	Install insulation, fiberglass, 1″ wall - 2″ diam.				.208	3.56	11		14.56	17.90	22
	Total				.312	3.56	16.48		20.04	**24.90**	**30.70**
1160	**Repair damaged pipe insulation, fiberglass 3″**	5	Q-14	Ea.							
	Remove 2′ length old insulation				.116		6.12		6.12	7.80	9.70
	Install insulation, fiberglass, 1″ wall - 3″ diam.				.232	3.88	12.20		16.08	19.85	24.50
	Total				.348	3.88	18.32		22.20	**27.65**	**34.20**
1170	**Repair damaged pipe insulation, fiberglass 4″**	5	Q-14	Ea.							
	Remove 2′ length old insulation				.139		7.34		7.34	9.35	11.60
	Install insulation, fiberglass, 1″ wall - 4″ diam.				.278	5.16	14.70		19.86	24.50	30
	Total				.417	5.16	22.04		27.20	**33.85**	**41.60**
1180	**Repair damaged pipe insulation, fiberglass 6″**	5	Q-14	Ea.							
	Remove 2′ length old insulation				.173		9.12		9.12	11.65	14.50
	Install insulation, fiberglass, 1″ wall - 6″ diam.				.347	6.14	18.30		24.44	30	36.50
	Total				.520	6.14	27.42		33.56	**41.65**	**51**

D3043 550 Pipe Insulation

	System Description	Freq. (Years)	Crew	Unit	Labor Hours	2020 Bare Costs				Total In-House	Total w/O&P
						Material	Labor	Equipment	Total		
1220	**Replace pipe insulation, fiberglass 1/2″**	5	Q-14	M.L.F.							
	Remove 1000′ length old insulation				33.330		1,760		1,760	2,250	2,800
	Install insulation, fiberglass, 1″ wall - 1/2″ diam.				66.670	920	3,520		4,440	5,500	6,750
	Total				100.000	920	5,280		6,200	**7,750**	**9,550**
1230	**Replace pipe insulation, fiberglass 3/4″**	5	Q-14	M.L.F.							
	Remove 1000′ length old insulation				34.780		1,840		1,840	2,350	2,900
	Install insulation, fiberglass, 1″ wall - 3/4″ diam.				69.570	980	3,670		4,650	5,750	7,025
	Total				104.350	980	5,510		6,490	**8,100**	**9,925**
1240	**Replace pipe insulation, fiberglass 1″**	5	Q-14	M.L.F.							
	Remove 1000′ length old insulation				36.360		1,920		1,920	2,450	3,050
	Install insulation, fiberglass, 1″ wall - 1″ diam.				72.730	1,080	3,840		4,920	6,075	7,450
	Total				109.090	1,080	5,760		6,840	**8,525**	**10,500**
1250	**Replace pipe insulation, fiberglass 1-1/4″**	5	Q-14	M.L.F.							
	Remove 1000′ length old insulation				38.100		2,010		2,010	2,550	3,200
	Install insulation, fiberglass, 1″ wall - 1-1/4″ diam.				76.190	1,160	4,020		5,180	6,400	7,850
	Total				114.290	1,160	6,030		7,190	**8,950**	**11,050**
1260	**Replace pipe insulation, fiberglass 1-1/2″**	5	Q-14	M.L.F.							
	Remove 1000′ length old insulation				38.100		2,010		2,010	2,550	3,200
	Install insulation, fiberglass, 1″ wall - 1-1/2″ diam.				76.190	1,250	4,020		5,270	6,500	7,975
	Total				114.290	1,250	6,030		7,280	**9,050**	**11,175**
1270	**Replace pipe insulation, fiberglass 2″**	5	Q-14	M.L.F.							
	Remove 1000′ length old insulation				40.000		2,110		2,110	2,700	3,350
	Install insulation, fiberglass, 1″ wall - 2″ diam.				80.000	1,780	4,220		6,000	7,350	8,925
	Total				120.000	1,780	6,330		8,110	**10,050**	**12,275**
1280	**Replace pipe insulation, fiberglass 3″**	5	Q-14	M.L.F.							
	Remove 1000′ length old insulation				44.440		2,350		2,350	3,000	3,725
	Install insulation, fiberglass, 1″ wall - 3″ diam.				88.890	1,940	4,690		6,630	8,100	9,875
	Total				133.330	1,940	7,040		8,980	**11,100**	**13,600**

D3043 550 Pipe Insulation

	System Description	Freq. (Years)	Crew	Unit	Labor Hours	2020 Bare Costs				Total In-House	Total w/O&P
						Material	Labor	Equipment	Total		
1290	**Replace pipe insulation, fiberglass 4″**	5	Q-14	M.L.F.							
	Remove 1000′ length old insulation				53.330		2,810		2,810	3,600	4,450
	Install insulation, fiberglass, 1″ wall - 4″ diam.				106.670	2,580	5,650		8,230	10,000	12,200
	Total				160.000	2,580	8,460		11,040	**13,600**	**16,650**
1300	**Replace pipe insulation, fiberglass 6″**	5	Q-14	M.L.F.							
	Remove 1000′ length old insulation				66.670		3,520		3,520	4,500	5,600
	Install insulation, fiberglass, 1″ wall - 6″ diam.				133.330	3,070	7,050		10,120	12,400	15,000
	Total				200.000	3,070	10,570		13,640	**16,900**	**20,600**
1410	**Repair damaged pipe insulation rubber 1/2″**	5	1 ASBE	Ea.							
	Remove 2′ length old insulation				.117		6.84		6.84	8.75	10.90
	Install insulation, foam rubber, 3/4″ wall - 1/2″ diam.				.234	2.26	13.70		15.96	19.95	24.50
	Total				.350	2.26	20.54		22.80	**28.70**	**35.40**
1420	**Repair damaged pipe insulation rubber 3/4″**	5	1 ASBE	Ea.							
	Remove 2′ length old insulation				.117		6.84		6.84	8.75	10.90
	Install insulation, foam rubber, 3/4″ wall - 3/4″ diam.				.234	4.26	13.70		17.96	22	27
	Total				.350	4.26	20.54		24.80	**30.75**	**37.90**
1430	**Repair damaged pipe insulation rubber 1″**	5	1 ASBE	Ea.							
	Remove 2′ length old insulation				.118		6.94		6.94	8.85	11
	Install insulation, foam rubber, 3/4″ wall - 1″ diam.				.236	4.20	13.90		18.10	22.50	27.50
	Total				.355	4.20	20.84		25.04	**31.35**	**38.50**
1440	**Repair damaged pipe insulation rubber 1-1/4″**	5	1 ASBE	Ea.							
	Remove 2′ length old insulation				.120		7.02		7.02	8.95	11.10
	Install insulation, foam rubber, 3/4″ wall - 1-1/4″ diam.				.239	4.92	14		18.92	23.50	28.50
	Total				.359	4.92	21.02		25.94	**32.45**	**39.60**
1450	**Repair damaged pipe insulation rubber 1-1/2″**	5	1 ASBE	Ea.							
	Remove 2′ length old insulation				.120		7.02		7.02	8.95	11.10
	Install insulation, foam rubber, 3/4″ wall - 1-1/2″ diam.				.239	6.14	14		20.14	24.50	30
	Total				.359	6.14	21.02		27.16	**33.45**	**41.10**

D3043 550 Pipe Insulation

	System Description	Freq. (Years)	Crew	Unit	Labor Hours	2020 Bare Costs				Total In-House	Total w/O&P
						Material	Labor	Equipment	Total		
1460	**Repair damaged pipe insulation rubber 2″**	5	1 ASBE	Ea.							
	Remove 2′ length old insulation				.121		7.10		7.10	9.05	11.30
	Install insulation, foam rubber, 3/4″ wall - 2″ diam.				.242	8.54	14.20		22.74	27.50	33
	Total				.364	8.54	21.30		29.84	**36.55**	**44.30**
1470	**Repair damaged pipe insulation rubber 3″**	5	1 ASBE	Ea.							
	Remove 2′ length old insulation				.122		7.18		7.18	9.15	11.40
	Install insulation, foam rubber, 3/4″ wall - 3″ diam.				.245	11	14.30		25.30	30.50	36.50
	Total				.367	11	21.48		32.48	**39.65**	**47.90**
1480	**Repair damaged pipe insulation rubber 4″**	5	1 ASBE	Ea.							
	Remove 2′ length old insulation				.130		7.62		7.62	9.70	12.10
	Install insulation, foam rubber, 3/4″ wall - 4″ diam.				.260	13.60	15.30		28.90	34.50	41
	Total				.390	13.60	22.92		36.52	**44.20**	**53.10**
1490	**Repair damaged pipe insulation rubber 6″**	5	1 ASBE	Ea.							
	Remove 2′ length old insulation				.130		7.62		7.62	9.70	12.10
	Install insulation, foam rubber, 3/4″ wall - 6″ diam.				.260	19.80	15.30		35.10	41	49
	Total				.390	19.80	22.92		42.72	**50.70**	**61.10**
1510	**Replace pipe insulation foam rubber 1/2″**	5	1 ASBE	L.F.							
	Remove 1000′ length old insulation				.090		5.28		5.28	6.70	8.35
	Install insulation, foam rubber, 3/4″ wall - 1/2″ diam.				.180	2.26	10.50		12.76	15.90	19.55
	Total				.270	2.26	15.78		18.04	**22.60**	**27.90**
1520	**Replace pipe insulation foam rubber 3/4″**	5	1 ASBE	L.F.							
	Remove 1000′ length old insulation				.090		5.28		5.28	6.70	8.35
	Install insulation, foam rubber, 3/4″ wall - 3/4″ diam.				.180	4.26	10.50		14.76	18.10	22
	Total				.270	4.26	15.78		20.04	**24.80**	**30.35**
1530	**Replace pipe insulation foam rubber 1″**	5	1 ASBE	L.F.							
	Remove 1000′ length old insulation				.091		5.34		5.34	6.80	8.45
	Install insulation, foam rubber, 3/4″ wall - 1″ diam.				.182	4.20	10.70		14.90	18.20	22
	Total				.273	4.20	16.04		20.24	**25**	**30.45**

D3043 550 Pipe Insulation

	System Description	Freq. (Years)	Crew	Unit	Labor Hours	2020 Bare Costs				Total In-House	Total w/O&P
						Material	Labor	Equipment	Total		
1540	**Replace pipe insulation foam rubber 1-1/4″**	5	1 ASBE	L.F.							
	Remove 1000′ length old insulation				.092		5.40		5.40	6.90	8.55
	Install insulation, foam rubber, 3/4″ wall - 1-1/4″ diam.				.184	4.92	10.80		15.72	19.20	23.50
	Total				.276	4.92	16.20		21.12	**26.10**	**32.05**
1550	**Replace pipe insulation foam rubber 1-1/2″**	5	1 ASBE	L.F.							
	Remove 1000′ length old insulation				.092		5.40		5.40	6.90	8.55
	Install insulation, foam rubber, 3/4″ wall - 1-1/2″ diam.				.184	6.14	10.80		16.94	20.50	25
	Total				.276	6.14	16.20		22.34	**27.40**	**33.55**
1560	**Replace pipe insulation foam rubber 2″**	5	1 ASBE	L.F.							
	Remove 1000′ length old insulation				.093		5.46		5.46	6.95	8.65
	Install insulation, foam rubber, 3/4″ wall - 2″ diam.				.186	8.54	10.90		19.44	23.50	28
	Total				.279	8.54	16.36		24.90	**30.45**	**36.65**
1570	**Replace pipe insulation foam rubber 3″**	5	1 ASBE	L.F.							
	Remove 1000′ length old insulation				.094		5.52		5.52	7.05	8.75
	Install insulation, foam rubber, 3/4″ wall - 3″ diam.				.188	11	11		22	26	31.50
	Total				.282	11	16.52		27.52	**33.05**	**40.25**
1580	**Replace pipe insulation foam rubber 4″**	5	1 ASBE	L.F.							
	Remove 1000′ length old insulation				.100		5.86		5.86	7.50	9.30
	Install insulation, foam rubber, 3/4″ wall - 4″ diam.				.200	13.60	11.70		25.30	30	35.50
	Total				.300	13.60	17.56		31.16	**37.50**	**44.80**
1590	**Replace pipe insulation foam rubber 6″**	5	1 ASBE	L.F.							
	Remove 1000′ length old insulation				.100		5.86		5.86	7.50	9.30
	Install insulation, foam rubber, 3/4″ wall - 6″ diam.				.200	19.80	11.70		31.50	36.50	43.50
	Total				.300	19.80	17.56		37.36	**44**	**52.80**

D3053 110 Unit Heater

	System Description	Freq. (Years)	Crew	Unit	Labor Hours	2020 Bare Costs Material	Labor	Equipment	Total	Total In-House	Total w/O&P
1010	**Repair unit heater, 12 MBH, 2 PSI steam**	10	1 STPI	Ea.							
	Repair controls				.200		13.10		13.10	16.20	20.50
	Remove fan motor				.976		64		64	79	99
	Replace fan motor				1.951	207	120		327	375	445
	Total				3.127	207	197.10		404.10	**470.20**	**564.50**
1030	**Replace unit heater, 12 MBH, 2 PSI steam**	15	Q-5	Ea.							
	Remove unit heater				.867		51		51	63	79
	Replace unit heater, 12 MBH, 2 PSI steam				1.733	415	102		517	585	675
	Total				2.600	415	153		568	**648**	**754**
2010	**Repair unit heater, 36 MBH, 2 PSI steam**	10	1 STPI	Ea.							
	Repair controls				.200		13.10		13.10	16.20	20.50
	Remove fan motor				.976		64		64	79	99
	Replace fan motor				1.951	247	120		367	420	495
	Total				3.127	247	197.10		444.10	**515.20**	**614.50**
2030	**Replace unit heater, 36 MBH, 2 PSI steam**	15	Q-5	Ea.							
	Remove unit heater				1.300		76.50		76.50	95	119
	Replace unit heater, 36 MBH, 2 PSI steam				2.600	625	153		778	875	1,025
	Total				3.900	625	229.50		854.50	**970**	**1,144**
3010	**Repair unit heater, 85 MBH, 2 PSI steam**	10	1 STPI	Ea.							
	Repair controls				.200		13.10		13.10	16.20	20.50
	Remove fan motor				.976		64		64	79	99
	Replace fan motor				1.951	320	120		440	500	585
	Total				3.127	320	197.10		517.10	**595.20**	**704.50**
3030	**Replace unit heater, 85 MBH, 2 PSI steam**	15	Q-5	Ea.							
	Remove unit heater				1.600		94.50		94.50	117	146
	Replace unit heater, 85 MBH, 2 PSI steam				3.200	750	189		939	1,050	1,225
	Total				4.800	750	283.50		1,033.50	**1,167**	**1,371**
4010	**Repair unit heater, 250 MBH, 2 PSI steam**	10	1 STPI	Ea.							
	Repair controls				.200		13.10		13.10	16.20	20.50
	Remove fan motor				1.156		71		71	87	109
	Replace fan motor				2.311	345	142		487	555	650
	Total				3.667	345	226.10		571.10	**658.20**	**779.50**

D3053 110 Unit Heater

	System Description	Freq. (Years)	Crew	Unit	Labor Hours	2020 Bare Costs				Total In-House	Total w/O&P
						Material	Labor	Equipment	Total		
4030	**Replace unit heater, 250 MBH, 2 PSI steam**	15	Q-5	Ea.							
	Remove unit heater				4.160		245		245	305	380
	Replace unit heater, 250 MBH, 2 PSI steam				8.320	1,600	490		2,090	2,375	2,750
	Total				12.481	1,600	735		2,335	**2,680**	**3,130**
5010	**Repair unit heater, 400 MBH, 2 PSI steam**	10	1 STPI	Ea.							
	Repair controls				.200		13.10		13.10	16.20	20.50
	Remove fan motor				1.156		71		71	87	109
	Replace fan motor				2.311	370	142		512	580	680
	Total				3.667	370	226.10		596.10	**683.20**	**809.50**
5020	**Replace unit heater, 400 MBH, 2 PSI steam**	15	Q-5	Ea.							
	Remove unit heater				6.499		385		385	475	595
	Replace unit heater, 400 MBH, 2 PSI steam				13.008	2,375	765		3,140	3,550	4,150
	Total				19.507	2,375	1,150		3,525	**4,025**	**4,745**

D3053 112 Infrared Heater Suspended, Commercial

	System Description	Freq. (Years)	Crew	Unit	Labor Hours	2020 Bare Costs				Total In-House	Total w/O&P
						Material	Labor	Equipment	Total		
0010	**Maintenance and repair**	1	1 ELEC	Ea.							
	Repair wiring connections				.615		38		38	46.50	58
	Total				.615		38		38	**46.50**	**58**
0020	**Maintenance and inspection**	0.50	1 ELEC	Ea.							
	Inspect and clean infrared heater				1.143		70		70	86	108
	Total				1.143		70		70	**86**	**108**
0030	**Replace infrared heater**	15	1 ELEC	Ea.							
	Remove infrared heater				.667		41		41	50.50	63
	Infrared heater 240 V, 1500 W				2.105	405	129		534	605	705
	Total				2.772	405	170		575	**655.50**	**768**

D3053 114 Standard Suspended Heater

	System Description	Freq. (Years)	Crew	Unit	Labor Hours	2020 Bare Costs Material	Labor	Equipment	Total	Total In-House	Total w/O&P
0010	**Maintenance and repair**	2	1 ELEC	Ea.							
	Remove unit heater component				.151		9.25		9.25	11.40	14.25
	Unit heater component				.444	81	27.50		108.50	123	143
	Total				.595	81	36.75		117.75	**134.40**	**157.25**
0020	**Maintenance and inspection**	0.50	1 ELEC	Ea.							
	Inspect and clean unit heater				1.143		70		70	86	108
	Total				1.143		70		70	**86**	**108**
0030	**Replace heater**	15	1 ELEC	Ea.							
	Remove unit heater				.667		41		41	50.50	63
	Unit heater, heavy duty, 480 V, 3 KW				1.333	470	82		552	620	715
	Total				2.000	470	123		593	**670.50**	**778**

D3053 116 Explosionproof Industrial Heater

	System Description	Freq. (Years)	Crew	Unit	Labor Hours	2020 Bare Costs Material	Labor	Equipment	Total	Total In-House	Total w/O&P
0010	**Maintenance and repair**	2	1 ELEC	Ea.							
	Remove explosionproof heater part				.200		12.25		12.25	15.10	18.90
	Explosionproof heater part				.615	110	38		148	167	196
	Total				.815	110	50.25		160.25	**182.10**	**214.90**
0020	**Maintenance and inspection**	0.50	1 ELEC	Ea.							
	Inspect / clean explosionproof heater				1.143		70		70	86	108
	Total				1.143		70		70	**86**	**108**
0030	**Replace heater**	15	1 ELEC	Ea.							
	Remove explosionproof heater				.800		49		49	60.50	75.50
	Explosionproof heater, 3 KW				3.810	5,025	234		5,259	5,825	6,650
	Total				4.610	5,025	283		5,308	**5,885.50**	**6,725.50**

D3053 150 Wall Mounted/Recessed Heater, With Fan

	System Description	Freq. (Years)	Crew	Unit	Labor Hours	2020 Bare Costs				Total In-House	Total w/O&P
						Material	Labor	Equipment	Total		
0010	**Maintenance and repair**	5	1 ELEC	Ea.							
	Remove heater fan				.250		15.35		15.35	18.85	23.50
	Heater fan				.727	102	44.50		146.50	167	196
	Total				.977	102	59.85		161.85	**185.85**	**219.50**
0020	**Maintenance and inspection**	1	1 ELEC	Ea.							
	Inspect and clean heater with fan				1.143		70		70	86	108
	Total				1.143		70		70	**86**	**108**
0030	**Replace heater**	20	1 ELEC	Ea.							
	Remove heater with fan				.667		41		41	50.50	63
	Heater with fan, commercial 2000 W				2.667	211	164		375	435	515
	Total				3.333	211	205		416	**485.50**	**578**

D3053 160 Convector Suspended, Commercial

	System Description	Freq. (Years)	Crew	Unit	Labor Hours	2020 Bare Costs				Total In-House	Total w/O&P
						Material	Labor	Equipment	Total		
0010	**Maintenance and repair**	2	1 ELEC	Ea.							
	Repair wiring connections				.615		38		38	46.50	58
	Total				.615		38		38	**46.50**	**58**
0020	**Maintenance and inspection**	0.50	1 ELEC	Ea.							
	Inspect and clean convector heater				1.143		70		70	86	108
	Total				1.143		70		70	**86**	**108**
0030	**Replace heater**	15	1 ELEC	Ea.							
	Remove cabinet convector heater				.667		41		41	50.50	63
	Cabinet convector heater, 2′ long				2.000	2,300	123		2,423	2,675	3,075
	Total				2.667	2,300	164		2,464	**2,725.50**	**3,138**

D3053 170 Terminal Reheat

	System Description	Freq. (Years)	Crew	Unit	Labor Hours	2020 Bare Costs				Total In-House	Total w/O&P
						Material	Labor	Equipment	Total		
1010	**Repair terminal reheat, 12″ x 24″ coil**	10	1 STPI	Ea.							
	Fix leak				1.200		78.50		78.50	97.50	122
	Total				1.200		78.50		78.50	**97.50**	**122**
1040	**Replace terminal reheat, 12″ x 24″ coil**	15	Q-5	Ea.							
	Remove terminal reheat unit				.858		50.50		50.50	62.50	78.50
	Replace terminal reheat unit, 12″ x 24″ coil				1.716	1,450	101		1,551	1,725	1,975
	Total				2.574	1,450	151.50		1,601.50	**1,787.50**	**2,053.50**
2010	**Repair terminal reheat, 18″ x 24″ coil**	10	1 STPI	Ea.							
	Fix leak				1.400		92		92	113	142
	Total				1.400		92		92	**113**	**142**
2040	**Replace terminal reheat, 18″ x 24″ coil**	15	Q-5	Ea.							
	Remove terminal reheat unit				1.284		76		76	93.50	117
	Replace terminal reheat unit, 18″ x 24″ coil				2.568	1,600	152		1,752	1,950	2,225
	Total				3.852	1,600	228		1,828	**2,043.50**	**2,342**
3010	**Repair terminal reheat, 36″ x 36″ coil**	10	1 STPI	Ea.							
	Fix leak				2.200		144		144	178	223
	Total				2.200		144		144	**178**	**223**
3040	**Replace terminal reheat, 36″ x 36″ coil**	15	Q-5	Ea.							
	Remove terminal reheat unit				3.852		227		227	281	350
	Replace terminal reheat unit, 36″ x 36″ coil				7.703	2,425	455		2,880	3,225	3,725
	Total				11.555	2,425	682		3,107	**3,506**	**4,075**
4010	**Repair terminal reheat, 48″ x 126″ coil**	10	1 STPI	Ea.							
	Fix leak				4.000		262		262	325	405
	Total				4.000		262		262	**325**	**405**
4040	**Replace terminal reheat, 48″ x 126″ coil**	15	Q-5	Ea.							
	Remove terminal reheat unit				17.937		1,050		1,050	1,300	1,625
	Replace terminal reheat unit, 48″ x 126″ coil				35.874	6,750	2,125		8,875	10,000	11,700
	Total				53.812	6,750	3,175		9,925	**11,300**	**13,325**

D3053 245 Heat Pump

	System Description	Freq. (Years)	Crew	Unit	Labor Hours	2020 Bare Costs				Total In-House	Total w/O&P
						Material	Labor	Equipment	Total		
1010	**Repair heat pump, 1.5 ton, air to air split**	10	1 STPI	Ea.							
	Repair controls				.600		39.50		39.50	48.50	61
	Remove / replace supply fan				.650		42.50		42.50	52.50	66
	Remove supply fan motor				.700		46		46	56.50	71
	Replace supply fan motor				1.951	280	128		408	465	550
	Remove compressor				1.860		122		122	151	189
	Replace compressor				3.714	276	243		519	605	720
	Remove / replace condenser fan				1.800		118		118	146	182
	Remove condenser fan motor				1.800		118		118	146	182
	Replace condenser fan motor				1.951	247	120		367	420	495
	Replace refrigerant				.133	24	8.75		32.75	37	43.50
	Remove / replace heater				.500		33		33	40.50	50.50
	Total				15.661	827	1,018.75		1,845.75	**2,168**	**2,610**
1030	**Replace heat pump, 1.5 ton, air to air split**	20	Q-5	Ea.							
	Remove heat pump				4.336		256		256	315	395
	Replace heat pump, 1.5 ton, air to air split				8.649	1,625	510		2,135	2,425	2,825
	Total				12.985	1,625	766		2,391	**2,740**	**3,220**
2010	**Repair heat pump, 5 ton, air to air split**	10	1 STPI	Ea.							
	Repair controls				.600		39.50		39.50	48.50	61
	Remove / replace supply fan				.650		42.50		42.50	52.50	66
	Remove supply fan motor				1.156		71		71	87	109
	Replace supply fan motor				1.951	247	120		367	420	495
	Remove compressor				2.477		162		162	201	251
	Replace compressor				4.954	520	325		845	975	1,150
	Remove / replace condenser fan				1.800		118		118	146	182
	Remove condenser fan motor				.976		64		64	79	99
	Replace condenser fan motor				1.951	207	120		327	375	445
	Replace refrigerant				.667	120	43.75		163.75	186	218
	Remove / replace heater				.500		33		33	40.50	50.50
	Total				17.681	1,094	1,138.75		2,232.75	**2,610.50**	**3,126.50**
2030	**Replace heat pump, 5 ton, air to air split**	20	Q-5	Ea.							
	Remove heat pump				20.806		1,225		1,225	1,525	1,900
	Replace heat pump, 5 ton, air to air split				41.558	2,700	2,450		5,150	6,000	7,175
	Total				62.365	2,700	3,675		6,375	**7,525**	**9,075**

D3053 245 Heat Pump

	System Description	Freq. (Years)	Crew	Unit	Labor Hours	2020 Bare Costs				Total In-House	Total w/O&P
						Material	Labor	Equipment	Total		
3010	**Repair heat pump, 10 ton, air to air split**	10	Q-5	Ea.							
	Repair controls				.600		39.50		39.50	48.50	61
	Remove / replace supply fan				.650		42.50		42.50	52.50	66
	Remove supply fan motor				.700		46		46	56.50	71
	Replace supply fan motor				2.311	370	142		512	580	680
	Remove compressor				4.000		236		236	292	365
	Replace compressor				8.000	1,375	470		1,845	2,100	2,450
	Remove / replace condenser fan				1.800		118		118	146	182
	Remove condenser fan motor				1.800		118		118	146	182
	Replace condenser fan motor				2.311	345	142		487	555	650
	Replace refrigerant				1.067	192	70		262	298	350
	Remove / replace heater				.500		33		33	40.50	50.50
	Total				23.739	2,282	1,457		3,739	**4,315**	**5,107.50**
3030	**Replace heat pump, 10 ton, air to air split**	20	Q-6	Ea.							
	Remove heat pump				24.490		1,500		1,500	1,850	2,325
	Replace heat pump, 10 ton, air to air split				48.980	6,575	3,000		9,575	10,900	12,800
	Total				73.469	6,575	4,500		11,075	**12,750**	**15,125**
4010	**Repair heat pump, 25 ton, air to air split**	10	Q-5	Ea.							
	Repair controls				.600		39.50		39.50	48.50	61
	Remove / replace supply fan				1.000		65.50		65.50	81	101
	Remove supply fan motor				1.156		71		71	87	109
	Replace supply fan motor				2.311	595	142		737	830	960
	Remove compressor				10.000		590		590	730	910
	Replace compressor				20.000	3,437.50	1,175		4,612.50	5,250	6,125
	Remove / replace condenser fan				1.800		118		118	146	182
	Remove condenser fan motor				1.800		118		118	146	182
	Replace condenser fan motor				2.312	805	142		947	1,050	1,225
	Replace refrigerant				2.000	360	131.25		491.25	560	655
	Remove / replace heater				.500		33		33	40.50	50.50
	Total				43.479	5,197.50	2,625.25		7,822.75	**8,969**	**10,560.50**
4030	**Replace heat pump, 25 ton, air to air split**	20	Q-7	Ea.							
	Remove heat pump				61.538		3,850		3,850	4,750	5,950
	Replace heat pump, 25 ton, air to air split				123.000	21,800	7,675		29,475	33,500	39,200
	Total				184.538	21,800	11,525		33,325	**38,250**	**45,150**

D3053 245 Heat Pump

	System Description	Freq. (Years)	Crew	Unit	Labor Hours	2020 Bare Costs				Total In-House	Total w/O&P
						Material	Labor	Equipment	Total		
5010	**Repair heat pump, 50 ton, air to air split**	10	Q-6	Ea.							
	Repair controls				.600		39.50		39.50	48.50	61
	Remove / replace supply fan				1.000		65.50		65.50	81	101
	Remove supply fan motor				1.156		71		71	87	109
	Replace supply fan motor				4.167	2,575	256		2,831	3,150	3,625
	Remove compressor				77.922		4,775		4,775	5,900	7,375
	Replace compressor				156.000	34,300	9,525		43,825	49,500	57,500
	Remove / replace condenser fan				2.400		157		157	195	243
	Remove condenser fan motor				1.800		118		118	146	182
	Replace condenser fan motor				4.000	2,125	245		2,370	2,650	3,025
	Replace refrigerant				4.667	840	306.25		1,146.25	1,300	1,525
	Remove / replace heater				1.200		78.50		78.50	97.50	122
	Total				254.912	39,840	15,636.75		55,476.75	**63,155**	**73,868**
5030	**Replace heat pump, 50 ton, air to air split**	20	Q-7	Ea.							
	Remove heat pump				96.970		6,050		6,050	7,475	9,350
	Replace heat pump, 50 ton, air to air split				208.000	54,000	13,000		67,000	75,500	87,500
	Total				304.970	54,000	19,050		73,050	**82,975**	**96,850**
6010	**Repair heat pump, thru-wall unit, 1.5 ton**	10	1 STPI	Ea.							
	Repair controls				.600		39.50		39.50	48.50	61
	Remove / replace supply fan				.650		42.50		42.50	52.50	66
	Remove supply fan motor				.700		46		46	56.50	71
	Replace supply fan motor				1.951	280	128		408	465	550
	Remove compressor				1.860		122		122	151	189
	Replace compressor				3.714	276	243		519	605	720
	Remove / replace condenser fan				1.800		118		118	146	182
	Remove condenser fan motor				1.800		118		118	146	182
	Replace condenser fan motor				1.951	247	120		367	420	495
	Replace refrigerant				.133	24	8.75		32.75	37	43.50
	Remove / replace heater				.500		33		33	40.50	50.50
	Total				15.661	827	1,018.75		1,845.75	**2,168**	**2,610**
6030	**Replace heat pump, thru-wall unit, 1.5 ton**	20	Q-5	Ea.							
	Remove heat pump				6.709		395		395	490	610
	Replace heat pump, thru-wall, 1.5 ton				13.423	3,375	790		4,165	4,700	5,450
	Total				20.131	3,375	1,185		4,560	**5,190**	**6,060**

D3053 245 Heat Pump

	System Description	Freq. (Years)	Crew	Unit	Labor Hours	2020 Bare Costs				Total In-House	Total w/O&P
						Material	Labor	Equipment	Total		
7010	**Repair heat pump, thru-wall unit, 5 ton**	10	1 STPI	Ea.							
	Repair controls				.600		39.50		39.50	48.50	61
	Remove / replace supply fan				.650		42.50		42.50	52.50	66
	Remove supply fan motor				1.156		71		71	87	109
	Replace supply fan motor				1.951	247	120		367	420	495
	Remove compressor				2.477		162		162	201	251
	Replace compressor				4.954	520	325		845	975	1,150
	Remove / replace condenser fan				1.800		118		118	146	182
	Remove condenser fan motor				.976		64		64	79	99
	Replace condenser fan motor				1.951	207	120		327	375	445
	Replace refrigerant				.667	120	43.75		163.75	186	218
	Remove / replace heater				.500		33		33	40.50	50.50
	Total				17.681	1,094	1,138.75		2,232.75	**2,610.50**	**3,126.50**
7030	**Replace heat pump, thru-wall unit, 5 ton**	20	Q-5	Ea.							
	Remove heat pump				16.000		945		945	1,175	1,450
	Replace heat pump, thru-wall, 5 ton				32.000	5,125	1,900		7,025	7,975	9,325
	Total				48.000	5,125	2,845		7,970	**9,150**	**10,775**

D3053 265 Air Conditioner, Window, 1 Ton

	System Description	Freq. (Years)	Crew	Unit	Labor Hours	2020 Bare Costs				Total In-House	Total w/O&P
						Material	Labor	Equipment	Total		
1010	**Repair air conditioner, window, 1 ton**	8	1 STPI	Ea.							
	Find / fix leak				1.000		65.50		65.50	81	101
	Replace refrigerant				.067	12	4.38		16.38	18.60	22
	Total				1.067	12	69.88		81.88	**99.60**	**123**
1030	**Replace air conditioner, window, 1 ton**	10	L-2	Ea.							
	Remove air conditioner				1.300		60.50		60.50	79	97.50
	Replace air conditioner, 1 ton				2.000	2,000	93		2,093	2,325	2,650
	Total				3.300	2,000	153.50		2,153.50	**2,404**	**2,747.50**

D3053 266 Air Conditioner, Thru-The-Wall

	System Description	Freq. (Years)	Crew	Unit	Labor Hours	2020 Bare Costs				Total In-House	Total w/O&P
						Material	Labor	Equipment	Total		
2010	**Repair air conditioner, thru-the-wall, 2 ton**	8	1 STPI	Ea.							
	Find / fix leak				1.000		65.50		65.50	81	101
	Replace refrigerant				.133	24	8.75		32.75	37	43.50
	Total				1.133	24	74.25		98.25	**118**	**144.50**
2030	**Replace air conditioner, thru-the-wall, 2 ton**	10	L-2	Ea.							
	Remove air conditioner				2.600		121		121	158	195
	Replace air conditioner, 2 ton				5.200	1,075	242		1,317	1,500	1,725
	Total				7.800	1,075	363		1,438	**1,658**	**1,920**

D3053 272 Air Conditioner, DX Package

	System Description	Freq. (Years)	Crew	Unit	Labor Hours	2020 Bare Costs				Total In-House	Total w/O&P
						Material	Labor	Equipment	Total		
1010	**Repair air conditioner, DX, 5 ton**	10	1 STPI	Ea.							
	Repair controls				.600		39.50		39.50	48.50	61
	Remove / replace supply fan				.650		42.50		42.50	52.50	66
	Remove supply fan bearing				1.000		65.50		65.50	81	101
	Replace bearings				2.000	70	131		201	239	291
	Remove supply fan motor				1.156		71		71	87	109
	Replace supply fan motor				2.311	345	142		487	555	650
	Remove compressor				2.477		162		162	201	251
	Replace compressor				4.954	520	325		845	975	1,150
	Remove / replace condenser fan				1.800		118		118	146	182
	Remove condenser fan bearing				1.200		78.50		78.50	97.50	122
	Replace bearings				2.000	70	131		201	239	291
	Remove condenser fan motor				.976		64		64	79	99
	Replace condenser fan motor				1.951	320	120		440	500	585
	Replace refrigerant				.667	120	43.75		163.75	186	218
	Remove / replace heating coils				3.900	678.75	240		918.75	1,050	1,225
	Total				27.641	2,123.75	1,773.75		3,897.50	**4,536.50**	**5,401**
1030	**Replace air conditioner, DX, 5 ton**	20	Q-6	Ea.							
	Remove air conditioner				13.001		795		795	985	1,225
	Replace air conditioner, 5 ton				26.002	4,525	1,600		6,125	6,950	8,100
	Total				39.003	4,525	2,395		6,920	**7,935**	**9,325**

D3053 272 Air Conditioner, DX Package

	System Description	Freq. (Years)	Crew	Unit	Labor Hours	2020 Bare Costs				Total In-House	Total w/O&P
						Material	Labor	Equipment	Total		
2010	**Repair air conditioner, DX, 20 ton**	10	Q-6	Ea.							
	Repair controls				.800		52.50		52.50	65	81
	Remove / replace supply fan				.650		42.50		42.50	52.50	66
	Remove supply fan bearing				1.000		65.50		65.50	81	101
	Replace bearings				2.000	70	131		201	239	291
	Remove supply fan motor				1.156		71		71	87	109
	Replace supply fan motor				2.312	805	142		947	1,050	1,225
	Remove compressor				24.390		1,500		1,500	1,850	2,300
	Replace compressor				48.780	28,700	2,975		31,675	35,300	40,500
	Remove / replace condenser fan				1.800		118		118	146	182
	Remove condenser fan bearing				1.200		78.50		78.50	97.50	122
	Replace bearings				2.000	70	131		201	239	291
	Remove condenser fan motor				.976		64		64	79	99
	Replace condenser fan motor				2.311	345	142		487	555	650
	Replace refrigerant				1.333	240	87.50		327.50	370	435
	Remove / replace heating coils				6.936	2,070	431.25		2,501.25	2,800	3,250
	Total				97.645	32,300	6,031.75		38,331.75	**43,011**	**49,702**
2030	**Replace air conditioner, DX, 20 ton**	20	Q-7	Ea.							
	Remove air conditioner				23.105		1,450		1,450	1,775	2,225
	Replace air conditioner, 20 ton				46.243	13,800	2,875		16,675	18,700	21,700
	Total				69.347	13,800	4,325		18,125	**20,475**	**23,925**

D3053 272 Air Conditioner, DX Package

	System Description	Freq. (Years)	Crew	Unit	Labor Hours	2020 Bare Costs				Total In-House	Total w/O&P
						Material	Labor	Equipment	Total		
3010	**Repair air conditioner, DX, 50 ton**	10	Q-6	Ea.							
	Repair controls				.800		52.50		52.50	65	81
	Remove / replace supply fan				.650		42.50		42.50	52.50	66
	Remove supply fan bearing				1.200		78.50		78.50	97.50	122
	Replace bearings				2.222	78.50	146		224.50	266	325
	Remove supply fan motor				1.156		71		71	87	109
	Replace supply fan motor				2.597	1,175	159		1,334	1,500	1,725
	Remove compressor				77.922		4,775		4,775	5,900	7,375
	Replace compressor				156.000	34,300	9,525		43,825	49,500	57,500
	Remove / replace condenser fan				1.800		118		118	146	182
	Remove condenser fan bearing				1.200		78.50		78.50	97.50	122
	Replace bearings				2.222	78.50	146		224.50	266	325
	Remove condenser fan motor				.976		64		64	79	99
	Replace condenser fan motor				2.311	345	142		487	555	650
	Replace refrigerant				3.333	600	218.75		818.75	930	1,100
	Remove / replace heating coil				12.468	7,575	776.25		8,351.25	9,300	10,700
	Total				266.857	44,152	16,393		60,545	**68,841.50**	**80,481**
3030	**Replace air conditioner, DX, 50 ton**	20	Q-7	Ea.							
	Remove air conditioner				41.558		2,600		2,600	3,200	4,000
	Replace air conditioner, 50 ton				83.117	50,500	5,175		55,675	62,000	71,000
	Total				124.675	50,500	7,775		58,275	**65,200**	**75,000**

D3053 274 Computer Room A/C Units, Air Cooled

	System Description	Freq. (Years)	Crew	Unit	Labor Hours	2020 Bare Costs				Total In-House	Total w/O&P
						Material	Labor	Equipment	Total		
1010	**Repair computer room A/C, air cooled, 5 ton**	10	1 STPI	Ea.							
	Repair controls				.600		39.50		39.50	48.50	61
	Remove / replace supply fan				.650		42.50		42.50	52.50	66
	Remove supply fan bearing				1.000		65.50		65.50	81	101
	Replace bearings				2.000	70	131		201	239	291
	Remove supply fan motor				1.156		71		71	87	109
	Replace supply fan motor				2.311	345	142		487	555	650
	Replace fan belt				.400	17.75	26		43.75	52	62.50
	Remove compressor				2.477		162		162	201	251
	Replace compressor				4.954	520	325		845	975	1,150
	Remove / replace condenser fan				1.800		118		118	146	182
	Remove condenser fan bearing				1.200		78.50		78.50	97.50	122
	Replace bearings				2.000	70	131		201	239	291
	Remove condenser fan motor				.976		64		64	79	99
	Replace condenser fan motor				1.951	320	120		440	500	585
	Replace fan belt				.400	17.75	26		43.75	52	62.50
	Replace refrigerant				.667	120	43.75		163.75	186	218
	Remove / replace heating coils				3.900	678.75	240		918.75	1,050	1,225
	Total				28.441	2,159.25	1,825.75		3,985	**4,640.50**	**5,526**
1015	**Replace computer room A/C, air cooled, 5 ton**	20	Q-6	Ea.							
	Remove air conditioner and condenser				13.001		795		795	985	1,225
	Replace air conditioner and remote condenser, 5 ton				35.556	29,500	2,100		31,600	35,000	40,100
	Total				48.557	29,500	2,895		32,395	**35,985**	**41,325**

D3053 274 Computer Room A/C Units, Air Cooled

	System Description	Freq. (Years)	Crew	Unit	Labor Hours	2020 Bare Costs				Total In-House	Total w/O&P
						Material	Labor	Equipment	Total		
1020	**Repair computer room A/C, air cooled, 10 ton**	10	1 STPI	Ea.							
	Repair controls				.400		26		26	32.50	40.50
	Remove / replace supply fan				.650		42.50		42.50	52.50	66
	Remove supply fan bearing				1.200		78.50		78.50	97.50	122
	Replace bearings				2.000	70	131		201	239	291
	Remove supply fan motor				1.156		71		71	87	109
	Replace supply fan motor				2.311	370	142		512	580	680
	Replace fan belt				.400	17.75	26		43.75	52	62.50
	Remove compressor				14.513		887.75		887.75	1,100	1,375
	Replace compressor				28.714	15,879	1,758.75		17,637.75	19,600	22,600
	Remove / replace condenser fan				1.800		118		118	146	182
	Remove condenser fan bearing				1.200		78.50		78.50	97.50	122
	Replace bearings				2.000	70	131		201	239	291
	Remove condenser fan motor				.976		64		64	79	99
	Replace condenser fan motor				1.951	320	120		440	500	585
	Replace fan belt				.400	17.75	26		43.75	52	62.50
	Replace refrigerant				1.333	240	87.50		327.50	370	435
	Remove / replace heating coils				5.200	905	320		1,225	1,400	1,625
	Total				66.204	17,889.50	4,108.50		21,998	**24,724**	**28,747.50**
1025	**Replace computer room A/C, air cooled, 10 ton**	20	Q-6	Ea.							
	Remove air conditioner and condenser				33.898		2,075		2,075	2,575	3,200
	Replace air conditioner and remote condenser, 10 ton				64.000	57,500	3,775		61,275	68,000	77,500
	Total				97.898	57,500	5,850		63,350	**70,575**	**80,700**

D3053 274 Computer Room A/C Units, Air Cooled

	System Description	Freq. (Years)	Crew	Unit	Labor Hours	2020 Bare Costs				Total In-House	Total w/O&P
						Material	Labor	Equipment	Total		
1030	**Repair computer room A/C, air cooled, 15 ton**	10	1 STPI	Ea.							
	Repair controls				.500		33		33	40.50	50.50
	Remove / replace supply fan				.650		42.50		42.50	52.50	66
	Remove supply fan bearing				1.200		78.50		78.50	97.50	122
	Replace bearings				2.000	70	131		201	239	291
	Remove supply fan motor				1.156		71		71	87	109
	Replace supply fan motor				2.312	805	142		947	1,050	1,225
	Replace fan belt				.400	17.75	26		43.75	52	62.50
	Remove compressor				21.661		1,325		1,325	1,650	2,050
	Replace compressor				42.857	23,700	2,625		26,325	29,300	33,700
	Remove / replace condenser fan				1.800		118		118	146	182
	Remove condenser fan bearing				1.200		78.50		78.50	97.50	122
	Replace bearings				2.000	70	131		201	239	291
	Remove condenser fan motor				.976		64		64	79	99
	Replace condenser fan motor				1.951	320	120		440	500	585
	Replace fan belt				.400	17.75	26		43.75	52	62.50
	Replace refrigerant				2.000	360	131.25		491.25	560	655
	Remove / replace heating coils				6.501	1,131.25	400		1,531.25	1,725	2,025
	Total				89.563	26,491.75	5,542.75		32,034.50	**35,967**	**41,697.50**
1035	**Replace computer room A/C, air cooled, 15 ton**	20	Q-6	Ea.							
	Remove air conditioner and condenser				50.314		3,075		3,075	3,800	4,750
	Replace air conditioner and remote condenser, 15 ton				72.727	63,500	4,300		67,800	75,000	86,000
	Total				123.042	63,500	7,375		70,875	**78,800**	**90,750**

D3053 274 Computer Room A/C Units, Air Cooled

	System Description	Freq. (Years)	Crew	Unit	Labor Hours	2020 Bare Costs				Total In-House	Total w/O&P
						Material	Labor	Equipment	Total		
1040	**Repair computer room A/C, air cooled, 20 ton**	10	1 STPI	Ea.							
	Repair controls				.800		52.50		52.50	65	81
	Remove / replace supply fan				.650		42.50		42.50	52.50	66
	Remove supply fan bearing				1.000		65.50		65.50	81	101
	Replace bearings				2.000	70	131		201	239	291
	Remove supply fan motor				1.156		71		71	87	109
	Replace supply fan motor				2.312	805	142		947	1,050	1,225
	Replace fan belt				.400	17.75	26		43.75	52	62.50
	Remove compressor				24.390		1,500		1,500	1,850	2,300
	Replace compressor				48.780	28,700	2,975		31,675	35,300	40,500
	Remove / replace condenser fan				1.800		118		118	146	182
	Remove condenser fan bearing				1.200		78.50		78.50	97.50	122
	Replace bearings				2.000	70	131		201	239	291
	Remove condenser fan motor				.976		64		64	79	99
	Replace condenser fan motor				2.311	345	142		487	555	650
	Replace fan belt				.400	17.75	26		43.75	52	62.50
	Replace refrigerant				2.667	480	175		655	745	870
	Remove / replace heating coils				6.936	2,070	431.25		2,501.25	2,800	3,250
	Total				99.778	32,575.50	6,171.25		38,746.75	**43,490**	**50,262**
1045	**Replace computer room A/C, air cooled, 20 ton**	20	Q-6	Ea.							
	Remove air conditioner and condenser				23.105		1,450		1,450	1,775	2,225
	Replace air conditioner and remote condenser, 20 ton				100.000	77,000	6,125		83,125	92,500	105,500
	Total				123.105	77,000	7,575		84,575	**94,275**	**107,725**

D3053 276 Computer Room A/C Units, Chilled Water

	System Description	Freq. (Years)	Crew	Unit	Labor Hours	2020 Bare Costs				Total In-House	Total w/O&P
						Material	Labor	Equipment	Total		
1010	**Repair computer room A/C, chilled water, 5 ton**	10	1 STPI	Ea.							
	Repair controls				.600		39.50		39.50	48.50	61
	Remove / replace supply fan				.650		42.50		42.50	52.50	66
	Remove supply fan bearing				1.000		65.50		65.50	81	101
	Replace bearings				2.000	70	131		201	239	291
	Remove supply fan motor				1.156		71		71	87	109
	Replace supply fan motor				2.311	345	142		487	555	650
	Replace fan belt				.400	17.75	26		43.75	52	62.50
	Remove / replace heating coils				26.002	4,525	1,600		6,125	6,950	8,100
	Total				34.119	4,957.75	2,117.50		7,075.25	**8,065**	**9,440.50**
1015	**Replace computer room A/C, chilled water, 5 ton**	20	Q-6	Ea.							
	Remove air conditioner				9.456		560		560	690	865
	Replace air conditioner, 5 ton				28.070	20,000	1,650		21,650	24,000	27,600
	Total				37.526	20,000	2,210		22,210	**24,690**	**28,465**
1020	**Repair computer room A/C, chilled water, 10 ton**	10	1 STPI	Ea.							
	Repair controls				.400		26		26	32.50	40.50
	Remove / replace supply fan				.650		42.50		42.50	52.50	66
	Remove supply fan bearing				1.200		78.50		78.50	97.50	122
	Replace bearings				2.000	70	131		201	239	291
	Remove supply fan motor				1.156		71		71	87	109
	Replace supply fan motor				2.311	370	142		512	580	680
	Replace fan belt				.400	17.75	26		43.75	52	62.50
	Remove / replace heating coils				26.002	4,525	1,600		6,125	6,950	8,100
	Total				34.119	4,982.75	2,117		7,099.75	**8,090.50**	**9,471**
1025	**Replace computer room A/C, chilled water, 10 ton**	20	Q-6	Ea.							
	Remove air conditioner				13.001		795		795	985	1,225
	Replace air conditioner, 10 ton				32.653	20,700	1,925		22,625	25,200	28,900
	Total				45.654	20,700	2,720		23,420	**26,185**	**30,125**

D3053 276 Computer Room A/C Units, Chilled Water

	System Description	Freq. (Years)	Crew	Unit	Labor Hours	2020 Bare Costs				Total In-House	Total w/O&P
						Material	Labor	Equipment	Total		
1030	**Repair computer room A/C, chilled water, 15 ton**	10	1 STPI	Ea.							
	Repair controls				.500		33		33	40.50	50.50
	Remove / replace supply fan				.650		42.50		42.50	52.50	66
	Remove supply fan bearing				1.200		78.50		78.50	97.50	122
	Replace bearings				2.000	70	131		201	239	291
	Remove supply fan motor				1.156		71		71	87	109
	Replace supply fan motor				2.312	805	142		947	1,050	1,225
	Replace fan belt				.400	17.75	26		43.75	52	62.50
	Remove / replace heating coils				26.002	4,525	1,300		6,125	6,950	8,100
	Total				34.220	5,417.75	2,124		7,541.75	**8,568.50**	**10,026**
1035	**Replace computer room A/C, chilled water, 15 ton**	20	Q-6	Ea.							
	Remove air conditioner				23.105		1,450		1,450	1,775	2,225
	Replace air conditioner, 15 ton				44.444	22,600	2,625		25,225	28,100	32,300
	Total				67.549	22,600	4,075		26,675	**29,875**	**34,525**
1040	**Repair computer room A/C, chilled water, 20 ton**	10	1 STPI	Ea.							
	Repair controls				.500		33		33	40.50	50.50
	Remove / replace supply fan				.650		42.50		42.50	52.50	66
	Remove supply fan bearing				1.200		78.50		78.50	97.50	122
	Replace bearings				2.000	70	131		201	239	291
	Remove supply fan motor				1.156		71		71	87	109
	Replace supply fan motor				2.312	805	142		947	1,050	1,225
	Replace fan belt				.400	17.75	26		43.75	52	62.50
	Remove / replace heating coils				46.243	13,800	2,875		16,675	18,700	21,700
	Total				54.461	14,692.75	3,399		18,091.75	**20,318.50**	**23,626**
1045	**Replace computer room A/C, chilled water, 20 ton**	20	Q-6	Ea.							
	Remove air conditioner				41.558		2,600		2,600	3,200	4,000
	Replace air conditioner, 20 ton				47.059	24,100	2,775		26,875	29,900	34,400
	Total				88.617	24,100	5,375		29,475	**33,100**	**38,400**

D3053 278 Multi-Zone Air Conditioner

	System Description	Freq. (Years)	Crew	Unit	Labor Hours	2020 Bare Costs				Total In-House	Total w/O&P
						Material	Labor	Equipment	Total		
1010	**Repair multi-zone rooftop unit, 15 ton**	10	Q-6	Ea.							
	Repair controls				1.000		65.50		65.50	81	101
	Remove fan bearings				2.400		157		157	195	244
	Replace bearings				4.000	140	262		402	480	580
	Remove fan motor				1.156		71		71	87	109
	Replace fan motor				2.312	805	142		947	1,050	1,225
	Remove compressor				17.778		1,050		1,050	1,300	1,625
	Replace compressor				42.857	23,700	2,625		26,325	29,300	33,700
	Replace refrigerant				1.333	240	87.50		327.50	370	435
	Remove / replace heater igniter				.500	33	33		66	77	92
	Total				73.336	24,918	4,493		29,411	**32,940**	**38,111**
1040	**Replace multi-zone rooftop unit, 15 ton**	15	Q-7	Ea.							
	Remove multi-zone rooftop unit				94.675		5,900		5,900	7,300	9,125
	Replace multi-zone rooftop, 15 ton				70.022	66,500	4,375		70,875	78,500	90,000
	Total				164.696	66,500	10,275		76,775	**85,800**	**99,125**
2010	**Repair multi - zone rooftop unit, 25 ton**	10	Q-6	Ea.							
	Repair controls				1.000		65.50		65.50	81	101
	Remove fan bearings				2.400		157		157	195	244
	Replace bearings				4.000	140	262		402	480	580
	Remove fan motor				1.156		71		71	87	109
	Replace fan motor				2.311	595	142		737	830	960
	Remove compressor				22.222		1,312.50		1,312.50	1,625	2,025
	Replace compressor				60.976	35,875	3,718.75		39,593.75	44,100	50,500
	Replace refrigerant				2.000	360	131.25		491.25	560	655
	Remove / replace heater igniter				1.000	66	66		132	154	184
	Total				97.065	37,036	5,926		42,962	**48,112**	**55,358**
2040	**Replace multi-zone rooftop unit, 25 ton**	15	Q-7	Ea.							
	Remove multi-zone rooftop unit				116.000		7,200		7,200	8,925	11,100
	Replace multi-zone rooftop, 25 ton				95.808	87,500	5,975		93,475	103,500	118,500
	Total				211.808	87,500	13,175		100,675	**112,425**	**129,600**

D3053 278 Multi-Zone Air Conditioner

	System Description	Freq. (Years)	Crew	Unit	Labor Hours	2020 Bare Costs				Total In-House	Total w/O&P
						Material	Labor	Equipment	Total		
3010	**Repair multi-zone rooftop unit, 40 ton**	10	Q-6	Ea.							
	Repair controls				1.000		65.50		65.50	81	101
	Remove fan bearings				2.400		157		157	195	244
	Replace bearings				4.444	157	292		449	535	645
	Remove fan motor				2.000		123		123	151	189
	Replace fan motor				4.000	2,125	245		2,370	2,650	3,025
	Remove compressor				35.451		2,175		2,175	2,675	3,350
	Replace compressor				71.006	33,000	4,350		37,350	41,700	48,000
	Replace refrigerant				2.667	480	175		655	745	870
	Remove / replace heater igniter				1.000	66	66		132	154	184
	Total				123.967	35,828	7,648.50		43,476.50	**48,886**	**56,608**
3040	**Replace multi-zone rooftop unit, 40 ton**	15	Q-7	Ea.							
	Remove multi-zone				173.000		10,800		10,800	13,300	16,700
	Replace multi-zone rooftop, 40 ton				152.000	126,500	9,475		135,975	151,000	172,500
	Total				325.000	126,500	20,275		146,775	**164,300**	**189,200**
4010	**Repair multi-zone rooftop unit, 70 ton**	10	Q-7	Ea.							
	Repair controls				1.000		65.50		65.50	81	101
	Remove fan bearings				2.400		157		157	195	244
	Replace bearings				4.444	157	292		449	535	645
	Remove fan motor				2.601		160		160	196	246
	Replace fan motor				5.195	3,625	320		3,945	4,375	5,025
	Remove compressor				76.923		4 800		4,800	5,925	7,425
	Replace compressor				154.000	38,000	9 600		47,600	53,500	62,500
	Replace refrigerant				4.667	840	306.25		1,146.25	1,300	1,525
	Remove / replace heater igniter				1.000	66	66		132	154	184
	Total				252.230	42,688	15,766.75		58,454.75	**66,261**	**77,895**
4040	**Replace multi-zone rooftop unit, 70 ton**	15	Q-7	Ea.							
	Remove multi-zone rooftop unit				232.000		14,500		14,500	17,900	22,400
	Replace multi-zone rooftop, 70 ton				264.000	171,500	16,500		188,000	209,000	240,000
	Total				496.000	171,500	31,000		202,500	**226,900**	**262,400**

D3053 278 Multi-Zone Air Conditioner

	System Description	Freq. (Years)	Crew	Unit	Labor Hours	2020 Bare Costs				Total In-House	Total w/O&P
						Material	Labor	Equipment	Total		
5010	**Repair multi-zone rooftop unit, 105 ton**	10	Q-7	Ea.							
	Repair controls				1.000		65.50		65.50	81	101
	Remove fan bearings				4.800		314		314	390	490
	Replace fan bearings				9.412	744	616		1,360	1,575	1,875
	Remove fan motor				4.334		266		266	325	410
	Replace fan motor				8.696	5,825	535		6,360	7,075	8,100
	Remove compressor				79.012		4,940		4,940	6,100	7,625
	Replace compressor				158.400	34,720	9,840		44,560	50,500	58,500
	Replace refrigerant				6.667	1,200	437.50		1,637.50	1,850	2,175
	Remove / replace heater igniter				1.500	99	99		198	230	275
	Total				273.820	42,588	17,113		59,701	**68,126**	**79,551**
5040	**Replace multi-zone rooftop unit, 105 ton**	15	Q-7	Ea.							
	Remove multi-zone rooftop unit				348.000		21,700		21,700	26,800	33,600
	Replace multi-zone rooftop, 105 ton				390.000	223,000	24,400		247,400	275,500	316,500
	Total				738.000	223,000	46,100		269,100	**302,300**	**350,100**

D3053 280 Single Zone Air Conditioner

	System Description	Freq. (Years)	Crew	Unit	Labor Hours	2020 Bare Costs				Total In-House	Total w/O&P
						Material	Labor	Equipment	Total		
1001	**Repair single zone rooftop unit, 3 ton**	10	2 STPI	Ea.							
	Repair controls				.300		19.65		19.65	24.50	30.50
	Remove fan bearings				2.400		157		157	195	244
	Replace bearings				4.000	140	262		402	480	580
	Remove fan motor				1.156		71		71	87	109
	Replace fan motor				1.951	207	120		327	375	445
	Remove compressor				2.477		162		162	201	251
	Replace compressor				4.522	370	296		666	775	925
	Replace refrigerant				.533	96	35		131	149	174
	Remove/replace heater igniter				.500	33	33		66	77	92
	Total				17.840	846	1,155.65		2,001.65	**2,363.50**	**2,850.50**
1002	**Replace single zone rooftop unit, 3 ton**	15	Q-5	Ea.							
	Remove single zone				8.000		470		470	585	730
	Replace single zone rooftop, 3 ton				30.476	2,900	1,800		4,700	5,425	6,400
	Total				38.476	2,900	2,270		5,170	**6,010**	**7,130**

D3053 280 Single Zone Air Conditioner

	System Description	Freq. (Years)	Crew	Unit	Labor Hours	2020 Bare Costs				Total In-House	Total w/O&P
						Material	Labor	Equipment	Total		
1003	**Repair single zone rooftop unit, 5 ton**	10	2 STPI	Ea.							
	Repair controls				.300		19.65		19.65	24.50	30.50
	Remove fan bearings				2.400		157		157	195	244
	Replace bearings				4.000	140	262		402	480	580
	Remove fan motor				1.156		71		71	87	109
	Replace fan motor				1.951	207	120		327	375	445
	Remove compressor				2.477		162		162	201	251
	Replace compressor				4.954	520	325		845	975	1,150
	Replace refrigerant				.533	96	35		131	149	174
	Remove/replace heater igniter				.500	33	33		66	77	92
	Total				18.271	996	1,184.65		2,180.65	**2,563.50**	**3,075.50**
1004	**Replace single zone rooftop unit, 5 ton**	15	Q-5	Ea.							
	Remove single zone				18.561		1,100		1,100	1,350	1,700
	Replace single zone rooftop, 5 ton				38.005	4,475	2,250		6,725	7,700	9,075
	Total				56.566	4,475	3,350		7,825	**9,050**	**10,775**
1005	**Repair single zone rooftop unit, 7.5 ton**	10	2 STPI	Ea.							
	Repair controls				.400		26		26	32.50	40.50
	Remove fan bearings				2.400		157		157	195	244
	Replace bearings				4.000	140	262		402	480	580
	Remove fan motor				1.156		71		71	87	109
	Replace fan motor				2.311	345	142		487	555	650
	Remove compressor				3.467		205		205	253	315
	Replace compressor				6.932	1,175	410		1,585	1,800	2,100
	Replace refrigerant				.667	120	43.75		163.75	186	218
	Remove/replace heater igniter				.500	33	33		66	77	92
	Total				21.833	1,813	1,349.75		3,162.75	**3,665.50**	**4,348.50**
1006	**Replace single zone rooftop unit, 7.5 ton**	15	Q-5	Ea.							
	Remove single zone				26.016		1,525		1,525	1,900	2,375
	Replace single zone rooftop, 7.5 ton				43.011	5,725	2,550		8,275	9,425	11,100
	Total				69.027	5,725	4,075		9,800	**11,325**	**13,475**

D3053 280 Single Zone Air Conditioner

	System Description	Freq. (Years)	Crew	Unit	Labor Hours	2020 Bare Costs				Total In-House	Total w/O&P
						Material	Labor	Equipment	Total		
1007	**Repair single zone rooftop unit, 10 ton**	10	2 STPI	Ea.							
	Repair controls				.400		26		26	32.50	40.50
	Remove fan bearings				2.400		157		157	195	244
	Replace bearings				4.000	140	262		402	480	580
	Remove fan motor				1.156		71		71	87	109
	Replace fan motor				2.311	370	142		512	580	680
	Remove compressor				14.513		887.75		887.75	1,100	1,375
	Replace compressor				28.714	15,879	1,758.75		17,637.75	19,600	22,600
	Replace refrigerant				.800	144	52.50		196.50	223	261
	Remove/replace heater igniter				.500	33	33		66	77	92
	Total				54.794	16,566	3,390		19,956	**22,374.50**	**25,981.50**
1008	**Replace single zone rooftop unit, 10 ton**	15	Q-6	Ea.							
	Remove single zone				33.898		2,075		2,075	2,575	3,200
	Replace single zone rooftop, 10 ton				48.000	9,675	2,925		12,600	14,300	16,600
	Total				81.898	9,675	5,000		14,675	**16,875**	**19,800**
1010	**Repair single zone rooftop unit, 15 ton**	10	2 STPI	Ea.							
	Repair controls				.500		33		33	40.50	50.50
	Remove fan bearings				2.400		157		157	195	244
	Replace bearings				4.000	140	262		402	480	580
	Remove fan motor				1.156		71		71	87	109
	Replace fan motor				2.312	805	142		947	1,050	1,225
	Remove compressor				21.661		1,325		1,325	1,650	2,050
	Replace compressor				42.857	23,700	2,625		26,325	29,300	33,700
	Replace refrigerant				1.333	240	87.50		327.50	370	435
	Remove / replace heater igniter				.500	33	33		66	77	92
	Total				76.719	24,918	4,735.50		29,653.50	**33,249.50**	**38,485.50**
1040	**Replace single zone rooftop unit, 15 ton**	15	Q-6	Ea.							
	Remove single zone rooftop unit				50.314		3,075		3,075	3,800	4,750
	Replace single zone rooftop, 15 ton				56.075	13,300	3,425		16,725	18,900	21,900
	Total				106.389	13,300	6,500		19,800	**22,700**	**26,650**

D3053 280 Single Zone Air Conditioner

	System Description	Freq. (Years)	Crew	Unit	Labor Hours	2020 Bare Costs				Total In-House	Total w/O&P
						Material	Labor	Equipment	Total		
2010	**Repair single zone rooftop unit, 25 ton**	10	Q-6	Ea.							
	Repair controls				.500		33		33	40.50	50.50
	Remove fan bearings				2.400		157		157	195	244
	Replace bearings				4.000	140	262		402	480	580
	Remove fan motor				1.156		71		71	87	109
	Replace fan motor				2.311	595	142		737	830	960
	Remove compressor				22.222		1,312.50		1,312.50	1,625	2,025
	Replace compressor				60.976	35,875	3,718.75		39,593.75	44,100	50,500
	Replace refrigerant				2.000	360	131.25		491.25	560	655
	Remove / replace heater igniter				1.000	66	66		132	154	184
	Total				96.565	37,036	5,893.50		42,929.50	**48,071.50**	**55,307.50**
2040	**Replace single zone rooftop unit, 25 ton**	15	Q-7	Ea.							
	Remove single zone rooftop unit				76.923		4,800		4,800	5,925	7,425
	Replace single zone rooftop, 25 ton				76.739	34,800	4,800		39,600	44,200	51,000
	Total				153.662	34,800	9,600		44,400	**50,125**	**58,425**
3010	**Repair single zone rooftop unit, 60 ton**	10	Q-6	Ea.							
	Repair controls				.500		33		33	40.50	50.50
	Remove fan bearings				2.400		157		157	195	244
	Replace bearings				4.444	157	292		449	535	645
	Remove fan motor				2.000		123		123	151	189
	Replace fan motor				4.167	2,575	256		2,831	3,150	3,625
	Remove compressor				93.507		5,730		5,730	7,075	8,850
	Replace compressor				187.200	41,160	11,430		52,590	59,500	69,000
	Replace refrigerant				4.667	840	306.25		1,146.25	1,300	1,525
	Remove / replace heater igniter				1.000	66	66		132	154	184
	Total				299.884	44,798	18,393.25		63,191.25	**72,100.50**	**84,312.50**
3040	**Replace single zone rooftop unit, 60 ton**	15	Q-7	Ea.							
	Remove single zone rooftop unit				188.000		11,700		11,700	14,500	18,200
	Replace single zone rooftop, 60 ton				182.000	60,000	11,300		71,300	80,000	92,500
	Total				370.000	60,000	23,000		83,000	**94,500**	**110,700**

D3053 280 Single Zone Air Conditioner

	System Description	Freq. (Years)	Crew	Unit	Labor Hours	2020 Bare Costs Material	Labor	Equipment	Total	Total In-House	Total w/O&P
4010	**Repair single zone rooftop unit, 100 ton**	10	Q-7	Ea.							
	Repair controls				.500		33		33	40.50	50.50
	Remove fan bearings				4.800		314		314	390	490
	Replace fan bearings				9.412	744	616		1,360	1,575	1,875
	Remove fan motor				4.334		266		266	325	410
	Replace fan motor				7.407	4,725	455		5,180	5,750	6,600
	Remove compressor				98.765		6,175		6,175	7,625	9,525
	Replace compressor				198.000	43,400	12,300		55,700	63,000	73,500
	Replace refrigerant				6.667	1,200	437.50		1,637.50	1,850	2,175
	Remove / replace heater igniter				1.500	99	99		198	230	275
	Total				331.385	50,168	20,695.50		70,863.50	**80,785.50**	**94,900.50**
4040	**Replace single zone rooftop unit, 100 ton**	15	Q-7	Ea.							
	Remove single zone rooftop unit				296.000		18,500		18,500	22,900	28,600
	Replace single zone rooftop, 100 ton				305.000	125,000	19,000		144,000	161,000	185,500
	Total				601.000	125,000	37,500		162,500	**183,900**	**214,100**

D3053 282 Multi-Zone Variable Volume

	System Description	Freq. (Years)	Crew	Unit	Labor Hours	2020 Bare Costs Material	Labor	Equipment	Total	Total In-House	Total w/O&P
1010	**Repair multi-zone variable volume, 50 ton**	10	Q-6	Ea.							
	Repair controls				1.300		85		85	105	132
	Remove fan bearings				2.400		157		157	195	244
	Replace bearings				4.444	157	292		449	535	645
	Remove fan motor				2.000		123		123	151	189
	Replace fan motor				4.000	2,125	245		2,370	2,650	3,025
	Remove compressor				77.922		4,775		4,775	5,900	7,375
	Replace compressor				156.000	34,300	9,525		43,825	49,500	57,500
	Replace refrigerant				3.333	600	218.75		818.75	930	1,100
	Remove / replace heater igniter				1.000	66	66		132	154	184
	Total				252.400	37,248	15,486.75		52,734.75	**60,120**	**70,394**
1040	**Replace multi-zone variable volume, 50 ton**	15	Q-7	Ea.							
	Remove multi-zone variable volume unit				160.000		9,975		9,975	12,300	15,400
	Replace multi-zone variable volume unit, 50 ton				219.000	117,500	13,700		131,200	146,000	168,000
	Total				379.000	117,500	23,675		141,175	**158,300**	**183,400**

D3053 282 Multi-Zone Variable Volume

	System Description	Freq. (Years)	Crew	Unit	Labor Hours	2020 Bare Costs				Total In-House	Total w/O&P
						Material	Labor	Equipment	Total		
2010	**Repair multi-zone variable volume, 70 ton**	10	Q-7	Ea.							
	Repair controls				1.300		85		85	105	132
	Remove fan bearings				2.400		157		157	195	244
	Replace bearings				4.444	157	292		449	535	645
	Remove fan motor				2.601		160		160	196	246
	Replace fan motor				5.195	3,625	320		3,945	4,375	5,025
	Remove compressor				76.923		4,800		4,800	5,925	7,425
	Replace compressor				154.000	38,000	9,600		47,600	53,500	62,500
	Replace refrigerant				4.667	840	306.25		1,146.25	1,300	1,525
	Remove / replace heater igniter				1.000	66	66		132	154	184
	Total				252.530	42,688	15,786.25		58,474.25	**66,285**	**77,926**
2040	**Replace multi-zone variable volume, 70 ton**	15	Q-7	Ea.							
	Remove multi-zone variable volume unit				232.000		14,500		14,500	17,900	22,400
	Replace multi-zone variable volume unit, 70 ton				305.000	164,500	19,000		183,500	204,500	235,000
	Total				537.000	164,500	33,500		198,000	**222,400**	**257,400**
3010	**Repair multi-zone variable volume, 90 ton**	10	Q-7	Ea.							
	Repair controls				1.300		85		85	105	132
	Remove fan bearings				2.400		157		157	195	244
	Replace bearings				4.444	157	292		449	535	645
	Remove fan motor				2.601		160		160	196	246
	Replace fan motor				6.504	3,800	400		4,200	4,675	5,375
	Remove compressor				92.308		5,760		5,760	7,125	8,900
	Replace compressor				184.800	45,600	11,520		57,120	64,500	75,000
	Replace refrigerant				6.667	1,200	437.50		1,637.50	1,850	2,175
	Remove / replace heater igniter				1.500	99	99		198	230	275
	Total				302.523	50,856	18,910.50		69,766.50	**79,411**	**92,992**
3040	**Replace multi-zone variable volume, 90 ton**	15	Q-7	Ea.							
	Remove multi-zone variable volume unit				296.000		18,500		18,500	22,900	28,600
	Replace multi-zone variable volume unit, 90 ton				395.000	180,000	24,700		204,700	228,500	263,000
	Total				691.000	180,000	43,200		223,200	**251,400**	**291,600**

D3053 282 Multi-Zone Variable Volume

	System Description	Freq. (Years)	Crew	Unit	Labor Hours	2020 Bare Costs				Total In-House	Total w/O&P
						Material	Labor	Equipment	Total		
4010	**Repair multi-zone variable volume, 105 ton**	10	Q-7	Ea.							
	Repair controls				1.300		85		85	105	132
	Remove fan bearings				4.800		314		314	390	490
	Replace fan bearings				9.412	744	616		1,360	1,575	1,875
	Remove fan motor				4.334		266		266	325	410
	Replace fan motor				8.696	5,825	535		6,360	7,075	8,100
	Remove compressor				98.765		6,175		6,175	7,625	9,525
	Replace compressor				198.000	43,400	12,300		55,700	63,000	73,500
	Replace refrigerant				7.333	1,320	481.25		1,801.25	2,050	2,400
	Remove / replace heater igniter				1.500	99	99		198	230	275
	Total				334.140	51,388	20,871.25		72,259.25	**82,375**	**96,707**
4040	**Replace multi-zone variable volume, 105 ton**	15	Q-7	Ea.							
	Remove multi-zone variable volume unit				348.000		21,700		21,700	26,800	33,600
	Replace multi-zone variable volume unit, 105 ton				444.000	214,000	27,700		241,700	269,500	310,500
	Total				792.000	214,000	49,400		263,400	**296,300**	**344,100**
5010	**Repair multi-zone variable volume, 140 ton**	10	Q-7	Ea.							
	Repair controls				1.300		85		85	105	132
	Remove fan bearings				7.200		471		471	585	730
	Replace fan bearings				14.118	1,116	924		2,040	2,375	2,825
	Remove fan motor				4.334		266		266	325	410
	Replace fan motor				11.594	5,950	710		6,660	7,425	8,550
	Remove compressor				106.370		6,650.48		6,650.48	8,200	10,300
	Replace compressor				213.246	46,741.80	13,247.10		59,988.90	68,000	79,000
	Replace refrigerant				8.666	1,560	568.75		2,128.75	2,425	2,825
	Remove / replace heater igniter				2.000	132	132		264	305	365
	Total				368.828	55,499.80	23,054.33		78,554.13	**89,745**	**105,137**
5040	**Replace multi-zone variable volume, 140 ton**	15	Q-7	Ea.							
	Remove multi-zone variable volume unit				516.000		32,200		32,200	39,800	49,800
	Replace multi-zone variable volume unit, 140 ton				593.000	261,000	37,000		298,000	333,000	383,500
	Total				1109.000	261,000	69,200		330,200	**372,800**	**433,300**

D3053 284 Single Zone Variable Volume

	System Description	Freq. (Years)	Crew	Unit	Labor Hours	2020 Bare Costs				Total In-House	Total w/O&P
						Material	Labor	Equipment	Total		
1010	**Repair single zone variable volume, 20 ton**	10	Q-6	Ea.							
	Repair controls				1.000		65.50		65.50	81	101
	Remove fan bearings				2.400		157		157	195	244
	Replace bearings				4.000	140	262		402	480	580
	Remove fan motor				1.156		71		71	87	109
	Replace fan motor				2.312	805	142		947	1,050	1,225
	Remove compressor				17.778		1,050		1,050	1,300	1,625
	Replace compressor				42.857	23,700	2,625		26,325	29,300	33,700
	Replace refrigerant				1.333	240	87.50		327.50	370	435
	Remove / replace heater igniter				1.000	66	66		132	154	184
	Total				73.836	24,951	4,526		29,477	**33,017**	**38,203**
1040	**Replace single zone variable volume, 20 ton**	15	Q-7	Ea.							
	Remove single zone variable volume unit				71.545		4,455		4,455	5,525	6,900
	Replace single zone variable volume unit, 20 ton				64.945	25,960	4,042.50		30,002.50	33,600	38,700
	Total				136.489	25,960	8,497.50		34,457.50	**39,125**	**45,600**
2010	**Repair single zone variable volume, 30 ton**	10	Q-6	Ea.							
	Repair controls				1.000		65.50		65.50	81	101
	Remove fan bearings				2.400		157		157	195	244
	Replace bearings				4.000	140	262		402	480	580
	Remove fan motor				1.156		71		71	87	109
	Replace fan motor				2.476	985	152		1,137	1,275	1,475
	Remove compressor				17.778		1,050		1,050	1,300	1,625
	Replace compressor				48.780	28,700	2,975		31,675	35,300	40,500
	Replace refrigerant				2.667	480	175		655	745	870
	Remove / replace heater igniter				1.000	66	66		132	154	184
	Total				81.257	30,371	4,973.50		35,344.50	**39,617**	**45,688**
2040	**Replace single zone variable volume, 30 ton**	15	Q-7	Ea.							
	Remove single zone variable volume unit				104.142		6,490		6,490	8,025	10,000
	Replace single zone variable volume unit, 30 ton				92.147	33,110	5,747.50		38,857.50	43,500	50,500
	Total				196.289	33,110	12,237.50		45,347.50	**51,525**	**60,500**

D3053 284 Single Zone Variable Volume

	System Description	Freq. (Years)	Crew	Unit	Labor Hours	2020 Bare Costs				Total In-House	Total w/O&P
						Material	Labor	Equipment	Total		
3010	**Repair single zone variable volume, 40 ton**	10	Q-6	Ea.							
	Repair controls				1.000		65.50		65.50	81	101
	Remove fan bearings				2.400		157		157	195	244
	Replace bearings				4.000	140	262		402	480	580
	Remove fan motor				1.156		71		71	87	109
	Replace fan motor				4.000	2,125	245		2,370	2,650	3,025
	Remove compressor				35.451		2,175		2,175	2,675	3,350
	Replace compressor				71.006	33,000	4,350		37,350	41,700	48,000
	Replace refrigerant				3.333	600	218.75		818.75	930	1,100
	Remove / replace heater igniter				1.000	66	66		132	154	184
	Total				123.346	35,931	7,610.25		43,541.25	**48,952**	**56,693**
3040	**Replace single zone variable volume, 40 ton**	15	Q-7	Ea.							
	Remove single zone variable volume unit				143.000		8,937.50		8,937.50	11,000	13,900
	Replace single zone variable volume unit, 40 ton				123.200	38,170	7,672.50		45,842.50	51,500	59,500
	Total				266.200	38,170	16,610		54,780	**62,500**	**73,400**
4010	**Repair single zone variable volume, 60 ton**	10	Q-6	Ea.							
	Repair controls				.500		33		33	40.50	50.50
	Remove fan bearings				2.400		157		157	195	244
	Replace bearings				4.444	157	292		449	535	645
	Remove fan motor				2.000		123		123	151	189
	Replace fan motor				4.167	2,575	256		2,831	3,150	3,625
	Remove compressor				93.507		5,730		5,730	7,075	8,850
	Replace compressor				187.200	41,160	11,430		52,590	59,500	69,000
	Replace refrigerant				4.000	720	262.50		982.50	1,125	1,300
	Remove / replace heater igniter				1.000	66	66		132	154	184
	Total				299.218	44,678	18,349.50		63,027.50	**71,925.50**	**84,087.50**
4040	**Replace single zone variable volume, 60 ton**	15	Q-7	Ea.							
	Remove single zone variable volume unit				190.300		11,880		11,880	14,700	18,400
	Replace single zone variable volume unit, 60 ton				184.800	54,670	11,550		66,220	74,500	86,500
	Total				375.100	54,670	23,430		78,100	**89,200**	**104,900**

D3053 286 Central Station Air Conditioning Air Handling Unit

	System Description	Freq. (Years)	Crew	Unit	Labor Hours	2020 Bare Costs				Total In-House	Total w/O&P
						Material	Labor	Equipment	Total		
1010	**Repair central station A.H.U., 1300 CFM**	10	1 STPI	Ea.							
	Repair controls				.300		19.65		19.65	24.50	30.50
	Remove blower motor				1.200		78.50		78.50	97.50	122
	Replace blower motor				1.951	320	120		440	500	585
	Total				3.451	320	218.15		538.15	**622**	**737.50**
1040	**Replace central station A.H.U., 1300 CFM**	15	Q-5	Ea.							
	Remove central station A.H.U.				8.667		510		510	630	790
	Replace central station A.H.U., 1300 CFM				17.335	7,100	1,025		8,125	9,075	10,500
	Total				26.002	7,100	1,535		8,635	**9,705**	**11,290**
2010	**Repair central station A.H.U., 1900 CFM**	10	1 STPI	Ea.							
	Repair controls				.300		19.65		19.65	24.50	30.50
	Remove blower motor				1.200		78.50		78.50	97.50	122
	Replace blower motor				2.311	345	142		487	555	650
	Total				3.811	345	240.15		585.15	**677**	**802.50**
2040	**Replace central station A.H.U., 1900 CFM**	15	Q-5	Ea.							
	Remove central station A.H.U.				9.456		560		560	690	865
	Replace central station A.H.U., 1900 CFM				18.913	10,900	1,125		12,025	13,400	15,400
	Total				28.369	10,900	1,685		12,585	**14,090**	**16,265**
3010	**Repair central station A.H.U., 5400 CFM**	10	1 STPI	Ea.							
	Repair controls				.300		19.65		19.65	24.50	30.50
	Remove blower motor				1.200		78.50		78.50	97.50	122
	Replace blower motor				2.312	805	142		947	1,050	1,225
	Total				3.812	805	240.15		1,045.15	**1,172**	**1,377.50**
3040	**Replace central station A.H.U., 5400 CFM**	15	Q-6	Ea.							
	Remove central station A.H.U.				19.496		1,200		1,200	1,475	1,850
	Replace central station A.H.U., 5400 CFM				39.024	17,700	2,400		20,100	22,400	25,800
	Total				58.521	17,700	3,600		21,300	**23,875**	**27,650**
4010	**Repair central station A.H.U., 8000 CFM**	10	1 STPI	Ea.							
	Repair controls				.300		19.65		19.65	24.50	30.50
	Remove blower motor				1.600		105		105	130	162
	Replace blower motor				2.476	985	152		1,137	1,275	1,475
	Total				4.376	985	276.65		1,261.65	**1,429.50**	**1,667.50**

D3053 286 Central Station Air Conditioning Air Handling Unit

	System Description	Freq. (Years)	Crew	Unit	Labor Hours	2020 Bare Costs				Total In-House	Total w/O&P
						Material	Labor	Equipment	Total		
4040	**Replace central station A.H.U., 8000 CFM**	15	Q-6	Ea.							
	Remove central station A.H.U.				25.974		1,600		1,600	1,975	2,450
	Replace central station A.H.U., 8000 CFM				51.948	28,300	3,175		31,475	35,100	40,300
	Total				77.922	28,300	4,775		33,075	**37,075**	**42,750**
5010	**Repair central station A.H.U., 16,000 CFM**	10	1 STPI	Ea.							
	Repair controls				.300		19.65		19.65	24.50	30.50
	Remove blower motor				1.800		118		118	146	182
	Replace blower motor				2.597	1,175	159		1,334	1,500	1,725
	Total				4.698	1,175	296.65		1,471.65	**1,670.50**	**1,937.50**
5040	**Replace central station A.H.U., 16,000 CFM**	15	Q-6	Ea.							
	Remove central station A.H.U.				41.026		2,500		2,500	3,100	3,875
	Replace central station A.H.U., 16,000 CFM				82.192	53,000	5,025		58,025	64,500	74,000
	Total				123.217	53,000	7,525		60,525	**67,600**	**77,875**
6010	**Repair central station A.H.U., 33,500 CFM**	10	1 STPI	Ea.							
	Repair controls				.300		19.65		19.65	24.50	30.50
	Remove blower motor				2.601		160		160	196	246
	Replace blower motor				5.195	3,625	320		3,945	4,375	5,025
	Total				8.096	3,625	499.65		4,124.65	**4,595.50**	**5,301.50**
6040	**Replace central station A.H.U., 33,500 CFM**	15	Q-6	Ea.							
	Remove central station A.H.U.				82.192		5,025		5,025	6,225	7,775
	Replace central station A.H.U., 33,500 CFM				164.000	113,000	10,100		123,100	136,500	157,000
	Total				246.192	113,000	15,125		128,125	**142,725**	**164,775**
7010	**Repair central station A.H.U., 63,000 CFM**	10	1 STPI	Ea.							
	Repair controls				.300		19.65		19.65	24.50	30.50
	Remove blower motor				4.334		266		266	325	410
	Replace blower motor				8.696	5,825	535		6,360	7,075	8,100
	Total				13.329	5,825	820.65		6,645.65	**7,424.50**	**8,540.50**
7040	**Replace central station A.H.U., 63,000 CFM**	15	Q-7	Ea.							
	Remove central station A.H.U.				160.000		9,975		9,975	12,300	15,400
	Replace central station A.H.U., 63,000 CFM				320.000	213,000	20,000		233,000	259,000	297,000
	Total				480.000	213,000	29,975		242,975	**271,300**	**312,400**

D3053 310 Residential Furnace, Gas

	System Description	Freq. (Years)	Crew	Unit	Labor Hours	2020 Bare Costs				Total In-House	Total w/O&P
						Material	Labor	Equipment	Total		
1010	**Repair furnace, gas, 25 MBH**	10	Q-1	Ea.							
	Repair controls				.500		33		33	40.50	50.50
	Remove fan motor				.976		64		64	79	99
	Replace fan motor				1.951	247	120		367	420	495
	Remove auto vent damper				.274		17.05		17.05	21.50	26.50
	Replace auto vent damper				.547	115	34		149	169	197
	Remove burner				1.000		58		58	71.50	89.50
	Replace burner				6.400	1,025	370		1,395	1,575	1,850
	Total				11.648	1,387	696.05		2,083.05	**2,376.50**	**2,807.50**
1030	**Replace furnace, gas, 25 MBH**	15	Q-9	Ea.							
	Remove furnace				2.000		112		112	141	176
	Replace furnace				4.000	700	224		924	1,050	1,225
	Total				6.000	700	336		1,036	**1,191**	**1,401**
2010	**Repair furnace, gas, 100 MBH**	10	Q-1	Ea.							
	Repair controls				.500		33		33	40.50	50.50
	Remove fan motor				.976		64		64	79	99
	Replace fan motor				1.951	207	120		327	375	445
	Remove auto vent damper				.289		18		18	22.50	28
	Replace auto vent damper				.578	148	36		184	208	242
	Remove burner				1.200		69.50		69.50	86	108
	Replace burner				8.000	1,725	465		2,190	2,475	2,875
	Total				13.494	2,080	805.50		2,885.50	**3,286**	**3,847.50**
2030	**Replace furnace, gas, 100 MBH**	15	Q-9	Ea.							
	Remove furnace				3.250		182		182	229	286
	Replace furnace				6.499	825	365		1,190	1,375	1,600
	Total				9.749	825	547		1,372	**1,604**	**1,886**

D3053 310 Residential Furnace, Gas

	System Description	Freq. (Years)	Crew	Unit	Labor Hours	2020 Bare Costs				Total In-House	Total w/O&P
						Material	Labor	Equipment	Total		
3010	**Repair furnace, gas, 200 MBH**	10	Q-1	Ea.							
	Repair controls				.500		33		33	40.50	50.50
	Remove fan motor				.976		64		64	79	99
	Replace fan motor				1.951	320	120		440	500	585
	Remove auto vent damper				.325		20.50		20.50	25.50	31.50
	Replace auto vent damper				.650	164	40.50		204.50	231	269
	Remove burner				3.077		178		178	221	276
	Replace burner				12.298	5,125	715		5,840	6,525	7,500
	Total				19.777	5,609	1,171		6,780	**7,622**	**8,811**
3030	**Replace furnace, gas, 200 MBH**	15	Q-9	Ea.							
	Remove furnace				4.000		224		224	282	350
	Replace furnace				8.000	3,525	450		3,975	4,450	5,100
	Total				12.000	3,525	674		4,199	**4,732**	**5,450**

D3053 320 Residential Furnace, Oil

	System Description	Freq. (Years)	Crew	Unit	Labor Hours	2020 Bare Costs				Total In-House	Total w/O&P
						Material	Labor	Equipment	Total		
1010	**Repair furnace, oil, 55 MBH**	10	Q-1	Ea.							
	Repair controls				.500		33		33	40.50	50.50
	Remove fan motor				.976		64		64	79	99
	Replace fan motor				1.951	207	120		327	375	445
	Remove auto vent damper				.473		26.50		26.50	33.50	41.50
	Replace auto vent damper				.990	115	55.50		170.50	196	231
	Remove burner				4.334		251		251	310	390
	Replace burner				8.667	360	505		865	1,025	1,225
	Total				17.891	682	1,055		1,737	**2,059**	**2,482**
1030	**Replace furnace, oil, 55 MBH**	15	Q-9	Ea.							
	Remove furnace				2.889		162		162	204	254
	Replace furnace, oil, 55 MBH				5.778	3,275	325		3,600	4,000	4,600
	Total				8.667	3,275	487		3,762	**4,204**	**4,854**

D3053 320 Residential Furnace, Oil

	System Description	Freq. (Years)	Crew	Unit	Labor Hours	2020 Bare Costs				Total In-House	Total w/O&P
						Material	Labor	Equipment	Total		
2010	**Repair furnace, oil, 100 MBH**	10	Q-1	Ea.							
	Repair controls				.500		33		33	40.50	50.50
	Remove fan motor				.976		64		64	79	99
	Replace fan motor				1.951	207	120		327	375	445
	Remove auto vent damper				.495		28		28	35	43.50
	Replace auto vent damper				.990	115	55.50		170.50	196	231
	Remove burner				4.334		251		251	310	390
	Replace burner				8.667	360	505		865	1,025	1,225
	Total				17.914	682	1,056.50		1,738.50	**2,060.50**	**2,484**
2030	**Replace furnace, oil, 100 MBH**	15	Q-1	Ea.							
	Remove furnace				3.059		172		172	216	269
	Replace furnace, oil, 100 MBH				6.119	3,400	345		3,745	4,175	4,800
	Total				9.177	3,400	517		3,917	**4,391**	**5,069**
3010	**Repair furnace, oil, 200 MBH**	10	Q-1	Ea.							
	Repair controls				.500		33		33	40.50	50.50
	Remove fan motor				.976		64		64	79	99
	Replace fan motor				1.951	320	120		440	500	585
	Remove auto vent damper				.520		29		29	36.50	45.50
	Replace auto vent damper				1.040	148	58.50		206.50	236	277
	Remove burner				4.334		251		251	310	390
	Replace burner				8.667	360	505		865	1,025	1,225
	Total				17.988	828	1,060.50		1,888.50	**2,227**	**2,672**
3030	**Replace furnace, oil, 200 MBH**	15	Q-1	Ea.							
	Remove furnace				4.000		224		224	282	350
	Replace furnace, oil, 200 MBH				8.000	4,175	450		4,625	5,150	5,925
	Total				12.000	4,175	674		4,849	**5,432**	**6,275**

D3053 330 Residential Furnace, Electric

	System Description	Freq. (Years)	Crew	Unit	Labor Hours	2020 Bare Costs Material	Labor	Equipment	Total	Total In-House	Total w/O&P
1010	**Repair furnace, electric, 25 MBH**	10	Q-20	Ea.							
	Repair controls				.500		33		33	40.50	50.50
	Remove fan motor				.976		64		64	79	99
	Replace fan motor				1.951	247	120		367	420	495
	Remove heat element				.867		49.50		49.50	62	77.50
	Replace heat element				1.733	1,375	99		1,474	1,625	1,875
	Total				6.027	1,622	365.50		1,987.50	**2,226.50**	**2,597**
1030	**Replace furnace, electric, 25 MBH**	15	Q-20	Ea.							
	Remove furnace				2.826		161		161	202	252
	Replace furnace, electric, 25 MBH				4.916	734.78	282.61		1,017.39	1,150	1,350
	Total				7.742	734.78	443.61		1,178.39	**1,352**	**1,602**
2010	**Repair furnace, electric, 50 MBH**	10	Q-20	Ea.							
	Repair controls				.500		33		33	40.50	50.50
	Remove fan motor				.976		64		64	79	99
	Replace fan motor				1.951	247	120		367	420	495
	Remove heat element				.929		53		53	66.50	83
	Replace heat element				1.857	1,425	106		1,531	1,700	1,950
	Total				6.212	1,672	376		2,048	**2,306**	**2,677.50**
2030	**Replace furnace , electric, 50 MBH**	15	Q-20	Ea.							
	Remove furnace				3.095		177		177	221	276
	Replace furnace, electric, 50 MBH				6.190	590	355		945	1,100	1,300
	Total				9.285	590	532		1,122	**1,321**	**1,576**
3010	**Repair furnace, electric, 85 MBH**	10	Q-20	Ea.							
	Repair controls				.500		33		33	40.50	50.50
	Remove fan motor				.976		64		64	79	99
	Replace fan motor				1.951	207	120		327	375	445
	Remove heat element				1.083		62		62	77.50	96.50
	Replace heat element				2.167	2,550	124		2,674	2,950	3,375
	Total				6.677	2,757	403		3,160	**3,522**	**4,066**
3030	**Replace furnace, electric, 85 MBH**	15	Q-20	Ea.							
	Remove furnace				4.000		224		224	282	350
	Replace furnace, electric, 85 MBH				6.842	650	390		1,040	1,200	1,425
	Total				10.842	650	614		1,264	**1,482**	**1,775**

D3053 410 Electric Baseboard Heating Units

	System Description	Freq. (Years)	Crew	Unit	Labor Hours	2020 Bare Costs Material	Labor	Equipment	Total	Total In-House	Total w/O&P
0010	**Maintenance and repair**	2	1 ELEC	Ea.							
	Repair wiring connections				.615		38		38	46.50	58
	Total				.615		38		38	**46.50**	**58**
0020	**Maintenance and inspection**	0.50	1 ELEC	Ea.							
	Inspect and clean baseboard heater				1.143		70		70	86	108
	Total				1.143		70		70	**86**	**108**
0030	**Replace baseboard heater**	20	1 ELEC	Ea.							
	Remove baseboard heater				.667		41		41	50.50	63
	Baseboard heater, 5′ long				1.860	44	114		158	189	231
	Total				2.527	44	155		199	**239.50**	**294**

D3053 420 Cast Iron Radiator, 10′ section, 1 side

	System Description	Freq. (Years)	Crew	Unit	Labor Hours	2020 Bare Costs Material	Labor	Equipment	Total	Total In-House	Total w/O&P
0020	**Refinish C.I. radiator, 1 side**	5	1 PORD	Ea.							
	Prepare surface				.296		13.20		13.20	16.80	21
	Prime and paint 1 coat, brushwork				.471	4.40	20.80		25.20	31.50	38.50
	Total				.767	4.40	34		38.40	**48.30**	**59.50**

D3053 710 Fireplaces, Firebrick

	System Description	Freq. (Years)	Crew	Unit	Labor Hours	2020 Bare Costs Material	Labor	Equipment	Total	Total In-House	Total w/O&P
0020	**Replace firebrick**	75	2 BRIC	S.F.							
	Masonry demolition, fireplace, brick, 30″ x 24″				1.155		48.62		48.62	63	77.50
	Install new firebrick				.208	10.05	9.75		19.80	24	28.50
	Total				1.363	10.05	58.37		68.42	**87**	**106**

D4013 110 Backflow Preventer

	System Description	Freq. (Years)	Crew	Unit	Labor Hours	2020 Bare Costs				Total In-House	Total w/O&P
						Material	Labor	Equipment	Total		
3010	**Rebuild 4″ diam. reduced pressure backflow preventer**	10	1 PLUM	Ea.							
	Notify proper authorities prior to closing valves				.250		16.10		16.10	19.95	25
	Close valves at 4″ diameter BFP and open by-pass				.167		10.75		10.75	13.30	16.60
	Rebuild 4″ diameter reduced pressure BFP				2.000	385	129		514	585	680
	Test backflow preventer				.500		32		32	40	50
	Total				2.917	385	187.85		572.85	**658.25**	**771.60**
3020	**Rebuild 6″ diam. reduced pressure backflow preventer**	10	1 PLUM	Ea.							
	Notify proper authorities prior to closing valves				.250		16.10		16.10	19.95	25
	Close valves at 6″ diameter BFP and open by-pass				.182		11.70		11.70	14.50	18.10
	Rebuild 6″ diameter reduced pressure BFP				3.008	405	194		599	685	805
	Test backflow preventer				.500		32		32	40	50
	Total				3.939	405	253.80		658.80	**759.45**	**898.10**
3030	**Rebuild 8″ diam. reduced pressure backflow preventer**	10	1 PLUM	Ea.							
	Notify proper authorities prior to closing valves				.250		16.10		16.10	19.95	25
	Close valves at 8″ diameter BFP and open by-pass				.200		12.90		12.90	15.95	19.95
	Rebuild 8″ diameter reduced pressure BFP				4.000	505	258		763	875	1,025
	Test backflow preventer				.500		32		32	40	50
	Total				4.950	505	319		824	**950.90**	**1,119.95**
3040	**Rebuild 10″ diam. reduced pressure backflow preventer**	10	1 PLUM	Ea.							
	Notify proper authorities prior to closing valves				.250		16.10		16.10	19.95	25
	Close valves at 10″ diameter BFP and open by-pass				.222		14.30		14.30	17.70	22
	Rebuild 10″ diameter reduced pressure BFP				5.000	585	320		905	1,050	1,225
	Test backflow preventer				.500		32		32	40	50
	Total				5.972	585	382.40		967.40	**1,127.65**	**1,322**

D4013 310 Sprinkler System, Fire Suppression

	System Description	Freq. (Years)	Crew	Unit	Labor Hours	2020 Bare Costs				Total In-House	Total w/O&P
						Material	Labor	Equipment	Total		
1020	**Inspect sprinkler system**	1	1 PLUM	Ea.							
	Inspect water pressure gauges				.044		2.80		2.80	3.50	4.40
	Test fire alarm (wet or dry)				.030		1.90		1.90	2.40	3
	Motor, exposed, inspect				.333		21.50		21.50	26.50	33
	Inspect, visual				.060		3.90		3.90	4.80	6
	Total				.467		30.10		30.10	**37.20**	**46.40**
1030	**Replace sprinkler head**	20	1 PLUM	Ea.							
	Remove old sprinkler head				.325		20.50		20.50	25.50	32
	Install new head				.650	13.45	41		54.45	65.50	80.50
	Total				.975	13.45	61.50		74.95	**91**	**112.50**
1040	**Rebuild double check 3″ backflow preventer**	1	1 PLUM	Ea.							
	Rebuild 3″ backflow preventer				3.910	272	252		524	610	730
	Test system				1.300		84		84	104	130
	Total				5.210	272	336		608	**714**	**860**
1050	**Rebuild double check 4″ backflow preventer**	1	1 PLUM	Ea.							
	Rebuild 4″ backflow preventer				4.561	290	294		584	685	820
	Test system				1.300		84		84	104	130
	Total				5.861	290	378		668	**789**	**950**
1060	**Rebuild double check 6″ backflow preventer**	1	1 PLUM	Ea.							
	Rebuild 6″ backflow preventer				5.202	350	335		685	800	960
	Test system				1.300		84		84	104	130
	Total				6.502	350	419		769	**904**	**1,090**
1070	**Rebuild reduced pressure backflow preventer**	1	1 PLUM	Ea.							
	Rebuild 3″ backflow preventer				3.910	272	252		524	610	730
	Test system				1.300		84		84	104	130
	Total				5.210	272	336		608	**714**	**860**

D4013 410 Fire Pump

	System Description	Freq. (Years)	Crew	Unit	Labor Hours	2020 Bare Costs				Total In-House	Total w/O&P
						Material	Labor	Equipment	Total		
1030	**Replace fire pump / electric motor assembly 100 H.P.**	25	Q-13	Ea.							
	Remove 100 H.P. electric-drive fire pump				31.373		1,900		1,900	2,350	2,925
	Install new electric-drive fire pump and controls, 100 H.P.				64.000	25,000	3,850		28,850	32,300	37,200
	Total				95.373	25,000	5,750		30,750	**34,650**	**40,125**

D5013 110 Primary Transformer, Liquid Filled

	System Description	Freq. (Years)	Crew	Unit	Labor Hours	2020 Bare Costs Material	Labor	Equipment	Total	Total In-House	Total w/O&P
0010	**Repair 500 kVA transformer**	10	1 ELEC	Ea.							
	Remove transformer component				1.000		61.50		61.50	75.50	94.50
	Transformer component				2.963	2,200	182		2,382	2,650	3,025
	Total				3.963	2,200	243.50		2,443.50	**2,725.50**	**3,119.50**
0020	**Maintenance and inspection**	0.50	1 ELEC	Ea.							
	Test energized transformer				.500		30.50		30.50	37.50	47
	Total				.500		30.50		30.50	**37.50**	**47**
0030	**Replace transformer**	30	R-3	Ea.							
	Remove transformer				13.793		845	130	975	1,175	1,450
	Transformer 500 kVA, 5 KV primary, 277/480 V secondary				50.000	21,100	3,050	475	24,625	27,500	31,600
	Total				63.793	21,100	3,895	605	25,600	**28,675**	**33,050**

D5013 120 Primary Transformer, Dry

	System Description	Freq. (Years)	Crew	Unit	Labor Hours	2020 Bare Costs Material	Labor	Equipment	Total	Total In-House	Total w/O&P
0010	**Repair 15 KV primary transformer**	15	1 ELEC	Ea.							
	Repair transformer terminal				2.000	52.50	123		175.50	209	255
	Total				2.000	52.50	123		175.50	**209**	**255**
0020	**Maintenance and inspection**	0.50	1 ELEC	Ea.							
	Test energized transformer				.500		30.50		30.50	37.50	47
	Total				.500		30.50		30.50	**37.50**	**47**
0030	**Replace transformer**	30	R-3	Ea.							
	Remove transformer				13.793		845	130	975	1,175	1,450
	Transformer 500 kVA, 15 KV primary, 277/480 V secondary				57.143	54,000	3,500	540	58,040	64,500	73,500
	Total				70.936	54,000	4,345	670	59,015	**65,675**	**74,950**

D5013 210 Switchgear, Mainframe

	System Description	Freq. (Years)	Crew	Unit	Labor Hours	2020 Bare Costs				Total In-House	Total w/O&P
						Material	Labor	Equipment	Total		
0010	**Repair switchgear 1200 A**	5	1 ELEC	Ea.							
	Remove relay				1.702		104		104	128	161
	Relay				5.000	1,175	305		1,480	1,675	1,950
	Total				6.702	1,175	409		1,584	**1,803**	**2,111**
0020	**Maintenance and inspection**	1	1 ELEC	Ea.							
	Check & retighten switchgear connections				.800		49		49	60.50	75.50
	Total				.800		49		49	**60.50**	**75.50**
0030	**Replace switchgear 1200 A**	20	3 ELEC	Ea.							
	Remove switchgear 1200 A				6.000		370		370	450	565
	Switchgear 1200 A				21.818	1,700	1,350		3,050	3,525	4,175
	Test switchgear				2.000		123		123	151	189
	Total				29.818	1,700	1,843		3,543	**4,126**	**4,929**

D5013 216 Fuses

	System Description	Freq. (Years)	Crew	Unit	Labor Hours	2020 Bare Costs				Total In-House	Total w/O&P
						Material	Labor	Equipment	Total		
0010	**Replace fuse**	25	1 ELEC	Ea.							
	Remove fuse				.151		9.25		9.25	11.40	14.25
	Fuse 600 A				.400	310	24.50		334.50	370	425
	Total				.551	310	33.75		343.75	**381.40**	**439.25**

D5013 220 Switchgear, Indoor, Less Than 600 V

	System Description	Freq. (Years)	Crew	Unit	Labor Hours	2020 Bare Costs				Total In-House	Total w/O&P
						Material	Labor	Equipment	Total		
0010	**Repair switchgear, - (5% of total C.B.)**	10	1 ELEC	Ea.							
	Remove circuit breaker				.400		24.50		24.50	30	38
	Circuit breaker				1.143	345	70		415	465	540
	Total				1.543	345	94.50		439.50	**495**	**578**
0020	**Maintenance and inspection**	3	1 ELEC	Ea.							
	Inspect and clean panel				.500		30.50		30.50	37.50	47
	Total				.500		30.50		30.50	**37.50**	**47**

D5013 220 Switchgear, Indoor, Less Than 600 V

	System Description	Freq. (Years)	Crew	Unit	Labor Hours	2020 Bare Costs				Total In-House	Total w/O&P
						Material	Labor	Equipment	Total		
0030	**Replace switchgear, 225 A**	30	1 ELEC	Ea.							
	Remove panelboard, 4 wire, 120/208 V, 225 A				4.000		245		245	300	380
	Panelboard, 4 wire, 120/208 V, 225 A main circuit breaker				11.940	4,125	735		4,860	5,450	6,275
	Test indoor switchgear				1.509		92.50		92.50	114	143
	Total				17.450	4,125	1,072.50		5,197.50	**5,864**	**6,798**
0200	**Replace switchgear, 400 A**	30	2 ELEC	Ea.							
	Turn power off, later turn on				.200		12.30		12.30	15.10	18.90
	Remove panelboard, 4 wire, 120/208 V, 100 A				3.333		205		205	251	315
	Remove panelboard, 4 wire, 120/208 V, 400 A				8.333		510		510	630	785
	Panelboard, 4 wire, 120/208 V, 100 A main circuit breaker				17.021	1,825	1,050		2,875	3,300	3,875
	Panelboard, 4 wire, 120/208 V, 400 A main circuit breaker				33.333	5,225	2,050		7,275	8,250	9,675
	Test circuits				.145		8.92		8.92	10.95	13.70
	Total				62.367	7,050	3,836.22		10,886.22	**12,457.05**	**14,682.60**
0240	**Replace switchgear, 600 A**	30	2 ELEC	Ea.							
	Turn power off, later turn on				.300		18.45		18.45	22.50	28.50
	Remove panelboard, 4 wire, 120/208 V, 200 A				6.667		410		410	505	630
	Remove switchboard, incoming section, 600 A				4.848		297		297	365	460
	Remove switchboard, distribution section, 600 A				4.000		245		245	300	380
	Panelboard, 4 wire, 120/208 V, 225 A main circuit breaker				28.571	4,125	1,750		5,875	6,700	7,850
	Switchboard, incoming section, 600 A				16.000	4,625	980		5,605	6,300	7,275
	Switchboard, distribution section, 600 A				16.000	1,700	980		2,680	3,075	3,625
	Test circuits				.218		13.38		13.38	16.45	20.50
	Total				76.605	10,450	4,693.83		15,143.83	**17,283.95**	**20,269**
0280	**Replace switchgear, 800 A**	30	2 ELEC	Ea.							
	Turn power off, later turn on				.400		24.60		24.60	30	38
	Remove panelboard, 4 wire, 120/208 V, 200 A (2 panelboards)				13.333		820		820	1,000	1,250
	Remove switchboard, incoming section, 800 A				5.517		340		340	415	520
	Remove switchboard, distribution section, 800 A				4.444		273		273	335	420
	Panelboard, 4 wire, 120/208 V, 225 A main lugs (2 panelboards)				47.059	4,400	2,900		7,300	8,400	9,950
	Switchboard, incoming section, 800 A				18.182	4,625	1,125		5,750	6,450	7,500
	Switchboard, distribution section, 800 A				18.182	2,175	1,125		3,300	3,775	4,450
	Test circuits				.291		17.84		17.84	22	27.50
	Total				107.408	11,200	6,625.44		17,825.44	**20,427**	**24,155.50**

D5013 220 Switchgear, Indoor, Less Than 600 V

	System Description	Freq. (Years)	Crew	Unit	Labor Hours	2020 Bare Costs				Total In-House	Total w/O&P
						Material	Labor	Equipment	Total		
0320	**Replace switchgear, 1200 A**	30	3 ELEC	Ea.							
	Turn power off, later turn on				.600		36.90		36.90	45	56.50
	Remove panelboard, 4 wire, 120/208 V, 200 A (4 panelboards)				26.667		1,640		1,640	2,000	2,525
	Remove switchboard, incoming section, 1200 A				6.667		410		410	505	630
	Remove switchboard, distribution section, 1200 A				5.161		315		315	390	485
	Panelboard, 4 wire, 120/208 V, 225 A main lugs (4 panelboards)				94.118	8,800	5,800		14,600	16,800	19,900
	Switchboard, incoming section, 1200 A				22.222	5,550	1,375		6,925	7,775	9,050
	Switchboard, distribution section, 1200 A				22.222	2,950	1,375		4,325	4,925	5,800
	Test circuits				.436		26.76		26.76	33	41
	Total				178.093	17,300	10,978.66		28,278.66	**32,473**	**38,487.50**
0360	**Replace switchgear, 1600 A**	30	3 ELEC	Ea.							
	Turn power off, later turn on				.800		49.20		49.20	60.50	75.50
	Remove panelboard, 4 wire, 120/208 V, 200 A (6 panelboards)				40.000		2,460		2,460	3,025	3,775
	Remove switchboard, incoming section, 1600 A				7.273		445		445	550	685
	Remove switchboard, distribution section, 1600 A				5.517		340		340	415	520
	Panelboard, 4 wire, 120/208 V, 225 A main lugs (6 panelboards)				141.176	13,200	8,700		21,900	25,200	29,900
	Switchboard, incoming section, 1600 A				24.242	5,550	1,475		7,025	7,925	9,250
	Switchboard, distribution section, 1600 A				24.242	4,400	1,475		5,875	6,675	7,800
	Test circuits				.582		35.68		35.68	44	55
	Total				243.833	23,150	14,979.88		38,129.88	**43,894.50**	**52,060.50**
0400	**Replace switchgear, 2000 A**	30	3 ELEC	Ea.							
	Turn power off, later turn on				1.000		61.50		61.50	75.50	94.50
	Remove panelboard, 4 wire, 120/208 V, 200 A (8 panelboards)				53.333		3,280		3,280	4,025	5,050
	Remove switchboard, incoming section, 2000 A				7.619		465		465	575	720
	Remove switchboard, distribution section, 2000 A				5.926		365		365	445	560
	Panelboard, 4 wire, 120/208 V, 225 A main lugs (8 panelboards)				188.235	17,600	11,600		29,200	33,600	39,800
	Switchboard, incoming section, 2000 A				25.806	6,000	1,575		7,575	8,550	9,925
	Switchboard, distribution section, 2000 A				25.806	5,500	1,575		7,075	8,000	9,300
	Test circuits				.727		44.60		44.60	55	68.50
	Total				308.454	29,100	18,966.10		48,066.10	**55,325.50**	**65,518**

D5013 224 Switchgear, Indoor, 600 V

	System Description	Freq. (Years)	Crew	Unit	Labor Hours	2020 Bare Costs				Total In-House	Total w/O&P
						Material	Labor	Equipment	Total		
0010	**Maintenance and repair, - (5% of total fuses)**	10	1 ELEC	Ea.							
	Remove fuse				.151		9.25		9.25	11.40	14.25
	Fuse				.400	310	24.50		334.50	370	425
	Total				.551	310	33.75		343.75	**381.40**	**439.25**
0020	**Maintenance and inspection**	3	1 ELEC	Ea.							
	Inspect and clean panel				.500		30.50		30.50	37.50	47
	Total				.500		30.50		30.50	**37.50**	**47**
0030	**Replace switchgear**	30	2 ELEC	Ea.							
	Remove indoor switchgear				4.000		245		245	300	380
	Switchgear				14.035	1,300	860		2,160	2,500	2,950
	Test indoor switchgear				2.000		123		123	151	189
	Total				20.035	1,300	1,228		2,528	**2,951**	**3,519**

D5013 230 Meters

	System Description	Freq. (Years)	Crew	Unit	Labor Hours	2020 Bare Costs				Total In-House	Total w/O&P
						Material	Labor	Equipment	Total		
0010	**Repair switchboard meter**	10	1 ELEC	Ea.							
	Repair meter				5.000	1,025	305		1,330	1,500	1,750
	Total				5.000	1,025	305		1,330	**1,500**	**1,750**
0020	**Replace switchboard meter**	20	1 ELEC	Ea.							
	Remove meter				.500		30.50		30.50	37.50	47
	Meter				1.600	5,175	98		5,273	5,825	6,625
	Total				2.100	5,175	128.50		5,303.50	**5,862.50**	**6,672**

D5013 240 Inverter

	System Description	Freq. (Years)	Crew	Unit	Labor Hours	2020 Bare Costs				Total In-House	Total w/O&P
						Material	Labor	Equipment	Total		
0010	**Maintenance and repair**	1	2 ELEC	Ea.							
	Repair inverter component				5.000	355	305		660	770	915
	Total				5.000	355	305		660	**770**	**915**

D5013 240 Inverter

	System Description	Freq. (Years)	Crew	Unit	Labor Hours	2020 Bare Costs				Total In-House	Total w/O&P
						Material	Labor	Equipment	Total		
0020	**Maintenance and inspection**	0.25	1 ELEC	Ea.							
	Check connections, controls & relays				1.600		98		98	121	151
	Total				1.600		98		98	**121**	**151**
0030	**Replace 1 kVA inverter**	20	2 ELEC	Ea.							
	Remove inverter				.500		30.50		30.50	37.50	47
	Inverter				1.600	3,550	98		3,648	4,025	4,600
	Total				2.100	3,550	128.50		3,678.50	**4,062.50**	**4,647**

D5013 250 Rectifier, Up To 600 V

	System Description	Freq. (Years)	Crew	Unit	Labor Hours	2020 Bare Costs				Total In-House	Total w/O&P
						Material	Labor	Equipment	Total		
0010	**Maintenance and repair**	2	1 ELEC	Ea.							
	Repair rectifier component				4.000	305	245		550	635	760
	Total				4.000	305	245		550	**635**	**760**
0020	**Maintenance and inspection**	0.33	1 ELEC	Ea.							
	Check connections, controls & relays				1.600		98		98	121	151
	Total				1.600		98		98	**121**	**151**
0030	**Replace rectifier**	20	2 ELEC	Ea.							
	Remove rectifier				.500		30.50		30.50	37.50	47
	Rectifier				1.600	925	98		1,023	1,150	1,300
	Total				2.100	925	128.50		1,053.50	**1,187.50**	**1,347**

D5013 264 Motor Starter, Up To 600 V

	System Description	Freq. (Years)	Crew	Unit	Labor Hours	2020 Bare Costs				Total In-House	Total w/O&P
						Material	Labor	Equipment	Total		
0010	**Maintenance and repair**	5	1 ELEC	Ea.							
	Remove motor starter coil				.500		30.50		30.50	37.50	47
	Motor starter coil				1.600	37	98		135	161	197
	Total				2.100	37	128.50		165.50	**198.50**	**244**

D5013 264 Motor Starter, Up To 600 V

	System Description	Freq. (Years)	Crew	Unit	Labor Hours	2020 Bare Costs Material	Labor	Equipment	Total	Total In-House	Total w/O&P
0020	**Maintenance and inspection**	0.50	1 ELEC	Ea.							
	Inspect & clean motor starter				.667		41		41	50.50	63
	Total				.667		41		41	**50.50**	**63**
0030	**Replace starter**	18	1 ELEC	Ea.							
	Remove motor starter				.899		55		55	68	85
	Magnetic motor starter 5 H.P., size 0				3.478	345	213		558	640	760
	Cut, form, align & connect wires				.151		9.25		9.25	11.40	14.25
	Check operation of motor starter				.018		1.12		1.12	1.38	1.73
	Total				4.546	345	278.37		623.37	**720.78**	**860.98**

D5013 266 Motor Starter, 600 V

	System Description	Freq. (Years)	Crew	Unit	Labor Hours	2020 Bare Costs Material	Labor	Equipment	Total	Total In-House	Total w/O&P
0010	**Maintenance and repair**	3	1 ELEC	Ea.							
	Remove motor starter coil				1.702		104		104	128	161
	Motor starter coil				5.000	125	305		430	515	625
	Total				6.702	125	409		534	**643**	**786**
0020	**Maintenance and inspection**	0.25	1 ELEC	Ea.							
	Inspect & clean motor starter				.667		41		41	50.50	63
	Total				.667		41		41	**50.50**	**63**
0030	**Replace starter**	18	2 ELEC	Ea.							
	Remove motor starter				3.333		205		205	251	315
	Motor starter FVNR 600 V, 3 P, 100 H.P. motor				13.333	2,550	820		3,370	3,800	4,450
	Cut, form, align & connect wires				.170		10.45		10.45	12.85	16.10
	Check operation of motor starter				.018		1.12		1.12	1.38	1.73
	Total				16.855	2,550	1,036.57		3,586.57	**4,065.23**	**4,782.83**

D5013 272 Secondary Transformer, Liquid Filled

	System Description	Freq. (Years)	Crew	Unit	Labor Hours	2020 Bare Costs Material	Labor	Equipment	Total	Total In-House	Total w/O&P
0010	**Maintenance and repair**	25	1 ELEC	Ea.							
	Remove transformer terminal				.348		21.50		21.50	26	33
	Transformer terminal				1.000	13.05	61.50		74.55	90	111
	Total				1.348	13.05	83		96.05	**116**	**144**
0020	**Maintenance and inspection**	0.50	1 ELEC	Ea.							
	Check connections & fluid				.667		41		41	50.50	63
	Total				.667		41		41	**50.50**	**63**

D5013 274 Secondary Transformer, Dry

	System Description	Freq. (Years)	Crew	Unit	Labor Hours	2020 Bare Costs Material	Labor	Equipment	Total	Total In-House	Total w/O&P
0010	**Maintenance and repair**	10	1 ELEC	Ea.							
	Remove transformer terminal				.500		30.50		30.50	37.50	47
	Transformer terminal				1.000	44	61.50		105.50	124	150
	Total				1.500	44	92		136	**161.50**	**197**
0020	**Maintenance and inspection**	0.50	1 ELEC	Ea.							
	Inspect connections & tighten				1.000		61.50		61.50	75.50	94.50
	Total				1.000		61.50		61.50	**75.50**	**94.50**
0030	**Replace 15 kVA transformer**	30	2 ELEC	Ea.							
	Remove transformer 15 KVA, incl. supp., wire & conduit terminations				4.360		267		267	330	410
	Transformer 15 KVA, 240/480 V primary, 120/208 V secondary				14.545	1,900	890		2,790	3,175	3,750
	Total				18.905	1,900	1,157		3,057	**3,505**	**4,160**
0040	**Replace 112.5 kVA transformer**	30	R-3	Ea.							
	Remove transformer 112.5 KVA, incl. support, wire, conduit termination				6.897		420	[illegible]	485	590	720
	Transformer 112.5 KVA, 240/480 V primary, 120/208 V secondary				22.222	4,075	1,350	[illegible]	5,635	6,400	7,425
	Total				29.119	4,075	1,770	[illegible]	6,120	**6,990**	**8,145**
0050	**Replace 500 kVA transformer**	30	R-3	Ea.							
	Remove transformer 500 KVA, incl. support,wire & conduit terminations				14.286		875	[illegible]	1,010	1,225	1,500
	Transformer 500 KVA, 240/480 V primary, 120/208 V secondary				44.444	18,200	2,725	[illegible]	21,345	23,800	27,400
	Total				58.730	18,200	3,600	[illegible]	22,355	**25,025**	**28,900**

D5013 280 Lighting Panel, Indoor

	System Description	Freq. (Years)	Crew	Unit	Labor Hours	2020 Bare Costs				Total In-House	Total w/O&P
						Material	Labor	Equipment	Total		
0020	**Maintenance and inspection**	3	1 ELEC	Ea.							
	Inspect and clean panels				.500		30.50		30.50	37.50	47
	Total				.500		30.50		30.50	**37.50**	**47**
0030	**Replace load center, 100 A**	20	2 ELEC	Ea.							
	Turn power off, later turn on				.100		6.15		6.15	7.55	9.45
	Remove load center, 3 wire, 120/240 V				3.077		89		189	232	291
	Load centers, 3 Wire, 120/240V, 100 A				6.667	159	410		569	680	830
	Test circuits				.073		4.46		4.46	5.50	6.85
	Total				9.916	159	609.61		768.61	**925.05**	**1,137.30**

D5023 110 Wireway

	System Description	Freq. (Years)	Crew	Unit	Labor Hours	2020 Bare Costs Material	Labor	Equipment	Total	Total In-House	Total w/O&P
0030	**Replace 8″ x 8″ wireway**	20	2 ELEC	L.F.							
	Remove and reinstall supply box				.050		3.05		3.05	3.77	4.70
	Remove wireway, fittings & support				.100		6.15		6.15	7.55	9.45
	Wireway 8″ x 8″ , fittings & support				.400	28	24.50		52.50	61	73
	Total				.550	28	33.70		61.70	**72.32**	**87.15**

D5023 112 Conduit EMT

	System Description	Freq. (Years)	Crew	Unit	Labor Hours	2020 Bare Costs Material	Labor	Equipment	Total	Total In-House	Total w/O&P
0010	**Replace 1″ EMT conduit**	50	2 ELEC	M.L.F.							
	Remove & reinstall cover plate				1.290		79		79	97.50	122
	Remove conduit w/ fittings & hangers				20.300		1,250		1,250	1,525	1,925
	Remove junction box				1.000		61.50		61.50	75.50	94.50
	1″ EMT with couplings				38.650	1,500	2,370		3,870	4,550	5,525
	Outlet box square				4.000	68.50	245		313.50	375	465
	Box connector				.889	11.10	54.50		65.60	79	98
	Beam clamp				5.000	338	322		660	770	920
	Conduit hanger				5.926	630	364		994	1,150	1,350
	Total				77.055	2,547.60	4,746		7,293.60	**8,622**	**10,499.50**

D5023 120 Cable, Non-Metallic (NM) Sheathed

	System Description	Freq. (Years)	Crew	Unit	Labor Hours	2020 Bare Costs Material	Labor	Equipment	Total	Total In-House	Total w/O&P
0010	**Replace NM cable**	50	2 ELEC	M.L.F.							
	Shut off power, later turn on				.100		6.15		6.15	7.55	9.45
	Remove & reinstall cover plate				1.290		79		79	97.50	122
	Remove junction box				1.000		61.50		61.50	75.50	94.50
	Remove NM cable #12, 2 wire				12.720		780		780	960	1,200
	NM cable #12, 2 wire				32.000	265	1,960		2,225	2,700	3,325
	Staples, for Romex or BX cable				4.000	4	246		250	305	385
	NM #12-2 wire cable connectors				.602	17.10	36.90		54	64	78.50
	Outlet box square 4″ for Romex or BX				4.000	91	245		336	400	495
	Total				55.712	377.10	3,414.55		3,791.65	**4,609.55**	**5,709.45**

D5023 122 Cable, Service

	System Description	Freq. (Years)	Crew	Unit	Labor Hours	2020 Bare Costs				Total In-House	Total w/O&P
						Material	Labor	Equipment	Total		
0010	**Replace service cable**	50	2 ELEC	M.L.F.							
	Shut off power, later turn on				.100		6.15		6.15	7.55	9.45
	Remove & reinstall cover plate				1.290		79		79	97.50	122
	Remove junction box				1.000		61.50		61.50	75.50	94.50
	Remove wire #6 XLPE				2.000		122.50		122.50	151	189
	600 V, copper type XLPE-USE				12.308	565	755		1,320	1,550	1,875
	Staples, for service cable				5.000	8	306		314	385	480
	SER cable connectors				3.333	39.40	205		244.40	295	365
	Outlet box square 4″				4.000	68.50	245		313.50	375	465
	Total				29.031	680.90	1,780.15		2,461.05	**2,936.55**	**3,599.95**

D5023 124 Cable, Armored

	System Description	Freq. (Years)	Crew	Unit	Labor Hours	2020 Bare Costs				Total In-House	Total w/O&P
						Material	Labor	Equipment	Total		
0010	**Replace armored cable**	60	2 ELEC	M.L.F.							
	Shut off power, later turn on				.100		6.15		6.15	7.55	9.45
	Remove & reinstall cover plate				1.290		79		79	97.50	122
	Remove junction box				1.000		61.50		61.50	75.50	94.50
	Remove armored cable				13.220		810		810	1,000	1,250
	BX #12, 2 wire				34.783	410	2,130		2,540	3,075	3,825
	Staples, for Romex or BX cable				4.000	4	246		250	305	385
	Cable connector				2.000	7.90	122.50		130.40	160	199
	Outlet box square 4″ for Romex or BX				4.000	91	245		336	400	495
	Total				60.393	512.90	3,700.15		4,213.05	**5,120.55**	**6,379.95**

D5023 126 Branch Wiring, With Junction Box

	System Description	Freq. (Years)	Crew	Unit	Labor Hours	2020 Bare Costs				Total In-House	Total w/O&P
						Material	Labor	Equipment	Total		
0010	**Replace branch wiring**	50	2 ELEC	M.L.F.							
	Shut off power, later turn on				.100		6.15		6.15	7.55	9.45
	Remove & reinstall cover plate				1.290		79		79	97.50	122
	Remove junction box				1.000		61.50		61.50	75.50	94.50
	Remove #12 wire				1.455		89		89	110	138
	Conductor THW-THHN, copper, stranded, #12				7.273	106	445		551	665	820
	Outlet box square 4"				4.000	68.50	245		313.50	375	465
	Wire connectors				1.000	2.10	61.20		63.30	77.50	97
	Total				16.117	176.60	986.35		1,163.45	**1,408.05**	**1,745.95**

D5023 128 Branch Wiring, 600 V

	System Description	Freq. (Years)	Crew	Unit	Labor Hours	2020 Bare Costs				Total In-House	Total w/O&P
						Material	Labor	Equipment	Total		
0010	**Replace branch wiring**	50	2 ELEC	M.L.F.							
	Shut off power, later turn on				.100		6.15		6.15	7.55	9.45
	Remove #12 wire				1.455		89		89	110	138
	Conductor THW-THHN, copper, stranded, #12				7.273	106	445		551	665	820
	Cut, form and align 4 wires				.410		25		25	31	39
	Splice, solder, insulate 2 wires				2.667		164		164	201	252
	Total				11.904	106	729.15		835.15	**1,014.55**	**1,258.45**

D5023 130 Circuit Breaker

	System Description	Freq. (Years)	Crew	Unit	Labor Hours	2020 Bare Costs				Total In-House	Total w/O&P
						Material	Labor	Equipment	Total		
0010	**Maintenance and repair breaker, molded case, 480 V, 1 pole**	20	1 ELEC	Ea.							
	Tighten connections				.889		54.50		54.50	67	84
	Total				.889		54.50		54.50	**67**	**84**
0020	**Maintenance and inspection C.B., molded case, 480 V, 1 pole**	0.50	1 ELEC	Ea.							
	Inspect, tighten C.B. wiring				.400		24.50		24.50	30	38
	Total				.400		24.50		24.50	**30**	**38**

D5023 130 Circuit Breaker

	System Description	Freq. (Years)	Crew	Unit	Labor Hours	2020 Bare Costs				Total In-House	Total w/O&P
						Material	Labor	Equipment	Total		
0030	**Replace C.B. molded case, 480 V, 1 pole**	50	1 ELEC	Ea.							
	Remove circuit breaker				.348		21.50		21.50	26	33
	Circuit breaker molded case, 480 V, 70 to 100 A, 1 P				1.143	345	70		415	465	540
	Cut, form, align & connect wires				.151		9.25		9.25	11.40	14.25
	Check operation of breaker				.018		1.12		1.12	1.38	1.73
	Total				1.660	345	101.87		446.87	**503.78**	**588.98**
1010	**Maintenance and repair breaker, molded case, 480 V, 2 pole**	20	1 ELEC	Ea.							
	Tighten connections				.889		54.50		54.50	67	84
	Total				.889		54.50		54.50	**67**	**84**
1020	**Maintenance and inspection C.B., molded case, 480 V, 2 pole**	0.50	1 ELEC	Ea.							
	Inspect, tighten C.B. wiring				.400		24.50		24.50	30	38
	Total				.400		24.50		24.50	**30**	**38**
1030	**Replace C.B. molded case, 480 V, 2 pole**	50	1 ELEC	Ea.							
	Remove circuit breaker				.500		30.50		30.50	37.50	47
	Circuit breaker molded case, 480 V, 70 to 100 A, 2 P				1.600	585	98		683	765	880
	Cut, form, align & connect wires				.151		9.25		9.25	11.40	14.25
	Check operation of breaker				.018		1.12		1.12	1.38	1.73
	Total				2.269	585	138.87		723.87	**815.28**	**942.98**
1040	**Maintenance and repair breaker, molded case, 480 V, 3 pole**	20	1 ELEC	Ea.							
	Tighten connections				.889		54.50		54.50	67	84
	Total				.889		54.50		54.50	**67**	**84**
1050	**Maintenance and inspection C.B., molded case, 480 V, 3 pole**	0.50	1 ELEC	Ea.							
	Inspect & tighten C.B. wiring				.400		24.50		24.50	30	38
	Total				.400		24.50		24.50	**30**	**38**
1060	**Replace C.B. molded case, 480 V, 3 pole**	50	1 ELEC	Ea.							
	Remove circuit breaker				.500		30.50		30.50	37.50	47
	Circuit breaker molded case, 480 V, 70 to 100 A, 3 P				2.000	690	123		813	910	1,050
	Cut, form, align & connect wires				.151		9.25		9.25	11.40	14.25
	Check operation of breaker				.018		1.12		1.12	1.38	1.73
	Total				2.669	690	163.87		853.87	**960.28**	**1,112.98**

D5023 130 Circuit Breaker

	System Description	Freq. (Years)	Crew	Unit	Labor Hours	2020 Bare Costs				Total In-House	Total w/O&P
						Material	Labor	Equipment	Total		
2010	**Repair failed breaker, molded case, 600 V, 2 pole**	10	1 ELEC	Ea.							
	Remove trip switch				.500		30.50		30.50	37.50	47
	Trip switch				1.600	164	98		262	300	355
	Total				2.100	164	128.50		292.50	**337.50**	**402**
2020	**Maintenance and inspection C.B., molded case, 600 V, 2 pole**	0.33	1 ELEC	Ea.							
	Inspect & tighten C.B. wiring				.400		24.50		24.50	30	38
	Total				.400		24.50		24.50	**30**	**38**
2030	**Replace C.B. molded case, 600 V, 2 pole**	50	1 ELEC	Ea.							
	Remove circuit breaker				.500		30.50		30.50	37.50	47
	Circuit breaker molded case, 600 V, 70 to 100 A, 2 P				1.600	665	98		763	850	980
	Cut, form, align & connect wires				.170		10.45		10.45	12.85	16.10
	Check operation of breakers				.018		1.12		1.12	1.38	1.73
	Total				2.289	665	140.07		805.07	**901.73**	**1,044.83**
3010	**Repair failed breaker, molded case, 600 V, 3 pole**	10	1 ELEC	Ea.							
	Remove trip switch				.500		30.50		30.50	37.50	47
	Trip switch				1.600	164	98		262	300	355
	Total				2.100	164	128.50		292.50	**337.50**	**402**
3020	**Maintenance and inspection C.B., molded case, 600 V, 3 pole**	0.33	1 ELEC	Ea.							
	Inspect & tighten C.B. wiring				.400		24.50		24.50	30	38
	Total				.400		24.50		24.50	**30**	**38**
3030	**Replace C.B. molded case, 600 V, 3 pole**	50	1 ELEC	Ea.							
	Remove circuit breaker				.889		54.50		54.50	67	84
	Circuit breaker molded case, 600 V, 125 to 400 A, 3 P				3.478	3,900	213		4,113	4,550	5,200
	Cut, form, align & connect wires				.170		10.45		10.45	12.85	16.10
	Check operation of breakers				.018		1.12		1.12	1.38	1.73
	Total				4.556	3,900	279.07		4,179.07	**4,631.23**	**5,301.83**
3040	**Maintenance and repair breaker, enclosed, 240 V, 1 pole**	25	1 ELEC	Ea.							
	Tighten connections				.889		54.50		54.50	67	84
	Total				.889		54.50		54.50	**67**	**84**

D5023 130 Circuit Breaker

	System Description	Freq. (Years)	Crew	Unit	Labor Hours	2020 Bare Costs				Total In-House	Total w/O&P
						Material	Labor	Equipment	Total		
3050	**Maintenance and inspection C.B., enclosed, 240 V, 1 pole**	1	1 ELEC	Ea.							
	Inspect & tighten C.B. wiring				.400		24.50		24.50	30	38
	Total				.400		24.50		24.50	**30**	**38**
3060	**Replace C.B. enclosed, 240 V, 1 pole**	50	1 ELEC	Ea.							
	Remove circuit breaker				.500		30.50		30.50	37.50	47
	Circuit breaker enclosed, 240 V, 15 to 60 A, 1 P				2.105	275	129		404	460	545
	Cut, form, align & connect wires				.151		9.25		9.25	11.40	14.25
	Check operation of breaker				.018		1.12		1.12	1.38	1.73
	Total				2.775	275	169.87		444.87	**510.28**	**607.98**
4010	**Maintenance and repair breaker, enclosed, 240 V, 2 pole**	25	1 ELEC	Ea.							
	Tighten connections				.889		54.50		54.50	67	84
	Total				.889		54.50		54.50	**67**	**84**
4020	**Maintenance and inspection C.B., enclosed, 240 V, 2 pole**	1	1 ELEC	Ea.							
	Inspect & tighten C.B. wiring				.400		24.50		24.50	30	38
	Total				.400		24.50		24.50	**30**	**38**
4030	**Replace C.B. enclosed, 240 V, 2 pole**	50	1 ELEC	Ea.							
	Remove circuit breaker				.500		30.50		30.50	37.50	47
	Circuit breaker enclosed, 240 V, 15 to 60 A, 2 P				2.286	525	140		665	750	870
	Cut, form, align & connect wires				.151		9.25		9.25	11.40	14.25
	Check operation of breaker				.018		1.12		1.12	1.38	1.73
	Total				2.955	525	180.87		705.87	**800.28**	**932.98**
4040	**Maintenance and repair breaker, enclosed, 240 V, 3 pole**	25	1 ELEC	Ea.							
	Tighten connections				.889		54.50		54.50	67	84
	Total				.889		54.50		54.50	**67**	**84**
4050	**Maintenance and inspection C.B., enclosed, 240 V, 3 pole**	1	1 ELEC	Ea.							
	Inspect & tighten C.B. wiring				.400		24.50		24.50	30	38
	Total				.400		24.50		24.50	**30**	**38**

D5023 130 Circuit Breaker

	System Description	Freq. (Years)	Crew	Unit	Labor Hours	2020 Bare Costs				Total In-House	Total w/O&P
						Material	Labor	Equipment	Total		
4060	**Replace C.B. enclosed, 240 V, 3 pole**	50	1 ELEC	Ea.							
	Remove circuit breaker				.500		30.50		30.50	37.50	47
	Circuit breaker enclosed, 240 V, 15 to 60 A, 3 P				2.500	725	153		878	985	1,150
	Cut, form, align & connect wires				.151		9.25		9.25	11.40	14.25
	Check operation of breaker				.018		1.12		1.12	1.38	1.73
	Total				3.169	725	193.87		918.87	**1,035.28**	**1,212.98**
5010	**Repair failed breaker, enclosed, 600 V, 2 pole**	4	1 ELEC	Ea.							
	Repair circuit breaker component				2.105	550	129		679	765	885
	Total				2.105	550	129		679	**765**	**885**
5020	**Maintenance and inspection C.B., enclosed, 600 V, 2 pole**	0.33	1 ELEC	Ea.							
	Inspect & tighten C.B. wiring				.400		24.50		24.50	30	38
	Total				.400		24.50		24.50	**30**	**38**
5030	**Replace C.B. enclosed, 600 V, 2 pole**	50	1 ELEC	Ea.							
	Remove circuit breaker				.800		49		49	60.50	75.50
	Circuit breaker enclosed, 600 V, 100 A, 2 P				3.200	665	196		861	975	1,125
	Cut, form, align & connect wires				.170		10.45		10.45	12.85	16.10
	Check operation of breaker				.018		1.12		1.12	1.38	1.73
	Total				4.189	665	256.57		921.57	**1,049.73**	**1,218.33**
6010	**Repair failed breaker, enclosed, 600 V, 3 pole**	4	1 ELEC	Ea.							
	Repair circuit breaker component				2.105	475	129		604	680	795
	Total				2.105	475	129		604	**680**	**795**
6020	**Maintenance and inspection C.B., enclosed, 600 V, 3 pole**	0.33	1 ELEC	Ea.							
	Inspect & tighten C.B. wiring				.400		24.50		24.50	30	38
	Total				.400		24.50		24.50	**30**	**38**
6030	**Replace C.B. enclosed, 600 V, 3 pole**	50	1 ELEC	Ea.							
	Remove circuit breaker				.889		54.50		54.50	67	84
	Circuit breaker enclosed, 600 V, 100 A, 3 P				3.478	735	213		948	1,075	1,250
	Cut, form, align & connect wires				.170		10.45		10.45	12.85	16.10
	Check operation of breaker				.018		1.12		1.12	1.38	1.73
	Total				4.556	735	279.07		1,014.07	**1,156.23**	**1,351.83**

D5023 130 Circuit Breaker

	System Description	Freq. (Years)	Crew	Unit	Labor Hours	2020 Bare Costs				Total In-House	Total w/O&P
						Material	Labor	Equipment	Total		
6830	**Replace C.B. enclosed, 600 V, 3 pole 800 A**	50	2 ELEC	Ea.							
	Remove circuit breaker				4.444		273		273	335	420
	Circuit breaker enclosed, 600 V, 800 A, 3 P				17.021	5,475	1,050		6,525	7,300	8,450
	Cut, form, align & connect wires				.340		20.90		20.90	25.50	32
	Check operation of breaker				.037		2.24		2.24	2.76	3.46
	Total				21.843	5,475	1,346.14		6,821.14	**7,663.26**	**8,905.46**
6930	**Replace C.B. enclosed, 600 V, 3 pole 1000 A**	50	2 ELEC	Ea.							
	Remove circuit breaker				5.161		315		315	390	485
	Circuit breaker enclosed, 600 V, 1000 A, 3 P				19.048	6,925	1,175		8,100	9,050	10,500
	Cut, form, align & connect wires				.516		31.67		31.67	39	49
	Check operation of breaker				.055		3.39		3.39	4.18	5.25
	Total				24.780	6,925	1,525.06		8,450.06	**9,483.18**	**11,039.25**
7130	**Replace C.B. enclosed, 600 V, 3 pole 1600 A**	50	2 ELEC	Ea.							
	Remove circuit breaker				5.714		350		350	430	540
	Circuit breaker enclosed, 600 V, 1600 A, 3 P				22.222	16,100	1,375		17,475	19,400	22,200
	Cut, form, align & connect wires				.681		41.80		41.80	51.50	64.50
	Check operation of breaker				.073		4.48		4.48	5.50	6.90
	Total				28.691	16,100	1,771.28		17,871.28	**19,887**	**22,811.40**
7230	**Replace C.B. enclosed, 600 V, 3 pole 2000 A**	50	2 ELEC	Ea.							
	Remove circuit breaker				6.400		395		395	485	605
	Circuit breaker enclosed, 600 V, 2000 A, 3 P				25.000	17,400	1,525		18,925	21,000	24,100
	Cut, form, align & connect wires				1.418		87.08		87.08	107	134
	Check operation of breaker				.153		9.33		9.33	11.50	14.40
	Total				32.971	17,400	2,016.41		19,416.41	**21,603.50**	**24,853.40**

D5023 132 Safety Switch, Heavy Duty

	System Description	Freq. (Years)	Crew	Unit	Labor Hours	2020 Bare Costs				Total In-House	Total w/O&P
						Material	Labor	Equipment	Total		
0010	**Maintenance and repair safety switch, H.D., 3 pole**	8	1 ELEC	Ea.							
	Resecure lugs				.500		30.50		30.50	37.50	47
	Total				.500		30.50		30.50	**37.50**	**47**

D5023 132 Safety Switch, Heavy Duty

	System Description	Freq. (Years)	Crew	Unit	Labor Hours	2020 Bare Costs				Total In-House	Total w/O&P
						Material	Labor	Equipment	Total		
0020	**Maintenance and inspection safety switch, 3 pole**	1	1 ELEC	Ea.							
	Inspect & clean safety switch				.500		30.50		30.50	37.50	47
	Total				.500		30.50		30.50	**37.50**	**47**
0030	**Replace safety switch, H.D. 30 A**	25	2 ELEC	Ea.							
	Remove safety switch				.650		40		40	49	61.50
	Fused safety switch, heavy duty, 600 V, 3 P, 30 A				2.500	166	153		319	370	445
	Fuses				.600	64.50	36.75		101.25	116	137
	Check operation of switch				.018		1.12		1.12	1.38	1.73
	Total				3.769	230.50	230.87		461.37	**536.38**	**645.23**
0130	**Replace safety switch, H.D. 100 A**	25	2 ELEC	Ea.							
	Remove safety switch				1.096		67		67	82.50	104
	Fused safety switch, heavy duty, 600 V, 3 P, 100 A				4.211	365	258		623	720	855
	Fuses				.667	199.50	40.95		240.45	270	310
	Check operation of switch				.018		1.12		1.12	1.38	1.73
	Total				5.991	564.50	367.07		931.57	**1,073.88**	**1,270.73**
0230	**Replace safety switch, H.D. 200 A**	25	2 ELEC	Ea.							
	Remove safety switch				1.600		98		98	121	151
	Fused safety switch, heavy duty, 600 V, 3 P, 200 A				6.154	520	380		900	1,025	1,225
	Fuses				.800	393	49.05		442.05	495	565
	Check operation of switch				.018		1.12		1.12	1.38	1.73
	Total				8.572	913	528.17		1,441.17	**1,642.38**	**1,942.73**
0430	**Replace safety switch, H.D. 400 A**	25	2 ELEC	Ea.							
	Remove safety switch				2.353		144		144	177	222
	Fused safety switch, heavy duty, 600 V, 3 P, 400 A				8.889	1,400	545		1,945	2,200	2,600
	Fuses				1.000	849	61.50		910.50	1,000	1,150
	Check operation of switch				.037		2.24		2.24	2.76	3.46
	Total				12.278	2,249	752.74		3,001.74	**3,379.76**	**3,975.46**
0630	**Replace safety switch, H.D. 600 A**	25	2 ELEC	Ea.							
	Remove safety switch				3.478		213		213	262	330
	Fused safety switch, heavy duty, 600 V, 3 P, 600 A				13.333	2,475	820		3,295	3,725	4,350
	Fuses				1.200	1,080	73.50		1,153.50	1,275	1,475
	Check operation of switch				.037		2.24		2.24	2.76	3.46
	Total				18.048	3,555	1,108.74		4,663.74	**5,264.76**	**6,158.46**

D5023 140 Safety Switch, General Duty

	System Description	Freq. (Years)	Crew	Unit	Labor Hours	2020 Bare Costs				Total In-House	Total w/O&P
						Material	Labor	Equipment	Total		
2010	**Maintenance and repair safety switch gen., 2 pole**	8	1 ELEC	Ea.							
	Resecure lugs				.500		30.50		30.50	37.50	47
	Total				.500		30.50		30.50	**37.50**	**47**
2020	**Maintenance and inspection safety switch, 2 pole**	1	1 ELEC	Ea.							
	Inspect & clean safety switch				.500		30.50		30.50	37.50	47
	Total				.500		30.50		30.50	**37.50**	**47**
2030	**Replace safety switch 240 V, 2 pole**	25	2 ELEC	Ea.							
	Remove safety switch				.650		40		40	49	61.50
	Fused safety switch, general duty, 240 V, 2 P, 30 A				2.299	77.50	141		218.50	259	315
	Fuse, dual element, time delay, 250 V, 30 A				.320	19.80	19.60		39.40	46	55
	Cut, form, align & connect wires				.151		9.25		9.25	11.40	14.25
	Check operation of switch				.018		1.12		1.12	1.38	1.73
	Total				3.439	97.30	210.97		308.27	**366.78**	**447.48**
3010	**Maintenance and repair safety switch gen., 3 pole**	8	1 ELEC	Ea.							
	Resecure lugs				.500		30.50		30.50	37.50	47
	Total				.500		30.50		30.50	**37.50**	**47**
3020	**Maintenance and inspection safety switch, 3 pole**	1	1 ELEC	Ea.							
	Inspect & clean safety switch				.500		30.50		30.50	37.50	47
	Total				.500		30.50		30.50	**37.50**	**47**
3030	**Replace safety switch 240 V, 3 pole**	25	2 ELEC	Ea.							
	Remove safety switch				.650		40		40	49	61.50
	Fused safety switch, general duty, 240 V, 3 P, 30 A				2.500	102	153		255	300	365
	Fuse, dual element, time delay, 250 V, 30 A				.480	29.70	29.40		59.10	69	82.50
	Cut, form, align & connect wires				.151		9.25		9.25	11.40	14.25
	Check operation of switch				.018		1.12		1.12	1.38	1.73
	Total				3.800	131.70	232.77		364.47	**430.78**	**524.98**
4010	**Replace low voltage cartridge**	50	1 ELEC	Ea.							
	Remove cartridge fuse cutouts				.100		6.15		6.15	7.55	9.45
	Cartridge fuse cutouts 3 P				.296	20	18.20		38.20	44.50	53
	Total				.396	20	24.35		44.35	**52.05**	**62.45**

D5023 140 Safety Switch, General Duty

	System Description	Freq. (Years)	Crew	Unit	Labor Hours	2020 Bare Costs				Total In-House	Total w/O&P
						Material	Labor	Equipment	Total		
5010	**Replace plug fuse**	25	1 ELEC	Ea.							
	Remove plug fuse				.080		4.91		4.91	6.05	7.55
	Fuse, plug, 120 V, 15 to 30 A				.160	4.17	9.80		13.97	16.65	20.50
	Total				.240	4.17	14.71		18.88	**22.70**	**28.05**

D5023 150 Receptacles And Plugs

	System Description	Freq. (Years)	Crew	Unit	Labor Hours	2020 Bare Costs				Total In-House	Total w/O&P
						Material	Labor	Equipment	Total		
0010	**Maintenance and repair**	20	1 ELEC	Ea.							
	Repair receptacle component (cover)				.500	2.57	30.50		33.07	40.50	50
	Total				.500	2.57	30.50		33.07	**40.50**	**50**
0020	**Replace receptacle/plug**	20	1 ELEC	Ea.							
	Turn power off, later turn on				.100		6.15		6.15	7.55	9.45
	Remove & install cover plates				.170		10.45		10.45	12.85	16.10
	Remove receptacle				.040		2.45		2.45	3.02	3.78
	Wiring devices, receptacle				.381	10.65	23.50		34.15	40.50	49.50
	Cut, form, align & connect wires				.026		1.59		1.59	1.96	2.45
	Test for operation				.015		.90		.90	1.10	1.38
	Total				.732	10.65	45.04		55.69	**66.98**	**82.66**

D5023 154 4-Pin Receptacle

	System Description	Freq. (Years)	Crew	Unit	Labor Hours	2020 Bare Costs				Total In-House	Total w/O&P
						Material	Labor	Equipment	Total		
0010	**Repair 4-pin receptacle cover**	10	1 ELEC	Ea.							
	Repair receptacle(cover)				.615	6.70	38		44.70	54	66.50
	Total				.615	6.70	38		44.70	**54**	**66.50**
0020	**Replace 4 - pin receptacle**	20	1 ELEC	Ea.							
	Remove 4-pin receptacle				.200		12.25		12.25	15.10	18.90
	Receptacle locking, 4-pin				.615	76	38		114	130	153
	Total				.815	76	50.25		126.25	**145.10**	**171.90**

D5023 210 Contactors And Relays

	System Description	Freq. (Years)	Crew	Unit	Labor Hours	2020 Bare Costs Material	Labor	Equipment	Total	Total In-House	Total w/O&P
0010	**Maintenance and repair**	3	1 ELEC	Ea.							
	Remove contactor coil				.602		37		37	45.50	57
	Contactor coil				2.000	37	123		160	192	235
	Total				2.602	37	160		197	**237.50**	**292**
0020	**Maintenance and inspection**	0.50	1 ELEC	Ea.							
	Inspect and clean contacts				.250		15.35		15.35	18.85	23.50
	Total				.250		15.35		15.35	**18.85**	**23.50**
0030	**Replace contactor**	18	1 ELEC	Ea.							
	Remove magnetic contactor				.800		49		49	60.50	75.50
	Magnetic contactor 600 V, 3 P, 25 H.P.				2.667	440	164		604	685	800
	Cut, form, align & connect wires				.151		9.25		9.25	11.40	14.25
	Check operation of contactor				.018		1.12		1.12	1.38	1.73
	Total				3.636	440	223.37		663.37	**758.28**	**891.48**

D5023 220 Wiring Devices, Switches

	System Description	Freq. (Years)	Crew	Unit	Labor Hours	2020 Bare Costs Material	Labor	Equipment	Total	Total In-House	Total w/O&P
0010	**Maintenance and repair**	10	1 ELEC	Ea.							
	Repair switch component (cover)				.500	2.57	30.50		33.07	40.50	50
	Total				.500	2.57	30.50		33.07	**40.50**	**50**
0020	**Replace switch**	15	1 ELEC	Ea.							
	Turn power off, later turn on				.100		6.15		6.15	7.55	9.45
	Remove & install cover plates				.170		10.45		10.45	12.85	16.10
	Remove switch				.040		2.45		2.45	3.02	3.78
	Wiring devices, switch				.381	3.81	23.50		27.31	33	41
	Cut, form, align & connect wires				.026		1.59		1.59	1.96	2.45
	Test for operation				.015		.90		.90	1.10	1.38
	Total				.732	3.81	45.04		48.85	**59.48**	**74.16**

D5023 222 Switch, Pull Cord

	System Description	Freq. (Years)	Crew	Unit	Labor Hours	2020 Bare Costs				Total In-House	Total w/O&P
						Material	Labor	Equipment	Total		
0010	**Maintenance and repair**	5	1 ELEC	Ea.							
	Repair switch component (cover)				.500	2.83	30.50		33.33	41	50.50
	Total				.500	2.83	30.50		33.33	**41**	**50.50**
0020	**Replace switch**	15	1 ELEC	Ea.							
	Turn power off, later turn on				.100		6.15		6.15	7.55	9.45
	Remove & install cover plates				.170		10.45		10.45	12.85	16.10
	Remove switch, pull cord				.040		2.45		2.45	3.02	3.78
	Switch, pull cord				.381	12.40	23.50		35.90	42.50	51.50
	Cut, form, align & connect wires				.026		1.59		1.59	1.96	2.45
	Test for operation				.015		.90		.90	1.10	1.38
	Total				.732	12.40	45.04		57.44	**68.98**	**84.66**

D5023 230 Light Dimming Panel

	System Description	Freq. (Years)	Crew	Unit	Labor Hours	2020 Bare Costs				Total In-House	Total w/O&P
						Material	Labor	Equipment	Total		
0010	**Minor repairs to light dimming panel**	5	1 ELEC	Ea.							
	Check and test components				.400		24.50		24.50	30	38
	Remove dimming panel component				.222		13.65		13.65	16.75	21
	Dimming panel component				.667	92	41		133	151	178
	Total				1.289	92	79.15		171.15	**197.75**	**237**
0020	**Maintenance and inspection**	1	1 ELEC	Ea.							
	Check light dimming panel				.533		32.50		32.50	40	50.50
	Total				.533		32.50		32.50	**40**	**50.50**
0030	**Replace light dimming panel**	15	1 ELEC	Ea.							
	Turn power off, later turn on				.100		6.15		6.15	7.55	9.45
	Remove dimming panel				1.600		98		98	121	151
	Dimming panel				8.000	720	490		1,210	1,400	1,650
	Test circuits				.073		4.46		4.46	5.50	6.85
	Total				9.773	720	598.31		1,318.61	**1,534.05**	**1,817.30**

D5023 240 Incandescent Lighting Fixtures

	System Description	Freq. (Years)	Crew	Unit	Labor Hours	2020 Bare Costs				Total In-House	Total w/O&P
						Material	Labor	Equipment	Total		
0010	**Maintenance and repair**	10	1 ELEC	Ea.							
	Remove glass globe				.100		6.15		6.15	7.55	9.45
	Glass globe				.296	39	18.20		57.20	65	77
	Total				.396	39	24.35		63.35	**72.55**	**86.45**
0020	**Replace lamp**	5	1 ELEC	Ea.							
	Remove incand. lighting fixture lamp				.022		1.35		1.35	1.66	2.08
	Incand. lighting fixture lamp				.065	27	3.99		30.99	34.50	40
	Total				.087	27	5.34		32.34	**36.16**	**42.08**
0030	**Replace lighting fixture**	20	1 ELEC	Ea.							
	Turn circuit off and on for fixture				.018		1.12		1.12	1.38	1.73
	Remove incand. lighting fixture				.258		15.85		15.85	19.45	24.50
	Incand. lighting fixture, 75 W				.800	51.50	49		100.50	117	140
	Total				1.076	51.50	65.97		117.47	**137.83**	**166.23**
0120	**Replace lamp, 200 W**	5	1 ELEC	Ea.							
	Remove incand. lighting fixture lamp				.022		1.36		1.36	1.68	2.10
	Incand. lighting fixture lamp, 200 W				.065	3.22	3.99		7.21	8.45	10.20
	Total				.087	3.22	5.35		8.57	**10.13**	**12.30**
0220	**Replace lamp for explosion proof fixture**	5	1 ELEC	Ea.							
	Remove incand. lighting fixture lamp				.044		2.72		2.72	3.36	4.20
	Incand. lighting fixture lamp for explosion proof fixture				.130	6.44	7.98		14.42	16.90	20.50
	Total				.175	6.44	10.70		17.14	**20.26**	**24.70**
0230	**Replace incan. lighting fixt., explosion proof, ceiling mtd., 200 W**	20	1 ELEC	Ea.							
	Turn branch circuit off and on				.018		1.12		1.12	1.38	1.73
	Remove incandescent lighting fixture, explosion proof				.920		56.50		56.50	69.50	87
	Incan. lighting fixture, explosion proof, ceiling mtd, 200 W				2.000	1,650	123		1,773	1,975	2,250
	Total				2.938	1,650	180.62		1,830.62	**2,045.88**	**2,338.73**
1230	**Replace incan. lighting fixt., high hat can type, ceiling mtd., 200 W**	20	1 ELEC	Ea.							
	Turn branch circuit off and on				.018		1.12		1.12	1.38	1.73
	Remove incandescent lighting fixture, can type				.258		15.85		15.85	19.45	24.50
	Incan. lighting fixture, high hat can type, ceiling mtd, 200 W				1.194	97.50	73.50		171	197	235
	Total				1.470	97.50	90.47		187.97	**217.83**	**261.23**

D5023 250 Quartz Fixture

	System Description	Freq. (Years)	Crew	Unit	Labor Hours	2020 Bare Costs				Total In-House	Total w/O&P
						Material	Labor	Equipment	Total		
0010	**Maintenance and repair**	10	1 ELEC	Ea.							
	Repair glass lens				.400	27	24.50		51.50	60	72
	Total				.400	27	24.50		51.50	**60**	**72**
0020	**Replace 1500 W quartz lamp**	10	1 ELEC	Ea.							
	Remove quartz lamp				.087		5.35		5.35	6.55	8.20
	Quartz lamp				.258	21	15.85		36.85	42.50	51
	Total				.345	21	21.20		42.20	**49.05**	**59.20**
0030	**Replace fixture**	20	1 ELEC	Ea.							
	Turn branch circuit off and on				.018		1.12		1.12	1.38	1.73
	Remove exterior fixt., quartz, 1500 W				.593		36.50		36.50	44.50	56
	Exterior fixt. quartz, 1500 W				1.905	108	117		225	262	315
	Total				2.516	108	154.62		262.62	**307.88**	**372.73**

D5023 260 Fluorescent Lighting Fixture

	System Description	Freq. (Years)	Crew	Unit	Labor Hours	2020 Bare Costs				Total In-House	Total w/O&P
						Material	Labor	Equipment	Total		
0010	**Replace fluor. ballast**	10	1 ELEC	Ea.							
	Remove indoor fluor., ballast				.333		20.50		20.50	25	31.50
	Fluorescent, electronic ballast				.667	41.50	41		82.50	96	115
	Test fixture				.018		1.12		1.12	1.38	1.73
	Total				1.018	41.50	62.62		104.12	**122.38**	**148.23**
0020	**Replace lamps (2 lamps), 4′, 34 W energy saver**	10	1 ELEC	Ea.							
	Remove fluor. lamps in fixture				.078		4.78		4.78	5.90	7.40
	Fluor. lamp, 4′ energy saver				.232	15.80	14.20		30	35	41.50
	Total				.310	15.80	18.98		34.78	**40.90**	**48.90**
0030	**Replace fixture, lay-in, recess mtd., 2′ x 4′, two 40 W**	20	1 ELEC	Ea.							
	Turn branch circuit off and on				.018		1.12		1.12	1.38	1.73
	Remove fluor. lighting fixture				.485		30		30	36.50	46
	Fluor. 2′ x 4′, recess mounted, two 40 W				1.509	56	92.50		148.50	175	213
	Total				2.013	56	123.62		179.62	**212.88**	**260.73**

D5023 260 Fluorescent Lighting Fixture

	System Description	Freq. (Years)	Crew	Unit	Labor Hours	2020 Bare Costs				Total In-House	Total w/O&P
						Material	Labor	Equipment	Total		
0040	**Replace fixture, lay-in, recess mtd., 2′ x 4′, four 32 W**	20	1 ELEC	Ea.							
	Turn branch circuit off and on				.018		1.12		1.12	1.38	1.73
	Remove fluor. lighting fixture				.533		32.50		32.50	40	50.50
	Fluor, lay-in, recess mtd., 2′ x 4′, four 32 W				1.702	75	104		179	211	255
	Total				2.254	75	137.62		212.62	**252.38**	**307.23**
0120	**Replace lamps (2 lamps), 8′, 60 W energy saver**	10	1 ELEC	Ea.							
	Remove fluor. lamps in fixture				.078		4.78		4.78	5.90	7.40
	Fluor. lamps, 8′ energy saver				.200	7.90	12.30		20.20	24	29
	Total				.278	7.90	17.08		24.98	**29.90**	**36.40**
0130	**Replace fixture, strip, surface mtd., 8′, two 75 W**	20	2 ELEC	Ea.							
	Turn branch circuit off and on				.018		1.12		1.12	1.38	1.73
	Remove fluor. lighting fixture				.400		24.50		24.50	30	38
	Fluor. strip, surface mounted, 8′, two 75 W				1.290	71	79		150	175	211
	Total				1.709	71	104.62		175.62	**206.38**	**250.73**
0140	**Replace fixture, strip, pendent mtd, 8′, two 75 W**	20	2 ELEC	Ea.							
	Turn branch circuit off and on				.018		1.12		1.12	1.38	1.73
	Remove fluor. lighting fixture				.593		36.50		36.50	44.50	56
	Fluor, strip, pendent mtd, 8′, two 75 W				1.818	101	112		213	248	298
	Total				2.429	101	149.62		250.62	**293.88**	**355.73**

D5023 270 Metal Halide Fixture

	System Description	Freq. (Years)	Crew	Unit	Labor Hours	2020 Bare Costs				Total In-House	Total w/O&P
						Material	Labor	Equipment	Total		
2010	**Replace M.H. ballast, 175 W**	10	1 ELEC	Ea.							
	Remove metal halide fixt. ballast				.333		20.50		20.50	25	31.50
	Metal halide fixt. 175 W, ballast				.667	97	41		138	157	184
	Test fixture				.018		1.12		1.12	1.38	1.73
	Total				1.018	97	62.62		159.62	**183.38**	**217.23**
2020	**Replace lamp, 175 W**	5	1 ELEC	Ea.							
	Remove metal halide fixt. lamp				.116		7.10		7.10	8.75	10.95
	Metal halide lamp, 175 W				.348	10.45	21.50		31.95	37.50	46
	Total				.464	10.45	28.60		39.05	**46.25**	**56.95**

D5023 270 Metal Halide Fixture

	System Description	Freq. (Years)	Crew	Unit	Labor Hours	2020 Bare Costs				Total In-House	Total w/O&P
						Material	Labor	Equipment	Total		
2030	**Replace fixture, 175 W**	20	1 ELEC	Ea.							
	Turn branch circuit off and on				.018		1.12		1.12	1.38	1.73
	Remove metal halide fixt.				.800		49		49	60.50	75.50
	Metal halide, recessed, square 175 W				2.353	460	144		604	685	795
	Total				3.171	460	194.12		654.12	**746.88**	**872.23**
2210	**Replace M.H. ballast, 400 W**	10	1 ELEC	Ea.							
	Remove metal halide fixt. ballast				.333		20.50		20.50	25	31.50
	Metal halide fixt. 400 W, ballast				.762	158	46.50		204.50	231	270
	Test fixture				.018		1.12		1.12	1.38	1.73
	Total				1.114	158	68.12		226.12	**257.38**	**303.23**
2220	**Replace lamp, 400 W**	5	1 ELEC	Ea.							
	Remove metal halide fixt. lamp				.118		7.20		7.20	8.85	11.10
	Metal halide lamp, 400 W				.348	20	21.50		41.50	48	58
	Total				.465	20	28.70		48.70	**56.85**	**69.10**
2230	**Replace fixture, 400 W**	20	2 ELEC	Ea.							
	Turn branch circuit off and on				.018		1.12		1.12	1.38	1.73
	Remove metal halide fixt.				1.067		65.50		65.50	80.50	101
	Metal halide, surface mtd, high bay, 400 W				3.478	425	213		638	730	860
	Total				4.563	425	279.62		704.62	**811.88**	**962.73**
2310	**Replace M.H. ballast, 1000 W**	10	1 ELEC	Ea.							
	Remove metal halide fixt. ballast				.333		20.50		20.50	25	31.50
	Metal halide fixt. 1000 W, ballast				.889	258	54.50		312.50	350	405
	Test fixture				.018		1.12		1.12	1.38	1.73
	Total				1.241	258	76.12		334.12	**376.38**	**438.23**
2320	**Replace lamp, 1000 W**	5	1 ELEC	Ea.							
	Remove metal halide fixt. lamp				.118		7.20		7.20	8.85	11.10
	Metal halide lamp, 1000 W				.523	39	32		71	82.50	98.50
	Total				.641	39	39.20		78.20	**91.35**	**109.60**

D5023 270 Metal Halide Fixture

	System Description	Freq. (Years)	Crew	Unit	Labor Hours	2020 Bare Costs				Total In-House	Total w/O&P
						Material	Labor	Equipment	Total		
2330	**Replace fixture, 1000 W**	20	2 ELEC	Ea.							
	Turn branch circuit off and on				.018		1.12		1.12	1.38	1.73
	Remove metal halide fixt.				1.333		82		82	101	126
	Metal halide, surface mtd, high bay, 1000 W				4.000	610	245		855	975	1,150
	Total				5.352	610	328.12		938.12	**1,077.38**	**1,277.73**

D5023 280 H.P. Sodium Fixture, 250 W

	System Description	Freq. (Years)	Crew	Unit	Labor Hours	2020 Bare Costs				Total In-House	Total w/O&P
						Material	Labor	Equipment	Total		
1010	**Replace H.P. sodium ballast**	10	2 ELEC	Ea.							
	Remove H.P. Sodium fixt. ballast				.333		20.50		20.50	25	31.50
	H.P. Sodium fixt., 250 W, ballast				.667	305	41		346	385	445
	Test fixture				.018		1.12		1.12	1.38	1.73
	Total				1.018	305	62.62		367.62	**411.38**	**478.23**
1020	**Replace lamp**	10	2 ELEC	Ea.							
	Remove H.P. Sodium fixt. lamp				.116		7.10		7.10	8.75	10.95
	H.P. Sodium fixt. lamp, 250 W				.348	19.55	21.50		41.05	47.50	57.50
	Total				.464	19.55	28.60		48.15	**56.25**	**68.45**
1030	**Replace fixture**	20	2 ELEC	Ea.							
	Turn branch circuit off and on				.018		1.12		1.12	1.38	1.73
	Remove H.P. Sodium fixt.				.889		54.50		54.50	67	84
	H.P. Sodium fixt., recessed, square 250 W				2.667	890	164		1,054	1,175	1,375
	Total				3.574	890	219.62		1,109.62	**1,243.38**	**1,460.73**

D5033 310 Telephone Cable

	System Description	Freq. (Years)	Crew	Unit	Labor Hours	2020 Bare Costs				Total In-House	Total w/O&P
						Material	Labor	Equipment	Total		
0010	**Repair cable, 22 A.W.G., 4 pair**	8	1 ELEC	M.L.F.							
	Cut telephone cable				.080		4.91		4.91	6.05	7.55
	Splice and insulate telephone cable				.471	12.55	29		41.55	49.50	60
	Total				.551	12.55	33.91		46.46	**55.55**	**67.55**
0020	**Replace telephone cable , #22-4 conductor**	50	2 ELEC	M.L.F.							
	Turn power off, later turn on				.100		6.15		6.15	7.55	9.45
	Remove & reinstall cover plates				1.290		79		79	97.50	122
	Remove telephone cable				2.581		158		158	195	244
	Telephone twisted, #22-4 conductor				10.000	140	615		755	910	1,125
	Total				13.971	140	858.15		998.15	**1,210.05**	**1,500.45**
1020	**Replace telephone jack**	20	1 ELEC	Ea.							
	Remove telephone jack				.059		3.34		3.64	4.47	5.60
	Telephone jack				.250	3.49	15.35		18.84	22.50	28
	Total				.309	3.49	18.39		22.48	**26.97**	**33.60**

D5033 410 Master Clock Control

	System Description	Freq. (Years)	Crew	Unit	Labor Hours	2020 Bare Costs				Total In-House	Total w/O&P
						Material	Labor	Equipment	Total		
0010	**Maintenance and repair**	10	1 ELEC	Ea.							
	Remove time control clock motor				.296		18.20		18.20	22.50	28
	Time control clock motor				1.000	76.50	61.50		138	160	190
	Total				1.296	76.50	79.70		156.20	**182.50**	**218**
0020	**Check operation**	1	1 ELEC	Ea.							
	Check time control clock operation				.533		32.50		32.50	40	50.50
	Total				.533		32.50		32.50	**40**	**50.50**
0030	**Replace time control clock**	15	1 ELEC	Ea.							
	Remove time control clock				.296		18.20		18.20	22.50	28
	Time control clock				1.333	109	82		191	220	262
	Total				1.630	109	100.20		209.20	**242.50**	**290**

D5033 410 Master Clock Control

	System Description	Freq. (Years)	Crew	Unit	Labor Hours	2020 Bare Costs Material	Labor	Equipment	Total	Total In-House	Total w/O&P
1120	**Maintenance and inspection**	1	2 ELEC	Ea.							
	Check operation of program bell				.320		19.65		19.65	24	30
	Total				.320		19.65		19.65	**24**	**30**
1130	**Replace program bell**	15	2 ELEC	Ea.							
	Remove program bell				.667		41		41	50.50	63
	Program bell				1.000	114	61.50		175.50	201	237
	Total				1.667	114	102.50		216.50	**251.50**	**300**

D5033 510 TV Cable Outlet

	System Description	Freq. (Years)	Crew	Unit	Labor Hours	2020 Bare Costs Material	Labor	Equipment	Total	Total In-House	Total w/O&P
0010	**Maintenance and repair**	10	1 ELEC	Ea.							
	Repair TV cable outlet (cover)				.615	2.83	38		40.83	49.50	61.50
	Total				.615	2.83	38		40.83	**49.50**	**61.50**
0020	**Replace TV cable outlet**	20	1 ELEC	Ea.							
	Remove TV cable outlet				.200		12.25		12.25	15.10	18.90
	TV cable outlet				.615	4.60	38		42.60	51.50	64
	Total				.815	4.60	50.25		54.85	**66.60**	**82.90**

D5033 610 Door Bell

	System Description	Freq. (Years)	Crew	Unit	Labor Hours	2020 Bare Costs Material	Labor	Equipment	Total	Total In-House	Total w/O&P
0010	**Maintenance and repair**	10	1 ELEC	Ea.							
	Remove door bell transformer				.200		12.25		12.25	15.10	18.90
	Door bell transformer				.667	24	41		65	76.50	93
	Total				.867	24	53.25		77.25	**91.60**	**111.90**
0020	**Maintenance and inspection**	1	1 ELEC	Ea.							
	Check operation of door bell				.320		19.65		19.65	24	30
	Total				.320		19.65		19.65	**24**	**30**

D5033 610 Door Bell

	System Description	Freq. (Years)	Crew	Unit	Labor Hours	2020 Bare Costs				Total In-House	Total w/O&P
						Material	Labor	Equipment	Total		
0030	**Replace door bell**	15	1 ELEC	Ea.							
	Remove door bell system				.667		41		41	50.50	63
	Door bell system				2.667	158	164		322	375	450
	Total				3.333	158	205		363	**425.50**	**513**

D5033 620 Sound System Components

	System Description	Freq. (Years)	Crew	Unit	Labor Hours	2020 Bare Costs				Total In-House	Total w/O&P
						Material	Labor	Equipment	Total		
0030	**Replace speaker**	20	1 ELEC	Ea.							
	Turn power off, later turn on				.100		6.15		6.15	7.55	9.45
	Remove speaker				.333		20.50		20.50	25	31.50
	Speaker				1.000	157	61.50		218.50	248	291
	Test for operation				.015		.90		.90	1.10	1.38
	Total				1.448	157	89.05		246.05	**281.65**	**333.33**
0120	**Inspect monitor panel**	0.50	1 ELEC	Ea.							
	Check monitor panel				.533		32.50		32.50	40	50.50
	Total				.533		32.50		32.50	**40**	**50.50**
0130	**Replace monitor panel**	15	1 ELEC	Ea.							
	Turn power off, later turn on				.100		6.15		6.15	7.55	9.45
	Remove monitor panel				1.667		102.50		102.50	126	158
	Monitor panel				2.000	520	123		643	725	840
	Test circuits				.073		4.46		4.46	5.50	6.85
	Total				3.839	520	236.11		756.11	**864.05**	**1,014.30**
0220	**Inspect volume control**	1	1 ELEC	Ea.							
	Check operation of volume control				.320		19.35		19.65	24	30
	Total				.320		19.35		19.65	**24**	**30**
0230	**Replace volume control**	15	2 ELEC	Ea.							
	Remove volume control				.333		20.50		20.50	25	31.50
	Volume control				1.000	60	61.50		121.50	141	170
	Total				1.333	60	82		142	**166**	**201.50**

D5033 620 Sound System Components

	System Description	Freq. (Years)	Crew	Unit	Labor Hours	2020 Bare Costs Material	Labor	Equipment	Total	Total In-House	Total w/O&P
0330	**Replace amplifier**	15	2 ELEC	Ea.							
	Turn power off, later turn on				.100		6.15		6.15	7.55	9.45
	Remove amplifier				1.667		102.50		102.50	126	158
	Amplifier				8.000	1,400	490		1,890	2,150	2,500
	Test for operation				.015		.90		.90	1.10	1.38
	Total				9.781	1,400	599.55		1,999.55	**2,284.65**	**2,668.83**

D5033 630 Intercom System Components

	System Description	Freq. (Years)	Crew	Unit	Labor Hours	2020 Bare Costs Material	Labor	Equipment	Total	Total In-House	Total w/O&P
0120	**Inspect intercom master station**	0.50	1 ELEC	Ea.							
	Check intercom master station				2.133		130		130	161	202
	Total				2.133		130		130	**161**	**202**
0130	**Replace intercom master station**	15	2 ELEC	Ea.							
	Turn power off, later turn on				.100		6.15		6.15	7.55	9.45
	Remove intercom master station				1.333		82		82	101	126
	Intercom master station				8.000	2,400	490		2,890	3,250	3,750
	Test circuits				.073		4.46		4.46	5.50	6.85
	Total				9.506	2,400	582.61		2,982.61	**3,364.05**	**3,892.30**
0220	**Inspect intercom remote station**	1	1 ELEC	Ea.							
	Check operation of intercom remote station				.320		19.65		19.65	24	30
	Total				.320		19.65		19.65	**24**	**30**
0230	**Replace intercom remote station**	15	1 ELEC	Ea.							
	Remove intercom remote station				.333		20.50		20.50	25	31.50
	Intercom remote station				1.000	217	61.50		278.50	315	365
	Total				1.333	217	82		299	**340**	**396.50**

D5033 640 Security System Components

	System Description	Freq. (Years)	Crew	Unit	Labor Hours	2020 Bare Costs				Total In-House	Total w/O&P
						Material	Labor	Equipment	Total		
0020	**Inspect camera and monitor**	0.50	1 ELEC	Ea.							
	Check camera and monitor				.808		49.24		49.24	61	76.50
	Total				.808		49.24		49.24	**61**	**76.50**
0030	**Replace camera and monitor**	12	2 ELEC	Ea.							
	Turn power off, later turn on				.100		6.15		6.15	7.55	9.45
	Remove closed circuit TV, one camera & one monitor				2.051		126		126	155	194
	Closed circuit TV, one camera & one monitor				6.154	665	380		1,045	1,200	1,400
	Test for operation				.015		.90		.90	1.10	1.38
	Total				8.320	665	513.05		1,178.05	**1,363.65**	**1,604.83**
0120	**Inspect camera**	0.50	1 ELEC	Ea.							
	Check camera				.533		32.50		32.50	40	50.50
	Total				.533		32.50		32.50	**40**	**50.50**
0130	**Replace camera**	12	1 ELEC	Ea.							
	Turn power off, later turn on				.100		6.15		6.15	7.55	9.45
	Remove closed circuit TV, one camera				1.000		61.50		61.50	75.50	94.50
	Closed circuit TV, one camera				2.963	320	182		502	575	680
	Test for operation				.015		.90		.90	1.10	1.38
	Total				4.078	320	250.55		570.55	**659.15**	**785.33**

D5033 710 Smoke Detector

	System Description	Freq. (Years)	Crew	Unit	Labor Hours	2020 Bare Costs				Total In-House	Total w/O&P
						Material	Labor	Equipment	Total		
0010	**Repair smoke detector**	10	1 ELEC	Ea.							
	Remove smoke detector lamp				.151		9.25		9.25	11.40	14.25
	Smoke detector lamp				.444	7	27.50		34.50	41	51
	Total				.595	7	36.75		43.75	**52.40**	**65.25**
0020	**Check operation**	1	1 ELEC	Ea.							
	Check smoke detector operation				.200		12.25		12.25	15.10	18.90
	Total				.200		12.25		12.25	**15.10**	**18.90**

D5033 710 Smoke Detector

	System Description	Freq. (Years)	Crew	Unit	Labor Hours	2020 Bare Costs Material	Labor	Equipment	Total	Total In-House	Total w/O&P
0030	**Replace smoke detector**	15	1 ELEC	Ea.							
	Turn power off, later turn on				.100		6.15		6.15	7.55	9.45
	Remove smoke detector				.348		21.50		21.50	26	33
	Smoke detector, ceiling type				1.290	120	79		199	229	272
	Test for operation				.015		.90		.90	1.10	1.38
	Total				1.753	120	107.55		227.55	**263.65**	**315.83**

D5033 712 Heat Detector

	System Description	Freq. (Years)	Crew	Unit	Labor Hours	2020 Bare Costs Material	Labor	Equipment	Total	Total In-House	Total w/O&P
0010	**Repair heat detector**	10	1 ELEC	Ea.							
	Remove heat detector element				.151		9.25		9.25	11.40	14.25
	Heat detector element				.444	10.20	27.50		37.70	44.50	55
	Total				.595	10.20	36.75		46.95	**55.90**	**69.25**
0020	**Check operation**	1	1 ELEC	Ea.							
	Check heat detector operation				.200		12.25		12.25	15.10	18.90
	Total				.200		12.25		12.25	**15.10**	**18.90**
0030	**Replace heat detector**	15	1 ELEC	Ea.							
	Turn power off, later turn on				.100		6.15		6.15	7.55	9.45
	Remove heat detector				.348		21.50		21.50	26	33
	Heat detector				1.103	42	67.50		109.50	129	157
	Test for operation				.015		.90		.90	1.10	1.38
	Total				1.566	42	96.05		138.05	**163.65**	**200.83**

D5033 720 Manual Pull Station

	System Description	Freq. (Years)	Crew	Unit	Labor Hours	2020 Bare Costs Material	Labor	Equipment	Total	Total In-House	Total w/O&P
0010	**Check and repair manual pull station**	10	1 ELEC	Ea.							
	Remove pull station component				.200		12.25		12.25	15.10	18.90
	Manual pull station component				.667	17.30	41		58.30	69.50	84.50
	Total				.867	17.30	53.25		70.55	**84.60**	**103.40**

D5033 720 Manual Pull Station

	System Description	Freq. (Years)	Crew	Unit	Labor Hours	2020 Bare Costs				Total In-House	Total w/O&P
						Material	Labor	Equipment	Total		
0020	**Replace manual pull station**	15	1 ELEC	Ea.							
	Turn power off, later turn on				.100		6.15		6.15	7.55	9.45
	Remove manual pull station				.348		21.50		21.50	26	33
	Manual pull station				1.000	88.50	61.50		150	173	205
	Test for operation				.015		.90		.90	1.10	1.38
	Total				1.462	88.50	90.05		178.55	**207.65**	**248.83**

D5033 760 Fire Alarm Control Panel

	System Description	Freq. (Years)	Crew	Unit	Labor Hours	2020 Bare Costs				Total In-House	Total w/O&P
						Material	Labor	Equipment	Total		
0010	**Minor repairs to fire alarm**	5	1 ELEC	Ea.							
	Check and test components				.400		24.50		24.50	30	38
	Remove relay				.222		13.65		13.65	16.75	21
	Relay				.667	44	4[illegible]		85	98.50	118
	Total				1.289	44	79.15		123.15	**145.25**	**177**
0020	**Maintenance and inspection**	0.50	1 ELEC	Ea.							
	Check control panel				.533		32.50		32.50	40	50.50
	Total				.533		32.50		32.50	**40**	**50.50**
0030	**Replace fire alarm panel**	15	1 ELEC	Ea.							
	Turn power off, later turn on				.100		6.15		6.15	7.55	9.45
	Remove fire alarm control panel				2.000		123		123	151	189
	Fire alarm control panel				16.000	875	980		1,855	2,175	2,600
	Test circuits				.073		4.46		4.46	5.50	6.85
	Total				18.173	875	1,113.61		1,988.61	**2,339.05**	**2,805.30**

D5033 766 Annunciation Panel

	System Description	Freq. (Years)	Crew	Unit	Labor Hours	2020 Bare Costs				Total In-House	Total w/O&P
						Material	Labor	Equipment	Total		
0010	**Minor repairs to annunciation panel**	5	1 ELEC	Ea.							
	Check and test components				.400		24.50		24.50	30	38
	Remove relay				.222		13.65		13.65	16.75	21
	Relay				.667	44	41		85	98.50	118
	Total				1.289	44	79.15		123.15	**145.25**	**177**
0020	**Maintenance and inspection**	0.50	1 ELEC	Ea.							
	Check annunciator panel				.533		32.50		32.50	40	50.50
	Total				.533		32.50		32.50	**40**	**50.50**
0030	**Replace annunciation panel**	15	1 ELEC	Ea.							
	Turn power off, later turn on				.100		6.15		6.15	7.55	9.45
	Remove annunciation panel				1.600		98		98	121	151
	Annunciation panel				6.154	415	380		795	920	1,100
	Test circuits				.073		4.46		4.46	5.50	6.85
	Total				7.927	415	488.61		903.61	**1,054.05**	**1,267.30**

D5033 770 Fire Alarm Bell

	System Description	Freq. (Years)	Crew	Unit	Labor Hours	2020 Bare Costs				Total In-House	Total w/O&P
						Material	Labor	Equipment	Total		
0010	**Replace fire alarm bell, 6″**	20	1 ELEC	Ea.							
	Turn power off, later turn on				.100		6.15		6.15	7.55	9.45
	Remove fire alarm bell				.348		21.50		21.50	26	33
	Fire alarm bell				1.000	66.50	61.50		128	149	178
	Test for operation				.015		.90		.90	1.10	1.38
	Total				1.462	66.50	90.05		156.55	**183.65**	**221.83**

D5093 110 Electrical Service Ground

	System Description	Freq. (Years)	Crew	Unit	Labor Hours	2020 Bare Costs				Total In-House	Total w/O&P
						Material	Labor	Equipment	Total		
0010	**Maintenance and repair**	25	1 ELEC	M.L.F.							
	Repair service ground (1 termination)				1.053	4.64	64.50		69.14	84.50	105
	Total				1.053	4.64	64.50		69.14	**84.50**	**105**
0020	**Replace electrical service ground**	50	1 ELEC	M.L.F.							
	Remove ground wire #6				2.000		122.50		122.50	151	189
	Excavate hole for ground rod				2.222		136.50		136.50	168	210
	Ground rod 5/8″, 10′				17.391	225	1,070		1,295	1,550	1,925
	Ground clamp				2.500	46.40	153.50		199.90	240	293
	Backfill over top of ground rod				1.053		64.50		64.50	79.50	99.50
	Insulated copper wire #6				12.308	440	755		1,195	1,400	1,700
	Total				37.474	711.40	2,302		3,013.40	**3,588.50**	**4,416.50**

D5093 120 Building Structure Ground

	System Description	Freq. (Years)	Crew	Unit	Labor Hours	2020 Bare Costs				Total In-House	Total w/O&P
						Material	Labor	Equipment	Total		
0010	**Maintenance and repair**	7	1 ELEC	M.L.F.							
	Repair bldg structure ground (1 termination)				1.053	4.64	64.50		69.14	84.50	105
	Total				1.053	4.64	64.50		69.14	**84.50**	**105**
0020	**Replace building structure ground**	50	1 ELEC	M.L.F.							
	Remove ground wire #4				3.077		189		189	232	290
	Excavate hole for ground rod				2.222		136.50		136.50	168	210
	Ground rod 5/8″, 10′				17.391	225	1,070		1,295	1,550	1,925
	Ground clamp				2.500	46.40	153.50		199.90	240	293
	Backfill over top of ground rod				1.053		64.50		64.50	79.50	99.50
	Insulated copper wire #4				15.094	700	925		1,625	1,900	2,300
	Total				41.337	971.40	2,538.50		3,509.90	**4,169.50**	**5,117.50**

D5093 130 Lightning Protection System

	System Description	Freq. (Years)	Crew	Unit	Labor Hours	2020 Bare Costs Material	Labor	Equipment	Total	Total In-House	Total w/O&P
1010	**Maintenance and repair of general wiring**	1	1 ELEC	M.L.F.							
	Repair lightning protection (1 termination)				1.053	10.45	64.50		74.95	91	113
	Total				1.053	10.45	64.50		74.95	**91**	**113**
1020	**Replace lightning protection general wiring**	25	2 ELEC	M.L.F.							
	Remove lightning protection cable				4.000		245		245	300	380
	Excavate hole for ground rod				2.222		136.50		136.50	168	210
	Ground rod 5/8″, 10′				17.391	225	1,070		1,295	1,550	1,925
	Ground clamp				2.500	46.40	153.50		199.90	240	293
	Backfill over top of ground rod				1.053		64.50		64.50	79.50	99.50
	Lightning protection cable				25.000	3,180	1,530		4,710	5,400	6,325
	Air terminals, copper 3/8″ x 10″				10.000	260	615		875	1,050	1,275
	Total				62.166	3,711.40	3,814.50		7,525.90	**8,787.50**	**10,507.50**

D5093 140 Lightning Ground Rod

	System Description	Freq. (Years)	Crew	Unit	Labor Hours	2020 Bare Costs Material	Labor	Equipment	Total	Total In-House	Total w/O&P
1010	**Maintenance and repair**	1	1 ELEC	Ea.							
	Repair ground rod connector				1.053	4.64	64.50		69.14	84.50	105
	Total				1.053	4.64	64.50		69.14	**84.50**	**105**
1020	**Replace lightning ground rod**	25	2 ELEC	Ea.							
	Excavate hole for ground rod				.333		20.50		20.50	25	31.50
	Ground rod 5/8″, 10′				1.739	22.50	107		129.50	156	192
	Ground clamp				.250	4.64	15.35		19.99	24	29.50
	Backfill over top of ground rod				.157		9.60		9.60	11.85	14.80
	Total				2.479	27.14	152.45		179.59	**216.85**	**267.80**

D5093 150 Computer Ground System

	System Description	Freq. (Years)	Crew	Unit	Labor Hours	2020 Bare Costs Material	Labor	Equipment	Total	Total In-House	Total w/O&P
0010	**Maintenance and repair**	4	1 ELEC	Ea.							
	Test bonding each connection				.267		16.35		16.35	20	25
	Total				.267		16.35		16.35	**20**	**25**

D5093 150 Computer Ground System

	System Description	Freq. (Years)	Crew	Unit	Labor Hours	2020 Bare Costs				Total In-House	Total w/O&P
						Material	Labor	Equipment	Total		
0020	**Replace computer ground system**	50	1 ELEC	M.L.F.							
	Remove ground wire #6				2.000		122.50		122.50	151	189
	Insulated ground wire #6				12.308	440	755		1,195	1,400	1,700
	Grounding connection				2.500	46.40	153.50		199.90	240	293
	Total				16.808	486.40	1,031		1,517.40	**1,791**	**2,182**

D5093 190 Special Ground System

	System Description	Freq. (Years)	Crew	Unit	Labor Hours	2020 Bare Costs				Total In-House	Total w/O&P
						Material	Labor	Equipment	Total		
0010	**Maintenance and repair**	4	1 ELEC	Ea.							
	Test bonding each connection				.267		16.35		16.35	20	25
	Total				.267		16.35		16.35	**20**	**25**
0020	**Replace special ground system**	50	2 ELEC	M.L.F.							
	Remove ground wire #6				2.000		122.50		122.50	151	189
	Insulated ground wire #6				12.308	440	755		1,195	1,400	1,700
	Grounding connection				2.500	46.40	153.50		199.90	240	293
	Total				16.808	486.40	1,031		1,517.40	**1,791**	**2,182**

D5093 210 Generator, Gasoline, 175 KW

	System Description	Freq. (Years)	Crew	Unit	Labor Hours	2020 Bare Costs				Total In-House	Total w/O&P
						Material	Labor	Equipment	Total		
0010	**Maintenance and inspection**	0.08	1 ELEC	Ea.							
	Run test gasoline generator				.800		49		49	60.50	75.50
	Total				.800		49		49	**60.50**	**75.50**
0020	**Replace generator component**	25	2 ELEC	Ea.							
	Remove gasoline generator				36.364		2,225		2,225	2,750	3,425
	Gasoline generator, 175 KW				64.000	84,500	3,925		88,425	98,000	111,500
	Total				100.364	84,500	6,150		90,650	**100,750**	**114,925**

D5093 220 Generator, Diesel, 750 KW

	System Description	Freq. (Years)	Crew	Unit	Labor Hours	2020 Bare Costs				Total In-House	Total w/O&P
						Material	Labor	Equipment	Total		
0010	**Maintenance and inspection**	0.08	1 ELEC	Ea.							
	Run test diesel generator				.800		49		49	60.50	75.50
	Total				.800		49		49	**60.50**	**75.50**
0020	**Replace diesel generator component**	25	4 ELEC	Ea.							
	Remove diesel generator				57.143		3,500		3,500	4,300	5,400
	Diesel generator, 750 KW				160.000	160,000	9,825		169,825	188,000	215,000
	Total				217.143	160,000	13,325		173,325	**192,300**	**220,400**

D5093 230 Transfer Switch

	System Description	Freq. (Years)	Crew	Unit	Labor Hours	2020 Bare Costs				Total In-House	Total w/O&P
						Material	Labor	Equipment	Total		
0010	**Maintenance and repair**	5	1 ELEC	Ea.							
	Remove transfer switch coil				.667		41		41	50.50	63
	Transfer switch coil				2.000	111	123		234	273	330
	Total				2.667	111	164		275	**323.50**	**393**
0020	**Maintenance and inspection**	0.50	1 ELEC	Ea.							
	Inspect and clean transfer switch				.500		30.50		30.50	37.50	47
	Total				.500		30.50		30.50	**37.50**	**47**
0030	**Replace transfer switch**	18	1 ELEC	Ea.							
	Remove transfer switch				.667		41		41	50.50	63
	Transfer switch, 480 V, 3 P, 1200 A				22.857	25,400	1,400		26,800	29,700	33,900
	Cut, form, align & connect wires				.151		9.25		9.25	11.40	14.25
	Check operation of switch				.018		1.12		1.12	1.38	1.73
	Total				23.693	25,400	1,451.37		26,851.37	**29,763.28**	**33,978.98**

D5093 240 Emergency Lighting Fixture

	System Description	Freq. (Years)	Crew	Unit	Labor Hours	2020 Bare Costs				Total In-House	Total w/O&P
						Material	Labor	Equipment	Total		
0020	**Replace lamp**	2	1 ELEC	Ea.							
	Remove emergency remote lamp				.100		6.15		6.15	7.55	9.45
	Emergency remote lamp				.296	29	18.20		47.20	54	64.50
	Total				.396	29	24.35		53.35	**61.55**	**73.95**
0030	**Replace fixture**	20	1 ELEC	Ea.							
	Turn branch circuit off and on				.018		1.12		1.12	1.38	1.73
	Remove emergency light units				.667		41		41	50.50	63
	Emergency light units, battery operated				2.000	330	123		453	515	600
	Total				2.685	330	165.12		495.12	**566.88**	**664.73**

D5093 250 Exit Light

	System Description	Freq. (Years)	Crew	Unit	Labor Hours	2020 Bare Costs				Total In-House	Total w/O&P
						Material	Labor	Equipment	Total		
0010	**Maintenance and repair**	20	1 ELEC	Ea.							
	Remove exit light face plate				.100		6.15		6.15	7.55	9.45
	Exit light face plate				.296	5.20	18.20		23.40	28	34.50
	Total				.396	5.20	24.35		29.55	**35.55**	**43.95**
0020	**Replace lamp**	5	1 ELEC	Ea.							
	Remove exit light lamp				.022		1.35		1.35	1.66	2.08
	Exit light lamp				.065	7.05	3.99		11.04	12.65	14.95
	Total				.087	7.05	5.34		12.39	**14.31**	**17.03**
0030	**Replace lighting fixture**	20	1 ELEC	Ea.							
	Turn circuit off and on for fixture				.018		1.12		1.12	1.38	1.73
	Remove exit light				.333		20.50		20.50	25	31.50
	Exit light				1.000	73	61.50		134.50	156	186
	Total				1.352	73	83.12		156.12	**182.38**	**219.23**

D5093 255 Exit Light L.E.D.

	System Description	Freq. (Years)	Crew	Unit	Labor Hours	2020 Bare Costs				Total In-House	Total w/O&P
						Material	Labor	Equipment	Total		
0020	**Replace lamp with exit light L.E.D. retrofit kits**	15	1 ELEC	Ea.							
	Remove exit light lamp				.022		1.35		1.35	1.66	2.08
	Exit light L.E.D. retrofit kits				.133	52.50	8.20		60.70	68	78
	Total				.155	52.50	9.55		62.05	**69.66**	**80.08**
0030	**Replace lighting fixture with exit light L.E.D. std.**	20	1 ELEC	Ea.							
	Turn circuit off and on for fixture				.018		1.12		1.12	1.38	1.73
	Remove exit light				.333		20.50		20.50	25	31.50
	Exit light, L.E.D. standard, double face				1.194	52	73.50		125.50	147	178
	Total				1.546	52	95.12		147.12	**173.38**	**211.23**
0040	**Replace lighting fixture with exit light L.E.D. w/battery unit**	20	1 ELEC	Ea.							
	Turn circuit off and on for fixture				.018		1.12		1.12	1.38	1.73
	Remove exit light				.333		20.50		20.50	25	31.50
	Exit light, L.E.D. with battery unit, double face				2.000	223	123		346	395	470
	Total				2.352	223	144.62		367.62	**421.38**	**503.23**

D5093 260 Battery, Wet

	System Description	Freq. (Years)	Crew	Unit	Labor Hours	2020 Bare Costs				Total In-House	Total w/O&P
						Material	Labor	Equipment	Total		
0010	**Maintenance and inspection**	0.02	1 ELEC	Ea.							
	Clean & maintain wet cell battery				.500		30.50		30.50	37.50	47
	Total				.500		30.50		30.50	**37.50**	**47**
0020	**Replace battery**	10	1 ELEC	Ea.							
	Remove wet cell battery				.100		6.15		6.15	7.55	9.45
	Wet cell battery				.296	610	18.20		628.20	695	790
	Total				.396	610	24.35		634.35	**702.55**	**799.45**

D5093 265 Battery, Dry

	System Description	Freq. (Years)	Crew	Unit	Labor Hours	2020 Bare Costs				Total In-House	Total w/O&P
						Material	Labor	Equipment	Total		
0010	**Maintenance and inspection**	0.08	1 ELEC	Ea.							
	Clean & maintain dry cell battery				.500		30.50		30.50	37.50	47
	Total				.500		30.50		30.50	**37.50**	**47**
0020	**Replace battery**	5	1 ELEC	Ea.							
	Remove dry cell battery				.100		6.15		6.15	7.55	9.45
	Dry cell battery				.296	150	18.20		168.20	187	216
	Total				.396	150	24.35		174.35	**194.55**	**225.45**

D5093 270 Battery Charger

	System Description	Freq. (Years)	Crew	Unit	Labor Hours	2020 Bare Costs				Total In-House	Total w/O&P
						Material	Labor	Equipment	Total		
0010	**Maintenance and repair**	2	1 ELEC	Ea.							
	Repair battery charger component				3.077	46.50	189		235.50	283	350
	Total				3.077	46.50	189		235.50	**283**	**350**
0020	**Maintenance and inspection**	0.25	1 ELEC	Ea.							
	Check battery charger				1.600		98		98	121	151
	Total				1.600		98		98	**121**	**151**
0030	**Replace charger**	20	1 ELEC	Ea.							
	Remove battery charger				.615		38		38	46.50	58
	Battery charger				1.600	790	98		888	990	1,150
	Total				2.215	790	136		926	**1,036.50**	**1,208**

D5093 280 UPS Battery

	System Description	Freq. (Years)	Crew	Unit	Labor Hours	2020 Bare Costs				Total In-House	Total w/O&P
						Material	Labor	Equipment	Total		
0020	**Maintenance and inspection**	0.17	1 ELEC	Ea.							
	UPS battery electrical test				.800		49		49	60.50	75.50
	Total				.800		49		49	**60.50**	**75.50**

D5093 280 UPS Battery

	System Description	Freq. (Years)	Crew	Unit	Labor Hours	2020 Bare Costs				Total In-House	Total w/O&P
						Material	Labor	Equipment	Total		
0030	**Replace motor generator UPS battery**	15	1 ELEC	Ea.							
	Remove UPS battery				1.333		82		82	101	126
	UPS battery				4.000	535	245		780	890	1,050
	Total				5.333	535	327		862	**991**	**1,176**

D5093 920 Communications Components

	System Description	Freq. (Years)	Crew	Unit	Labor Hours	2020 Bare Costs				Total In-House	Total w/O&P
						Material	Labor	Equipment	Total		
0020	**Maintenance and repair voice/data outlet**	10	1 ELEC	Ea.							
	Repair voice/data outlet				.615	2.83	38		40.83	49.50	61.50
	Total				.615	2.83	38		40.83	**49.50**	**61.50**
0030	**Replace voice/data outlet**	20	1 ELEC	Ea.							
	Remove voice/data outlet				.057		3.51		3.51	4.31	5.40
	Voice/data outlet, single opening, excludes voice/data device				.167	7.60	10.25		17.85	21	25.50
	Total				.224	7.60	13.76		21.36	**25.31**	**30.90**
0120	**Maintenance and inspection patch panel**	0.50	1 ELEC	Ea.							
	Check patch panel				1.067		65		65	80.50	101
	Total				1.067		65		65	**80.50**	**101**
0130	**Replace patch panel**	15	1 ELEC	Ea.							
	Remove patch panel				1.010		62.50		62.50	76	95.50
	Patch panel				4.000	320	245		565	655	780
	Test circuits				.996		61.10		61.10	75	94
	Total				6.006	320	368.60		688.60	**806**	**969.50**

E1023 710 Laboratory Equipment

	System Description	Freq. (Years)	Crew	Unit	Labor Hours	2020 Bare Costs				Total In-House	Total w/O&P
						Material	Labor	Equipment	Total		
0020	**Replace glove box gloves**	5	1 SKWK	Ea.							
	Laboratory equipment, remove bacteriological glove box gloves				.500		27.50		27.50	35	43.50
	Lab equipment, replace fiberglass, bacteriological glove box gloves				1.000	220	55		275	310	365
	Laboratory equipment, test replacement gloves				1.000		55		55	70.50	87.50
	Total				2.500	220	137.50		357.50	**415.50**	**496**
0080	**Replace fume hood sash**	20	2 SKWK	Ea.							
	Laboratory equipment, remove fume hood sash				1.000		55		55	70.50	87.50
	Laboratory equipment, replace fume hood sash				2.000	890	110		1,000	1,125	1,300
	Total				3.000	890	165		1,055	**1,195.50**	**1,387.50**

E1033 110 Automotive Equipment

	System Description	Freq. (Years)	Crew	Unit	Labor Hours	2020 Bare Costs				Total In-House	Total w/O&P
						Material	Labor	Equipment	Total		
0020	**Remove and replace vehicle lift hydraulic pump**	15	1 SKWK	Ea.							
	Automotive hoist, two posts, adj. frame, remove and replace pump				4.000	3,175	219		3,394	3,775	4,325
	Total				4.000	3,175	219		3,394	**3,775**	**4,325**
0100	**Remove and replace compressor, electric, 5 H.P.**	10	1 ELEC	Ea.							
	Automotive compressors, remove and replace 5 H.P. motor				2.000	750	123		873	975	1,125
	Total				2.000	750	123		873	**975**	**1,125**

E1093 310 Loading Dock Equipment

	System Description	Freq. (Years)	Crew	Unit	Labor Hours	2020 Bare Costs Material	Labor	Equipment	Total	Total In-House	Total w/O&P
0020	**Remove and replace hydraulic dock leveler lift cylinder**	15	2 SKWK	Ea.							
	Loading dock, dock levelers, 10 ton cap., replace lift cylinder				4.000	6,975	219		7,194	7,950	9,075
	Total				4.000	6,975	219		7,194	**7,950**	**9,075**
0040	**Remove and replace hydraulic dock leveler hydraulic pump**	20	1 SKWK	Ea.							
	Loading dock, dock levelers, 10 ton cap, replace lift cylinder				3.077	1,300	169		1,469	1,650	1,900
	Total				3.077	1,300	169		1,469	**1,650**	**1,900**

E1093 315 Dishwasher

	System Description	Freq. (Years)	Crew	Unit	Labor Hours	2020 Bare Costs Material	Labor	Equipment	Total	Total In-House	Total w/O&P
0020	**Replace commercial dishwasher, 10 to 12 racks per hour**	10	2 PLUM	Ea.							
	Cleaning and disposal, dishwasher, commercial, rack type				12.000		760		760	940	1,175
	Install new dishwasher				6.499	3,450	375		3,825	4,250	4,900
	Total				18.499	3,450	1,135		4,585	**5,190**	**6,075**
0040	**Remove and replace dishwasher pump**	15	1 PLUM	Ea.							
	Laboratory equip., glassware washer, remove and replace pump				1.778	885	115		1,000	1,125	1,275
	Total				1.778	885	115		1,000	**1,125**	**1,275**
0080	**Replace commercial dishwasher, to 375 racks per hour**	20	L-4	Ea.	76.000						
	Commercial dishwasher, to 275 racks per hour, selective demo				24.000		1,200		1,200	1,550	1,925
	Dishwasher, automatic, 235 to 275 racks/hr.				60.000	28,000	3,800		31,800	35,500	40,900
	Total				84.000	28,000	5,000		33,000	**37,050**	**42,825**

E1093 316 Waste Disposal, Residential

	System Description	Freq. (Years)	Crew	Unit	Labor Hours	2020 Bare Costs Material	Labor	Equipment	Total	Total In-House	Total w/O&P
0010	**Unstop disposal**	1	1 PLUM	Ea.							
	Unstop clogged unit				.889		57.50		57.50	71	88.50
	Total				.889		57.50		57.50	**71**	**88.50**

E1093 316 Waste Disposal, Residential

	System Description	Freq. (Years)	Crew	Unit	Labor Hours	2020 Bare Costs				Total In-House	Total w/O&P
						Material	Labor	Equipment	Total		
0020	**Replace waste disposal unit, residential**	8	1 PLUM	Ea.							
	Remove unit				1.040		65.50		65.50	80.50	101
	Install new disposal				2.080	216	131		347	400	470
	Total				3.120	216	196.50		412.50	**480.50**	**571**

E1093 320 Waste Handling Equipment

	System Description	Freq. (Years)	Crew	Unit	Labor Hours	2020 Bare Costs				Total In-House	Total w/O&P
						Material	Labor	Equipment	Total		
0020	**Remove and replace waste compactor hydraulic cylinder**	15	2 SKWK	Ea.							
	Replace ram, 5 C.Y. capacity				4.000	5,650	219		5,869	6,500	7,425
	Total				4.000	5,650	219		5,869	**6,500**	**7,425**
0040	**Remove and replace waste compactor hydraulic pump**	20	1 SKWK	Ea.							
	Replace hydraulic pump, 5 C.Y. capacity				3.200	1,100	176		1,276	1,425	1,650
	Total				3.200	1,100	176		1,276	**1,425**	**1,650**

E1093 510 Darkroom Dryer

	System Description	Freq. (Years)	Crew	Unit	Labor Hours	2020 Bare Costs				Total In-House	Total w/O&P
						Material	Labor	Equipment	Total		
0020	**Remove and replace darkroom dryer fan**	10	1 SKWK	Ea.							
	48″ x 25″ x 68″ high				1.600	163	88		251	292	345
	Total				1.600	163	88		251	**292**	**345**
0040	**Remove and replace darkroom dryer heating element**	15	1 SKWK	Ea.							
	Darkroom equip., dryers, dehumidifier, replace compressor				1.333	89	73		162	192	228
	Total				1.333	89	73		162	**192**	**228**

E1093 610 Dust Collector

	System Description	Freq. (Years)	Crew	Unit	Labor Hours	2020 Bare Costs				Total In-House	Total w/O&P
						Material	Labor	Equipment	Total		
0100	**Remove and replace 20″ dia dust collector bag**	5	1 SKWK	Ea.							
	Dust collector bag, 20″ diameter				1.600	600	99.50		699.50	785	905
	Total				1.600	600	99.50		699.50	**785**	**905**

E1093 910 Pump Systems

	System Description	Freq. (Years)	Crew	Unit	Labor Hours	2020 Bare Costs				Total In-House	Total w/O&P
						Material	Labor	Equipment	Total		
0500	**Remove and replace 50 HP pump motor**	25	2 ELEC	Ea.							
	Electrical dml, motors, 230/460 V, 60 Hz, 50 HP				1.667		102		102	126	157
	Install 50 HP motor, 230/460 V, totly encld, 1800 RPM				8.000	4,775	490		5,265	5,850	6,725
	Total				9.667	4,775	592		5,367	**5,976**	**6,882**

G1023 210 Underground Storage Tank Removal

	System Description	Freq. (Years)	Crew	Unit	Labor Hours	2020 Bare Costs				Total In-House	Total w/O&P
						Material	Labor	Equipment	Total		
1010	**Remove 500 gal. underground storage tank(non-leaking)**	20	B-34P	Ea.							
	Saw cut asphalt				2.700	10	135.50	174.50	320	375	420
	Disconnect and remove piping				1.500		96.60		96.60	120	149
	Transfer liquids, 10% of volume				.250		16		16	20	25
	Insert solid carbon dioxide, 1.5 lbs./100 gal.				.150	9.15	6.30		15.45	18.25	21.50
	Cut accessway to inside of tank				1.501		63		63	81.50	101
	Remove sludge, wash and wipe tank				1.000		64.50		64.50	79.50	99.50
	Properly dispose of sludge/water					36.50			36.50	40	45.50
	Excavate, pull, & load tank, backfill hole				16.000		790	214	1,004	1,250	1,475
	Select structural backfill					196			196	216	245
	Haul tank to certified dump, 100 miles round trip				8.000		425	196	621	750	880
	Total				31.101	251.65	1,596.90	584.50	2,433.05	**2,950.25**	**3,461.50**
1020	**Remove 3000 gal. underground storage tank(non-leaking)**	20	B-34R	Ea.							
	Saw cut asphalt				5.400	20	271	349	640	750	835
	Disconnect and remove piping				2.500		161		161	200	249
	Transfer liquids, 10% of volume				1.500		96		96	120	150
	Insert solid carbon dioxide, 1.5 lbs./100 gal.				.900	54.90	37.80		92.70	109	129
	Cut accessway to inside of tank				1.501		63		63	81.50	101
	Remove sludge, wash and wipe tank				1.199		77.50		77.50	95.50	120
	Properly dispose of sludge/water					328.50			328.50	360	410
	Excavate, pull, & load tank, backfill hole				32.000		1,575	465	2,040	2,525	3,000
	Select structural backfill					735			735	810	920
	Haul tank to certified dump, 100 miles round trip				8.000		425	820	1,245	1,425	1,575
	Total				53.000	1,138.40	2,706.30	1,634	5,478.70	**6,476**	**7,489**
1030	**Remove 5000 gal. underground storage tank(non-leaking)**	20	B-34R	Ea.							
	Saw cut asphalt				5.940	22	298.10	383.90	704	825	920
	Disconnect and remove piping				2.500		161		161	200	249
	Transfer liquids, 10% of volume				2.500		160		160	200	250
	Insert solid carbon dioxide, 1.5 lbs./100 gal.				1.500	91.50	63		154.50	182	216
	Cut accessway to inside of tank				1.501		63		63	81.50	101
	Remove sludge, wash and wipe tank				1.301		84		84	104	130
	Properly dispose of sludge/water					547.50			547.50	600	685
	Excavate, pull, & load tank, backfill hole				32.000		1,575	465	2,040	2,525	3,000
	Select structural backfill					1,225			1,225	1,350	1,525
	Haul tank to certified dump, 100 miles round trip				8.000		425	820	1,245	1,425	1,575
	Total				55.242	1,886	2,829.10	1,668.90	6,384	**7,492.50**	**8,651**

G1023 210 Underground Storage Tank Removal

	System Description	Freq. (Years)	Crew	Unit	Labor Hours	2020 Bare Costs				Total In-House	Total w/O&P
						Material	Labor	Equipment	Total		
1040	**Remove 8000 gal. underground storage tank(non-leaking)**	20	B-34R	Ea.							
	Saw cut asphalt				6.480	24	325.20	[illegible]	768	900	1,000
	Disconnect and remove piping				2.500		161		161	200	249
	Transfer liquids, 10% of volume				4.000		256		256	320	400
	Insert solid carbon dioxide, 1.5 lbs./100 gal.				2.400	146.40	100.80		247.20	292	345
	Cut accessway to inside of tank				1.501		63		63	81.50	101
	Remove sludge, wash and wipe tank				1.501		96.50		96.50	120	150
	Properly dispose of sludge/water					876			876	965	1,100
	Excavate, pull, & load tank, backfill hole				32.000		1,625	[illegible]	3,525	4,150	4,650
	Select structural backfill					1,715			1,715	1,875	2,150
	Haul tank to certified dump, 100 miles round trip				8.000		390	[illegible]	1,245	1,450	1,575
	Total				58.382	2,761.40	3,017.50	[illegible]	8,952.70	**10,353.50**	**11,720**
1050	**Remove 10000 gal. ugnd. storage tank(non-leaking)**	20	B-34S	Ea.							
	Saw cut asphalt				10.800	40	542	[illegible]	1,280	1,500	1,675
	Disconnect and remove piping				3.750		241.50		241.50	299	375
	Transfer liquids, 10% of volume				5.000		320		320	400	500
	Insert solid carbon dioxide, 1.5 lbs./100 gal.				3.000	183	126		309	365	430
	Cut accessway to inside of tank				1.501		63		63	81.50	101
	Remove sludge, wash and wipe tank				1.751		113		113	140	174
	Properly dispose of sludge/water					1,095			1,095	1,200	1,375
	Excavate, pull, & load tank, backfill hole				32.000		1,625	[illegible]	3,525	4,150	4,650
	Select structural backfill					2,205			2,205	2,425	2,750
	Haul tank to certified dump, 100 miles round trip				8.000		390	[illegible]	1,245	1,450	1,575
	Total				65.801	3,523	3,420.50	[illegible]	10,396.50	**12,010.50**	**13,605**
1060	**Remove 12000 gal. ugnd. storage tank(non-leaking)**	20	B-34S	Ea.							
	Saw cut asphalt				12.150	45	609.75	[illegible]	1,440	1,700	1,875
	Disconnect and remove piping				5.000		322		322	400	500
	Transfer liquids, 10% of volume				6.000		384		384	480	600
	Insert solid carbon dioxide, 1.5 lbs./100 gal.				3.600	219.60	151.20		370.80	435	520
	Cut accessway to inside of tank				1.501		63		63	81.50	101
	Remove sludge, wash and wipe tank				1.900		122		122	151	189
	Properly dispose of sludge/water					1,314			1,314	1,450	1,650
	Excavate, pull, & load tank, backfill hole				32.000		1,625	[illegible]	3,525	4,150	4,650
	Select structural backfill					2,450			2,450	2,700	3,075
	Haul tank to certified dump, 100 miles round trip				8.000		390	[illegible]	1,245	1,450	1,575
	Total				70.151	4,028.60	3,666.95	[illegible]	11,235.80	**12,997.50**	**14,735**

G2023 210 Parking Lot Repairs

	System Description	Freq. (Years)	Crew	Unit	Labor Hours	2020 Bare Costs				Total In-House	Total w/O&P
						Material	Labor	Equipment	Total		
1010	**Parking lot repair and sealcoating**	5	B-1	M.S.F.							
	Thoroughly clean surface				.312		13.32	2.22	15.54	21	25.50
	Patch holes				.240	22.41	10.40	.95	33.76	39	45.50
	Fill cracks				.715	82.50	31	18.50	132	151	173
	Install 2 coat petroleum resistant emulsion				3.330	155.40	142.08		297.48	355	425
	Restripe lot				.291	20.20	12.60	6.40	39.20	45.50	52.50
	Total				4.887	280.51	209.40	28.07	517.98	**611.50**	**721.50**
1020	**Parking lot repair and resurface**	10	B-25B	M.S.F.							
	Thoroughly clean surface				.312		13.32	2.22	15.54	21	25.50
	Patch holes				.240	22.41	10.40	.95	33.76	39	45.50
	Fill cracks				.715	82.50	31	18.50	132	151	173
	Emulsion tack coat, .05 gal. per S.Y.				.710	38.85	37.74	36.63	113.22	131	147
	Install 1″ thick asphaltic concrete wearing course				1.008	406.26	47.73	31.08	485.07	545	620
	Restripe lot				.291	20.20	12.60	6.40	39.20	45.50	52.50
	Total				3.276	570.22	152.79	95.78	818.79	**932.50**	**1,063.50**

G2023 310 General

	System Description	Freq. (Years)	Crew	Unit	Labor Hours	2020 Bare Costs				Total In-House	Total w/O&P
						Material	Labor	Equipment	Total		
1000	**Remove and replace steel guard rail**	7	B-80	L.F.							
	Remove old metal guide/guard rail and posts				.213		9.95	6.80	16.75	20	23.50
	Install metal guide/guard rail including posts 6'3″				.107	27.50	4.97	3.40	35.87	40.50	46
	Total				.320	27.50	14.92	10.20	52.62	**60.50**	**69.50**
1100	**Raise MH or catch basin frame and cover**	10	D-1	Ea.							
	Demolish pavement around frame				4.800		212	35.50	247.50	310	380
	Raise MH or catch basin frame with brick				1.332	9.32	62.60		71.92	92	113
	Total				6.132	9.32	274.60	35.50	319.42	**402**	**493**

G2033 130 Asphalt Sidewalk & Curb

	System Description	Freq. (Years)	Crew	Unit	Labor Hours	2020 Bare Costs				Total In-House	Total w/O&P
						Material	Labor	Equipment	Total		
0100	**Remove & replace asphalt sidewalk, 4′ wide**	15	B-37	L.F.							
	Remove existing asphalt pavement				.080		3.72	1.60	5.32	6.55	7.65
	Bank run gravel, 3″ avg.				.002	.63	.11	.21	.95	1.06	1.19
	Vibratory plate compaction				.003		.11	.01	.12	.15	.19
	Fine grade gravel base				.026		1.29	1.73	3.02	3.53	3.93
	Place asphalt sidewalk, 2-1/2″ thick				.027	2.54	1.18	.14	3.86	4.48	5.20
	Total				.138	3.17	6.41	3.69	13.27	**15.77**	**18.16**
1000	**Remove and replace asphalt curb or berm**	10	B-37	L.F.							
	Demolish existing curb or berm				.096		4.25	.71	4.96	6.25	7.55
	Remove broken curb				.032		1.49	.64	2.13	2.62	3.08
	Asphalt curb, 8″ wide, 6″ high machine formed				.042	.05	1.77	.35	2.17	2.74	3.29
	Total				.170	.05	7.51	1.70	9.26	**11.61**	**13.92**

G2033 140 Concrete Sidewalk & Curb

	System Description	Freq. (Years)	Crew	Unit	Labor Hours	2020 Bare Costs				Total In-House	Total w/O&P
						Material	Labor	Equipment	Total		
0100	**Remove and replace concrete sidewalk, 4′ wide**	25	B-37	L.F.							
	Remove existing concrete pavement				.107		4.96	2.14	7.10	8.70	10.25
	Bank run gravel, 3″ avg.				.002	.63	.11	.21	.95	1.06	1.19
	Vibratory plate compaction				.003		.11	.01	.12	.15	.19
	Fine grade gravel				.026		1.29	1.73	3.02	3.53	3.93
	Broom finished concrete sidewalk, 4″ thick				.209	8.60	10.12		18.72	22.50	27
	Total				.346	9.23	16.59	4.09	29.91	**35.94**	**42.56**
2000	**Remove and replace concrete curb or berm**	25	B-37	L.F.							
	Demolish existing curb or berm				.240		10.60	1.79	12.39	15.60	18.90
	Remove broken curb				.080		3.72	1.60	5.32	6.55	7.65
	Concrete curb, 6″ x 18″, steel forms				.091	8.90	4.68		13.58	15.85	18.65
	Total				.411	8.90	19	3.39	31.29	**38**	**45.20**

G2033 150 Patios

	System Description	Freq. (Years)	Crew	Unit	Labor Hours	2020 Bare Costs				Total In-House	Total w/O&P
						Material	Labor	Equipment	Total		
1030	**Refinish Concrete Patio**	3	1 PORD	S.F.							
	Prepare surface				.003		.14	.03	.17	.21	.25
	Paint 1 coat, roller & brush				.008	.08	.36		.44	.56	.68
	Total				.012	.08	.50	.03	.61	**.77**	**.93**
2030	**Refinish Masonry Patio**	3	1 PORD	S.F.							
	Prepare surface				.003		.14	.03	.17	.21	.25
	Apply 1 coat of sealer, roller & brush				.005	.28	.21		.49	.58	.69
	Total				.008	.28	.35	.03	.66	**.79**	**.94**
3030	**Refinish Stone Patio**	3	1 PORD	S.F.							
	Prepare surface				.003		.14	.03	.17	.21	.25
	Apply 1 coat of sealer, roller & brush				.005	.28	.21		.49	.58	.69
	Total				.008	.28	.35	.03	.66	**.79**	**.94**
4030	**Refinish Wood Patio**	3	1 PORD	S.F.							
	Prepare surface				.015		.66		.66	.84	1.04
	Primer + 1 top coat, roller & brush				.022	.15	.99		1.14	1.43	1.76
	Total				.037	.15	1.65		1.80	**2.27**	**2.80**

G2033 250 Handicap Ramp

	System Description	Freq. (Years)	Crew	Unit	Labor Hours	2020 Bare Costs				Total In-House	Total w/O&P
						Material	Labor	Equipment	Total		
1030	**Refinish Metal Handicap Ramp**	3	1 PORD	S.F.							
	Prepare surface				.015		.66		.66	.84	1.04
	Prime & paint 1 coat, roller & brush				.015	.31	.65		.96	1.17	1.42
	Total				.029	.31	1.31		1.62	**2.01**	**2.46**
2030	**Refinish Wood Handicap Ramp**	3	1 PORD	S.F.							
	Prepare surface				.015		.66		.66	.84	1.04
	Prime & paint 1 coat, roller & brush				.022	.15	.99		1.14	1.43	1.76
	Total				.037	.15	1.65		1.80	**2.27**	**2.80**

G2043 105 Chain Link Fence and Gate Repairs

	System Description	Freq. (Years)	Crew	Unit	Labor Hours	2020 Bare Costs				Total In-House	Total w/O&P
						Material	Labor	Equipment	Total		
1010	**Minor chain link fence repairs (per 10 L.F.)**	1	2 CLAB	Ea.							
	Straighten bent 2″ line post				.167		7		7	9.05	11.25
	Re-tie fence fabric				.167	1.75	7		8.75	11	13.40
	Refasten loose barbed wire arm				.083		3.51		3.51	4.53	5.60
	Refasten loose barbed wire, 3 strands				.083		3.51		3.51	4.53	5.60
	Total				.500	1.75	21.02		22.77	**29.11**	**35.85**
1110	**Replace bent 1-5/8″ top rail (per 20 L.F.)**	2	2 CLAB	Ea.							
	Remove fabric ties, top rail & couplings				.333		14.05		14.05	18.15	22.50
	Install new top rail, couplings & fabric ties				.667	38.50	28.05		66.55	78.50	93
	Total				1.000	38.50	42.10		80.60	**96.65**	**115.50**
1120	**Replace broken barbed wire arm**	2	2 CLAB	Ea.							
	Remove fabric ties, top rail & couplings				.333		14.05		14.05	18.15	22.50
	Release barbed wires & remove old arm				.067		2.81		2.81	3.63	4.50
	Install new arm & refasten barbed wires				.167	.53	7.05		7.58	9.70	11.90
	Install old top rail, couplings & fabric ties				.667		28		28	36.50	45
	Total				1.234	.53	51.91		52.44	**67.98**	**83.90**
1130	**Replace barbed wire, 3 strands (per 100 L.F.)**	5	2 CLAB	Ea.							
	Remove old barbed wire				3.002		126		126	163	202
	Install new barbed wire				3.000	105	126		231	279	330
	Total				6.002	105	252		357	**442**	**532**
1140	**Replace fence fabric, 6′ high, 9 ga. (per 10 L.F.)**	10	2 CLAB	Ea.							
	Remove ties from top rail, bottom tension wire & posts				.167		7		7	9.05	11.25
	Unstitch & remove section of fabric				.667		28		28	36.50	45
	Hang & stretch new fabric, stitch into existing fabric				1.333	70	56.20		126.20	150	178
	Re-tie fence fabric				.167	1.75	7		8.75	11	13.40
	Total				2.333	71.75	98.20		169.95	**206.55**	**247.65**
1150	**Replace double swing gates, 6′ high, 20′ opng.**	5	2 CLAB	Opng.							
	Remove 2 old gates (6′x10′ ea.) & 4 hinges				1.111		47		47	60.50	75
	Supply 4 new gate hinges					61.10			61.10	67	76.50
	Install 2 new gates (6′x10′ ea.), 4 hinges, adjust alignment				2.412	561.18	101.49		662.67	750	865
	Remove & replace gate latch				.914	15.04	38.40		53.44	66.50	80.50
	Remove & replace gate cane bolt				.883	15.04	37.20		52.24	64.50	78.50
	Total				5.320	652.36	224.09		876.45	**1,008.50**	**1,175.50**

G2043 105 Chain Link Fence and Gate Repairs

	System Description	Freq. (Years)	Crew	Unit	Labor Hours	2020 Bare Costs Material	Labor	Equipment	Total	Total In-House	Total w/O&P
1160	**Replace 6′x18′ cantilever slide gate**	5	2 CLAB	Opng.							
	Remove old gate & 4 rollers				.926		39		39	50.50	62.50
	Supply 4 new cantilever gate rollers				.003	412.50	.05		412.55	455	515
	Install new 6′x18′ cantilever slide gate, 4 rollers, adjust				2.148	2,052.27	90.18		2,142.45	2,375	2,700
	Remove & replace gate latch				.914	15.04	38.40		53.44	66.50	80.50
	Total				3.991	2,479.81	167.63		2,647.44	**2,947**	**3,358**
1210	**Replace 2″ line post**	20	B-55	Ea.							
	Remove fabric ties, top rail & couplings				.333		14.05		14.05	18.15	22.50
	Release barbed wires & remove old arm				.067		2.81		2.81	3.63	4.50
	Unstitch & roll up section of fabric				1.000		44	50.50	94.50	112	126
	Remove 2″ post & concrete base from ground				.750		33	38	71	84	94
	Install new 2″ line post with concrete in old hole				1.500	61.25	65.80	75.25	202.30	235	266
	Install old top rail, couplings & new fabric ties				.667		28		28	36.50	45
	Unroll & stretch fabric, stitch into existing fabric				2.000		87.50	101	188.50	224	251
	Re-tie fence fabric				.333	3.50	14		17.50	22	27
	Total				6.650	64.75	289.16	264.75	618.66	**735.28**	**836**
1220	**Replace 3″ corner post**	10	B-55	Ea.							
	Remove fabric ties, retainer bars & ring fittings				.500		22	25	47	56	63
	Roll back fabric				.250		10.95	12.60	23.55	28	31.50
	Remove top rail & ring fittings				.500		22	25	47	56	63
	Remove barbed wires & ring fittings				.750		33	38	71	84	94
	Remove diagonal & horizontal braces, ring fittings				.500		22	25	47	56	63
	Remove 3″ post & concrete base from ground				1.000		44	50.50	94.50	112	126
	Install new 3″ post with concrete in old hole				2.000	90	87.50	101	278.50	325	365
	Install old diagonal & horizontal braces, ring fittings				.500		22	25	47	56	63
	Install old barbed wires & ring fittings				.750		33	38	71	84	94
	Install old top rail & ring fittings				.500		22	25	47	56	63
	Hang & stretch fabric, insert retainer bars & ring fittings				1.500		65.50	75.50	141	168	188
	Re-tie fence fabric				.333	3.50	14		17.50	22	27
	Total				9.083	93.50	397.95	440.60	932.05	**1,103**	**1,240.50**

G2043 105 Chain Link Fence and Gate Repairs

	System Description	Freq. (Years)	Crew	Unit	Labor Hours	2020 Bare Costs				Total In-House	Total w/O&P
						Material	Labor	Equipment	Total		
1230	**Replace 3″ gate post**	5	B-55	Ea.							
	Remove gate & hinges				.833		36.50	42	78.50	93.50	105
	Remove fabric ties, retainer bars & ring fittings				.500		22	25	47	56	63
	Roll back fabric				.250		10.95	12.60	23.55	28	31.50
	Remove top rail & ring fittings				.500		22	25	47	56	63
	Remove barbed wires & ring fittings				.750		33	38	71	84	94
	Remove diagonal & horizontal braces, ring fittings				.500		22	25	47	56	63
	Remove 3″ post & concrete base from ground				1.000		44	50.50	94.50	112	126
	Install new 3″ post with concrete in old hole				2.000	90	87.50	101	278.50	325	365
	Install old diagonal & horizontal braces, ring fittings				.500		22	25	47	56	63
	Install old barbed wires & ring fittings				.750		33	38	71	84	94
	Install old top rail & ring fittings				.500		22	25	47	56	63
	Hang & stretch fabric, insert retainer bars & ring fittings				1.500		65.50	75.50	141	168	188
	Re-tie fence fabric				.333	3.50	14		17.50	22	27
	Install old hinges & gate				1.807		79	91	170	203	227
	Total				11.724	93.50	513.45	573.60	1,180.55	**1,399.50**	**1,572.50**

G2043 110 Wood Fence

	System Description	Freq. (Years)	Crew	Unit	Labor Hours	2020 Bare Costs				Total In-House	Total w/O&P
						Material	Labor	Equipment	Total		
1020	**Refinish Wood Fence, 6′ High**	3	1 PORD	L.F.							
	Prepare surface				.089		3.96		3.96	5.05	6.25
	Prime & 1 top coat, roller & brush				.046	.54	2.04		2.58	3.24	3.92
	Total				.135	.54	6		6.54	**8.29**	**10.17**

G2043 710 Bleachers, Exterior

	System Description	Freq. (Years)	Crew	Unit	Labor Hours	2020 Bare Costs				Total In-House	Total w/O&P
						Material	Labor	Equipment	Total		
1020	**Refinish Wood Bleachers**	3	1 PORD	S.F.							
	Prepare surface				.015		.66		.66	.84	1.04
	Prime & 1 top coat, roller & brush				.022	.15	.99		1.14	1.43	1.76
	Total				.037	.15	1.65		1.80	**2.27**	**2.80**

G2043 750 Tennis Court Resurfacing

	System Description	Freq. (Years)	Crew	Unit	Labor Hours	2020 Bare Costs Material	Labor	Equipment	Total	Total In-House	Total w/O&P
1010	**Resurface asphalt tennis court**	7	2 SKWK	Ea.							
	Remove/reinstall tennis net				1.000		42		42	54.50	67.50
	Thoroughly clean surface				7.760		328		328	425	520
	Install 3 coat acrylic emulsion sealcoat				16.000	5,640	672		6,312	7,075	8,125
	Total				24.760	5,640	1,042		6,682	**7,554.50**	**8,712.50**
1020	**Resurface cushioned asphalt tennis court**	7	2 SKWK	Ea.							
	Remove/reinstall tennis net				1.000		42		42	54.50	67.50
	Thoroughly clean surface				7.760		328		328	425	520
	Two cushion coats				16.000	2,625	675		3,300	3,750	4,350
	Install 3 coat acrylic emulsion sealcoat				16.000	5,640	672		6,312	7,075	8,125
	Total				40.760	8,265	1,717		9,982	**11,304.50**	**13,062.50**
2010	**Clay tennis court preparation**	1	2 SKWK	Ea.							
	Remove/reinstall tennis net				1.000		42		42	54.50	67.50
	Remove and reinstall tapes				6.015		253		253	325	405
	Remove weeds and deteriorated surface material				2.000		84		84	109	135
	Check court for level				1.000		42		42	54.50	67.50
	Clay court top dressing, 1.5 tons				7.500	720	316.50		1,036.50	1,200	1,400
	Clay court, roll after dressing				1.500		78	21	99	121	146
	Total				19.015	720	815.50	21	1,556.50	**1,864**	**2,221**

G2043 810 Flag Pole

	System Description	Freq. (Years)	Crew	Unit	Labor Hours	2020 Bare Costs Material	Labor	Equipment	Total	Total In-House	Total w/O&P
1020	**Refinish 25′ Wood Flag Pole**	5	1 PORD	Ea.	.295						
	Set up, secure and take down ladder				.258		11.45		11.45	14.65	18.20
	Prepare surface				.370		16.50		16.50	21	26
	Prime & 1 top coat, roller & brush				.556	3.75	24.75		28.50	36	44
	Total				1.184	3.75	52.70		56.45	**71.65**	**88.20**

G3013 400 Ground Level Water Storage Tank

	System Description	Freq. (Years)	Crew	Unit	Labor Hours	2020 Bare Costs				Total In-House	Total w/O&P
						Material	Labor	Equipment	Total		
0100	**Prep & paint 100k gal. ground level water stor tank, 30′ dia x 19′ tall**	10	E-11	Ea.							
	Scaffolding, outer ring				72.000		3,816		3,816	4,950	6,125
	Scaffolding planks				35.666		1,893.9[illegible]		1,893.90	2,450	3,050
	Sandblast exterior, near white blast (SSPC-SP10), loose scale, fine rust				177.775	2,400	8,300	[illegible],675	12,375	15,600	18,600
	Epoxy primer, spray exterior				13.325	650	600		1,250	1,600	1,875
	Epoxy topcoat, 2 coats, spray exterior				28.550	1,450	1,300		2,750	3,450	4,075
	Clean up				10.000		450		450	675	800
	Total				337.316	4,500	16,359.9[illegible]	1,675	22,534.90	**28,725**	**34,525**
0200	**Prep & paint 250k gal. ground level water stor tank, 40′ dia x 27′ tall**	10	E-11	Ea.							
	Scaffolding, outer ring				126.000		6,678		6,678	8,650	10,700
	Scaffolding planks				63.333		3,363		3,363	4,350	5,425
	Sandblast exterior, near white blast (SSPC-SP10), loose scale, fine rust				327.106	4,416	15,272	3,08[illegible]	22,770	28,700	34,200
	Epoxy primer, spray exterior				24.518	1,196	1,104		2,300	2,950	3,425
	Epoxy topcoat, 2 coats, spray exterior				52.532	2,668	2,392		5,060	6,350	7,475
	Clean up				18.400		828		828	1,250	1,475
	Total				611.889	8,280	29,637	3,08[illegible]	40,999	**52,250**	**62,700**
0300	**Prep & paint 500k gal. ground level water stor tank, 50′ dia x 34′ tall**	10	E-11	Ea.							
	Scaffolding, outer ring				192.000		10,176		10,176	13,200	16,300
	Scaffolding planks				96.666		5,133		5,133	6,650	8,275
	Sandblast exterior, near white blast (SSPC-SP10), loose scale, fine rust				519.103	7,008	24,236	4,891	36,135	45,600	54,000
	Epoxy primer, spray exterior				38.909	1,898	1,752		3,650	4,675	5,450
	Epoxy topcoat, 2 coats, spray exterior				83.366	4,234	3,796		8,030	10,100	11,900
	Clean up				29.200		1,314		1,314	1,975	2,325
	Total				959.244	13,140	46,407	4,89[illegible]	64,438	**82,200**	**98,250**
0400	**Prep & paint 750k gal. ground level water stor tank, 60′ dia x 36′ tall**	10	E-11	Ea.							
	Scaffolding, outer ring				234.000		12,402		12,402	16,100	19,900
	Scaffolding planks				117.332		6,230.4[illegible]		6,230.40	8,050	10,000
	Sandblast exterior, near white blast (SSPC-SP10), loose scale, fine rust				675.545	9,120	31,540	6,3[illegible]	47,025	59,500	70,500
	Epoxy primer, spray exterior				50.635	2,470	2,280		4,750	6,075	7,075
	Epoxy topcoat, 2 coats, spray exterior				108.490	5,510	4,940		10,450	13,100	15,400
	Clean up				38.000		1,710		1,710	2,575	3,050
	Total				1224.002	17,100	59,102.4[illegible]	6,3[illegible]	82,567.40	**105,400**	**125,925**

G3013 400 Ground Level Water Storage Tank

	System Description	Freq. (Years)	Crew	Unit	Labor Hours	2020 Bare Costs				Total In-House	Total w/O&P
						Material	Labor	Equipment	Total		
0500	**Prep & paint 1M gal. ground level water stor tank, 70′ dia x 35′ tall**	10	E-11	Ea.							
	Scaffolding, outer ring				261.000		13,833		13,833	17,900	22,200
	Scaffolding planks				131.332		6,973.80		6,973.80	9,025	11,200
	Sandblast exterior, near white blast (SSPC-SP10), loose scale, fine rust				817.765	11,040	38,180	7,705	56,925	72,000	85,500
	Epoxy primer, spray exterior				61.295	2,990	2,760		5,750	7,350	8,575
	Epoxy topcoat, 2 coats, spray exterior				131.330	6,670	5,980		12,650	15,900	18,700
	Clean up				46.000		2,070		2,070	3,100	3,675
	Total				1448.722	20,700	69,796.80	7,705	98,201.80	**125,275**	**149,850**
0600	**Prep & paint 2M gal. ground level water stor tank, 100′ dia x 34′ tall**	10	E-11	Ea.							
	Scaffolding, outer ring				354.000		18,762		18,762	24,300	30,100
	Scaffolding planks				176.998		9,398.70		9,398.70	12,200	15,100
	Sandblast exterior, near white blast (SSPC-SP10), loose scale, fine rust				1315.535	17,760	61,420	12,395	91,575	115,500	137,500
	Epoxy primer, spray exterior				98.605	4,810	4,440		9,250	11,800	13,800
	Epoxy topcoat, 2 coats, spray exterior				211.270	10,730	9,620		20,350	25,500	30,100
	Clean up				74.000		3,330		3,330	5,000	5,925
	Total				2230.408	33,300	106,970.70	12,395	152,665.70	**194,300**	**232,525**
0700	**Prep & paint 4M gal. ground level water stor tank, 130′ dia x 40′ tall**	10	E-11	Ea.							
	Scaffolding, outer ring				531.000		28,143		28,143	36,500	45,100
	Scaffolding planks				266.664		14,160		14,160	18,300	22,800
	Sandblast exterior, near white blast (SSPC-SP10), loose scale, fine rust				2133.300	28,800	99,600	20,100	148,500	187,000	223,000
	Epoxy primer, spray exterior				159.900	7,800	7,200		15,000	19,200	22,400
	Epoxy topcoat, 2 coats, spray exterior				342.600	17,400	15,600		33,000	41,400	48,800
	Clean up				120.000		5,400		5,400	8,100	9,600
	Total				3553.464	54,000	170,103	20,100	244,203	**310,500**	**371,700**

G3013 405 Ground Level Water Storage Standpipe

	System Description	Freq. (Years)	Crew	Unit	Labor Hours	2020 Bare Costs				Total In-House	Total w/O&P
						Material	Labor	Equipment	Total		
0100	**Prep & paint 100k gal. grnd. level water standpipe, 24′ dia x 30′ tall**	10	E-11	Ea.							
	Scaffolding, outer ring				96.000		5,088		5,088	6,600	8,150
	Scaffolding planks				47.666		2,531.10		2,531.10	3,275	4,075
	Sandblast exterior, near white blast (SSPC-SP10), loose scale, fine rust				191.997	2,592	8,964	1,809	13,365	16,800	20,100
	Epoxy primer, spray exterior				14.391	702	648		1,350	1,725	2,000
	Epoxy topcoat, 2 coats, spray exterior				30.834	1,566	1,404		2,970	3,725	4,400
	Clean up				10.800		486		486	730	865
	Total				391.688	4,860	19,121.10	1,809	25,790.10	**32,855**	**39,590**
0200	**Prep & paint 250k gal. grnd. level water standpipe, 32′ dia x 42′ tall**	10	E-11	Ea.							
	Scaffolding, outer ring				165.000		8,745		8,745	11,300	14,000
	Scaffolding planks				82.666		4,389.60		4,389.60	5,675	7,075
	Sandblast exterior, near white blast (SSPC-SP10), loose scale, fine rust				355.550	4,800	16,600	3,35[illegible]	24,750	31,200	37,200
	Epoxy primer, spray exterior				26.650	1,300	1,200		2,500	3,200	3,725
	Epoxy topcoat, 2 coats, spray exterior				57.100	2,900	2,600		5,500	6,900	8,125
	Clean up				20.000		900		900	1,350	1,600
	Total				706.966	9,000	34,434.60	3,35[illegible]	46,784.60	**59,625**	**71,725**
0300	**Prep & paint 500k gal. grnd. level water standpipe, 40′ dia x 53′ tall**	10	E-11	Ea.							
	Scaffolding, outer ring				252.000		13,356		13,356	17,300	21,400
	Scaffolding planks				125.665		6,672.90		6,672.90	8,625	10,700
	Sandblast exterior, near white blast (SSPC-SP10), loose scale, fine rust				568.880	7,680	26,560	5,36[illegible]	39,600	49,900	59,500
	Epoxy primer, spray exterior				42.640	2,080	1,920		4,000	5,125	5,950
	Epoxy topcoat, 2 coats, spray exterior				91.360	4,640	4,160		8,800	11,000	13,000
	Clean up				32.000		1,440		1,440	2,150	2,550
	Total				1112.545	14,400	54,108.90	5,36[illegible]	73,868.90	**94,100**	**113,100**
0400	**Prep & paint 750k gal. grnd. level water standpipe, 46′ dia x 60′ tall**	10	E-11	Ea.							
	Scaffolding, outer ring				318.000		16,854		16,854	21,800	27,000
	Scaffolding planks				159.665		8,478.30		8,478.30	11,000	13,700
	Sandblast exterior, near white blast (SSPC-SP10), loose scale, fine rust				739.544	9,984	34,528	6,96[illegible]	51,480	65,000	77,500
	Epoxy primer, spray exterior				55.432	2,704	2,496		5,200	6,650	7,750
	Epoxy topcoat, 2 coats, spray exterior				118.768	6,032	5,408		11,440	14,400	16,900
	Clean up				41.600		1,872		1,872	2,800	3,325
	Total				1433.009	18,720	69,636.30	6,96[illegible]	95,324.30	**121,650**	**146,175**

G3013 405 — Ground Level Water Storage Standpipe

	System Description	Freq. (Years)	Crew	Unit	Labor Hours	2020 Bare Costs				Total In-House	Total w/O&P
						Material	Labor	Equipment	Total		
0500	**Prep & paint 1M gal. grnd. level water standpipe, 50′ dia x 68′ tall**	10	E-11	Ea.							
	Scaffolding, outer ring				384.000		20,352		20,352	26,400	32,600
	Scaffolding planks				192.998		10,248.30		10,248.30	13,300	16,500
	Sandblast exterior, near white blast (SSPC-SP10), loose scale, fine rust				903.097	12,192	42,164	8,509	62,865	79,000	94,500
	Epoxy primer, spray exterior				67.691	3,302	3,048		6,350	8,125	9,450
	Epoxy topcoat, 2 coats, spray exterior				145.034	7,366	6,604		13,970	17,500	20,600
	Clean up				50.800		2,286		2,286	3,425	4,075
	Total				1743.620	22,860	84,702.30	8,509	116,071.30	**147,750**	**177,725**
0600	**Prep & paint 2M gal. grnd. level water standpipe, 64′ dia x 83′ tall**	10	E-11	Ea.							
	Scaffolding, outer ring				579.000		30,687		30,687	39,800	49,200
	Scaffolding planks				290.664		15,434.40		15,434.40	20,000	24,900
	Sandblast exterior, near white blast (SSPC-SP10), loose scale, fine rust				1422.200	19,200	66,400	13,400	99,000	125,000	148,500
	Epoxy primer, spray exterior				106.600	5,200	4,800		10,000	12,800	14,900
	Epoxy topcoat, 2 coats, spray exterior				228.400	11,600	10,400		22,000	27,600	32,500
	Clean up				80.000		3,600		3,600	5,400	6,400
	Total				2706.864	36,000	131,321.40	13,400	180,721.40	**230,600**	**276,400**

G3013 410 — Fire Hydrants

	System Description	Freq. (Years)	Crew	Unit	Labor Hours	2020 Bare Costs				Total In-House	Total w/O&P
						Material	Labor	Equipment	Total		
1010	**Remove and replace fire hydrant**	25	B-21	Ea.							
	Isolate hydrant from system				1.000		42		42	54.50	67.50
	Excavate to expose base and feeder pipe				.974		48.30	13.02	61.32	75.50	90.50
	Disconnect and remove hydrant and associated piping				4.000		168		168	218	270
	Install pipe to new hydrant				3.200	510	162		672	765	895
	Install hydrant, including lower barrel and shoe				4.000	3,300	195	27	3,522	3,900	4,475
	Install thrust blocks				1.200	60.90	60.30	.29	121.49	145	172
	Backfill excavation				5.091		213.50		213.50	277	345
	Fine grade, hand				8.000		336		336	435	540
	Total				27.465	3,870.90	1,225.10	40.31	5,136.31	**5,870**	**6,855**

G3013 470 Post Indicator Valve

	System Description	Freq. (Years)	Crew	Unit	Labor Hours	2020 Bare Costs Material	Labor	Equipment	Total	Total In-House	Total w/O&P
1010	**Remove and replace post indicator valve**	35	B-21	Ea.							
	Isolate PIV from system				.126		8.10		8.10	10.05	12.55
	Excavate to expose base of valve				.139		6.90	1.86	8.76	10.80	12.90
	Disconnect and remove post indicator and valve				4.000		168		168	218	270
	Install new 4″ diameter valve				4.000	2,000	183	35.50	2,218.50	2,475	2,825
	Install new post indicator				3.636	1,475	177	24.50	1,676.50	1,875	2,150
	Backfill excavation				.727		30.50		30.50	39.50	49
	Fine grade, hand				8.000		336		336	435	540
	Total				20.629	3,475	909.50	61.86	4,446.36	**5,063.35**	**5,859.45**

G3013 510 Elevated Water Storage Tank

	System Description	Freq. (Years)	Crew	Unit	Labor Hours	2020 Bare Costs Material	Labor	Equipment	Total	Total In-House	Total w/O&P
0100	**Prep & paint 50K gal. stg tank, 30′ dia x 10′ tall, 50′ above grnd.**	10	E-11	Ea.							
	Scaffolding, outer ring				225.000		11,925		11,925	15,500	19,100
	Scaffolding, center core under tank				80.000		4,260		4,260	5,500	6,825
	Scaffolding planks				58.333		3,097.50		3,097.50	4,000	5,000
	Sandblast exterior, near white blast (SSPC-SP10), loose scale, fine rust				167.464	2,260.80	7,818.60	1,577.85	11,657.25	14,700	17,500
	Epoxy primer, spray exterior				12.552	612.30	565.20		1,177.50	1,500	1,750
	Epoxy topcoat, 2 coats, spray exterior				26.894	1,365.90	1,224.60		2,590.50	3,250	3,825
	Clean up				9.420		423.90		423.90	635	755
	Total				579.663	4,239	29,314.80	1,577.85	35,131.65	**45,085**	**54,755**
0200	**Prep & paint 100K gal. stg tank, 30′ dia x 20′ tall, 50′ above grnd.**	10	E-11	Ea.							
	Scaffolding, outer ring				261.000		13,833		13,833	17,900	22,200
	Scaffolding, center core under tank				80.000		4,260		4,260	5,500	6,825
	Scaffolding planks				153.332		8,142		8,142	10,500	13,100
	Sandblast exterior, near white blast (SSPC-SP10), loose scale, fine rust				227.552	3,072	10,624	2,144	15,840	20,000	23,800
	Epoxy primer, spray exterior				17.056	832	768		1,600	2,050	2,375
	Epoxy topcoat, 2 coats, spray exterior				36.544	1,856	1,664		3,520	4,425	5,200
	Clean up				12.800		576		576	865	1,025
	Total				788.284	5,760	39,867	2,144	47,771	**61,240**	**74,525**

G3013 510 Elevated Water Storage Tank

	System Description	Freq. (Years)	Crew	Unit	Labor Hours	2020 Bare Costs				Total In-House	Total w/O&P
						Material	Labor	Equipment	Total		
0300	**Prep & paint 250K gal. stg tank, 40′ dia x 27′ tall, 50′ above grnd.**	10	E-11	Ea.							
	Scaffolding, outer ring				360.000		19,080		19,080	24,700	30,600
	Scaffolding, center core under tank				134.000		7,135.50		7,135.50	9,200	11,400
	Scaffolding planks				219.998		11,682		11,682	15,100	18,800
	Sandblast exterior, near white blast (SSPC-SP10), loose scale, fine rust				419.549	5,664	19,588	3,953	29,205	36,800	43,800
	Epoxy primer, spray exterior				31.447	1,534	1,416		2,950	3,775	4,400
	Epoxy topcoat, 2 coats, spray exterior				67.378	3,422	3,068		6,490	8,150	9,600
	Clean up				23.600		1,062		1,062	1,600	1,900
	Total				1255.972	10,620	63,031.50	3,953	77,604.50	**99,325**	**120,500**
0400	**Prep & paint 500K gal. stg tank, 50′ dia x 34′ tall, 50′ above grnd.**	10	E-11	Ea.							
	Scaffolding, outer ring				480.000		25,440		25,440	33,000	40,800
	Scaffolding, center core under tank				202.000		10,756.50		10,756.50	13,900	17,200
	Scaffolding planks				296.664		15,753		15,753	20,400	25,400
	Sandblast exterior, near white blast (SSPC-SP10), loose scale, fine rust				661.323	8,928	30,876	6,231	46,035	58,000	69,000
	Epoxy primer, spray exterior				49.569	2,418	2,232		4,650	5,950	6,925
	Epoxy topcoat, 2 coats, spray exterior				106.206	5,394	4,836		10,230	12,800	15,100
	Clean up				37.200		1,674		1,674	2,500	2,975
	Total				1832.962	16,740	91,567.50	6,231	114,538.50	**146,550**	**177,400**
0500	**Prep & paint 750K gal. stg tank, 60′ dia x 36′ tall, 50′ above grnd.**	10	E-11	Ea.							
	Scaffolding, outer ring				570.000		30,210		30,210	39,200	48,500
	Scaffolding, center core under tank				283.000		15,069.75		15,069.75	19,400	24,100
	Scaffolding planks				366.663		19,470		19,470	25,200	31,400
	Sandblast exterior, near white blast (SSPC-SP10), loose scale, fine rust				881.764	11,904	41,168	8,308	61,380	77,500	92,000
	Epoxy primer, spray exterior				66.092	3,224	2,976		6,200	7,925	9,250
	Epoxy topcoat, 2 coats, spray exterior				141.608	7,192	6,448		13,640	17,100	20,200
	Clean up				49.600		2,232		2,232	3,350	3,975
	Total				2358.727	22,320	117,573.75	8,308	148,201.75	**189,675**	**229,425**
0600	**Prep & paint 1M gal. stg tank, 65′ dia x 40′ tall, 50′ above grnd.**	10	E-11	Ea.							
	Scaffolding, outer ring				645.000		34,185		34,185	44,300	55,000
	Scaffolding, center core under tank				329.000		17,519.25		17,519.25	22,600	28,000
	Scaffolding planks				413.329		21,948		21,948	28,400	35,300
	Sandblast exterior, near white blast (SSPC-SP10), loose scale, fine rust				1059.539	14,304	49,468	9,983	73,755	93,000	110,500
	Epoxy primer, spray exterior				79.417	3,874	3,576		7,450	9,525	11,100
	Epoxy topcoat, 2 coats, spray exterior				170.158	8,642	7,748		16,390	20,600	24,200
	Clean up				59.600		2,682		2,682	4,025	4,775
	Total				2756.043	26,820	137,126.25	9,983	173,929.25	**222,450**	**268,875**

G3063 410 Ground Level Fuel Storage Tank

	System Description	Freq. (Years)	Crew	Unit	Labor Hours	2020 Bare Costs				Total In-House	Total w/O&P
						Material	Labor	Equipment	Total		
0100	**Prep & paint 100k gal. ground level fuel storage tank, 24′ dia x 30′ tall**	25	E-11	Ea.							
	Scaffolding, outer ring				96.000		5,088		5,088	6,600	8,150
	Scaffolding planks				47.333		2,513.40		2,513.40	3,250	4,050
	Sandblast exterior, near white blast (SSPC-SP10), loose scale, fine rust				191.997	2,592	8,964	1,309	13,365	16,800	20,100
	Epoxy primer, spray exterior				14.391	702	648		1,350	1,725	2,000
	Epoxy topcoat, 2 coats, spray exterior				30.834	1,566	1,404		2,970	3,725	4,400
	Clean up				10.800		486		486	730	865
	Total				391.355	4,860	19,103.40	1,309	25,772.40	**32,830**	**39,565**

G4013 210 Overhead Service Cables

	System Description	Freq. (Years)	Crew	Unit	Labor Hours	2020 Bare Costs				Total In-House	Total w/O&P
						Material	Labor	Equipment	Total		
0010	**Repair cable splice**	12	1 ELEC	M.L.F.							
	Repair splice cable, polyethylene jacket				7.619	14.50	465		479.50	590	740
	Total				7.619	14.50	465		479.50	**590**	**740**
0020	**Cable inspection**	5	1 ELEC	M.L.F.							
	Check cables for damage				.667		41		41	50.50	63
	Total				.667		41		41	**50.50**	**63**
0030	**Replace service cable**	30	2 ELEC	M.L.F.							
	Remove overhead service cables				1.860		114		114	140	176
	Overhead service 3 ACSR cable				16.000	475	980		1,455	1,725	2,100
	Install 3 conductors at terminations				.941		57.50		57.50	71	89
	Install conductors to insulators and clipping				2.500		153		153	189	236
	Install and remove sag gauge				1.000		61.50		61.50	75.50	94.50
	Make up & install conductor jumper				2.000		123		123	151	189
	Total				24.302	475	1,489		1,964	**2,351.50**	**2,884.50**

G40 SITE ELEC. UTILITIES | G4023 | Site Lighting

G4023 210 | Outdoor Pole Lights

	System Description	Freq. (Years)	Crew	Unit	Labor Hours	2020 Bare Costs Material	Labor	Equipment	Total	Total In-House	Total w/O&P
1010	**Replace 400W H.P.S. lamp, pole-mounted fixture**	10	R-26	Ea.							
	Position truck, raise and lower boom bucket				.289		17.70	4.1[illegible]	21.84	26.50	32
	Remove & install 400W H.P.S. lamp, pole-mounted fixture				.533	33	32.50		65.50	76.50	92
	Total				.822	33	50.20	4.1[illegible]	87.34	**103**	**124**
1020	**Replace 400W H.P.S. ballast, pole-mounted fixture**	10	R-26	Ea.							
	Position truck, raise and lower boom bucket				.289		17.70	4.1[illegible]	21.84	26.50	32
	Remove & install 400W H.P.S. multi-tap ballast, pole-mounted fixt.				2.974	269	182	[illegible]	493.50	565	665
	Test pole-mounted H.I.D. fixture				.074		4.54	1.0[illegible]	5.60	6.75	8.15
	Total				3.337	269	204.24	[illegible]	520.94	**598.25**	**705.15**
1030	**Replace 400W H.P.S. pole-mounted fixt. w/ lamp & blst.**	20	R-26	Ea.							
	Turn branch circuit off and on, pole light				.370		22.50	5.3[illegible]	27.80	34	41
	Turn branch circuit off and on, pole light				.370		22.50	5.3[illegible]	27.80	34	41
	Position truck, raise and lower boom bucket				.289		17.70	4.1[illegible]	21.84	26.50	32
	Remove & install pole-mounted 400W H.P.S. fixture w/ lamp & blst.				9.456	320	580	13[illegible]	1,036	1,225	1,450
	Test pole-mounted H.I.D. fixture				.074		4.54	1.0[illegible]	5.60	6.75	8.15
	Total				10.560	320	647.24	[illegible]	1,119.04	**1,326.25**	**1,572.15**
1040	**Replace light pole, 2 fixtures (conc. base not incl.)**	10	R-3	Ea.							
	Turn branch circuit off and on, pole light				.370		22.50	5.3[illegible]	27.80	34	41
	Remove parking lot light pole with fixtures				2.500		153	23.5[illegible]	176.50	215	262
	Install 2 brackets arms onto aluminum light pole				2.602	560	160		720	810	945
	Install two 400W H.P.S. fixtures w/ lamps & ballasts onto light pole				9.467	640	580		1,220	1,425	1,700
	Install 30′ aluminum light pole (incl. arms & fixtures) onto base				11.111	2,075	680	105	2,860	3,225	3,750
	Pull new wires in existing underground conduits (3#10 500′)				15.605	234	960		1,194	1,425	1,775
	Test pole-mounted H.I.D. fixture				.074		4.54	1.0[illegible]	5.60	6.75	8.15
	Total				41.729	3,509	2,560.04	134.5[illegible]	6,203.90	**7,140.75**	**8,481.15**

G4023 210 Outdoor Pole Lights

	System Description	Freq. (Years)	Crew	Unit	Labor Hours	2020 Bare Costs				Total In-House	Total w/O&P
						Material	Labor	Equipment	Total		
1050	**Replace concrete base for parking lot light pole**	20	B-18	Ea.							
	Sawcut asphalt paving				1.296	4.80	65.04	83.76	153.60	180	201
	Remove concrete light pole base				4.000		236	520	756	870	945
	Excavate by hand				12.987		545		545	705	875
	Install 1-1/4″ PVC electrical conduit				.612	10.50	37.50		48	57.50	71
	Install 1-1/4″ PVC electrical sweep				.433	8.55	26.50		35.05	42	51.50
	Install new concrete light pole base, 24″ dia. x 8′				12.030	325	605		930	1,150	1,375
	Backfill excavation by hand				4.728		200		200	257	320
	Compact backfill				.578		24.35	2.30	26.65	34	41.50
	Install asphalt patch				.960	94	42.40	5.12	141.52	164	191
	Total				37.624	442.85	1,781.79	611.18	2,835.82	**3,459.50**	**4,071**

H1043 200 Temporary Cranes

	System Description	Freq. (Years)	Crew	Unit	Labor Hours	2020 Bare Costs				Total In-House	Total w/O&P
						Material	Labor	Equipment	Total		
1010	**Daily use of crane, portal to portal, 12-ton**	1	A-3H	Day							
	12-ton truck-mounted hydraulic crane, daily use for small jobs				8.000		475	725	1,200	1,400	1,525
	Total				8.000		475	725	1,200	**1,400**	**1,525**
1020	**Daily use of crane, portal to portal, 25-ton**	1	A-3I	Day							
	25-ton truck-mounted hydraulic crane, daily use for small jobs				8.000		475	800	1,275	1,475	1,625
	Total				8.000		475	800	1,275	**1,475**	**1,625**
1030	**Daily use of crane, portal to portal, 40-ton**	1	A-3J	Day							
	40-ton truck-mounted hydraulic crane, daily use for small jobs				8.000		475	1,275	1,750	2,000	2,150
	Total				8.000		475	1,275	1,750	**2,000**	**2,150**
1040	**Daily use of crane, portal to portal, 55-ton**	1	A-3K	Day							
	55-ton truck-mounted hydraulic crane, daily use for small jobs				16.000		880	1,500	2,380	2,750	3,025
	Total				16.000		880	1,500	2,380	**2,750**	**3,025**
1050	**Daily use of crane, portal to portal, 80-ton**	1	A-3L	Day							
	80-ton truck-mounted hydraulic crane, daily use for small jobs				16.000		880	2,250	3,130	3,575	3,850
	Total				16.000		880	2,250	3,130	**3,575**	**3,850**
1060	**Daily use of crane, portal to portal, 100-ton**	1	A-3M	Day							
	100-ton truck-mounted hydraulic crane, daily use for small jobs				16.000		880	2,375	3,255	3,700	3,975
	Total				16.000		880	2,375	3,255	**3,700**	**3,975**

Preventive Maintenance

Table of Contents

Table of Contents (cont.)

E1095 Other Equipment

F10 Special Construction

F1045 Special Facilities

G20 Site Improvements

G2015 Roadways

G2045 Site Development

G30 Site Mechanical Utilities

G3015 Water Supply

G3025 Sanitary Sewer

G3065 Fuel Distribution

G40 Site Electrical Utilities

G4095 Other Site Electrical Utilities

How to Use the Preventive Maintenance Cost Tables

The following is a detailed explanation of a sample Preventive Maintenance Cost Table. The Preventive Maintenance Tables are separated into two parts: 1) the components and related labor-hours and frequencies of a typical system and 2) the costs for systems on an annual or annualized basis. Next to each bold number that follows is the described item with the appropriate component of the sample entry in parentheses.

D30 HVAC | **D3025 130** (1) | **Boiler, Hot Water, Oil/Gas/Comb.**

PM Components (2)	Labor-hrs. (3)	W	M (4)	Q	S	A
PM System D3025 130 1950						
Boiler, hot water; oil, gas or combination fired, up to 120 MBH						
1 Check combustion chamber for air or gas leaks.	.077					√
2 Inspect and clean oil burner gun and ignition assembly, where applicable.	.658					√
3 Inspect fuel system for leaks and change fuel filter element, where applicable.	.098					√
4 Check fuel lines and connections for damage.	.023		√	√	√	√
5 Check for proper operational response of burner to thermostat controls.	.133			√	√	√
6 Check and lubricate burner and blower motors.	.079			√	√	√
7 Check main flame failure protection and main flame detection scanner on boiler equipped with spark ignition (oil burner).	.124		√	√	√	√
8 Check electrical wiring to burner controls and blower.	.079					√
9 Clean firebox (sweep and vacuum).	.577					√
10 Check operation of mercury control switches (i.e., steam pressure, hot water temperature limit, atomizing or combustion air proving, etc.).	.143		√	√	√	√
11 Check operation and condition of safety pressure relief valve.	.030		√	√	√	√
12 Check operation of boiler low water cut off devices.	.056		√	√	√	√
13 Check hot water pressure gauges.	.073		√	√	√	√
14 Inspect and clean water column sight glass (or replace).	.127		√	√	√	√
15 Clean fire side of water jacket boiler.	.433					√
16 Check condition of flue pipe, damper and exhaust stack.	.147			√	√	√
17 Check boiler operation through complete cycle, up to 30 minutes.	.650					√
18 Check fuel level with gauge pole, add as required.	.046		√	√	√	√
19 Clean area around boiler.	.066		√	√	√	√
20 Fill out maintenance checklist and report deficiencies.	.022		√	√	√	√
Total labor-hours/period (5)			.710	1.069	1.069	3.641
Total labor-hours/year			5.678	2.138	1.069	3.641

	Description (6)	Labor-hrs. (7)	Material	Labor	Equip.	Total	Total In-House (9)	Total w/O&P (10)
			Cost Each — 2020 Bare Costs (8)					
1900	Boiler, hot water, O/G/C, up to 120 MBH, annually	3.641	57	239		296	358.48	440
1950	Annualized	12.528	66.50	820		886.50	1,089.40	1,350

1 System/Line Numbers (D3025 130 1950)

Each Preventive Maintenance Assembly has been assigned a unique identification number based on the UNIFORMAT II classification system.

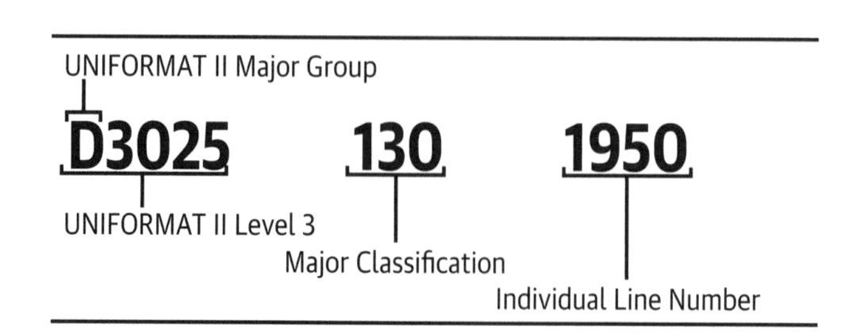

2 PM Components

The individual preventive maintenance operations required to be performed annually are listed separately to show what has been included in the development of the total system price. The cost table below the PM Components listing contains prices for the system on an annual or annualized basis.

3 Labor Hours

The "Labor-hrs." figure represents the number of labor-hours required to perform the individual operations one time.

4 Frequency

The columns marked with "W" for weekly, "M" for monthly, "Q" for quarterly, "S" for semi-annually, and "A" for annually indicate the recommended frequency for performing a task. For example, a check in the "M" column indicates the recommended frequency of the operation is monthly. Note that, as a result, checks also appear in the "Q," "S," and "A" columns, as they would be part of a monthly frequency.

5 Total Labor-Hour

"Total Labor-hours" are provided on both a per-period and per-year basis. The per-period totals are derived by adding the hours for each operation recommended to be performed (indicated with a check mark) in a given period. In the example given here:

M	= .023 + .124 + .143 + .030 + .056 + .073 + .127 + .046 + .066 + .022	= .710
Similarly		
Q	= .023 + .133 + .079 + . . . + .046 + .066 + .022	= 1.069
S	= .023 + .133 + .079 + . . . + .046 + .066 + .022	= 1.069
A	= .077 + .658 + .098 + .023 + . . . + .650 + .046 + .066 + .022	= 3.641

The per-year totals are derived by multiplying the per-period total by the net frequency for each time period. The net frequency per period is based on the following table:

Total Labor-hours		Labor-hours		
Annual	=	Annual	X	1
Semi-Annual	=	Semi-Annual	X	1
Quarterly	=	Quarterly	X	2
Monthly	=	Monthly	X	8
Weekly	=	Weekly	X	40

Based on these values, the total labor-hours/year are calculated for the given example as follows:

M	–	(.710)	X	8	=	5.68
[illegible]	=	[illegible]	X	[illegible]	=	[illegible]
S	=	(1.069)	X	1	=	1.069
A	–	(3.641)	X	1	–	3.641
						12.528 hrs.

6 System Description (Boiler, Hot Water, O/G/C, etc.)

The "System Description" corresponds to what's given at the beginning of the PM Components list. A description of the cost basis, annual or annualized, is given as well. Annual costs are based on performance of each operation listed in the PM Components one time in a given year. Annualized costs are based on performance of each operation at the recommended periodic frequencies in a given year.

7 Labor-Hours: Annual (3.641)/ Annualized (12.528)

Annual labor-hours are based on performance of each task once during a given year and are equal to the total labor-hours/period under the "A" column in the PM Components list. Annualized labor-hours are based on the recommended periodic frequencies and are equivalent to the sum of all the values in the total labor-hours/year row at the bottom of the PM Components.

8 Bare Costs

Annualized Material (66.50)

This figure for the system material cost is the "bare" material cost with no overhead and profit allowances included. *Costs shown reflect national average material prices for the current year and generally include delivery to the facility. Small purchases may require an additional delivery charge. No sales taxes are included.*

Labor (820.00)

The labor costs are derived by multiplying bare labor-hour costs by labor-hour units. (The hourly labor costs are listed in the Crew Listings, or for a single trade, the wage rate is found on the inside back cover of the cost data.)

Equip. (Equipment) (0.00)

The unit equipment cost is derived by multiplying the bare equipment hourly cost by the labor-hour units. Equipment costs for each crew are listed in the description of each crew.

Total (886.50)

The total of the bare costs is the arithmetic total of the three previous columns: material, labor, and equipment.

Material	+	Labor	+	Equip.	=	Total
$66.50	+	$820.00	+	$0.00	=	$886.50

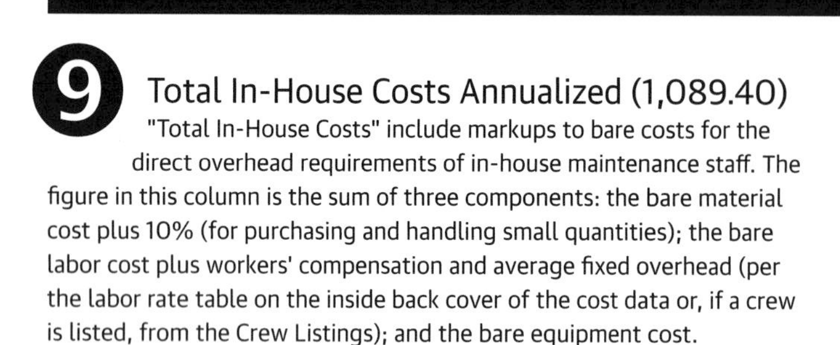

9 Total In-House Costs Annualized (1,089.40)

"Total In-House Costs" include markups to bare costs for the direct overhead requirements of in-house maintenance staff. The figure in this column is the sum of three components: the bare material cost plus 10% (for purchasing and handling small quantities); the bare labor cost plus workers' compensation and average fixed overhead (per the labor rate table on the inside back cover of the cost data or, if a crew is listed, from the Crew Listings); and the bare equipment cost.

10 Total Costs Including O&P Annualized (1,350)

"Total Costs Including O&P" include suggested markups to bare costs for an outside subcontractor's overhead and profit requirements. The figure in this column is the sum of three components: the bare material cost plus 25% for profit; the bare labor cost plus total overhead and profit (per the labor rate table on the inside back cover of the printed product or the Reference Section of the electronic product, or, if a crew is listed, from the Crew Listings); and the bare equipment cost plus 10% for profit.

PM Components	Labor-hrs.	W	M	Q	S	A
PM System B2035 110 1950						
Door, sliding, electric						
1 Check with operating or area personnel for deficiencies.	.035				√	√
2 Check for proper operation, binding or misalignment; adjust as necessary.	.062				√	√
3 Check and lubricate door guides, pulleys and hinges.	.143				√	√
4 Inspect and lubricate motor gearbox, drive chain (or belt), and motor; adjust as necessary.	.200				√	√
5 Check operation of limit switch; adjust as necessary.	.222				√	√
6 Check electrical operator, wiring, connections and contacts; adjust as necessary.	.471				√	√
7 Clean area around door.	.000				√	√
8 Fill out maintenance checklist and report deficiencies.	.022				√	√
Total labor-hours/period					1.221	1.221
Total labor-hours/year					1.221	1.221

	Description	Labor-hrs.	Cost Each					
			2020 Bare Costs				Total	Total
			Material	Labor	Equip.	Total	In-House	w/O&P
1900	Door, sliding, electric, annually	1.221	22	65		87	108.47	132
1950	Annualized	2.448	44	131		175	216.38	263

PM Components	Labor-hrs.	W	M	Q	S	A
PM System B2035 110 2950						
Hanger doors, sliding						
1 Check with door operating personnel for deficiencies.	.027			√	√	√
2 Remove debris from door track.	.040			√	√	√
3 Operate door.	.050			√	√	√
4 Check alignment of hanger door, door guides and lubricate.	.650					√
5 Inspect and lubricate wheel drive chain.	.030			√	√	√
6 Adjust brake shoes and check for wear.	.151					√
7 Inspect and lubricate drive wheels, guides, stops and rollers.	.200			√	√	√
8 Fill out maintenance report.	.017			√	√	√
Total labor-hours/period				.364	.364	1.165
Total labor-hours/year				.728	.364	1.165

	Description	Labor-hrs.	Cost Each					
			2020 Bare Costs				Total	Total
			Material	Labor	Equip.	Total	In-House	w/O&P
2900	Hanger doors, sliding, annually	1.165	33	62		95	116.26	141
2950	Annualized	2.257	77	120		197	240.17	288

B2035 225 Door, Emergency Egress

PM Components	Labor-hrs.	W	M	Q	S	A
PM System B2035 225 1950						
Door, emergency egress, swinging						
1 Remove obstructions that restrict full movement/swing of door.	.013			√	√	√
2 Check swing of door; door must latch on normal closing.	.013			√	√	√
3 Test operation of panic hardware and local alarm battery.	.013			√	√	√
4 Lubricate hardware.	.013			√	√	√
5 Fill out maintenance checklist and report deficiencies.	.013			√	√	√
Total labor-hours/period				.065	.065	.065
Total labor-hours/year				.130	.065	.065

	Description	Labor-hrs.	Cost Each					
			2020 Bare Costs				Total	Total
			Material	Labor	Equip.	Total	In-House	w/O&P
1900	Door, emergency egress, swinging, annually	.065	4.77	3.45		8.22	9.71	11.50
1950	Annualized	.260	14.30	13.80		28.10	33.58	40

PM Components	Labor-hrs.	W	M	Q	S	A
PM System B2035 300 1950						
Door, revolving, manual						
1 Check with operating or area personnel for deficiencies.	.043				√	√
2 Check for proper operation, binding or misalignment; adjust as necessary.	.043				√	√
3 Check for proper emergency breakaway operation of wings.	.043				√	√
4 Check top, bottom, edge and shaft weatherstripping.	.217				√	√
5 Check and lubricate pivots and bearings.	.217				√	√
6 Clean area around door.	.065				√	√
7 Fill out maintenance checklist and report deficiencies.	.022				√	√
Total labor-hours/period					[illegible]	[illegible]
Total labor-hours/year					.650	.650

	Description	Labor-hrs.	Material	Labor	Equip.	Total	Total In-House	Total w/O&P
			Cost Each					
			2020 Bare Costs					
1900	Door, revolving, manual, annually	.650	11.05	34.50		45.55	56.73	69.50
1950	Annualized	1.300	22	69		91	113.67	139

PM Components	Labor-hrs.	W	M	Q	S	A
PM System B2035 300 2950						
Door, revolving, electric						
1 Check with operating or area personnel for deficiencies.	.043				√	√
2 Check for proper operation, binding or misalignment; adjust as necessary.	.043				√	√
3 Check for proper emergency breakaway operation of wings.	.043				√	√
4 Check top, bottom, edge and shaft weatherstripping.	.217				√	√
5 Check elec. operator, safety devices and activators for proper operation.	.433				√	√
6 Check and lubricate pivots and bearings and operator gearbox.	.325				√	√
7 Clean area around door.	.065				√	√
8 Fill out maintenance checklist and report deficiencies.	.022				√	√
Total labor-hours/period					1.191	1.191
Total labor-hours/year					1.191	1.191

	Description	Labor-hrs.	Material	Labor	Equip.	Total	Total In-House	Total w/O&P
			Cost Each					
			2020 Bare Costs					
2900	Door, revolving, electric, annually	1.191	11.05	63.50		74.55	93.73	115
2950	Annualized	2.382	22	126		148	187.67	231

B20 EXTERIOR CLOSURE | B2035 400 | Door, Overhead, Manual

PM Components	Labor-hrs.	W	M	Q	S	A
PM System B2035 400 1950						
Door, overhead, manual, up to 24′ high x 25′ wide						
1 Check with operating or area personnel for deficiencies.	.035				√	√
2 Check for proper operation, binding or misalignment; adjust as necessary.	.136				√	√
3 Check and lubricate door guides, pulleys and hinges.	.889				√	√
4 Clean area around door.	.026				√	√
5 Fill out maintenance checklist and report deficiencies.	.022				√	√
Total labor-hours/period					1.107	1.107
Total labor-hours/year					1.107	1.107

	Description	Labor-hrs.	Cost Each					
			2020 Bare Costs				Total	Total
			Material	Labor	Equip.	Total	In-House	w/O&P
1900	Dr., overhead, manual, up to 24′H x 25′W, annually	1.107	11.05	58.50		69.55	88	108
1950	Annualized	2.196	22	117		139	174.95	215

B20 EXTERIOR CLOSURE | B2035 410 | Door, Overhead, Roll-up, Electric

PM Components	Labor-hrs.	W	M	Q	S	A
PM System B2035 410 1950						
Door, overhead, roll-up, electric, up to 24′ high x 25′ wide						
1 Check with operating or area personnel for deficiencies.	.035				√	√
2 Check for proper operation, binding or misalignment; adjust as necessary.	.136				√	√
3 Check and lubricate door guides, pulleys and hinges.	.889				√	√
4 Inspect and lubricate motor gearbox, drive chain (or belt), and motor; adjust as necessary.	.200				√	√
5 Check operation of limit switch; adjust as necessary.	.222				√	√
6 Check electrical operator, wiring, connections and contacts; adjust as necessary.	.471				√	√
7 Clean area around door.	.066				√	√
8 Fill out maintenance checklist and report deficiencies.	.022				√	√
Total labor-hours/period					2.040	2.040
Total labor-hours/year					2.040	2.040

	Description	Labor-hrs.	Cost Each					
			2020 Bare Costs				Total	Total
			Material	Labor	Equip.	Total	In-House	w/O&P
1900	Dr., overhead, electric, to 24′H x 25′W, annually	2.040	52	108		160	197.50	238
1950	Annualized	4.070	104	217		321	393.93	475

PM Components	Labor-hrs.	W	M	Q	S	A
PM System B2035 450 1950						
Shutter, roll-up, electric						
1 Check with operating or area personnel for deficiencies.	.035					√
2 Check for proper operation, binding or misalignment; adjust as necessary.	.136					√
3 Check and lubricate door guides, pulleys and hinges.	.889					√
4 Inspect and lubricate motor gearbox, drive chain (or belt), and motor; adjust as necessary.	.200					√
5 Check operation of limit switch; adjust as necessary	.222					√
6 Check electrical operator wiring, connections and contactor; adjust as necessary.	.471					√
7 Clean area around door.	.074					√
8 Fill out maintenance checklist and report deficiencies.	.022					√
Total labor-hours/period						2.048
Total labor-hours/year						2.048

	Description	Labor-hrs.	Cost Each					
			2020 Bare Costs				Total	Total
			Material	Labor	Equip.	Total	In-House	w/O&P
1900	Shutter, roll up, electric, annually	2.048	22	109		131	165.06	202
1950	Annualized	2.048	22	109		131	165.06	202

PM Components	Labor-hrs.	W	M	Q	S	A
PM System C1025 110 1950						
Fire Door, Swinging						
1 Remove fusible link hold open devices.	.026			√	√	√
2 Remove obstructions that retard full movement/swing of door.	.013			√	√	√
3 Check swing of door. Door must latch on normal closing.	.013			√	√	√
4 Test operation of panic hardware.	.007			√	√	√
5 Check operation of special devices such as smoke detectors or magnetic door releases.	.013			√	√	√
6 Lubricate hardware.	.013			√	√	√
7 Fill out maintenance checklist and report deficiencies.	.013			√	√	√
Total labor-hours/period				.098	.098	.098
Total labor-hours/year				.196	.098	.098

	Description	Labor-hrs.	Cost Each 2020 Bare Costs Material	Labor	Equip.	Total	Total In-House	Total w/O&P
1900	Fire doors, swinging, annually	.098	4.77	5.35		10.12	12.15	14.55
1950	Annualized	.392	14.30	21.50		35.80	43.31	52.50

PM Components	Labor-hrs.	W	M	Q	S	A
PM System C1025 110 2950						
Fire Door, Sliding						
1 Clean track.	.013			√	√	√
2 Lubricate pulleys.	.013			√	√	√
3 Inspect cable or chain for wear or damage and proper threading through pulleys.	.078			√	√	√
4 Replace fusible links or other heat activated devices that have been painted. Check operation of other heat activated devices.	.130			√	√	√
5 Permanently remove obstructions that prohibit movement of door.	.013			√	√	√
6 Check operation of door by disconnecting or lifting counterweight or other appropriate means.	.129			√	√	√
7 Check fit of door in binders and fit against stay roll.	.013			√	√	√
8 Check door for breaks in covering and examine for dry rot.	.013			√	√	√
9 Fill out maintenance checklist and report deficiencies.	.013			√	√	√
Total labor-hours/period				.415	.415	.415
Total labor-hours/year				.830	.415	.415

	Description	Labor-hrs.	Cost Each 2020 Bare Costs Material	Labor	Equip.	Total	Total In-House	Total w/O&P
2900	Fire doors, sliding, annually	.415	18.80	22		40.80	49.20	59
2950	Annualized	1.659	56.50	88		144.50	176.28	212

PM Components	Labor-hrs.	W	M	Q	S	A
PM System C1025 110 3950						
Fire Door, Rollup						
1 Check with door operating personnel for deficiencies.	.027			√	√	√
2 Drop test heat activated or fusible link roll up fire door.	.070			√	√	√
3 Adjust rate of descent governor to desired closing speed.	.200					√
4 Inspect, clean and lubricate door gear assembly.	.200			√	√	√
5 Check alignment of overhead door guides, clean and lubricate.	.100			√	√	√
6 Check and adjust door counter balance assembly.	.250					√
7 Check electrical wiring and contacts for wear.	.200					√
8 Check and adjust limit switches.	.170					√
9 Inspect general condition of door for need of paint and repair.	.005			√	√	√
10 Replace missing or tighten loose nuts and bolts.	.040			√	√	√
11 After maintenance clean around door and ensure access.	.050			√	√	√
12 Fill out maintenance report.	.017			√	√	√
Total labor-hours/period				.509	.509	1.329
Total labor-hours/year				1.018	.509	1.329

	Description	Labor-hrs.	Cost Each					
			2020 Bare Costs				Total	Total
			Material	Labor	Equip.	Total	In-House	w/O&P
3900	Fire doors, roll-up, annually	1.329	34	70.50		104.50	129.28	157
3950	Annualized	2.856	79.50	152		231.50	284.09	340

PM Components	Labor-hrs.	W	M	Q	S	A
PM System D1015 100 1950						
Elevator, cable, electric, passenger/freight						
1 Ride car, check for any unusual noise or operation.	.411		√	√	√	√
2 Inspect machine room equipment.	.286		√	√	√	√
3 Motor room:						
A) visually inspect controllers and starters.	.012		√	√	√	√
B) visually inspect selector.	.012		√	√	√	√
C) lubricate and adjust tension of selector.	.113		√	√	√	√
D) inspect sleeve bearings of hoist motor.	.012		√	√	√	√
E) inspect brushes and commutator of hoist motor.	.012		√	√	√	√
F) inspect sleeve bearings of motor generator.	.012		√	√	√	√
G) inspect brushes and commutator of motor generator.	.014		√	√	√	√
H) inspect sleeve bearings of exciter.	.022		√	√	√	√
I) inspect brushes and commutator of exciter.	.022		√	√	√	√
J) inspect sleeve bearings of regulator dampening motors and tach generators.	.044		√	√	√	√
K) inspect brushes and commutator of regulator dampening motors and tach generators.	.047		√	√	√	√
L) inspect and grease sleeve bearings of geared machines.	.044		√	√	√	√
M) inspect oil level in worm and gear of geared machines.	.044		√	√	√	√
N) visually inspect the brakes.	.022		√	√	√	√
O) inspect and grease, if necessary, sleeve bearings of drive deflectors and secondary sheaves.	.186		√	√	√	√
P) visually inspect and adjust contacts of controllers and starters.	.338			√	√	√
Q) inspect and adjust main operating contactor and switches of controllers and starters.	.338			√	√	√
R) inspect selector, adjusting contacts, brushes and cams if needed.	.103			√	√	√
S) inspect selsyn and advancer, motor brushes and commutators of selector.	.022			√	√	√
T) clean hoist motor brush rigging and commutator.	.068			√	√	√
U) clean motor generator and tighten brush rigging and clean commutator.	.068			√	√	√
V) clean exciter brush rigging and commutator.	.066			√	√	√
W) clean brush rigging and commutator of regulator dampening motors and tach generators.	.099			√	√	√
X) inspect machine worm and gear thrust and backlash and spider bolts; replace oil.	.047			√	√	√
Y) clean and lubricate brakes, pins and linkage.	.099			√	√	√
Z) inspect governors; lubricate sleeve bearings.	.099			√	√	√
Aa) inspect controllers and starters; adjust setting of overloads.	.169					√
Ab) tighten connections on controllers and starters; clean fuses and holders.	.169					√
Ac) set controllers and starter timers.	.169					√
Ad) clean controllers and starters.	.169					√
Ae) clean and inspect selector components.	.068					√
Af) inspect group operation controllers; clean and lubricate contacts, timers, steppers of operation controller.	.099					√
Ag) change oil and/or grease roller bearings of the hoist motor.	.094					√
Ah) vacuum and blowout the hoist motor.	.094					√
Ai) tighten all connections on the hoist motor.	.022					√

	PM Components	Labor-hrs.	W	M	Q	S	A
	Aj) change oil and/or grease generator's roller bearings.	.069					√
	Ak) vacuum and blowout motor generator.	.094					√
	Al) inspect all connections on the motor generator and tighten if necessary.	.022					√
	Am) inspect roller bearings of the exciter and change oil and/or grease.	.069					√
	An) vacuum and blowout exciter.	.094					√
	Ao) change oil and/or grease roller bearings of regulator dampening motors and tach generators.	.094					√
	Ap) vacuum/blowout regulator dampening motors and tach generators.	.044					√
	Aq) inspect all regulator dampening motors and tach generator connections.	.047					√
	Ar) change oil and/or grease roller bearings of the geared machines.	[illegible]					√
	As) clean and lubricate brake cores.	.099					√
	At) inspect drive deflectors and secondary sheaves; change oil and grease roller bearings if necessary.	.170					√
	Au) inspect drive deflectors and secondary sheaves' cable grooves for cracks.	.012					√
	Av) inspect governors; change oil and grease roller bearings, if necessary.	.094					√
4	Hatch:						
	A) lubricate rails of hoistway.	.733		√	√	√	√
	B) inspect car top and grease sleeve bearings if needed.	.151		√	√	√	√
	C) inspect counterweight and grease sleeve bearings if needed.	.068		√	√	√	√
	D) visually inspect cables, chains, hoist, compensating governor for wear and equalization.	.047		√	√	√	√
	E) lubricate cables, chain hoist and compensating governor if necessary.	.186		√	√	√	√
	F) inspect door operator; clean and lubricate chain and belt tension.	.177			√	√	√
	G) inspect door, clean and adjust safety edge, light ray and cables.	.052			√	√	√
	H) clean and adjust proximity devices on door.	.103			√	√	√
	I) inspect and lubricate governor tape and compensating sheave in pit.	.068			√	√	√
	J) clean and lubricate compensating tie down in pit.	.047			√	√	√
	K) inspect pit's run-by on counter weight.	.142			√	√	√
	L) inspect governor and compensating sheaves or chains for clearance to pit floor.	.022			√	√	√
	M) inspect/lubricate overhead hatch switches and cams.	.096					√
	N) clean and lubricate rollers, cables, sight guards, chains, motors, closures, limit and zone switches.	.068					√
	O) inspect door gibs and fastening.	.022					√
5	Inspect car top; change oil and grease roller bearings, if necessary.	.044					√
6	Check safety switches, indicators, leveling devices, selector tape, switches and hatches.	.044					√
7	Check and lubricate fan motor.	.072					√
8	Inspect counterweight; change oil and grease roller bearings, if necessary.	.094					√
9	Check safety linkage safety and cwt guides.	.098					√
10	Inspect compensation hatch.	.047					√
11	Inspect chains, hoist and compensating governor.	.094					√
12	Inspect traveling cables for tracking and wear.	.070					√

	PM Components	Labor-hrs.	W	M	Q	S	A
13	Inspect door operator commutator and brushes.	.012					√
14	Inspect/clean/lubricate door operator clutch, retiring cam, door gib, rollers, tracks, upthrusts and relating cables.	.103					√
15	Under car:						
	A) inspect/clean/lubricate safety devices, linkages, car guides, selector tape, switches and hitches.	.020					√
	B) inspect cable hitch and loops; load weighing devices.	.027					√
16	Inspect/lubricate pit hatch switches, cams and pit oil buffer.	.083					√
17	In car:						
	A) inspect and clean fixtures and signal in operating panel and car position and direction indicator.	.103		√	√	√	√
	B) check operation of emergency lights and bells.	.229		√	√	√	√
	C) check handrails, ceiling panels and hang on panels for tightness.	.027		√	√	√	√
	D) check for tripping hazards.	.044		√	√	√	√
18	Hallway corridor:						
	A) inspect hall buttons, signal lamps, lanterns and hall position indicator.	.046		√	√	√	√
	B) inspect starter station, key operation and lamps.	.069		√	√	√	√
19	Clean equipment and surrounding area.	.066		√	√	√	√
20	Fill out maintenance checklist.	.022		√	√	√	√
	Total labor-hours/period			3.106	5.063	5.063	8.112
	Total labor-hours/year			24.848	10.126	5.063	8.112

	Description	Labor-hrs.	Cost Each					
			2020 Bare Costs				Total	Total
			Material	Labor	Equip.	Total	In-House	w/O&P
1900	Elevator, cable, electric, passenger / freight, annually	8.112	1,525	695		2,220	2,516.02	2,975
1950	Annualized	47.895	4,025	4,100		8,125	9,438.01	11,300

PM Components	Labor-hrs.	W	M	Q	S	A
PM System D1015 110 1950						
Elevator, hydraulic, passenger/freight						
1 Ride car, checking for any unusual noise or operation.	.052		√	√	√	√
2 In car:						
A) inspect and clean fixtures and signal in operating panel and car position and direction indicator.	.103		√	√	√	√
B) check operation of emergency lights and bell.	.020		√	√	√	√
C) check handrails, ceiling panels and hang on panels for tightness.	.027		√	√	√	√
D) check for tripping hazards.	.012		√	√	√	√
3 Inspect and lubricate rails of hoistway.	.094		√	√	√	√
4 Hallway corridor:						
A) inspect hall buttons, signal lamps, lanterns and hall position indicator.	.010		√	√	√	√
B) inspect starter station, key operation and lamps.	.060		√	√	√	√
5 Motor room:						
A) inspect machine room equipment.	.077		√	√	√	√
B) lockout and log record.	.022		√	√	√	√
C) inspect tank oil level.	.012		√	√	√	√
D) inspect and adjust controller contacts; main operating contactors and switches.	.195			√	√	√
E) inspect pump and valve unit for leaks.	.051			√	√	√
F) inspect and adjust controller overloads; set timers.	.049					√
G) tighten connections and clean controller fuses and holders.	.020					√
H) inspect and lubricate pump motor bearings.	.099					√
6 Hatch:						
A) check hoistway car rails, brackets and fish plates.	.103					√
B) inspect/lubricate overhead hatch switches and cams.	.070					√
C) inspect/clean/lubricate hatch doors locks, rollers, tracks, upthrusts, relating cables, racks, sight guards and closers, motors, gear boxes, limit and zone switches.	.096					√
D) inspect door gibs and fastening.	.022					√
E) inspect/clean/lubricate cab top guides, steading devices, safety switches, inductors, leveling devices, selector tape, switches, hitches and fan motor.	.068					√
F) check traveling cables for wear.	.070					√
G) inspect/clean/lubricate door operator roller tracks, upthrusts, related cables, clutch, retiring cam and door gib.	.078					√
H) inspect door operator; clean and lubricate chain and belt tension.	.052			√	√	√
I) inspect door, clean and adjust safety edge, light ray and cables.	.103			√	√	√
J) clean and adjust proximity devices on door.	.068			√	√	√
7 Inspect pit gland packing.	.010			√	√	√
8 Inspect/clean/lubricate under car guides, selector tape, traveling cable, switches and platen plate assembly.	.068					√
9 Clean equipment and surrounding area.	.074		√	√	√	√
10 Fill out maintenance checklist.	.022		√	√	√	√
Total labor-hours/period			.630	1.109	1.109	1.852
Total labor-hours/year			5.042	2.217	1.109	1.852

	Description	Labor-hrs.	Cost Each: 2020 Bare Costs Material	Labor	Equip.	Total	Total In-House	Total w/O&P
1900	Elevator, hydraulic, passenger / freight, annually	1.852	1,425	158		1,583	1,753.25	2,025
1950	Annualized	10.224	1,375	875		2,250	2,578.43	3,075

PM Components	Labor-hrs.	W	M	Q	S	A
PM System D1015 310 1950						
Wheelchair Lift						
1 Check with operating personnel for any deficiencies.	.033				√	√
2 Check voltage and amperage under load.	.091				√	√
3 Check bolts securing drive cabinet and base, tighten accordingly.	.033				√	√
4 Check belt tension.	.033				√	√
5 Check lift nut assembly.	.026				√	√
6 Check cam rollers.	.026				√	√
7 Check wear pads for excessive wear.	.026				√	√
8 Inspect motor and shaft pulleys.	.039				√	√
9 Check Acme screw alignment.	.033				√	√
10 Inspect and lubricate upper and lower bearings.	.039				√	√
11 Check fastening of cable harness.	.026				√	√
12 Check alignment of platform and doors.	.033				√	√
13 Check door interlock switch for proper operation.	.039				√	√
14 Check operation of final limit switch and emergency stop/alarm.	.033				√	√
15 Check call/send controls at each station and on platform.	.039				√	√
16 Lube hinge of flip up ramp.	.033				√	√
17 Fill out maintenance checklist and report deficiencies.	.022				√	√
Total labor-hours/period					.604	.604
Total labor-hours/year					.604	.604

	Description	Labor-hrs.	Cost Each					
			2020 Bare Costs				Total	Total
			Material	Labor	Equip.	Total	In-House	w/O&P
1900	Wheelchair lift, annually	.604	22	51.50		73.50	87.56	107
1950	Annualized	1.208	44	103		147	174.31	214

PM Components	Labor-hrs.	W	M	Q	S	A
PM System D1025 100 1950 **Escalator**						
1 Ride escalator; check operation for smoothness, unusual vibration or noise, condition of handrails, etc.	.104	√	√	√	√	√
2 Deenergize, tag, and lockout the electrical circuit.	.017	√	√	√	√	√
3 Inspect comb plates at both ends of escalator for broken teeth and check for proper clearance between combs and step teeth and check for broken step treads.	.039	√	√	√	√	√
4 Check clearance between steps and skirt panel; look for loose trim, screws or bolts, that could snag or damage clothing or cause injury; [illegible]	.087	√	√	√	√	√
5 Clean escalator machine space.	.113	√	√	√	√	√
6 Lubricate step rollers, step chain, drive gears or chains, handrail drive chains, etc., according to manufacturers' instructions; observe gears and chains for signs of wear, misalignment, etc.; adjust as required.	.007	√	√	√	√	√
7 Check motor for signs of overheating.	.004	√	√	√	√	√
8 Inspect controller for loose leads, burned contacts, etc.; repair as required.	.022	√	√	√	√	√
9 Clean handrails as required.	.217	√	√	√	√	√
10 Check escalator lighting; replace bulbs as required.	.099	√	√	√	√	√
11 Operate each emergency stop button and note that the escalator stops; if the escalator has the capabilities of running in both directions, stop buttons should function properly for each direction of travel; observe the stopping distance.	.013	√	√	√	√	√
12 Clean surrounding area.	.066	√	√	√	√	√
13 Fill out maintenance checklist and report deficiencies.	.022	√	√	√	√	√
Total labor-hours/period		1.070	1.070	1.070	1.070	1.070
Total labor-hours/year		40.642	8.556	2.139	1.070	1.070

	Description	Labor-hrs.	Cost Each					
			2020 Bare Costs				Total	Total
			Material	Labor	Equip.	Total	In-House	w/O&P
1900	Escalator, annually	1.070	2,300	91.50		2,391.50	2,642	3,025
1950	Annualized	53.509	3,225	4,575		7,800	9,126.50	11,100

D10 CONVEYING | D1095 100 | Dumbwaiter, Electric

PM Components	Labor-hrs.	W	M	Q	S	A
PM System D1095 100 1950						
Dumbwaiter, electric						
1 Check with operating or area personnel for any obvious deficiencies.	.086					√
2 Operate dumbwaiter through complete cycle and check operation for unusual noises or problems.	.052					√
3 Inspect and clean operating panel, starter station, key operation and lamps.	.072					√
4 Check operation of emergency lights and alarms.	.052					√
5 Inspect, clean and lubricate motor.	.072					√
6 Check operation of indicator lights and buttons.	.052					√
7 Inspect and lubricate rails of guides.	.148					√
8 Clean equipment and surrounding area.	.077					√
9 Fill out maintenance checklist and report deficiencies.	.022					√
Total labor-hours/period						.633
Total labor-hours/year						.633

	Description	Labor-hrs.	Cost Each					
			2020 Bare Costs				Total	Total
			Material	Labor	Equip.	Total	In-House	w/O&P
1900	Dumbwaiter, electric, annually	.633	54	54		108	125.95	151
1950	Annualized	.633	54	54		108	125.95	151

D10 CONVEYING | D1095 110 | Dumbwaiter, Hydraulic

PM Components	Labor-hrs.	W	M	Q	S	A
PM System D1095 110 1950						
Dumbwaiter, hydraulic						
1 Check with operating or area personnel for any obvious deficiencies.	.086				√	√
2 Operate dumbwaiter through complete cycle and check operation for unusual noises or problems.	.052				√	√
3 Inspect and clean operating panel, starter station, key operation and lamps.	.072				√	√
4 Check operation of emergency lights and alarms.	.052				√	√
5 Inspect pump and valve unit for leaks.	.072				√	√
6 Inspect tank oil level; add as required.	.039				√	√
7 Inspect and lubricate rails of guides.	.148				√	√
8 Inspect jack seals.	.051				√	√
9 Clean equipment and surrounding area.	.077				√	√
10 Fill out maintenance checklist and report deficiencies.	.022				√	√
Total labor-hours/period					.671	.671
Total labor-hours/year					.671	.671

	Description	Labor-hrs.	Cost Each					
			2020 Bare Costs				Total	Total
			Material	Labor	Equip.	Total	In-House	w/O&P
1900	Dumbwaiter, hydraulic, annually	.671	69	57.50		126.50	146.20	174
1950	Annualized	1.342	138	115		253	291.46	350

Pneumatic Tube System

PM Components	Labor-hrs.	W	M	Q	S	A
PM System D1095 200 1950						
Pneumatic Tube System						
1 Check with operating personnel for any deficiencies.	.039			√	√	√
2 Clean air intake screen.	.039			√	√	√
3 Check seal at doors.	.129			√	√	√
4 Inspect check valves	.026			√	√	√
5 Check blower motor and lubricate.	.065			√	√	√
6 Inspect carrier ends, latches and air washers	.026			√	√	√
7 Fill out maintenance checklist and report deficiencies.	.026			√	√	√
Total labor-hours/period				.350	.350	.350
Total labor-hours/year				.700	.350	.350

	Description	Labor-hrs.	Cost Each					
			2020 Bare Costs				Total	Total
			Material	Labor	Equip.	Total	In-House	w/O&P
1900	Pneumatic Tube System, annually	.350	23	23		46	53.62	64.50
1950	Annualized	1.404	69	92		161	189.70	228

PM Components	Labor-hrs.	W	M	Q	S	A
PM System D2015 100 1950						
Urinals (Time is per fixture)						
1 Urinals - Flush and adjust water flow if required.	.020			√	√	√
2 Inspect for missing or damaged parts/caps and replace.	.020			√	√	√
3 Fill out maintenance checklist and report deficiencies.	.017			√	√	√
Total labor-hours/period				.057	.057	.057
Total labor-hours/year				.114	.057	.057

	Description	Labor-hrs.	Cost Each					
			2020 Bare Costs				Total	Total
			Material	Labor	Equip.	Total	In-House	w/O&P
1900	Urinals, annually	.057	6.10	3.13		9.23	10.71	12.60
1950	Annualized	.228	6.10	12.50		18.60	22.78	27.50

PM Components	Labor-hrs.	W	M	Q	S	A
PM System D2015 100 2950						
Toilet, vacuum breaker type (Time is per fixture)						
1 Toilet (vacuum breaker type) - Flush and adjust water flow if required.	.020			√	√	√
2 Inspect for missing or damaged parts/caps, seat supports, and replace.	.020			√	√	√
3 Fill out maintenance checklist and report deficiencies.	.017			√	√	√
Total labor-hours/period				.057	.057	.057
Total labor-hours/year				.114	.057	.057

	Description	Labor-hrs.	Cost Each					
			2020 Bare Costs				Total	Total
			Material	Labor	Equip.	Total	In-House	w/O&P
2900	Toilet (vacuum breaker type), annually	.057	4.17	3.13		7.30	8.61	10.20
2950	Annualized	.177	8.35	9.70		18.05	21.70	26

PM Components	Labor-hrs.	W	M	Q	S	A
PM System D2015 100 3950						
Toilet, tank type (Time is per fixture)						
1 Toilet - Clean flapper seat if leaking, adjust fill level if required.	.050			√	√	√
2 Inspect for missing or damaged parts/caps, seat supports, and replace.	.030			√	√	√
3 Fill out maintenance checklist and report deficiencies.	.017			√	√	√
Total labor-hours/period				.097	.097	.097
Total labor-hours/year				.194	.097	.097

	Description	Labor-hrs.	Cost Each					
			2020 Bare Costs				Total	Total
			Material	Labor	Equip.	Total	In-House	w/O&P
3900	Toilet (tank type), annually	.097	7.55	5.30		12.85	15.12	17.95
3950	Annualized	.388	7.55	21.50		29.05	35.58	43.50

PM Components	Labor-hrs.	W	M	Q	S	A
PM System D2015 100 4950						
Lavatories (Time is per fixture)						
1 Lavatories - Operate faucets, replace washers/"O" Rings as necessary.	.040			√	√	√
2 Observe drain flow, clean trap if flow is obstructed.	.030			√	√	√
3 Fill out maintenance checklist and report deficiencies.	.017			√	√	√
Total labor-hours/period				.087	.087	.087
Total labor-hours/year				.174	.087	.087

D20 PLUMBING — D2015 100 Facility Plumbing Fixture Service

	Description	Labor-hrs.	2020 Bare Costs Material	Labor	Equip.	Total	Total In-House	Total w/O&P
4900	Lavatories, annually	.087	9.40	5.60		15	17.29	20.50
4950	Annualized	.348	9.40	22.50		31.90	37.95	46.50

PM Components	Labor-hrs.	W	M	Q	S	A
PM System D2015 100 5950						
Showers (Time is per fixture)						
1 Showers: Check for damaged, or missing shower heads/handles and replace if required.	.040			√	√	√
2 Fill out maintenance checklist and report deficiencies.	.017			√	√	√
Total labor-hours/period				.057	.057	.057
Total labor-hours/year				.114	.057	.057

	Description	Labor-hrs.	2020 Bare Costs Material	Labor	Equip.	Total	Total In-House	Total w/O&P
5900	Showers, annually	.057	12.80	3.68		16.48	18.65	21.50
5950	Annualized	.228	12.80	14.70		27.50	32.40	39

D20 PLUMBING — D2015 800 Drinking Fountain

PM Components	Labor-hrs.	W	M	Q	S	A
PM System D2015 800 1950						
Drinking fountain						
1 Check unit for proper operation, excessive noise or vibration.	.035					√
2 Clean condenser coils and fan as required.	.222					√
3 Check for water leaks in supply line and drain.	.077					√
4 Check water flow; adjust as necessary.	.039					√
5 Check drinking water temperature to ensure the unit is operating properly.	.039					√
6 Check electrical connections and cord; tighten and repair as necessary.	.120					√
7 Clean area around fountain.	.066					√
8 Fill out maintenance checklist and report deficiencies.	.022					√
Total labor-hours/period						.620
Total labor-hours/year						.620

	Description	Labor-hrs.	2020 Bare Costs Material	Labor	Equip.	Total	Total In-House	Total w/O&P
1900	Drinking fountain, annually	.620	20	34		54	65.71	79
1950	Annualized	.620	20	34		54	65.71	79

D20 PLUMBING | D2025 120 | Valve, Butterfly

PM Components	Labor-hrs.	W	M	Q	S	A
PM System D2025 120 1950						
Valve, butterfly, above 4″						
1 Lubricate valves that have grease fittings.	.022					√
2 Open and close valve using handle, wrench, or hand wheel to check operation.	.049					√
3 Check for leaks.	.007					√
4 Clean valve exterior and area around valve.	.066					√
5 Fill out maintenance report and report deficiencies.	.022					√
Total labor-hours/period						.166
Total labor-hours/year						.166

	Description	Labor-hrs.	Cost Each 2020 Bare Costs Material	Labor	Equip.	Total	Total In-House	Total w/O&P
1900	Valve, butterfly, above 4″, annually	.166	11.05	9.10		20.15	23.84	28.50
1950	Annualized	.166	11.05	9.10		20.15	23.84	28.50

PM Components	Labor-hrs.	W	M	Q	S	A
PM System D2025 120 2950						
Valve, butterfly, auto, above 4″						
1 Lubricate valve actuator stem and valve stem, where possible.	.022					√
2 Check automatic valve for proper operation.	.062					√
3 Check packing gland for leaks; adjust as required.	.160					√
4 Check pneumatic operator and tubing for air leaks.	.047					√
5 Clean valve exterior and area around valve.	.034					√
6 Fill out maintenance report and report deficiencies.	.022					√
Total labor-hours/period						.347
Total labor-hours/year						.347

	Description	Labor-hrs.	Cost Each 2020 Bare Costs Material	Labor	Equip.	Total	Total In-House	Total w/O&P
2900	Valve, butterfly, auto, above 4″, annually	.347	11.05	19.05		30.10	36.56	44.50
2950	Annualized	.347	11.05	19.05		30.10	36.56	44.50

D20 PLUMBING | D2025 125 | Valve, Check

PM Components	Labor-hrs.	W	M	Q	S	A
PM System D2025 125 1950						
Valve, check, above 4″						
1 Inspect valve for leaks; repair as necessary.	.009					√
2 Adjust system pressure/flow to verify that valve is opening and closing properly, if applicable.	.160					√
3 Clean valve and area around valve.	.066					√
4 Fill out maintenance report and report deficiencies.	.022					√
Total labor-hours/period						.257
Total labor-hours/year						.257

	Description	Labor-hrs.	Cost Each 2020 Bare Costs Material	Labor	Equip.	Total	Total In-House	Total w/O&P
1900	Valve, check, above 4″, annually	.257	11.05	14.15		25.20	30.25	36.50
1950	Annualized	.257	11.05	14.15		25.20	30.25	36.50

D20 PLUMBING | D2025 130 | Valve, Cock

PM Components	Labor-hrs.	W	M	Q	S	A
PM System D2025 130 1950						
Valve, ball, above 4″						
1 Lubricate valves that have grease fittings.	.022					√
2 Open and close valve using handle, wrench, or hand wheel to check operation.	.049					√
3 Check for leaks.	.007					√
4 Clean valve and area around valve.	.066					√
5 Fill out maintenance report and report deficiencies.	.022					√
Total labor-hours/period						.166
Total labor-hours/year						.166

	Description	Labor-hrs.	Cost Each					
			2020 Bare Costs				Total	Total
			Material	Labor	Equip.	Total	In-House	w/O&P
1900	Valve, ball, above 4″, annually	.166	11.05	9.10		20.15	23.84	28.50
1950	Annualized	.166	11.05	9.10		20.15	23.84	28.50

D20 PLUMBING | D2025 135 | Valve, Diaphragm

PM Components	Labor-hrs.	W	M	Q	S	A
PM System D2025 135 1950						
Valve, diaphragm, above 4″						
1 Lubricate valve stem, close and open valve to check operation.	.022					√
2 Check valve for proper operation and leaks; tighten packing and flange bolts as required.	.044					√
3 Clean valve exterior and area around valve.	.034					√
4 Fill out maintenance report and report deficiencies.	.022					√
Total labor-hours/period						.122
Total labor-hours/year						.122

	Description	Labor-hrs.	Cost Each					
			2020 Bare Costs				Total	Total
			Material	Labor	Equip.	Total	In-House	w/O&P
1900	Valve, diaphragm, above 4″, annually	.122	11.05	6.70		17.75	20.73	24.50
1950	Annualized	.122	11.05	6.70		17.75	20.73	24.50

D20 PLUMBING | D2025 140 | Valve, Gate

PM Components	Labor-hrs.	W	M	Q	S	A
PM System D2025 140 1950						
Valve, gate, above 4″						
1 Lubricate valve stem; close and open valve to check operation.	.049					√
2 Check packing gland for leaks; tighten packing and flange bolts as req.	.022					√
3 Clean valve exterior and area around valve.	.066					√
4 Fill out maintenance report and report deficiencies.	.022					√
Total labor-hours/period						.159
Total labor-hours/year						.159

	Description	Labor-hrs.	Cost Each					
			2020 Bare Costs				Total	Total
			Material	Labor	Equip.	Total	In-House	w/O&P
1900	Valve, gate, above 4″, annually	.159	11.05	8.75		19.80	23.35	28
1950	Annualized	.159	11.05	8.75		19.80	23.35	28

D20 PLUMBING | D2025 145 | Valve, Globe

PM Components	Labor-hrs.	W	M	Q	S	A
PM System D2025 145 1950						
Valve, globe, above 4″						
1 Lubricate stem, close and open valve to check operation.	.049					√
2 Check packing gland for leaks; tighten packing and flange bolts as req.	.022					√
3 Clean valve exterior.	.066					√
4 Fill out maintenance report and report deficiencies.	.022					√
Total labor-hours/period						.159
Total labor-hours/year						.159

	Description	Labor-hrs.	Cost Each					
			2020 Bare Costs				Total	Total
			Material	Labor	Equip.	Total	In-House	w/O&P
1900	Valve, globe, above 4′′, annually	.159	11.05	8.75		19.80	23.35	28
1950	Annualized	.159	11.05	8.75		19.80	23.35	28

PM Components	Labor-hrs.	W	M	Q	S	A
PM System D2025 145 2950						
Valve, globe, auto, above 4″						
1 Lubricate valve actuator stem and valve stem, where possible.	.022					√
2 Check automatic valve for proper operation.	.062					√
3 Check packing gland for leaks; adjust as required.	.160					√
4 Check penumatic operator and tubing for air leaks.	.025					√
5 Clean valve exterior and area around valve.	.034					√
6 Fill out maintenance report and report deficiencies.	.022					√
Total labor-hours/period						.325
Total labor-hours/year						.325

	Description	Labor-hrs.	Cost Each					
			2020 Bare Costs				Total	Total
			Material	Labor	Equip.	Total	In-House	w/O&P
2900	Valve, globe, auto, above 4′′, annually	.325	11.05	17.85		28.90	35.01	42.50
2950	Annualized	.325	11.05	17.85		28.90	35.01	42.50

D20 PLUMBING | D2025 150 | Valve, Motor Operated

PM Components	Labor-hrs.	W	M	Q	S	A
PM System D2025 150 1950						
Valve, motor operated, above 4″						
1 Lubricate valve actuator stem and valve stem, where possible.	.091				√	√
2 Check motor and valve for proper operation, including limit switch; adjust as required.	.022				√	√
3 Check packing gland for leaks; adjust as required.	.113				√	√
4 Check electrical wiring, connections and contacts; repair as necessary.	.119				√	√
5 Inspect and lubricate motor gearbox as required.	.099				√	√
6 Clean valve exterior and area around valve.	.034				√	√
7 Fill out maintenance report and report deficiencies.	.022				√	√
Total labor-hours/period					.500	.500
Total labor-hours/year					.500	.500

	Description	Labor-hrs.	Cost Each					
			2020 Bare Costs				Total	Total
			Material	Labor	Equip.	Total	In-House	w/O&P
1900	Valve, motor operated, above 4′′, annually	.500	22	27.50		49.50	59.49	71
1950	Annualized	1.002	44	55		99	118.78	143

PM Components	Labor-hrs.	W	M	Q	S	A
PM System D2025 155 1950 **Valve, OS&Y, above 4″**						
1 Lubricate stem; open and close valve to check operation.	.049					√
2 Check packing gland for leaks; tighten packing and flange bolts as required.	.022					√
3 Clean valve exterior and area around valve.	.066					√
4 Fill out maintenance report and report deficiencies	.022					√
Total labor-hours/period						.159
Total labor-hours/year						.159

	Description	Labor-hrs.	Cost Each					
			2020 Bare Costs				Total	Total
			Material	Labor	Equip.	Total	In-House	w/O&P
1900	Valve, OS&Y, above 4″, annually	.160	11.05	8.75		19.80	[illegible]	[illegible]
1950	Annualized	.159	11.05	8.75		19.80	23.35	28

D2025 190 Water Heater, Solar

PM Components	Labor-hrs.	W	M	Q	S	A
PM System D2025 190 1950						
Solar, closed loop hot water heating system, up to 6 panels						
1 Check with operating or area personnel for deficiencies.	.035				√	√
2 Inspect interior piping and connections for leaks and damaged insulation; tighten connections and repair damaged insulation as necessary.	.125				√	√
3 Check zone and circulating pump motors for excessive overheating; lubricate motor bearings.	.077				√	√
4 Check pressure and air relief valves for proper operation.	.030				√	√
5 Check control panel and differential thermostat for proper operation.	.094				√	√
6 Clean sight glasses, controls, pumps, and flow indicators on tanks.	.127				√	√
7 Check system pressure on closed loop for loss of fluid.	.046				√	√
8 Check fluid level on drain-back systems; add fluid as necessary.	.029				√	√
9 Test glycol strength in closed systems, as applicable; if required, drain system and replace with new fluid mixture.	.222					√
10 Check heat exchanger for exterior leaks.	.077				√	√
11 Clean strainers and traps.	.181					√
12 Check storage and expansion tanks; for leaks and deteriorated insulation.	.077					√
13 Inspect all collector piping for leaks and damaged insulation; tighten connections and repair as required.	.133				√	√
14 Inspect collector glazing for cracks and seals for tightness; tighten or replace seals as necessary.	.124				√	√
15 Wash/clean glazing on collector panels.	.585					√
16 Inspect ferrule around pipe flashing where solar piping runs through roof; repair as necessary.	.086				√	√
17 Check collector mounting brackets and bolts; tighten as required.	.094				√	√
18 Clean area.	.066				√	√
19 Fill out maintenance checklist and report deficiencies.	.022				√	√
Total labor-hours/period					1.165	2.230
Total labor-hours/year					1.165	2.230

	Description	Labor-hrs.	Cost Each					
			2020 Bare Costs				Total	Total
			Material	Labor	Equip.	Total	In-House	w/O&P
1900	Wtr. htng. sys., solar clsd. lp., up to 6 panels, annually	2.230	355	122		477	547.54	640
1950	Annualized	3.395	480	186		666	768.72	895

PM Components	Labor-hrs.	W	M	Q	S	A
PM System D2025 260 1950						
Water heater, gas, to 120 gallon						
1 Check with operating or area personnel for deficiencies.	.035				√	√
2 Check for water leaks to tank and piping. check for fuel system leaks.	.077				√	√
3 Check gas burner and pilot for proper flame; adjust if required.	.118				√	√
4 Check operation and condition of pressure relief valve.	.010				√	√
5 Check automatic controls for proper operation (temperature regulators, thermostatic devices, automatic fuel shut off valve, etc.).	.094				√	√
6 Check draft diverter and clear openings, if clogged.	.027				√	√
7 Check electrical wiring for fraying and loose connections on oil burner.	.072				√	√
8 Check for proper water temperature setting; adjust as required.	.029				√	√
9 Check condition of flue pipe, and chimney.	.148				√	√
10 Drain sediment from tank	.327					√
11 Clean up area around unit	.066				√	√
12 Fill out maintenance checklist and report deficiencies	.022				√	√
Total labor-hours/period					.698	1.024
Total labor-hours/year					.698	1.024

	Description	Labor-hrs.	Cost Each					
			2020 Bare Costs				Total	Total
			Material	Labor	Equip.	Total	In-House	w/O&P
1900	Water heater, gas, to 120 gal., annually	1.024	48	56		104	124.85	150
1950	Annualized	1.721	119	94.50		213.50	251.55	299

PM Components	Labor-hrs.	W	M	Q	S	A
PM System D2025 260 2950						
Water heater, oil fired, to 100 gallon						
1 Check with operating or area personnel for deficiencies.	.035				√	√
2 Check for water leaks to tank and piping; check for fuel system leaks.	.077				√	√
3 Check burner flame and pilot on oil burner; adjust if required.	.073				√	√
4 Check operation and condition of pressure relief valve.	.010				√	√
5 Check automatic controls for proper operation (temperature regulators, thermostatic devices, automatic fuel shut off valve, etc.).	.094				√	√
6 Check fuel filter element on oil burner.	.068				√	√
7 Check fuel level in tank; check tank, fill pipe and fuel lines and connections for damage.	.022				√	√
8 Inspect, clean, and adjust electrodes and nozzles on oil burners; inspect fire box and flame detection scanner.	.254				√	√
9 Check electrical wiring for fraying and loose connections on oil burner.	.072				√	√
10 Check for proper water temperature setting; adjust as required.	.029				√	√
11 Clean fire box.	.577					√
12 Check for proper draft adjustment; adjust draft meter if necessary.	.005				√	√
13 Check condition of flue pipe, damper and chimney.	.147				√	√
14 Drain sediment from tank.	.325					√
15 Clean up area around unit.	.066				√	√
16 Fill out maintenance checklist and report deficiencies.	.022				√	√
Total labor-hours/period					.973	1.875
Total labor-hours/year					.973	1.875

	Description	Labor-hrs.	Cost Each					
			2020 Bare Costs				Total	Total
			Material	Labor	Equip.	Total	In-House	w/O&P
2900	Water heater, oil fired, to 100 gal., annually	1.875	53	103		156	189.87	230
2950	Annualized	2.850	129	156		285	341.25	410

PM Components	Labor-hrs.	W	M	Q	S	A
PM System D2025 260 3950						
Water heater, steam coil, to 2500 gallon						
1 Check with operating or area personnel for deficiencies.	.035				√	√
2 Check for water leaks to tank and piping; check steam lines for leaks.	.077				√	√
3 Check operation and condition of pressure relief valve.	.010				√	√
4 Check steam modulating valve and steam condensate trap for proper operation.	.068				√	√
5 Check electrical wiring connections on controls and switches.	.094				√	√
6 Check for proper water temperature setting; adjust as required.	.029				√	√
7 Clean, test and inspect sight gauges, valves and drains.	.040				√	√
8 Check automatic controls for proper operation including temperature regulators and thermostatic devices.	.094				√	√
9 Check insulation on heater; repair as necessary.	.077				√	√
10 Drain sediment from tank.	.325					√
11 Clean up area around unit.	.066				√	√
12 Fill out maintenance checklist and report deficiencies.	.022				√	√
Total labor-hours/period					.612	.937
Total labor-hours/year					.612	.937

	Description	Labor-hrs.	Cost Each					
			2020 Bare Costs				Total	Total
			Material	Labor	Equip.	Total	In-House	w/O&P
3900	Water heater, steam coil, to 2500 gal., annually	.937	142	51.50		193.50	221.58	260
3950	Annualized	1.549	142	85		227	265.36	315

D20 PLUMBING — D2025 262 Valve, Pressure Relief

PM Components	Labor-hrs.	W	M	Q	S	A
PM System D2025 262 1950						
Valve, pressure relief, above 4″						
1 Inspect valve for leaks, tighten fittings as necessary.	.022					√
2 Manually operate to check operation.	.038					√
3 Clean valve and area around valve.	.066					√
4 Fill out maintenance report and report deficiencies.	.022					√
Total labor-hours/period						.140
Total labor-hours/year						.148

	Description	Labor-hrs.	Cost Each: 2020 Bare Costs: Material	Labor	Equip.	Total	Total In-House	Total w/O&P
1900	Valve, pressure relief, above 4″, annually	.148	5.55	8.15		13.70	16.82	[illegible]
1950	Annualized	.148	5.55	8.15		13.70	16.82	[illegible]

D20 PLUMBING — D2025 265 Valve, Pressure Regulator

PM Components	Labor-hrs.	W	M	Q	S	A
PM System D2025 265 1950						
Valve, pressure regulator, above 4″						
1 Inspect valve for leaks; repair as necessary.	.160					√
2 Manually operate to check operation; adjust as required.	.049					√
3 Check pressure mechanism for proper opening and closing action.	.062					√
4 Clean valve and area around regulator.	.066					√
5 Fill out maintenance report and report deficiencies.	.022					√
Total labor-hours/period						.359
Total labor-hours/year						.359

	Description	Labor-hrs.	Cost Each: 2020 Bare Costs: Material	Labor	Equip.	Total	Total In-House	Total w/O&P
1900	Valve, pressure regular, above 4″, annually	.359	5.55	19.75		25.30	31.36	38.50
1950	Annualized	.359	5.55	19.75		25.30	31.36	38.50

D20 PLUMBING — D2025 270 Valve, Sediment Strainer

PM Components	Labor-hrs.	W	M	Q	S	A
PM System D2025 270 1950						
Valve, sediment strainer, above 4″						
1 Inspect valve for leaks, tighten fittings as necessary.	.160					√
2 Open valve drain to remove collected sediment.	.065					√
3 Clean valve exterior and around strainer.	.066					√
4 Fill out maintenance checklist and report deficiencies.	.022					√
Total labor-hours/period						.313
Total labor-hours/year						.313

	Description	Labor-hrs.	Cost Each: 2020 Bare Costs: Material	Labor	Equip.	Total	Total In-House	Total w/O&P
1900	Valve, sediment strainer, above 4″, annually	.313	5.55	17.20		22.75	28.13	34.50
1950	Annualized	.313	5.55	17.20		22.75	28.13	34.50

PM Components	Labor-hrs.	W	M	Q	S	A
PM System D2025 310 1950						
Valve, automatic, above 4″						
1 Lubricate valve actuator stem and valve stem, where possible.	.022					√
2 Check automatic valve for proper operation.	.062					√
3 Check packing gland for leaks; adjust as required.	.022					√
4 Check pneumatic operator and tubing for air leaks.	.025					√
5 Clean valve exterior and area around valve.	.034					√
6 Fill out maintenance report and report deficiencies.	.022					√
Total labor-hours/period						.187
Total labor-hours/year						.187

Description		Labor-hrs.	Cost Each					
			2020 Bare Costs				Total	Total
			Material	Labor	Equip.	Total	In-House	w/O&P
1900	Valve, automatic, above 4″, annually	.187	11.05	10.25		21.30	25.31	30
1950	Annualized	.187	11.05	10.25		21.30	25.31	30

PM Components	Labor-hrs.	W	M	Q	S	A
PM System D2025 310 2950						
Valve, auto diaphragm, above 4″						
1 Lubricate valve actuator mechanism and valve stem.	.022					√
2 Check valve for proper operation and leaks; tighten packing and flange bolts as required.	.044					√
3 Check pneumatic tubing and operator mechanism for proper alignment and damage; adjust as necessary and soap solution test for leaks after maintenance.	.053					√
4 Clean valve exterior and area around valve.	.034					√
5 Fill out maintenance report and report deficiencies.	.022					√
Total labor-hours/period						.175
Total labor-hours/year						.175

Description		Labor-hrs.	Cost Each					
			2020 Bare Costs				Total	Total
			Material	Labor	Equip.	Total	In-House	w/O&P
2900	Valve, auto diaphragm, above 4″, annually	.175	11.05	9.60		20.65	24.46	29
2950	Annualized	.175	11.05	9.60		20.65	24.46	29

D20 PLUMBING | D2095 905 | Duplex Sump

PM Components	Labor-hrs.	W	M	Q	S	A
PM System D2095 905 1950						
Duplex sump pump						
1 Check electrical cords, plugs and connections.	.239				√	√
2 Activate float switches and check pumps for proper operation.	.078				√	√
3 Lubricate pumps as required.	.094				√	√
4 Inspect packing and tighten as required.	.062				√	√
5 Check pumps for misalignment and bearings for overheating.	.258				√	√
6 Clean out trash from sump bottom.	.072				√	√
7 Fill out maintenance checklist and report deficiencies.	.022				√	√
Total labor-hours/period					.825	.825
Total labor-hours/year					.825	.825

	Description	Labor-hrs.	Cost Each					
			2020 Bare Costs				Total	Total
			Material	Labor	Equip.	Total	In-House	w/O&P
1900	Duplex sump pump, annually	.825	23	45		68	83.36	101
1950	Annualized	1.654	45.50	90.50		136	166.40	202

D20 PLUMBING | D2095 910 | Pump, Submersible

PM Components	Labor-hrs.	W	M	Q	S	A
PM System D2095 910 1950						
Submersible pump, 1 H.P. and over						
1 Check with operating personnel for any deficiencies.	.035				√	√
2 Remove pump from pit.	.585				√	√
3 Clean out trash from pump intake.	.338				√	√
4 Check electrical plug, cord and connection.	.119				√	√
5 Inspect pump body for corrosion; prime and paint as necessary.	.053				√	√
6 Check pump and motor operation for excessive vibration, noise and overheating.	.022				√	√
7 Lubricate pump and motor, where applicable.	.099				√	√
8 Return pump to pit; reset and check float switch for proper operation.	.585				√	√
9 Clean area.	.066				√	√
10 Fill out maintenance checklist and report deficiencies.	.022				√	√
Total labor-hours/period					1.924	1.924
Total labor-hours/year					1.924	1.924

	Description	Labor-hrs.	Cost Each					
			2020 Bare Costs				Total	Total
			Material	Labor	Equip.	Total	In-House	w/O&P
1900	Submersible pump, 1 H.P. and over, annually	1.924	19.75	105		124.75	157.01	193
1950	Annualized	3.850	30.50	211		241.50	304.82	375

D2095 930 Oxygen Monitor

PM Components	Labor-hrs.	W	M	Q	S	A
PM System D2095 930 1950						
Oxygen monitor.						
1 Check alarm set points.	.167			√	√	√
2 Check alarm lights for replacement.	.167			√	√	√
3 Check audible annunciator for proper operation.	.167			√	√	√
4 Check battery backup for proper operation.	.167			√	√	√
5 Check and test oxygen sensor for proper operation.	.276			√	√	√
6 Replace oxygen sensor annually.	1.000					√
7 Clean area.	.034			√	√	√
8 Fill out maintenance checklist and report deficiencies.	.022			√	√	√
Total labor-hours/period				.999	.999	1.999
Total labor-hours/year				1.997	.999	1.999

	Description	Labor-hrs.	Cost Each					
			2020 Bare Costs				Total	Total
			Material	Labor	Equip.	Total	In-House	w/O&P
1900	Oxygen monitor annually	1.999	208	110		318	370.84	435
1950	Annualized	5.000	395	274		669	788.25	935

D3025 110 Boiler, Electric

PM Components	Labor-hrs.	W	M	Q	S	A
PM System D3025 110 1950						
Boiler, electric, to 1500 gallon						
1 Check with operating or area personnel for deficiencies.	.035			√	√	√
2 Check hot water pressure gauges.	.073			√	√	√
3 Check operation and condition of pressure relief valve.	.030			√	√	√
4 Check for proper operation of primary controls for resistance-type or electrode-type heating elements; check and adjust thermostat.	.133			√	√	√
5 Check electrical wiring to heating elements, blower, motors, overcurrent protective devices, grounding system and other electrical components as required.	.238			√	√	√
6 Check over temperature and over-pressure limit controls for proper operation.	.140			√	√	√
7 Check furnace operation through complete cycle or up to 10 minutes.	.216			√	√	√
8 [illegible]	[illegible]			√	√	√
9 [illegible]	[illegible]			√	√	√
Total labor-hours/period				.956	.956	.956
Total labor-hours/year				1.912	.956	.956

	Description	Labor-hrs.	Cost Each					
			2020 Bare Costs				Total	Total
			Material	Labor	Equip.	Total	In-House	w/O&P
1900	Boiler, electric, to 1500 gal., annually	.956	48	62.50		110.50	130.56	157
1950	Annualized	3.828	48	251		299	363.50	450

D3025 130 Boiler, Hot Water, Oil/Gas/Comb.

PM Components	Labor-hrs.	W	M	Q	S	A
PM System D3025 130 1950						
Boiler, hot water; oil, gas or combination fired, up to 120 MBH						
1 Check combustion chamber for air or gas leaks.	.077					√
2 Inspect and clean oil burner gun and ignition assembly, where applicable.	.658					√
3 Inspect fuel system for leaks and change fuel filter element, where applicable.	.098					√
4 Check fuel lines and connections for damage.	.023		√	√	√	√
5 Check for proper operational response of burner to thermostat controls.	.133			√	√	√
6 Check and lubricate burner and blower motors.	.079			√	√	√
7 Check main flame failure protection and main flame detection scanner on boiler equipped with spark ignition (oil burner).	.124		√	√	√	√
8 Check electrical wiring to burner controls and blower.	.079					√
9 Clean firebox (sweep and vacuum).	.577					√
10 Check operation of mercury control switches (i.e., steam pressure, hot water temperature limit, atomizing or combustion air proving, etc.).	.143		√	√	√	√
11 Check operation and condition of safety pressure relief valve.	.030		√	√	√	√
12 Check operation of boiler low water cut off devices.	.056		√	√	√	√
13 Check hot water pressure gauges.	.073		√	√	√	√
14 Inspect and clean water column sight glass (or replace).	.127		√	√	√	√
15 Clean fire side of water jacket boiler.	.433					√
16 Check condition of flue pipe, damper and exhaust stack.	.147			√	√	√
17 Check boiler operation through complete cycle, up to 30 minutes.	.650					√
18 Check fuel level with gauge pole, add as required.	.046		√	√	√	√
19 Clean area around boiler.	.066		√	√	√	√
20 Fill out maintenance checklist and report deficiencies.	.022		√	√	√	√
Total labor-hours/period			.710	1.069	1.069	3.641
Total labor-hours/year			5.678	2.138	1.069	3.641

	Description	Labor-hrs.	Cost Each					
			2020 Bare Costs				Total	Total
			Material	Labor	Equip.	Total	In-House	w/O&P
1900	Boiler, hot water, O/G/C, up to 120 MBH, annually	3.641	57	239		296	358.48	440
1950	Annualized	12.528	66.50	820		886.50	1,089.40	1,350

PM Components	Labor-hrs.	W	M	Q	S	A
PM System D3025 130 2950						
Boiler, hot water; oil, gas or combination fired, 120 to 500 MBH						
1 Check combustion chamber for air or gas leaks.	.077					√
2 Inspect and clean oil burner gun and ignition assembly where applicable.	.835					√
3 Inspect fuel system for leaks and change fuel filter element, where applicable.	.125					√
4 Check fuel lines and connections for damage.	.023		√	√	√	√
5 Check for proper operational response of burner to thermostat controls.	.170			√	√	√
6 Check and lubricate burner and blower motors.	.099			√	√	√
7 Check main flame failure protection and main flame detection scanner on boiler equipped with spark ignition (oil burner).	.155		√	√	√	√
8 Check electrical wiring to burner controls and blower.	.100					√
9 Clean firebox (sweep and vacuum).	.700					√
10 Check operation of mercury control switches (i.e., steam pressure, hot water temperature limit, atomizing or combustion air proving, etc.).	.185		√	√	√	√
11 Check operation and condition of safety pressure relief valve.	.038		√	√	√	√
12 Check operation of boiler low water cut off devices.	.070		√	√	√	√
13 Check hot water pressure gauges.	.073		√	√	√	√
14 Inspect and clean water column sight glass (or replace).	.160		√	√	√	√
15 Clean fire side of water jacket	.433					√
16 Check condition of flue pipe, damper and exhaust stack.	.184			√	√	√
17 Check boiler operation through complete cycle, up to 30 minutes.	.806					√
18 Check fuel level with gauge pole, add as required.	.046		√	√	√	√
19 Clean area around boiler.	.138		√	√	√	√
20 Fill out maintenance checklist and report deficiencies.	.022		√	√	√	√
Total labor-hours/period			.910	1.363	1.363	4.532
Total labor-hours/year			7.278	2.725	1.363	4.532

	Description	Labor-hrs.	Cost Each: 2020 Bare Costs: Material	Labor	Equip.	Total	Total In-House	Total w/O&P
2900	Boiler, hot water, O/G/C, 120 to 500 MBH, annually	4.532	57	297		354	430.54	530
2950	Annualized	15.881	66.50	1,050		1,116.50	1,361.80	1,675

PM Components	Labor-hrs.	W	M	Q	S	A
PM System D3025 130 3950						
Boiler, hot water; oil, gas or combination fired, 500 to 1000 MBH						
1 Check combustion chamber for air or gas leaks.	.086					√
2 Inspect and clean oil burner gun and ignition assembly where applicable.	.910					√
3 Inspect fuel system for leaks and change fuel filter element, where applicable.	.140					√
4 Check fuel lines and connections for damage.	.026		√	√	√	√
5 Check for proper operational response of burner to thermostat controls.	.186			√	√	√
6 Check and lubricate burner and blower motors.	.110			√	√	√
7 Check main flame failure protection and main flame detection scanner on boiler equipped with spark ignition (oil burner).	.169		√	√	√	√
8 Check electrical wiring to burner controls and blower.	.111					√
9 Clean firebox (sweep and vacuum).	.889					√
10 Check operation of mercury control switches (i.e., steam pressure, hot water temperature limit, atomizing or combustion air proving, etc.).	.203		√	√	√	√
11 Check operation and condition of safety pressure relief valve.	.420		√	√	√	√
12 Check operation of boiler low water cut off devices.	.077		√	√	√	√
13 Check hot water pressure gauges.	.081		√	√	√	√
14 Inspect and clean water column sight glass (or replace).	.176		√	√	√	√
15 Clean fire side of water jacket	.432					√
16 Check condition of flue pipe, damper and exhaust stack.	.203			√	√	√
17 Check boiler operation through complete cycle, up to 30 minutes.	.887					√
18 Check fuel level with gauge pole, add as required.	.049		√	√	√	√
19 Clean area around boiler.	.151		√	√	√	√
20 Fill out maintenance checklist and report deficiencies.	.022		√	√	√	√
Total labor-hours/period			1.374	1.872	1.872	5.327
Total labor-hours/year			10.988	3.743	1.872	5.327

	Description	Labor-hrs.	Cost Each					
			2020 Bare Costs				Total	Total
			Material	Labor	Equip.	Total	In-House	w/O&P
3900	Boiler, hot water, O/G/C, 500 to 1000 MBH, annually	5.327	57	350		407	495.31	610
3950	Annualized	17.378	66.50	1,150		1,216.50	1,482.60	1,825

D30 HVAC | D3025 130 | Boiler, Hot Water, Oil/Gas/Comb.

PM Components	Labor-hrs.	W	M	Q	S	A
PM System D3025 130 4950						
Boiler, hot water; oil, gas or combination fired, over 1000 MBH						
1 Check combustion chamber for air or gas leaks.	.117					√
2 Inspect and clean oil burner gun and ignition assembly where applicable.	.987					√
3 Inspect fuel system for leaks and change fuel filter element, where applicable.	.147					√
4 Check fuel lines and connections for damage.	.035		√	√	√	√
5 Check for proper operational response of burner to thermostat controls.	.200			√	√	√
6 Check and lubricate burner and blower motors.	.119			√	√	√
7 Check main flame failure protection and main flame detection scanner on boiler equipped with spark ignition (oil burner).	.186		√	√	√	√
8 Check electrical wiring to burner controls and blower.	.120					√
9 [illegible]	.010					√
10 Check operation of mercury control switches (i.e., steam pressure, hot water temperature limit, atomizing or combustion air proving, etc.).	.215		√	√	√	√
11 Check operation and condition of safety pressure relief valve.	.046		√	√	√	√
12 Check operation of boiler low water cut off devices.	.085		√	√	√	√
13 Check hot water pressure gauges.	.109		√	√	√	√
14 Inspect and clean water column sight glass (or replace).	.190		√	√	√	√
15 Clean fire side of water jacket	.433					√
16 Check condition of flue pipe, damper and exhaust stack.	.222			√	√	√
17 Check boiler operation through complete cycle, up to 30 minutes.	.887					√
18 Check fuel level with gauge pole, add as required.	.098		√	√	√	√
19 Clean area around boiler.	.182		√	√	√	√
20 Fill out maintenance checklist and report deficiencies.	.022		√	√	√	√
Total labor-hours/period			1.169	1.710	1.710	5.220
Total labor-hours/year			9.350	3.421	1.710	5.220

	Description	Labor-hrs.	Cost Each					
			2020 Bare Costs				Total	Total
			Material	Labor	Equip.	Total	In-House	w/O&P
4900	Boiler, hot water, O/G/C, over 1000 MBH, annually	5.220	57	340		397	486.43	600
4950	Annualized	19.698	66.50	1,300		1,366.50	1,670.45	2,075

	PM Components	Labor-hrs.	W	M	Q	S	A
	PM System D3025 140 1950						
	Boiler, steam; oil, gas or combination fired, up to 120 MBH						
1	Inspect fuel system for leaks or damage.	.098		√	√	√	√
2	Change fuel filter element and clean strainers; repair leaks, where applicable.	.581					√
3	Check main flame failure protection, positive fuel shutoff and main flame detection scanner on boiler equipped with spark ignition (oil burner).	.124		√	√	√	√
4	Check for proper operational response of burner to thermostat controls.	.133			√	√	√
5	Inspect all gas, steam and water lines, valves, connections for leaks or damage; repair as necessary.	.195			√	√	√
6	Check feedwater system and feedwater makeup control and pump.	.056		√	√	√	√
7	Check and lubricate burner and blower motors as required.	.083			√	√	√
8	Check operation and condition of safety pressure relief valve.	.030		√	√	√	√
9	Check combustion controls, combustion blower and damper modulation control.	.133					√
10	Check all indicator lamps and water/steam pressure gauges.	.073		√	√	√	√
11	Check electrical panels and wiring to burner, blowers and other components.	.079			√	√	√
12	Clean blower air-intake dampers, if required.	.055			√	√	√
13	Check condition of flue pipe, damper and exhaust stack.	.147		√	√	√	√
14	Check boiler operation through complete cycle, up to 30 minutes.	.640			√	√	√
15	Check water column sight glass and water level system; clean or replace sight glass, if required.	.127		√	√	√	√
16	Clean firebox (sweep and vacuum).	.577					√
17	Check fuel level with gauge pole for oil burning boilers.	.046		√	√	√	√
18	Inspect and clean oil burner gun and ignition assembly where applicable.	.650					√
19	Clean area around boiler.	.066		√	√	√	√
20	Fill out maintenance checklist and report deficiencies.	.022		√	√	√	√
	Total labor-hours/period			.788	1.974	1.974	3.915
	Total labor-hours/year			6.304	3.948	1.974	3.915

	Description	Labor-hrs.	Cost Each					
			2020 Bare Costs				Total	Total
			Material	Labor	Equip.	Total	In-House	w/O&P
1900	Boiler, steam, O/G/C, up to 120 MBH, annually	3.915	64.50	257		321.50	389.14	475
1950	Annualized	16.189	78	1,050		1,128	1,399.15	1,750

D3025 140 Boiler, Steam, Oil/Gas/Comb.

	PM Components	Labor-hrs.	W	M	Q	S	A
	PM System D3025 140 2950						
	Boiler, steam; oil, gas or combination fired, 120 to 500 MBH						
1	Inspect fuel system for leaks or damage.	.098		√	√	√	√
2	Change fuel filter element and clean strainers; repair leaks, where applicable.	.835					√
3	Check main flame failure protection, positive fuel shutoff and main flame detection scanner on boiler equipped with spark ignition (oil burner).	.125		√	√	√	√
4	Check for proper operational response of burner to thermostat controls.	.168			√	√	√
5	Inspect all gas, steam and water lines, valves, connections for leaks or damage, repair as necessary.	.195			√	√	√
6	Check feedwater system and feedwater makeup control and pump.	.069		√	√	√	√
7	Check and lubricate burner blowers and motors as required.	.100			√	√	√
8	Check operation and condition of safety pressure relief valve.	.038		√	√	√	√
9	Check combustion controls, combustion blower and damper modulation control.	.169					√
10	Check all indicator lamps and water/steam pressure gauges.	.073		√	√	√	√
11	Check electrical panels and wiring to burner, blowers and other components.	.104			√	√	√
12	Clean blower air-intake dampers, if required.	.069			√	√	√
13	Check condition of flue pipe, damper and exhaust stack.	.147		√	√	√	√
14	Check boiler operation through complete cycle, up to 30 minutes.	.800			√	√	√
15	Check water column sight glass and water level system; clean or replace sight glass, if required.	.127		√	√	√	√
16	Clean firebox (sweep and vacuum).	.793					√
17	Check fuel level with gauge pole for oil burning boilers.	.046		√	√	√	√
18	Inspect and clean oil burner gun and ignition assembly where applicable.	.819					√
19	Clean area around boiler.	.137		√	√	√	√
20	Fill out maintenance checklist and report deficiencies.	.022		√	√	√	√
	Total labor-hours/period			.882	2.318	2.318	4.934
	Total labor-hours/year			7.059	4.635	2.318	4.934

	Description	Labor-hrs.	Cost Each					
			2020 Bare Costs				Total	Total
			Material	Labor	Equip.	Total	In-House	w/O&P
2900	Boiler, steam, O/G/C, 120 to 500 MBH, annually	4.934	64.50	325		389.50	471.97	580
2950	Annualized	18.981	78	1,250		1,328	1,625.20	2,025

	PM Components	Labor-hrs.	W	M	Q	S	A
	PM System D3025 140 3950						
	Boiler, steam; oil, gas or combination fired, 500 to 1000 MBH						
1	Inspect fuel system for leaks or damage.	.098		√	√	√	√
2	Change fuel filter element and clean strainers; repair leaks, where applicable.	.910					√
3	Check main flame failure protection, positive fuel shutoff and main flame detection scanner on boiler equipped with spark ignition (oil burner).	.140		√	√	√	√
4	Check for proper operational response of burner to thermostat controls.	.186			√	√	√
5	Inspect all gas, steam and water lines, valves, connections for leaks or damage; repair as necessary.	.195			√	√	√
6	Check feedwater system and feedwater makeup control and pump.	.075		√	√	√	√
7	Check and lubricate burner, blowers and motors as required.	.109			√	√	√
8	Check operation and condition of safety pressure relief valve.	.052		√	√	√	√
9	Check combustion controls, combustion blower and damper modulation control.	.185					√
10	Check all indicator lamps and water/steam pressure gauges.	.091		√	√	√	√
11	Check electrical panels and wiring to burner, blowers and other components.	.117			√	√	√
12	Clean blower air-intake dampers, if required.	.069			√	√	√
13	Check condition of flue pipe, damper and exhaust stack.	.147		√	√	√	√
14	Check boiler operation through complete cycle, up to 30 minutes.	.889			√	√	√
15	Check water column sight glass and water level system; clean or replace sight glass, if required.	.127		√	√	√	√
16	Clean firebox (sweep and vacuum).	1.053					√
17	Check fuel level with gauge pole for oil burning boilers.	.046		√	√	√	√
18	Inspect and clean oil burner gun and ignition assembly where applicable.	.923					√
19	Clean area around boiler.	.151		√	√	√	√
20	Fill out maintenance checklist and report deficiencies.	.022		√	√	√	√
	Total labor-hours/period			.949	2.514	2.514	5.585
	Total labor-hours/year			7.593	5.028	2.514	5.585

Description		Labor-hrs.	Cost Each					
			2020 Bare Costs				Total	Total
			Material	Labor	Equip.	Total	In-House	w/O&P
3900	Boiler, steam, O/G/C, 500 to 1000 MBH, annually	5.585	64.50	365		429.50	524.92	645
3950	Annualized	20.696	78	1,350		1,428	1,765.50	2,200

PM Components	Labor-hrs.	W	M	Q	S	A
PM System D3025 140 4950						
Boiler, steam; oil, gas or combination fired, over 1000 MBH						
1 Inspect fuel system for leaks or damage.	.098		√	√	√	√
2 Change fuel filter element and clean strainers; repair leaks, where applicable.	1.027					√
3 Check main flame failure protection, positive fuel shutoff and main flame detection scanner on boiler equipped with spark ignition (oil burner).	.147		√	√	√	√
4 Check for proper operational response of burner to thermostat controls.	.200			√	√	√
5 Inspect all gas, steam and water lines, valves, connections for leaks and/or damage; repair as necessary.	.158			√	√	√
6 Check feedwater system and feedwater makeup control and pump.	.091		√	√	√	√
7 [illegible]	[illegible]			√	√	√
8 [illegible]	[illegible]		√	√	√	√
9 Check combustion controls, combustion blower and damper modulation control.	.199					√
10 Check all indicator lamps and water/steam pressure gauges.	.109		√	√	√	√
11 Check electrical panels and wiring to burner, blowers and other components.	.140			√	√	√
12 Clean blower air-intake dampers, if required.	.069			√	√	√
13 Check condition of flue pipe, damper and exhaust stack.	.147		√	√	√	√
14 Check boiler operation through complete cycle, up to 30 minutes.	.889			√	√	√
15 Check water column sight glass and water level system; clean or replace sight glass, if required.	.127		√	√	√	√
16 Clean firebox (sweep and vacuum).	1.144					√
17 Check fuel level with gauge pole for oil burning boilers.	.046		√	√	√	√
18 Inspect and clean oil burner gun and ignition assembly where applicable.	1.196					√
19 Clean area around boiler.	.182		√	√	√	√
20 Fill out maintenance checklist and report deficiencies.	.022		√	√	√	√
Total labor-hours/period			1.037	2.651	2.651	6.217
Total labor-hours/year			8.299	5.302	2.651	6.217

	Description	Labor-hrs.	Cost Each					
			2020 Bare Costs				Total	Total
			Material	Labor	Equip.	Total	In-House	w/O&P
4900	Boiler, steam, O/G/C, over 1000 MBH, annually	6.217	64.50	410		474.50	574.71	710
4950	Annualized	22.451	78	1,475		1,553	1,906.65	2,375

D3025 210 Deaerator Tank

PM Components	Labor-hrs.	W	M	Q	S	A
PM System D3025 210 1950						
Deaerator tank						
1 Check tank and associated piping for leaks.	.013				√	√
2 Bottom - blow deaerator tank.	.022				√	√
3 Perform sulfite test on deaerator water sample.	.020				√	√
4 Check low and high float levels for proper operation and respective water level alarms.	.340				√	√
5 Clean steam and feedwater strainers.	.143				√	√
6 Check steam pressure regulating valve operation.	.007				√	√
7 Check all indicator lights.	.066				√	√
8 Clean unit and surrounding area.	.022				√	√
9 Fill out maintenance checklist and report deficiencies.	.130				√	√
Total labor-hours/period					.763	.763
Total labor-hours/year					.763	.763

	Description	Labor-hrs.	Cost Each					
			2020 Bare Costs				Total	Total
			Material	Labor	Equip.	Total	In-House	w/O&P
1900	Deaerator tank, annually	.763	13.50	50		63.50	76.63	94.50
1950	Annualized	1.506	23.50	98.50		122	147.08	182

PM Components	Labor-hrs.	W	M	Q	S	A
PM System D3025 310 1950						
Pump, boiler fuel oil						
1 Check with operating or area personnel for deficiencies.	.035				√	√
2 Check for leaks on discharge piping and seals, etc.	.077				√	√
3 Check pump and motor operation for vibration, noise, overheating, etc.	.022				√	√
4 Check alignment and clearances of shaft and coupler.	.291				√	√
5 Tighten or replace loose, missing, or damaged nuts, bolts, and screws.	.005				√	√
6 Lubricate pump and motor as required.	[illegible]				√	√
7 Clean pump, motor and surrounding area	[illegible]				√	√
8 Fill out maintenance checklist and report deficiencies.	.022				√	√
Total labor-hours/period					.617	.617
Total labor-hours/year					.617	.617

	Description	Labor-hrs.	Cost Each					
			2020 Bare Costs				Total	Total
			Material	Labor	Equip.	Total	In-House	w/O&P
1900	Pump, boiler fuel oil, annually	.617	67	40.50		107.50	123.65	146
1950	Annualized	1.232	107	81		188	217.30	259

PM Components	Labor-hrs.	W	M	Q	S	A
PM System D3025 310 2950						
Pump, condensate return, over 1 H.P.						
1 Check for proper operation of pump.	.022				√	√
2 Check for leaks on suction and discharge piping, seals, packing glands, etc.; make minor adjustments as required.	.077				√	√
3 Check pump and motor operation for vibration, noise, overheating, etc.	.022				√	√
4 Check alignment pump and motor; adjust as necessary.	.291				√	√
5 Lubricate pump and motor.	.099				√	√
6 Clean exterior of pump, motor and surrounding area.	.030				√	√
7 Fill out maintenance checklist and report deficiencies.	.022				√	√
Total labor-hours/period					.563	.563
Total labor-hours/year					.563	.563

	Description	Labor-hrs.	Cost Each					
			2020 Bare Costs				Total	Total
			Material	Labor	Equip.	Total	In-House	w/O&P
2900	Pump, steam condensate return, over 1 H.P., annually	.563	78	37		115	131.27	155
2950	Annualized	1.124	67	73.50		140.50	164.57	198

PM Components	Labor-hrs.	W	M	Q	S	A
PM System D3025 310 3950						
Pump, steam condensate return, duplex						
1 Check with operating or area personnel for deficiencies.	.035				√	√
2 Check for proper operation.	.022				√	√
3 Check for leaks on suction and discharge piping, seals, packing glands, etc.; make minor adjustments as required.	.154				√	√
4 Check pumps and motors operation for excessive vibration, noise and overheating.	.044				√	√
5 Check pump controller for proper operation.	.130				√	√
6 Lubricate pumps and motors.	.099				√	√
7 Clean condensate return unit and surrounding area.	.066				√	√
8 Fill out maintenance checklist and report deficiencies.	.022				√	√
Total labor-hours/period					.572	.572
Total labor-hours/year					.572	.572

	Description	Labor-hrs.	Cost Each					
			2020 Bare Costs				Total	Total
			Material	Labor	Equip.	Total	In-House	w/O&P
3900	Pump, steam condensate return, duplex, annually	.572	68.50	37.50		106	121.85	144
3950	Annualized	1.142	79.50	75		154.50	180.14	215

PM Components	Labor-hrs.	W	M	Q	S	A
PM System D3035 110 1950						
Water cooling tower, forced draft, up to 50 tons						
1 Check with operating or area personnel for deficiencies.	.035				√	√
2 Check operation of unit for water leaks, noise or vibration.	.077				√	√
3 Clean and inspect hot water basin.	.155				√	√
4 Remove access panel.	.048				√	√
5 Check electrical wiring and connections; make appropriate adjustments.	.120				√	√
6 Lubricate all motor and fan bearings.	.047				√	√
7 Check fan blades or blowers for imbalance and tip clearance.	[illegible]				√	√
8 Check belt for wear, tension and alignment; adjust as required.	[illegible]				√	√
9 Drain and flush cold water sump and clean strainer.	.381				√	√
10 Clean inside of water tower using water hose; scrape, brush and wipe as required; heavy deposits of scale should be removed with scale removing compound.	.598				√	√
11 Refill with water, chk. make-up water asemb. for leakage, adj. float if nec.	.211				√	√
12 Replace access panel.	.039				√	√
13 Remove, clean and reinstall conductivity and pH electrodes in chemical water treatment system.	.390				√	√
14 Inspect and clean around cooling tower.	.066				√	√
15 Fill out maintenance checklist and report deficiencies.	.022				√	√
Total labor-hours/period					2.273	2.273
Total labor-hours/year					2.273	2.273

	Description	Labor-hrs.	Cost Each					
			2020 Bare Costs				Total	Total
			Material	Labor	Equip.	Total	In-House	w/O&P
1900	Cooling tower, forced draft, up to 50 tons, annually	2.273	25	149		174	211.71	261
1950	Annualized	4.550	36	298		334	407.87	505

PM Components	Labor-hrs.	W	M	Q	S	A
PM System D3035 110 2950						
Water cooling tower, forced draft, 51 tons through 500 tons						
1 Check with operating or area personnel for deficiencies.	.074				√	√
2 Check operation of unit for water leaks, noise or vibration.	.163				√	√
3 Clean and inspect hot water basin.	.333				√	√
4 Remove access panel.	.103				√	√
5 Check electrical wiring and connections; make appropriate adjustments.	.254				√	√
6 Lubricate all motor and fan bearings.	.100				√	√
7 Check fan blades or blowers for imbalance and tip clearance.	.246				√	√
8 Check belt for wear, tension and alignment; adjust as required.	.096				√	√
9 Drain and flush cold water sump and clean strainer.	.800				√	√
10 Clean inside of water tower using water hose; scrape, brush and wipe as required; heavy deposits of scale should be removed with scale removing compound.	1.271				√	√
11 Refill with water, chk. make-up water asemb. for leakage, adj. float if nec.	.444				√	√
12 Replace access panel.	.078				√	√
13 Remove, clean and reinstall conductivity and pH electrodes in chemical water treatment system.	.842				√	√
14 Inspect and clean around cooling tower.	.117				√	√
15 Fill out maintenance checklist and report deficiencies.	.022				√	√
Total labor-hours/period					4.944	4.944
Total labor-hours/year					4.944	4.944

	Description	Labor-hrs.	Cost Each					
			2020 Bare Costs				Total	Total
			Material	Labor	Equip.	Total	In-House	w/O&P
2900	Cooling tower, forced draft, 50 thru 500 tons, annually	4.944	120	325		445	532.63	650
2950	Annualized	9.912	166	650		816	985.87	1,200

PM Components	Labor-hrs.	W	M	Q	S	A
PM System D3035 110 3950						
Water cooling tower, forced draft, 500 tons through 1000 tons						
1 Check with operating or area personnel for deficiencies.	.109				√	√
2 Check operation of unit for water leaks, noise or vibration.	.239				√	√
3 Clean and inspect hot water basin.	.485				√	√
4 Remove access panel.	.151				√	√
5 Check electrical wiring and connections; make appropriate adjustments.	.372				√	√
6 Lubricate all motor and fan bearings.	.147				√	√
7 Check fan blades or blowers for imbalance and tip clearance.	.302				√	√
8 Check belt for wear, tension and alignment; adjust as required.	.219				√	√
9 Drain and flush cold water sump and clean strainer.	1.455				√	√
10 Clean inside of water tower using water hose; scrape, brush and wipe as required; heavy deposits of scale should be removed with scale removing compound.	2.259				√	√
11 Refill with water, check make-up water assembly for leakage, adjust float if necessary.	1.578				√	√
12 Replace access panel.	.117				√	√
13 Remove, clean and reinstall conductivity and pH electrodes in chemical water treatment system.	1.231				√	√
14 Inspect and clean around cooling tower.	.172				√	√
15 Fill out maintenance checklist and report deficiencies.	.022				√	√
Total labor-hours/period					8.857	8.857
Total labor-hours/year					8.857	8.857

	Description	Labor-hrs.	Cost Each					
			2020 Bare Costs				Total	Total
			Material	Labor	Equip.	Total	In-House	w/O&P
3900	Cooling tower, forced draft, 500 to 1000 tons, annually	8.857	120	580		700	850.13	1,050
3950	Annualized	17.728	166	1,150		1,316	1,617.67	2,000

D3035 130 Chiller, Recip., Air Cooled

PM Components	Labor-hrs.	W	M	Q	S	A
PM System D3035 130 1950						
Chiller, reciprocating, air cooled, up to 25 tons						
1 Check unit for proper operation, excessive noise or vibration.	.033		√	√	√	√
2 Run system diagnostics test.	.325		√	√	√	√
3 Check oil level in sight glass of lead compressor only, add oil as necessary.	.042		√	√	√	√
4 Check superheat and subcooling temperatures.	.325					√
5 Check liquid line sight glass, oil and refrigerant pressures.	.036		√	√	√	√
6 Check contactors, sensors and mechanical safety limits.	.094					√
7 Check electrical wiring and connections; tighten loose connections.	.120					√
8 Clean intake side of condenser coils, fans and intake screens.	1.282					√
9 Inspect fan(s) or blower(s) for bent blades or imbalance.	.086					√
10 Lubricate shaft bearings and motor bearings as required.	.291					√
11 Inspect plumbing and valves for leaks, adjust as necessary.	.077		√	√	√	√
12 Check evaporator and condenser for corrosion.	.026		√	√	√	√
13 Clean chiller and surrounding area.	.066		√	√	√	√
14 Fill out maintenance checklist and report deficiencies.	.022		√	√	√	√
Total labor-hours/period			.627	.627	.627	2.825
Total labor-hours/year			5.016	1.254	.627	2.825

	Description	Labor-hrs.	Cost Each					
			2020 Bare Costs				Total	Total
			Material	Labor	Equip.	Total	In-House	w/O&P
1900	Chiller, recip., air cooled, up to 25 tons, annually	2.825	29.50	186		215.50	261.63	325
1950	Annualized	9.723	48	640		688	839.80	1,050

PM Components	Labor-hrs.	W	M	Q	S	A
PM System D3035 130 2950						
Chiller, reciprocating, air cooled, over 25 tons						
1 Check unit for proper operation, excessive noise or vibration.	.033		√	√	√	√
2 Run system diagnostics test.	.455		√	√	√	√
3 Check oil level in sight glass of lead compressor only, add oil as necessary.	.042		√	√	√	√
4 Check superheat and subcooling temperatures.	.325					√
5 Check liquid line sight glass, oil and refrigerant pressures.	.036		√	√	√	√
6 Check contactors, sensors and mechanical safety limits.	.094					√
7 Check electrical wiring and connections; tighten loose connections.	.120					√
8 Clean intake side of condenser coils, fans and intake screens.	1.282					√
9 Inspect fan(s) or blower(s) for bent blades of imbalance.	.237					√
10 Lubricate shaft bearings and motor bearings as required.	.341					√
11 Inspect plumbing and valves for leaks, adjust as necessary.	.117		√	√	√	√
12 Check evaporator and condenser for corrosion.	.052		√	√	√	√
13 Clean chiller and surrounding area.	.117		√	√	√	√
14 Fill out maintenance checklist and report deficiencies.	.022		√	√	√	√
Total labor-hours/period			.874	.874	.874	3.273
Total labor-hours/year			6.990	1.747	.874	3.273

	Description	Labor-hrs.	Cost Each					
			2020 Bare Costs				Total	Total
			Material	Labor	Equip.	Total	In-House	w/O&P
2900	Chiller, recip., air cooled, over 25 tons, annually	3.273	29.50	215		244.50	297.78	365
2950	Annualized	12.888	40	845		885	1,091.55	1,350

D3035 135 Chiller, Recip., Water Cooled

PM Components	Labor-hrs.	W	M	Q	S	A
PM System D3035 135 1950						
Chiller, reciprocating, water cooled, up to 50 tons						
1 Check unit for proper operation, excessive noise or vibration.	.033		√	√	√	√
2 Run system diagnostics test.	.327		√	√	√	√
3 Check oil level in sight glass of lead compressor only, add oil as necessary.	.042		√	√	√	√
4 Check superheat and subcooling temperatures.	.325					√
5 Check liquid line sight glass, oil and refrigerant pressures.	.036		√	√	√	√
6 Check contactors, sensors and mechanical limits, adjust as necessary.	.094					√
7 Inspect plumbing and valves for leaks, tighten connections as necessary.	.077		√	√	√	√
8 Check condenser and evaporator for corrosion.	.026		√	√	√	√
9 Clean chiller and surrounding area.	.066		√	√	√	√
10 Fill out maintenance checklist and report deficiencies.	.022		√	√	√	√
Total labor-hours/period			.629	.629	.629	1.048
Total labor-hours/year			5.030	1.258	.629	1.048

	Description	Labor-hrs.	Cost Each					
			2020 Bare Costs				Total	Total
			Material	Labor	Equip.	Total	In-House	w/O&P
1900	Chiller, recip., water cooled, up to 50 tons, annually	1.048	10.65	69		79.65	96.83	119
1950	Annualized	7.944	13.30	520		533.30	658.10	820

PM Components	Labor-hrs.	W	M	Q	S	A
PM System D3035 135 2950						
Chiller, reciprocating, water cooled, over 50 tons						
1 Check unit for proper operation, excessive noise or vibration.	.033		√	√	√	√
2 Run system diagnostics test.	.457		√	√	√	√
3 Check oil level in sight glass of lead compressor only, add oil as necessary.	.042		√	√	√	√
4 Check superheat and subcooling temperatures.	.325					√
5 Check liquid line sight glass, oil and refrigerant pressures.	.036		√	√	√	√
6 Check contactors, sensors and mechanical limits, adjust as necessary.	.094					√
7 Inspect plumbing and valves for leaks, tighten connections as necessary.	.117		√	√	√	√
8 Check condenser and evaporator for corrosion.	.052		√	√	√	√
9 Clean chiller and surrounding area.	.117		√	√	√	√
10 Fill out maintenance checklist and report deficiencies.	.022		√	√	√	√
Total labor-hours/period			.876	.876	.876	1.295
Total labor-hours/year			7.007	1.752	.876	1.295

	Description	Labor-hrs.	Cost Each					
			2020 Bare Costs				Total	Total
			Material	Labor	Equip.	Total	In-House	w/O&P
2900	Chiller, recip., water cooled, over 50 tons, annually	1.295	13.30	85		98.30	119.64	149
2950	Annualized	10.908	10.65	715		725.65	899.10	1,125

D3035 140 Chiller, Centrif., Water Cooled

PM Components	Labor-hrs.	W	M	Q	S	A
PM System D3035 140 1950						
Chiller, centrifugal water cooled, up to 100 tons						
1 Check unit for proper operation.	.035	√	√	√	√	√
2 Check oil level; add oil as necessary.	.022	√	√	√	√	√
3 Check oil temperature.	.038	√	√	√	√	√
4 Check dehydrator or purge system; remove water if observed in sight glass.	.046	√	√	√	√	√
5 Run system control tests.	.327		√	√	√	√
6 Check refrigerant charge/level, add as necessary.	.271		√	√	√	√
7 [illegible]	[illegible]		√	√	√	√
8 Check sensor and mechanical safety limits; replace as necessary.	.094				√	√
9 Clean dehydrator float valve.	.19[illegible]					√
10 Perform spectrochemical analysis of compressor oil; replace oil as necessary.	.039					√
11 Replace oil filters and add oil as necessary.	.081					√
12 Inspect cooler and condenser tubes for leaks; clean screens as necessary.	5.202					√
13 Inspect utility vessel vent piping and safety relief valve; replace as necessary.	.195					√
14 Inspect/clean the economizer (vane) gas line,damper valve and actuator arm.	.650					√
15 Run an insulation test on the centrifugal motor.	1.300					√
16 Clean area around equipment.	.066	√	√	√	√	√
17 Document all maintenance and cleaning procedures.	.022	√	√	√	√	√
Total labor-hours/period		.229	.852	.852	.946	8.607
Total labor-hours/year		8.701	6.814	1.703	.946	8.607

	Description	Labor-hrs.	Cost Each					
			2020 Bare Costs				Total	Total
			Material	Labor	Equip.	Total	In-House	w/O&P
1900	Chiller, centrif., water cooled, up to 100 tons, annually	8.607	204	565		769	920.21	1,125
1950	Annualized	26.771	220	1,750		1,970	2,408.35	2,975

	PM Components	Labor-hrs.	W	M	Q	S	A
	PM System D3035 140 2950						
	Chiller, centrifugal water cooled, over 100 tons						
1	Check unit for proper operation.	.035	√	√	√	√	√
2	Check oil level; add oil as necessary.	.022	√	√	√	√	√
3	Check oil temperature.	.038	√	√	√	√	√
4	Check dehydrator or purge system; remove water if observed in sight glass.	.046	√	√	√	√	√
5	Run system control tests.	.455		√	√	√	√
6	Check refrigerant charge/level, add as necessary.	.381		√	√	√	√
7	Check compressor for excessive noise/vibration.	.039		√	√	√	√
8	Check sensor and mechanical safety limits; replace as necessary.	.133				√	√
9	Clean dehydrator float valve.	.351					√
10	Perform spectrochemical analysis of compressor oil; replace oil as necessary.	.039					√
11	Replace oil filters and add oil as necessary.	.161					√
12	Inspect cooler and condenser tubes for leaks; clean screens as necessary.	5.202					√
13	Inspect utility vessel vent piping and safety relief valve; replace as necessary.	.247					√
14	Inspect/clean the economizer (vane), gas line damper valve and actuator arm.	.650					√
15	Run an insulation test on the centrifugal motor.	1.950					√
16	Clean area around equipment.	.117	√	√	√	√	√
17	Document all maintenance and cleaning procedures.	.022	√	√	√	√	√
	Total labor-hours/period		.280	1.155	1.155	1.288	9.887
	Total labor-hours/year		10.631	9.238	2.310	1.288	9.887

			Cost Each					
			2020 Bare Costs				Total	Total
Description		Labor-hrs.	Material	Labor	Equip.	Total	In-House	w/O&P
2900	Chiller, centrif., water cooled, over 100 tons, annually	9.887	236	645		881	1,059.05	1,300
2950	Annualized	33.364	212	2,200		2,412	2,937.55	3,650

PM Components	Labor-hrs.	W	M	Q	S	A
PM System D3035 150 1950						
Chiller, absorption unit, up to 500 tons						
1 Check with operating personnel for deficiencies; check operating log sheets for indications of increased temperature trends.	.035			√	√	√
2 Check unit for proper operation, excessive noise or vibration.	.026			√	√	√
3 Check and clean strainers in all lines as required.	1.300					√
4 Check pulley alignment and belts for condition, proper tension and misalignment on external purge pump system, if applicable; adjust for proper tension and or alignment.	.043			√	√	√
5 Check purge pump vacuum oil level, as required; add/change oil as necessary.	.650			√	√	√
6 Lubricate pump shaft bearings and motor bearings.	.281			√	√	√
7 Check and service system controls, wirings and connections; tighten loose connections.	.119					√
8 Inspect cooling and system water piping circuits for leakage.	.077			√	√	√
9 Clean area around equipment.	.066			√	√	√
10 Fill out maintenance checklist.	.022			√	√	√
Total labor-hours/period				1.200	1.200	2.619
Total labor-hours/year				2.399	1.200	2.619

	Description	Labor-hrs.	Cost Each					
			2020 Bare Costs				Total	Total
			Material	Labor	Equip.	Total	In-House	w/O&P
1900	Chiller, absorption unit, up to 500 tons, annually	2.619	40.50	171		211.50	256.82	315
1950	Annualized	6.219	57.50	405		462.50	568.65	700

PM Components	Labor-hrs.	W	M	Q	S	A
PM System D3035 150 2950						
Chiller, absorption unit, 500 to 5000 tons						
1 Check with operating personnel for deficiencies; check operating log sheets for indications of increased temperature trends.	.035			√	√	√
2 Check unit for proper operation, excessive noise or vibration.	.026			√	√	√
3 Check and clean strainers in all lines as required.	2.286					√
4 Check pulley alignment and belts for condition, proper tension and misalignment on external purge pump system, if applicable; adjust for proper tension and or alignment.	.086			√	√	√
5 Check purge pump vacuum oil level, as required; add/change oil as necessary.	.975			√	√	√
6 Lubricate pump shaft bearings and motor bearings.	.320			√	√	√
7 Check and service system controls, wirings and connections; tighten loose connections.	.139					√
8 Inspect cooling and system water piping circuits for leakage.	.152			√	√	√
9 Clean area around equipment.	.117			√	√	√
10 Fill out maintenance checklist.	.022			√	√	√
Total labor-hours/period				1.733	1.733	4.158
Total labor-hours/year				3.466	1.733	4.158

	Description	Labor-hrs.	Cost Each					
			2020 Bare Costs				Total	Total
			Material	Labor	Equip.	Total	In-House	w/O&P
2900	Chiller, absorption unit, 500 to 5000 tons, annually	4.158	35.50	273		308.50	375.82	465
2950	Annualized	9.361	55.50	615		670.50	817.75	1,025

D3035 160 Chiller, Screw, Water Cooled

PM Components	Labor-hrs.	W	M	Q	S	A
PM System D3035 160 1950						
Chiller, screw, water cooled, up to 100 tons						
1 Check unit for proper operation, excessive noise or vibration.	.033		√	√	√	√
2 Run system diagnostics test.	.327		√	√	√	√
3 Check oil level and oil temperature; add oil as necessary.	.022		√	√	√	√
4 Check refrigerant pressures; add as necessary.	.271		√	√	√	√
5 Replace oil filters and oil, if applicable.	.081				√	√
6 Check contactors, sensors and mechanical safety limits.	.094				√	√
7 Perform spectrochemical analysis of compressor oil.	.039				√	√
8 Check electrical wiring and connections; tighten loose connections.	.119				√	√
9 Inspect cooler and condenser tubes for leaks.	5.202					√
10 Check evaporator and condenser for corrosion.	.026					√
11 Clean chiller and surrounding area.	.066		√	√	√	√
12 Fill out maintenance checklist and report deficiencies.	.022		√	√	√	√
Total labor-hours/period			.741	.741	1.074	6.302
Total labor-hours/year			5.926	1.482	1.074	6.302

	Description	Labor-hrs.	Cost Each					
			2020 Bare Costs				Total	Total
			Material	Labor	Equip.	Total	In-House	w/O&P
1900	Chiller, screw, water cooled, up to 100 tons, annually	6.302	97	410		507	616.06	755
1950	Annualized	14.776	105	970		1,075	1,309.95	1,625

PM Components	Labor-hrs.	W	M	Q	S	A
PM System D3035 160 2950						
Chiller, screw, water cooled, over 100 tons						
1 Check unit for proper operation, excessive noise or vibration.	.033		√	√	√	√
2 Run system diagnostics test.	.325		√	√	√	√
3 Check oil level and oil temperature; add oil as necessary.	.022		√	√	√	√
4 Check refrigerant pressures; add as necessary.	.381		√	√	√	√
5 Replace oil filters and oil, if applicable.	.081				√	√
6 Check contactors, sensors and mechanical safety limits.	.094				√	√
7 Perform spectrochemical analysis of compressor oil.	.039				√	√
8 Check electrical wiring and connections; tighten loose connections.	.119				√	√
9 Inspect cooler and condenser tubes for leaks.	5.202					√
10 Check evaporator and condenser for corrosion.	.052					√
11 Clean chiller and surrounding area.	.117		√	√	√	√
12 Fill out maintenance checklist and report deficiencies.	.022		√	√	√	√
Total labor-hours/period			.900	.900	1.233	6.487
Total labor-hours/year			7.200	1.800	1.233	6.487

	Description	Labor-hrs.	Cost Each					
			2020 Bare Costs				Total	Total
			Material	Labor	Equip.	Total	In-House	w/O&P
2900	Chiller, screw, water cooled, over 100 tons, annually	6.487	252	425		677	803.76	970
2950	Annualized	16.722	114	1,100		1,214	1,478.55	1,850

D30 HVAC | D3035 170 | Evaporative Cooler

PM Components	Labor-hrs.	W	M	Q	S	A
PM System D3035 170 1950						
Evaporative cooler						
1 Check with operating or area personnel for deficiencies.	.035					√
2 Check unit for proper operation, noise and vibration.	.016					√
3 Clean evaporation louver panels.	.083					√
4 Lubricate fan, bearings and motor.	.052					√
5 Check belt(s) for excessive wear and deterioration.	.095					√
6 Visually inspect wiring for damage or loose connections; tighten loose connections.	.119					√
7 Clean sump.	[illegible]					√
8 Clean and adjust float.	.258					√
9 Check interior surfaces and check components for loose paint/lime deposits.	[illegible]					√
10 Check/clean, adjust nozzles (6).	.039					√
11 Remove/clean and reinstall evaporative pads.	[illegible]					√
12 Start unit and check for proper operation.	.055					√
13 Remove debris from surrounding area.	.066					√
14 Fill out maintenance checklist.	.022					√
Total labor-hours/period						1.251
Total labor-hours/year						1.251

	Description	Labor-hrs.	Cost Each					
			2020 Bare Costs				Total	Total
			Material	Labor	Equip.	Total	In-House	w/O&P
1900	Evaporative cooler, annually	1.251	41	82		123	146.94	178
1950	Annualized	1.251	41	82		123	146.94	178

D30 HVAC | D3035 180 | Evaporative Cooler, Rotating Drum

PM Components	Labor-hrs.	W	M	Q	S	A
PM System D3035 180 1950						
Evaporative cooler, rotating drum						
1 Check with operating or area personnel for deficiencies.	.035					√
2 Check unit for proper operation, noise and vibration.	.016					√
3 Clean evaporation louver panels.	.083					√
4 Lubricate rotary and fan, bearings and motor.	.078					√
5 Check belt(s) for excessive wear and deterioration.	.095					√
6 Visually inspect wiring for damage or loose connections; tighten loose connections.	.119					√
7 Clean sump and pump intake.	.039					√
8 Clean and adjust float.	.258					√
9 Check interior surfaces and check components for loose paint/lime deposits.	.055					√
10 Check/clean, adjust nozzles.	.039					√
11 Remove/clean and reinstall evaporative pads.	.356					√
12 Start unit and check for proper operation.	.055					√
13 Remove debris from surrounding area.	.066					√
14 Fill out maintenance checklist.	.022					√
Total labor-hours/period						1.316
Total labor-hours/year						1.316

	Description	Labor-hrs.	Cost Each					
			2020 Bare Costs				Total	Total
			Material	Labor	Equip.	Total	In-House	w/O&P
1900	Evaporative cooler, rotating drum, annually	1.316	41	86.50		127.50	152.25	184
1950	Annualized	1.316	41	86.50		127.50	152.25	184

PM Components	Labor-hrs.	W	M	Q	S	A
PM System D3035 210 1950						
Condenser, air cooled, 3 tons to 25 tons						
1 Check with operating or area personnel for deficiencies.	.027			√	√	√
2 Check unit for proper operation, excessive noise or vibration.	.025			√	√	√
3 Pressure wash coils and fans with coil cleaning solution.	.471					√
4 Check electrical wiring and connections; tighten loose connections.	.092					√
5 Lubricate shaft bearings and motor bearings.	.036			√	√	√
6 Inspect fan(s) or blower(s) for bent blades or imbalance; adjust as necessary.	.024			√	√	√
7 Check belt(s) for condition, proper tension, and misalignment; adjust for proper tension and/or alignment, if applicable.	.030			√	√	√
8 Inspect piping and valves for leaks; tighten connections as necessary.	.059			√	√	√
9 Clean area around equipment.	.051			√	√	√
10 Fill out maintenance checklist and report deficiencies.	.017			√	√	√
Total labor-hours/period				.269	.269	.832
Total labor-hours/year				.538	.269	.832

	Description	Labor-hrs.	Cost Each					
			2020 Bare Costs				Total	Total
			Material	Labor	Equip.	Total	In-House	w/O&P
1900	Condenser, air cooled, 3 tons to 25 tons, annually	.832	33	54.50		87.50	103.72	125
1950	Annualized	1.708	69	112		181	214.68	259

PM Components	Labor-hrs.	W	M	Q	S	A
PM System D3035 210 2950						
Condenser, air cooled, 26 tons through 100 tons						
1 Check with operating or area personnel for deficiencies.	.027			√	√	√
2 Check unit for proper operation, excessive noise or vibration.	.025			√	√	√
3 Pressure wash coils and fans with coil cleaning solution.	.571					√
4 Check electrical wiring and connections; tighten loose connections.	.092					√
5 Lubricate shaft bearings and motor bearings.	.042			√	√	√
6 Inspect fan(s) or blower(s) for bent blades or imbalance; adjust as necessary.	.030			√	√	√
7 Check belt(s) for condition, proper tension, and misalignment; adjust for proper tension and/or alignment, if applicable.	.080			√	√	√
8 Inspect piping and valves for leaks; tighten connections as necessary.	.059			√	√	√
9 Clean area around equipment.	.051			√	√	√
10 Fill out maintenance checklist and report deficiencies.	.017			√	√	√
Total labor-hours/period				.331	.331	.994
Total labor-hours/year				.662	.331	.994

	Description	Labor-hrs.	Cost Each					
			2020 Bare Costs				Total	Total
			Material	Labor	Equip.	Total	In-House	w/O&P
2900	Condenser, air cooled, 26 tons to 100 tons, annually	.994	57	65.50		122.50	143.45	172
2950	Annualized	1.697	79	111		190	224.43	271

D3035 210 Condenser, Air Cooled

PM Components	Labor-hrs.	W	M	Q	S	A
PM System D3035 210 3950						
Condenser, air cooled, over 100 tons						
1 Check with operating or area personnel for deficiencies.	.027			√	√	√
2 Check unit for proper operation, excessive noise or vibration.	.031			√	√	√
3 Pressure wash coils and fans with coil cleaning solution.	.667					√
4 Check electrical wiring and connections; tighten loose connections.	.092					√
5 Lubricate shaft bearings and motor bearings.	.042			√	√	√
6 Inspect fan(s) or blower(s) for bent blades or imbalance; adjust as necessary.	.030			√	√	√
7 Check belt(s) for condition, proper tension and misalignment; adjust for proper tension and/or alignment, if applicable.	.100			√	√	√
8 Inspect piping and valves for leaks; tighten connections as necessary.	.059			√	√	√
9 Clean area around equipment.	.051			√	√	√
10 Fill out maintenance checklist and report deficiencies.	.017			√	√	√
Total labor-hours/period				[illegible]	[illegible]	[illegible]
Total labor-hours/year				.714	.357	1.116

	Description	Labor-hrs.	Cost Each					
			2020 Bare Costs				Total	Total
			Material	Labor	Equip.	Total	In-House	w/O&P
3900	Condenser, air cooled, over 100 tons, annually	1.116	89.50	73		162.50	188.93	225
3950	Annualized	2.187	126	143		269	315.70	380

PM Components	Labor-hrs.	W	M	Q	S	A
PM System D3035 220 1950						
Condensing unit, air cooled, 3 tons to 25 tons						
1 Check with operating or area personnel for deficiencies.	.035			√	√	√
2 Check unit for proper operation, excessive noise or vibration.	.033			√	√	√
3 Pressure wash coils and fans with coil cleaning solution.	.615					√
4 Check electrical wiring and connections; tighten loose connections.	.119					√
5 Lubricate shaft bearings and motor bearings.	.047			√	√	√
6 Inspect fan(s) or blower(s) for bent blades or imbalance; adjust as necessary.	.031			√	√	√
7 Check belt(s) for condition, proper tension, and misalignment; adjust for proper tension and/or alignment, if applicable.	.078			√	√	√
8 Inspect piping and valves for leaks; tighten connections as necessary.	.077			√	√	√
9 Check refrigerant pressure; add refrigerant, if necessary.	.271					√
10 Clean area around equipment.	.066			√	√	√
11 Fill out maintenance checklist and report deficiencies.	.022			√	√	√
Total labor-hours/period				.389	.389	1.395
Total labor-hours/year				.778	.389	1.395

	Description	Labor-hrs.	Cost Each					
			2020 Bare Costs				Total	Total
			Material	Labor	Equip.	Total	In-House	w/O&P
1900	Condensing unit, air cooled, 3 to 25 tons, annually	1.395	88	91.50		179.50	209.89	252
1950	Annualized	2.562	110	169		279	329.45	400

PM Components	Labor-hrs.	W	M	Q	S	A
PM System D3035 220 2950						
Condensing unit, air cooled, 26 tons through 100 tons						
1 Check with operating or area personnel for deficiencies.	.035			√	√	√
2 Check unit for proper operation, excessive noise or vibration.	.033			√	√	√
3 Pressure wash coils and fans with coil cleaning solution.	.727					√
4 Check electrical wiring and connections; tighten loose connections.	.119					√
5 Lubricate shaft bearings and motor bearings.	.055			√	√	√
6 Inspect fan(s) or blower(s) for bent blades or imbalance; adjust as necessary.	.039			√	√	√
7 Check belt(s) for condition, proper tension, and misalignment; adjust for proper tension and/or alignment, if applicable.	.104			√	√	√
8 Inspect piping and valves for leaks; tighten connections as necessary.	.077			√	√	√
9 Check refrigerant pressure; add refrigerant, if necessary.	.390					√
10 Clean area around equipment.	.066			√	√	√
11 Fill out maintenance checklist and report deficiencies.	.022			√	√	√
Total labor-hours/period				.431	.431	1.668
Total labor-hours/year				.862	.431	1.668

	Description	Labor-hrs.	Cost Each					
			2020 Bare Costs				Total	Total
			Material	Labor	Equip.	Total	In-House	w/O&P
2900	Condensing unit, air cooled, 26 to 100 tons, annually	1.668	135	109		244	283.69	340
2950	Annualized	2.961	189	194		383	448.55	535

PM Components	Labor-hrs.	W	M	Q	S	A
PM System D3035 220 3950						
Condensing unit, air cooled, over 100 tons						
1 Check with operating or area personnel for deficiencies.	.035			√	√	√
2 Pressure wash coils and fans with coil cleaning solution.	.871					√
3 Clean intake side of condenser coils, fans and intake screens.	.091			√	√	√
4 Check electrical wiring and connections; tighten loose connections.	.120					√
5 Lubricate shaft bearings and motor bearings.	.055			√	√	√
6 Inspect fan(s) or blower(s) for bent blades or imbalance; adjust as necessary.	.039			√	√	√
7 [illegible] for proper tension and/or alignment, if applicable.	[illegible]			√	√	√
8 Inspect piping and valves for leaks; tighten connections as necessary.	.077			√	√	√
9 Check refrigerant pressure; add refrigerant, if necessary.	.455					√
10 Clean area around equipment.	.066			√	√	√
11 Fill out maintenance checklist and report deficiencies.	.022			√	√	√
Total labor-hours/period				.515	.515	1.961
Total labor-hours/year				1.030	.515	1.961

	Description	Labor-hrs.	Cost Each					
			2020 Bare Costs				Total	Total
			Material	Labor	Equip.	Total	In-House	w/O&P
3900	Condensing unit, air cooled, over 100 tons, annually	1.961	135	129		264	307.92	370
3950	Annualized	3.440	189	225		414	488.03	585

PM Components	Labor-hrs.	W	M	Q	S	A
PM System D3035 240 1950						
Condensing unit, water cooled, 3 tons to 24 tons						
1 Check with operating or area personnel for deficiencies.	.035			√	√	√
2 Check unit for proper operation, excessive noise or vibration.	.033			√	√	√
3 Check electrical wiring and connections; tighten loose connections.	.119					√
4 Inspect piping and valves for leaks; tighten connections as necessary.	.077			√	√	√
5 Check refrigerant pressure; add refrigerant, if necessary.	.271					√
6 Clean area around equipment.	.066			√	√	√
7 Fill out maintenance checklist and report deficiencies.	.022			√	√	√
Total labor-hours/period				.233	.233	.624
Total labor-hours/year				.466	.233	.624

	Description	Labor-hrs.	Cost Each					
			2020 Bare Costs				Total	Total
			Material	Labor	Equip.	Total	In-House	w/O&P
1900	Condensing unit, water cooled, 3 to 24 tons, annually	.624	31	41		72	84.94	102
1950	Annualized	1.323	36.50	86.50		123	147.40	180

PM Components	Labor-hrs.	W	M	Q	S	A
PM System D3035 240 2950						
Condensing unit, water cooled, 25 tons through 100 tons						
1 Check with operating or area personnel for deficiencies.	.035			√	√	√
2 Check unit for proper operation, excessive noise or vibration.	.033			√	√	√
3 Run system diagnostics test.	.216			√	√	√
4 Check electrical wiring and connections; tighten loose connections.	.119			√	√	√
5 Check oil level in sight glass of lead compressor only; add oil as necessary.	.029			√	√	√
6 Check contactors, sensors and mechanical limits, adjust as necessary.	.157					√
7 Check condenser for corrosion.	.026					√
8 Inspect piping and valves for leaks; tighten connections as necessary.	.077			√	√	√
9 Check refrigerant pressure; add refrigerant, if necessary.	.271					√
10 Clean area around equipment.	.066			√	√	√
11 Fill out maintenance checklist and report deficiencies.	.022			√	√	√
Total labor-hours/period				.598	.598	1.052
Total labor-hours/year				1.195	.598	1.052

	Description	Labor-hrs.	Cost Each					
			2020 Bare Costs				Total	Total
			Material	Labor	Equip.	Total	In-House	w/O&P
2900	Condensing unit, WC, 25 to 100 tons, annually	1.052	41.50	69		110.50	131.09	159
2950	Annualized	2.846	57.50	186		243.50	294	360

D3035 240 Condensing Unit, Water Cooled

	PM Components	Labor-hrs.	W	M	Q	S	A
	PM System D3035 240 3950						
	Condensing unit, water cooled, over 100 tons						
1	Check with operating or area personnel for deficiencies.	.035			√	√	√
2	Check unit for proper operation, excessive noise or vibration.	.033			√	√	√
3	Run system diagnostics test.	.216			√	√	√
4	Check electrical wiring and connections; tighten loose connections.	.119			√	√	√
5	Check oil level in sight glass of lead compressor only; add oil as necessary.	.029			√	√	√
6	Check contactors, sensors and mechanical limits, adjust as necessary.	.157					√
7	[illegible]	[illegible]					√
8	Inspect piping and valves for leaks; tighten connections as necessary.	.077			√	√	√
9	Check refrigerant pressure; add refrigerant, if necessary.	.271					√
10	Clean area around equipment.	.066			√	√	√
11	Fill out maintenance checklist and report deficiencies.	.022			√	√	√
	Total labor-hours/period				[illegible]	[illegible]	1.052
	Total labor-hours/year				1.195	.598	1.052

	Description	Labor-hrs.	Cost Each					
			2020 Bare Costs				Total	Total
			Material	Labor	Equip.	Total	In-House	w/O&P
3900	Condensing unit, WC, over 100 tons, annually	1.052	41.50	69		110.50	131.09	159
3950	Annualized	2.846	52	186		238	288	355

PM Components	Labor-hrs.	W	M	Q	S	A
PM System D3035 260 1950						
Compressor, DX Refrigeration, to 25 tons						
1 Remove/replace access panel/cover.	.070			√	√	√
2 Check unit for proper operation, excessive noise or vibration.	.030			√	√	√
3 Run systems diagnostics test, if applicable.	.060			√	√	√
4 Check oil level in compressor, if possible, and add oil as required.	.030			√	√	√
5 Check refrigerant pressures; add refrigerant as necessary.	.200			√	√	√
6 Check contactors, sensors and mechanical limits, adjust as necessary.	.100			√	√	√
7 Inspect refrigerant piping and insulation.	.050			√	√	√
8 Clean compressor and surrounding area.	.050			√	√	√
9 Fill out maintenance checklist and report deficiencies.	.020			√	√	√
Total labor-hours/period				.610	.610	.610
Total labor-hours/year				1.220	.610	.610

	Description	Labor-hrs.	Cost Each					
			2020 Bare Costs				Total	Total
			Material	Labor	Equip.	Total	In-House	w/O&P
1900	Compressor, DX refrigeration, to 25 tons, annually	.610	39.50	40		79.50	92.98	111
1950	Annualized	2.440	51.50	160		211.50	255.10	310

PM Components	Labor-hrs.	W	M	Q	S	A
PM System D3035 260 2950						
Compressor, DX refrigeration, 25 tons to 100 tons						
1 Check unit for proper operation, excessive noise or vibration.	.030			√	√	√
2 Run system diagnostics test, if applicable.	.098			√	√	√
3 Check oil level in compressor; add oil as necessary.	.033			√	√	√
4 Check refrigerant pressures; add refrigerant as necessary.	.271			√	√	√
5 Check contactors, sensors and mechanical limits, adjust as necessary.	.216			√	√	√
6 Inspect refrigerant piping and insulation.	.077			√	√	√
7 Clean compressor and surrounding area.	.066			√	√	√
8 Fill out maintenance checklist and report deficiencies.	.022			√	√	√
Total labor-hours/period				.814	.814	.814
Total labor-hours/year				1.627	.814	.814

	Description	Labor-hrs.	Cost Each					
			2020 Bare Costs				Total	Total
			Material	Labor	Equip.	Total	In-House	w/O&P
2900	Compressor, DX refrigeration, 25 to 100 tons, annually	.814	39.50	53.50		93	109.66	132
2950	Annualized	3.256	51.50	213		264.50	321.40	395

PM Components	Labor-hrs.	W	M	Q	S	A
PM System D3035 290 1950						
Fluid cooler, 2 fans (no compressor)						
1 Check with operating or area personnel for deficiencies.	.035					√
2 Check unit for proper operation, excessive noise or vibration.	.159					√
3 Clean intake side of condenser coils, fans and intake screens.	.473					√
4 Check electrical wiring and connections; tighten loose connections.	.120					√
5 Inspect fan(s) for bent blades or imbalance; adjust as necessary.	.040					√
6 Check belts for condition, proper tension and misalignment; adjust for proper tension and/or alignment, if required.	.029					√
7 Lubricate shaft bearings and motor bearings.	.047					√
8 Inspect piping and valves for leaks; tighten connections as necessary.	.077					√
9 Lubricate and check operation of dampers, if applicable.	.055					√
10 Clean area around fluid cooler.	.066					√
11 Fill out maintenance checklist and report deficiencies.	.022					√
Total labor-hours/period						1.123
Total labor-hours/year						1.123

	Description	Labor-hrs.	Cost Each					
			2020 Bare Costs				Total	Total
			Material	Labor	Equip.	Total	In-House	w/O&P
1900	Fluid cooler, 2 fans (no compressor), annually	1.123	44	73.50		117.50	139.76	169
1950	Annualized	1.123	44	73.50		117.50	139.76	169

D3045 110 Air Handling Unit

PM Components	Labor-hrs.	W	M	Q	S	A
PM System D3045 110 1950						
Air handling unit, 3 tons through 24 tons						
1 Check with operating or area personnel for deficiencies.	.035			√	√	√
2 Check controls and unit for proper operation.	.033			√	√	√
3 Check for unusual noise or vibration.	.033			√	√	√
4 Check tension, condition and alignment of belts, adjust as necessary.	.029			√	√	√
5 Clean coils, evaporator drain pan, blower, motor and drain piping, as required.	.381					√
6 Lubricate shaft and motor bearings.	.047			√	√	√
7 Replace air filters.	.078			√	√	√
8 Inspect exterior piping and valves for leaks; tighten connections as required.	.077			√	√	√
9 Clean area around equipment.	.066			√	√	√
10 Fill out maintenance checklist and report deficiencies.	.022			√	√	√
Total labor-hours/period				.421	.421	.802
Total labor-hours/year				.841	.421	.802

	Description	Labor-hrs.	Cost Each					
			2020 Bare Costs				Total	Total
			Material	Labor	Equip.	Total	In-House	w/O&P
1900	Air handling unit, 3 thru 24 tons, annually	.802	104	44		148	170.67	200
1950	Annualized	2.061	238	113		351	407.70	480

PM Components	Labor-hrs.	W	M	Q	S	A
PM System D3045 110 2950						
Air handling unit, 25 tons through 50 tons						
1 Check with operating or area personnel for deficiencies.	.035			√	√	√
2 Check controls and unit for proper operation.	.033			√	√	√
3 Check for unusual noise or vibration.	.033			√	√	√
4 Clean coils, evaporator drain pan, blower, motor and condensate drain piping, as required.	.381					√
5 Lubricate shaft and motor bearings.	.047			√	√	√
6 Check belts for wear, proper tension, and alignment; adjust as necessary.	.029			√	√	√
7 Inspect exterior piping and valves for leaks; tighten connections as required.	.077			√	√	√
8 Check operation and clean dampers, louvers and shutters; lubricate all pivot points and linkages.	.078					√
9 Replace air filters.	.078			√	√	√
10 Clean area around equipment.	.066			√	√	√
11 Fill out maintenance checklist and report deficiencies.	.022			√	√	√
Total labor-hours/period				.421	.421	.880
Total labor-hours/year				.841	.421	.880

	Description	Labor-hrs.	Cost Each					
			2020 Bare Costs				Total	Total
			Material	Labor	Equip.	Total	In-House	w/O&P
2900	Air handling unit, 25 thru 50 tons, annually	.880	225	48.50		273.50	310.92	360
2950	Annualized	1.940	435	107		542	610.05	715

D30 HVAC — D3045 110 Air Handling Unit

PM Components	Labor-hrs.	W	M	Q	S	A
PM System D3045 110 3950						
Air handling unit, over 50 tons						
1 Check with operating or area personnel for deficiencies.	.035			√	√	√
2 Check controls and unit for proper operation.	.033			√	√	√
3 Check for unusual noise or vibration.	.033			√	√	√
4 Clean coils, evaporator drain pan, blower, motor and drain piping, as required.	.516					√
5 Lubricate shaft and motor bearings.	.047			√	√	√
6 Check belts for wear, proper tension, and alignment; adjust as necessary.	.055			√	√	√
7 Inspect exterior piping and valves for leaks; tighten connections as required.	.077			√	√	√
8 Check operation and clean dampers, louvers and shutters; lubricate all pivot points and linkages.	.078					√
9 Clean centrifugal fan	.065					√
10 Replace air filters.	.200			√	√	√
11 Clean area around equipment.	.066			√	√	√
12 Fill out maintenance checklist and report deficiencies	.022			√	√	√
Total labor-hours/period				.654	.654	1.313
Total labor-hours/year				1.307	.654	1.313

	Description	Labor-hrs.	Cost Each: 2020 Bare Costs: Material	Labor	Equip.	Total	Total In-House	Total w/O&P
3900	Air handling unit, over 50 tons, annually	1.313	246	72		318	364	425
3950	Annualized	3.276	455	180		635	729.53	855

D30 HVAC — D3045 112 A.H.U., Computer Room

PM Components	Labor-hrs.	W	M	Q	S	A
PM System D3045 112 1950						
Air handling unit, computer room						
1 Check with operating or area personnel for deficiencies.	.035			√	√	√
2 Run microprocessor check, if available, or check controls and unit for proper operation.	.216			√	√	√
3 Check for unusual noise or vibration.	.033			√	√	√
4 Clean coils, evaporator drain pan, blower, motor and drain piping, as required.	.380					√
5 Lubricate shaft and motor bearings.	.047			√	√	√
6 Check belts for wear, proper tension, and alignment; adjust as necessary.	.029			√	√	√
7 Check humidity lamp, replace if necessary.	.157			√	√	√
8 Inspect exterior piping and valves for leaks; tighten connections as required.	.077			√	√	√
9 Replace air filters.	.078			√	√	√
10 Clean area around equipment.	.066			√	√	√
11 Fill out maintenance checklist and report deficiencies.	.022			√	√	√
Total labor-hours/period				.761	.761	1.141
Total labor-hours/year				1.521	.761	1.141

	Description	Labor-hrs.	Cost Each: 2020 Bare Costs: Material	Labor	Equip.	Total	Total In-House	Total w/O&P
1900	Air handling unit, computer room, annually	1.141	63	62.50		125.50	149.54	178
1950	Annualized	3.416	141	187		328	396.40	475

D3045 120 Fan Coil Unit

PM Components	Labor-hrs.	W	M	Q	S	A
PM System D3045 120 1950						
Fan coil unit						
1 Check with operating or area personnel for deficiencies.	.035			√	√	√
2 Check coil unit while operating.	.120				√	√
3 Remove access panel and vacuum inside of unit and coils.	.471				√	√
4 Check coils and piping for leaks, damage and corrosion; repair as necessary.	.077			√	√	√
5 Lubricate blower shaft and fan motor bearings.	.047				√	√
6 Clean coil, drip pan, and drain line with solvent.	.471				√	√
7 Replace filters as required.	.009			√	√	√
8 Replace access panel.	.023				√	√
9 Check operation after repairs.	.120				√	√
10 Clean area.	.066			√	√	√
11 Fill out maintenance checklist and report deficiencies.	.022			√	√	√
Total labor-hours/period				.209	1.461	1.461
Total labor-hours/year				.418	1.461	1.461

	Description	Labor-hrs.	Cost Each					
			2020 Bare Costs				Total	Total
			Material	Labor	Equip.	Total	In-House	w/O&P
1900	Fan coil unit, annually	1.461	47	80.50		127.50	154.54	186
1950	Annualized	3.338	123	183		306	369.14	445

D3045 150 Air Filters, Electrostatic

PM Components	Labor-hrs.	W	M	Q	S	A
PM System D3045 150 1950						
Filters, Electrostatic						
1 Review manufacturers instructions.	.091			√	√	√
2 Check indicators for defective tubes or broken ionizing wires.	.130			√	√	√
3 De-energize power supply,tag and lockout disconnect switch.	.130			√	√	√
4 Ground bus trips, top and bottom.	.065			√	√	√
5 Secure filter and fan unit.	.065			√	√	√
6 Wash each manifold until clean, approximately 4 minutes with hot water, 7 minutes with cold water.	.390			√	√	√
7 Clean or replace dry filters as necessary.	.258			√	√	√
8 Inspect for broken on hum suppressors and wipe insulators with soft dry cloth.	.258			√	√	√
9 Disassemble unit as required, check it thoroughly, clean and adjust.	.640			√	√	√
10 Restore to service and check for shorts.	.130			√	√	√
11 Fill out maintenance checklist and report deficiencies.	.022			√	√	√
Total labor-hours/period				2.180	2.180	2.180
Total labor-hours/year				4.359	2.180	2.180

	Description	Labor-hrs.	Cost Each					
			2020 Bare Costs				Total	Total
			Material	Labor	Equip.	Total	In-House	w/O&P
1900	Air filter, electrostatic, annually	2.180	10.10	120		130.10	164.21	203
1950	Annualized	7.602	10.10	415		425.10	545.30	680

D3045 160 VAV Boxes

PM Components	Labor-hrs.	W	M	Q	S	A
PM System D3045 160 1950						
VAV Boxes						
1 Open/close VAV control box access.	.060				√	√
2 Check that pneumatic tubing/electrical connections are in place and tight.	.040				√	√
3 Tighten arm on motor output shaft.	.040				√	√
4 Cycle actuator while watching for proper operation.	.200				√	√
Verify that blades fully open and close.	.040				√	√
5 Lubricate actuator linkage and damper blade pivot points.	.070				√	√
6 Fill out maintenance checklist and report deficiencies.	.017				√	√
Total labor-hours/period					.467	.467
Total labor-hours/year					.467	.467

	Description	Labor-hrs.	Cost Each					
			2020 Bare Costs				Total	Total
			Material	Labor	Equip.	Total	In-House	w/O&P
1900	VAV Boxes, annually	.467	4.77	30.50		35.27	43.13	53.50
1950	Annualized	.934	9.55	61		70.55	86.51	106

D3045 170 Fire Dampers

PM Components	Labor-hrs.	W	M	Q	S	A
PM System D3045 170 1950						
Fire dampers						
1 Remove/replace access door.	.800					√
2 Clean out debris/dirt blown against damper.	.200					√
3 Remove fusible link and check that blades operate freely.	.070					√
4 Lubricate pivot points.	.050					√
5 Replace fusible link.	.020					√
6 Fill out maintenance checklist and report deficiencies.	.017					√
Total labor-hours/period						1.157
Total labor-hours/year						1.157

	Description	Labor-hrs.	Cost Each					
			2020 Bare Costs				Total	Total
			Material	Labor	Equip.	Total	In-House	w/O&P
1900	Fire dampers, annually	1.157	12.50	76		88.50	107.73	133
1950	Annualized	1.157	12.50	76		88.50	107.73	133

D3045 210 Fan, Axial

PM Components	Labor-hrs.	W	M	Q	S	A
PM System D3045 210 1950						
Fan, axial, up to 5,000 CFM						
1 Start and stop fan with local switch.	.012				√	√
2 Check fan for noise and vibration.	.091				√	√
3 Check electrical wiring and connections; tighten loose connections.	.029				√	√
4 Check motor and fan shaft bearings for noise, vibration, overheating; lubricate as required.	.327				√	√
5 Clean area around fan.	.143				√	√
6 Fill out maintenance checklist and report deficiencies.	.022				√	√
Total labor-hours/period					.623	.623
Total labor-hours/year					.623	.623

	Description	Labor-hrs.	Cost Each					
			2020 Bare Costs				Total	Total
			Material	Labor	Equip.	Total	In-House	w/O&P
1900	Fan, axial, up to 5,000 CFM, annually	.623	11.05	34		45.05	55.88	68.50
1950	Annualized	1.244	22	68		90	111.17	137

PM Components	Labor-hrs.	W	M	Q	S	A
PM System D3045 210 2950						
Fan, axial, 5,000 to 10,000 CFM						
1 Start and stop fan with local switch.	.012				√	√
2 Check motor and fan shaft bearings for noise, vibration, overheating; lubricate bearings.	.327				√	√
3 Check belts for wear, tension, and alignment, if applicable; adjust as required.	.057				√	√
4 Check fan pitch operator, lubricate; if applicable.	.029				√	√
5 Check electrical wiring and connections; tighten loose connections.	.057				√	√
6 Clean fan and surrounding fan.	.143				√	√
7 Fill out maintenance checklist and report deficiencies.	.022				√	√
Total labor-hours/period					.647	.647
Total labor-hours/year					.647	.647

	Description	Labor-hrs.	Cost Each					
			2020 Bare Costs				Total	Total
			Material	Labor	Equip.	Total	In-House	w/O&P
2900	Fan, axial, 5,000 to 10,000 CFM, annually	.647	29.50	35.50		65	78.06	93.50
2950	Annualized	1.290	45.50	70.50		116	140.39	170

D3045 210 Fan, Axial

PM Components	Labor-hrs.	W	M	Q	S	A
PM System D3045 210 3950						
Fan, axial, 36″ to 48″ dia (over 10,000 CFM)						
1 Start and stop fan with local switch.	.012				√	√
2 Check motor and fan shaft bearings for noise, vibration, overheating; lubricate bearings.	.327				√	√
3 Check belts for wear, tension, and alignment, if applicable; adjust as required.	.086				√	√
4 Check fan pitch operator, lubricate; if applicable.	.029				√	√
5 Check electrical wiring and connections; tighten loose connections.	.078				√	√
6 Clean fan and surrounding area.	.143				√	√
7 Fill out maintenance checklist and report deficiencies.	[illegible]				√	√
Total labor-hours/period					.696	.696
Total labor-hours/year					.696	.696

	Description	Labor-hrs.	Cost Each					
			2020 Bare Costs				Total	Total
			Material	Labor	Equip.	Total	In-House	w/O&P
3900	Fan, axial, 36″ to 48″ dia (over 10,000 CFM), annually	.696	29.50	38		67.50	81.69	98
3950	Annualized	1.390	45.50	76		121.50	147.34	179

PM Components	Labor-hrs.	W	M	Q	S	A
PM System D3045 220 1950						
Fan, centrifugal, up to 5,000 CFM						
1 Start and stop fan with local switch.	.012				√	√
2 Check motor and fan shaft bearings for noise, vibration, overheating; lubricate bearings.	.327				√	√
3 Check belts for wear, tension, and alignment, if applicable; adjust as required.	.057				√	√
4 Check blower intake dampers, lubricate; if applicable.	.029				√	√
5 Check electrical wiring and connections; tighten loose connections.	.029				√	√
6 Clean fan and surrounding area.	.066				√	√
7 Fill out maintenance checklist and report deficiencies.	.022				√	√
Total labor-hours/period					.542	.542
Total labor-hours/year					.542	.542

	Description	Labor-hrs.	Cost Each					
			2020 Bare Costs				Total	Total
			Material	Labor	Equip.	Total	In-House	w/O&P
1900	Fan, centrifugal, up to 5,000 CFM, annually	.542	29.50	29.50		59	70.68	84.50
1950	Annualized	1.080	45.50	59		104.50	125.77	151

PM Components	Labor-hrs.	W	M	Q	S	A
PM System D3045 220 2950						
Fan, centrifugal, 5,000 to 10,000 CFM						
1 Start and stop fan with local switch.	.012				√	√
2 Check motor and fan shaft bearings for noise, vibration, overheating; lubricate bearings.	.327				√	√
3 Check belts for wear, tension, and alignment, if applicable; adjust as required.	.057				√	√
4 Check blower intake dampers, lubricate; if applicable.	.029				√	√
5 Check electrical wiring and connections; tighten loose connections.	.057				√	√
6 Clean fan and surrounding area.	.066				√	√
7 Fill out maintenance checklist and report deficiencies.	.022				√	√
Total labor-hours/period					.570	.570
Total labor-hours/year					.570	.570

	Description	Labor-hrs.	Cost Each					
			2020 Bare Costs				Total	Total
			Material	Labor	Equip.	Total	In-House	w/O&P
2900	Fan, centrifugal, 5,000 to 10,000 CFM, annually	.570	29.50	31.50		61	72.66	87
2950	Annualized	1.136	45.50	62		107.50	129.69	156

D3045 220 Fan, Centrifugal

PM Components	Labor-hrs.	W	M	Q	S	A
PM System D3045 220 3950 **Fan, centrifugal, over 10,000 CFM**						
1 Start and stop fan with local switch.	.007				√	√
2 Check motor and fan shaft bearings for noise, vibration, overheating; lubricate bearings.	.327				√	√
3 Check belts for wear, tension, and alignment, if applicable; adjust as required.	.086				√	√
4 Check blower intake dampers, lubricate; if applicable.	.029				√	√
5 Check electrical wiring and connections; tighten loose connections.	.057				√	√
6 Clean fan and surrounding area.	.066				√	√
7 Fill out maintenance checklist and report deficiencies.	.022				√	√
Total labor-hours/period					.594	.594
Total labor-hours/year					.594	.594

	Description	Labor-hrs.	Cost Each					
			2020 Bare Costs				Total	Total
			Material	Labor	Equip.	Total	In-House	w/O&P
3900	Fan, centrifugal, over 10,000 CFM, annually	.594	29.50	32.50		62	74.51	89
3950	Annualized	1.184	45.50	65		110.50	132.98	161

D30 HVAC

D3045 250 Hood And Blower

PM Components	Labor-hrs.	W	M	Q	S	A
PM System D3045 250 1950 **Hood and blower**						
1 Check with operating or area personnel for any deficiencies.	.044			√	√	√
2 Check unit for proper operation, including switches, controls and thermostat; calibrate thermostat and repair components as required.	.109			√	√	√
3 Check operation of spray nozzles for spray coverage and drainage, if applicable.	.022					√
4 Inspect soap and spray solution feeder lines.	.007			√	√	√
5 Check components for automatic cleaning system and automatic fire protection system - solenoid valve, line strainer, shut-off valve, detergent tank, etc.	.055			√	√	√
6 Clean spray nozzles.	1.104					√
7 Tighten or replace loose, missing or damaged nuts, bolts or screws.	.005			√	√	√
8 Check operation of exhaust blower on roof; lubricate bearings; adjust tension of fan belts and clean fan blades as required.	.056			√	√	√
9 Fill out maintenance checklist and report deficiencies.	.022			√	√	√
Total labor-hours/period				.298	.298	1.424
Total labor-hours/year				.595	.298	1.424

	Description	Labor-hrs.	Cost Each					
			2020 Bare Costs				Total	Total
			Material	Labor	Equip.	Total	In-House	w/O&P
1900	Hood and blower, annually	1.424	68.50	78		146.50	175.38	210
1950	Annualized	2.318	90.50	127		217.50	262.53	315

PM Components	Labor-hrs.	W	M	Q	S	A
PM System D3045 410 1950						
Centrifugal pump over 1 H.P.						
1 Check for proper operation of pump.	.022				√	√
2 Check for leaks on suction and discharge piping, seals, packing glands, etc.; make minor adjustments as required.	.077				√	√
3 Check pump and motor operation for excessive vibration, noise and overheating.	.022				√	√
4 Check alignment of pump and motor; adjust as necessary.	.258				√	√
5 Lubricate pump and motor.	.099				√	√
6 Clean exterior of pump, motor and surrounding area.	.096				√	√
7 Fill out maintenance checklist and report deficiencies.	.022				√	√
Total labor-hours/period					.596	.596
Total labor-hours/year					.596	.596

	Description	Labor-hrs.	Cost Each 2020 Bare Costs Material	Labor	Equip.	Total	Total In-House	Total w/O&P
1900	Centrifugal pump, over 1 H.P., annually	.596	11.05	32.50		43.55	54.05	66
1950	Annualized	1.196	22	65.50		87.50	108.15	133

PM Components	Labor-hrs.	W	M	Q	S	A
PM System D3045 410 2950						
Centrifugal pump w/reduction gear, over 1 H.P.						
1 Check with operating or area personnel for deficiencies.	.035				√	√
2 Clean pump exterior and check for corrosion on pump exterior and base plate.	.030				√	√
3 Check for leaks on suction and discharge piping, seals, packing glands, etc.	.077				√	√
4 Check pump, gear and motor operation for vibration, noise, overheating, etc.	.022				√	√
5 Check alignment and clearances of shaft reduction gear and coupler.	.258				√	√
6 Tighten or replace loose, missing, or damaged nuts, bolts and screws.	.005				√	√
7 Lubricate pump and motor as required.	.099				√	√
8 When available, check suction or discharge, pressure gauge readings and flow rate.	.022				√	√
9 Clean area around pump.	.066				√	√
10 Fill out maintenance checklist and report deficiencies.	.022				√	√
Total labor-hours/period					.636	.636
Total labor-hours/year					.636	.636

	Description	Labor-hrs.	Cost Each 2020 Bare Costs Material	Labor	Equip.	Total	Total In-House	Total w/O&P
2900	Centrifugal pump, w/ red. gear, over 1H.P., annually	.636	11.05	35		46.05	56.87	69.50
2950	Annualized	1.276	22	70		92	113.79	140

D3045 420 Pump w/Oil Reservoir, Electric

PM Components	Labor-hrs.	W	M	Q	S	A
PM System D3045 420 1950						
Pump w/oil reservoir, over 1 H.P.						
1 Check for proper operation of pump.	.022				√	√
2 Check for leaks on suction and discharge piping, seals, packing glands, etc.; make minor adjustments as required.	.077				√	√
3 Check pump and motor operation for excessive vibration, noise and overheating.	.022				√	√
4 Check alignment of pump and motor; adjust as necessary.	.258				√	√
5 Lubricate motor, check oil level in pump [illegible]	[illegible]				√	√
6 [illegible]	[illegible]				√	√
7 Fill out maintenance checklist and report deficiencies.	.022				√	√
Total labor-hours/period					[illegible]	[illegible]
Total labor-hours/year					.596	.596

Description		Labor-hrs.	Cost Each					
			2020 Bare Costs				Total	Total
			Material	Labor	Equip.	Total	In-House	w/O&P
1900	Pump w/ oil reservoir, electric, annually	.596	24	32.50		56.50	68.45	82
1950	Annualized	1.196	12.95	65.50		78.45	98.15	121

D3045 600 Heat Exchanger, Steam

PM Components	Labor-hrs.	W	M	Q	S	A
PM System D3045 600 1950 **Heat exchanger, steam**						
1 Check with operating or area personnel for deficiencies.	.035				√	√
2 Check temperature gauges for proper operating temperatures.	.091				√	√
3 Check steam modulating valve and steam condensate trap for proper operation.	.101				√	√
4 Inspect heat exchanger and adjacent piping for torn or deteriorated insulation.	.147				√	√
5 Clean heat exchanger and surrounding area.	.066				√	√
6 Fill out maintenance checklist and report deficiencies.	.022				√	√
Total labor-hours/period					.462	.462
Total labor-hours/year					.462	.462

	Description	Labor-hrs.	Cost Each					
			2020 Bare Costs				Total	Total
			Material	Labor	Equip.	Total	In-House	w/O&P
1900	Heat exchanger, steam, annually	.462	22	25.50		47.50	56.41	68
1950	Annualized	.924	22	51		73	88.62	108

D3055 110 Space Heater

PM Components	Labor-hrs.	W	M	Q	S	A
PM System D3055 110 1950						
Unit heater, gas radiant						
1 Check with operating or area personnel for deficiencies.	.035					√
2 Inspect, clean and adjust control valves and thermo sensing bulbs on gas burners.	.254					√
3 Inspect fuel system for leaks.	.016					√
4 Check for proper operation of burner controls, check and adjust thermostat.	.133					√
5 Check fan and motor for vibration and noise, lubricate bearings.	[illegible]					√
6 Check electrical wiring to blower motor.	[illegible]					√
7 Check condition of flue pipe, damper and stack.	.117					√
8 Check unit heater operation through complete cycle or up to ten minutes.	.133					√
9 Clean area around unit heater.	.133					√
10 Fill out maintenance checklist and report deficiencies.	.022					√
Total labor-hours/period						1.009
Total labor-hours/year						1.009

	Description	Labor-hrs.	Cost Each					
			2020 Bare Costs				Total	Total
			Material	Labor	Equip.	Total	In-House	w/O&P
1900	Unit heater, gas radiant, annually	1.009	2.77	55.50		58.27	74.09	91.50
1950	Annualized	1.009	2.77	55.50		58.27	74.09	91.50

PM Components	Labor-hrs.	W	M	Q	S	A
PM System D3055 110 2950						
Unit heater, gas infrared						
1 Check with operating or area personnel for deficiencies.	.035					√
2 Inspect fuel system for leaks around unit.	.016					√
3 Replace primary air intake filter.	.048					√
4 Blow out burner.	.130					√
5 Clean spark electrode and reset gap, replace if necessary.	.254					√
6 Clean pilot and thermo-sensor bulb, replace if necessary.	.133					√
7 Check alignment of thermo-sensor bulb.	.133					√
8 Check wiring and connections; tighten any loose connections.	.079					√
9 Check and clean pilot sight glass.	.091					√
10 Vacuum pump (fan):						
A) check motor bearings for overheating; adjust or repair as required.	.022					√
B) lubricate motor bearings.	.099					√
C) check motor mounting; adjust as required.	.030					√
D) check fan blade clearance; adjust as required.	.042					√
E) check wiring, connections, switches, etc.; tighten any loose connections.	.079					√
11 Operate unit to ensure that it is in proper working condition.	.225					√
12 Clean equipment and surrounding area.	.066					√
13 Fill out maintenance checklist and report deficiencies.	.022					√
Total labor-hours/period						1.505
Total labor-hours/year						1.505

	Description	Labor-hrs.	Cost Each					
			2020 Bare Costs				Total	Total
			Material	Labor	Equip.	Total	In-House	w/O&P
2900	Unit heater, gas infrared, annually	1.505	76	82.50		158.50	190.41	226
2950	Annualized	1.505	76	82.50		158.50	190.41	226

D3055 110 Space Heater

PM Components	Labor-hrs.	W	M	Q	S	A
PM System D3055 110 3950						
Unit heater, steam						
1 Check with operating or area personnel for deficiencies.	.035					√
2 Inspect, clean and adjust control valves and thermostat.	.254					√
3 Inspect coils, connections, trap and steam piping for leaks; repair as necessary.	.195					√
4 Check fan and motor for vibration and noise; lubricate bearings.	.056					√
5 Check electrical wiring to motor.	.079					√
6 Check unit heater operation through complete cycle or up to ten minutes.	.133					√
7 Clean equipment and surrounding area.	.066					√
8 Fill out maintenance checklist and report deficiencies.	.022					√
Total labor-hours/period						.841
Total labor-hours/year						.841

	Description	Labor-hrs.	Cost Each					
			2020 Bare Costs				Total	Total
			Material	Labor	Equip.	Total	In-House	w/O&P
3900	Unit heater, steam, annually	.841	52	46		98	116.46	139
3950	Annualized	.841	52	46		98	116.46	139

D3055 122 Forced Air Heater, Oil/Gas Fired

	PM Components	Labor-hrs.	W	M	Q	S	A
	PM System D3055 122 1950						
	Forced air heater, oil or gas fired, up to 120 MBH						
1	Check with operating or area personnel for deficiencies.	.035			√	√	√
2	Inspect, clean and adjust electrodes and nozzles on oil burners or controls, valves and thermo-sensing bulbs on gas burners; lubricate oil burner motor bearings as applicable.	.254					√
3	Inspect fuel system for leaks.	.016			√	√	√
4	Change fuel filter element on oil burner, where applicable.						
5	Check for proper operation of burner primary controls. check and adjust thermostat.	.133			√	√	√
6	Replace air filters in air handler.	.009			√	√	√
7	Check blower and motor for vibration and noise. lubricate bearings.	.042			√	√	√
8	Check belts for wear and proper tension, tighten if required.	.029			√	√	√
9	Check electrical wiring to burner controls and blower.	.079					√
10	Inspect and clean firebox.	.571					√
11	Clean blower and air plenum.	.290					√
12	Check condition of flue pipe, damper and stack.	.147			√	√	√
13	Check furnace operation through complete cycle or up to 10 minutes.	.640			√	√	√
14	Clean area around furnace.	.066			√	√	√
15	Fill out maintenance checklist and report deficiencies.	.022			√	√	√
	Total labor-hours/period				1.139	1.139	2.340
	Total labor-hours/year				2.278	1.139	2.340

	Description	Labor-hrs.	Cost Each					
			2020 Bare Costs				Total	Total
			Material	Labor	Equip.	Total	In-House	w/O&P
1900	Forced air heater, oil/gas, up to 120 MBH, annually	2.340	69	153		222	266.09	325
1950	Annualized	5.628	58	370		428	520.27	645

	PM Components	Labor-hrs.	W	M	Q	S	A
	PM System D3055 122 2950						
	Forced air heater, oil or gas fired, over 120 MBH						
1	Check with operating or area personnel for deficiencies.	.035			√	√	√
2	Inspect, clean and adjust electrodes and nozzles on oil burners or controls valves and thermo-sensing bulbs on gas burners; lubricate oil burner motor bearings as applicable.	.358					√
3	Inspect fuel system for leaks.	.155			√	√	√
4	Change fuel filter element on oil burner, where applicable.	.119					√
5	Check for proper operation of burner primary controls. check and adjust thermostat.	.133			√	√	√
6	Replace air filters in air handler.	.182			√	√	√
7	Check blower and motor for vibration and noise. lubricate bearings.	.047			√	√	√
8	Check belts for wear and proper tension, tighten if required.	.057			√	√	√
9	Check electrical wiring to burner controls and blower.	.079					√
10	Inspect and clean firebox.	.577					√
11	Clean blower and air plenum.	.294					√
12	Check condition of flue pipe, damper and stack.	.147			√	√	√
13	Check furnace operation through complete cycle or up to 10 minutes.	.640			√	√	√
14	Clean area around furnace.	.066			√	√	√
15	Fill out maintenance checklist and report deficiencies.	.022			√	√	√
	Total labor-hours/period				1.484	1.484	2.912
	Total labor-hours/year				2.968	1.484	2.912

	Description	Labor-hrs.	Cost Each					
			2020 Bare Costs				Total	Total
			Material	Labor	Equip.	Total	In-House	w/O&P
2900	Forced air heater, oil/gas, over 120 MBH, annually	2.912	121	191		312	369.52	445
2950	Annualized	7.408	235	485		720	860.40	1,050

D3055 210 Package Unit, Air Cooled

	PM Components	Labor-hrs.	W	M	Q	S	A
	PM System D3055 210 1950						
	Package unit, air cooled, 3 tons through 24 tons						
1	Check with operating or area personnel for deficiencies.	.035			√	√	√
2	Check tension, condition, and alignment of belts; adjust as necessary.	.029			√	√	√
3	Lubricate shaft and motor bearings.	.047			√	√	√
4	Replace air filters.	.055			√	√	√
5	Clean electrical wiring and connections; tighten loose connections.	.119					√
6	Clean coils, evaporator drain pan, blowers, fans, motors and drain piping as required.	.381					√
7	Perform operational check of unit; make adjustments on controls and other components as required.	.077			√	√	√
8	During operation of unit, check refrigerant pressure; add refrigerant as necessary.	.134			√	√	√
9	Check compressor oil level; add oil as required.	.033					√
10	Clean area around unit.	.066			√	√	√
11	Fill out maintenance checklist and report deficiencies.	.022			√	√	√
	Total labor-hours/period				.465	.465	.999
	Total labor-hours/year				.930	.465	.999

	Description	Labor-hrs.	Cost Each					
			2020 Bare Costs				Total	Total
			Material	Labor	Equip.	Total	In-House	w/O&P
1900	Package unit, air cooled, 3 thru 24 ton, annually	.999	99	65.50		164.50	190.71	225
1950	Annualized	2.397	177	157		334	390.90	465

	PM Components	Labor-hrs.	W	M	Q	S	A
	PM System D3055 210 2950						
	Package unit, air cooled, 25 tons through 50 tons						
1	Check with operating or area personnel for deficiencies.	.035			√	√	√
2	Check tension, condition, and alignment of belts; adjust as necessary.	.029			√	√	√
3	Lubricate shaft and motor bearings.	.047			√	√	√
4	Replace air filters.	.078			√	√	√
5	Clean electrical wiring and connections; tighten loose connections.	.119					√
6	Clean coils, evaporator drain pan, blowers, fans, motors and drain piping as required.	.381					√
7	Perform operational check of unit; make adjustments on controls and other components as required.	.130			√	√	√
8	During operation of unit, check refrigerant pressure; add refrigerant as necessary.	.271			√	√	√
9	Check compressor oil level; add oil as required.	.033					√
10	Clean area around unit.	.066			√	√	√
11	Fill out maintenance checklist and report deficiencies.	.022			√	√	√
	Total labor-hours/period				.679	.679	1.212
	Total labor-hours/year				1.358	.679	1.212

	Description	Labor-hrs.	Cost Each					
			2020 Bare Costs				Total	Total
			Material	Labor	Equip.	Total	In-House	w/O&P
2900	Package unit, air cooled, 25 thru 50 ton, annually	1.212	99	79.50		178.50	208.01	247
2950	Annualized	3.249	177	213		390	460.40	550

D3055 220 Package Unit, Water Cooled

	PM Components	Labor-hrs.	W	M	Q	S	A
	PM System D3055 220 1950						
	Package unit, water cooled, 3 tons through 24 tons						
1	Check with operating or area personnel for deficiencies.	.035			√	√	√
2	Check tension, condition, and alignment of belts; adjust as necessary.	.029			√	√	√
3	Lubricate shaft and motor bearings.	.047			√	√	√
4	Replace air filters.	.055			√	√	√
5	Clean electrical wiring and connections; tighten loose connections.	.120					√
6	Clean coils, evaporator drain pan, blowers, fans, motors and drain piping as required.	.385					√
7	Perform operational check of unit; make adjustments on controls and other components as required.	.077			√	√	√
8	During operation of unit, check refrigerant pressure; add refrigerant as necessary.	.134			√	√	√
9	Check compressor oil level; add oil as required.	.033					√
10	Clean area around unit.	.066			√	√	√
11	Fill out maintenance checklist and report deficiencies.	.022			√	√	√
	Total labor-hours/period				.465	.465	1.003
	Total labor-hours/year				.931	.465	1.003

	Description	Labor-hrs.	Cost Each					
			2020 Bare Costs				Total	Total
			Material	Labor	Equip.	Total	In-House	w/O&P
1900	Package unit, water cooled, 3 thru 24 ton, annually	1.003	99	65.50		164.50	190.76	226
1950	Annualized	2.402	177	157		334	390.95	465

	PM Components	Labor-hrs.	W	M	Q	S	A
	PM System D3055 220 2950						
	Package unit, water cooled, 25 tons through 50 tons						
1	Check with operating or area personnel for deficiencies.	.035			√	√	√
2	Check tension, condition, and alignment of belts; adjust as necessary.	.029			√	√	√
3	Lubricate shaft and motor bearings.	.047			√	√	√
4	Replace air filters.	.078			√	√	√
5	Clean electrical wiring and connections; tighten loose connections.	.119					√
6	Clean coils, evaporator drain pan, blowers, fans, motors and drain piping as required.	.381					√
7	Perform operational check of unit; make adjustments on controls and other components as required.	.130			√	√	√
8	During operation of unit, check refrigerant pressure; add refrigerant as necessary.	.271			√	√	√
9	Check compressor oil level; add oil as required.	.033					√
10	Clean area around unit.	.066			√	√	√
11	Fill out maintenance checklist and report deficiencies.	.022			√	√	√
	Total labor-hours/period				.679	.679	1.212
	Total labor-hours/year				1.358	.679	1.212

	Description	Labor-hrs.	Cost Each					
			2020 Bare Costs				Total	Total
			Material	Labor	Equip.	Total	In-House	w/O&P
2900	Package unit, water cooled, 25 thru 50 ton, annually	1.212	99	79.50		178.50	208.01	247
2950	Annualized	3.249	177	213		390	460.40	550

D3055 230 Package Unit, Computer Room

PM Components	Labor-hrs.	W	M	Q	S	A
PM System D3055 230 1950						
Package unit, computer room						
1 Check with operating or area personnel for deficiencies.	.035			√	√	√
2 Run microprocessor check, if available, or check controls and unit for proper operation.	.216			√	√	√
3 Check for unusual noise or vibration.	.033			√	√	√
4 Clean coils, evaporator drain pan, humidifier pan, blower, motor and drain piping as required.	.381				√	√
5 Replace air filters.	.078			√	√	√
6 Lubricate shaft and motor bearings.	.047			√	√	√
7 Check belts for wear, proper tension, and alignment; adjust as necessary.	.029			√	√	√
8 Check humidity lamp, replace if necessary.	.155			√	√	√
9 During operation of unit, check refrigerant pressures; add refrigerant as necessary.	.271				√	√
10 Inspect exterior piping and valves for leaks; tighten connections as required.	.077			√	√	√
11 Clean area around unit.	.066			√	√	√
12 Fill out maintenance checklist and report deficiencies.	.022			√	√	√
Total labor-hours/period				.759	1.411	1.411
Total labor-hours/year				1.518	1.411	1.411

	Description	Labor-hrs.	Cost Each					
			2020 Bare Costs				Total	Total
			Material	Labor	Equip.	Total	In-House	w/O&P
1900	Package unit, computer room, annually	1.411	90.50	92.50		183	214.69	256
1950	Annualized	4.336	177	284		461	547.20	660

D3055 240 Package Terminal Unit

PM Components	Labor-hrs.	W	M	Q	S	A
PM System D3055 240 1950						
Package unit, with duct gas heater						
1 Check with operating or area personnel for deficiencies.	.035			√	√	√
2 Check tension, condition and alignment of belts; adjust as necessary.	.029			√	√	√
3 Lubricate shaft and motor bearings.	.047			√	√	√
4 Replace air filters.	.078			√	√	√
5 Check electrical wiring and connections; tighten loose connections.	.119					√
6 Clean coils, evaporator drain pan, blowers, fans, motors and drain piping as required.	.385					√
7 Perform operational check of unit; make adjustments on controls and other components as required.	.077			√	√	√
8 During operation of unit, check refrigerant pressures, add refrigerant as necessary.	.271			√	√	√
9 Check compressor oil level; add oil as required.	.033					√
10 Inspect, clean and adjust control valves and thermo-sensing bulbs on gas burners.	[illegible]					√
11 Inspect fuel system for leaks.	.016			√	√	√
12 Check for proper operation of burner primary controls. Check and adjust thermostat.	.133					√
13 Check electrical wiring to burner controls.	.079					√
14 Inspect and clean firebox.	.571					√
15 Check condition of flue pipe, damper and stack.	.147			√	√	√
16 Check heater operation through complete cycle or up to 10 minutes.	.225					√
17 Clean area around entire unit.	.066			√	√	√
18 Fill out maintenance checklist and report deficiencies.	.022			√	√	√
Total labor-hours/period				.788	.788	2.588
Total labor-hours/year				1.577	.788	2.588

	Description	Labor-hrs.	Cost Each					
			2020 Bare Costs				Total	Total
			Material	Labor	Equip.	Total	In-House	w/O&P
1900	Package unit with duct gas heater, annually	2.588	100	170		270	320.31	385
1950	Annualized	4.956	178	325		503	599	725

PM Components	Labor-hrs.	W	M	Q	S	A
PM System D3055 250 1950						
Air conditioning split system, DX, air cooled, up to 10 tons						
1 Check with operating or area personnel for deficiencies.	.035			√	√	√
2 Clean intake side of condenser coils, fans and intake screens.	.055			√	√	√
3 Lubricate shaft and motor bearings.	.047			√	√	√
4 Pressure wash condenser coils with coil clean solution, as required.	.611					√
5 Replace air filters.	.078			√	√	√
6 Clean electrical wiring and connections; tighten loose connections.	.120					√
7 Clean evaporator coils, drain pan, blowers, fans, motors and drain piping as required.	.380					√
8 Perform operational check of unit; make adjustments on controls and other components as required.	.033			√	√	√
9 During operation of unit, check refrigerant pressure; add refrigerant as necessary.	.272			√	√	√
10 Clean area around equipment.	.066			√	√	√
11 Fill out maintenance checklist and report deficiencies.	.022			√	√	√
Total labor-hours/period				.608	.608	1.719
Total labor-hours/year				1.216	.608	1.719

	Description	Labor-hrs.	Cost Each					
			2020 Bare Costs				Total	Total
			Material	Labor	Equip.	Total	In-House	w/O&P
1900	A/C, split sys., DX, air cooled, to 10 tons, annually	1.719	82.50	113		195.50	231.14	277
1950	Annualized	3.543	177	232		409	484.45	580

PM Components	Labor-hrs.	W	M	Q	S	A
PM System D3055 250 2950						
Air conditioning split system, DX, air cooled, over 10 tons						
1 Check with operating or area personnel for deficiencies.	.035			√	√	√
2 Check tension, condition, and alignment of belts; adjust as necessary.	.029			√	√	√
3 Lubricate shaft and motor bearings.	.047			√	√	√
4 Pressure wash condenser coils with coil clean solution, as required.	.611					√
5 Replace air filters.	.078			√	√	√
6 Clean electrical wiring and connections; tighten loose connections.	.120					√
7 Clean evaporator coils, drain pan, blowers, fans, motors and drain piping as required.	.380					√
8 Perform operational check of unit; make adjustments on controls and other components as required.	.033			√	√	√
9 During operation of unit, check refrigerant pressure; add refrigerant as necessary.	.271			√	√	√
10 Check compressor oil level; add oil as required.	.033			√	√	√
11 Clean area around equipment.	.066			√	√	√
12 Fill out maintenance checklist and report deficiencies.	.022			√	√	√
Total labor-hours/period				.615	.615	1.726
Total labor-hours/year				1.230	.615	1.726

	Description	Labor-hrs.	Cost Each					
			2020 Bare Costs				Total	Total
			Material	Labor	Equip.	Total	In-House	w/O&P
2900	A/C, split sys., DX, air cooled, over 10 T, annually	1.726	99	113		212	249.69	299
2950	Annualized	3.571	177	234		411	486.45	580

D3055 310 Heat Pump, Air Cooled

PM Components	Labor-hrs.	W	M	Q	S	A
PM System D3055 310 1950						
Heat pump, air cooled, up to 5 tons						
1 Check with operating or area personnel for deficiencies.	.035			√	√	√
2 Check unit for proper operation, excessive noise or vibration.	.033			√	√	√
3 Clean intake side of condenser coils, fans and intake screens.	.055			√	√	√
4 Check electrical wiring and connections; tighten loose connections.	.119					√
5 Inspect fan(s) for bent blades or imbalance; adjust and clean as necessary.	.027					√
6 Check belts for condition, proper tension and misalignment; adjust as required.	.029			√	√	√
7 Lubricate shaft bearings and motor bearings.	.047			√	√	√
8 Inspect piping and valves for leaks; tighten connections as necessary.	.077			√	√	√
9 Replace air filters.	.078			√	√	√
10 Check refrigerant pressure; add refrigerant as necessary.	.134			√	√	√
11 Clean evaporative drain pan, and drain piping as required.	.390					√
12 Cycle the reverse cycle valve to insure proper operation.	.091			√	√	√
13 Clean area around equipment.	.066			√	√	√
14 Fill out maintenance checklist and report deficiencies.	.022			√	√	√
Total labor-hours/period				.667	.667	1.204
Total labor-hours/year				1.335	.667	1.204

	Description	Labor-hrs.	Cost Each					
			2020 Bare Costs				Total	Total
			Material	Labor	Equip.	Total	In-House	w/O&P
1900	Heat pump, air cooled, up to 5 ton, annually	1.204	80.50	79		159.50	186.70	223
1950	Annualized	3.002	187	197		384	450.07	540

PM Components	Labor-hrs.	W	M	Q	S	A
PM System D3055 310 2950						
Heat pump, air cooled, over 5 tons						
1 Check with operating or area personnel for deficiencies.	.035			√	√	√
2 Check unit for proper operation, excessive noise or vibration.	.033			√	√	√
3 Clean intake side of condenser coils, fans and intake screens.	.055			√	√	√
4 Check electrical wiring and connections; tighten loose connections.	.119					√
5 Inspect fan(s) for bent blades or imbalance; adjust and clean as necessary.	.055					√
6 Check belts for condition, proper tension and misalignment; adjust as required.	.057			√	√	√
7 Lubricate shaft bearings and motor bearings.	.047			√	√	√
8 Inspect piping and valves for leaks; tighten connections as necessary.	.077					√
9 Replace air filters.	.155			√	√	√
10 Check refrigerant pressure; add refrigerant as necessary.	.134			√	√	√
11 Lubricate and check operation of dampers, if applicable.	.029					√
12 Check compressor oil level and add oil, if required.	.033			√	√	√
13 Cycle the reverse cycle valve to insure proper operation.	.091			√	√	√
14 Clean evaporative drain pan, and drain piping as required.	.390					√
15 Clean area around equipment.	.066			√	√	√
16 Fill out maintenance checklist and report deficiencies.	.022			√	√	√
Total labor-hours/period				.729	.729	1.400
Total labor-hours/year				1.458	.729	1.400

	Description	Labor-hrs.	Cost Each					
			2020 Bare Costs				Total	Total
			Material	Labor	Equip.	Total	In-House	w/O&P
2900	Heat pump, air cooled, over 5 ton, annually	1.400	104	91.50		195.50	227.91	272
2950	Annualized	3.591	184	236		420	494.72	595

D3055 320 Heat Pump, Water Cooled

PM Components	Labor-hrs.	W	M	Q	S	A
PM System D3055 320 1950						
Heat pump, water cooled, up to 5 tons						
1 Check with operating or area personnel for deficiencies.	.035			√	√	√
2 Check unit for proper operation, excessive noise or vibration.	.033			√	√	√
3 Clean intake side of evaporator coil, fans and intake screens.	.055			√	√	√
4 Check electrical wiring and connections; tighten loose connections.	.119					√
5 Inspect fan(s) for bent blades or imbalance; adjust and clean as necessary.	.027					√
6 Check belts for condition, proper tension and misalignment; adjust as required.	.029			√	√	√
7 Lubricate shaft bearings and motor bearings.	.047			√	√	√
8 Inspect piping and valves for leaks; tighten connections as necessary.	.077			√	√	√
9 Replace air filters.	.078			√	√	√
10 Check refrigerant pressure; add refrigerant as necessary.	.136			√	√	√
11 Lubricate and check operation of dampers, if applicable.	.029					√
12 Clean evaporator drain pan and drain line with solvent.	.381					√
13 Cycle reverse cycle valve to insure proper operation.	.091			√	√	√
14 Backwash condenser coil to remove sediment.	.327					√
15 Clean area around equipment.	.066			√	√	√
16 Fill out maintenance checklist and report deficiencies.	.022			√	√	√
Total labor-hours/period				.668	.668	1.551
Total labor-hours/year				1.337	.668	1.551

	Description	Labor-hrs.	Cost Each					
			2020 Bare Costs				Total	Total
			Material	Labor	Equip.	Total	In-House	w/O&P
1900	Heat pump, water cooled, up to 5 ton, annually	1.551	93.50	102		195.50	229.13	274
1950	Annualized	3.555	188	233		421	496.69	595

PM Components	Labor-hrs.	W	M	Q	S	A
PM System D3055 320 2950						
Heat pump, water cooled, over 5 tons						
1 Check with operating or area personnel for deficiencies.	.035			√	√	√
2 Check unit for proper operation, excessive noise or vibration.	.033			√	√	√
3 Clean intake side of condenser coils, fans and intake screens.	.055			√	√	√
4 Check electrical wiring and connections; tighten loose connections.	.119					√
5 Inspect fan(s) for bent blades or imbalance; adjust and clean as necessary.	.027					√
6 Check belts for condition, proper tension and misalignment; adjust as required.	.029			√	√	√
7 Lubricate shaft bearings and motor bearings.	.047			√	√	√
8 Inspect piping and valves for leaks; tighten connections as necessary.	.077			√	√	√
9 Replace air filters.	.078			√	√	√
10 Check refrigerant pressure; add refrigerant as necessary.	.136			√	√	√
11 Lubricate and check operation of dampers, if applicable.	.029					√
12 Clean evaporator drain pan and drain line with solvent.	.381					√
13 Cycle reverse cycle valve to insure proper operation.	.091			√	√	√
14 Clean area around equipment.	.066			√	√	√
15 Fill out maintenance checklist and report deficiencies.	.022			√	√	√
Total labor-hours/period				.668	.668	1.225
Total labor-hours/year				1.337	.668	1.225

	Description	Labor-hrs.	Cost Each					
			2020 Bare Costs				Total	Total
			Material	Labor	Equip.	Total	In-House	w/O&P
2900	Heat pump, water cooled, over 5 ton, annually	1.225	93.50	80.50		174	202.63	241
2950	Annualized	3.228	188	212		400	470.19	560

D3065 100 Controls, Central System

PM Components	Labor-hrs.	W	M	Q	S	A
PM System D3065 100 1950						
Controls, central system, electro/pneumatic						
1 With panel disconnected from power source, clean patrol panel compartment with a vacuum.	.471					√
2 Inspect wiring/components for loose connections; tighten, as required.	.119					√
3 Check set point of controls temperature, humidity or pressure.	.033					√
4 Check unit over its range of control.	.033					√
5 Check for correct pressure differential on all two position controllers.	.195					√
6 Check source of the signal and its amplification on electronic controls.	.471					√
7 [illegible]	[illegible]					√
8 Check relays, pilot valves and pressure regulators for proper operation, repair or replace as necessary.	.327					√
9 Replace air filters in sensors, controllers, and thermostats as necessary.	.029					√
10 Clean area around equipment.	.066					√
11 Fill out maintenance checklist and report deficiencies.	.022					√
Total labor-hours/period						1.921
Total labor-hours/year						1.921

	Description	Labor-hrs.	Cost Each					
			2020 Bare Costs				Total	Total
			Material	Labor	Equip.	Total	In-House	w/O&P
1900	Controls, central system, electro/pneumatic, annually	1.921	66	126		192	228.09	277
1950	Annualized	1.921	66	126		192	228.09	277

D3095 110 Air Compressor, Gas Engine

PM Components	Labor-hrs.	W	M	Q	S	A
PM System D3095 110 1950						
Air compressor, gas engine						
1 Check with operating or area personnel for any obvious deficiencies.	.035		√	√	√	√
2 Check compressor oil level; add oil as required.	.022		√	√	√	√
3 Replace compressor oil.	.222				√	√
4 Clean air intake filter on air compressor(s); replace if necessary.	.178				√	√
5 Clean cylinder cooling fins and air cooler on compressor(s).	.155				√	√
6 Check tension, condition, and alignment of v-belts; adjust as necessary.	.030				√	√
7 Clean oil and water traps.	.178				√	√
8 Drain moisture from air storage tank and check discharge for indication of interior corrosion.	.046				√	√
9 Perform operation check of compressor; check the operation of low pressure cut-in and high pressure cut-out switches.	.221		√	√	√	√
10 Check operation of safety pressure relief valve.	.030				√	√
11 Check and tighten compressor foundation anchor bolts.	.022				√	√
12 Check radiator coolant; add if required.	.012		√	√	√	√
13 Check battery water; add if required.	.242		√	√	√	√
14 Check wiring, connections, switches, etc.; tighten loose connections.	.119				√	√
15 Check engine oil level; add if required.	.014		√	√	√	√
16 Change engine oil.	.225				√	√
17 Change engine oil filter.	.059				√	√
18 Check spark plug and reset cap.	.035				√	√
19 Check condition of engine air filter; replace if necessary.	.039				√	√
20 Test run engine for proper operation.	.410		√	√	√	√
21 Wipe dust and dirt from engine and compressor.	.109		√	√	√	√
22 Check muffler/exhaust system for corrosion.	.020				√	√
23 Clean area around equipment.	.066		√	√	√	√
24 Fill out maintenance checklist and report deficiencies.	.022		√	√	√	√
Total labor-hours/period			1.154	1.154	2.512	2.512
Total labor-hours/year			9.228	2.307	2.512	2.512

	Description	Labor-hrs.	Cost Each					
			2020 Bare Costs				Total	Total
			Material	Labor	Equip.	Total	In-House	w/O&P
1900	Air compressor, gas engine powered, annually	2.512	120	165		285	335.98	405
1950	Annualized	16.575	214	1,075		1,289	1,576.87	1,950

D3095 114 Air Compressor, Centrifugal

	PM Components	Labor-hrs.	W	M	Q	S	A
	PM System D3095 114 1950						
	Air compressor, centrifugal, to 40 H.P.						
1	Check compressor oil level; add oil as necessary.	.022			√	√	√
2	Perform operational check of compressor system and adjust as required.	.222			√	√	√
3	Check motor for excessive vibration, noise and overheating; lubricate.	.039			√	√	√
4	Check operation of pressure relief valve.	.030			√	√	√
5	Clean cooling fans and air cooler.	.023			√	√	√
6	Check tension, condition, and alignment of V-belts; adjust as necessary.	.030			√	√	√
7	Drain moisture from air storage tank and check low pressure cut-in; while draining, check discharge for indication of interior corrosion.	.046			√	√	√
8	Replace air intake filter, as needed.	.178			√	√	√
9	Clean oil and water trap.	.178			√	√	√
10	Clean compressor and surrounding area.	[illegible]			√	√	√
11	Fill out maintenance checklist and report deficiencies.	[illegible]			√	√	√
	Total labor-hours/period				.856	.856	.856
	Total labor-hours/year				1.712	.856	.856

	Description	Labor-hrs.	Cost Each					
			2020 Bare Costs				Total	Total
			Material	Labor	Equip.	Total	In-House	w/O&P
1900	Air compressor, centrifugal, to 40 H.P., annually	.856	54	56		110	128.75	154
1950	Annualized	3.412	76	224		300	360.25	440

D3095 114 Air Compressor, Centrifugal

PM Components	Labor-hrs.	W	M	Q	S	A
PM System D3095 114 2950 **Air compressor, centrifugal, over 40 H.P.**						
1 Check with operating or area personnel for any obvious deficiencies.	.035		√	√	√	√
2 Perform control system check.	.327				√	√
3 Lubricate main driver coupling if necessary.	.047				√	√
4 Lubricate prelube pump motor and pump coupling, if necessary.	.047				√	√
5 Perform operation check of air compressor, adjust as required.	.222		√	√	√	√
6 Check main driver coupling alignment; realign if necessary.	.327				√	√
7 Check compressor and motor operation for excessive vibration, noise and overheating; lubricate motor.	.039		√	√	√	√
8 Check compressor scrolls, piping, and impeller housing for all leaks or cracks.	.327		√	√	√	√
9 Check intercoolers and aftercoolers for high cooling water temperatures or cooling water leakage.	.195		√	√	√	√
10 Check oil level in compressor oil reservoir; add oil as necessary.	.022		√	√	√	√
11 Replace compressor oil and oil filters.	.282					√
12 Record oil pressure and oil temperature.	.013		√	√	√	√
13 Check compressor for oil leaks.	.026		√	√	√	√
14 Visually inspect oil mist arrestor, clean housing, lines, and replace element if saturated, if applicable.	.177				√	√
15 Check operation of pressure relief valve.	.030		√	√	√	√
16 Visually inspect discharge check valve.	.009				√	√
17 Check oil and water trap.	.022		√	√	√	√
18 Check indicating lamps or gauges for proper operation if appropriate; replace burned out lamps or repair/replace gauges.	.020					√
19 Visually check all air intake filter elements; replace if necessary.	.022		√	√	√	√
20 Replace all air intake filter elements.	.177					√
21 Lubricate motor.	.047				√	√
22 Clean compressor, motor and surrounding area.	.066		√	√	√	√
23 Fill out maintenance check and report deficiencies.	.022		√	√	√	√
Total labor-hours/period			1.041	1.041	2.021	2.500
Total labor-hours/year			8.325	2.081	2.021	2.500

	Description	Labor-hrs.	Cost Each					
			2020 Bare Costs				Total	Total
			Material	Labor	Equip.	Total	In-House	w/O&P
2900	Air compressor, centrifugal, over 40 H.P., annually	2.500	98.50	164		262.50	311.45	375
2950	Annualized	14.889	102	975		1,077	1,317.51	1,625

PM Components	Labor-hrs.	W	M	Q	S	A
PM System D3095 118 1950						
Air compressor, reciprocating, less than 5 H.P.						
1 Replace compressor oil.	.340			√	√	√
2 Perform operation check of compressor system and adjust as required.	.222			√	√	√
3 Check motor operation for excessive vibration, noise and overheating.	.042			√	√	√
4 Lubricate motor.	.047			√	√	√
5 Check operation of pressure relief valve.	.030			√	√	√
6 Check tension, condition, and alignment of V-belts; adjust as necessary.	.030			√	√	√
7 Drain moisture from air storage tank and check low pressure cut-in; while draining, check discharge for indication of interior corrosion.	.046			√	√	√
8 Clean air intake filter on compressor.	.178			√	√	√
9 Clean oil and water trap.	.178			√	√	√
10 Clean exterior of compressor, motor and surrounding area.	.066			√	√	√
11 Fill out maintenance checklist and report deficiencies.	.022			√	√	√
Total labor-hours/period				1.201	1.201	1.201
Total labor-hours/year				2.403	1.201	1.201

	Description	Labor-hrs.	Cost Each					
			2020 Bare Costs				Total	Total
			Material	Labor	Equip.	Total	In-House	w/O&P
1900	Air compressor, recip., less than 5 H.P., annually	1.201	46.50	79		125.50	148.20	180
1950	Annualized	4.796	74	315		389	471.40	580

PM Components	Labor-hrs.	W	M	Q	S	A
PM System D3095 118 2950						
Air compressor, reciprocating, 5 to 40 H.P.						
1 Replace compressor oil.	.340			√	√	√
2 Perform operation check of compressor system and adjust as required.	.222			√	√	√
3 Check motor operation for excessive vibration, noise and overheating; lubricate motor.	.042			√	√	√
4 Check operation of pressure relief valve.	.043			√	√	√
5 Clean cooling fans and air cooler on compressor.	.023			√	√	√
6 Check tension, condition, and alignment of V-belts; adjust as necessary.	.030			√	√	√
7 Drain moisture from air storage tank and check low pressure cut-in; while draining, check discharge for indication of interior corrosion.	.059			√	√	√
8 Clean air intake filter on compressor.	.178			√	√	√
9 Clean oil and water trap.	.190			√	√	√
10 Clean compressor and surrounding area.	.066			√	√	√
11 Fill out maintenance checklist and report deficiencies.	.022			√	√	√
Total labor-hours/period				1.216	1.216	1.216
Total labor-hours/year				2.432	1.216	1.216

	Description	Labor-hrs.	Cost Each					
			2020 Bare Costs				Total	Total
			Material	Labor	Equip.	Total	In-House	w/O&P
2900	Air compressor, reciprocating, 5 to 40 H.P., annually	1.216	46.50	80		126.50	149.41	181
2950	Annualized	4.856	111	320		431	516.70	630

D3095 118 Air Compressor, Reciprocating

PM Components	Labor-hrs.	W	M	Q	S	A
PM System D3095 118 3950						
Air compressor, reciprocating, over 40 H.P.						
1 Check with operating or area personnel for deficiencies.	.035			√	√	√
2 Perform operation check of compressor system and adjust as required.	.222			√	√	√
3 Replace compressor oil.	.340			√	√	√
4 Check motor(s) operation for excessive vibration, noise and overheating; lubricate motor(s).	.042			√	√	√
5 Clean cylinder cooling fins and air cooler on compressor.	.023			√	√	√
6 Check tension, condition, and alignment of V-belts; adjust as necessary.	.056			√	√	√
7 Check operation of pressure relief valve.	.030			√	√	√
8 Check low pressure cut in and high pressure cut out switches.	.120			√	√	√
9 Drain moisture from air storage tank and check low pressure cut-in; while draining, check discharge for indication of interior corrosion.	.072			√	√	√
10 Clean air intake filter on air compressor(s); replace if necessary.	.178			√	√	√
11 Clean oil and water trap.	.205			√	√	√
12 Check indicating lamps or gauges for proper operation if appropriate; replace burned out lamps or repair/replace gauges.	.020			√	√	√
13 Clean area around equipment.	.066			√	√	√
14 Fill out maintenance checklist and report deficiencies.	.022			√	√	√
Total labor-hours/period				1.432	1.432	1.432
Total labor-hours/year				2.864	1.432	1.432

	Description	Labor-hrs.	Cost Each					
			2020 Bare Costs				Total	Total
			Material	Labor	Equip.	Total	In-House	w/O&P
3900	Air compressor, reciprocating, over 40 H.P., annually	1.432	68.50	94		162.50	191.20	231
3950	Annualized	5.716	135	375		510	612.90	750

PM Components	Labor-hrs.	W	M	Q	S	A
PM System D3095 210 1950						
Steam Humidification System						
1 Operate humidistat through its throttling range to verify activation and deactivation.	.117				√	√
2 Inspect steam trap for proper operation.	.117				√	√
3 Turn off steam supply.	.065				√	√
4 Secure electrical service before servicing humidification unit.	.065				√	√
5 Clean strainer	.195				√	√
6 Clean and/or replace water/steam nozzle as necessary.	.457				√	√
7 Inspect pneumatic controller for air leaks.	[illegible]				√	√
8 Inspect steam lines for leaks and corrosion and repair leaks.	.106				√	√
9 Fill out maintenance checklist and report deficiencies.	.022				√	√
Total labor-hours/period					1.272	1.272
Total labor-hours/year					1.272	1.272

	Description	Labor-hrs.	Cost Each					
			2020 Bare Costs				Total	Total
			Material	Labor	Equip.	Total	In-House	w/O&P
1900	Steam humidification system, annually	1.272	25.50	83.50		109	130.94	161
1950	Annualized	2.540	25.50	166		191.50	233.87	289

PM Components	Labor-hrs.	W	M	Q	S	A
PM System D3095 210 2950						
Evaporative Pan with Heating Coil Humidification System						
1 Operate humidistat through its throttling range to verify activation and deactivation.	.117				√	√
2 Inspect steam trap for proper operation.	.117				√	√
3 Turn off water and steam supply.	.065				√	√
4 Secure electrical service before servicing humidification unit.	.065				√	√
5 Drain and flush water pans, clean drains, etc.	.258				√	√
6 Check condition of heating element/steam coils and clean.	.065				√	√
7 Inspect pneumatic controller for air leaks.	.039				√	√
8 Inspect steam lines for leaks and corrosion and repair leaks.	.195				√	√
9 Fill out maintenance checklist and report deficiencies.	.022				√	√
Total labor-hours/period					.943	.943
Total labor-hours/year					.943	.943

	Description	Labor-hrs.	Cost Each					
			2020 Bare Costs				Total	Total
			Material	Labor	Equip.	Total	In-House	w/O&P
2900	Evap. pan with heating coil humidif. system, annually	.943	29	62		91	108.64	132
2950	Annualized	1.890	29	124		153	184.87	227

D3095 220 Dehumidifier, Desiccant Wheel

PM Components	Labor-hrs.	W	M	Q	S	A
PM System D3095 220 1950						
Dehumidifier, desiccant wheel						
1 Check with operating or area personnel for deficiencies.	.035		√	√	√	√
2 Check filters for any blockage or fouling, remove and clean as required.	.036		√	√	√	√
3 Check wheel seals for tears or punctures.	.055		√	√	√	√
4 Check the control valve, thermo sensor bulb and burner, if it has heaters; or check for scaling or leaking on steam heating coils, as applicable.	.094				√	√
5 Check that the outlet air temperature is within the proper heat range.	.105		√	√	√	√
6 Clean the desiccant wheel and check for softening of wheel faces.	.031				√	√
7 Check desiccant wheel and motor for vibration and noise, adjust as required.	.109				√	√
8 Check gear reducer oil level; add as required.	.035				√	√
9 Check wheel belt(s) for wear, proper tension and alignment; adjust as required.	.029				√	√
10 Check blower and motor for excessive vibration and noise; adjust as required.	.109				√	√
11 Check blower belt(s) for wear, proper tension and alignment; adjust as required.	.029				√	√
12 Lubricate wheel, blower and motor bearings.	.047				√	√
13 Check electrical wiring and connections; make appropriate adjustments.	.120				√	√
14 Check reactivation ductwork for condensation and air leaks.	.013				√	√
15 Clean the equipment and the surrounding area.	.066		√	√	√	√
16 Fill out maintenance checklist and report deficiencies.	.022		√	√	√	√
Total labor-hours/period			.320	.320	.935	.935
Total labor-hours/year			2.556	.639	.935	.935

	Description	Labor-hrs.	Cost Each					
			2020 Bare Costs				Total	Total
			Material	Labor	Equip.	Total	In-House	w/O&P
1900	Dehumidifier, desiccant wheel, annually	.935	72	51.50		123.50	145.18	172
1950	Annualized	5.060	77	277		354	440.79	535

PM Components	Labor-hrs.	W	M	Q	S	A
PM System D4015 100 1950						
Backflow prevention device, up to 4″						
NOTE: Test frequency may vary depending on local regulations and application.						
1 Test and calibrate check valve operation of backflow prevention device with test set.	.190					√
2 Bleed air from backflow preventer.	.047					√
3 Inspect for leaks under pressure.	.007					√
4 Clean backflow preventer and surrounding area.	[illegible]					√
5 Fill out maintenance checklist and report deficiencies.	.022					√
Total labor-hours/period						.333
Total labor-hours/year						.333

	Description	Labor-hrs.	Cost Each: 2020 Bare Costs Material	Labor	Equip.	Total	Total In-House	Total w/O&P
1900	Backflow prevention device, up to 4″, annually	.333	13.30	21.50		34.80	41.15	49.50
1950	Annualized	.333	13.30	21.50		34.80	41.15	49.50

PM Components	Labor-hrs.	W	M	Q	S	A
PM System D4015 100 2950						
Backflow prevention device, over 4″						
NOTE: Test frequency may vary depending on local regulations and application.						
1 Test and calibrate check valve operation of backflow prevention device with test set.	.333					√
2 Bleed air from backflow preventer.	.065					√
3 Inspect for leaks under pressure.	.007					√
4 Clean backflow preventer and surrounding area.	.066					√
5 Fill out maintenance checklist and report deficiencies.	.022					√
Total labor-hours/period						.493
Total labor-hours/year						.493

	Description	Labor-hrs.	Cost Each: 2020 Bare Costs Material	Labor	Equip.	Total	Total In-House	Total w/O&P
2900	Backflow prevention device, over 4″, annually	.493	13.30	32		45.30	53.90	65.50
2950	Annualized	.493	13.30	32		45.30	53.90	65.50

PM Components	Labor-hrs.	W	M	Q	S	A
PM System D4015 150 1950						
Extinguishing system, wet pipe						
1 Notify proper authorities prior to testing any alarm systems.	.130		√	√	√	√
2 Open and close post indicator valve (PIV) to check operation; make minor adjustments such as lubricating valve stem, cleaning and/or replacing target windows as required.	.176					√
3 Open and close OS&Y (outside stem and yoke) cut-off valve to check operation; make minor repairs such as lubricating stems and tightening packing glands as required.	.176					√
4 Perform operational test of water flow detectors; make minor adjustments and restore system to proper operating condition.	.148					√
5 Check to ensure that alarm drain is open; clean drain line if necessary.	.081		√	√	√	√
6 Open water motor alarm test valve and ensure that outside alarm operates; lubricate alarm, make adjustments as required.	.176		√	√	√	√
7 Conduct main drain test by opening 2″ test valve; maintain a continuous record of drain tests; make minor adjustments if applicable; restore system to proper operating condition.	.333			√	√	√
8 Check general condition of sprinklers and sprinkler system; make minor adjustments as required.	.229					√
9 Check equipment gaskets, piping, packing glands, and valves for leaks; tighten flange bolts and loose connections to stop all leaks.	.013					√
10 Check condition of fire department connections; replace missing or broken covers as required.	.103					√
11 Trip test wet pipe system using test valve furthest from wet pipe valve (control valve); make minor adjustments as necessary and restore system to proper operating condition.	1.733					√
12 Inspect OS&Y and PIV cut-off valves for open position.	.066		√	√	√	√
13 Clean area around system components.	.432			√	√	√
14 Fill out maintenance checklist and report deficiencies.	.022		√	√	√	√
Total labor-hours/period			.475	1.241	1.241	3.818
Total labor-hours/year			3.798	2.481	1.241	3.818

	Description	Labor-hrs.	Cost Each					
			2020 Bare Costs				Total	Total
			Material	Labor	Equip.	Total	In-House	w/O&P
1900	Extinguishing system, wet pipe, annually	3.818	105	247		352	419.59	510
1950	Annualized	11.342	119	730		849	1,034.04	1,275

PM Components	Labor-hrs.	W	M	Q	S	A
PM System D4015 180 1950 **Extinguishing system, deluge/preaction**						
1 Notify proper authorites prior to testing any alarm systems.	.130		√	√	√	√
2 Open and close post indicator valve (PIV) to check operation; make minor repairs such as lubricating valve stem, cleaning and/or replacing target windows as required.	.176					√
3 Open and close outside stem and yoke (OS&Y) cut-off valve to check operation; make minor repairs such as lubricating stems, tightening packing glands as required.	.176					√
4 Perform operational test of supervisory initiating devices and water flow detectors; make minor adjustments and restore system to proper operating condition.	.148					√
5 Check to ensure that alarm drain is open, clean drain if necessary.	.081		√	√	√	√
6 Open water motor alarm test valve and ensure that outside alarm operates; lubricate alarm and adjust as required.	.176		√	√	√	√
7 Visually check water pressure to ensure adequate operating pressure is available; make adjustments as required.	.004		√	√	√	√
8 Conduct main drain test by opening 2″ test valve and observing drop in water pressure on gauge; pressure drop should not exceed 20 PSI; maintain a continuous record of drain tests; make minor adjustments; restore system to proper operating condition.	.333			√	√	√
9 Check general condition of sprinklers and sprinkler system; make minor adjustments as required.	.228					√
10 Check equipment gaskets, piping, packing glands, and valves for leaks; tighten flange bolts and loose connections to stop all leaks.	.025					√
11 Check condition of fire department connections; replace missing or broken covers as required.	.103					√
12 Check and inspect pneumatic system for physical damage and proper operation; make minor adjustments as required.	.320					√
13 Trip test deluge system (control valve closed); make minor adjustments as necessary and restore system to fully operational condition.	1.956					√
14 Inspect OS&Y and PIV cut-off valves for open position.	.066		√	√	√	√
15 Clean area around system components.	.066					√
16 Ensure that system is restored to proper operating condition.	.109		√	√	√	√
17 Fill out maintenance checklist and report deficiencies.	.022		√	√	√	√
Total labor-hours/period			.588	.921	.921	4.119
Total labor-hours/year			4.700	1.842	.921	4.119

	Description	Labor-hrs.	Cost Each: 2020 Bare Costs: Material	Labor	Equip.	Total	Total In-House	Total w/O&P
1900	Extinguishing system, deluge / preaction, annually	4.119	105	265		370	443.24	540
1950	Annualized	11.586	119	745		864	1,053.05	1,300

PM Components	Labor-hrs.	W	M	Q	S	A
PM System D4015 210 1950						
Fire pump, electric motor driven						
1 Check control panel and wiring for loose connections; tighten connections as required.	.109			√	√	√
2 Ensure all valves relating to water system are in correct position.	.008	√	√	√	√	√
3 Open and close OS&Y(outside steam and yoke) cut-off valve to check operation; make minor repairs such as lubricating stems and tightening packing glands as required.	.176			√	√	√
4 Centrifugal pump:						
A) perform 10 minute pump test run; check for proper operation and adjust if required.	.216	√	√	√	√	√
B) check for leaks on suction and discharge piping, seals, packing glands, etc.	.077	√	√	√	√	√
C) check for excessive vibration, noise, overheating, etc.	.022	√	√	√	√	√
D) check alignment, clearances, and rotation of shaft and coupler (includes removing and reinstalling safety cover).	.160	√	√	√	√	√
E) tighten or replace loose, missing or damaged nuts, bolts, or screws.	.005	√	√	√	√	√
F) lubricate pump and motor as required.	.099			√	√	√
G) check suction or discharge pressure gauge readings and flow rate.	.078	√	√	√	√	√
H) check packing glands and tighten or repack as required; note that slight dripping is required for proper lubrication of shaft.	.113			√	√	√
5 Inspect and clean strainers after each use and flow test.	.258	√	√	√	√	√
6 Clean equipment and surrounding area.	.066	√	√	√	√	√
7 Fill out maintenance checklist and report deficiencies.	.022	√	√	√	√	√
Total labor-hours/period		.912	.912	1.408	1.408	1.408
Total labor-hours/year		34.654	7.296	2.816	1.408	1.408

	Description	Labor-hrs.	Cost Each					
			2020 Bare Costs				Total	Total
			Material	Labor	Equip.	Total	In-House	w/O&P
1900	Fire pump, electric motor driven, annually	1.408	33	91		124	148.29	181
1950	Annualized	47.752	99	3,075		3,174	3,902.95	4,875

D4015 250 Fire Pump, Motor/Engine Driven

PM Components	Labor-hrs.	W	M	Q	S	A
[illegible]						
Fire pump, engine driven						
1 Ensure all valves relating to water system are in correct position.	.008	✓	✓	✓	✓	✓
2 Open and close OS&Y (outside steam and yoke) cut-off valve to check operation; make minor repairs such as lubricating stems and tightening packing glands as required.	.176			✓	✓	✓
3 Controller panel:						
A) check controller panel for proper operation in accordance with [illegible] out bulbs and fuses; make other repairs as necessary.	.030			✓	✓	✓
B) check wiring for loose connections and fraying; tighten or replace as required.	.100			✓	✓	✓
4 Diesel engine:						
A) check oil level in crankcase with dipstick; add oil as necessary.	.014	✓	✓	✓	✓	✓
B) drain and replace engine oil.	[illegible]					✓
C) replace engine oil filter.	.059					✓
D) check radiator coolant level; add water if low.	.012	✓	✓	✓	✓	✓
E) drain and replace radiator coolant.	.511					✓
F) check battery water level; add water if low.	.012			✓	✓	✓
G) check battery terminals for corrosion; clean if necessary.	.124			✓	✓	✓
H) check belts for tension and wear; adjust or replace if required.	.012			✓	✓	✓
I) check fuel level in tank; refill as required.	.046			✓	✓	✓
J) check electrical wiring, connections, switches, etc.; adjust or tighten as necessary.	.120			✓	✓	✓
K) perform 10 minute pump check for proper operation and adjust as required.	.216	✓	✓	✓	✓	✓
5 Centrifugal pump:						
A) check for leaks on suction and discharge piping, seals, packing glands, etc.	.077	✓	✓	✓	✓	✓
B) check pump operation; vibration, noise, overheating, etc.	.022	✓	✓	✓	✓	✓
C) check alignment, clearances, and rotation of shaft and coupler (includes removing and reinstalling safety cover).	.160	✓	✓	✓	✓	✓
D) tighten or replace loose, missing or damaged nuts, bolts, or screws.	.005	✓	✓	✓	✓	✓
E) lubricate pump as required.	.099			✓	✓	✓
F) check suction or discharge pressure gauge readings and flow rate.	.078	✓	✓	✓	✓	✓
G) check packing glands and tighten or repack as required; note that slight dripping is required for proper lubrication of shaft.	.113			✓	✓	✓
6 Inspect and clean strainers after each use and flow test.	.262	✓	✓	✓	✓	✓
7 Clean equipment and surrounding area.	.432	✓	✓	✓	✓	✓
8 Fill out maintenance checklist and report deficiencies.	.022	✓	✓	✓	✓	✓
Total labor-hours/period		1.309	1.309	2.158	2.158	3.239
Total labor-hours/year		49.723	10.468	4.316	2.158	3.239

	Description	Labor-hrs.	Cost Each					
			2020 Bare Costs				Total	Total
			Material	Labor	Equip.	Total	In-House	w/O&P
1950	Annualized	69.866	276	4,500		4,776	5,862.83	7,300

PM Components	Labor-hrs.	W	M	Q	S	A
PM System D4015 310 1950						
Extinguishing system, dry pipe						
1 Notify proper authorities prior to testing any alarm systems.	.130		√	√	√	√
2 Open and close post indicator valve to check operation; make minor repairs such as lubricating valve stem, cleaning and/or replacing target windows as required.	.176					√
3 Open and close OS&Y (outside stem and yoke) cut-off valve to check operation; make minor repairs such as lubricating stems and tightening packing glands as required.	.176					√
4 Perform operational test of water flow detectors; make minor adjustments and restore system to proper operating condition.	.148					√
5 Check to ensure that alarm drain is open; clean drain line if necessary.	.081		√	√	√	√
6 Open water motor alarm test valve and ensure that outside alarm operates; lubricate alarm, make adjustments as required.	.176		√	√	√	√
7 Visually check water pressure to ensure that adequate operating pressure is available; make adjustments as required.	.003		√	√	√	√
8 Conduct main drain test by opening 2″ test valve; maintain a continuous record of drain tests; make minor adjustments if applicable; restore system to proper operating condition.	.333			√	√	√
9 Check general condition of sprinklers and sprinkler system; make minor adjustments as required.	.228					√
10 Check equipment gaskets, piping, packing glands, and valves for leaks; tighten flange bolts and loose connections to stop all leaks.	.013					√
11 Check condition of fire department connections; replace missing or broken covers as required.	.103					√
12 Visually check system air pressure; pump up system and make minor adjustments as required.	.135		√	√	√	√
13 Trip test dry pipe system using test valve furthest from dry pipe valve (control valve partially open); make minor adjustments as necessary and restore system to fully operational condition.	1.733					√
14 Check operation of quick opening device.	.022					√
15 Inspect OS&Y and PIV cut-off valves for open position.	.066		√	√	√	√
16 Clean area around system components.	.433			√	√	√
17 Fill out maintenance checklist and report deficiencies.	.022		√	√	√	√
Total labor-hours/period			.613	1.379	1.379	3.978
Total labor-hours/year			4.902	2.758	1.379	3.978

Description		Labor-hrs.	Cost Each					
			2020 Bare Costs				Total	Total
			Material	Labor	Equip.	Total	In-House	w/O&P
1900	Extinguishing system, dry pipe, annually	3.978	144	257		401	475.03	575
1950	Annualized	13.019	144	840		984	1,194.61	1,475

D4095 100 Extinguishing System, CO_2

PM Components	Labor-hrs.	W	M	Q	S	A
PM System D4095 100 1950						
Extinguishing system, CO_2						
1 Check that nozzles, heads and hand hose lines are clear from obstructions, have not been damaged and in proper position.	.098		√	√	√	√
2 Check to ensure that all operating controls are properly set.	.066		√	√	√	√
3 Clean nozzles as required.	.098		√	√	√	√
4 Visually inspect the control panel for obstructions or physical damage; clean dirt and dust from interior and exterior; make sure that all cards are plugged in tightly, tighten loose connections and make other minor adjustments as necessary.	.222		√	√	√	√
5 Check battery voltages where installed, recharge or replace as required.	.013		√	√	√	√
6 Weigh CO_2 cylinders, replace any that show a weight loss greater than 10%.	.040				√	√
7 Blow out entire system to make sure that the system is not plugged or restricted.	.927					√
8 Run out hose reels and check hoses.	.216					√
9 Conduct operational test of actuating devices, both automatic, manual and alarm system. Prior to conducting tests, close valves, disable CO_2dumping control and manually override computer shutdown feature if applicable, all in accordance with the manufacturer's specifications.	.432					√
10 Check the operation and timing of the time delay control and check operation of the abort station.	.108					√
11 Restore system to proper operating condition and notify personnel upon completion of tests.	.079					√
12 Clean up around system.	.066		√	√	√	√
13 Fill out maintenance checklist and report deficiencies.	.022		√	√	√	√
Total labor-hours/period			.585	.585	1.225	2.387
Total labor-hours/year			4.678	1.169	1.225	2.387

	Description	Labor-hrs.	Cost Each					
			2020 Bare Costs				Total	Total
			Material	Labor	Equip.	Total	In-House	w/O&P
1900	Extinguishing system, CO_2, annually	2.387	202	131		333	390.15	460
1950	Annualized	9.470	705	520		1,225	1,439.40	1,700

D4095 200 Exting. System, Foam Bottle

PM Components	Labor-hrs.	W	M	Q	S	A
PM System D4095 200 1950						
Extinguishing system, foam bottle						
1 Check that all nozzles, heads and hand held lines are clear from obstructions, have not been damaged and are in proper position.	.195		√	√	√	√
2 Check to ensure that all operating controls are properly set.	.022		√	√	√	√
3 Observe pressure on system to ensure that proper pressure is being maintained.	.004		√	√	√	√
4 Clean area around system.	.066		√	√	√	√
5 Fill out maintenance checklist and report deficiencies.	.022		√	√	√	√
Total labor-hours/period			.309	.309	.309	.309
Total labor-hours/year			2.473	.618	.309	.309

	Description	Labor-hrs.	Cost Each					
			2020 Bare Costs				Total	Total
			Material	Labor	Equip.	Total	In-House	w/O&P
1900	Extinguishing system, foam bottle, annually	.309	4.53	16.95		21.48	26.73	32.50
1950	Annualized	3.708	18.10	203		221.10	280.98	350

PM Components	Labor-hrs.	W	M	Q	S	A
PM System D4095 210 1950						
Extinguishing system, foam, electric pump, deluge system						
1 Check foam concentrate level in tank; add concentrate as required to maintain proper level.	.043		√	√	√	√
2 Ensure all valves relating to foam/water system are in correct position.	.008		√	√	√	√
3 Visually check proportioning devices, pumps, and foam nozzles; correct any observed deficiencies.	.386		√	√	√	√
4 Check water supply pressure; adjust as necessary.	.029		√	√	√	√
5 Open and close OS&Y (outside stem and yoke) cut-off valve to check operation; make minor repairs such as lubricating stems and tightening packing glands as required.	.059			√	√	√
6 Centrifugal pump:						
A) perform 10 minute pump test run, check for proper operation and adjust as required.	.217			√	√	√
B) test start pump without foam discharger.	.022		√	√	√	√
C) check for leaks on suction and discharge piping, seals, packing glands, etc.	.077			√	√	√
D) check for excessive vibration, noise, overheating, etc.	.022			√	√	√
E) check alignment, clearances, and rotation of shaft and coupler (includes removing and reinstalling safety cover).	.160			√	√	√
F) tighten or replace loose, missing or damaged nuts, bolts, or screws.	.005			√	√	√
G) lubricate pump and motor as required.	.099			√	√	√
H) check suction or discharge pressure gauge readings and flow rate.	.078			√	√	√
I) check packing glands and tighten or repack as required; note that slight dripping is required for proper lubrication of shaft.	.113			√	√	√
7 Inspect and clean strainers after each use and flow test.	.258		√	√	√	√
8 Clean equipment and surrounding area.	.432		√	√	√	√
9 Fill out maintenance checklist.	.022		√	√	√	√
Total labor-hours/period			1.200	2.029	2.029	2.029
Total labor-hours/year			9.604	4.059	2.029	2.029

	Description	Labor-hrs.	Cost Each					
			2020 Bare Costs				Total	Total
			Material	Labor	Equip.	Total	In-House	w/O&P
1900	Extinguishing system, foam, electric pump, annually	2.029	94.50	111		205.50	246.95	295
1950	Annualized	17.721	355	970		1,325	1,637.96	2,000

D4095 220 Exting. System, Foam, Diesel

PM Components	Labor-hrs.	W	M	Q	S	A
PM System D4095 220 1950						
Extinguishing system, foam, diesel pump, deluge system						
1 Check foam concentrate level in tank; add concentrate as required to maintain proper level.	.043		√	√	√	√
2 Ensure all valves relating to foam/water system are in correct position.	.008		√	√	√	√
3 Visually check proportioning devices, pumps, and foam nozzles; correct any observed deficiencies.	.386		√	√	√	√
4 Check water supply pressure; adjust as necessary.	.003		√	√	√	√
5 Open and close OS&Y (outside stem and yoke) cut-off valve to check operation; make minor repairs such as lubricating stems and tightening packing glands as required.	.059			√	√	√
6 Controller panel:						
A) check controller panel for proper operation in accordance with standard procedures for the type of panel installed, replace burned out bulbs and fuses; make other repairs as necessary.	.039			√	√	√
B) check wiring for loose connections and fraying; tighten or replace as required.	.109			√	√	√
7 Diesel engine:						
A) check oil level in crankcase with dipstick; add oil as necessary.	.014		√	√	√	√
B) drain and replace engine oil.	.511					√
C) replace engine oil filter.	.059					√
D) check radiator coolant level; add water if low.	.012		√	√	√	√
E) drain and replace radiator coolant.	.511					√
F) check battery water level; add water if low.	.012			√	√	√
G) check battery terminals for corrosion; clean if necessary.	.124			√	√	√
H) check belts for tension and wear; adjust or replace if required.	.012			√	√	√
I) check fuel level in tank; refill as required.	.046			√	√	√
J) check electrical wiring, connections, switches, etc.; adjust or tighten as necessary.	.120			√	√	√
K) perform 10 minute pump check for proper operation and adjust as required, start pump without foam discharger.	.216		√	√	√	√
8 Centrifugal pump:						
A) check for leaks on suction and discharge piping, seals, packing glands, etc.	.077			√	√	√
B) check pump operation; vibration, noise, overheating, etc.	.022			√	√	√
C) check alignment, clearances, and rotation of shaft and coupler (includes removing and reinstalling safety cover).	.160			√	√	√
D) tighten or replace loose, missing or damaged nuts, bolts, or screws.	.005			√	√	√
E) lubricate pump as required.	.099			√	√	√
F) check suction or discharge pressure gauge readings and flow rate.	.078			√	√	√
G) check packing glands and tighten or repack as required; note that slight dripping is required for proper lubrication of shaft.	.113			√	√	√
9 Inspect and clean strainers after each use and flow test.	.262		√	√	√	√
10 Clean equipment and surrounding area.	.432		√	√	√	√
11 Fill out maintenance checklist.	.022		√	√	√	√
Total labor-hours/period			1.399	2.473	2.473	3.554
Total labor-hours/year			11.192	4.946	2.473	3.554

	Description	Labor-hrs.	Cost Each					
			2020 Bare Costs				Total	Total
			Material	Labor	Equip.	Total	In-House	w/O&P
1900	Extinguishing system, foam, diesel pump, annually	3.554	320	195		515	604.75	710
1950	Annualized	22.156	880	1,225		2,105	2,532.74	3,050

	PM Components	Labor-hrs.	W	M	Q	S	A
	PM System D4095 400 1950 **Extinguishing system, dry chemical**						
1	Check nozzles, heads, and fusible links or heat detectors to ensure that they are clean, have adequate clearance from obstructions and have not been damaged.	.098		√	√	√	√
2	Check hand hose lines, if applicable, to ensure that they are clear, in proper position and in good condition.	.098		√	√	√	√
3	Check to ensure that all operating controls are properly set.	.066		√	√	√	√
4	Observe pressure on system to ensure that proper pressure is maintained.	.004		√	√	√	√
5	Check expellant gas cylinders of gas cartridge systems for proper pressure or weight to ensure that sufficient gas is available.	.258				√	√
6	Test all actuating and operating devices; turn valve on dry chemical bottle or cylinder to off position in accordance with manufacturers specifications, and test operation of control head by removing fusible link, activating heat detectors or pulling test control. Restore system to normal operation and make minor adjustments as applicable.	.133					√
7	Open and check dry chemical in cylinder and stored pressure systems to ensure it is free flowing and without lumps.	.327					√
8	When an alarm system is reactivated, ensure that it functions in accordance with manufacturers specifications.	.079					√
9	Clean area around system.	.066		√	√	√	√
10	Fill out maintenance checklist and report deficiencies.	.022		√	√	√	√
	Total labor-hours/period			.355	.355	.613	1.451
	Total labor-hours/year			2.836	.709	.613	1.451

	Description	Labor-hrs.	Cost Each					
			2020 Bare Costs				Total	Total
			Material	Labor	Equip.	Total	In-House	w/O&P
1900	Extinguishing system, dry chemical, annually	1.451	52	80		132	158.98	192
1950	Annualized	5.606	104	310		414	507.98	620

PM Components	Labor-hrs.	W	M	Q	S	A
PM System D4095 450 1950						
Extinguishing system, FM200						
1 Check that nozzles, heads and hand hose lines are clear from obstructions, have not been damaged and in proper position.	.098		√	√	√	√
2 Check to ensure that all operating controls are properly set.	.066		√	√	√	√
3 Clean nozzles as required.	.098		√	√	√	√
4 Visually inspect the control panel for obstructions or physical damage; clean dirt and dust from interior and exterior; make sure that all cards are plugged in tightly, tighten loose connections and make other minor adjustments as necessary.	.222		√	√	√	√
5 Check battery voltages where installed, recharge or replace as required.	.012		√	√	√	√
6 Weigh cylinders; replace any that show a weight loss greater than 5% or a pressure loss greater than 10%.	.640				√	√
7 Blow out entire system to make sure that the system is not plugged or restricted.	.333					√
8 Conduct operational test of actuating devices, both automatic, manual and alarm system; prior to conducting tests, close valves disable CO_2 dumping control and manually override computer shutdown feature if applicable, all in accordance with manufacturers specifications.	.432					√
9 Check the operation and timing of the time delay control and check operation of abort station.	.108					√
10 Restore system to proper operating condition and notify cognizant personnel upon completion of tests.	.079					√
11 Clean out around system.	.066		√	√	√	√
12 Fill out maintenance checklist and report deficiencies.	.022		√	√	√	√
Total labor-hours/period			.585	.585	1.225	2.178
Total labor-hours/year			4.678	1.169	1.225	2.178

	Description	Labor-hrs.	Cost Each					
			2020 Bare Costs				Total	Total
			Material	Labor	Equip.	Total	In-House	w/O&P
1900	Extinguishing system, FM200, annually	2.178	176	119		295	346.95	410
1950	Annualized	9.261	650	510		1,160	1,367.70	1,625

D5015 210 Switchboard, Electrical

PM Components	Labor-hrs.	W	M	Q	S	A
PM System D5015 210 1950						
Switchboard, electrical						
1 Check with operating or area personnel for deficiencies.	.044					√
2 Check indicating lamps for proper operation, if appropriate; replace burned out lamps.	.018					√
3 Remove and reinstall cover.	.195					√
4 Check for discolorations, hot spots, odors and charred insulation.	.356					√
5 Clean switchboard exterior and surrounding area.	.066					√
6 Fill out maintenance checklist and report deficiencies.	.022					√
Total labor-hours/period						.701
Total labor-hours/year						.701

	Description	Labor-hrs.	Cost Each					
			2020 Bare Costs				Total	Total
			Material	Labor	Equip.	Total	In-House	w/O&P
1900	Switchboard, electrical, annually	.701	3.79	43		46.79	57.21	70.50
1950	Annualized	.701	3.79	43		46.79	57.21	70.50

D5015 214 Switchboard, w/Air Circ. Breaker

PM Components	Labor-hrs.	W	M	Q	S	A
PM System D5015 214 1950						
Switchboard w/air circuit breaker						
1 Check with operating or area personnel for deficiencies.	.035					√
2 Check indicating lamps for proper operation, if appropriate; replace burned out lamps.	.020					√
3 Remove switchboard from service, coordinate with appropriate authority.	.038					√
4 Remove and reinstall cover.	.051					√
5 Check for discolorations, hot spots, odors and charred insulation.	.016					√
6 Check the records on the circuit breaker for any full current rated trips which may indicate a potential problem.	.020					√
7 Examine the exterior of the switchboard for any damage or moisture.	.020					√
8 Check operation of circuit breaker tripping mechanism and relays.	.096					√
9 Check the operation of the circuit breaker charging motor.	.020					√
10 Check hardware for tightness.	.120					√
11 Check interlock for proper operation.	.025					√
12 Check that the charger is operating properly.	.022		√	√	√	√
13 Check for any signs of corrosion on the battery terminals or wires.	.020		√	√	√	√
14 Check the electrolyte level in the batteries; add if required.	.624		√	√	√	√
15 Check the specific gravity of the electrolyte in a 10% sample of the batteries.	.069		√	√	√	√
16 Check 25% of terminal-to-cell connection resistance; rehabilitate connections as required; add anti-corrosion grease to battery terminals and connections.	.113					√
17 Measure and record individual cell and string float voltages.	2.808					√
18 Place switchboard back in service, notify appropriate authority.	.057					√
19 Clean switchboard, charger and the surrounding area.	.066		√	√	√	√
20 Fill out maintenance checklist and report deficiencies.	.022		√	√	√	√
Total labor-hours/period			.823	.823	.823	4.262
Total labor-hours/year			6.585	1.646	.823	4.262

	Description	Labor-hrs.	Cost Each					
			2020 Bare Costs				Total	Total
			Material	Labor	Equip.	Total	In-House	w/O&P
1900	Switchboard, with air circuit breaker, annually	4.262	16.40	261		277.40	339.98	420
1950	Annualized	13.318	16.40	815		831.40	1,022.36	1,275

PM Components	Labor-hrs.	W	M	Q	S	A
PM System D5015 217 1950						
Switchboard w/air circuit breaker and tie switch						
1 Check with operating or area personnel for deficiencies.	.035					√
2 Check indicating lamps for proper operation, if appropriate; replace burned out lamps.	.020					√
3 Remove switchboard from service, coordinate with appropriate authority.	.038					√
4 Remove and reinstall cover.	.051					√
5 Check for discolorations, hot spots, odors and charred insulation.	.016					√
6 Check the records on the circuit breaker for any full current rated trips which may indicate a potential problem.	.020					√
7 Examine the exterior of the switchboard for any damage or moisture.	.020					√
8 Check operation of circuit breaker tripping mechanism and relays.	.096					√
9 Check the operation of the circuit breaker charging motor.	.020					√
10 Check hardware for tightness.	.120					√
11 Check interlock for proper operation.	.013					√
12 Operate the switching handle and check the locking mechanism.	.052					√
13 Check that the charger is operating properly.	.022		√	√	√	√
14 Check for any signs of corrosion on the battery terminals or wires.	.020		√	√	√	√
15 Check the electrolyte level in the batteries; add if required.	.624		√	√	√	√
16 Check the specific gravity of the electrolyte in a 10% sample of the batteries.	.069		√	√	√	√
17 Check 25% of terminal-to-cell connection resistance; rehabilitate connections as required; add anti-corrosion grease to battery terminals and connections.	.113					√
18 Measure and record individual cell and string float voltages.	2.808					√
19 Place switchboard back in service, notify appropriate authority.	.057					√
20 Clean switchboard, charger and the surrounding area.	.066		√	√	√	√
21 Fill out maintenance checklist and report deficiencies.	.022		√	√	√	√
Total labor-hours/period			.823	.823	.823	4.302
Total labor-hours/year			6.585	1.646	.823	4.302

	Description	Labor-hrs.	Cost Each					
			2020 Bare Costs				Total	Total
			Material	Labor	Equip.	Total	In-House	w/O&P
1900	Switchboard, with air C.B. and tie switch, annually	4.302	20	264		284	347.12	430
1950	Annualized	13.358	20	820		840	1,029.50	1,275

D5015 220 Circuit Breaker, HV Air

PM Components	Labor-hrs.	W	M	Q	S	A
PM System D5015 220 1950						
Circuit breaker, HV air						
1 Check the records on the circuit breaker for any full current rated trips which may indicate a potential problem.	.066					√
2 Examine the exterior of the circuit breaker for any damage or moisture.	.060					√
3 Remove the circuit breaker from service, coordinate with appropriate authority.	.038					√
4 Check operation of tripping mechanisms and relays.	.096					√
5 Check the operation of the charging motor, if applicable.	.038					√
6 Check hardware for tightness.	[illegible]					√
7 Check interlock for proper operation.	[illegible]					√
8 Clean area around switch.	.066					√
9 Place the circuit breaker back in service, coordinate with appropriate authority.	.040					√
10 Fill out maintenance checklist.	.022					√
Total labor-hours/period						.470
Total labor-hours/year						.470

	Description	Labor-hrs.	Cost Each					
			2020 Bare Costs				Total	Total
			Material	Labor	Equip.	Total	In-House	w/O&P
1900	Circuit breaker, high voltage air, annually	.470	14.90	29		43.90	51.90	63
1950	Annualized	.470	14.90	29		43.90	51.90	63

D50 ELECTRICAL

D5015 222 Circuit Breaker, HV Oil

PM Components	Labor-hrs.	W	M	Q	S	A
PM System D5015 222 1950						
Circuit breaker, HV oil						
1 Check the records on the circuit breaker for any full current rated trips which may indicate a potential problem.	.066					√
2 Examine the exterior of the circuit breaker for any damage, cracks, rust or leaks; check around the gaskets and the drain valves.	.074					√
3 Check the entrance bushings for cracks in the porcelain.	.320					√
4 Check the condition of the ground system.	.170					√
5 Draw oil sample from bottom of circuit breaker; have sample tested; replace removed oil to the proper level.	.042					√
6 Clean the exterior of the circuit breaker and operating mechanism.	.094					√
7 Clean the area around the circuit breaker.	.066					√
8 Fill out maintenance checklist.	.022					√
Total labor-hours/period						.855
Total labor-hours/year						.855

	Description	Labor-hrs.	Cost Each					
			2020 Bare Costs				Total	Total
			Material	Labor	Equip.	Total	In-House	w/O&P
1900	Circuit breaker, high voltage oil, annually	.855	33.50	52.50		86	101.19	122
1950	Annualized	.855	33.50	52.50		86	101.19	122

D5015 230 Switch, Selector, HV Air

PM Components	Labor-hrs.	W	M	Q	S	A
PM System D5015 230 1950						
Switch, selector, HV, air						
1 Examine the exterior of the selector switch for any damage.	.060					√
2 Check that the switching handle is locked.	.044					√
3 Check the condition of the ground system.	.170					√
4 Clean the exterior of the selector switch and operating mechanism.	.046					√
5 Clean the area around the selector switch.	.066					√
6 Fill out maintenance checklist and report deficiencies.	.022					√
Total labor-hours/period						.408
Total labor-hours/year						.408

	Description	Labor-hrs.	Cost Each: 2020 Bare Costs: Material	Labor	Equip.	Total	Total In-House	Total w/O&P
1900	Switch, selector, high voltage, air, annually	.408	14.90	25		39.90	47.20	57
1950	Annualized	.408	14.90	25		39.90	47.20	57

D5015 232 Switch, Selector, HV Oil

PM Components	Labor-hrs.	W	M	Q	S	A
PM System D5015 232 1950						
Switch, selector, HV, oil						
1 Examine the exterior of the interrupt switch for any damage, cracks, rust or leaks; check around the gaskets and the drain valves.	.060					√
2 Check that the switching handle is locked.	.044					√
3 Check the condition of the ground system.	.170					√
4 Draw oil sample from bottom of selector switch; have sample tested; replace removed oil to the proper level.	.042					√
5 Clean the exterior of the selector switch and operating mechanism.	.046					√
6 Clean the area around the selector switch.	.066					√
7 Fill out maintenance checklist and report deficiencies.	.022					√
Total labor-hours/period						.450
Total labor-hours/year						.450

	Description	Labor-hrs.	Cost Each: 2020 Bare Costs: Material	Labor	Equip.	Total	Total In-House	Total w/O&P
1900	Switch, selector, high voltage, oil, annually	.450	33.50	27.50		61	70.82	84.50
1950	Annualized	.450	33.50	27.50		61	70.82	84.50

D5015 234 Switch, Automatic Transfer

PM Components	Labor-hrs.	W	M	Q	S	A
PM System D5015 234 1950						
Switch, automatic transfer						
1 Check with operating or area personnel for deficiencies.	.044		√	√	√	√
2 Inspect wiring, wiring connections and fuse blocks for looseness, charring evidence of short circuiting, overheating and tighten all connections.	.263		√	√	√	√
3 Inspect gen. cond. of transf. switch and clean exter. and surrounding area.	.114		√	√	√	√
4 Fill out maintenance checklist and report deficiencies.	.022		√	√	√	√
Total labor-hours/period			.443	.443	.443	.443
Total labor-hours/year			3.546	.886	.443	.443

	Description	Labor-hrs.	Cost Each					
			2020 Bare Costs				Total	Total
			Material	Labor	Equip.	Total	In-House	w/O&P
1900	Switch, automatic transfer, annually	.443	14.90	27		41.90	49.82	60.50
1950	Annualized	5.316	14.90	325		339.90	416.90	520

D50 ELECTRICAL

D5015 236 Switch, Interrupt, HV, Fused Air

PM Components	Labor-hrs.	W	M	Q	S	A
PM System D5015 236 1950						
Switch, interrupt, HV, fused air						
1 Examine the exterior of the interrupt switch for any damage.	.060					√
2 Check the condition of the ground system.	.170					√
3 Clean the exterior of the interrupt switch and operating mechanism.	.046					√
4 Clean the area around the interrupt switch.	.066					√
5 Fill out maintenance checklist and report deficiencies.	.022					√
Total labor-hours/period						.364
Total labor-hours/year						.364

	Description	Labor-hrs.	Cost Each					
			2020 Bare Costs				Total	Total
			Material	Labor	Equip.	Total	In-House	w/O&P
1900	Switch, interrupt, high voltage, fused air, annually	.364	14.90	22.50		37.40	43.89	53
1950	Annualized	.364	14.90	22.50		37.40	43.89	53

D50 ELECTRICAL

D5015 238 Switch, HV w/Aux Fuses, Air

PM Components	Labor-hrs.	W	M	Q	S	A
PM System D5015 238 1950						
Switch, interrupt, HV, w/auxiliary fuses, air						
1 Examine the exterior of the interrupt switch for any damage.	.060					√
2 Check the condition of the ground system.	.170					√
3 Clean the exterior of the interrupt switch and operating mechanism.	.046					√
4 Clean the area around the interrupt switch.	.066					√
5 Fill out maintenance checklist and report deficiencies.	.022					√
Total labor-hours/period						.364
Total labor-hours/year						.364

	Description	Labor-hrs.	Cost Each					
			2020 Bare Costs				Total	Total
			Material	Labor	Equip.	Total	In-House	w/O&P
1900	Switch, interrupt, HV, w/ aux fuses, air, annually	.364	14.90	22.50		37.40	43.89	53
1950	Annualized	.364	14.90	22.50		37.40	43.89	53

D5015 240 Transformer

PM Components	Labor-hrs.	W	M	Q	S	A
PM System D5015 240 1950						
Transformer, dry type 500 KVA and over						
1 Examine the exterior of the transformer for any damage.	.060					√
2 Remove the covers or open the doors.	.195					√
3 Check for signs of moisture or overheating.	.059					√
4 Check for voltage creeping over insulated surfaces, such as evidenced by tracking or carbonization.	.018					√
5 Check fans, motors and other auxiliary devices for proper operation; where applicable.	.049					√
6 Check the condition of the ground system.	.170					√
7 Replace the covers or close the doors.	.195					√
8 Fill out maintenance checklist and report deficiencies.	.022					√
Total labor-hours/period						.769
Total labor-hours/year						.769

	Description	Labor-hrs.	Cost Each					
			2020 Bare Costs				Total	Total
			Material	Labor	Equip.	Total	In-House	w/O&P
1900	Transformer, dry type 500 KVA and over, annually	.769	14.90	47		61.90	74.25	91
1950	Annualized	.769	14.90	47		61.90	74.25	91

PM Components	Labor-hrs.	W	M	Q	S	A
PM System D5015 240 2950						
Transformer, oil, pad mounted						
1 Examine the exterior of the transformer for any damage, cracks, rust or leaks; check around bushings, gaskets and pressure relief device.	.014					√
2 Check the condition of the ground system.	.170					√
3 Check and record the oil level, pressure and temperature readings.	.263					√
4 Check fans, motors and other auxiliary devices for proper operation; where applicable.	.022					√
5 Draw oil sample from top of transformer and have sample tested for dielectric strength; replace removed oil.	.525					√
6 Clean transformer exterior and the surrounding area.	.066					√
7 Fill out maintenance checklist and report deficiencies.	.022					√
Total labor-hours/period						1.082
Total labor-hours/year						1.082

	Description	Labor-hrs.	Cost Each					
			2020 Bare Costs				Total	Total
			Material	Labor	Equip.	Total	In-House	w/O&P
2900	Transformer, oil, pad mounted, annually	1.082	22	66		88	105.51	130
2950	Annualized	1.082	22	66		88	105.51	130

PM Components	Labor-hrs.	W	M	Q	S	A
PM System D5015 240 3950						
Transformer, oil, pad mounted, PCB						
1 Check with operating or area personnel for deficiencies.	.040			√	√	√
2 Examine the exterior of the transformer for any damage, cracks, rust or leaks; check around the gaskets, the pressure relief device and the cooling fins.	.014			√	√	√
3 Check and record the oil levels, pressure and temperature readings: level ____; pressure ____; temperature ____.	.263					√
4 Check fans, motors and other auxiliary devices for proper operation, where applicable.	.061					√
5 Check the exterior bushings for leaking oil.	.321			√	√	√
6 Check the condition of the ground system.	.170					√
7 Draw oil sample from top of transformer and have sample tested; replace removed oil.	.525					√
8 Clean transformer exterior and the surrounding area.	.066					√
9 Verify the PCB warning markings are intact and clearly visible.	.022					√
10 Fill out maintenance checklist and report deficiencies. Note: the EPA PCB regulations under 40 CFR 716.30 shall apply to all inspections, samples, disposal, cleanups, decontamination, reporting and record-keeping.	.022			√	√	√
Total labor-hours/period				.397	.397	1.494
Total labor-hours/year				.794	.397	1.494

	Description	Labor-hrs.	Cost Each					
			2020 Bare Costs				Total	Total
			Material	Labor	Equip.	Total	In-House	w/O&P
3900	Transformer, oil, pad mounted, PCB, annually	1.494	66.50	91.50		158	185.88	224
3950	Annualized	2.685	66.50	165		231.50	276.06	335

D50 ELECTRICAL | D5015 260 | Panelboard, 225 Amps and Above

PM Components	Labor-hrs.	W	M	Q	S	A
PM System D5015 260 1950						
Panelboard, 225 amps and above						
1 Check with operating or area personnel for deficiencies.	.066					√
2 Check for excessive heat, odors, noise, and vibration.	.239					√
3 Clean and check general condition of panel.	.114					√
4 Fill out maintenance checklist and report deficiencies.	.022					√
Total labor-hours/period						.441
Total labor-hours/year						.441

	Description	Labor-hrs.	Cost Each					
			2020 Bare Costs				Total	Total
			Material	Labor	Equip.	Total	In-House	w/O&P
1900	Panelboard, 225 A and above, annually	.441	22.50	27		49.50	58.15	69.50
1950	Annualized	.441	22.50	27		49.50	58.15	69.50

D50 ELECTRICAL | D5015 280 | Motor Control Center

PM Components	Labor-hrs.	W	M	Q	S	A
PM System D5015 280 1950						
Motor control center, over 400 amps						
1 Check with operating or area personnel for deficiencies.	.044					√
2 Check starter lights, replace if required.	.018					√
3 Check for excessive heat, odors, noise, and vibration.	.239					√
4 Clean motor control center exterior and surrounding area.	.066					√
5 Fill out maintenance checklist and report deficiencies.	.022					√
Total labor-hours/period						.389
Total labor-hours/year						.389

	Description	Labor-hrs.	Cost Each					
			2020 Bare Costs				Total	Total
			Material	Labor	Equip.	Total	In-House	w/O&P
1900	Motor control center, over 400 A, annually	.389	22.50	24		46.50	54.17	64.50
1950	Annualized	.389	22.50	24		46.50	54.17	64.50

PM Components	Labor-hrs.	W	M	Q	S	A
PM System D5035 610 1950						
Clocks, Central System						
1 Clean dirt and dust from interior and exterior of cabinet.	.280				√	√
2 Adjust relays and check transmission of signal.	.117				√	√
3 Tighten contacts and terminal screws.	.065				√	√
4 Burnish contacts.	.195				√	√
5 Fill out maintenance checklist and report deficiencies.	.022				√	√
Total labor-hours/period					.659	.659
Total labor-hours/year					.659	.659

	Description	Labor-hrs.	Cost Each					
			2020 Bare Costs				Total	Total
			Material	Labor	Equip.	Total	In-House	w/O&P
1000	Central clock systems, annually	.659	5.95	36		41.95	53.00	65
1950	Annualized	1.318	11.90	72.50		84.40	105.70	130

D50 ELECTRICAL

D5035 710 Fire Alarm Annunciator System

PM Components	Labor-hrs.	W	M	Q	S	A
PM System D5035 710 1950						
Fire alarm annunciator system						
1 Visually inspect all alarm equipment for obstructions or physical damage, clean dirt and dust from interior and exterior of panel/pull boxes, tighten loose connections.	.222		√	√	√	√
2 Notify cognizant personnel prior to testing.	.112		√	√	√	√
3 Conduct operational test of initiating and signal transmitting devices in populated buildings by building zone/area; for those circuits which do not operate properly, check detectors, control units, and annunciators for dust on defective components; make minor adjustments as required.	.112		√	√	√	√
4 Check battery voltages where installed; replace as required.	.012		√	√	√	√
5 Conduct operational test of 10% of total number of spot type heat detectors and all smoke detectors; for those circuits which do not operate properly, check to determine if problem relates to circuit, device, or control unit, and make minor adjustments as required; if detector is defective but no replacement is immediately available, remove detector, re-establish Initiating circuit and tag location until a replacement detector is installed.	2.172				√	√
6 Restore system to proper operating condition and notify personnel upon completion of tests.	.079		√	√	√	√
7 Fill out maintenance checklist and report deficiencies.	.022		√	√	√	√
Total labor-hours/period			.559	.559	2.731	2.731
Total labor-hours/year			4.474	1.119	2.731	2.731

	Description	Labor-hrs.	Cost Each					
			2020 Bare Costs				Total	Total
			Material	Labor	Equip.	Total	In-House	w/O&P
1900	Fire alarm annunciator system, annually	2.731	165	150		315	373.85	445
1950	Annualized	11.051	174	605		779	968.05	1,175

D50 ELECTRICAL

D5035 810 Security/Intrusion Alarm Ctl. Panel

PM Components	Labor-hrs.	W	M	Q	S	A
PM System D5035 810 1950						
Security / intrusion alarm system						
1 Check in and out with area security officer, notify operating/facility personnel, obtain necessary alarm keys, alarm codes and escort when required.	.520			√	√	√
2 Inspect alarm control panel and conduct operational test of initiating and signal transmitting devices; make minor adjustments as required.	.241			√	√	√
3 Check indicating lamps for proper operation, replace if necessary.	.018			√	√	√
4 Check battery voltages where installed, replace as required.	.012			√	√	√
5 Restore system to proper operating condition and notify personnel upon completion of tests.	.079			√	√	√
6 Clean exterior of cabinet and surrounding area.	.066			√	√	√
7 Fill out maintenance checklist and report deficiencies.	.022			√	√	√
Total labor-hours/period				.958	.958	.958
Total labor-hours/year				1.916	.958	.958

	Description	Labor-hrs.	Cost Each					
			2020 Bare Costs				Total	Total
			Material	Labor	Equip.	Total	In-House	w/O&P
1900	Security, intrusion alarm system, annually	.958	158	52.50		210.50	240.80	281
1950	Annualized	3.832	158	210		368	442.15	535

	PM Components	Labor-hrs.	W	M	Q	S	A
	PM System D5095 210 1950						
	Emergency generator up to 15 KVA						
1	Check with the operating or area personnel for any obvious deficiencies.	.044		√	√	√	√
2	Check engine oil level; add as required.	.010		√	√	√	√
3	Change engine oil and oil filter.	.170					√
4	Check battery charge and electrolyte specific gravity, add water as required; check terminals for corrosion, clean as required.	.120		√	√	√	√
5	Check belt tension and wear; adjust as required, if applicable.	.012		√	√	√	√
6	Check engine air filter; change as required.	.042					√
7	Check spark plug or injector nozzle condition; service or replace as required.	.211					√
8	Check wiring, connections, switches, etc.; adjust as required.	.036		√	√	√	√
9	Perform 30 minute generator test run; check for proper operation.	.650		√	√	√	√
10	Check fuel level; add as required.	.046		√	√	√	√
11	Wipe dust and dirt from engine and generator.	.056		√	√	√	√
12	Clean area around generator.	.066		√	√	√	√
13	Fill out maintenance checklist and report deficiencies.	.022		√	√	√	√
	Total labor-hours/period			1.062	1.062	1.062	1.486
	Total labor-hours/year			8.499	2.125	1.062	1.486

	Description	Labor-hrs.	Cost Each					
			2020 Bare Costs				Total	Total
			Material	Labor	Equip.	Total	In-House	w/O&P
1900	Emergency diesel generator, up to 15 KVA, annually	1.486	123	81.50		204.50	240.55	284
1950	Annualized	13.164	114	725		839	1,053.20	1,300

	PM Components	Labor-hrs.	W	M	Q	S	A
	PM System D5095 210 2950						
	Emergency diesel generator, over 15 KVA						
1	Check with the operating or area personnel for any obvious deficiencies.	.044		√	√	√	√
2	Check engine oil level; add as required.	.010		√	√	√	√
3	Change engine oil and oil filter.	.506					√
4	Check battery charge and electrolyte specific gravity, add water as required; check terminals for corrosion, clean as required.	.241		√	√	√	√
5	Check belts for wear and proper tension; adjust as necessary.	.012		√	√	√	√
6	Check that crank case heater is operating.	.038		√	√	√	√
7	Check engine air filter; change as required.	.042					√
8	Check wiring, connections, switches, etc.; adjust as required.	.036		√	√	√	√
9	Check spark plug or injector nozzle condition; service or replace as required.	.281					√
10	Perform 30 minute generator test run; check for proper operation.	.650		√	√	√	√
11	Check fuel level with gauge pole, add as required.	.046		√	√	√	√
12	Wipe dust and dirt from engine and generator.	.109		√	√	√	√
13	Clean area around generator.	.066		√	√	√	√
14	Fill out maintenance checklist and report deficiencies.	.025		√	√	√	√
	Total labor-hours/period			1.277	1.277	1.277	2.106
	Total labor-hours/year			10.216	2.554	1.277	2.106

	Description	Labor-hrs.	Cost Each					
			2020 Bare Costs				Total	Total
			Material	Labor	Equip.	Total	In-House	w/O&P
2900	Emergency diesel generator, over 15 KVA, annually	2.106	134	116		250	295.72	350
2950	Annualized	16.150	134	890		1,024	1,284.65	1,575

PM Components	Labor-hrs.	W	M	Q	S	A
PM System D5095 210 3950						
Emergency diesel generator, turbine						
1 Check with the operating or area personnel for any obvious deficiencies.	.044		√	√	√	√
2 Check turbine oil level; add oil as required.	.036		√	√	√	√
3 Change turbine oil and oil filter; check transmission oil level.	.569					√
4 Check that the crankcase heater is operating properly.	.022		√	√	√	√
5 Replace turbine air filter.	.224					√
6 Check wiring, connections, switches, etc.; adjust as required.	.060					√
7 Check starter for proper operation; lubricate as necessary.	.047		√	√	√	√
8 Check fuel nozzles, fuel regulator and ignition device condition; service or replace as required.	.813					√
9 Perform 30 minute generator test run; check for proper operation.	7.797		√	√	√	√
10 Check and record transmission oil pressure and temperature, and natural gas pressure: oil press:_____ oil temp:_____ gas press:_____.	.021					√
11 Record running time: beginning _____ hours; ending _____ hours.	.010					√
12 Check that the charger is operating properly.	.127		√	√	√	√
13 Check for any signs of corrosion on battery terminals or wires.	.117		√	√	√	√
14 Check the electrolyte level in the batteries; add if required.	.049		√	√	√	√
15 Check the specific gravity of the electrolyte in a 10% sample of the batteries.	.069					√
16 Check 25% of terminal-to-cell connection resistance; rehabilitate connections as required; add anti-corrosion grease to battery terminals and connections.	.038					√
17 Measure and record individual cell and string float voltages.	.140					√
18 Clean area around generator.	.066		√	√	√	√
19 Fill out maintenance checklist and report deficiencies.	.022		√	√	√	√
Total labor-hours/period			8.327	8.327	8.327	10.271
Total labor-hours/year			66.618	16.654	8.327	10.271

	Description	Labor-hrs.	Cost Each					
			2020 Bare Costs				Total	Total
			Material	Labor	Equip.	Total	In-House	w/O&P
3900	Emergency diesel generator, turbine, annually	10.271	170	565		735	911.27	1,100
3950	Annualized	16.633	155	915		1,070	1,341.42	1,650

D5095 220 Power Stabilizer

PM Components	Labor-hrs.	W	M	Q	S	A
PM System D5095 220 1950						
Power stabilizer						
1 Remove the covers or open the doors.	[illegible]					√
2 Check for signs of moisture or overheating.	.022					√
3 Check fans and other auxiliary devices for proper operation.	.027					√
4 Check for voltage creeping over transformer insulated surfaces, such as evidenced by tracking or carbonization.	.018					√
5 Check condition of the ground system.	.170					√
6 Check wiring connections, relays, etc.; tighten and adjust as necessary.	.060					√
7 Replace the covers or close the doors.	.022					√
8 Clean the surrounding area.	[illegible]					√
9 Fill out maintenance checklist and report deficiencies.	.022					√
Total labor-hours/period						.625
Total labor-hours/year						.625

	Description	Labor-hrs.	Cost Each					
			2020 Bare Costs				Total	Total
			Material	Labor	Equip.	Total	In-House	w/O&P
1900	Power stabilizer, annually	.625	9.05	34.50		43.55	53.94	66
1950	Annualized	.625	9.05	34.50		43.55	53.94	66

D5095 230 Uninterruptible Power System

PM Components	Labor-hrs.	W	M	Q	S	A
PM System D5095 230 1950						
Uninterruptible power system, up to 200 KVA						
1 Check with operating or area personnel for any obvious deficiencies.	.044		√	√	√	√
2 Check electrolyte level of batteries; add water as required; check terminals for corrosion, clean as required.	.878		√	√	√	√
3 Check 25% of the batteries for charge and electrolyte specific gravity.	.777			√	√	√
4 Check batteries for cracks or leaks.	.059		√	√	√	√
5 Check 25% of the terminal-to-cell connection resistances; rehabilitate connections as required; add anti-corrosion grease to battery terminals and connections.	.143			√	√	√
6 Measure and record individual cell and string float voltages.	.618			√	√	√
7 Check integrity of battery rack.	.176		√	√	√	√
8 Check battery room temperature and ventilation systems.	.010		√	√	√	√
9 Replace air filters on UPS modules.	.036		√	√	√	√
10 Check output voltage and amperages, from control panel.	.007		√	√	√	√
11 Check UPS room temperature and ventilation system.	.062		√	√	√	√
12 Notify personnel and test UPS fault alarm system.	.112			√	√	√
13 Clean around batteries and UPS modules.	.066		√	√	√	√
14 Fill out maintenance checklist and report deficiencies.	.022		√	√	√	√
Total labor-hours/period			1.360	3.010	3.010	3.010
Total labor-hours/year			10.880	6.020	3.010	3.010

	Description	Labor-hrs.	Cost Each					
			2020 Bare Costs				Total	Total
			Material	Labor	Equip.	Total	In-House	w/O&P
1900	Uninter. power system, up to 200 KVA, annually	3.010	191	165		356	421.75	500
1950	Annualized	22.924	272	1,250		1,522	1,911.40	2,350

PM Components	Labor-hrs.	W	M	Q	S	A
PM System D5095 230 2950						
Uninterrupted power system, 200 KVA to 800 KVA						
1 Check with operating or area personnel for any obvious deficiencies.	.044		√	√	√	√
2 Check electrolyte level of batteries; add water as required; check terminals for corrosion, clean as required.	3.510		√	√	√	√
3 Check 25% of the batteries for charge and electrolyte specific gravity.	3.200			√	√	√
4 Check batteries for cracks or leaks.	.593		√	√	√	√
5 Check 25% of the terminal-to-cell connection resistances; rehabilitate connections as required; add anti-corrosion grease to battery terminals and connections.	.234			√	√	√
6 Measure and record individual cell and string float voltages.	.941			√	√	√
7 Check integrity of battery rack.	.696		√	√	√	√
8 Check battery room temperature and ventilation systems.	.010		√	√	√	√
9 Replace air filters on UPS modules.	.036		√	√	√	√
10 Check output voltage and amperages, from control panel.	.007		√	√	√	√
11 Check UPS room temperature and ventilation system.	.062		√	√	√	√
12 Notify personnel and test UPS fault alarm system.	.112			√	√	√
13 Clean around batteries and UPS modules.	.066		√	√	√	√
14 Fill out maintenance checklist and report deficiencies.	.022		√	√	√	√
Total labor-hours/period			5.046	9.533	9.533	9.533
Total labor-hours/year			40.365	19.065	9.533	9.533

	Description	Labor-hrs.	Cost Each					
			2020 Bare Costs				Total	Total
			Material	Labor	Equip.	Total	In-House	w/O&P
2900	UPS, 200 KVA to 800 KVA, annually	9.533	191	525		716	880.19	1,075
2950	Annualized	78.058	282	4,275		4,557	5,779.40	7,150

D5095 240 Wet Battery System and Charger

PM Components	Labor-hrs.	W	M	Q	S	A
PM System D5095 240 1950						
Battery system and charger						
1 Check that the charger is operating properly.	.061		√	√	√	√
2 Check for any signs of corrosion on the battery terminals or wires.	.117		√	√	√	√
3 Check the electrolyte level in the batteries; add if required.	.062		√	√	√	√
4 Check the specific gravity of the electrolyte in a 10% sample of the batteries.	.069		√	√	√	√
5 Check 25% of terminal-to-cell connection resistance; rehabilitate connections as required; add anti corrosion grease to battery terminals and connections.	.741			√	√	√
6 Measure and record individual cell and string float voltages.	.281			√	√	√
7 Clean area around batteries.	.066		√	√	√	√
8 Fill out maintenance checklist and report deficiencies.	.012		√	√	√	√
Total labor-hours/period			.387	1.409	1.409	1.409
Total labor-hours/year			3.097	2.818	1.409	1.409

	Description	Labor-hrs.	Cost Each					
			2020 Bare Costs				Total	Total
			Material	Labor	Equip.	Total	In-House	w/O&P
1900	Battery system and charger, annually	1.409	12.60	77		89.60	112.74	139
1950	Annualized	8.732	19.20	480		499.20	636.55	790

PM Components	Labor-hrs.	W	M	Q	S	A
PM System D5095 250 1950						
Light, emergency, hardwired system						
1 Test for proper operation, replace lamps as required.	.013				√	√
2 Clean exterior of cabinet and light heads.	.025				√	√
3 Check wiring for obvious defects.	.026				√	√
4 Adjust lamp heads for maximum illumination of area.	.039				√	√
5 Fill out maintenance checklist and report deficiencies.	.022				√	√
Total labor-hours/period					.125	.125
Total labor-hours/year					.125	.125

	Description	Labor-hrs.	Cost Each					
			2020 Bare Costs				Total	Total
			Material	Labor	Equip.	Total	In-House	w/O&P
1900	Light, emergency, hardwired system, annually	.125	5.55	6.85		12.40	14.95	17.90
1950	Annualized	.250	7.35	13.70		21.05	25.71	31

PM Components	Labor-hrs.	W	M	Q	S	A
PM System D5095 250 2950						
Light, emergency, dry cell						
1 Test for proper operation, replace lamps as required.	.013				√	√
2 Clean exterior of emergency light cabinet and lamp heads.	.025				√	√
3 Check condition of batteries and change as required.	.012				√	√
4 Clean interior of cabinet, top of battery, and battery terminal.	.009				√	√
5 Inspect cabinet, relay, relay contacts, pilot light, wiring, and gen. cond.	.065				√	√
6 Adjust lamp heads for maximum illumination of area.	.039				√	√
7 Fill out maintenance checklist and report deficiencies.	.022				√	√
Total labor-hours/period					.185	.185
Total labor-hours/year					.185	.185

	Description	Labor-hrs.	Cost Each					
			2020 Bare Costs				Total	Total
			Material	Labor	Equip.	Total	In-House	w/O&P
2900	Light, emergency, dry cell, annually	.185	41.50	10.15		51.65	58.95	68
2950	Annualized	.356	33.50	19.60		53.10	62.04	73

PM Components	Labor-hrs.	W	M	Q	S	A
PM System D5095 250 3950						
Light, emergency, wet cell						
1 Test for proper operation; correct deficiencies as required.	.013				√	√
2 Clean exterior of emergency light cabinet and lamp heads.	.025				√	√
3 Clean interior of cabinet, top of battery, and battery terminal.	.009				√	√
4 Check battery charge and electrolyte specific gravity; add water if nec.	.012				√	√
5 Inspect cabinet, relay, relay contacts, pilot light, wiring, and gen. cond.	.065				√	√
6 Apply anti-corrosion coating to battery terminals.	.010				√	√
7 Adjust lamp heads for maximum illumination of area.	.039				√	√
8 Fill out maintenance checklist and report deficiencies.	.022				√	√
Total labor-hours/period					.195	.195
Total labor-hours/year					.195	.195

	Description	Labor-hrs.	Cost Each					
			2020 Bare Costs				Total	Total
			Material	Labor	Equip.	Total	In-House	w/O&P
3900	Light, emergency, wet cell, annually	.195	22	10.70		32.70	38.21	44.50
3950	Annualized	.390	33	21.50		54.50	63.88	75.50

PM Components	Labor-hrs.	W	M	Q	S	A
PM System E1015 810 1950						
Personal Computers						
1 Check with operating personnel for any deficiencies.	[illegible]				√	√
2 Check all cables for fraying, cracking, or exposed wires; ensure connections are secure.	.065				√	√
3 Remove metal casing and vacuum interior, power supply vents, keyboard, monitor, and surrounding work area.	.130				√	√
4 Check for secure connections of all interior wires and cables. Secure all peripheral cards.	.065				√	√
5 Replace PC cover and clean case, keyboard, monitor screen and housing with an anti static spray cleaner. Apply cleaner to towel and not the components.	.130				√	√
6 Disassemble the mouse; clean the tracking ball, interior rollers and housing. Blow out the interior of the mouse with clean compressed air.	.065				√	√
7 Turn on PC and check for unusual noises.	[illegible]				√	√
8 Clean floppy drive heads with cleaning diskette and solution.	.039				√	√
9 Defragment the hard drive using a defragmentation utility.	.234				√	√
10 Run diagnostics software to test all the PC's components.	.234				√	√
11 Fill out maintenance checklist and report deficiencies.	.029				√	√
Total labor-hours/period					1.082	1.082
Total labor-hours/year					1.082	1.082

	Description	Labor-hrs.	Cost Each					
			2020 Bare Costs				Total	Total
			Material	Labor	Equip.	Total	In-House	w/O&P
1900	Personal Computers, annually	1.082	3.13	59.50		62.63	79.53	98.50
1950	Annualized	2.164	6.30	118		124.30	159.14	197

E1035 100 Hydraulic Lift, Automotive

PM Components	Labor-hrs.	W	M	Q	S	A
PM System E1035 100 1950						
Hydraulic lift						
1 Check for proper operation of pump.	.035				√	√
2 Check for leaks on suction and discharge piping, seals, packing glands, etc.; make minor adjustments as required.	.077				√	√
3 Check pump and motor operation for excessive vibration, noise and overheating.	.022				√	√
4 Check alignment of pump and motor; adjust as necessary.	.258				√	√
5 Lubricate pump and motor.	.099				√	√
6 Inspect hydraulic lift post(s) for wear or leaks.	.065				√	√
7 Inspect, clean and tighten valves.	.034				√	√
8 Inspect and clean motor contactors.	.026				√	√
9 Inspect and test control relays and check wiring terminals.	.034				√	√
10 Clean pump unit and surrounding area.	.099				√	√
11 Fill out maintenance checklist and report deficiencies.	.022				√	√
Total labor-hours/period					.771	.771
Total labor-hours/year					.771	.771

	Description	Labor-hrs.	Cost Each					
			2020 Bare Costs				Total	Total
			Material	Labor	Equip.	Total	In-House	w/O&P
1900	Hydraulic lift, annually	.771	280	42		322	362.25	420
1950	Annualized	1.546	560	85		645	724.79	835

E10 EQUIPMENT

E1035 310 Hydraulic Lift, Loading Dock

PM Components	Labor-hrs.	W	M	Q	S	A
PM System E1035 310 1950						
Hydraulic lift, loading dock						
1 Check for proper operation of pump.	.022				√	√
2 Check for leaks on suction and discharge piping, seals, packing glands, etc.; make minor adjustments as required.	.077				√	√
3 Check pump and motor operation for excessive vibration, noise and overheating.	.022				√	√
4 Check alignment of pump and motor; adjust as necessary.	.100				√	√
5 Lubricate pump and motor.	.099				√	√
6 Inspect hydraulic lift post(s) for wear or leaks.	.051				√	√
7 Inspect, clean and tighten valves.	.072				√	√
8 Inspect and clean motor contactors.	.148				√	√
9 Inspect and test control relays and check wiring terminals.	.137				√	√
10 Clean pump unit and surrounding area.	.066				√	√
11 Fill out maintenance checklist and report deficiencies.	.022				√	√
Total labor-hours/period					.816	.816
Total labor-hours/year					.816	.816

	Description	Labor-hrs.	Cost Each					
			2020 Bare Costs				Total	Total
			Material	Labor	Equip.	Total	In-House	w/O&P
1900	Hydraulic lift, loading dock, annually	.816	41	45		86	102.39	123
1950	Annualized	1.632	61.50	89.50		151	182.65	220

PM Components	Labor-hrs.	W	M	Q	S	A
PM System E1095 122 1950						
Crane, electric bridge, up to 5 ton						
1 Check with operating or area personnel for deficiencies.	.035					√
2 Inspect crane for damaged structural members, alignment, worn pins, column fasteners, stops and capacity markings.	.235					√
3 Inspect and lubricate power cable tension reel and inspect condition of cable.	.485					√
4 Check hoist operation - travel function, up, down, speed stopping, upper and lower travel limit switches, etc.	.072					√
5 Inspect, clean and lubricate trolley assembly and reduction gear bearings.	.112					√
6 Inspect and clean hook assembly and chain or wire rope sheaves.	.457					√
7 Inspect, clean and lubricate hoist motor.	.094					√
8 Inspect hoist brake mechanism.	.444					√
9 Lubricate rails.	.114					√
10 Weight capacity test hoisting equipment at 125% of rated capacity in accordance with 29 CFR 1910.179.	.780					√
11 Clean work area.	.066					√
12 Fill out maintenance report and report deficiencies.	.022					√
Total labor-hours/period						2.917
Total labor-hours/year						2.917

	Description	Labor-hrs.	Cost Each					
			2020 Bare Costs				Total	Total
			Material	Labor	Equip.	Total	In-House	w/O&P
1900	Crane, electric bridge, up to 5 ton, annually	2.917	160	163		323	377.26	455
1950	Annualized	2.917	160	163		323	377.26	455

PM Components	Labor-hrs.	W	M	Q	S	A
PM System E1095 122 2950						
Crane, electric bridge 5 to 15 ton						
1 Check with operating or area personnel for deficiencies.	.035			√	√	√
2 Inspect crane for damaged structural members, alignment, worn pins, column fasteners, stops and capacity markings.	.235			√	√	√
3 Inspect and lubricate power cable tension reel and inspect condition of cable.	.491			√	√	√
4 Check operation of upper and lower travel limit switches.	.072			√	√	√
5 Inspect brake mechanism.	.250					√
6 Inspect, clean and lubricate trolley assembly and reduction gear bearings.	.099			√	√	√
7 Drain and replace reduction gear housing oil.	.170					√
8 Inspect and clean hook assembly and chain or wire rope sheaves.	.099			√	√	√
9 Inspect, clean and lubricate hoist motor.	.094			√	√	√
10 Check pendent control for hoist operation - travel function, up, down, speed stepping, etc.	.088			√	√	√
11 Lubricate rails.	.325			√	√	√
12 Weight capacity test hoisting equipment at 125% of rated capacity in accordance with 29 CFR 1910.179.	1.040					√
13 Clean work area.	.066			√	√	√
14 Fill out maintenance report and report deficiencies.	.022			√	√	√
Total labor-hours/period				1.626	1.626	3.086
Total labor-hours/year				3.252	1.626	3.086

	Description	Labor-hrs.	Cost Each					
			2020 Bare Costs				Total	Total
			Material	Labor	Equip.	Total	In-House	w/O&P
2900	Crane, electric bridge, 5 to 15 ton, annually	3.086	186	172		358	417.50	500
2950	Annualized	7.667	410	430		840	985.35	1,175

Crane, Electric Bridge

PM Components	Labor-hrs.	W	M	Q	S	A
PM System E1095 122 3950						
Crane, electric bridge, over 15 tons						
1 Check with operating or area personnel for deficiencies.	.035			√	√	√
2 Inspect crane for damaged structural members, alignment, worn pins, column fasteners, stops and capacity markings.	.235			√	√	√
3 Inspect and lubricate power cable tension reel and inspect condition of cable.	.491			√	√	√
4 Check operation of upper and lower travel limit switches.	.072			√	√	√
5 Inspect brake mechanism.	.250					√
6 Inspect, clean and lubricate trolley assembly and reduction gear bearings.	.099			√	√	√
7 Drain and replace reduction gear housing oil.	.170					√
8 Inspect and clean hook assembly and chain or wire rope sheaves.	.099			√	√	√
9 Inspect, clean and lubricate hoist motor.	.094			√	√	√
10 Check pendent control for hoist operation - travel function, up, down, speed stepping, etc.	.088			√	√	√
11 Lubricate rails.	.325			√	√	√
12 Inspect control cabinet.	.238					√
13 Weight capacity test hoisting equipment, at 125% of rated capacity in accordance with 29 CFR 1910.179.	1.170					√
14 Clean work area.	.066			√	√	√
15 Fill out maintenance report and report deficiencies.	.022			√	√	√
Total labor-hours/period				1.626	1.626	3.454
Total labor-hours/year				3.252	1.626	3.454

	Description	Labor-hrs.	Cost Each					
			2020 Bare Costs				Total	Total
			Material	Labor	Equip.	Total	In-House	w/O&P
3900	Crane, electric bridge, over 15 tons, annually	3.454	93.50	193		286.50	342.15	415
3950	Annualized	8.266	320	460		780	926.77	1,125

PM Components	Labor-hrs.	W	M	Q	S	A
PM System E1095 123 1950						
Crane, manual bridge, up to 5 tons						
1 Check with operating or area personnel for deficiencies.	[illegible]					√
2 Operate crane and hoist, inspect underhung hoist monorail for damaged structural members, braces, stops and capacity markings.	.075					√
3 Inspect and lubricate trolley wheels and bearings.	.096					√
4 Inspect, clean and lubricate hooks and chain sheaves.	.099					√
5 Check operation of crane and hoist brakes.	.016					√
6 Make minor adjustments as required for operational integrity; tighten bolts, etc.	.134					√
7 Weight capacity test hoisting equipment at 125% of rated capacity in accordance with 29 CFR 1910.179.	.300					√
8 Clean work area.	.066					√
9 Fill out maintenance report and report deficiencies.	.022					√
Total labor-hours/period						.933
Total labor-hours/year						.933

	Description	Labor-hrs.	Cost Each					
			2020 Bare Costs				Total	Total
			Material	Labor	Equip.	Total	In-House	w/O&P
1900	Crane, manual bridge, up to 5 ton, annually	.933	92	52.50		144.50	165.63	196
1950	Annualized	.933	92	52.50		144.50	165.63	196

PM Components	Labor-hrs.	W	M	Q	S	A
PM System E1095 123 2950						
Crane, manual bridge, 5 to 15 tons						
1 Check with operating or area personnel for deficiencies.	.035					√
2 Operate crane and hoist, inspect underhung hoist monorail for damaged structural members, braces, stops and capacity markings.	.130					√
3 Inspect and lubricate trolley wheels and bearings.	.096					√
4 Inspect, clean and lubricate hooks and chain sheaves.	.099					√
5 Check operation of crane and hoist brakes.	.016					√
6 Make minor adjustments as required for operational integrity; tighten bolts, etc.	.134					√
7 Weight capacity test hoisting equipment at 125% of rated capacity in accordance with 29 CFR 1910.179.	.772					√
8 Clean work area.	.066					√
9 Fill out maintenance report and report deficiencies.	.022					√
Total labor-hours/period						1.370
Total labor-hours/year						1.370

	Description	Labor-hrs.	Cost Each					
			2020 Bare Costs				Total	Total
			Material	Labor	Equip.	Total	In-House	w/O&P
2900	Crane, manual bridge, 5 to 15 ton, annually	1.370	43	76.50		119.50	141.93	173
2950	Annualized	1.370	43	76.50		119.50	141.93	173

E1095 123 Crane, Manual Bridge

PM Components	Labor-hrs.	W	M	Q	S	A
PM System E1095 123 3950						
Crane, manual bridge, over 15 tons						
1 Check with operating or area personnel for deficiencies.	.035					√
2 Operate crane and hoist, inspect for damaged structural members, alignment, worn pins, fasteners, stops and capacity markings.	.235					√
3 Inspect and lubricate trolley wheels and bearings.	.096					√
4 Inspect, clean and lubricate hooks and chain sheaves.	.099					√
5 Check operation of crane and hoist brakes.	.016					√
6 Make minor adjustments as required for operational integrity; tighten bolts, etc.	.134					√
7 Weight capacity test hoisting equipment at 125% of rated capacity in accordance with 29 CFR 1910.179.	1.162					√
8 Clean work area.	.066					√
9 Fill out maintenance report and report deficiencies.	.022					√
Total labor-hours/period						1.865
Total labor-hours/year						1.865

	Description	Labor-hrs.	Cost Each					
			2020 Bare Costs				Total	Total
			Material	Labor	Equip.	Total	In-House	w/O&P
3900	Crane, manual bridge, over 15 tons, annually	1.865	43	104		147	176.18	216
3950	Annualized	1.865	43	104		147	176.18	216

E1095 125 Hoist, Pneumatic

PM Components	Labor-hrs.	W	M	Q	S	A
PM System E1095 125 1950						
Hoist, pneumatic						
1 Check with operating or area personnel for deficiencies.	.035					√
2 Inspect for damaged structural members, alignment, worn pins, column fasteners, stops and capacity markings.	.235					√
3 Inspect and lubricate cable tension reel and inspect condition of cable.	.452					√
4 Check pendent control for hoist operation - travel function, up, down, speed stepping, upper and lower travel limit switches, etc.	.088					√
5 Inspect, clean and lubricate trolley assembly.	.099					√
6 Drain and replace reduction gear housing oil.	.170					√
7 Inspect brake mechanism.	.016					√
8 Inspect and clean hook assembly and chain or wire rope sheaves.	.099					√
9 Inspect, clean and lubricate hoist motor.	.029					√
10 Clean work area.	.066					√
11 Fill out maintenance report and report deficiencies.	.022					√
Total labor-hours/period						1.311
Total labor-hours/year						1.311

	Description	Labor-hrs.	Cost Each					
			2020 Bare Costs				Total	Total
			Material	Labor	Equip.	Total	In-House	w/O&P
1900	Hoist, pneumatic, annually	1.311	164	73.50		237.50	271.32	320
1950	Annualized	1.311	164	73.50		237.50	271.32	320

E10 EQUIPMENT | E1095 126 | Hoist/Winch, Chain/Cable

PM Components	Labor-hrs.	W	M	Q	S	A
PM System E1095 126 1950						
Hoist/winch, chain fall/cable, electric						
1 Check with operating or area personnel for deficiencies.	.035					√
2 Inspect for damaged structural members, alignment, worn pins, column fasteners, stops and capacity markings.	.235					√
3 Inspect and lubricate power cable tension reel; inspect condition of cable.	.491					√
4 Check hoist operation - travel function, up, down, speed stepping, upper and lower travel limit switches, etc.	.088					√
5 Inspect, clean and lubricate trolley assembly and reduction gear bearings.	.099					√
6 Drain and replace reduction gear housing oil.	.170					√
7 Inspect brake mechanism.	.250					√
8 Inspect and clean hook assembly and chain or wire rope sheaves.	.099					√
9 Inspect, clean and lubricate hoist motor.	.094					√
10 Clean work area.	.066					√
11 Fill out maintenance report and report deficiencies.	.022					√
Total labor-hours/period						1.649
Total labor-hours/year						1.649

	Description	Labor-hrs.	Cost Each					
			2020 Bare Costs				Total	Total
			Material	Labor	Equip.	Total	In-House	w/O&P
1900	Hoist / winch, chain / cable, electric, annually	1.649	151	92		243	280.02	330
1950	Annualized	1.649	151	92		243	280.02	330

E10 EQUIPMENT | E1095 302 | Beverage Dispensing Unit

PM Components	Labor-hrs.	W	M	Q	S	A
PM System E1095 302 1950						
Beverage dispensing unit						
1 Check with operating or area personnel for deficiencies.	.044			√	√	√
2 Remove access panels and clean coils, fans, fins, and other areas; vacuum, blow down with compressed air or CO_2, and/or wipe off.	.517			√	√	√
3 Lubricate motor bearings.	.008			√	√	√
4 Check electrical wiring and connections; tighten loose connections.	.120			√	√	√
5 Visually check for refrigerant, water, and other liquid leaks.	.153			√	√	√
6 Check door gaskets for proper sealing; adjust door catch as required.	.033			√	√	√
7 Inspect dispensing valves; adjust as necessary.	.108			√	√	√
8 Perform operational check of unit; make adjustments as necessary.	.040			√	√	√
9 Clean area around equipment.	.066			√	√	√
10 Fill out maintenance checklist and report deficiencies.	.022			√	√	√
Total labor-hours/period				1.111	1.111	1.111
Total labor-hours/year				2.223	1.111	1.111

	Description	Labor-hrs.	Cost Each					
			2020 Bare Costs				Total	Total
			Material	Labor	Equip.	Total	In-House	w/O&P
1900	Beverage dispensing unit, annually	1.111	41	61		102	123.22	148
1950	Annualized	4.444	53.50	243		296.50	372.70	455

E10 EQUIPMENT | E1095 304 | Braising Pan

PM Components	Labor-hrs.	W	M	Q	S	A
PM System E1095 304 1950						
Braising pan, tilting, gas/electric						
1 Check with operating or area personnel for deficiencies.	.044			√	√	√
2 Check electric transformer, heating elements, insulators, connections and wiring; tighten connections and repair as required.	.239			√	√	√
3 Check surface temperature with meter for hot spots.	.073			√	√	√
4 Check piping and valves on gas operated units for leaks.	.077			√	√	√
5 Check pilot/igniter, burner and combustion chamber for proper operation; adjust as required.	.079			√	√	√
6 Check temperature control system for proper operation; adjust as required.	.458			√	√	√
7 Lubricate tilting gear mechanism and trunnion bearings.	.047			√	√	√
8 Check nuts, bolts and screws for tightness; tighten as required.	.007			√	√	√
9 Check complete operation of unit.	.055			√	√	√
10 Fill out maintenance checklist and report deficiencies.	.022			√	√	√
Total labor-hours/period				1.101	1.101	1.101
Total labor-hours/year				2.202	1.101	1.101

	Description	Labor-hrs.	Cost Each					
			2020 Bare Costs				Total	Total
			Material	Labor	Equip.	Total	In-House	w/O&P
1900	Braising pan, tilting, gas / electric, annually	1.101	41	60.50		101.50	122.36	148
1950	Annualized	4.404	53.50	241		294.50	369.07	450

E10 EQUIPMENT | E1095 306 | Bread Slicer

PM Components	Labor-hrs.	W	M	Q	S	A
PM System E1095 306 1950						
Bread slicer, electric						
1 Check with operating or area personnel for deficiencies.	.044			√	√	√
2 Check electric motor, controls, wiring connections, insulation and contacts for damage and/or proper operation; adjust as required.	.338			√	√	√
3 Check V-belt for wear, tension and alignment; change/adjust belt as necessary.	.029			√	√	√
4 Lubricate bushings and check oil level in gear drive mechanism; add oil if required.	.065			√	√	√
5 Clean machine of dust, grease and food particles.	.074			√	√	√
6 Check motor and motor bearings for overheating.	.039			√	√	√
7 Check for loose, missing or damaged nuts, bolts, or screws; tighten or replace as necessary.	.005			√	√	√
8 Fill out maintenance checklist and report deficiencies.	.022			√	√	√
Total labor-hours/period				.616	.616	.616
Total labor-hours/year				1.232	.616	.616

	Description	Labor-hrs.	Cost Each					
			2020 Bare Costs				Total	Total
			Material	Labor	Equip.	Total	In-House	w/O&P
1900	Bread slicer, electric, annually	.616	24.50	34		58.50	70.39	84.50
1950	Annualized	2.464	75	135		210	255.96	310

PM Components	Labor-hrs.	W	M	Q	S	A
PM System E1095 310 1950						
Broiler, conveyor type, gas						
1 Check with operating or area personnel for deficiencies.	.044			√	√	√
2 Check broiler pilot/igniters and jets for uniform flame; adjust as required.	.166			√	√	√
3 Check air filter; change as required.	.059			√	√	√
4 Check operation of chains and counterweights.	.012			√	√	√
5 Check piping and valves for leaks.	.077			√	√	√
6 Check electrical connections and wiring for defects; tighten loose connections.	.120			√	√	√
7 Check thermocouple and thermostat; calibrate thermostat as required.	.640			√	√	√
8 Tighten and replace any loose nuts, bolts, or screws.	.005			√	√	√
9 Check operation of unit.	.055			√	√	√
10 Fill out maintenance checklist and report deficiencies.	.022			√	√	√
Total labor-hours/period				1.200	1.200	1.200
Total labor-hours/year				2.400	1.200	1.200

	Description	Labor-hrs.	Cost Each					
			2020 Bare Costs				Total	Total
			Material	Labor	Equip.	Total	In-House	w/O&P
1900	Broiler, conveyor type, gas, annually	1.200	23	65.50		88.50	109.36	134
1950	Annualized	4.824	68.50	265		333.50	414.68	505

PM Components	Labor-hrs.	W	M	Q	S	A
PM System E1095 310 2950						
Broiler, hot dog, electric						
1 Check with operating or area personnel for deficiencies.	.044			√	√	√
2 Check electric motor, controls, contacts, wiring connections for damage and proper operation; adjust as required.	.152			√	√	√
3 Lubricate bushings.	.007			√	√	√
4 Inspect lines and unit for mechanical defects; adjust or repair as required.	.021			√	√	√
5 Clean machine of dust and grease.	.109			√	√	√
6 Check unit for proper operation.	.082			√	√	√
7 Fill out maintenance checklist and report deficiencies.	.022			√	√	√
Total labor-hours/period				.437	.437	.437
Total labor-hours/year				.874	.437	.437

	Description	Labor-hrs.	Cost Each					
			2020 Bare Costs				Total	Total
			Material	Labor	Equip.	Total	In-House	w/O&P
2900	Broiler, hot dog type, electric, annually	.437	9.40	24		33.40	41.09	50
2950	Annualized	1.748	28	96		124	153.80	187

E1095 310 Broiler

PM Components	Labor-hrs.	W	M	Q	S	A
PM System E1095 310 3950						
Broiler/oven (salamander), gas/electric						
1 Check with operating or area personnel for deficiencies.	.044			√	√	√
2 Check broiler pilot and oven for uniform flame; adjust as required.	.167			√	√	√
3 Check piping and valves for leaks.	.077			√	√	√
4 Check doors, shelves for warping, alignment, and seals; adjust as required.	.148			√	√	√
5 Check electrical connections and wiring for defects; tighten loose connections.	.242			√	√	√
6 Tighten and replace any loose nuts, bolts and/or screws.	.012			√	√	√
7 Check operation of broiler arm, adjust as required.	.007			√	√	√
8 Check thermocouple and thermostat; calibrate thermostat as required.	.571			√	√	√
9 Fill out maintenance checklist and report deficiencies.	.022			√	√	√
Total labor-hours/period				1.291	1.291	1.291
Total labor-hours/year				2.581	1.291	1.291

			Cost Each					
			2020 Bare Costs				Total	Total
	Description	Labor-hrs.	Material	Labor	Equip.	Total	In-House	w/O&P
3900	Broiler / oven, (salamander), gas / electric, annually	1.291	4.64	71		75.64	95.68	119
3950	Annualized	5.144	13.90	283		296.90	376.77	465

E10 EQUIPMENT

E1095 312 Chopper, Electric

PM Components	Labor-hrs.	W	M	Q	S	A
PM System E1095 312 1950						
Chopper, electric						
1 Check with operating or area personnel for deficiencies.	.044			√	√	√
2 Check cutting surfaces, clutch assembly and guards for damage and proper alignment; adjust as required.	.065			√	√	√
3 Check nuts, bolts, and screws for tightness; tighten or replace as required.	.005			√	√	√
4 Check switch and controls for proper operation; adjust as required.	.010			√	√	√
5 Clean machine of dust, grease and food particles.	.014			√	√	√
6 Check operation of motor and check motor bearings for overheating; lubricate as required.	.069			√	√	√
7 Check belt tension; replace belt and/or adjust if required.	.314			√	√	√
8 Fill out maintenance checklist and report deficiencies.	.022			√	√	√
Total labor-hours/period				.543	.543	.543
Total labor-hours/year				1.085	.543	.543

			Cost Each					
			2020 Bare Costs				Total	Total
	Description	Labor-hrs.	Material	Labor	Equip.	Total	In-House	w/O&P
1900	Chopper, electric, annually	.543	41.50	30		71.50	83.96	99.50
1950	Annualized	2.176	87	119		206	249.11	299

E1095 314 Coffee Maker/Urn

PM Components	Labor-hrs.	W	M	Q	S	A
PM System E1095 314 1950						
Coffee maker/urn, gas/electric/steam						
1 Check thermostat, switch, and temperature gauge; calibrate if required.	.465			√	√	√
2 Check pilots and flame on gas operated units; adjust as required.	.038			√	√	√
3 Check working pressure on steam operated unit.	.005			√	√	√
4 Check electrical connections and wiring for defects; tighten connections.	.120			√	√	√
5 Check for clogged or defective steam trap.	.100			√	√	√
6 Inspect and clean steam strainer.	.207			√	√	√
7 Examine equipment, valves, and piping for leaks.	.077			√	√	√
8 Inspect for leaks at water gauge glasses and at valves; repack valves if necessary.	.046			√	√	√
9 Tighten or replace loose, missing, or damaged nuts, bolts, or screws.	.005			√	√	√
10 Lubricate water filter valve, shaft rings; tighten as required.	[illegible]			√	√	√
11 Check operation of lights.	.027			√	√	√
12 Check timer mechanism and operation of unit.	.018			√	√	√
13 Fill out maintenance checklist and report deficiencies.	.022			√	√	√
Total labor-hours/period				1.163	1.163	1.163
Total labor-hours/year				2.326	1.163	1.163

	Description	Labor-hrs.	Cost Each					
			2020 Bare Costs				Total	Total
			Material	Labor	Equip.	Total	In-House	w/O&P
1900	Coffee maker / urn, gas / electric / steam, annually	1.163	21.50	64		85.50	105.09	129
1950	Annualized	4.652	64	255		319	398.16	485

PM Components	Labor-hrs.	W	M	Q	S	A
PM System E1095 316 1950						
Cooker, gas						
1 Check with operating or area personnel for any deficiencies.	.044			√	√	√
2 Examine doors and door gaskets; make necessary adjustments.	.021			√	√	√
3 Visually inspect hinges and latches; lubricate hinges.	.091			√	√	√
4 Tighten or replace any loose nuts, bolts, screws; replace broken knobs.	.007			√	√	√
5 Check gas piping and valves for leaks.	.077			√	√	√
6 Check pilot and thermosensor bulb.	.079			√	√	√
7 Check gas burner for uniform flame.	.007			√	√	√
8 Check calibration of thermostats.	.457			√	√	√
9 Operate unit to ensure that it is in proper working condition.	.130			√	√	√
10 Fill out maintenance checklist and report deficiencies.	.022			√	√	√
Total labor-hours/period				.935	.935	.935
Total labor-hours/year				1.870	.935	.935

	Description	Labor-hrs.	Cost Each						
			2020 Bare Costs				Total	Total
			Material	Labor	Equip.	Total	In-House	w/O&P
1900	Cooker, gas, annually	.935	9.40	51		60.40	75.96	93.50
1950	Annualized	3.744	28	205		233	294.22	360

PM Components	Labor-hrs.	W	M	Q	S	A
PM System E1095 316 2950						
Cooker, vegetable steamer						
1 Check with operating or area personnel for any deficiencies.	.044			√	√	√
2 Examine doors and door gaskets; make necessary adjustments.	.021			√	√	√
3 Visually inspect hinges and latches; lubricate hinges.	.091			√	√	√
4 Tighten or replace any loose nuts, bolts, screws; replace broken knobs.	.005			√	√	√
5 Check working pressure on steam gauge.	.005			√	√	√
6 Inspect piping and valves for leaks; tighten as necessary.	.077			√	√	√
7 Check operation of low water cut-off; adjust as required.	.030			√	√	√
8 Check water level sight glass; adjust as required.	.007			√	√	√
9 Fill out maintenance checklist and report deficiencies.	.022			√	√	√
Total labor-hours/period				.302	.302	.302
Total labor-hours/year				.603	.302	.302

	Description	Labor-hrs.	Cost Each						
			2020 Bare Costs				Total	Total
			Material	Labor	Equip.	Total	In-House	w/O&P
2900	Cooker, vegetable steamer, annually	.302	9.40	16.55		25.95	31.57	38.50
2950	Annualized	1.208	28	66.50		94.50	116.03	141

E1095 316 Cooker

PM Components	Labor-hrs.	W	M	Q	S	A
PM System E1095 316 3950						
Cooker, steam						
1 Check with operating or area personnel for any deficiencies.	.044			√	√	√
2 Examine doors and door gaskets; make necessary adjustments.	.021			√	√	√
3 Visually inspect hinges and latches; lubricate hinges.	.091			√	√	√
4 Tighten or replace any loose nuts, bolts, screws; replace broken knobs.	.005			√	√	√
5 Check for clogged or defective steam hose; clean hose.	[illegible]			√	√	√
6 Inspect and clean steam strainer.	.207			√	√	√
7 Check working pressure on steam gauge.	.005			√	√	√
8 Turn unit on, allow pressure to build, inspect piping, valve and doors for leaks; adjust as required.	.077			√	√	√
9 Check operation of pressure relief valve; adjust as required.	.010			√	√	√
10 Check operation of low water cut-off; adjust as required.	.030			√	√	√
11 Check water level sight glass; adjust as required.	.007			√	√	√
12 Check timer mechanism and operation, adjust as required.	.018			√	√	√
13 Fill out maintenance checklist and report deficiencies.	.022			√	√	√
Total labor-hours/period				.635	.635	.635
Total labor-hours/year				1.270	.635	.635

	Description	Labor-hrs.	2020 Bare Costs Material	Labor	Equip.	Total	Total In-House	Total w/O&P
3900	Cooker, steam, annually	.635	9.40	35		44.40	54.99	67.50
3950	Annualized	2.540	28	139		167	209.89	258

E10 EQUIPMENT

E1095 318 Cookie Maker

PM Components	Labor-hrs.	W	M	Q	S	A
PM System E1095 318 1950						
Cookie maker, electric						
1 Check with operating or area personnel for any deficiencies.	.044			√	√	√
2 Check electrical panel switches, connections, timing device for loose, frayed wiring; tighten connections as required.	.120			√	√	√
3 Check operation of electric motor and bearings for overheating; lubricate motor bearings.	.086			√	√	√
4 Check belt; adjust tension or replace belt as required.	.372			√	√	√
5 Check conveyor and rollers by observing movement; lubricate rollers and adjust as required.	.127			√	√	√
6 Tighten or replace loose, missing or damaged nuts, bolts or screws.	.012			√	√	√
7 Wipe off and clean unit.	.096			√	√	√
8 Fill out maintenance checklist and report deficiencies.	.022			√	√	√
Total labor-hours/period				.879	.879	.879
Total labor-hours/year				1.758	.879	.879

	Description	Labor-hrs.	2020 Bare Costs Material	Labor	Equip.	Total	Total In-House	Total w/O&P
1900	Cookie maker, electric, annually	.879	28	48.50		76.50	92.79	112
1950	Annualized	3.504	84	192		276	339.80	410

PM Components	Labor-hrs.	W	M	Q	S	A
PM System E1095 320 1950						
Deep fat fryer, pressurized broaster, gas/electric						
1 Check with operating or area personnel for any deficiencies.	.044			√	√	√
2 Check piping and valves for leaks; tighten fittings as required.	.125			√	√	√
3 Check pilot and flame on gas operated unit; adjust as required.	.009			√	√	√
4 Check elements, switches, controls, contacts, and wiring on electrically heated units; repair or adjust as required.	.025			√	√	√
5 Check thermostat; calibrate if necessary.	.444			√	√	√
6 Check basket/racks for bends, breaks or defects; straighten bends or repair as necessary.	.013			√	√	√
7 Check cooking oil filter on circulating system; change as required.	.068			√	√	√
8 Check operation of unit.	.163			√	√	√
9 Check operation of motor for excessive noise and overheating.	.022			√	√	√
10 Check nuts, bolts, and screws for tightness; tighten or replace as necessary.	.005			√	√	√
11 Fill out maintenance checklist and report deficiencies.	.022			√	√	√
Total labor-hours/period				.940	.940	.940
Total labor-hours/year				1.881	.940	.940

	Description	Labor-hrs.	Cost Each					
			2020 Bare Costs				Total	Total
			Material	Labor	Equip.	Total	In-House	w/O&P
1900	Fryer, pressurized broaster, gas / electric, annually	.940	13	51.50		64.50	80.70	99
1950	Annualized	3.744	39	206		245	306.24	375

PM Components	Labor-hrs.	W	M	Q	S	A
PM System E1095 320 2950						
Deep fat fryer, automatic conveyor belt type, gas/electric						
1 Check with operating or area personnel for any deficiencies.	.044			√	√	√
2 Check compartments, valves and piping for leaks; tighten as required.	.077			√	√	√
3 Check pilot and flame on gas operated fryers; adjust as required.	.168			√	√	√
4 Check elements, switches, controls, contacts and wiring on electrically heated units for defects; repair or adjust as necessary.	.051			√	√	√
5 Check thermostats; calibrate as necessary.	.444			√	√	√
6 Check cooking oil filter on circulating system; change as required.	.068			√	√	√
7 Check operation of motor for excessive noise and overheating; lubricate bearings.	.022			√	√	√
8 Check operation of belt hoist; lubricate and adjust as required.	.052			√	√	√
9 Check nuts, bolts and screws for tightness; replace or tighten as necessary.	.017			√	√	√
10 Clean machine of dust, grease and food particles.	.044			√	√	√
11 Clean area around unit.	.066			√	√	√
12 Fill out maintenance checklist and report deficiencies.	.022			√	√	√
Total labor-hours/period				1.076	1.076	1.076
Total labor-hours/year				2.151	1.076	1.076

	Description	Labor-hrs.	Cost Each					
			2020 Bare Costs				Total	Total
			Material	Labor	Equip.	Total	In-House	w/O&P
2900	Fryer, conveyor belt type, gas / electric, annually	1.076	22.50	59		81.50	100.67	122
2950	Annualized	4.280	67.50	235		302.50	375.90	460

E1095 320 Deep Fat Fryer

PM Components	Labor-hrs.	W	M	Q	S	A
PM System E1095 320 3950						
Deep fat fryer, conventional type, gas/electric						
1 Check with operating or area personnel for any deficiencies.	.044			√	√	√
2 Clean machine thoroughly of dust, grease, and food particles.	.014			√	√	√
3 Check compartments, valves, and piping for leaks; tighten as required.	.077			√	√	√
4 Check pilot and flame on gas operated deep fat fryer; adjust as required.	.005			√	√	√
5 Check thermostat; calibrate if necessary.	.439			√	√	√
6 On electrically heated units check elements, switches, controls, contacts, and wiring for defects; repair or adjust as necessary.	.025			√	√	√
7 Check basket or rack for bends, breaks or defects, straighten bends or repair as necessary.	.013			√	√	√
8 Check nuts, bolts and screws for tightness; tighten or replace as necessary.	.005			√	√	√
9 Fill out maintenance checklist and report deficiencies.	.022			√	√	√
Total labor-hours/period				.644	.644	.644
Total labor-hours/year				1.288	.644	.644

	Description	Labor-hrs.	Cost Each: 2020 Bare Costs: Material	Labor	Equip.	Total	Total In-House	Total w/O&P
3900	Fryer, conventional type, gas / electric, annually	.644	4.64	35		39.64	50.50	62.50
3950	Annualized	2.576	13.90	142		155.90	196.86	242

PM Components	Labor-hrs.	W	M	Q	S	A
PM System E1095 322 1950						
Dishwasher, electric						
1 Check with operating or area personnel for any deficiencies.	.044			√	√	√
2 Check electric insulators, connections, and wiring (includes removing access panels).	.120			√	√	√
3 Check motor and bearings for excessive noise, vibration and overheating.	.039			√	√	√
4 Check dish conveyor by turning on switch and observing conveyor movement; adjust chain if necessary.	.070			√	√	√
5 Check operation of wash and rinse spray mechanism for spray coverage and drainage.	.022			√	√	√
6 Inspect soap and spray solution feeder lines; clean as necessary.	.013			√	√	√
7 Inspect water lines and fittings for leaks; tighten fittings as necessary.	.077			√	√	√
8 Lubricate conveyor drive bushings and chain drive.	.055			√	√	√
9 Check belt tension; adjust if required.	.100			√	√	√
10 Check doors for: operation of chains and counterweights, warping, alignment and water tightness; adjust if necessary.	.022			√	√	√
11 Check packing glands on wash, rinse, and drain valves; add or replace packing as required.	.147			√	√	√
12 Check lubricant in gear case; add oil if required.	.049			√	√	√
13 Inspect splash curtain for tears, clearance and water tightness; adjust if required.	.016			√	√	√
14 Check proper operation of solenoid valve and float in fill tank; adjust as required.	.096			√	√	√
15 Check pumps for leakage and obstructions; adjust as required.	.120			√	√	√
16 Check proper operation of micro-switch.	.010			√	√	√
17 Check water for proper temperature.	.004			√	√	√
18 Check operation of safety stop and clutch on conveyor belt motor; adjust cut-off switch mechanism as required.	.049			√	√	√
19 Check temperature regulator and adjust if necessary.	.017			√	√	√
20 Clean lime off thermostatic probe and heating elements.	.020			√	√	√
21 Clean area around equipment.	.066			√	√	√
22 Fill out maintenance checklist and report deficiencies.	.022			√	√	√
Total labor-hours/period				1.179	1.179	1.179
Total labor-hours/year				2.357	1.179	1.179

	Description	Labor-hrs.	Cost Each					
			2020 Bare Costs				Total	Total
			Material	Labor	Equip.	Total	In-House	w/O&P
1900	Dishwasher, electric, annually	1.179	134	64.50		198.50	230.78	271
1950	Annualized	4.712	405	259		664	775.03	915

E1095 322 Dishwasher, Electric/Steam

	PM Components	Labor-hrs.	W	M	Q	S	A
	PM System E1095 322 2950						
	Dishwasher, steam						
1	Check with operating or area personnel for any deficiencies.	.044			√	√	√
2	Check electric insulators, connections, and wiring (includes removing access panels).	.120			√	√	√
3	Check motor and bearings for excessive noise, vibration and overheating	.039			√	√	√
4	Check dish conveyor by turning on switch and observing conveyor movement; adjust chain if necessary.	.070			√	√	√
5	Check operation of wash and rinse spray mechanism for spray coverage and drainage.	.022			√	√	√
6	Inspect soap and spray solution feeder lines; clean as necessary.	.013			√	√	√
7	Inspect steam and water lines and fittings for leaks; tighten fittings as necessary.	.077			√	√	√
8	Lubricate conveyor drive bushings and chain drive.	.055			√	√	√
9	Check belt tension; adjust if required.	.100			√	√	√
10	Check doors for: operation of chains and counterweights, warping, alignment and water tightness; adjust if necessary.	.022			√	√	√
11	Tighten loose nuts, bolts and screws.	.005			√	√	√
12	Check packing glands on wash, rinse, and drain valves; add or replace packing as required.	.147			√	√	√
13	Check lubricant in gear case; add oil if required.	.049			√	√	√
14	Inspect splash curtain for tears, clearance and water tightness; adjust if required.	.016			√	√	√
15	Check proper operation of solenoid valve and float in fill tank; adjust as required.	.096			√	√	√
16	Check pumps for leakage and obstructions; adjust as required.	.120			√	√	√
17	Check proper operation of micro-switch.	.010			√	√	√
18	Check water for proper temperature readings.	.004			√	√	√
19	Check operation of safety stop and clutch on conveyor belt motor; adjust cut-off switch mechanism as required.	.049			√	√	√
20	Check temperature regulator and adjust if necessary.	.017			√	√	√
21	Clean lime off thermostatic probe.	.020			√	√	√
22	Clean area around equipment.	.066			√	√	√
23	Fill out maintenance checklist and report deficiencies.	.022			√	√	√
	Total labor-hours/period				1.184	1.184	1.184
	Total labor-hours/year				2.367	1.184	1.184

	Description	Labor-hrs.	Cost Each					
			2020 Bare Costs				Total	Total
			Material	Labor	Equip.	Total	In-House	w/O&P
2900	Dishwasher, steam, annually	1.184	134	65		199	231.12	271
2950	Annualized	4.732	405	260		665	776.44	920

E10 EQUIPMENT | E1095 328 | Dispenser, Beverage

PM Components	Labor-hrs.	W	M	Q	S	A
PM System E1095 328 1950						
Dispenser, carbonated beverage						
1 Check with operating or area personnel for any deficiencies.	.044			√	√	√
2 Check all lines, connections and valves for leaks.	.077			√	√	√
3 Check pressure gauges; adjust as required.	.018			√	√	√
4 Check water strainer screen on carbonator; clean as required.	.158			√	√	√
5 Check wiring, connectors and switches; repair as required.	.120			√	√	√
6 Check nuts, bolts and screws for tightness; replace or tighten as required.	.005			√	√	√
7 Wipe and clean unit.	.056			√	√	√
8 Fill out maintenance checklist and report deficiencies.	.022			√	√	√
Total labor-hours/period				.501	.501	.501
Total labor-hours/year				1.001	.501	.501

	Description	Labor-hrs.	Cost Each					
			2020 Bare Costs				Total	Total
			Material	Labor	Equip.	Total	In-House	w/O&P
1900	Dispenser, carbonated beverage, annually	.501	24.50	27.50		52	62.35	74
1950	Annualized	2.004	74	110		184	222.10	268

E10 EQUIPMENT | E1095 332 | Disposal, Garbage

PM Components	Labor-hrs.	W	M	Q	S	A
PM System E1095 332 1950						
Disposal, garbage, electric						
1 Check with operating or area personnel for any deficiencies.	.044			√	√	√
2 Check motor and drive shaft for excessive noise, vibration, overheating, etc.	.044			√	√	√
3 Inspect electrical wiring.	.014			√	√	√
4 Visually examine grinder; check for obstructions; adjust cutters if required.	.074			√	√	√
5 Check drive belt(s) for wear and tension; adjust if required.	.014			√	√	√
6 Check for leaks to supply and drain connections; tighten if required.	.103			√	√	√
7 Check electrical switch for proper operation.	.014			√	√	√
8 Lubricate grinder, drive, and motor.	.022			√	√	√
9 Check water sprayer operation.	.030			√	√	√
10 Fill out maintenance checklist and report deficiencies.	.022			√	√	√
Total labor-hours/period				.381	.381	.381
Total labor-hours/year				.762	.381	.381

	Description	Labor-hrs.	Cost Each					
			2020 Bare Costs				Total	Total
			Material	Labor	Equip.	Total	In-House	w/O&P
1900	Disposal, garbage, electric, annually	.381	4.77	21		25.77	32.06	39.50
1950	Annualized	1.524	14.30	83.50		97.80	123.27	151

E1095 334 Dough, Divider/Roller

PM Components	Labor-hrs.	W	M	Q	S	A
PM System E1095 334 1950						
Dough divider						
1 Check with operating or area personnel for any deficiencies.	.044			√	√	√
2 Check motor and drive shaft for excessive noise, vibration, overheating, etc.	.039			√	√	√
3 Inspect electrical wiring.	.120			√	√	√
4 Visually examine plates; check for obstructions; adjust cutters if required.	.074			√	√	√
5 Check electrical switch for proper operation.	.014			√	√	√
6 Lubricate press, drive, and motor.	.219			√	√	√
7 Fill out maintenance checklist and report deficiencies.	.022			√	√	√
Total labor-hours/period				.500	.500	.500
Total labor-hours/year				1.065	.533	.533

	Description	Labor-hrs.	Cost Each: 2020 Bare Costs: Material	Labor	Equip.	Total	Total In-House	Total w/O&P
1900	Dough divider, annually	.533	4.77	29		33.77	42.53	52.50
1950	Annualized	2.124	14.30	117		131.30	165.99	204

PM Components	Labor-hrs.	W	M	Q	S	A
PM System E1095 334 2950						
Dough roller						
1 Check with operating or area personnel for any deficiencies.	.044			√	√	√
2 Check electric motor, switches, controls, wiring, connections and insulation for defective materials; repair as necessary.	.241			√	√	√
3 Check belt tension; adjust or replace as required.	.444			√	√	√
4 Check operation of conveyor-roller clearance; adjust as necessary.	.109			√	√	√
5 Lubricate all moving parts, motors, pivot points and chain drives.	.017			√	√	√
6 Clean unit and treat for rust and/or corrosion.	.083			√	√	√
7 Tighten or replace loose, missing or damaged nuts, bolts and screws.	.005			√	√	√
8 Fill out maintenance checklist and report deficiencies.	.022			√	√	√
Total labor-hours/period				.965	.965	.965
Total labor-hours/year				1.930	.965	.965

	Description	Labor-hrs.	Cost Each: 2020 Bare Costs: Material	Labor	Equip.	Total	Total In-House	Total w/O&P
2900	Dough roller, annually	.965	23.50	53		76.50	93.99	114
2950	Annualized	3.868	69.50	213		282.50	349.80	425

E10 EQUIPMENT | E1095 336 | Drink Cooler

PM Components	Labor-hrs.	W	M	Q	S	A
PM System E1095 336 1950						
Drink cooler, with external condenser						
1 Check with operating or area personnel for deficiencies.	.035				√	√
2 Clean condenser coils, fans, and intake screens; lubricate motor.	.500				√	√
3 Inspect door gaskets for damage and proper fit; adjust as necessary.	.033				√	√
4 Check starter panels and controls for proper operation, burned or loose contacts, and loose connections.	.094				√	√
5 Clean coils, evaporator drain pan, blowers, fans, motors and drain piping as required; lubricate motor(s).	.473				√	√
6 During operation of unit, check refrigerant pressures and compressor oil level; add refrigerant and/or oil as necessary.	.066				√	√
7 Check operation of low pressure cut-out; adjust or replace as necessary.	.057				√	√
8 Inspect defrost systems for proper operation; adjust as required.	.027				√	√
9 Clean area around equipment.	.066				√	√
10 Fill out maintenance checklist and report deficiencies.	.022				√	√
Total labor-hours/period					1.373	1.373
Total labor-hours/year					1.373	1.373

	Description	Labor-hrs.	Cost Each					
			2020 Bare Costs				Total	Total
			Material	Labor	Equip.	Total	In-House	w/O&P
1900	Drink cooler with external condenser, annually	1.373	104	75.50		179.50	210.99	250
1950	Annualized	2.754	147	151		298	355.51	425

E10 EQUIPMENT | E1095 338 | Fluid Cooler

PM Components	Labor-hrs.	W	M	Q	S	A
PM System E1095 338 1950						
Fluid cooler, air cooled condenser						
1 Check with operating or area personnel for deficiencies.	.035					√
2 Check unit for proper operation, excessive noise or vibration.	.158					√
3 Clean intake side of condenser coils, fans and intake screens.	.471					√
4 Check electrical wiring and connections; tighten loose connections.	.120					√
5 Inspect fan(s) for bent blades or imbalance; adjust as necessary.	.014					√
6 Check belts for condition, proper tension and misalignment, if required.	.029					√
7 Lubricate shaft bearings and motor bearings.	.047					√
8 Inspect piping and valves for leaks; tighten connections as necessary.	.077					√
9 Lubricate and check operation of dampers, if applicable.	.055					√
10 Clean area around fluid cooler.	.066					√
11 Fill out maintenance checklist and report deficiencies.	.022					√
Total labor-hours/period						1.094
Total labor-hours/year						1.094

	Description	Labor-hrs.	Cost Each					
			2020 Bare Costs				Total	Total
			Material	Labor	Equip.	Total	In-House	w/O&P
1900	Fluid cooler, annually	1.094	9.55	60.50		70.05	87.39	107
1950	Annualized	1.094	9.55	60.50		70.05	87.39	107

PM Components	Labor-hrs.	W	M	Q	S	A
PM System E1095 340 1950						
Food saw, electric						
1 Check with operating or area personnel for any deficiencies.	.044			√	√	√
2 Check operation of saw for excessive vibration, blade tension and proper adjustment of safety guards; repair or adjust as required.	.095			√	√	√
3 Check motor and bearings for overheating; lubricate motor bearings.	.039			√	√	√
4 Check switches, motor, controls and wiring for damage and proper operation; repair as required.	.120			√	√	√
5 Clean machine of dust and grease.	.044			√	√	√
6 Check nuts, bolts and screws for tightness; tighten as required.	.005			√	√	√
7 Check belt tension and alignment; adjust as required.	.029			√	√	√
8 Fill out maintenance checklist and report deficiencies.	.022			√	√	√
Total labor-hours/period				.398	.398	.398
Total labor-hours/year				.796	.398	.398

	Description	Labor-hrs.	Cost Each					
			2020 Bare Costs				Total	Total
			Material	Labor	Equip.	Total	In-House	w/O&P
1900	Food saw, electric, annually	.398	25	22		47	55.32	66.50
1950	Annualized	1.592	74.50	87.50		162	193.56	232

PM Components	Labor-hrs.	W	M	Q	S	A
PM System E1095 340 2950						
Food slicer, electric						
1 Check with operating or area personnel for any deficiencies.	.044			√	√	√
2 Check nuts, bolts and screws for tightness; tighten or replace as required.	.005			√	√	√
3 Lubricate food slicer and add lubricant to gear case if required.	.085			√	√	√
4 Check condition of blade, blade guard, guides, and controls; adjust as required.	.047			√	√	√
5 Check operation of motor and check motor bearings for overheating; lubricate motor bearings.	.039			√	√	√
6 Check sharpening stones; install new stones as required.	.013			√	√	√
7 Clean machine of dust, grease, and food particles.	.014			√	√	√
8 Fill out maintenance checklist and report deficiencies.	.022			√	√	√
Total labor-hours/period				.269	.269	.269
Total labor-hours/year				.538	.269	.269

	Description	Labor-hrs.	Cost Each					
			2020 Bare Costs				Total	Total
			Material	Labor	Equip.	Total	In-House	w/O&P
2900	Food slicer, electric, annually	.269	151	14.75		165.75	185.64	212
2950	Annualized	1.076	455	59		514	573.99	665

E1095 346 Grill, Gas/Electric

PM Components	Labor-hrs.	W	M	Q	S	A
PM System E1095 346 1950						
Grill, gas/electric						
1 Check with operating or area personnel for any deficiencies.	.044			√	√	√
2 Check nuts, bolts and screws for tightness; tighten or replace as required.	.005			√	√	√
3 On gas operated units, check piping and valves for leaks.	.077			√	√	√
4 On gas operated units, check pilot and gas burners for uniform flame; adjust as required.	.079			√	√	√
5 On electrically operated units, check switches, connections, and wiring for proper operation; adjust as required.	.134			√	√	√
6 Check calibration of thermostats; adjust as required.	.457			√	√	√
7 Fill out maintenance checklist and report deficiencies.	.022			√	√	√
Total labor-hours/period				.819	.819	.819
Total labor-hours/year				1.637	.819	.819

	Description	Labor-hrs.	Cost Each					
			2020 Bare Costs				Total	Total
			Material	Labor	Equip.	Total	In-House	w/O&P
1900	Grill, gas / electric, annually	.819	24.50	45		69.50	84.54	102
1950	Annualized	3.280	74	179		253	311.80	380

E10 EQUIPMENT

E1095 348 Ice Machine

PM Components	Labor-hrs.	W	M	Q	S	A
PM System E1095 348 1950						
Ice machine, flake or cube						
1 Check with operating or area personnel for any deficiencies.	.044			√	√	√
2 Remove and install access panel.	.044			√	√	√
3 Lubricate all moving parts, pivot points and fan motor(s).	.023			√	√	√
4 Visually check for refrigerant, oil or water leaks.	.022			√	√	√
5 Open and close water valve.	.007			√	√	√
6 Replace in-line water filter.	.009			√	√	√
7 Check and clear ice machine draining system (drain vent and trap).	.198			√	√	√
8 Clean motor, compressor and condenser coil.	.400			√	√	√
9 Check and tighten any loose screw-type electrical connections.	.005			√	√	√
10 Inspect door(s) hinges, gaskets, handles; lubricate as required.	.053			√	√	√
11 Clean area around equipment.	.066			√	√	√
12 Fill out maintenance checklist and report deficiencies.	.022			√	√	√
Total labor-hours/period				.893	.893	.893
Total labor-hours/year				1.785	.893	.893

	Description	Labor-hrs.	Cost Each					
			2020 Bare Costs				Total	Total
			Material	Labor	Equip.	Total	In-House	w/O&P
1900	Ice machine, flake or cube, annually	.893	90	49		139	161.52	191
1950	Annualized	3.584	271	197		468	549.13	655

PM Components	Labor-hrs.	W	M	Q	S	A
PM System E1095 350 1950						
Kettle, steam, fixed or tilt						
1 Check with operating or area personnel for any deficiencies.	.044			√	√	√
2 Check piping and fittings for leaks; tighten as required.	.085			√	√	√
3 Check operation of electric water valve.	.030			√	√	√
4 Inspect cover, hinges, and seals on units so equipped; lubricate hinge.	.053			√	√	√
5 Lubricate tilting gear mechanism and trunnion bearings, if applicable.	.047			√	√	√
6 Fill out maintenance checklist and report deficiencies.	.022			√	√	√
Total labor-hours/period				.281	.281	.281
Total labor-hours/year				.562	.281	.281

	Description	Labor-hrs.	Cost Each: 2020 Bare Costs: Material	Labor	Equip.	Total	Total In-House	Total w/O&P
1900	Kettle, steam, fixed or tilt, annually	.281	9.55	15.40		24.95	30.30	36.50
1950	Annualized	1.124	28.50	61.50		90	110.55	134

PM Components	Labor-hrs.	W	M	Q	S	A
PM System E1095 354 1950						
Mixer, counter, electric						
1 Check with operating or area personnel for any deficiencies.	.044			√	√	√
2 Check operation of mixer at varying speeds for excessive noise and vibrations; align or adjust as required.	.033			√	√	√
3 Check belt for proper tension and alignment; adjust as required.	.068			√	√	√
4 Lubricate mixer gears.	.042			√	√	√
5 Clean machine thoroughly of dust, grease, and food particles.	.014			√	√	√
6 Check motor, switches, controls and motor bearings for overheating; lubricate motor bearings.	.039			√	√	√
7 Check anchor bolts for tightness; tighten as required.	.010			√	√	√
8 Fill out maintenance checklist and report deficiencies.	.022			√	√	√
Total labor-hours/period				.272	.272	.272
Total labor-hours/year				.544	.272	.272

	Description	Labor-hrs.	Cost Each					
			2020 Bare Costs				Total	Total
			Material	Labor	Equip.	Total	In-House	w/O&P
1900	Mixer, counter, electric, annually	.272	9.55	14.90		24.45	29.63	36
1950	Annualized	1.088	28.50	59.50		88	107.80	131

PM Components	Labor-hrs.	W	M	Q	S	A
PM System E1095 354 2950						
Mixer, floor, electric						
1 Check with operating or area personnel for any deficiencies.	.044			√	√	√
2 Check operation of mixer at varying speeds for excessive noise and vibrations; align or adjust as required.	.120			√	√	√
3 Tighten loose bolts, nuts, and screws.	.005			√	√	√
4 Check oil level in transmission; add or replace oil if required.	.158			√	√	√
5 Lubricate mixer gears and bowl lift mechanism.	.042			√	√	√
6 Check switches and control for damage and proper operation; adjust as required.	.068			√	√	√
7 Check belt for proper tension and alignment; adjust as required.	.029			√	√	√
8 Clean machine of dust and grease.	.030			√	√	√
9 Check anchor bolts for tightness; tighten as required.	.010			√	√	√
10 Clean area around equipment.	.066			√	√	√
11 Fill out maintenance checklist and report deficiencies.	.022			√	√	√
Total labor-hours/period				.595	.595	.595
Total labor-hours/year				1.189	.595	.595

	Description	Labor-hrs.	Cost Each					
			2020 Bare Costs				Total	Total
			Material	Labor	Equip.	Total	In-House	w/O&P
2900	Mixer, floor, electric, annually	.595	17.70	32.50		50.20	61.41	74
2950	Annualized	2.380	53.50	131		184.50	226.12	275

PM Components	Labor-hrs.	W	M	Q	S	A
PM System E1095 356 1950						
Oven, convection, gas/electric						
1 Check with operating or area personnel for any deficiencies.	.044		√	√	√	√
2 Check doors and seals for warping and misalignment; lubricate hinges and repair as necessary.	.027		√	√	√	√
3 Check piping and valves for leaks.	.077		√	√	√	√
4 Check nuts, bolts, and screws for tightness; replace or tighten as required.	.005		√	√	√	√
5 Check pilot and gas burner for uniform flame, adjust as required.	.083		√	√	√	√
6 Check element, switches, controls and wiring on electrically heated units for defects; repair as required.	.120		√	√	√	√
7 Check fan blades and fan motor for proper operation.	.120		√	√	√	√
8 Check operation of thermostat; calibrate as required.	.440		√	√	√	√
9 Fill out maintenance checklist and report deficiencies.	.022		√	√	√	√
Total labor-hours/period			.938	.938	.938	.938
Total labor-hours/year			7.501	1.875	.938	.938

			Cost Each					
			2020 Bare Costs				Total	Total
	Description	Labor-hrs.	Material	Labor	Equip.	Total	In-House	w/O&P
1900	Oven, convection, gas / electric, annually	.938	9.40	51.50		60.90	76.39	94
1950	Annualized	11.242	37.50	615		652.50	831.05	1,025

PM Components	Labor-hrs.	W	M	Q	S	A
PM System E1095 356 2950						
Oven, rotary, electric						
1 Check with operating or area personnel for any deficiencies.	.044		√	√	√	√
2 Remove and reinstall access panel; lubricate main bearing bushings.	.064		√	√	√	√
3 Turn motor on and check operation of chain drive mechanism; lubricate chain.	.064		√	√	√	√
4 Inspect motor and wiring; sensory inspect motor and bearings for overheating; lubricate motor bearings.	.055		√	√	√	√
5 Check V-belt; adjust tension and/or pulley as required.	.029		√	√	√	√
6 Inspect shelves for defects and level if required.	.142		√	√	√	√
7 Tighten or replace any loose nuts, bolts, screws.	.005		√	√	√	√
8 Check thermostat with thermometer; adjust as required.	.444		√	√	√	√
9 Check operation of stop and reverse switch.	.008		√	√	√	√
10 Check timer mechanism and operation.	.018		√	√	√	√
11 Check lubricant in gear case; add as required.	.055		√	√	√	√
12 Check ventilator for proper operation.	.160		√	√	√	√
13 Clean area around equipment.	.066		√	√	√	√
14 Fill out maintenance checklist and report deficiencies.	.022		√	√	√	√
Total labor-hours/period			1.176	1.176	1.176	1.176
Total labor-hours/year			9.410	2.353	1.176	1.176

			Cost Each					
			2020 Bare Costs				Total	Total
	Description	Labor-hrs.	Material	Labor	Equip.	Total	In-House	w/O&P
2900	Oven, rotary, electric, annually	1.176	27	64.50		91.50	112.83	137
2950	Annualized	14.050	109	770		879	1,107.50	1,350

E1095 356 Oven, Convection/Rotary

PM Components	Labor-hrs.	W	M	Q	S	A
PM System E1095 356 3950						
Oven, rotary, gas						
1 Check with operating or area personnel for any deficiencies.	.044		√	√	√	√
2 Examine doors and door gaskets; make necessary adjustments.	.021		√	√	√	√
3 Visually inspect hinges and latches; lubricate hinges.	.091		√	√	√	√
4 Remove and reinstall access panel; lubricate main bearing bushings.	.025		√	√	√	√
5 Turn motor on and chk. operation of chain drive mechanism; lubricate chain.	.064		√	√	√	√
6 Inspect motor and wiring; sensory inspect motor and bearings for overheating; lubricate motor bearings.	.055		√	√	√	√
7 Check V-belt; adjust tension and/or pulley as required.	.029		√	√	√	√
8 Check pilot and thermosensor bulb.	.042		√	√	√	√
9 Check gas burner for uniform flame.	.042		√	√	√	√
10 Check piping and valves for leaks.	.077		√	√	√	√
11 Inspect shelves for defects and level if required.	.142		√	√	√	√
12 Tighten or replace any loose nuts, bolts, screws.	.005		√	√	√	√
13 Check thermostat with thermometer; calibrate if required.	.444		√	√	√	√
14 Check operation of stop and reverse switch.	.008		√	√	√	√
15 Check timer mechanism and operation.	.018		√	√	√	√
16 Check lubricant in gear case, add as required.	.055		√	√	√	√
17 Check ventilator for proper operation.	.160		√	√	√	√
18 Clean area around equipment.	.066		√	√	√	√
19 Fill out maintenance checklist and report deficiencies.	.022		√	√	√	√
Total labor-hours/period			1.410	1.410	1.410	1.410
Total labor-hours/year			11.283	2.821	1.410	1.410

	Description	Labor-hrs.	Cost Each 2020 Bare Costs Material	Labor	Equip.	Total	Total In-House	Total w/O&P
3900	Oven, rotary, gas, annually	1.410	32	77.50		109.50	134.53	163
3950	Annualized	16.858	128	925		1,053	1,326.25	1,625

E10 EQUIPMENT

E1095 366 Peeler, Vegetable

PM Components	Labor-hrs.	W	M	Q	S	A
PM System E1095 366 1950						
Peeler, vegetable, electric						
1 Check with operating or area personnel for any deficiencies.	.044			√	√	√
2 Check vegetable peeler for proper operation, including switches and controls; lubricate motor bearings.	.156			√	√	√
3 Lubricate vegetable peeler bushings, as required.	.062			√	√	√
4 Check door for loose hinges and latch fittings; adjust if required.	.007			√	√	√
5 Clean peeler of dust, grease, and food particles.	.014			√	√	√
6 Check piping and valves for leaks.	.077			√	√	√
7 Inspect abrasive disk.	.018			√	√	√
8 Tighten or replace loose, missing or damaged nuts, bolts, screws.	.005			√	√	√
9 Check belt for proper tension and alignment; adjust or replace as required.	.029			√	√	√
10 Fill out maintenance checklist and report deficiencies.	.022			√	√	√
Total labor-hours/period				.434	.434	.434
Total labor-hours/year				.868	.434	.434

	Description	Labor-hrs.	Cost Each 2020 Bare Costs Material	Labor	Equip.	Total	Total In-House	Total w/O&P
1900	Peeler, vegetable, electric, annually	.434	23.50	24		47.50	56.19	67.50
1950	Annualized	1.736	69.50	95		164.50	198.26	239

E1095 368 Pie Maker, Electric

PM Components	Labor-hrs.	W	M	Q	S	A
PM System E1095 368 1950						
Pie maker, electric						
1 Check with operating or area personnel for any deficiencies.	.044			√	√	√
2 Check electrical system for loose connections and frayed wiring; tighten connections as required.	.120			√	√	√
3 Check operation of motors and bearings for overheating; lubricate motor bearings.	.168			√	√	√
4 Check V-belt for alignment and proper tension; adjust or replace belt as required.	.348			√	√	√
5 Check drain drive and sprocket for excessive wear; lubricate if applicable.	.087			√	√	√
6 Tighten or replace loose, missing or damaged nuts, bolts or screws.	.005			√	√	√
7 Wipe clean the machine.	.096			√	√	√
8 Fill out maintenance checklist and report deficiencies.	.022			√	√	√
Total labor-hours/period				.891	.891	.891
Total labor-hours/year				1.781	.891	.891

	Description	Labor-hrs.	Cost Each					
			2020 Bare Costs				Total	Total
			Material	Labor	Equip.	Total	In-House	w/O&P
1900	Pie maker, electric, annually	.891	28	49		77	93.79	113
1950	Annualized	3.552	84	195		279	343.30	415

E10 EQUIPMENT

E1095 370 Proofer, Automatic

PM Components	Labor-hrs.	W	M	Q	S	A
PM System E1095 370 1950						
Proofer, automatic, electric						
1 Check with operating or area personnel for any deficiencies.	.044			√	√	√
2 Check operation of electric motor, switches, controls and bearings; adjust, repair and lubricate as required.	.281			√	√	√
3 Check belt tension; adjust as required.	.029			√	√	√
4 Check chain drive operation for proper alignment and tension; repair as required.	.129			√	√	√
5 Check action of flour sifter mechanism, connecting linkage and conveyor; lubricate bushings as required.	.291			√	√	√
6 Check roller conveyor and connecting drive mechanism.	.113			√	√	√
7 Check door latch; adjust if needed.	.014			√	√	√
8 Check nuts, bolts and screws for tightness; replace or tighten as required.	.029			√	√	√
9 Fill out maintenance checklist and report deficiencies.	.022			√	√	√
Total labor-hours/period				.951	.951	.951
Total labor-hours/year				1.903	.951	.951

	Description	Labor-hrs.	Cost Each					
			2020 Bare Costs				Total	Total
			Material	Labor	Equip.	Total	In-House	w/O&P
1900	Proofer, automatic, electric, annually	.951	14.20	52		66.20	82.87	101
1950	Annualized	3.800	42.50	209		251.50	315.19	385

PM Components	Labor-hrs.	W	M	Q	S	A
PM System E1095 380 1950						
Refrigerator, domestic						
1 Check with operating or area personnel for deficiencies.	.035					√
2 Clean coils, fans, fan motors, drip pan and other areas with vacuum, brush or wipe as necessary.	.066					√
3 Inspect door gaskets for damage and proper fit; adjust gaskets as required and lubricate hinges.	.022					√
4 Check door latch and adjust as necessary.	.023					√
5 Clean area around equipment.	.066					√
6 Fill out maintenance checklist and report deficiencies.	.022					√
Total labor-hours/period						.234
Total labor-hours/year						.234

	Description	Labor-hrs.	Cost Each					
			2020 Bare Costs				Total	Total
			Material	Labor	Equip.	Total	In-House	w/O&P
1900	Refrigerator, domestic, annually	.234	4.77	12.85		17.62	21.72	26.50
1950	Annualized	.234	4.77	12.85		17.62	21.72	26.50

PM Components	Labor-hrs.	W	M	Q	S	A
PM System E1095 380 2950						
Refrigerator, display case						
1 Check with operating or area personnel for deficiencies.	.035					√
2 Clean coils, fans, fan motors, drip pan and other areas with vacuum, brush or wipe as necessary.	.066					√
3 Inspect door gaskets for damage and proper fit; adjust gaskets as required and lubricate hinges.	.044					√
4 Check door latch and adjust as necessary.	.047					√
5 Clean area around equipment.	.066					√
6 Fill out maintenance checklist and report deficiencies.	.022					√
Total labor-hours/period						.280
Total labor-hours/year						.280

	Description	Labor-hrs.	Cost Each					
			2020 Bare Costs				Total	Total
			Material	Labor	Equip.	Total	In-House	w/O&P
2900	Refrigerator, display case, annually	.280	4.77	15.35		20.12	24.96	30.50
2950	Annualized	.280	4.77	15.35		20.12	24.96	30.50

PM Components	Labor-hrs.	W	M	Q	S	A
PM System E1095 380 3950						
Refrigerator/display, walk-in w/external condenser						
1 Check with operating or area personnel for deficiencies.	.035				√	√
2 Clean condenser coils, fans, and intake screens; lubricate motor.	.066				√	√
3 Inspect door gaskets for damage and proper fit; adjust gaskets as required and lubricate hinges.	.137				√	√
4 Check starter panels and controls for proper operation, burned or loose contacts, and loose connections.	.182				√	√
5 Clean coils, evaporator drain pan, blowers, fans, motors and drain piping as required; lubricate motor(s).	.473				√	√
6 During operation of unit, check refrigerant pressures and compressor oil level; add refrigerant and/or oil as necessary.	.066				√	√
7 Check operation of low pressure cut-out; adjust or replace as required.	.047				√	√
8 Inspect defrost systems for proper operation, adjust as required.	[illegible]				√	√
9 Clean area around equipment.	.066				√	√
10 Fill out maintenance checklist and report deficiencies.	.022				√	√
Total labor-hours/period					1.188	1.188
Total labor-hours/year					1.188	1.188

	Description	Labor-hrs.	Cost Each					
			2020 Bare Costs				Total	Total
			Material	Labor	Equip.	Total	In-House	w/O&P
3900	Refrig. display, walk-in w/ ext. condenser, annually	1.188	98	65		163	191.81	227
3950	Annualized	2.376	196	130		326	383.37	455

PM Components	Labor-hrs.	W	M	Q	S	A
PM System E1095 382 1950						
Refrigerator/freezer, walk-in box w/external condenser						
1 Check with operating or area personnel for deficiencies.	.035				√	√
2 Clean condenser coils, fans, and intake screens; lubricate motor.	.036				√	√
3 Inspect door gaskets for damage and proper fit; adjust gaskets as required and lubricate hinges.	.022				√	√
4 Check starter panels and controls for proper operation, burned or loose contacts, and loose connections.	.182				√	√
5 Clean coils, evaporator drain pan, blowers, fans, motors and drain piping as required; lubricate motor(s).	.114				√	√
6 During operation of unit, check refrigerant pressures and compressor oil level; add refrigerant and/or oil as necessary.	.083				√	√
7 Check operation of low pressure cut-out; adjust or replace as required.	.079				√	√
8 Inspect defrost systems for proper operation; adjust as required.	.094				√	√
9 Clean area around equipment.	.066				√	√
10 Fill out maintenance checklist and report deficiencies.	.022				√	√
Total labor-hours/period					.733	.733
Total labor-hours/year					.733	.733

	Description	Labor-hrs.	Cost Each					
			2020 Bare Costs				Total	Total
			Material	Labor	Equip.	Total	In-House	w/O&P
1900	Refrig. freezer, walk-in box w/ext. condenser, annually	.733	98	40		138	159.66	187
1950	Annualized	1.466	196	80.50		276.50	318.27	375

PM Components	Labor-hrs.	W	M	Q	S	A
PM System E1095 382 2950						
Refrigerator unit/display case/freezer w/external condenser						
1 Check with operating or area personnel for deficiencies.	.035				√	√
2 Clean condenser coils, fans, and intake screens; lubricate motor.	.500				√	√
3 Inspect door gaskets for damage and proper fit; adjust gaskets as required and lubricate hinges.	.033				√	√
4 Check starter panels and controls for proper operation, burned or loose contacts, and loose connections.	.213				√	√
5 Clean coils, evaporator drain pan, blowers, fans, motors and drain piping as required; lubricate motor(s).	.473				√	√
6 During operation of unit, check refrigerant pressures and compressor oil level; add refrigerant and/or oil as necessary.	.066				√	√
7 Check operation of low pressure cut-out; adjust or replace as required.	.057				√	√
8 Inspect defrost systems for proper operation; adjust as required.	.027				√	√
9 Clean area around equipment.	.066				√	√
10 Fill out maintenance checklist and report deficiencies.	.022				√	√
Total labor-hours/period					1.492	1.492
Total labor-hours/year					1.492	1.492

	Description	Labor-hrs.	Cost Each					
			2020 Bare Costs				Total	Total
			Material	Labor	Equip.	Total	In-House	w/O&P
2900	Refrig., display case, freezer w/ ext. cond., annually	1.492	98	82		180	213.16	253
2950	Annualized	2.992	196	165		361	426.27	505

E10 EQUIPMENT — E1095 386 — Steam Table

PM Components	Labor-hrs.	W	M	Q	S	A
PM System E1095 386 1950						
Steam table						
1 Check with operating or area personnel for any deficiencies.	.044			√	√	√
2 Inspect water compartment, steam coil, valves, and piping for leaks.	.117			√	√	√
3 Check steam trap and strainer; clean as required.	.308			√	√	√
4 Check operation of pressure regulating valve and gauge.	.035			√	√	√
5 Check pilots and flame on gas burner units; adjust as required.	.085			√	√	√
6 Check insulators, connections, and wiring, if applicable, tighten connections.	.120			√	√	√
7 Check condition of covers and receptacles; adjust as required.	.025			√	√	√
8 Check thermostat and temperature gauge; calibrate thermostat if necessary.	.152			√	√	√
9 Fill out maintenance checklist and report deficiencies.	.022			√	√	√
Total labor-hours/period				[illegible]	[illegible]	[illegible]
Total labor-hours/year				1.816	.908	.908

	Description	Labor-hrs.	Cost Each					
			2020 Bare Costs				Total	Total
			Material	Labor	Equip.	Total	In-House	w/O&P
1900	Steam table, annually	.908		50		50	63.71	79.50
1950	Annualized	3.624		199		199	255.50	315

E10 EQUIPMENT — E1095 388 — Steamer, Vegetable

PM Components	Labor-hrs.	W	M	Q	S	A
PM System E1095 388 1950						
Steamer, vegetable, direct connected units						
1 Check with operating or area personnel for any deficiencies.	.044			√	√	√
2 Inspect doors, door hardware and gaskets; lubricate hinges.	.112			√	√	√
3 Tighten or replace loose, missing, or damaged nuts, bolts, screws.	.005			√	√	√
4 Check for clogged or defective steam and strainers; clean as required.	.302			√	√	√
5 Check working pressure on steam gauge and inspect piping, valves, and doors for leaks; make necessary repairs as required.	.082			√	√	√
6 Check operation of pressure relief valve, low water cut-off, timer operation and water level in sight glass.	.065			√	√	√
7 Fill out maintenance checklist and report deficiencies.	.022			√	√	√
Total labor-hours/period				.632	.632	.632
Total labor-hours/year				1.264	.632	.632

	Description	Labor-hrs.	Cost Each					
			2020 Bare Costs				Total	Total
			Material	Labor	Equip.	Total	In-House	w/O&P
1900	Steamer, vegetable, annually	.632	9.40	34.50		43.90	54.52	67.50
1950	Annualized	2.536	28	139		167	209.60	256

E10 EQUIPMENT | E1095 390 | Toaster, Rotary

PM Components	Labor-hrs.	W	M	Q	S	A
PM System E1095 390 1950						
Toaster, rotary, gas/electric						
1 Check with operating or area personnel for any deficiencies.	.044			√	√	√
2 Check operation of toaster by toasting sample slice of bread.	.034			√	√	√
3 Clean toaster.	.014			√	√	√
4 Check valves and piping for leaks on gas units.	.077			√	√	√
5 Check pilot light and burner flame on gas units; adjust when required.	.017			√	√	√
6 Check insulators, heating elements, connections, and wiring on electric units; tighten connections and repair as necessary.	.134			√	√	√
7 Inspect alignment of baskets and conveyor chains; lubricate chains and align as necessary.	.044			√	√	√
8 Check motor and motor bearings for overheating; lubricate motor bearings.	.086			√	√	√
9 Tighten or replace loose, missing or damaged nuts, bolts, screws.	.005			√	√	√
10 Fill out maintenance checklist and report deficiencies.	.022			√	√	√
Total labor-hours/period				.477	.477	.477
Total labor-hours/year				.954	.477	.477

	Description	Labor-hrs.	Cost Each					
			2020 Bare Costs				Total	Total
			Material	Labor	Equip.	Total	In-House	w/O&P
1900	Toaster, rotary, gas / electric, annually	.477	14.20	26		40.20	49.17	59.50
1950	Annualized	1.908	42.50	105		147.50	180.57	220

E10 EQUIPMENT | E1095 930 | Pump, Vacuum

PM Components	Labor-hrs.	W	M	Q	S	A
PM System E1095 930 1950						
Vacuum						
1 Check with operating or area personnel for deficiencies.	.035				√	√
2 Check for leaks on suction and discharge piping, seals, etc.	.077				√	√
3 Check pump and motor operation for excessive vibration, noise and overheating.	.022				√	√
4 Check alignment and clearances of shaft and coupler.	.258				√	√
5 Tighten or replace loose, missing, or damaged nuts, bolts and screws.	.005				√	√
6 Lubricate pump and motor as required.	.099				√	√
7 Clean pump, motor and surrounding area.	.096				√	√
8 Fill out maintenance checklist and report deficiencies.	.022				√	√
Total labor-hours/period					.613	.613
Total labor-hours/year					.613	.613

	Description	Labor-hrs.	Cost Each					
			2020 Bare Costs				Total	Total
			Material	Labor	Equip.	Total	In-House	w/O&P
1900	Vacuum, annually	.613	9.40	33.50		42.90	53.51	65.50
1950	Annualized	1.232	18.85	67.50		86.35	107.42	132

	PM Components	Labor-hrs.	W	M	Q	S	A
	PM System F1045 110 1950						
	Swimming pool						
	Quantities of chemicals used vary significantly by climate and pool usage, therefore the price of these materials are not provided.						
1	Clean strainer basket.	.050	√	√	√	√	√
2	Backwash pool water filter.	.120	√	√	√	√	√
3	Add soda ash to feeder.	.080	√	√	√	√	√
4	Clean/flush soda ash pump head.	.400		√	√	√	√
5	Check chlorine, change bottles and check for leaks.	.120	√	√	√	√	√
6	Clean chlorine pump, flush with acid, and check for signs of deterioration	.200		√	√	√	√
7	Check acid source, add acid, check for leaks.	.100	√	√	√	√	√
8	Clean acid pump head.	.400		√	√	√	√
9	Check circulating pump for leaks and unusual noises. Lubricate as required	.050	√	√	√	√	√
10	Check diving board, tighten or replace missing hardware. Inspect for structural defects or deterioration.	.080	√	√	√	√	√
11	Wash exposed stainless steel ladders and scum gutters with tap water to remove oil and dirt.	.302	√	√	√	√	√
12	Fill out maintenance report.	.017	√	√	√	√	√
	Total labor-hours/period		.919	1.919	1.919	1.919	1.919
	Total labor-hours/year		34.940	15.356	3.839	1.919	1.919

	Description	Labor-hrs.	Cost Each					
			2020 Bare Costs				Total	Total
			Material	Labor	Equip.	Total	In-House	w/O&P
1900	Swimming pool outdoor, annually	1.919	4.77	105		109.77	139.77	174
1950	Annualized	57.859	24	3,175		3,199	4,089.50	5,075

G2015 610 Traffic Signal Light

PM Components	Labor-hrs.	W	M	Q	S	A
PM System G2015 610 1950						
Traffic signal light						
1 Inspect and clean control cabinet.	.030					√
2 Check wiring for obvious defects and tighten control electrical connections.	.090					√
3 Check operation of cabinet exhaust fan and inspect vents.	.020					√
4 Inspect and tighten power line connections to control cabinet.	.034					√
5 Check operation of cabinet heater.	.037					√
6 Perform the following operational checks of traffic controller.	.082			√	√	√
A) Check controller phase timing with stop watch.						
B) Check controller pedestrian phase timing with stop watch.						
C) Check controller clocks for correct time and log settings.						
7 Walk a circuit of the intersection to ensure all lights are functioning, replace bulbs as required.	.070			√	√	√
8 Check operation of pedestrian push buttons, average two per intersection.	.084			√	√	√
9 Check operation of all informational signs.	.042			√	√	√
10 Check operation of phase detectors (loop amplifiers).	.144			√	√	√
11 Visually inspect road sensors.	.074			√	√	√
12 Check operation of flashers.	.018					√
13 Fill out maintenance checklist and report deficiencies.	.017			√	√	√
Total labor-hours/period				.514	.514	.743
Total labor-hours/year				1.027	.514	.743

	Description	Labor-hrs.	Cost Each					
			2020 Bare Costs				Total	Total
			Material	Labor	Equip.	Total	In-House	w/O&P
1900	Traffic signal light, annually	.743	10.10	40.50		50.60	63.24	77.50
1950	Annualized	2.281	21	125		146	183.66	226

PM Components	Labor-hrs.	W	M	Q	S	A
PM System G2045 150 1950						
Manual swing gate						
1 Check gate and hinge alignment.	.327				√	√
2 Tighten diagonal brace turnbuckle.	.108				√	√
3 Check cane bolt alignment.	.108				√	√
4 Grease hinges.	.216				√	√
5 Oil latch.	.108				√	√
Total labor-hours/period					.867	.867
Total labor-hours/year					.867	.867

	Description	Labor-hrs.	Cost Each					
			2020 Bare Costs				Total	Total
			Material	Labor	Equip.	Total	In-House	w/O&P
1900	Manual swing gate, annually	.867	5.15	47.50		52.65	66.65	82.50
1950	Annualized	1.732	10.30	95		105.30	132.90	165

PM Components	Labor-hrs.	W	M	Q	S	A
PM System G2045 150 2950						
Electric swing gate						
1 Check gate and hinge alignment.	.327				√	√
2 Tighten diagonal brace turnbuckle.	.108				√	√
3 Check cane bolt alignment.	.108				√	√
4 Grease hinges.	.216				√	√
5 Oil latch.	.108				√	√
6 Check electric motor and connections.	.327				√	√
7 Grease gearbox.	.216				√	√
Total labor-hours/period					1.410	1.410
Total labor-hours/year					1.410	1.410

	Description	Labor-hrs.	Cost Each					
			2020 Bare Costs				Total	Total
			Material	Labor	Equip.	Total	In-House	w/O&P
2900	Electric swing gate, annually	1.410	10.70	77.50		88.20	111.15	136
2950	Annualized	2.841	21.50	156		177.50	222.90	276

PM Components	Labor-hrs.	W	M	Q	S	A
PM System G2045 150 3950						
Manual slide gate						
1 Check alignment of rollers and gate.	.327				√	√
2 Grease rollers.	.327				√	√
3 Oil latch.	.108				√	√
Total labor-hours/period					.761	.761
Total labor-hours/year					.761	.761

	Description	Labor-hrs.	Cost Each					
			2020 Bare Costs				Total	Total
			Material	Labor	Equip.	Total	In-House	w/O&P
3900	Manual slide gate, annually	.761	5.15	42		47.15	59.20	73
3950	Annualized	1.516	10.30	83		93.30	117.50	146

PM Components	Labor-hrs.	W	M	Q	S	A
PM System G2045 150 4950						
Electric slide gate						
1 Check alignment of rollers and gate.	.327				√	√
2 Grease rollers.	.327				√	√
3 Oil latch.	.108				√	√
4 Check electric motor and connections.	.327				√	√
5 Clean and grease chain.	.640			√	√	√
Total labor-hours/period				.640	1.728	1.728
Total labor-hours/year				1.280	1.728	1.728

	Description	Labor-hrs.	Cost Each					
			2020 Bare Costs				Total	Total
			Material	Labor	Equip.	Total	In-House	w/O&P
4900	Electric slide gate, annually	1.728	7.90	94.50		102.40	130.20	161
4950	Annualized	4.766	18.60	261		279.60	355	440

PM Components	Labor-hrs.	W	M	Q	S	A
PM System G3015 108 1950						
Storage tank, tower						
1 Inspect exterior of tank for leaks or damage, including tank base and base plates.	.151					√
2 Inspect condition of ladders, sway bracing hardware and structural forms.	.624					√
3 Tighten and lubricate altitude valve and inspect valve vault.	.039					√
4 Check operation of storage tank lighting.	.121					√
5 Check cathodic protection system.	.123					√
6 Operate elevated tank heat system, if installed, and inspect insulation and coils.	.107					√
7 Clean up around tank area.	.134					√
8 Fill out maintenance checklist and report deficiencies.	.022					√
Total labor-hours/period						1.321
Total labor-hours/year						1.321

	Description	Labor-hrs.	Cost Each					
			2020 Bare Costs				Total	Total
			Material	Labor	Equip.	Total	In-House	w/O&P
1900	Storage tank, tower, annually	1.321	18.65	72		90.65	113.50	138
1950	Annualized	1.321	18.65	72		90.65	113.50	138

PM Components	Labor-hrs.	W	M	Q	S	A
PM System G3015 108 2950						
Storage tank, ground level						
1 Inspect exterior of tank for leaks or damage.	.053					√
2 Check condition of roof and roof hatches including locks; remove trash from roof.	.147					√
3 Inspect condition of ladders and tighten or replace missing hardware.	.088					√
4 Inspect valves, fittings, drains and controls for leakage or damage.	.039					√
5 Check for deterioration of tank foundation.	.077					√
6 Inspect ground drainage slope, and vent screens; where applicable.	.043					√
7 Inspect tank lighting system; replace bulbs as needed.	.121					√
8 Clean up around tank.	.134					√
9 Fill out maintenance checklist and report deficiencies.	.022					√
Total labor-hours/period						.724
Total labor-hours/year						.724

	Description	Labor-hrs.	Cost Each					
			2020 Bare Costs				Total	Total
			Material	Labor	Equip.	Total	In-House	w/O&P
2900	Storage tank, ground level, annually	.724	12.35	39.50		51.85	64.61	79
2950	Annualized	.724	12.35	39.50		51.85	64.61	79

G30 SITE MECH. UTILITIES | G3015 116 | Water Flow Meter

PM Components	Labor-hrs.	W	M	Q	S	A
PM System G3015 116 1950						
Water flow meter, turbine						
1 Inspect meter for leaks, corrosion or broken sight glass; replace broken glass as required.	.035				√	√
2 Check meter operation for unusual noise.	.034				√	√
3 Clean meter to remove mineral buildup.	.039				√	√
4 Clean out trash or sludge from bottom of pit.	.165				√	√
5 Fill out maintenance report and report deficiencies.	.022				√	√
Total labor-hours/period					.295	.295
Total labor-hours/year					.295	.295

	Description	Labor-hrs.	Cost Each					
			2020 Bare Costs				Total	Total
			Material	Labor	Equip.	Total	In-House	w/O&P
1900	Water flow meter, turbine, annually	.295	12.95	16.20		29.15	34.99	42
1950	Annualized	.590	12.95	32.50		45.45	55.53	67.50

G30 SITE MECH. UTILITIES | G3015 118 | Reservoir Controls

PM Components	Labor-hrs.	W	M	Q	S	A
PM System G3015 118 1950						
Fresh water distribution reservoir controls						
1 Check with reservoir personnel for any known deficiencies.	.086			√	√	√
2 Visually inspect and clean interior and exterior of controls cabinet.	.061			√	√	√
3 Inspect electrical system for frayed wires or loose connections; repair as required.	.120			√	√	√
4 Inspect water level receiver at reservoir.	.193			√	√	√
5 Inspect and adjust chronoflo transmitter.	.471			√	√	√
6 Perform operational check of relays, safety switches and electrical contactors.	.542			√	√	√
7 Perform operational check of controls system.	.659			√	√	√
8 Clean up controls area and dispose of debris.	.066			√	√	√
9 Fill out maintenance checklist and report deficiencies.	.022			√	√	√
Total labor-hours/period				2.220	2.220	2.220
Total labor-hours/year				4.440	2.220	2.220

	Description	Labor-hrs.	Cost Each					
			2020 Bare Costs				Total	Total
			Material	Labor	Equip.	Total	In-House	w/O&P
1900	Reservoir controls, fresh water distribution, annually	2.220	19.45	122		141.45	177.55	218
1950	Annualized	8.860	28.50	485		513.50	653.75	810

PM Components	Labor-hrs.	W	M	Q	S	A
PM System G3015 126 0950						
Pump, air lift, well						
1 Check with pump operating personnel for any known deficiencies.	.035				√	√
2 Check operation of compressor and pump	.033				√	√
3 Check water sample for air or oil contamination.	.009				√	√
4 Check compressor oil level; add oil as necessary and lubricate pump and compressor shaft bearings as applicable.	.121				√	√
5 Inspect electrical system for frayed wires or loose connections, repair as required.	.120				√	√
6 Check and cycle high and low shut off valves.	.130				√	√
7 Inspect and clean air intake filter.	.094				√	√
8 Clean pump body and tighten or replace loose hardware.	.086				√	√
9 Calibrate and adjust pressure gauge.	.190				√	√
10 Clean surrounding area.	.066				√	√
11 Fill out maintenance checklist and report deficiencies.	.022				√	√
Total labor-hours/period					.907	.907
Total labor-hours/year					.907	.907

	Description	Labor-hrs.	Cost Each					
			2020 Bare Costs				Total	Total
			Material	Labor	Equip.	Total	In-House	w/O&P
0900	Pump, air lift, well, annually	.907	10.75	50		60.75	75.77	93
0950	Annualized	1.814	10.75	99.50		110.25	139.78	171

PM Components	Labor-hrs.	W	M	Q	S	A
PM System G3015 126 1950						
Pump, centrifugal ejector						
1 Check for proper operation of pump.	.022				√	√
2 Check for leaks on suction and discharge piping, seals, packing glands, etc.; make minor adjustments as required.	.077				√	√
3 Check pump and motor operation for excessive vibration, noise and overheating.	.022				√	√
4 Check alignment of pump and motor; adjust as necessary.	.258				√	√
5 Lubricate pump and motor.	.099				√	√
6 Clean exterior of pump, motor and surrounding area.	.096				√	√
7 Fill out maintenance checklist and report deficiencies.	.022				√	√
Total labor-hours/period					.596	.596
Total labor-hours/year					.596	.596

	Description	Labor-hrs.	Cost Each					
			2020 Bare Costs				Total	Total
			Material	Labor	Equip.	Total	In-House	w/O&P
1900	Pump, centrifugal ejector, annually	.596	22	32.50		54.50	65.95	79.50
1950	Annualized	1.196	22	65.50		87.50	108.15	133

PM Components	Labor-hrs.	W	M	Q	S	A
PM System G3015 126 3950						
Pump, metering (slurry)						
1 Check with pump operating personnel for any obvious deficiencies.	.035				√	√
2 Obtain and put on safety equipment for working with fluoride: goggles, apron, gloves and respirator.	.014				√	√
3 Check oil level in pump with dipstick; add oil as necessary.	.009				√	√
4 Inspect belt condition and tension on pump; adjust as required.	.029				√	√
5 Flush slurry pump with fresh water.	.139				√	√
6 Pressure test pump.	.012				√	√
7 Inspect pump motor.	.039				√	√
8 Operate pump and check for leaks, excessive noise and vibration.	.099				√	√
9 Clean up area.	.066				√	√
10 Fill out maintenance checklist and report deficiencies.	.022				√	√
Total labor-hours/period					.464	.464
Total labor-hours/year					.464	.464

	Description	Labor-hrs.	Cost Each					
			2020 Bare Costs				Total	Total
			Material	Labor	Equip.	Total	In-House	w/O&P
3900	Pump, metering (slurry), annually	.464	27	25.50		52.50	62.24	74.50
3950	Annualized	.928	27	51		78	94.88	115

PM Components	Labor-hrs.	W	M	Q	S	A
PM System G3015 126 4950						
Pump, mixed or axial flow velocity						
1 Check with pump operating personnel for any obvious deficiencies.	.035				√	√
2 Clean pump exterior.	.014				√	√
3 Check impeller and shaft for correct alignment and clearances.	.031				√	√
4 Check oil level; add oil as needed.	.009				√	√
5 Inspect packing for leaks and tighten as needed.	.031				√	√
6 Inspect bearings for wear, damage or overheating; lubricate as required.	.053				√	√
7 Inspect float operation and adjust as needed.	.098				√	√
8 Inspect V-belt tension and condition; adjust as required.	.029				√	√
9 Operate pump and check for leaks, excessive noise and vibration.	.099				√	√
10 Tighten loose nuts and bolts and replace missing hardware.	.005				√	√
11 Check electrical system for frayed wires and loose connections.	.120				√	√
12 Clean up area.	.066				√	√
13 Fill out maintenance checklist and report deficiencies.	.022				√	√
Total labor-hours/period					.612	.612
Total labor-hours/year					.612	.612

	Description	Labor-hrs.	Cost Each					
			2020 Bare Costs				Total	Total
			Material	Labor	Equip.	Total	In-House	w/O&P
4900	Pump, mixed or axial flow, annually	.612	38	33.50		71.50	84.81	101
4950	Annualized	1.224	49	67		116	139.76	168

PM Components	Labor-hrs.	W	M	Q	S	A
PM System G3015 126 5950						
Pump, reciprocating positive displacement						
1 Check for proper operation of pump.	.022				√	√
2 Check for leaks on suction and discharge piping, seals, packing glands, etc.; make minor adjustments as required.	.077				√	√
3 Check pump and motor operation for excessive vibration, noise and overheating	.022				√	√
4 Check operation of pressure controls	.133				√	√
5 Check alignment of pump and motor; adjust as necessary.	.258				√	√
6 Lubricate pump and motor.	.099				√	√
7 Clean exterior of pump, motor and surrounding area.	.094				√	√
8 Fill out maintenance checklist and report deficiencies.	.022				√	√
Total labor-hours/period					.727	.727
Total labor-hours/year					.727	.727

	Description	Labor-hrs.	Cost Each					
			2020 Bare Costs				Total	Total
			Material	Labor	Equip.	Total	In-House	w/O&P
5900	Pump, reciprocating displacement, annually	.727	11.05	40		51.05	63.30	77.50
5950	Annualized	1.458	22	80		102	126.60	156

PM Components	Labor-hrs.	W	M	Q	S	A
PM System G3015 126 6950						
Pump, rotary positive displacement						
1 Check with operating personnel for obvious defects.	.035				√	√
2 Clean surface of pump with solvent.	.014				√	√
3 Operate pump and check for leaks at pipes and connections and for excessive noise and vibration.	.099				√	√
4 Lubricate pump.	.047				√	√
5 Check shaft alignment and clearances and pump rotation.	.031				√	√
6 Inspect packing for leaks and tighten.	.031				√	√
7 Tighten loose bolts and nuts; replace loose and missing hardware as needed.	.009				√	√
8 Inspect electrical system for loose connections and frayed wires; repair as necessary.	.120				√	√
9 Operate pump upon completion of maintenance, read and record pressure gauge readings and check discharge pressure.	.221				√	√
10 Clean up area around pump after repairs.	.066				√	√
11 Fill out maintenance checklist and report deficiencies.	.022				√	√
Total labor-hours/period					.695	.695
Total labor-hours/year					.695	.695

	Description	Labor-hrs.	Cost Each					
			2020 Bare Costs				Total	Total
			Material	Labor	Equip.	Total	In-House	w/O&P
6900	Pump, rotary displacement, annually	.695	16.60	38		54.60	67.15	82
6950	Annualized	1.390	33	76		109	133.81	162

PM Components	Labor-hrs.	W	M	Q	S	A
PM System G3015 126 7950						
Pump, sump, up to 1 H.P.						
1 Check electrical plug, cord and connections.	.120				√	√
2 Activate float switch and check pump for proper operation.	.039				√	√
3 Lubricate pump as required.	.047				√	√
4 Inspect packing and tighten as required.	.031				√	√
5 Check pump for misalignment and bearings for overheating.	.199				√	√
6 Clean out trash from sump.	.065				√	√
7 Fill out maintenance checklist and report deficiencies.	.022				√	√
Total labor-hours/period					.523	.523
Total labor-hours/year					.523	.523

	Description	Labor-hrs.	Material	Labor	Equip.	Total	Total In-House	Total w/O&P
			Cost Each					
			2020 Bare Costs					
7900	Pump, sump, up to 1 H.P., annually	.523	11.05	28.50		39.55	48.96	60
7950	Annualized	1.046	22	57.50		79.50	97.51	119

PM Components	Labor-hrs.	W	M	Q	S	A
PM System G3015 126 8950						
Pump, turbine, well						
1 Check with pump operating personnel for any known deficiencies.	.035				√	√
2 Remove turbine and pump from deep well.	.585				√	√
3 Inspect pump body, bowls, water passages and impeller for corrosion, wear or pitting; repair as necessary.	.057				√	√
4 Check packing for leaks; tighten or replace as necessary.	.327				√	√
5 Adjust impeller for optimum operation.	.229				√	√
6 Inspect wear ring clearances, thrust and pump bearings for wear; repair as necessary.	.142				√	√
7 Lubricate all pump parts as required.	.047				√	√
8 Clean suction strainer or replace if damaged.	.338				√	√
9 Inspect operation of discharge valve.	.026				√	√
10 Inspect electrical system for frayed wires or loose connections; repair as necessary.	.120				√	√
11 Check operation of pressure and thermal controls.	.052				√	√
12 Paint pump with underwater paint.	.259				√	√
13 Run pump and check for noise and vibration.	.022				√	√
14 Reinstall pump in deep well and test operation.	.585				√	√
15 Fill out maintenance checklist and report deficiencies.	.022				√	√
Total labor-hours/period					2.845	2.845
Total labor-hours/year					2.845	2.845

	Description	Labor-hrs.	Material	Labor	Equip.	Total	Total In-House	Total w/O&P
			Cost Each					
			2020 Bare Costs					
8900	Pump, turbine, well, annually	2.845	147	156		303	362.54	430
8950	Annualized	5.686	165	310		475	581.98	700

PM Components	Labor-hrs.	W	M	Q	S	A
PM System G3015 126 9950						
Pump, vacuum						
1 Check with pump operating personnel for any known deficiencies.	.035				√	√
2 Check for leaks on suction and discharge piping, seals, etc.	.077				√	√
3 Check pump and motor operation for vibration, noise, overheating, etc.	.022				√	√
4 Check alignment and clearances of shaft and coupler.	.258				√	√
5 Tighten or replace loose, missing, or damaged nuts, bolts, and screws.	.005				√	√
6 Lubricate pump and motor as required.	.099				√	√
7 Clean pump, motor and surrounding area.	.099				√	√
8 Fill out maintenance checklist and report deficiencies.	.022				√	√
Total labor-hours/period					.617	.617
Total labor-hours/year					.617	.617

	Description	Labor-hrs.	Cost Each					
			2020 Bare Costs				Total	Total
			Material	Labor	Equip.	Total	In-House	w/O&P
9900	Pump, vacuum, annually	.617	16.60	34		50.60	61.61	75
9950	Annualized	1.238	33	68		101	123.17	149

G30 SITE MECH. UTILITIES | G3015 310 | Pump, Turbine Well

PM Components	Labor-hrs.	W	M	Q	S	A
PM System G3015 310 1950						
Turbine well pump						
1 Check with operating personnel for any deficiencies.	.035				√	√
2 Remove turbine pump from deep well.	.585				√	√
3 Inspect pump body, bowls, water passages and impeller for corrosion, wear or pitting; repair as necessary.	.057				√	√
4 Check packing for leaks; tighten or repack as necessary.	.333				√	√
5 Adjust impeller for optimun operation.	.229				√	√
6 Inspect wear ring clearances, thrust and pump bearings for wear; repair as necessary.	.142				√	√
7 Lubricate all pump parts as required.	.047				√	√
8 Clean suction strainer or replace if damaged.	.338				√	√
9 Inspect operation of discharge valve.	.030				√	√
10 Inspect electrical system for frayed wires or loose connections; repair as necessary.	.120				√	√
11 Check operation of pressure and thermal controls.	.124				√	√
12 Paint pump with underwater paint.	.259				√	√
13 Run pump and check for noise and vibration.	.022				√	√
14 Reinstall pump in deep well and test operation.	.585				√	√
15 Fill out maintenance checklist and report deficiencies.	.022				√	√
Total labor-hours/period					2.928	2.928
Total labor-hours/year					2.928	2.928

	Description	Labor-hrs.	Cost Each					
			2020 Bare Costs				Total	Total
			Material	Labor	Equip.	Total	In-House	w/O&P
1900	Turbine well pump, annually	2.928	76	160		236	289.37	350
1950	Annualized	5.848	95.50	320		415.50	516.69	630

G30 SITE MECH. UTILITIES | G3015 320 | Pump, Vertical Lift

PM Components	Labor-hrs.	W	M	Q	S	A
PM System G3015 320 1950						
Vertical lift pump, over 1 H.P.						
1 Check for proper operation of pump.	.130				√	√
2 Check for leaks on suction and discharge piping, seals, packing glands, etc.; make minor adjustments as required.	.077				√	√
3 Check pump and motor operation for excessive vibration, noise and overheating.	.022				√	√
4 Check alignment of pump and motor; adjust as necessary.	.258				√	√
5 Lubricate pump and motor.	.099				√	√
6 When available, check and record suction or discharge gauge pressure and flow rate.	.022				√	√
7 Clean exterior of pump and surrounding area.	.096				√	√
8 Fill out maintenance checklist and report deficiencies.	.022				√	√
Total labor-hours/period					.726	.726
Total labor-hours/year					.726	.726

	Description	Labor-hrs.	Cost Each					
			2020 Bare Costs				Total	Total
			Material	Labor	Equip.	Total	In-House	w/O&P
1900	Vertical lift pump, over 1 H.P., annually	.726	30	40		70	83.90	101
1950	Annualized	1.456	46.50	80		126.50	153.10	185

G30 SITE MECH. UTILITIES | G3015 410 | Fire Hydrant

PM Components	Labor-hrs.	W	M	Q	S	A
PM System G3015 410 1950						
Fire hydrant						
1 Remove hydrant caps and check condition of gaskets; replace as required.	.242					√
2 Flush hydrant, lubricate cap threads and reinstall cap.	.072					√
3 Fill out maintenance checklist and report deficiencies.	.022					√
Total labor-hours/period						.636
Total labor-hours/year						.636

	Description	Labor-hrs.	Cost Each					
			2020 Bare Costs				Total	Total
			Material	Labor	Equip.	Total	In-House	w/O&P
1900	Fire hydrant, annually	[illegible]	[illegible]	[illegible]		[illegible]	[illegible]	78
1950	Annualized	.636	18.05	35		53.05	64.55	78

G30 SITE MECH. UTILITIES | G3015 420 | Valve, Post Indicator

PM Components	Labor-hrs.	W	M	Q	S	A
PM System G3015 420 1950						
Valve, post indicator						
1 Remove set screw, cap and wire seal.	.111				√	√
2 Apply lubricant to threads, open and close valve.	.291				√	√
3 Check operation of and sign and clean glass indicator windows.	.130				√	√
4 Install cap and tighten screw, install handle and wire seal with valve open.	.088				√	√
5 Clean valve exterior and area around valve.	.066				√	√
6 Fill out maintenance checklist and report deficiencies.	.022				√	√
Total labor-hours/period					.708	.708
Total labor-hours/year					.708	.708

	Description	Labor-hrs.	Cost Each					
			2020 Bare Costs				Total	Total
			Material	Labor	Equip.	Total	In-House	w/O&P
1900	Valve, post indicator, annually	.708	13.85	39		52.85	65.35	79.50
1950	Annualized	1.420	23	78		101	125.35	153

G30 SITE MECH. UTILITIES | G3015 612 | Water Treatment, Filtration Plant

PM Components	Labor-hrs.	W	M	Q	S	A
PM System G3015 612 1950						
Filter plant						
1 Check daily operating records for unusual occurances.	.035			√	√	√
2 Inspect plant sand or gravel filters.	.694			√	√	√
3 Inspect clear well for cleanliness.	.018			√	√	√
4 Inspect and lubricate flocculator chain and baffle mechanism and waste-line valves.	.163			√	√	√
5 Inspect and adjust rate-of-flow controller.	.036			√	√	√
6 Calibrate and adjust loss-of-head gauge.	.190			√	√	√
7 Inspect clear well vent screen and manhole.	.143			√	√	√
8 Inspect operation of pumps, water mixers, chemical feeders and flow measuring devices.	1.145			√	√	√
9 Check aerator nozzles for obstructions; clean as required.	.065			√	√	√
10 Drain, clean and inspect mixing chambers.	1.455					√
11 Drain sedimentation basins and clean with fire hose.	2.220					√
12 Drain, inspect and clean flocculator.	2.401					√
13 Clean up filter plant area.	.334			√	√	√
14 Fill out maintenance checklist and report deficiencies.	.022			√	√	√
Total labor-hours/period				2.845	2.845	8.921
Total labor-hours/year				5.691	2.845	8.921

	Description	Labor-hrs.	Cost Each					
			2020 Bare Costs				Total	Total
			Material	Labor	Equip.	Total	In-House	w/O&P
1900	Filter plant, annually	8.921	20	490		510	649.31	805
1950	Annualized	17.459	51	960		1,011	1,282.25	1,600

G30 SITE MECH. UTILITIES | G3015 630 | Water Treatment, De-Ionization

PM Components	Labor-hrs.	W	M	Q	S	A
PM System G3015 630 1950						
Water de-ionization station						
1 Check with operating or area personnel for deficiencies.	.035				√	√
2 Check bulk storage tanks, pumps, valves and piping for damage, corrosion or leaks.	.130				√	√
3 Lubricate pumps.	.047				√	√
4 Check venting of bulk storage tanks.	.043				√	√
5 Check operation of sodium hydroxide tank and piping heating system.	.124				√	√
6 Check operation of resin vent, blower and motor; lubricate as required.	.033				√	√
7 Check calibration of meters.	.571				√	√
8 Clean and check probes of metering system.	.040				√	√
9 Check condition of carbon filter bed; replace annually.	.696				√	√
10 Check water softener system for proper operation.	.195				√	√
11 Clean equipment and surrounding area.	.066				√	√
12 Fill out maintenance checklist and report deficiencies.	.022				√	√
Total labor-hours/period					2.002	2.002
Total labor-hours/year					2.002	2.002

	Description	Labor-hrs.	Cost Each					
			2020 Bare Costs				Total	Total
			Material	Labor	Equip.	Total	In-House	w/O&P
1900	Water de-ionization system, annually	2.002	390	110		500	574.20	665
1950	Annualized	4.012	780	220		1,000	1,143.92	1,325

G30 SITE MECH. UTILITIES | G3015 632 | Reverse Osmosis System

PM Components	Labor-hrs.	W	M	Q	S	A
PM System G3015 632 1950						
Reverse osmosis system, 750 gallons/month						
1 Check with operating personnel for report of treatment effectiveness.	.046			√	√	√
2 Inspect unit, piping and connections for leaks.	.077			√	√	√
3 Check operation of pumps and motors; check shafts for alignment and lubricate as required.	.333			√	√	√
4 Inspect electrical system for frayed wires or loose connections; repair as required.	.120			√	√	√
5 Check filters; replace as required.	.030			√	√	√
6 Check water level controls; adjust as required.	.030			√	√	√
7 Check holding tank temperature control valve operation.	.030			√	√	√
8 Replace charcoal filters as indicated by water test.	.143			√	√	√
9 Check to ensure that chemical supply level is properly maintained in the chemical feed system.	.120			√	√	√
Total labor-hours/period				.936	.936	.936
Total labor-hours/year				1.871	.936	.936

	Description	Labor-hrs.	Cost Each					
			2020 Bare Costs				Total	Total
			Material	Labor	Equip.	Total	In-House	w/O&P
1900	Reverse osmosis system, annually	.936	64	51.50		115.50	136.25	162
1950	Annualized	3.728	134	205		339	409.85	495

G30 SITE MECH. UTILITIES | G3015 634 | Water Ion Exchange System

PM Components	Labor-hrs.	W	M	Q	S	A
PM System G3015 634 1950						
Water ion exchange system						
1 Check with operating personnel for deficiencies.	.035			√	√	√
2 Clean unit and piping and check for leaks.	.077			√	√	√
3 Clean brine pump and check for proper operation.	.120			√	√	√
4 Clean and check strainers pilot valve, vents and orifices.	.593					√
5 Inspect electrical system including pump contactor and contacts; repair as required.	.120			√	√	√
6 Check regeneration cycle; make adjustments as required.	.327			√	√	√
7 Clean surrounding area.	.066			√	√	√
8 Fill out maintenance report and report deficiencies.	.022			√	√	√
Total labor-hours/period				.767	.767	1.360
Total labor-hours/year				1.534	.767	1.360

	Description	Labor-hrs.	Cost Each					
			2020 Bare Costs				Total	Total
			Material	Labor	Equip.	Total	In-House	w/O&P
1900	Water ion exchange system, annually	1.360	11.10	74.50		85.60	107.56	133
1950	Annualized	3.653	22	201		223	281.60	350

G30 SITE MECH. UTILITIES | G3015 650 | Water Softener

PM Components	Labor-hrs.	W	M	Q	S	A
PM System G3015 650 1950						
Water softener						
1 Check with building personnel for report of water softener effectiveness.	.086			√	√	√
2 Check pressure gauges for proper operation.	.190			√	√	√
3 Check density of brine solution in salt tank.	.124			√	√	√
4 Check operation of float control in brine.	.040			√	√	√
5 Inspect water softener piping, fittings and valves for leaks.	.077			√	√	√
6 Lubricate valves and motors.	.333			√	√	√
7 Make minor adjustments to water softener controls if required.	.014			√	√	√
8 Inspect softener base and brine tank for corrosion and repair as needed.	.258			√	√	√
9 Check operation of automatic fill valve in brine tank.	.030			√	√	√
10 Check softener electrical wiring and phasing.	.120			√	√	√
11 Clean up area around softener.	.066			√	√	√
12 Fill out maintenance report and report deficiencies.	.022			√	√	√
Total labor-hours/period				1.361	1.361	1.361
Total labor-hours/year				2.721	1.361	1.361

	Description	Labor-hrs.	Cost Each					
			2020 Bare Costs				Total	Total
			Material	Labor	Equip.	Total	In-House	w/O&P
1900	Water softener, annually	1.361	19.45	74.50		93.95	117.51	143
1950	Annualized	5.436	58.50	298		356.50	446.89	550

G30 SITE MECH. UTILITIES | G3015 660 | Fluoride Saturator

PM Components	Labor-hrs.	W	M	Q	S	A
PM System G3015 660 1950						
Fluoride saturator						
1 Check with operating personnel for any obvious discrepancies.	.035		√	√	√	√
2 Obtain and put on safety equipment such as goggles, apron, gloves and respirator prior to working on the fluoride saturator.	.014		√	√	√	√
3 Check oil level in gearbox, agitator and injector pump; add oil as necessary.	.018		√	√	√	√
4 Check water level in fluoride mixing tank.	.007		√	√	√	√
5 Check operation of fluoride flow regulator, float switch and solenoid.	.127		√	√	√	√
6 Check saturator for leaks.	.077		√	√	√	√
7 Check operation of unit.	.098		√	√	√	√
8 Clean fluoride tank with brush and detergent.	.085		√	√	√	√
9 Clean surrounding area.	.066		√	√	√	√
10 Fill out maintenance checklist and report deficiencies.	.022		√	√	√	√
Total labor-hours/period			.549	.549	.549	.549
Total labor-hours/year			4.394	1.098	.549	.549

	Description	Labor-hrs.	Cost Each					
			2020 Bare Costs				Total	Total
			Material	Labor	Equip.	Total	In-House	w/O&P
1900	Fluoride saturator, annually	.549	15.60	30		45.60	55.78	67.50
1950	Annualized	6.588	24.50	360		384.50	490.25	605

PM Components	Labor-hrs.	W	M	Q	S	A
PM System G3015 670 1950						
Chlorinator						
1 Check with operating or area personnel for deficiencies.	.035			√	√	√
2 Put on safety equipment prior to checking equipment.	.014			√	√	√
3 Clean chlorinator and check for leaks.	.057			√	√	√
4 Check water trap for proper level and bleed air.	.017			√	√	√
5 Check and clean water strainers.	.675			√	√	√
6 Grease fittings.	.023			√	√	√
7 Make minor adjustments to chlorinator as needed.	.094			√	√	√
8 Fill out maintenance checklist and report deficiencies.	.022			√	√	√
Total labor-hours/period				.937	.937	.937
Total labor-hours/year				1.874	.937	.937

	Description	Labor-hrs.	Cost Each					
			2020 Bare Costs				Total	Total
			Material	Labor	Equip.	Total	In-House	w/O&P
1900	Chlorinator, annually	.937	8.30	51.50		59.80	75.05	92.50
1950	Annualized	3.748	25	205		230	291.32	360

PM Components	Labor-hrs.	W	M	Q	S	A
PM System G3015 670 2950						
Chlorine detector						
1 Put on safety equipment prior to checking equipment.	.014			√	√	√
2 Test alarm and inspect fan and sensor wiring.	.254			√	√	√
3 Test detector with chlorine test kit.	.044			√	√	√
4 Flush sensor unit with fresh water and drain and fill reservoir.	.068			√	√	√
5 Check drip rate and reservoir level in sight glass.	.013			√	√	√
6 Check "O" rings; replace as required.	.029			√	√	√
7 Inspect sensor mounting grommet, fan operation, detector mounting bolts and drain tube connection.	.147			√	√	√
8 Check charcoal filter cartridge; replace as needed.	.009			√	√	√
9 Check basket for cracked or worn areas.	.059			√	√	√
10 Check detector for leaks under pressure.	.077			√	√	√
11 Clean unit and surrounding area.	.066			√	√	√
12 Fill out maintenance checklist and report deficiencies.	.022			√	√	√
Total labor-hours/period				.801	.801	.801
Total labor-hours/year				1.603	.801	.801

	Description	Labor-hrs.	Cost Each					
			2020 Bare Costs				Total	Total
			Material	Labor	Equip.	Total	In-House	w/O&P
2900	Chlorine detector, annually	.801	43	44		87	103.41	124
2950	Annualized	3.200	129	176		305	366.50	440

PM Components	Labor-hrs.	W	M	Q	S	A
PM System G3025 410 1950						
Ejector, pump, sewage						
1 Check for proper operation of pump.	.130				√	√
2 Check for leaks on suction and discharge piping, seals, packing glands, etc.; make minor adjustments as required.	.069				√	√
3 Check ejector and motor operation for excessive vibration, noise and overheating.	.022				√	√
4 Check float or pressure controls for proper operation.	.039				√	√
5 Lubricate ejector and motor.	.099				√	√
6 Clean pump unit and surrounding area.	.096				√	√
7 Fill out maintenance checklist and report deficiencies.	.022				√	√
Total labor-hours/period					.477	.477
Total labor-hours/year					.477	.477

	Description	Labor-hrs.	Cost Each					
			2020 Bare Costs				Total	Total
			Material	Labor	Equip.	Total	In-House	w/O&P
1900	Ejector pump, sewage, annually	.477	24.50	26		50.50	60.35	72
1950	Annualized	.954	48.50	52.50		101	120	144

PM Components	Labor-hrs.	W	M	Q	S	A
PM System G3025 412 1950						
Ejector, pump						
1 Check for proper operation of pump.	.130				√	√
2 Check for leaks on suction and discharge piping, seals, packing glands, etc., make minor adjustments as required.	.077				√	√
3 Check ejector and motor operation for excessive vibration, noise and overheating.	.022				√	√
4 Check float or pressure controls for proper operation.	.099				√	√
5 Lubricate ejector and motor.	.000				√	√
6 Clean pump unit and surrounding area.	.096				√	√
7 Fill out maintenance checklist and report deficiencies.	.022				√	√
Total labor-hours/period					.485	.485
Total labor-hours/year					.485	.485

	Description	Labor-hrs.	Cost Each					
			2020 Bare Costs				Total	Total
			Material	Labor	Equip.	Total	In-House	w/O&P
1900	Ejector pump, annually	.485	24.50	26.50		51	60.85	73
1950	Annualized	.970	48.50	53		101.50	121.50	145

PM Components	Labor-hrs.	W	M	Q	S	A
PM System G3025 412 2950						
Ejector pump, sump type						
1 Operate float switch, if applicable, and check for proper operation of pump.	.130				√	√
2 Check for leaks on suction and discharge piping, seals, packing glands, etc.; make minor adjustments as required.	.077				√	√
3 Check pump and motor operation for excessive vibration, noise and overheating.	.022				√	√
4 Check alignment of pump and motor; adjust as necessary.	.258				√	√
5 Lubricate motor.	.099				√	√
6 Check electrical wiring and connections and repair as applicable.	.120				√	√
7 Clean exterior of pump and surrounding area.	.096				√	√
8 Fill out maintenance checklist and report deficiencies.	.022				√	√
Total labor-hours/period					.824	.824
Total labor-hours/year					.824	.824

	Description	Labor-hrs.	Cost Each					
			2020 Bare Costs				Total	Total
			Material	Labor	Equip.	Total	In-House	w/O&P
2900	Ejector pump, sump type, annually	.824	24.50	45		69.50	84.70	103
2950	Annualized	1.652	48.50	90.50		139	169.40	205

G3025 420 Pump, Sewage Lift

PM Components	Labor-hrs.	W	M	Q	S	A
PM System G3025 420 1950						
Sewage lift pump, over 1 H.P.						
1 Check for proper operation of pump.	.130				√	√
2 Check for leaks on suction and discharge piping, seals, packing glands, etc.; make minor adjustments as required.	.077				√	√
3 Check pump and motor operation for excessive vibration, noise and overheating.	.022				√	√
4 Check alignment of pump and motor; adjust as necessary.	.258				√	√
5 Lubricate pump and motor.	.099				√	√
6 When available, check and record suction or discharge gauge pressure and flow rate.	.022				√	√
7 Clean exterior of pump, motor and surrounding area.	.096				√	√
8 Fill out maintenance checklist and report deficiencies.	.022				√	√
Total labor-hours/period					.726	.726
Total labor-hours/year					.726	.726

	Description	Labor-hrs.	Cost Each					
			2020 Bare Costs				Total	Total
			Material	Labor	Equip.	Total	In-House	w/O&P
1900	Sewage lift pump, over 1 H.P., annually	.726	24.50	40		64.50	78.05	94
1950	Annualized	1.456	35.50	80		115.50	141.15	171

G30 SITE MECH. UTILITIES G3025 530 Aerator, Floating

PM Components	Labor-hrs.	W	M	Q	S	A
PM System G3025 530 1950						
Aerator, floating						
1 Check with operating or area personnel for deficiencies.	.035			√	√	√
2 Remove aerator from pond.	.390			√	√	√
3 Clean exterior of pump and motor.	.066			√	√	√
4 Check pump and motor operation for excessive vibration, noise and overheating.	.022			√	√	√
5 Check alignment of pump and motor; adjust as necessary.	.050			√	√	√
6 Lubricate pump and motor.	.000			√	√	√
7 Position aerator back in pond.	.390			√	√	√
8 Fill out maintenance checklist and report deficiencies.	.022			√	√	√
Total labor-hours/period				1.282	1.282	1.282
Total labor-hours/year				2.565	1.282	1.282

	Description	Labor-hrs.	Cost Each					
			2020 Bare Costs				Total	Total
			Material	Labor	Equip.	Total	In-House	w/O&P
1900	Aerator, floating, annually	1.282	16.60	70.50		87.10	108.56	133
1950	Annualized	5.136	49.50	281		330.50	416.75	510

G30 SITE MECH. UTILITIES G3025 532 Barminutor

PM Components	Labor-hrs.	W	M	Q	S	A
PM System G3025 532 1950						
Barminutor						
1 Check with operating or area personnel for deficiencies.	.044		√	√	√	√
2 Check condition and clearance of cutting knives and inspect base seal.	.217		√	√	√	√
3 Check oil level in gearbox; add oil as necessary.	.035		√	√	√	√
4 Change oil in gearbox.	.681					√
5 Wire brush and lubricate directional flow valve stem.	.060		√	√	√	√
6 Check for rust and corrosion; scrape, wire brush and spot paint as necessary.	.105		√	√	√	√
7 Fill out maintenance checklist and report deficiencies.	.022		√	√	√	√
Total labor-hours/period			.483	.483	.483	1.164
Total labor-hours/year			3.864	.966	.483	1.164

	Description	Labor-hrs.	Cost Each					
			2020 Bare Costs				Total	Total
			Material	Labor	Equip.	Total	In-House	w/O&P
1900	Barminutor, annually	1.164	49.50	64		113.50	136.60	164
1950	Annualized	6.477	118	355		473	585.55	715

G30 SITE MECH. UTILITIES | G3025 534 | Blower, Aerator

PM Components	Labor-hrs.	W	M	Q	S	A
PM System G3025 534 1950						
Blower, aerator						
1 Check with operating or area personnel for deficiencies.	.035		√	√	√	√
2 Check blower oil level; add oil as necessary.	.046		√	√	√	√
3 Change blower oil.	.390				√	√
4 Check bolts and tighten as necessary: foundation, cylinder head, belt guard, etc.	.155		√	√	√	√
5 Check tension, condition, and alignment of V-belts on blower; adjust or replace as necessary.	.030		√	√	√	√
6 Check pump for excessive heat or vibration.	.042		√	√	√	√
7 Clean air intake filter on blower; replace as necessary.	.178		√	√	√	√
8 Perform operation check of air blower and adjust as required.	.222		√	√	√	√
9 Clean area around equipment.	.066		√	√	√	√
10 Fill out maintenance checklist and report deficiencies.	.022		√	√	√	√
Total labor-hours/period			.796	.796	1.187	1.187
Total labor-hours/year			6.371	1.593	1.187	1.187

	Description	Labor-hrs.	Cost Each					
			2020 Bare Costs				Total	Total
			Material	Labor	Equip.	Total	In-House	w/O&P
1900	Blower, aerator, annually	1.187	71	65		136	161.62	193
1950	Annualized	10.308	233	565		798	981.55	1,200

G30 SITE MECH. UTILITIES | G3025 536 | Clarifier

PM Components	Labor-hrs.	W	M	Q	S	A
PM System G3025 536 1950						
Clarifier						
1 Check with operating or area personnel for deficiencies.	.035		√	√	√	√
2 Check chain and sprocket for wear, where applicable.	.033		√	√	√	√
3 Grease drive gear fittings.	.150		√	√	√	√
4 Check gearbox oil level; add if necessary.	.046		√	√	√	√
5 Drain and refill gearbox oil reservoir.	.681					√
6 Check unit for rust and corrosion; scrape, wire brush, and paint as required.	.105					√
7 Clean clarifier drive and surrounding area.	.098		√	√	√	√
8 Fill out maintenance checklist and report deficiencies.	.022		√	√	√	√
Total labor-hours/period			.384	.384	.384	1.170
Total labor-hours/year			3.072	.768	.384	1.170

	Description	Labor-hrs.	Cost Each					
			2020 Bare Costs				Total	Total
			Material	Labor	Equip.	Total	In-House	w/O&P
1900	Clarifier, annually	1.170	54.50	64.50		119	141.98	170
1950	Annualized	5.394	127	296		423	519.30	630

G30 SITE MECH. UTILITIES | G3025 538 | Comminutor

PM Components	Labor-hrs.	W	M	Q	S	A
PM System G3025 538 1950						
Comminutor						
1 Check with operating or area personnel for deficiencies.	.035			√	√	√
2 Change gearbox oil and lubricate fittings.	.681				√	√
3 Check comminutor and motor for excessive noise, vibration and overheating.	.022			√	√	√
4 Lubricate fittings and check oil in gearbox.	.150			√	√	√
5 Check operation.	.108			√	√	√
6 Fill out maintenance checklist and report deficiencies.	.022			√	√	√
Total labor-hours/period				.337	1.018	1.018
Total labor-hours/year				.674	1.018	1.018

	Description	Labor-hrs.	Cost Each					
			2020 Bare Costs				Total	Total
			Material	Labor	Equip.	Total	In-House	w/O&P
1900	Comminutor, annually	1.018	37	56		93	112.16	135
1950	Annualized	2.710	85	148		233	284.25	345

G30 SITE MECH. UTILITIES | G3025 540 | Filter, Trickling

PM Components	Labor-hrs.	W	M	Q	S	A
PM System G3025 540 1950						
Filter, trickling						
1 Lubricate fittings and check oil level in oil reservoir.	.035				√	√
2 Check distribution arms, support cables, clevis pins, and turnbuckles for deterioration.	.327				√	√
3 Check for leaks.	.077				√	√
4 Check operation.	.108				√	√
5 Fill out maintenance checklist and report deficiencies.	.022				√	√
Total labor-hours/period					.569	.569
Total labor-hours/year					.569	.569

	Description	Labor-hrs.	Cost Each					
			2020 Bare Costs				Total	Total
			Material	Labor	Equip.	Total	In-House	w/O&P
1900	Filter, trickling, annually	.569	24	31		55	66.55	79.50
1950	Annualized	1.134	48	62		110	132.65	160

G3025 542 Grit Drive

PM Components	Labor-hrs.	W	M	Q	S	A
PM System G3025 542 1950						
Grit drive						
1 Check with operating personnel for obvious deficiencies.	.035				√	√
2 Lubricate fittings and check oil level in gearbox.	.047				√	√
3 Change oil in gearbox.	.681					√
4 Check chain linkage and operation.	.033				√	√
5 Clean area around drive mechanism.	.066				√	√
6 Fill out maintenance checklist and report deficiencies.	.022				√	√
Total labor-hours/period					.203	.884
Total labor-hours/year					.203	.884

	Description	Labor-hrs.	Cost Each					
			2020 Bare Costs				Total	Total
			Material	Labor	Equip.	Total	In-House	w/O&P
1900	Grit drive, annually	.884	50	48.50		98.50	117.47	140
1950	Annualized	1.087	74	60		134	157.96	188

G30 SITE MECH. UTILITIES

G3025 544 Mixer, Sewage

PM Components	Labor-hrs.	W	M	Q	S	A
PM System G3025 544 1950						
Mixer, sewage						
1 Check with operating or area personnel for deficiencies.	.035			√	√	√
2 Clean drive motor exterior and check for corrosion on motor exterior and base plate.	.053			√	√	√
3 Check motor operation for vibration, noise, overheating; etc.	.022			√	√	√
4 Check alignment and clearances of shaft and coupler.	.031			√	√	√
5 Tighten or replace loose, missing, or damaged nuts, bolts and screws.	.005			√	√	√
6 Lubricate motor and shaft bearings as required.	.047			√	√	√
7 Clean area around motor.	.066			√	√	√
8 Fill out maintenance checklist and report deficiencies.	.022			√	√	√
Total labor-hours/period				.281	.281	.281
Total labor-hours/year				.562	.281	.281

	Description	Labor-hrs.	Cost Each					
			2020 Bare Costs				Total	Total
			Material	Labor	Equip.	Total	In-House	w/O&P
1900	Mixer, sewage, annually	.281	11.05	15.45		26.50	31.92	38.50
1950	Annualized	1.124	33	61.50		94.50	115.81	140

G3065 310 Fuel Oil Tank

PM Components	Labor-hrs.	W	M	Q	S	A
PM System G3065 310 1950						
Fuel oil storage tank, above ground						
1 Inspect tank for corrosion.	.192					√
2 Inspect floating tank roof, filter or cone and automatic float gauge, if applicable.	.233					√
3 Inspect fuel tank fire protection devices.	.022					√
4 Inspect berms around fuel storage tank for erosion.	.047					√
5 Inspect non-freeze draw valves.	.060					√
6 Inspect storage tank vacuum and pressure vents.	.179					√
7 Inspect vapor recovery line for deterioration.	.082					√
8 Run pump and check operation.	.159					√
9 Calibrate and adjust fuel oil pressure.	.382					√
10 Inspect fuel strainer and clean as needed.	.415					√
11 Inspect tank electrical grounds for breaks or loose fittings.	1.202					√
12 Inspect liquid level gauge for proper operation.	.191					√
13 Fill out maintenance checklist.	.022					√
Total labor-hours/period						3.276
Total labor-hours/year						3.276

	Description	Labor-hrs.	Cost Each					
			2020 Bare Costs				Total	Total
			Material	Labor	Equip.	Total	In-House	w/O&P
1900	Fuel oil storage tank, above ground, annually	3.276	67	180		247	303.78	370
1950	Annualized	3.276	67	180		247	303.78	370

PM Components	Labor-hrs.	W	M	Q	S	A
PM System G4095 110 1950 **Cathodic protection system**						
1 Inspect exterior of rectifier cabinet, all exposed connecting conduit, and fittings for corrosion, deteriorated paint, mechanical integrity and cleanliness, make minor repairs and clean/touch-up paint as required.	.060				√	√
2 Open and inspect interior of rectifier cabinet, inspect wiring for fraying, signs of overheating, deterioration mechanical damage, and loose connections; make adjustments as required.	.138				√	√
3 Check electrical contacts for overheating and pitting; make adjustments.	.075				√	√
4 Clean interior of rectifier cabinet.	.008				√	√
5 Check panel voltmeter and ampmeter with standard test instruments.	.075				√	√
6 Read and record panel meter voltage and current values.	.007				√	√
7 Measure and record rectifier voltage and current with portable instrument.	.059				√	√
8 Check potential at cathode locations; tighten loose connections.	.286				√	√
9 Fill out maintenance checklist and record deficiencies.	.022				√	√
Total labor-hours/period					.729	.729
Total labor-hours/year					.729	.729

	Description	Labor-hrs.	Cost Each					
			2020 Bare Costs				Total	Total
			Material	Labor	Equip.	Total	In-House	w/O&P
1900	Cathodic protection system, annually	.729	8.80	44.50		53.30	64.63	80
1950	Annualized	1.460	13.20	89.50		102.70	124.59	155

General Maintenance

Table of Contents

Landscaping Maintenance

Exterior General Maintenance

Interior General Maintenance

Facility Maintenance Equipment

How to Use the General Maintenance Section

The following is a detailed explanation of a sample entry in the General Maintenance Section. Next to each bold number that follows is the described item with the appropriate component of the sample entry in parentheses.

1 Division Number/Title (01 93/Facility Maintenance)

Use the General Maintenance Section Table of Contents to locate specific items. This section is classified according to the CSI MasterFormat® 2018 system.

2 Line Numbers (01 93 04.40 0030)

Each General Maintenance unit price line item has been assigned a unique 12-digit code based on the CSI MasterFormat classification.

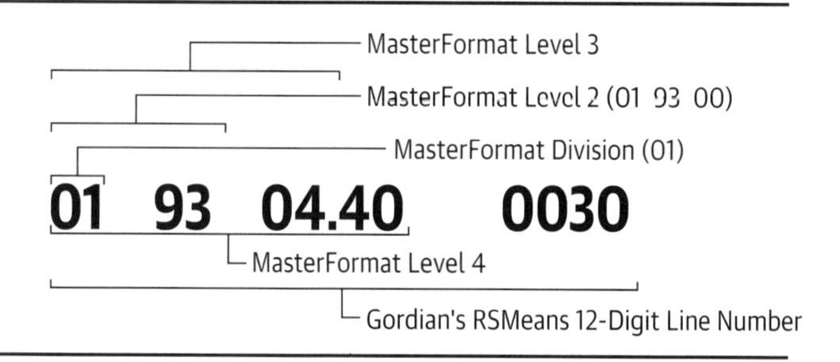

3 Description (Lawn Renovation, etc.)

Each line item is described in detail. Sub-items and additional sizes are indented beneath the appropriate line items. The first line or two after the main item (in boldface) may contain descriptive information that pertains to all line items beneath this boldface listing.

4 Crew (A-18)

The "Crew" column designates the typical trade or crew used for the task. If the task can be accomplished by one trade and requires no power equipment, that trade and the number of workers are listed (for example, "2 Clam"). If the task requires a composite crew, a crew code designation is listed (for example, "A-18"). You'll find full details on all composite crews in the Crew Listings.

- For a complete list of all trades utilized in this cost data and their abbreviations, see the inside back cover of the cost data.

Crews - Maintenance

Crew No.	Bare Costs		In-House Costs		Incl. Subs O&P		Cost Per Labor-Hour		
Crew A-18	Hr.	Daily	Hr.	Daily	Hr.	Daily	Bare Costs	In House	Incl. O&P
1 Maintenance Laborer	$31.60	$252.80	$40.85	$326.80	$50.65	$405.20	$31.60	$40.85	$50.65
1 Farm Tractor w/ Attachment		373.40		373.40		410.74	46.67	46.67	51.34
8 L.H., Daily Totals		$626.20		$700.20		$815.94	$78.28	$87.53	$101.99

5 Productivity: Daily Output (750)/Labor-Hours (.011)

The "Daily Output" represents the typical number of units the designated crew will perform in a normal eight-hour day. To find out the number of days the given crew would require to complete your task, divide your quantity by the daily output. See the second table for an example:

1

01 93 Facility Maintenance

01 93 04 – Landscaping Maintenance

2 01 93 04.15 Edging		Crew	Daily Output	Labor-Hours	Unit	Material	2020 Bare Costs Labor	Equipment	Total	Total In-House	Total Incl O&P
0010	**EDGING**										
0020	Hand edging, at walks	1 Clam	16	.500	C.L.F.		15.80		15.80	20.50	25.50
0030	At planting, mulch or stone beds		7	1.143			36		36	46.50	58
0040	Power edging, at walks		88	.091			2.87		2.87	3.71	4.60
0050	At planting, mulch or stone beds	↓	24	.333	↓		10.55		10.55	13.60	16.90
3 **01 93 04.40 Lawn Renovation**											
0010	**LAWN RENOVATION**	**4**	**5**		**6**		**7**			**8**	**9**
0020	Lawn renovations, aerating, 18" walk behind cultivator	1 Clam	95	.084	M.S.F.		2.66		2.66	3.44	4.27
0030	48" tractor drawn cultivator	A-18	750	.011			.34	.50	.84	.99	1.09
0040	72" tractor drawn cultivator	"	1100	.007			.23	.34	.57	.67	.74
0050	Fertilizing, dry granular, 4#/M.S.F., drop spreader	1 Clam	24	.333		2.32	10.55		12.87	16.15	19.80
0060	Rotary spreader	"	140	.057	↓	2.32	1.81		4.13	4.88	5.80

Quantity	÷	Daily Output	=	Duration
2000 M.S.F.	÷	750 M.S.F./ Crew Day	=	2.7 Crew Days

The "Labor-Hours" figure represents the number of labor-hours required to perform one unit of work. To find out the number of labor-hours required for your particular task, multiply the quantity of the item and the number of labor-hours shown. For example:

Quantity	x	Productivity Rate	=	Duration
2000 M.S.F.	x	.011 Labor-Hours/ MSF	=	22 Labor-Hours

6 Unit (M.S.F.)

The abbreviated designation indicates the unit of measure upon which the price, production, and crew are based (M.S.F. = Thousand Square Feet). For a complete listing of abbreviations, refer to the Abbreviations Listing in the Reference Section of this cost data.

7 Bare Costs:

The unit material cost is the "bare" material cost with no overhead and profit included. *Costs shown reflect national average material prices for January of the current year and include delivery to the facility. Small purchases may require an additional delivery charge. No sales taxes are included.*

Labor (.34)

The unit labor cost is derived by multiplying bare labor-hour costs for Crew A-18 by labor-hour units. The bare labor-hour cost is found in the Crew Section under A-18. (If a trade is listed, the hourly labor cost—the wage rate—is found on the inside back cover of the cost data.)

Labor-Hour Cost Crew A-18	X	Labor-Hour Units	=	Labor
$31.60	X	.011	=	$0.34

Equip. (Equipment) (.50)

Equipment costs for each crew are listed in the description of each crew. Tools or equipment whose value justifies purchase or ownership by a contractor are considered overhead as shown on the inside back cover of the cost data. The unit equipment cost is derived by multiplying the bare equipment hourly cost and the labor-hour units.

Equipment Cost Crew A-18	X	Labor-Hour Units	=	Equip.
$46.67	X	.011	=	$0.50

Total (.84)

The total of the bare costs is the arithmetic total of the three previous columns: material, labor, and equip.

Material	+	Labor	+	Equip.	=	Total
$0.00	+	$0.34	+	$0.50	=	$0.84

8 Total In-House Costs (.99)

"Total In-House Costs" include suggested markups to bare costs for the direct overhead requirements of in-house maintenance staff. The figure in this column is the sum of three components: the bare material cost plus 10% (for purchasing and handling small quantities); the bare labor cost plus workers' compensation and average fixed overhead (per the labor rate table on the inside back cover of the cost data or, if a crew is listed, from the Crew Listings); and the bare equipment cost.

Material is Bare Material Cost + 10% = $0.00 + $0.00	=	$0.00
Labor for Crew A-18 = Labor-Hour Cost ($40.85) x Labor-Hour Units (.011)	=	$0.44
Equip. is Bare Equip. Cost	=	$0.50
Total	=	$0.99

9 Total Costs Including O&P (1.09)

"Total Costs Including O&P" include suggested markups to bare costs for an outside subcontractor's overhead and profit requirements. The figure in this column is the sum of three components: the bare material cost plus 25% for profit; the bare labor cost plus total overhead and profit (per the labor rate table on the inside back cover of the printed product or the Reference Section of the electronic product, or, if a crew is listed, from the Crew Listings); and the bare equipment cost plus 10% for profit.

Material is Bare Material Cost + 25% = $0.00 + $0.00	=	$0.00
Labor for Crew A-18 = Labor-Hour Cost ($50.65) x Labor-Hour Units (.011)	=	$0.54
Equip. is Bare Equip. Cost + 10% = $0.50 + $0.05	=	$0.55
Total	=	$1.09

01 93 04 – Landscaping Maintenance

01 93 04.15 Edging		Crew	Daily Output	Labor-Hours	Unit	Material	2020 Bare Costs Labor	Equipment	Total	Total In-House	Total Incl O&P
0010	**EDGING**										
0020	Hand edging, at walks	1 Clam	16	.500	C.L.F.		15.80		15.80	20.50	25.50
0030	At planting, mulch or stone beds		7	1.143			36		36	46.50	58
0040	Power edging, at walks		88	.091			2.87		2.87	3.71	4.60
0050	At planting, mulch or stone beds		24	.333			10.55		10.55	13.60	16.90
01 93 04.20 Flower, Shrub and Tree Care											
0010	**FLOWER, SHRUB & TREE CARE**										
0020	Flower or shrub beds, bark mulch, 3" deep hand spreader	1 Clam	100	.080	S.Y.	4.12	2.53		6.65	7.80	9.20
0030	Peat moss, 1" deep hand spreader		900	.009		5.15	.28		5.43	6.05	6.90
0040	Wood chips, 2" deep hand spreader		220	.036		1.68	1.15		2.83	3.34	3.94
0050	Cleaning		1	8	M.S.F.		253		253	325.00	405
0060	Fertilizing, dry granular, 3#/M.S.F.		85	.094		1.23	2.97		4.20	5.20	6.30
0070	Weeding, mulched bed		20	.400			12.65		12.65	16.35	20.50
0080	Unmulched bed		8	1			31.50		31.50	41.00	50.50
0090	Trees, pruning from ground, 1-1/2" caliper		84	.095	Ea.		3.01		3.01	3.89	4.82
0100	2" caliper		70	.114			3.61		3.61	4.67	5.80
0110	2-1/2" caliper		50	.160			5.05		5.05	6.55	8.10
0120	3" caliper		30	.267			8.45		8.45	10.90	13.50
0130	4" caliper	2 Clam	21	.762			24		24	31.00	38.50
0140	6" caliper		12	1.333			42		42	54.50	67.50
0150	9" caliper		7.50	2.133			67.50		67.50	87.00	108
0160	12" caliper		6.50	2.462			78		78	101.00	125
0170	Fertilize, slow release tablets	1 Clam	100	.080		2.30	2.53		4.83	5.80	6.95
0180	Pest control, spray		24	.333		28.50	10.55		39.05	45.00	52.50
0190	Systemic		48	.167		29	5.25		34.25	38.50	44.50
0200	Watering, under 1-1/2" caliper		34	.235			7.45		7.45	9.60	11.90
0210	1-1/2" to 4" caliper		14.50	.552			17.45		17.45	22.50	28
0220	4" caliper and over		10	.800			25.50		25.50	32.50	40.50
0230	Shrubs, prune, entire bed		7	1.143	M.S.F.		36		36	46.50	58
0240	Per shrub, 3' height		190	.042	Ea.		1.33		1.33	1.72	2.13
0250	4' height		90	.089			2.81		2.81	3.63	4.50
0260	6' height		50	.160			5.05		5.05	6.55	8.10
0270	Fertilize, dry granular, 3#/M.S.F.		85	.094	M.S.F.	1.23	2.97		4.20	5.20	6.30
0280	Watering, entire bed		7	1.143	"		36		36	46.50	58
0290	Per shrub		32	.250	Ea.		7.90		7.90	10.20	12.65
01 93 04.35 Lawn Care											
0010	**LAWN CARE**										
0020	Mowing lawns, power mower, 18"-22"	1 Clam	80	.100	M.S.F.		3.16		3.16	4.09	5.05
0030	22"-30"		120	.067			2.11		2.11	2.72	3.38
0040	30"-32"		140	.057			1.81		1.81	2.33	2.89
0050	Self propelled or riding mower, 36"-44"	A-16	300	.027			.84	.35	1.19	1.48	1.74
0060	48"-58"		480	.017			.53	.22	.75	.92	1.08
0070	Tractor, 3 gang reel, 7' cut		930	.009			.27	.11	.38	.48	.57
0080	5 gang reel, 12' cut		1200	.007			.21	.09	.30	.37	.44
0090	Edge trimming with weed whacker	1 Clam	5760	.001	L.F.		.04		.04	.06	.07
01 93 04.40 Lawn Renovation											
0010	**LAWN RENOVATION**										
0020	Lawn renovations, aerating, 18" walk behind cultivator	1 Clam	95	.084	M.S.F.		2.66		2.66	3.44	4.27
0030	48" tractor drawn cultivator	A-18	750	.011			.34	.50	.84	.99	1.09
0040	72" tractor drawn cultivator	"	1100	.007			.23	.34	.57	.67	.74
0050	Fertilizing, dry granular, 4#/M.S.F., drop spreader	1 Clam	24	.333		2.32	10.55		12.87	16.15	19.80
0060	Rotary spreader	"	140	.057		2.32	1.81		4.13	4.88	5.80

01 93 Facility Maintenance

01 93 04 – Landscaping Maintenance

01 93 04.40 Lawn Renovation		Crew	Daily Output	Labor-Hours	Unit	2020 Bare Costs Material	Labor	Equipment	Total	Total In-House	Total Incl O&P
0070	Tractor drawn 8′ spreader	A-18	500	.016	M.S.F.	2.32	.51	.75	3.58	4.02	4.53
0080	Tractor drawn 12′ spreader	"	800	.010		2.32	.32	.47	3.11	3.47	3.92
0090	Overseeding, utility mix, 7#/M.S.F., drop spreader	1 Clam	10	.800		12.45	25.50		37.95	46.50	56
0100	Tractor drawn spreader	A-18	52	.154		12.45	4.86	7.20	24.51	28.00	31.50
0110	Watering, 1″ of water, applied by hand	1 Clam	21	.381			12.05		12.05	15.55	19.30
0120	Soaker hoses	"	82	.098			3.08		3.08	3.99	4.94

01 93 04.65 Raking		Crew	Daily Output	Labor-Hours	Unit	Material	Labor	Equipment	Total	Total In-House	Total Incl O&P
0010	**RAKING**										
0020	Raking, leaves, by hand	1 Clam	7.50	1.067	M.S.F.		33.50		33.50	43.50	54
0030	Power blower		45	.178			5.60		5.60	7.25	9
0040	Grass clippings, by hand		7.50	1.067			33.50		33.50	43.50	54

01 93 04.85 Walkways and Parking Care		Crew	Daily Output	Labor-Hours	Unit	Material	Labor	Equipment	Total	Total In-House	Total Incl O&P
0010	**WALKWAYS AND PARKING CARE**										
0020	General cleaning, parking area and walks, sweeping by hand	1 Clam	15	.533	M.S.F.		16.85		16.85	22.00	27
0030	Power vacuum		100	.080			2.53		2.53	3.27	4.05
0040	Power blower		100	.080			2.53		2.53	3.27	4.05
0050	Hose down		30	.267			8.45		8.45	10.90	13.50

01 93 04.90 Yard Waste Disposal		Crew	Daily Output	Labor-Hours	Unit	Material	Labor	Equipment	Total	Total In-House	Total Incl O&P
0010	**YARD WASTE DISPOSAL**										
0030	On site dump or compost heap	2 Clam	24	.667	C.Y.		21		21	27.00	34
0040	Into 6 C.Y. dump truck, 2 mile haul to compost heap	A-17	125	.064			2.02	1.57	3.59	4.34	4.97
0050	4 mile haul to compost heap, fees not included	"	85	.094			2.97	2.31	5.28	6.40	7.30

01 93 05 – Exterior General Maintenance

01 93 05.95 Window Washing		Crew	Daily Output	Labor-Hours	Unit	Material	Labor	Equipment	Total	Total In-House	Total Incl O&P
0010	**WINDOW WASHING**										
0030	Minimum productivity	1 Clam	1	8	M.S.F.	3.55	253		256.55	330.00	410
0040	Average productivity		2.50	3.200		3.55	101		104.55	135.00	166
0050	Maximum productivity		4	2		3.55	63		66.55	85.50	105

01 93 06 – Interior General Maintenance

01 93 06.20 Carpet Care		Crew	Daily Output	Labor-Hours	Unit	Material	Labor	Equipment	Total	Total In-House	Total Incl O&P
0010	**CARPET CARE**										
0100	Carpet cleaning, portable extractor with floor wand	1 Clam	20	.400	M.S.F.	8.50	12.65		21.15	25.50	31
0200	Deep carpet cleaning, self-contained extractor										
0210	24″ path @ 100 feet per minute	1 Clam	64	.125	M.S.F.	10.65	3.95		14.60	16.85	19.65
0220	20″ path		54	.148		10.65	4.68		15.33	17.75	21
0230	16″ path		43	.186		10.65	5.90		16.55	19.30	22.50
0240	12″ path		32	.250		10.65	7.90		18.55	22.00	26
0300	Carpet pile lifting, self-contained extractor										
0310	24″ path at 90 feet per minute	1 Clam	60	.133	M.S.F.	11.70	4.21		15.91	18.30	21.50
0320	20″ path		50	.160		11.70	5.05		16.75	19.40	22.50
0330	16″ path		40	.200		11.70	6.30		18	21.00	25
0340	12″ path		30	.267		11.70	8.45		20.15	24.00	28
0400	Carpet restoration cleaning, self-contained extractor										
0410	24″ path at 50 feet per minute	1 Clam	32	.250	M.S.F.	12.75	7.90		20.65	24.00	28.50
0420	20″ path		27	.296		12.75	9.35		22.10	26.00	31
0430	16″ path		21	.381		12.75	12.05		24.80	29.50	35
0440	12″ path		16	.500		12.75	15.80		28.55	34.50	41.50

01 93 06.25 Ceiling Care		Crew	Daily Output	Labor-Hours	Unit	Material	Labor	Equipment	Total	Total In-House	Total Incl O&P
0010	**CEILING CARE**										
0020	Washing hard ceiling from ladder, hand	1 Clam	2.05	3.902	M.S.F.	2.02	123		125.02	162.00	201

01 93 06 – Interior General Maintenance

01 93 06.25 Ceiling Care		Crew	Daily Output	Labor-Hours	Unit	2020 Bare Costs Material	Labor	Equipment	Total	Total In-House	Total Incl O&P
0030	Machine	1 Clam	2.52	3.175	M.S.F.	2.02	100		102.02	132.00	164
0035	Acoustic tile cleaning, chemical spray, including masking										
0040	Unobstructed floor	4 Clam	5400	.006	S.F.	.07	.19		.26	.32	.39
0045	Obstructed floor	"	4000	.008	"	.07	.25		.32	.41	.50
0040	Chemical spray cleaning and coating, including masking										
0050	Unobstructed floor	4 Clam	1800	.007	S.F.	.14	.21		.35	.42	.52
0055	Obstructed floor	"	3400	.009	"	.14	.30		.44	.53	.66

01 93 06.40 General Cleaning		Crew	Daily Output	Labor-Hours	Unit	Material	Labor	Equipment	Total	Total In-House	Total Incl O&P
0010	**GENERAL CLEANING**										
0020	Sweep stairs & landings, damp wipe handrails	1 Clam	80	.100	Floor	.04	3.16		3.20	4.13	5.10
0030	Dust mop stairs & landings, damp wipe handrails		62.50	.128		.04	4.04		4.08	5.25	6.55
0040	Damp mop stairs & landings, damp wipe handrails		45	.178		.24	5.60		5.84	7.50	9.30
0050	Vacuum stairs & landings, damp wipe handrails, canister		41	.195		.04	6.15		6.19	8.00	9.95
0060	Vacuum stairs & landings, damp wipe handrails, backpack		58.22	.137		.04	4.34		4.38	5.65	7
0070	Vacuum stairs & landings, damp wipe handrails, upright		31.57	.253		.04	8		8.04	10.40	12.90
0080	Clean carpeted elevators		48	.167	Ea.	.64	5.25		5.89	7.50	9.25
0090	Clean uncarpeted elevators		32	.250		.49	7.90		8.39	10.75	13.25
0100	Clean escalator		8	1		.40	31.50		31.90	41.50	51
0110	Restrooms, fixture cleaning, toilets		180	.044		.06	1.40		1.46	1.89	2.33
0120	Urinals		192	.042		.05	1.32		1.37	1.76	2.17
0130	Sinks		206	.039		.01	1.23		1.24	1.60	1.98
0140	Utility sink		175	.046		.02	1.44		1.46	1.89	2.35
0150	Bathtub		85	.094		.07	2.97		3.04	3.92	4.86
0160	Shower		85	.094		1.48	2.97		4.45	5.45	6.60
0170	Accessories cleaning, soap dispenser		1280	.006		.01	.20		.21	.27	.33
0180	Sanitary napkin dispenser		640	.013		.01	.40		.41	.52	.64
0190	Paper towel dispenser		1455	.006		.01	.17		.18	.23	.29
0200	Toilet tissue dispenser		1455	.006		.01	.17		.18	.23	.29
0210	Accessories, refilling/restocking, soap dispenser (1 bag)		640	.013		10.75	.40		11.15	12.35	14.05
0220	Sanitary napkin/tampon dispenser (25/50)		384	.021		19.40	.66		20.06	22.00	25.50
0230	Paper towel dispenser, folded (2 pkgs of 250)		545	.015		6.75	.46		7.21	8.05	9.20
0240	Paper towel dispenser, rolled (350′ roll)		340	.024		5.75	.74		6.49	7.30	8.40
0250	Toilet tissue dispenser (500 2-ply sheet roll)		1440	.006		1.35	.18		1.53	1.72	1.97
0260	Sanitary seat covers (pkg of 250)		960	.008		5.35	.26		5.61	6.25	7.10
0270	Toilet partition cleaning		96	.083		.15	2.63		2.78	3.57	4.41
0280	Ceramic tile wall cleaning		2400	.003	S.F.	.03	.11		.14	.17	.21
0290	General cleaning, empty to larger portable container, wastebaskets		960	.008	Ea.		.26		.26	.34	.42
0300	Ash trays		2880	.003			.09		.09	.11	.14
0310	Cigarette stands		960	.008			.26		.26	.34	.42
0320	Sand urns		640	.013			.40		.40	.51	.63
0330	Pencil sharpeners		2880	.003			.09		.09	.11	.14
0340	Remove trash to central location		290	.028	Lb.		.87		.87	1.13	1.40
0347	Bench		480	.017	Ea.	.02	.53		.55	.70	.87
0350	Chairs		960	.008		.01	.26		.27	.35	.43
0360	Chalkboard erasers		960	.008			.26		.26	.34	.42
0370	Chalkboards		16	.500	M.S.F.	2.02	15.80		17.82	22.50	28
0380	Desk tops		225	.036	Ea.	.03	1.12		1.15	1.48	1.84
0390	Glass doors		168	.048	"	.15	1.50		1.65	2.12	2.60
0400	Hard surface audience type seating		40	.200	M.S.F.	2.53	6.30		8.83	10.95	13.30
0409	Locker, metal		320	.025	Ea.	.06	.79		.85	1.09	1.35
0410	Microwave ovens		240	.033		.01	1.05		1.06	1.37	1.70
0420	Office partitions, to 48 S.F.		160	.050		.10	1.58		1.68	2.15	2.66

01 93 06 – Interior General Maintenance

01 93 06.40 General Cleaning		Crew	Daily Output	Labor-Hours	Unit	Material	2020 Bare Costs Labor	Equipment	Total	Total In-House	Total Incl O&P
0430	Overhead air vents	1 Clam	175	.046	Ea.	.01	1.44		1.45	1.88	2.33
0440	Refrigerators		35	.229		.16	7.20		7.36	9.50	11.80
0450	Revolving doors		35	.229		.89	7.20		8.09	10.30	12.70
0460	Tables		240	.033		.04	1.05		1.09	1.40	1.74
0470	Telephone		288	.028		.01	.88		.89	1.14	1.42
0480	Telephone booth		190	.042		.46	1.33		1.79	2.23	2.71
0490	Water fountain		480	.017		.01	.53		.54	.69	.85
0500	Wastebaskets		260	.031		.01	.97		.98	1.27	1.57
0510	Dusting, beds		360	.022			.70		.70	.91	1.13
0520	Bedside stand		960	.008			.26		.26	.34	.42
0530	Bookcase		960	.008			.26		.26	.34	.42
0540	Calculator		1920	.004			.13		.13	.17	.21
0550	Chair		1340	.006			.19		.19	.24	.30
0560	Coat rack		400	.020			.63		.63	.82	1.01
0570	Couch		320	.025			.79		.79	1.02	1.27
0580	Desk		550	.015			.46		.46	.59	.74
0590	Desk lamp		1200	.007			.21		.21	.27	.34
0600	Desk trays		1920	.004			.13		.13	.17	.21
0610	Dictaphone		1920	.004			.13		.13	.17	.21
0620	Dresser		480	.017			.53		.53	.68	.84
0630	File cabinets		960	.008			.26		.26	.34	.42
0640	Lockers		2400	.003			.11		.11	.14	.17
0650	Piano		425	.019			.59		.59	.77	.95
0660	Radio		960	.008			.26		.26	.34	.42
0665	Shelf, storage, 1′ x 20′		480	.017			.53		.53	.68	.84
0670	Table		670	.012			.38		.38	.49	.60
0680	Table lamp		960	.008			.26		.26	.34	.42
0690	Telephones		960	.008			.26		.26	.34	.42
0700	Television		960	.008			.26		.26	.34	.42
0710	Typewriter		960	.008			.26		.26	.34	.42
0720	Vacuuming, armless chair		825	.010			.31		.31	.40	.49
0730	Armchair		480	.017			.53		.53	.68	.84
0740	Couch		320	.025			.79		.79	1.02	1.27
0750	Entry mats		8	1	M.S.F.		31.50		31.50	41.00	50.50
0760	Overhead air vents		480	.017	Ea.		.53		.53	.68	.84
0770	Upholstered auditorium seating		5.27	1.518	M.S.F.		48		48	62.00	77

01 93 06.45 Hard Floor Care		Crew	Daily Output	Labor-Hours	Unit	Material	Labor	Equipment	Total	Total In-House	Total Incl O&P
0010	**HARD FLOOR CARE**										
0020	Damp mop w/bucket & wringer, 12 oz. mop head, unobstructed	1 Clam	25	.320	M.S.F.	5.70	10.10		15.80	19.35	23.50
0030	Obstructed		17	.471		5.70	14.85		20.55	25.50	31
0040	20 oz. mop head, unobstructed		29	.276		5.70	8.70		14.40	17.55	21
0050	Obstructed		20	.400		5.70	12.65		18.35	22.50	27.50
0060	24 oz. mop head, unobstructed		35	.229		5.70	7.20		12.90	15.60	18.75
0070	Obstructed		23	.348		5.70	11		16.70	20.50	24.50
0080	32 oz. mop head, unobstructed		43	.186		5.70	5.90		11.60	13.85	16.55
0090	Obstructed		29	.276		5.70	8.70		14.40	17.55	21
0100	Dust mop, 12″ mop, unobstructed		33	.242			7.65		7.65	9.90	12.30
0110	Obstructed		19	.421			13.30		13.30	17.20	21.50
0120	18″ mop, unobstructed		49	.163			5.15		5.15	6.65	8.25
0130	Obstructed		28	.286			9.05		9.05	11.65	14.45
0140	24″ mop, unobstructed		61	.131			4.14		4.14	5.35	6.65
0150	Obstructed		34	.235			7.45		7.45	9.60	11.90

01 93 Facility Maintenance

01 93 06 – Interior General Maintenance

01 93 06.45 Hard Floor Care		Crew	Daily Output	Labor-Hours	Unit	Material	2020 Bare Costs Labor	Equipment	Total	Total In-House	Total Incl O&P
0160	30″ mop, unobstructed	1 Clam	74	.108	M.S.F.		3.42		3.42	4.42	5.50
0170	Obstructed		42	.190			6		6	7.00	9.65
0180	36″ mop, unobstructed		92	.087			2.75		2.75	3.55	4.40
0190	Obstructed		52	.154			4.86		4.86	6.30	7.80
0200	42″ mop, unobstructed		124	.065			2.04		2.04	2.64	3.27
0210	Obstructed		70	.114			3.61		3.61	4.67	5.00
0220	48″ mop, unobstructed		185	.043			1.37		1.37	1.77	2.19
0230	Obstructed		104	.077			2.43		2.43	3.14	3.90
0240	60″ mop, unobstructed		246	.033			1.03		1.03	1.33	1.65
0250	Obstructed		138	.058			1.83		1.83	2.37	2.94
0260	72″ mop, unobstructed		369	.022			.69		.69	.89	1.10
0270	Obstructed		208	.038			1.22		1.22	1.57	[illegible]
0280	Sweeping, 8″ broom, light soil, unobstructed		19	.421			13.30		13.30	17.20	21.50
0290	Obstructed		14	.571			18.05		18.05	23.50	29
0300	Heavy soil, unobstructed		11	.727			23		23	29.50	37
0310	Obstructed		9	.889			28		28	36.50	45
0320	12″ push broom, light soil, unobstructed		32	.250			7.90		7.90	10.20	12.65
0330	Obstructed		24	.333			10.55		10.55	13.60	16.90
0340	Heavy soil, unobstructed		19	.421			13.30		13.30	17.20	21.50
0350	Obstructed		14	.571			18.05		18.05	23.50	29
0360	16″ push broom, light soil, unobstructed		40	.200			6.30		6.30	8.15	10.15
0370	Obstructed		30	.267			8.45		8.45	10.90	13.50
0380	Heavy soil, unobstructed		24	.333			10.55		10.55	13.60	16.90
0390	Obstructed		18	.444			14.05		14.05	18.15	22.50
0400	18″ push broom, light soil, unobstructed		44	.182			5.75		5.75	7.45	9.20
0410	Obstructed		33	.242			7.65		7.65	9.90	12.30
0420	Heavy soil, unobstructed		36	.222			7		7	9.10	11.25
0430	Obstructed		20	.400			12.65		12.65	16.35	20.50
0440	24″ push broom, light soil, unobstructed		57	.140			4.44		4.44	5.75	7.10
0450	Obstructed		43	.186			5.90		5.90	7.60	9.40
0460	Heavy soil, unobstructed		34	.235			7.45		7.45	9.60	11.90
0470	Obstructed		26	.308			9.70		9.70	12.55	15.60
0480	30″ push broom, light soil, unobstructed		81	.099			3.12		3.12	4.03	5
0490	Obstructed		60	.133			4.21		4.21	5.45	6.75
0500	Heavy soil, unobstructed		48	.167			5.25		5.25	6.80	8.45
0510	Obstructed		36	.222			7		7	9.10	11.25
0520	36″ push broom, light soil, unobstructed		100	.080			2.53		2.53	3.27	4.05
0530	Obstructed		75	.107			3.37		3.37	4.36	5.40
0540	Heavy soil, unobstructed		60	.133			4.21		4.21	5.45	6.75
0550	Obstructed		45	.178			5.60		5.60	7.25	9
0560	42″ push broom, light soil, unobstructed		133	.060			1.90		1.90	2.46	3.05
0570	Obstructed		100	.080			2.53		2.53	3.27	4.05
0580	Heavy soil, unobstructed		80	.100			3.16		3.16	4.09	5.05
0590	Obstructed		60	.133			4.21		4.21	5.45	6.75
0600	48″ push broom, light soil, unobstructed		200	.040			1.26		1.26	1.63	2.03
0610	Obstructed		150	.053			1.69		1.69	2.18	2.70
0620	Heavy soil, unobstructed		120	.067			2.11		2.11	2.72	3.38
0630	Obstructed		90	.089			2.81		2.81	3.63	4.50
0640	Wet mop/rinse w/bucket & wringer, 12 oz. mop, unobstructed		11	.727		2.02	23		25.02	32.00	39.50
0650	Obstructed		8	1		2.02	31.50		33.52	43.00	53
0660	20 oz. mop head, unobstructed		14	.571		2.02	18.05		20.07	25.50	31.50
0670	Obstructed		10	.800		2.02	25.50		27.52	35.00	43
0680	24 oz. mop head, unobstructed		20	.400		2.02	12.65		14.67	18.55	23

01 93 06 – Interior General Maintenance

01 93 06.45 Hard Floor Care		Crew	Daily Output	Labor-Hours	Unit	Material	2020 Bare Costs Labor	Equipment	Total	Total In-House	Total Incl O&P
0690	Obstructed	1 Clam	15	.533	M.S.F.	2.02	16.85		18.87	24.00	29.50
0700	32 oz. mop head, unobstructed		27	.296		2.02	9.35		11.37	14.30	17.55
0710	Obstructed		20	.400		2.02	12.65		14.67	18.55	23
0720	Dry buffing or polishing, 175 RPM, 12″ diameter machine		14.21	.563			17.80		17.80	23.00	28.50
0730	14″ diameter machine		16.51	.485			15.30		15.30	19.80	24.50
0740	17″ diameter machine		19.20	.417			13.15		13.15	17.00	21
0750	20″ diameter machine		22.85	.350			11.05		11.05	14.30	17.75
0760	24″ diameter machine		28.99	.276			8.70		8.70	11.25	14
0770	350 RPM, 17″ diameter machine		30.76	.260			8.20		8.20	10.60	13.15
0780	20″ diameter machine		40.60	.197			6.25		6.25	8.05	10
0790	24″ diameter machine		47.06	.170			5.35		5.35	6.95	8.60
0800	1000 RPM, 17″ diameter machine		62.09	.129			4.07		4.07	5.25	6.55
0810	20″ diameter machine		67.68	.118			3.74		3.74	4.83	6
0820	24″ diameter machine		88.79	.090			2.85		2.85	3.68	4.56
0830	27″ diameter machine		99.34	.081			2.54		2.54	3.29	4.08
0840	Spray buffing, 175 RPM, 12″ diameter		10.66	.750		.93	23.50		24.43	31.50	39
0850	14″ diameter machine		11.95	.669		.93	21		21.93	28.50	35
0860	17″ diameter machine		13.80	.580		.93	18.30		19.23	24.50	30.50
0870	20″ diameter machine		16	.500		.93	15.80		16.73	21.50	26.50
0880	24″ diameter machine		19.05	.420		.93	13.25		14.18	18.15	22.50
0890	350 RPM, 17″ diameter machine		17.39	.460		.93	14.55		15.48	19.80	24.50
0900	20″ diameter machine		22.13	.362		.93	11.40		12.33	15.80	19.45
0910	24″ diameter machine		29.22	.274		.93	8.65		9.58	12.20	15
0920	1000 RPM, 17″ diameter machine		41.02	.195		.93	6.15		7.08	9.00	11.05
0930	20″ diameter machine		44.14	.181		.93	5.75		6.68	8.40	10.35
0940	24″ diameter machine		52.21	.153		.93	4.84		5.77	7.30	8.90
0950	27″ diameter machine		57.43	.139		.93	4.40		5.33	6.70	8.20
0960	Burnish, 1000 RPM, 17″ diameter machine		60	.133			4.21		4.21	5.45	6.75
0970	2000 RPM, 17″ diameter machine		72	.111			3.51		3.51	4.54	5.65
0980	Scrubbing, 12″ brush by hand		2	4		5.70	126		131.70	170.00	210
0990	175 RPM, 12″ diameter machine		10.05	.796		5.70	25		30.70	39.00	47.50
1000	14″ diameter machine		11.62	.688		5.70	22		27.70	34.50	42
1010	17″ diameter machine		14	.571		5.70	18.05		23.75	29.50	36
1020	20″ diameter machine		17.61	.454		5.70	14.35		20.05	25.00	30
1030	24″ diameter machine		23.58	.339		5.70	10.70		16.40	20.00	24.50
1040	350 RPM, 17″ diameter machine		24.12	.332		5.70	10.50		16.20	19.80	24
1050	20″ diameter machine		28.91	.277		5.70	8.75		14.45	17.55	21
1060	24″ diameter machine		36.92	.217		5.70	6.85		12.55	15.10	18.10
1070	Automatic, 21″ machine		60.75	.132		5.70	4.16		9.86	11.65	13.80
1080	24″ machine		83.47	.096		5.70	3.03		8.73	10.20	12
1090	32″ machine		128.68	.062		5.70	1.96		7.66	8.80	10.30
1100	Stripping, separate wet pick-up required, 175 RPM, 17″ diameter		6.86	1.166		3.90	37		40.90	52.00	64
1110	20″ diameter machine		7.68	1.042		3.90	33		36.90	47.00	58
1120	350 RPM, 17″ diameter machine		8.80	.909		3.90	28.50		32.40	41.50	51
1130	20″ diameter machine		10.50	.762		3.90	24		27.90	35.50	43.50
1140	Wet pick-up, 12″ opening attachment		18	.444			14.05		14.05	18.15	22.50
1150	20″ opening attachment		24.12	.332			10.50		10.50	13.55	16.80
1160	Finish application, mop		14	.571		17.10	18.05		35.15	42.00	50.50
1170	Gravity fed applicator		21	.381		17.10	12.05		29.15	34.50	40.50
1180	Lambswool applicator		17	.471		17.10	14.85		31.95	38.00	45.50
1190	Sealer application, mop		14	.571		25	18.05		43.05	51.00	60.50
1200	Gravity fed applicator		21	.381		25	12.05		37.05	43.00	50.50
1210	Lambswool applicator	↓	17	.471	↓	25	14.85		39.85	46.50	55.50

01 93 06.80 Vacuuming		Crew	Daily Output	Labor-Hours	Unit	Material	2020 Bare Costs Labor	Equipment	Total	Total In-House	Total Incl O&P
0010	**VACUUMING**										
0020	Vacuum carpet, 12" upright, spot areas	1 Clam	73.68	.109	M.S.F.		3.43		3.43	4.44	5.50
0030	Traffic areas		44.21	.181			5.70		5.70	7.40	9.15
0040	All areas		21.05	.380			12		12	15.55	19.25
0050	14" upright, spot areas		79.98	.100			3.16		3.16	4.09	5.05
0060	Traffic areas		47.99	.167			5.25		5.25	6.80	8.45
0070	All areas		22.85	.350			11.05		11.05	14.30	17.75
0080	16" upright, spot areas		87.50	.091			2.89		2.89	3.73	4.63
0090	Traffic areas		52.50	.152			4.82		4.82	6.20	7.70
0100	All areas		25	.320			10.10		10.10	13.05	16.20
0110	18" upright, spot areas		96.50	.083			2.62		2.62	3.39	4.20
0120	Traffic areas		57.92	.138			4.36		4.36	5.65	7
0130	All areas		27.58	.290			9.15		9.15	11.85	14.70
0140	20" upright, spot areas		107.66	.074			2.35		2.35	3.04	3.76
0150	Traffic areas		64.60	.124			3.91		3.91	5.05	6.25
0160	All areas		30.76	.260			8.20		8.20	10.60	13.15
0170	22" upright, spot areas		121.73	.066			2.08		2.08	2.68	3.33
0180	Traffic areas		73.04	.110			3.46		3.46	4.47	5.55
0190	All areas		34.78	.230			7.25		7.25	9.40	11.65
0200	24" upright, spot areas		140	.057			1.81		1.81	2.33	2.89
0210	Traffic areas		84	.095			3.01		3.01	3.89	4.82
0220	All areas		40	.200			6.30		6.30	8.15	10.15
0230	14" twin motor upright, spot areas		98.81	.081			2.56		2.56	3.31	4.10
0240	Traffic areas		59.28	.135			4.26		4.26	5.50	6.85
0250	All areas		28.23	.283			8.95		8.95	11.60	14.35
0260	18" twin motor upright, spot areas		112	.071			2.26		2.26	2.92	3.62
0270	Traffic areas		67.20	.119			3.76		3.76	4.86	6.05
0280	All areas		32	.250			7.90		7.90	10.20	12.65
0290	Backpack, 12" opening carpet attachment, spot areas		77.77	.103			3.25		3.25	4.20	5.20
0300	Traffic areas		46.66	.171			5.40		5.40	7.00	8.70
0310	All areas		22.22	.360			11.40		11.40	14.70	18.25
0320	14" opening carpet attachment, spot areas		84.84	.094			2.98		2.98	3.85	4.78
0330	Traffic areas		50.90	.157			4.97		4.97	6.40	7.95
0340	All areas		24.24	.330			10.45		10.45	13.50	16.70
0350	16" opening carpet attachment, spot areas		93.35	.086			2.71		2.71	3.50	4.34
0360	Traffic areas		56.01	.143			4.51		4.51	5.85	7.25
0370	All areas		26.67	.300			9.50		9.50	12.25	15.20
0380	18" opening carpet attachment, spot areas		103.67	.077			2.44		2.44	3.15	3.91
0390	Traffic areas		62.20	.129			4.06		4.06	5.25	6.50
0400	All areas		29.62	.270			8.55		8.55	11.05	13.70
0410	20" opening carpet attachment, spot areas		116.66	.069			2.17		2.17	2.80	3.47
0420	Traffic areas		69.99	.114			3.61		3.61	4.67	5.80
0430	All areas		33.33	.240			7.60		7.60	9.80	12.15
0440	22" opening carpet attachment, spot areas		133.32	.060			1.90		1.90	2.45	3.04
0450	Traffic areas		79.99	.100			3.16		3.16	4.09	5.05
0460	All areas		38.09	.210			6.65		6.65	8.60	10.65
0470	24" opening carpet attachment, spot areas		155.54	.051			1.63		1.63	2.10	2.60
0480	Traffic areas		93.32	.086			2.71		2.71	3.50	4.34
0490	All areas		44.44	.180			5.70		5.70	7.35	9.10
0500	Tank/canister, 12" opening carpet attachment, spot areas		70	.114			3.61		3.61	4.67	5.80
0510	Traffic areas		42	.190			6		6	7.80	9.65
0520	All areas		20	.400			12.65		12.65	16.35	20.50

01 93 06 – Interior General Maintenance

01 93 06.80 Vacuuming

	01 93 06.80 Vacuuming	Crew	Daily Output	Labor-Hours	Unit	Material	2020 Bare Costs Labor	Equipment	Total	Total In-House	Total Incl O&P
0530	14" opening carpet attachment, spot areas	1 Clam	75.67	.106	M.S.F.		3.34		3.34	4.32	5.35
0540	Traffic areas		45.40	.176			5.55		5.55	7.20	8.95
0550	All areas		21.62	.370			11.70		11.70	15.10	18.75
0560	16" opening carpet attachment, spot areas		82.32	.097			3.07		3.07	3.97	4.92
0570	Traffic areas		49.39	.162			5.10		5.10	6.60	8.20
0580	All areas		23.52	.340			10.75		10.75	13.90	17.25
0590	18" opening carpet attachment, spot areas		90.30	.089			2.80		2.80	3.62	4.49
0600	Traffic areas		54.18	.148			4.67		4.67	6.05	7.50
0610	All areas		25.80	.310			9.80		9.80	12.65	15.70
0620	20" opening carpet attachment, spot areas		100	.080			2.53		2.53	3.27	4.05
0630	Traffic areas		60	.133			4.21		4.21	5.45	6.75
0640	All areas		28.57	.280			8.85		8.85	11.45	14.20
0650	22" opening carpet attachment, spot areas		112	.071			2.26		2.26	2.92	3.62
0660	Traffic areas		67.20	.119			3.76		3.76	4.86	6.05
0670	All areas		32	.250			7.90		7.90	10.20	12.65
0680	24" opening carpet attachment, spot areas		127.26	.063			1.99		1.99	2.57	3.18
0690	Traffic areas		76.36	.105			3.31		3.31	4.28	5.30
0700	All areas		36.36	.220			6.95		6.95	9.00	11.15
0710	Large area machine, no obstructions, 26" opening		44.44	.180			5.70		5.70	7.35	9.10
0720	28" opening		64	.125			3.95		3.95	5.10	6.35
0730	30" opening		80	.100			3.16		3.16	4.09	5.05
0740	32" opening		120	.067			2.11		2.11	2.72	3.38

01 93 06.90 Wall Care

	01 93 06.90 Wall Care	Crew	Daily Output	Labor-Hours	Unit	Material	Labor	Equipment	Total	Total In-House	Total Incl O&P
0010	**WALL CARE**										
0020	Dusting walls, hand duster	1 Clam	87	.092	M.S.F.		2.91		2.91	3.76	4.66
0030	Treated cloth		40	.200			6.30		6.30	8.15	10.15
0040	Hand held duster vacuum		53	.151			4.77		4.77	6.15	7.65
0050	Canister vacuum		32	.250			7.90		7.90	10.20	12.65
0060	Backpack vacuum		45	.178			5.60		5.60	7.25	9
0070	Washing walls, hand, painted surfaces		2.50	3.200		2.02	101		103.02	133.00	165
0080	Vinyl surfaces		3.25	2.462		2.02	78		80.02	103.00	128
0090	Machine, painted surfaces		4.80	1.667		2.02	52.50		54.52	70.50	87
0100	Washing walls from ladder, hand, painted surfaces		1.65	4.848		2.02	153		155.02	200.00	249
0110	Vinyl surfaces		2.15	3.721		2.02	118		120.02	154.00	191
0120	Machine, painted surfaces		3.17	2.524		2.02	80		82.02	105.00	131

01 93 06.95 Window and Accessory Care

	01 93 06.95 Window and Accessory Care	Crew	Daily Output	Labor-Hours	Unit	Material	Labor	Equipment	Total	Total In-House	Total Incl O&P
0010	**WINDOW AND ACCESSORY CARE**										
0025	Wash windows, trigger sprayer & wipe cloth, 12 S.F.	1 Clam	4.50	1.778	M.S.F.	3.55	56		59.55	76.50	94.50
0030	Over 12 S.F.		12.75	.627		3.55	19.85		23.40	29.50	36.50
0040	With squeegee & bucket, 12 S.F. and under		5.50	1.455		3.55	46		49.55	63.50	78
0050	Over 12 S.F.		15.50	.516		3.55	16.30		19.85	25.00	30.50
0060	Dust window sill		8640	.001	L.F.		.03		.03	.04	.05
0070	Damp wipe, venetian blinds		160	.050	Ea.	.04	1.58		1.62	2.08	2.58
0080	Vacuum, venetian blinds, horizontal		87.27	.092			2.90		2.90	3.74	4.64
0090	Draperies, in place, to 64 S.F.		160	.050			1.58		1.58	2.04	2.53

01 93 08 – Facilities Maintenance, Equipment

01 93 08.50 Equipment

	01 93 08.50 Equipment	Crew	Daily Output	Labor-Hours	Unit	Material	Labor	Equipment	Total	Total In-House	Total Incl O&P
0010	**EQUIPMENT**, Purchase										
0090	Carpet care equipment										
0110	Dual motor vac, 1 HP, 16" brush				Ea.	690			690	760.00	865
0120	Upright vacuum, 12" brush					227			227	250.00	284
0130	14" brush					455			455	500.00	570

01 93 08 – Facilities Maintenance, Equipment

01 93 08.50 Equipment		Crew	Daily Output	Labor-Hours	Unit	2020 Bare Costs Material	Labor	Equipment	Total	Total In-House	Total Incl O&P
0140	Soil extractor, hot water 5′ wand, 12″ head				Ea.	3,050			3,050	3,350.00	3,825
0150	Dry foam, 13″ brush					2,500			2,500	2,750.00	3,125
0160	24″ brush					4,250			4,250	4,675.00	5,325
0240	Floor care equipment										
0260	Polishing, buffing, waxing machine, 175 RPM										
0270	.33 HP, 20″ diam. brush				Ea.	870			870	955.00	1,100
0280	1 HP, 16″ diam. brush					1,225			1,225	1,350.00	1,525
0290	17″ diam. brush					1,150			1,150	1,275.00	1,450
0300	1.5 HP, 20″ diam. brush					1,400			1,400	1,550.00	1,750
0310	20″ diam. brush, 1500 RPM					1,625			1,625	1,800.00	2,025
0330	Scrubber, automatic, 2 stage, 1 HP vacuum motor										
0340	20″ diam. brush				Ea.	13,800			13,800	15,200.00	17,300
0350	28″ diam. brush				″	18,900			18,900	20,800.00	23,600
1200	Plumbing maintenance equipment										
1220	Kinetic water ram				Ea.	223			223	245.00	279
1240	Cable pipe snake, 104 ft., self feed, electric, .5 HP				″	2,700			2,700	2,975.00	3,375
1500	Specialty equipment										
1550	Litterbuggy, 9 HP gasoline				Ea.	4,800			4,800	5,275.00	6,000
1560	Propane					4,075			4,075	4,475.00	5,100
1570	Pressure cleaner, hot water, 1000 psi					3,375			3,375	3,725.00	4,225
1580	3500 psi					8,675			8,675	9,550.00	10,800
1590	Vacuum cleaners, steel canister										
1600	Dry only, two stage, 1 HP				Ea.	400			400	440.00	500
1610	Wet/dry, two stage, 2 HP					710			710	780.00	890
1620	4 HP					920			920	1,000.00	1,150
1630	Wet/dry, three stage, 1.5 HP					685			685	755.00	855
1670	Squeegee wet only, 2 stage, 1 HP					1,100			1,100	1,200.00	1,375
1680	2 HP					1,575			1,575	1,725.00	1,975
1690	Asbestos and hazardous waste dry vacs, 1 HP					925			925	1,025.00	1,150
1700	2 HP					790			790	870.00	990
1710	Three stage, 1 HP					630			630	695.00	790
1800	Upholstery soil extractor, dry-foam					3,625			3,625	4,000.00	4,525
1900	Snow removal equipment										
1920	Thrower, 3 HP, 20″, single stage, gas				Ea.	730			730	805.00	915
1930	Electric start					875			875	965.00	1,100
1940	5 HP, 24″, 2 stage, gas					1,225			1,225	1,350.00	1,525
1950	11 HP, 36″					2,400			2,400	2,650.00	3,000
2000	Tube cleaning equip., for heat exchangers, condensers, evap.										
2010	Tubes/pipes 1/4″-1″, 115 volt, .5 HP drive				Ea.	1,675			1,675	1,850.00	2,100
2020	Air powered					1,725			1,725	1,900.00	2,150
2040	1″ and up, 115 volt, 1 HP drive				Ea.	2,125			2,125	2,350.00	2,650

Facilities Audits

Table of Contents

Auditing the Facility

Introduction

Capital budget planning begins with compiling a complete inventory of the facility, including equipment, that identifies existing deficiencies. The Facilities Audit is a vehicle for producing such an inventory; it is a system for thoroughly assessing the existing physical condition and functional performance of buildings, grounds, utilities, and equipment. The results of this audit are used to address major and minor, urgent and long-term needs for corrective action, and for short- and long-term financial planning. The basic principles presented here can be applied to any scale operation, from a single structure to a facility consisting of multiple building complexes in dispersed locations.

The first step is determining the scope of the audit. Next, the audit team is selected, and a set of forms is designed. A survey team then records field observations on the forms for individual building components. Finally, the information from these forms is summarized, prioritized, presented, and used to:

- compare a building's condition and functional performance to that of other buildings within a facility
- define regular maintenance requirements
- define capital repair and replacement projects in order to eliminate deferred maintenance
- develop cost estimates for capital repair and replacement projects
- restore functionally obsolete facilities to a usable condition
- eliminate conditions that are either potentially damaging to property or present life safety hazards
- identify energy conservation measures

The audit ultimately produces a database on the condition of a facility's buildings, identifying the physical adequacy of construction, material, and equipment, and assessing the functional adequacy of the facility to perform its designated purpose. This comprehensive inventory can be used to guide property management decisions and strategic planning of capital assets, such as determining those expenditures necessary to extend the life of a facility.

The audit procedures and forms included here are for the purpose of identifying the condition of buildings and assessing the cost and priority of their repairs. While this is a comprehensive stand-alone process, it is for the limited purposes of assessing costs and establishing priorities for repairs. It represents only a portion of the requirements of a full facilities audit. For further information and guidance on conducting a full audit, contact the RSMeans engineering staff.

The Rating Forms

The audit approach involves conducting a thorough inspection of a facility, then employing a systematic process of completing individual forms for each component, following the order in which building components are assembled. This is a self-evaluation process designed to fit all types of structures. The audit procedure described is a "generic" approach that can be used as a starting point; the facilities manager should incorporate the special characteristics of an organization and its facilities into an individualized facilities audit.

Audits are best performed by the in-house personnel familiar with the building being audited. However, outside consultants may be required when staff is not available or when special expertise is required to further examine suspected deficiencies.

Four types of forms that can be used to conduct the building audit are described here. A complete set of the blank forms ready for photocopying follows.

1. *Building Data Summary*
   This form can be used to summarize all buildings to be covered by the audit.
2. *Facilities Inventory Form*
   This form is designed to collect the basic data about a specific building's characteristics (name, identification number, location, area), including land data. One form is prepared for each building to be audited as summarized on the *Building Data Summary* form.
3. *Standard Inspection Form*
   The standard form for recording observed deficiencies, it provides a uniform way to comment on, prioritize, and estimate the cost to correct those deficiencies. The *Standard Inspection Form* is designed to make use of cost data from the Maintenance & Repair and Preventive Maintenance sections of this publication. Cost data from publications such as *RSMeans Facilities Construction Cost Data* may be utilized as well.
4. *System Checklists*
   For each major building system from foundations through site work, there is a checklist to guide the audit and record the review process. The checklists are intended as general guides and may not be totally comprehensive. Be sure to make any necessary modifications (additions or deletions) required for the specific building being audited. The forms prompt the user on what to check and provide a place to check off or indicate by "Yes" or "No" the presence of deficiencies requiring further detailed analysis on the *Standard Inspection Form*.

Completing the Inspection Forms

The standard inspection forms are designed to be used in conjunction with the system checklists. A system checklist of typical deficiencies for each building system is completed as part of the inspection process. Then for each building deficiency, a standard inspection report is generated listing the identifying information as required on the form. With the deficiency defined and described, the specific work items required to correct the deficiency can be entered, prioritized, and costed in Part 3 of the form. The priority rating is further described below.

Priority Ratings

A priority rating is the inspector's judgment of the priority of an observed deficiency. The rating provides guidance for facilities managers who must review overall deficiencies and develop maintenance schedules and capital budgets. The following priority rating terms, commonly used in maintenance management, are a suggested guideline for priority values. The scale may be expanded as required to fit specific needs.

1. *Emergency*: This designation indicates work that demands immediate attention to repair an essential operating system or building component, or because life safety or property is endangered. A response should occur within two hours of notification.
2. *Urgent*: Work demanding prompt attention to alleviate temporary emergency repairs or a condition requiring correction within one to five days to prevent a subsequent emergency is classified as urgent.
3. *Routine*: A specific completion date can be requested or required for routine work. This includes work that has no short range effect on life safety, security of personnel, or property. Routine work can be planned in detail and incorporated into a trade backlog for scheduling within twelve months.
4. *Deferred*: Projects that can be deferred into the following year's work planning are classified as deferred.

Cost Estimates for Maintenance

A major characteristic of the audit described here is the application of detailed cost estimates. To prepare cost estimates for work required to correct deficiencies, detailed information about the nature and extent of the facility's condition is required at the beginning of the audit. When cost estimates are incorporated, the audit process generates highly useful summaries outlining major maintenance costs and priorities. Including cost estimates allows for comparisons to be made in order to prioritize repair and replacement needs on a project-by-project and building-by-building basis.

Other Considerations

A problem that may have to be settled by the audit leader is what to do with mixed-use buildings, those that have recent additions to the original construction, or buildings that have, under emergency circumstances, been forced to house new activities for which they were not designed. Such problems will have to be dealt with individually by the audit team.

The data gathered in a facilities audit must be kept current, gathered consistently, and regularly updated. An annual cycle of inspections is recommended. Each organization initiating a comprehensive audit and continuing with annual cycles of inspections will benefit from organizing data in a systematic format. Conducting annual audits in a consistent format provides the basis for monitoring the backlog of deferred maintenance and preparing effective long-term plans and budgets for maintenance and repair operations.

The forms are designed for ease of field entry of data and manual preparation of summaries and reports. The forms can also be created using basic word processing and spreadsheet software or database programming. Although the self-evaluation process is readily adaptable to data processing programs, not every organization has the capabilities or resources available. For those without computer capabilities, a manual method of storing and updating must be developed and used.

Summarizing the Audit Data

Information collected on the audit forms is compiled and used to determine the condition of the facility's buildings and what should be done to maintain and/or improve them. Summaries of a building audit can serve several purposes, some of which are listed below.

- *Routine Maintenance:* A current property owner may be interested in planning and estimating a program for the cost of routine operations and maintenance.
- *Major Maintenance:* A current property owner may want to evaluate major maintenance needs to plan and estimate the cost of a corrective program.
- *Deferred Maintenance:* A backlog of routine and major maintenance can be addressed using a survey of conditions to plan and budget a deferred maintenance program.
- *Renovations and/or Additions:* Prior to developing feasibility studies on alternatives to renovations and/or additions, a survey of existing conditions may be necessary.
- *Capital Budgeting and Planning:* A comprehensive audit of facilities conditions will incorporate major maintenance requirements into overall capital needs.

Conclusion

The information and forms for conducting an audit of the facility are included in *RSMeans Facilities Maintenance & Repair Cost Data* to provide a method for applying the cost data to the actual maintenance and repair needs of a specific facility. The facility audit presented here is a subset of a more comprehensive audit program. The more comprehensive approach includes assessment of the building's functional performance, as well as additional forms and procedures to incorporate the cost of maintenance and repair into a comprehensive multi-year facility management plan.

BUILDING DATA SUMMARY

Building Number	Building Name	Building Use*	Location / Address	Ownership** Own/Lea/O-L

* Principal and Special Uses

** Own = Owned

Lea = Leased

O-L = Owned/Leased

FACILITIES INVENTORY FORM

Building Data

1. Organization Name ______________________________
2. Approved by: Name ______________ Title ______________ Date __________
3. Building Number ______________ 4. Property Name ______________________
5. Property Address ______________________________
6. Property Main Use/Special Use ______________________________
7. Gross Area (Sq. Ft.) ______________ 8. Net Area (Sq. Ft.) ______________
9. Type of Ownership: Owned __________ Leased __________ Owned / Leased __________
10. Book Value $ ______________ Year Appraised __________
11. Replacement Value $______________
12. Year Built ______________ Year Acquired __________
13. Year Major Additions ______________________________

Land Data

14. Type of Ownership: Owned __________ Leased __________ Owned / Leased __________
15. Book Value $ ______________ Year Appraised __________
16. Year Acquired ______________
17. Size in Acres ______________
18. Market Value $ ______________

Notes

____

____

____

____

STANDARD INSPECTION FORM

1. Facility Inspection Data

Facility #: ____________________ Name: ____________________

Component #: ____________________ Name: ____________________

Inspector: ____________________ Date: ____________________

2. Component Description

____

____

____

3. Component Repair Evaluation

Defi-ciency #	Component Description	Line Number	Pri-ority	Unit	Quant.	Unit Cost	Total In-House	Unit Cost	Total W/O&P

A10 FOUNDATIONS INSPECTION CHECKLIST

	Deficiencies		Causes
_____	Settlement, alignment changes or cracks	_____	Soils - changes in load bearing capacity due to shrinkage, erosion, or compaction. Adjacent construction undermining foundations. Reduced soil cover resulting in frost exposure.
		_____	Design loads - building equipment loads exceeding design loads. Vibration from heavy equipment requiring isolated foundations.
		_____	Structural or occupancy changes - inadequate bearing capacities. Foundation settling.
		_____	Earthquake resistance non-functioning.
_____	Moisture penetration	_____	Water table changes - inadequate drainage.
		_____	Ineffective drains or sump pump/sump pits.
		_____	Roof drainage - storm sewer connections inadequate or defective. Installation of roof drain restrictors, gutters, and downspouts where required.
		_____	Surface drainage - exterior grades should slope away from building and structures.
		_____	Utilities - broken or improperly functioning utility service lines or drains.
		_____	Leakage - wall cracks, opening of construction joints, inadequate or defective waterproofing.
		_____	Condensation - inadequate ventilation, vapor barrier, and/or dehumidification.
_____	Temperature changes	_____	Insulation - improperly selected for insulating value, fire ratings, and vermin resistance.
_____	Surface material deterioration	_____	Concrete, masonry, or stucco - spalling, corrosion of reinforcing, moisture penetration, or chemical reaction between cement and soil.
		_____	Steel or other ferrous metals - corrosion due to moisture or contact with acid-bearing soils.
		_____	Wood - decay due to moisture or insect infestation.
_____	Openings deterioration	_____	Non-functioning of doors, windows, hatchways, and stairways.
		_____	Utilities penetration due to damage, weather, wear, or other cause.

A20 BASEMENT INSPECTION CHECKLIST

	Deficiencies		Causes
_____	Floors, concrete - cracking or arching	_____	Shrinkage, or settlement of subsoil.
		_____	Inadequate drainage.
		_____	Movement in exterior walls, or frost heave.
		_____	Improper compaction of base. Erosion of base.
		_____	Heaving from hydraulic pressure.
_____	Floors, wood - rotting or arching	_____	Excessive dampness or insect infestation.
		_____	Leak in building exterior.
		_____	Lack of ventilation.
_____	Wall deterioration	_____	Concrete or masonry (see *A10 Foundations*)
_____	Crawl space ventilation and maintenance	_____	Inadequate air circulation due to blockage of openings in foundation walls.
		_____	Moisture barrier ineffective.
		_____	Pest control, housekeeping, and proper drainage.

B10 SUPERSTRUCTURES INSPECTION CHECKLIST

The primary materials encountered in the superstructure inspection are concrete, steel, and wood. Typical observations of deficiencies will be observed by: failures in the exterior closure systems of exterior walls, openings, and roofs; cracks; movement of materials; moisture penetration; and discoloration. The exterior visual survey will detect failures of surface materials or at openings that will require further inspection to determine whether the cause was the superstructure system.

Deficiencies	Causes
Concrete (columns, walls, and floor and roof slabs)	
_____ Overall alignment	_____ Settlement due to design and construction techniques. _____ Under designed for loading conditions (see *A10 Foundations*).
_____ Deflection	_____ Expansion and/or contraction due to changes in design loads. _____ Original design deficient. _____ Original materials deficient.
_____ Surface Conditions	
_____ Cracks	_____ Inadequate design and/or construction due to changes in design loads. _____ Stress concentration. _____ Extreme temperature changes; secondary effects of freeze-thaw.
_____ Scaling, spalls, and pop-outs	_____ Extreme temperature changes. _____ Reinforcement corrosion. _____ Environmental conditions. _____ Mechanical damage. _____ Poor materials.
_____ Stains	_____ Chemical reaction of reinforcing. _____ Reaction of materials in concrete mixture. _____ Environmental conditions.
_____ Exposed reinforcing	_____ Corrosion of steel. _____ Insufficient cover. _____ Mechanical damage.

B10 SUPERSTRUCTURES INSPECTION CHECKLIST

	Deficiencies		Causes
Steel (structural members, stairs, and connections)			
_____	Overall alignment	_____	Settlement due to design and construction techniques; improper fabrication.
_____	Deflection or cracking	_____	Expansion and/or contraction.
		_____	Changes in design loads.
		_____	Fatigue due to vibration or impact.
_____	Corrosion	_____	Electrochemical reaction.
		_____	Failure of protective coating.
		_____	Excessive moisture exposure.
_____	Surface deterioration	_____	Excessive wear.
Wood (structural members and connections)			
_____	Overall alignment	_____	Settlement; design and construction techniques.
_____	Deflection or cracking	_____	Expansion and/or contraction.
		_____	Changes in design loads.
		_____	Fatigue due to vibration or impact.
		_____	Failure of compression members.
		_____	Poor construction techniques.
		_____	General material failure.
_____	Rot (Decay)	_____	Direct contact with moisture.
		_____	Condensation.
		_____	Omission or deterioration of vapor barrier.
		_____	Poor construction techniques.
		_____	Damage from rodents or insects.

B20 EXTERIOR CLOSURE INSPECTION CHECKLIST

General Inspection

- Overall appearance
- Displacement
- Paint conditions
- Caulking
- Window & door fit
- Flashing condition
- Material integrity
- Cracks
- Evidence of moisture
- Construction joints
- Hardware condition

Exterior Walls

_____ Wood (Shingles, weatherboard siding, plywood).

_____ Paint or surface treatment condition
_____ Rot or decay
_____ Moisture penetration
_____ Loose, cracked, warped, or broken boards and shingles.

_____ Concrete, Masonry, and Tile (Concrete, brick, concrete masonry units, structural tile, glazed tile, stucco, stone).

_____ Settlement
_____ Construction and expansion joints
_____ Surface deterioration
_____ Parapet movement
_____ Structural frame movement causing cracks
_____ Condition of caulking and mortar
_____ Efflorescence and staining
_____ Tightness of fasteners

_____ Metal (Corrugated iron or steel, aluminum, enamel coated steel, protected metals).

_____ Settlement
_____ Condition of bracing
_____ Tightness of fasteners
_____ Flashings
_____ Structural frame movement
_____ Surface damage due to impact
_____ Caulking
_____ Corrosion

_____ Finishes (Mineral products, fiberglass, polyester resins, and plastics).

_____ Settlement
_____ Surface damage due to impact
_____ Stains
_____ Adhesion to substrate
_____ Flashings
_____ Asbestos products
_____ Structural frame movement
_____ Cracks
_____ Fasteners
_____ Caulking
_____ Lead paint

_____ Insulation

_____ Insulation present
_____ Satisfactory condition

Windows and Doors

_____ Frame fitting
_____ Paint or surface finish
_____ Hardware and operating parts
_____ Cleanliness
_____ Rot or corrosion
_____ Frame and molding condition
_____ Putty and weatherstripping
_____ Security
_____ Material condition (glass, wood, & metal Panels)
_____ Screens and storm windows

Shading Devices

_____ Material conditions
_____ Cleanliness
_____ Operations

C10 PARTITIONS & DOORS INSPECTION CHECKLIST

General Inspection

- Strength & stability
- Acoustical quality
- Maintainability
- Physical condition
- Adaptability
- Code compliance

Wall Material

	Deficiencies		Causes
_____	Cracks	_____	Settlement
_____	Holes	_____	Defective material
_____	Looseness	_____	Operational abuse or vibrations
_____	Missing segments	_____	Environmental attack
_____	Water stains	_____	Moisture
_____	Joints	_____	Structural expansion or contraction
_____	Surface appearance	_____	Wind pressure

Hardware

_____	Overall condition	_____	Appearance
_____	Keying system	_____	Operations
_____	Fit	_____	Locksets
_____	Cylinders	_____	Panic devices
_____	Maintainability	_____	Security operations

C30 WALLS & FINISHES INSPECTION CHECKLIST

General Inspection

Code compliance ______________________________

Maintainability ______________________________

Means of egress ______________________________

Fire ratings ______________________________

Audible & visual device condition ______________________________

Extinguishing systems (see also *08 Mechanical/Plumbing*): ______________

Type ______________________________

Lighting system (see also *09 Electrical Lighting/Power*): ______________

Type ______________________________

Handicapped accessibility ______________________________

Building user comments ______________________________

____

Exterior Lighting

Adequacy:

Good _____ Fair _____ Poor_____

Condition ______________________________

Controls (type & location) ______________________________

Fire Alarm Systems

_____ Panel visible _____ Operational

_____ Pull station condition _____ Detector conditions

Stairs and Ramps

_____ Exits marked _____ Hardware operational

_____ Tripping hazards _____ Surface conditions

_____ Lighting adequate _____ Handrails

General Inspection

Overall Appearance

Good   Fair   Poor

- Evidence of moisture
- Irregular surface
- Handicapped hazards
- Visible settlement
- Tripping hazards
- Replacement necessary

Carpet (Tufted, tile)

Age
_____ Stains
_____ Holes, tears
_____ Excessive wear
_____ Discoloration
_____ Seam conditions

Resilient (Asphalt tile, cork tile, linoleum, rubber, vinyl)

_____ Broken tiles
_____ Shrinkage
_____ Fading
_____ Porosity
_____ Loose tiles
_____ Lifting, cupping
_____ Cuts, holes
_____ Asbestos present

Masonry (Stone, brick)

_____ Cracks
_____ Joints
_____ Porosity
_____ Deterioration
_____ Stains
_____ Sealing

Monolithic Topping (Concrete, granolithic, terrazzo, magnesite)

_____ Cracks
_____ Joints
_____ Porosity
_____ Sealing

Wood (Plank, strips, block, parquet)

_____ Shrinkage
_____ Excessive wear
_____ Decay
_____ Cupping, warpage
_____ Unevenness
_____ Sealing

C30 CEILINGS & FINISHES INSPECTION CHECKLIST

General Inspection

Overall appearance

Good ____ Fair ____ Poor ____

- Settlement or sagging
- Attachment
- Stains, discoloration
- Suitability
- Code compliance
- Alignment
- Evidence of moisture
- Missing units
- Acoustic quality

Exposed Systems (Unpainted, painted, spray-on, decorative)

_____ Cracks
_____ Surface deterioration
_____ Missing elements
_____ Adhesion

Applied to Structure & Suspended

General Condition:

Good ____ Fair ____ Poor ____

_____ Fasteners
_____ Trim condition

Openings:

_____ Panels
_____ Lighting fixtures
_____ Fire protection
_____ Inserts
_____ Air distribution
_____ Other

D10 CONVEYING INSPECTION CHECKLIST

General Inspection (Passenger Conveying)

Overall appearance (interior)

Good _____ Fair _____ Poor _____

Overall appearance (exterior)

Good _____ Fair _____ Poor _____

_____ Maintenance history available

- _____ Regular inspection frequency
- _____ Door operations
- _____ Control systems
- _____ Noise
- _____ Code compliance
- _____ Handicapped access
- _____ Major repairs necessary
- _____ Replacement necessary

D20 MECHANICAL/PLUMBING INSPECTION CHECKLIST

General Inspection

General appearance

Good ____ Fair ____ Poor ____

Leaks, dripping, running faucets and valves ____________

Maintenance history ____________

Supply adequacy ____________

Sanitation hazards ____________

Drain & waste connection ____________

Backflow protection ____________

Cross connections ____________

Fixture quantity ____________

Fixture types & conditions ____________

Handicapped fixtures ____________

Female facilities ____________

Metal pipe & fittings corrosion ____________

Pipe joints & sealing ____________

Pipe insulation ____________

Hanger supports & clamps ____________

Filters ____________

Building user comments ____________

_____ **Water System**

_____ Water pressure adequate
_____ Main cutoff operable
_____ Pump condition
_____ Odors, tastes
_____ Water heating temperature setting
_____ Insulation condition

_____ **Sanitary & Storm System**

_____ Flow adequate
_____ Floor drains
_____ Gradient
_____ Cleanouts access
_____ Chemical resistance
_____ On-site disposal system

_____ **Code Requirements**

_____ EPA/local permits
_____ Other

_____ **Fire protection system**

_____ Regular inspections
_____ Complies with code
_____ Hose cabinets functional
_____ Sprinkler heads operable
_____ Controls operable
_____ Water pressure sufficient

General Inspection

General appearance

Good ___ Fair ___ Poor ___

Lubrication, bearings and moving parts ___

Rust and corrosion ___

Motors, fans, drive assemblies, and pumps ___

Wiring and electrical controls ___

Thermostats and automatic temperature controls ___

Thermal insulation and protective coatings ___

Guards, casings, hangers, supports, platforms, and mounting bolts ___

Piping system identification ___

Solenoid valves ___

Burner assemblies ___

Combustion chambers, smokepipes, and breeching ___

Electrical heating units ___

Guards, casings, hangers, supports, platforms, and mounting bolts ___

Steam and hot water heating equipment ___

Accessible steam, water, and fuel piping ___

Traps ___

Humidifier assemblies and controls ___

Strainers ___

Water sprays, weirs, and similar devices ___

Shell-and tube-type condensers ___

Self-contained evaporative condensers ___

Air cooled condensers ___

Compressors ___

Liquid receivers ___

Refrigerant driers, strainers, valves, oil traps, and accessories ___

Building user comments ___

D30 MECHANICAL/HVAC INSPECTION CHECKLIST

Cleaning, maintenance, repair, and replacement

_____ Registers
_____ Dampers
_____ Plenum chambers
_____ Louvers
_____ Grills
_____ Draft diverters
_____ Supply and return ducts
_____ Fire dampers

Air filters

_____ Correct type
_____ Replacement schedule

Heating System Evaluation

_____ Heating capacity
_____ Temperature control
_____ Heating:
    _____ Seasonal
    _____ All year
_____ Noise level
_____ Air circulation & ventilation
_____ Humidity control
_____ Energy consumption
_____ Filtration

Cooling System Evaluation

_____ Cooling capacity
_____ Cooling all season
_____ Energy consumption
_____ Filtration
_____ Temperature and humidity control
_____ Noise level
_____ Air circulation & ventilation
_____ Reliability

Ventilation System Evaluation

_____ Air velocity
_____ Bag collection
_____ Steam and hot water coils
_____ Fire hazards
_____ Exhaust air systems
_____ Wet collectors
_____ Electrical heating units
_____ Fire protective devices

D50 ELEC./SERVICE & DIST. INSPECTION CHECKLIST

General Inspection

Safety conditions ______________________________

Service capacity, % used, and age ______________________________

Switchgear capacity, % used, and age ______________________________

Feeder capacity, % used, and age ______________________________

Panel capacity ______________________________

Thermo-scanning ______________________________

Maintenance records available ______________________________

Convenience outlets ______________________________

Building user comments ______________________________

Exterior Service

_____ Line drawing

_____ Feed source:

- _____ Utility/owned
- _____ Above/below ground

_____ Transformer:

- _____ Transformer tested
- _____ Transformer PCB's
- _____ Transformer arcing or burning
- _____ Ownership (facility or utility)

Interior Distribution System

- _____ Line drawing
- _____ Main circuit breaker marked
- _____ All wiring in conduit
- _____ Panels marked
- _____ Missing breakers
- _____ Incoming conduit marked
- _____ Panel boards, junction boxes covered
- _____ Conduit properly secured
- _____ Panel schedules

Emergency Circuits

Emergency generator(s):

- _____ Condition and age
- _____ Testing schedule
- _____ Service schedule
- _____ Circuits appropriate
- _____ Fuel storage (capacity)
- _____ Auto start and switchover
- _____ Test records available
- _____ Service schedule records available
- _____ Cooling & exhaust

Emergency lighting/power systems:

- _____ Battery operation
- _____ Exit signs
- _____ Elevators
- _____ HVAC
- _____ Separate power feed
- _____ Stairways/corridors
- _____ Interior areas
- _____ Exterior

G20 SITE WORK INSPECTION CHECKLIST

General Inspection

Overall appearance

Good _____ Fair _____ Poor _____

Maintainability ______________________________

Repairs/replacements ______________________________

Code compliance ______________________________

Roads, Walks, and Parking Lots

Surface conditions ______________________________

Subsurface conditions ______________________________

Settling and uplift ______________________________

Cracks, holes ______________________________

Drainage and slope ______________________________

Curbing

Alignment ______________________________

Erosion ______________________________

Repairs/replacements ______________________________

Drainage and Erosion Controls

Surface drainage ______________________________

Manholes, catch basins ______________________________

Vegetation ______________________________

Channels, dikes ______________________________

Retention, detention ______________________________

Drains ______________________________

Parking Lot Controls

Location ______________________________

Operation ______________________________

Repairs/replacements ______________________________

H10 SAFETY INSPECTION CHECKLIST

General Inspection

Code compliance ______________________________

Maintainability ______________________________

Means of egress ______________________________

Fire ratings ______________________________

Audible & visual device condition ______________________________

Extinguishing systems (see also *D20 Mechanical/Plumbing*): ______________

Type ______________________________

Lighting system (see also *D50 Electrical Lighting/Power*): ______________

Type ______________________________

Handicapped accessibility ______________________________

Building user comments ______________________________

____

Exterior Lighting

Adequacy:

Good _____ Fair _____ Poor_____

Condition ______________________________

Controls (type & location) ______________________________

Fire Alarm Systems

_____ Panel visible _____ Operational

_____ Pull station condition _____ Detector conditions

Stairs and Ramps

_____ Exits marked _____ Hardware operational

_____ Tripping hazards _____ Surface conditions

_____ Lighting adequate _____ Handrails

Reference Section

All the reference information is in one section, making it easy to find what you need to know . . . and easy to use the data set on a daily basis. This section is visually identified by a vertical black bar on the page edges.

In this Reference Section we've included information on Equipment Rental Costs, Crew Listings, Travel Costs, Reference Tables, Life Cycle Costing, and a listing of Abbreviations.

Table of Contents

Equipment Rental Costs

Estimating Tips

- This section contains the average costs to rent and operate hundreds of pieces of construction equipment. This is useful information when one is estimating the time and material requirements of any particular operation in order to establish a unit or total cost. Bare equipment costs shown on a unit cost line include, not only rental, but also operating costs for equipment under normal use.

Rental Costs

- Equipment rental rates are obtained from the following industry sources throughout North America: contractors, suppliers, dealers, manufacturers, and distributors.
- Rental rates vary throughout the country, with larger cities generally having lower rates. Lease plans for new equipment are available for periods in excess of six months, with a percentage of payments applying toward purchase.
- Monthly rental rates vary from 2% to 5% of the purchase price of the equipment depending on the anticipated life of the equipment and its wearing parts.
- Weekly rental rates are about 1/3 of the monthly rates, and daily rental rates are about 1/3 of the weekly rate.
- Rental rates can also be treated as reimbursement costs for contractor-owned equipment. Owned equipment costs include depreciation, loan payments, interest, taxes, insurance, storage, and major repairs.

Operating Costs

- The operating costs include parts and labor for routine servicing, such as the repair and replacement of pumps, filters, and worn lines. Normal operating expendables, such as fuel, lubricants, tires, and electricity (where applicable), are also included.
- Extraordinary operating expendables with highly variable wear patterns, such as diamond bits and blades, are excluded. These costs can be found as material costs in the Unit Price section.
- The hourly operating costs listed do not include the operator's wages.

Equipment Cost/Day

- Any power equipment required by a crew is shown in the Crew Listings with a daily cost.
- This daily cost of equipment needed by a crew includes both the rental cost and the operating cost and is based on dividing the weekly rental rate by 5 (the number of working days in the week), then adding the hourly operating cost multiplied by 8 (the number of hours in a day). This "Equipment Cost/Day" is shown in the far right column of the Equipment Rental section.
- If equipment is needed for only one or two days, it is best to develop your own cost by including components for daily rent and hourly operating costs. This is important when the listed Crew for a task does not contain the equipment needed, such as a crane for lifting mechanical heating/cooling equipment up onto a roof.
- If the quantity of work is less than the crew's Daily Output shown for a Unit Price line item that includes a bare unit equipment cost, the recommendation is to estimate one day's rental cost and operating cost for equipment shown in the Crew Listing for that line item.
- Please note, in some cases the equipment description in the crew is followed by a time period in parenthesis. For example: (daily) or (monthly). In these cases the equipment cost/day is calculated by adding the rental cost per time period to the hourly operating cost multiplied by 8.

Mobilization, Demobilization Costs

- The cost to move construction equipment from an equipment yard or rental company to the job site and back again is not included in equipment rental costs listed in the Reference Section. It is also not included in the bare equipment cost of any Unit Price line item or in any equipment costs shown in the Crew Listings.
- Mobilization (to the site) and demobilization (from the site) costs can be found in the Unit Price section.
- If a piece of equipment is already at the job site, it is not appropriate to utilize mobilization or demobilization costs again in an estimate. ■

01 54 33 | Equipment Rental

			UNIT	HOURLY OPER. COST	RENT PER DAY	RENT PER WEEK	RENT PER MONTH	EQUIPMENT COST/DAY
10	0010	**CONCRETE EQUIPMENT RENTAL** without operators R015433-10						
	0200	Bucket, concrete lightweight, 1/2 C.Y.	Ea.	.87	38.50	115	345	30
	0300	1 C.Y.		.98	63.50	190	570	45.80
	0400	1-1/2 C.Y.		1.23	60	180	540	45.85
	0500	2 C.Y.		1.34	73.50	220	660	54.70
	0580	8 C.Y.		6.47	93.50	280	840	107.75
	0600	Cart, concrete, self-propelled, operator walking, 10 C.F.		2.85	175	525	1,575	127.85
	0700	Operator riding, 18 C.F.		4.81	192	575	1,725	153.45
	0800	Conveyer for concrete, portable, gas, 16″ wide, 26′ long		10.61	160	480	1,450	180.85
	0900	46′ long		10.98	175	525	1,575	192.85
	1000	56′ long		11.15	192	575	1,725	204.15
	1100	Core drill, electric, 2-1/2 H.P., 1″ to 8″ bit diameter		1.56	83.50	250	750	62.50
	1150	11 H.P., 8″ to 18″ cores		5.38	119	356.32	1,075	114.30
	1200	Finisher, concrete floor, gas, riding trowel, 96″ wide		9.64	153	459.60	1,375	169
	1300	Gas, walk-behind, 3 blade, 36″ trowel		2.03	96	287.50	865	73.75
	1400	4 blade, 48″ trowel		3.06	104	312.50	940	87
	1500	Float, hand-operated (Bull float), 48″ wide		.08	12.35	37	111	8.05
	1570	Curb builder, 14 H.P., gas, single screw		14.00	253	760	2,275	263.95
	1590	Double screw		15.00	253	760	2,275	272
	1600	Floor grinder, concrete and terrazzo, electric, 22″ path		3.03	134	401.75	1,200	104.60
	1700	Edger, concrete, electric, 7″ path		1.18	57.50	172.50	520	43.95
	1750	Vacuum pick-up system for floor grinders, wet/dry		1.61	102	305.71	915	74.05
	1800	Mixer, powered, mortar and concrete, gas, 6 C.F., 18 H.P.		7.39	97	291	875	117.35
	1900	10 C.F., 25 H.P.		8.97	114	342.50	1,025	140.30
	2000	16 C.F.		9.33	144	432.50	1,300	161.15
	2100	Concrete, stationary, tilt drum, 2 C.Y.		7.21	80	240	720	105.70
	2120	Pump, concrete, truck mounted, 4″ line, 80′ boom		29.79	287	860	2,575	410.35
	2140	5″ line, 110′ boom		37.34	287	860	2,575	470.75
	2160	Mud jack, 50 C.F. per hr.		6.43	228	685	2,050	188.45
	2180	225 C.F. per hr.		8.52	293	880	2,650	244.15
	2190	Shotcrete pump rig, 12 C.Y./hr.		13.93	223	670	2,000	245.40
	2200	35 C.Y./hr.		15.75	287	860	2,575	298
	2600	Saw, concrete, manual, gas, 18 H.P.		5.52	112	337	1,000	111.55
	2650	Self-propelled, gas, 30 H.P.		7.87	81	242.71	730	111.50
	2675	V-groove crack chaser, manual, gas, 6 H.P.		1.64	100	300	900	73.10
	2700	Vibrators, concrete, electric, 60 cycle, 2 H.P.		.47	73	218.50	655	47.45
	2800	3 H.P.		.56	73.50	221	665	48.70
	2900	Gas engine, 5 H.P.		1.54	16.85	50.61	152	22.45
	3000	8 H.P.		2.08	17.05	51.12	153	26.85
	3050	Vibrating screed, gas engine, 8 H.P.		2.80	88	263.50	790	75.10
	3120	Concrete transit mixer, 6 x 4, 250 H.P., 8 C.Y., rear discharge		50.57	70	210	630	446.55
	3200	Front discharge		58.71	135	405	1,225	550.65
	3300	6 x 6, 285 H.P., 12 C.Y., rear discharge		57.97	150	450	1,350	553.80
	3400	Front discharge	↓	60.41	170	510	1,525	585.25
20	0010	**EARTHWORK EQUIPMENT RENTAL** without operators R015433-10						
	0040	Aggregate spreader, push type, 8′ to 12′ wide	Ea.	2.59	75	225	675	65.75
	0045	Tailgate type, 8′ wide		2.54	63.50	190	570	58.30
	0055	Earth auger, truck mounted, for fence & sign posts, utility poles		13.81	150	450	1,350	200.50
	0060	For borings and monitoring wells		42.52	83.50	250	750	390.20
	0070	Portable, trailer mounted		2.29	100	300	900	78.35
	0075	Truck mounted, for caissons, water wells		85.14	150	450	1,350	771.10
	0080	Horizontal boring machine, 12″ to 36″ diameter, 45 H.P.		22.70	104	312	935	244
	0090	12″ to 48″ diameter, 65 H.P.		31.16	108	325	975	314.25
	0095	Auger, for fence posts, gas engine, hand held		.45	84	251.50	755	53.85
	0100	Excavator, diesel hydraulic, crawler mounted, 1/2 C.Y. cap.		21.66	465	1,394.28	4,175	452.10
	0120	5/8 C.Y. capacity		28.95	610	1,833.22	5,500	598.25
	0140	3/4 C.Y. capacity		32.56	725	2,168.88	6,500	694.25
	0150	1 C.Y. capacity	↓	41.09	780	2,333	7,000	795.30

01 54 33 | Equipment Rental

		01 54 33 Equipment Rental	UNIT	HOURLY OPER. COST	RENT PER DAY	RENT PER WEEK	RENT PER MONTH	EQUIPMENT COST/DAY
20	0200	1-1/2 C.Y. capacity	Ea.	48.44	500	1,500	4,500	687.55
	0300	2 C.Y. capacity		56.41	835	2,500	7,500	951.30
	0320	2-1/2 C.Y. capacity		82.39	1,350	4,027.92	12,100	1,465
	0325	3-1/2 C.Y. capacity		119.76	2,000	6,000	18,000	2,158
	0330	4-1/2 C.Y. capacity		151.17	3,675	11,000	33,000	3,409
	0335	6 C.Y. capacity		191.81	3,225	9,650	29,000	3,464
	0340	7 C.Y. capacity		174.67	3,400	10,200	30,600	3,437
	0342	Excavator attachments, bucket thumbs		3.39	258	774.60	2,325	182.05
	0345	Grapples		3.13	222	666.16	2,000	158.30
	0346	Hydraulic hammer for boom mounting, 4000 ft lb.		13.44	890	2,670	8,000	641.50
	0347	5000 ft lb.		15.90	950	2,850	8,550	697.25
	0348	8000 ft lb.		23.47	1,200	3,600	10,800	907.75
	0349	12,000 ft lb.		25.64	1,100	3,333	10,000	871.75
	0350	Gradall type, truck mounted, 3 ton @ 15′ radius, 5/8 C.Y.		43.31	835	2,500	7,500	846.50
	0370	1 C.Y. capacity		59.22	835	2,500	7,500	973.80
	0400	Backhoe-loader, 40 to 45 H.P., 5/8 C.Y. capacity		11.86	244	732.50	2,200	241.40
	0450	45 H.P. to 60 H.P., 3/4 C.Y. capacity		17.97	117	350	1,050	213.75
	0460	80 H.P., 1-1/4 C.Y. capacity		20.30	117	350	1,050	232.40
	0470	112 H.P., 1-1/2 C.Y. capacity		32.89	610	1,833.22	5,500	629.75
	0482	Backhoe-loader attachment, compactor, 20,000 lb.		6.42	155	464.76	1,400	144.30
	0485	Hydraulic hammer, 750 ft lb.		3.67	107	320.17	960	93.40
	0486	Hydraulic hammer, 1200 ft lb.		6.53	205	614.52	1,850	175.15
	0500	Brush chipper, gas engine, 6″ cutter head, 35 H.P.		9.14	247	740	2,225	221.10
	0550	Diesel engine, 12″ cutter head, 130 H.P.		23.60	340	1,020	3,050	392.80
	0600	15″ cutter head, 165 H.P.		26.51	415	1,239.36	3,725	459.90
	0750	Bucket, clamshell, general purpose, 3/8 C.Y.		1.40	91.50	275	825	66.15
	0800	1/2 C.Y.		1.51	91.50	275	825	67.10
	0850	3/4 C.Y.		1.64	91.50	275	825	68.10
	0900	1 C.Y.		1.70	91.50	275	825	68.55
	0950	1-1/2 C.Y.		2.78	91.50	275	825	77.25
	1000	2 C.Y.		2.91	91.50	275	825	78.30
	1010	Bucket, dragline, medium duty, 1/2 C.Y.		.82	91.50	275	825	61.55
	1020	3/4 C.Y.		.78	91.50	275	825	61.25
	1030	1 C.Y.		.80	91.50	275	825	61.35
	1040	1-1/2 C.Y.		1.26	91.50	275	825	65.05
	1050	2 C.Y.		1.29	91.50	275	825	65.30
	1070	3 C.Y.		2.07	91.50	275	825	71.60
	1200	Compactor, manually guided 2-drum vibratory smooth roller, 7.5 H.P.		7.20	181	542.50	1,625	166.10
	1250	Rammer/tamper, gas, 8″		2.20	48	144.59	435	46.50
	1260	15″		2.62	55	165.25	495	54
	1300	Vibratory plate, gas, 18″ plate, 3000 lb. blow		2.12	24.50	72.81	218	31.55
	1350	21″ plate, 5000 lb. blow		2.61	241	722.50	2,175	165.35
	1370	Curb builder/extruder, 14 H.P., gas, single screw		13.99	253	760	2,275	263.95
	1390	Double screw		14.99	253	760	2,275	271.95
	1500	Disc harrow attachment, for tractor		.47	82.50	246.84	740	53.15
	1810	Feller buncher, shearing & accumulating trees, 100 H.P.		39.08	460	1,380	4,150	588.60
	1860	Grader, self-propelled, 25,000 lb.		33.25	1,100	3,333	10,000	932.60
	1910	30,000 lb.		32.76	1,325	4,000	12,000	1,062
	1920	40,000 lb.		51.73	1,550	4,667	14,000	1,347
	1930	55,000 lb.		66.73	1,775	5,333	16,000	1,600
	1950	Hammer, pavement breaker, self-propelled, diesel, 1000 to 1250 lb.		28.31	600	1,800	5,400	586.50
	2000	1300 to 1500 lb.		42.67	1,000	3,020.94	9,075	945.55
	2050	Pile driving hammer, steam or air, 4150 ft lb. @ 225 bpm		12.11	500	1,500	4,500	396.90
	2100	8750 ft lb. @ 145 bpm		14.30	700	2,100	6,300	534.45
	2150	15,000 ft lb. @ 60 bpm		14.63	835	2,500	7,500	617.05
	2200	24,450 ft lb. @ 111 bpm		15.64	965	2,900	8,700	705.15
	2250	Leads, 60′ high for pile driving hammers up to 20,000 ft lb.		3.66	300	900	2,700	209.25
	2300	90′ high for hammers over 20,000 ft lb.	↓	5.43	540	1,620	4,850	367.45

01 54 33 | Equipment Rental

			UNIT	HOURLY OPER. COST	RENT PER DAY	RENT PER WEEK	RENT PER MONTH	EQUIPMENT COST/DAY	
20	2350	Diesel type hammer, 22,400 ft lb.	Ea.	17.76	490	1,471.74	4,425	436.45	20
	2400	41,300 ft lb.		25.61	620	1,859.04	5,575	576.65	
	2450	141,000 ft lb.		41.20	980	2,943.48	8,825	918.35	
	2500	Vib. elec. hammer/extractor, 200 kW diesel generator, 34 H.P.		41.25	715	2,143.06	6,425	758.60	
	2550	80 H.P.		72.81	1,025	3,098.40	9,300	1,202	
	2600	150 H.P.		134.74	2,000	5,964.42	17,900	2,271	
	2700	Hydro Excavator w/EXT boom 12 C.Y., 1200 gallons		37.70	1,600	4,800	14,400	1,262	
	2800	Log chipper, up to 22" diameter, 600 H.P.		46.13	305	915	2,750	552	
	2850	Logger, for skidding & stacking logs, 150 H.P.		43.40	930	2,785	8,350	904.20	
	2860	Mulcher, diesel powered, trailer mounted		17.99	305	915	2,750	326.90	
	2900	Rake, spring tooth, with tractor		14.67	370	1,110.26	3,325	339.45	
	3000	Roller, vibratory, tandem, smooth drum, 20 H.P.		7.78	320	967	2,900	255.65	
	3050	35 H.P.		10.10	260	779.76	2,350	236.75	
	3100	Towed type vibratory compactor, smooth drum, 50 H.P.		25.20	520	1,566	4,700	514.75	
	3150	Sheepsfoot, 50 H.P.		25.56	385	1,161.90	3,475	436.90	
	3170	Landfill compactor, 220 H.P.		69.80	1,650	4,985	15,000	1,555	
	3200	Pneumatic tire roller, 80 H.P.		12.88	405	1,213.54	3,650	345.75	
	3250	120 H.P.		19.33	665	1,988.14	5,975	552.25	
	3300	Sheepsfoot vibratory roller, 240 H.P.		62.02	1,425	4,260.30	12,800	1,348	
	3320	340 H.P.		83.57	2,175	6,500	19,500	1,969	
	3350	Smooth drum vibratory roller, 75 H.P.		23.27	655	1,962.32	5,875	578.60	
	3400	125 H.P.		27.53	740	2,220.52	6,650	664.35	
	3410	Rotary mower, brush, 60", with tractor		18.73	360	1,084.44	3,250	366.75	
	3420	Rototiller, walk-behind, gas, 5 H.P.		2.13	60	180	540	53.05	
	3422	8 H.P.		2.80	132	395	1,175	101.40	
	3440	Scrapers, towed type, 7 C.Y. capacity		6.42	127	382.14	1,150	127.80	
	3450	10 C.Y. capacity		7.18	170	511.24	1,525	159.70	
	3500	15 C.Y. capacity		7.38	196	588.70	1,775	176.75	
	3525	Self-propelled, single engine, 14 C.Y. capacity		132.89	2,225	6,660	20,000	2,395	
	3550	Dual engine, 21 C.Y. capacity		140.95	2,500	7,500	22,500	2,628	
	3600	31 C.Y. capacity		187.28	3,625	10,844.40	32,500	3,667	
	3640	44 C.Y. capacity		231.98	4,650	13,942.80	41,800	4,644	
	3650	Elevating type, single engine, 11 C.Y. capacity		61.68	1,075	3,200	9,600	1,133	
	3700	22 C.Y. capacity		114.28	1,625	4,850	14,600	1,884	
	3710	Screening plant, 110 H.P. w/5' x 10' screen		21.07	645	1,933	5,800	555.15	
	3720	5' x 16' screen		26.60	1,325	4,000	12,000	1,013	
	3850	Shovel, crawler-mounted, front-loading, 7 C.Y. capacity		218.00	3,925	11,773.92	35,300	4,099	
	3855	12 C.Y. capacity		335.89	5,450	16,318.24	49,000	5,951	
	3860	Shovel/backhoe bucket, 1/2 C.Y.		2.68	73	218.95	655	65.25	
	3870	3/4 C.Y.		2.66	82	245.81	735	70.40	
	3880	1 C.Y.		2.75	91	272.66	820	76.50	
	3890	1-1/2 C.Y.		2.94	107	320.17	960	87.60	
	3910	3 C.Y.		3.43	145	433.78	1,300	114.15	
	3950	Stump chipper, 18" deep, 30 H.P.		6.88	232	697	2,100	194.45	
	4110	Dozer, crawler, torque converter, diesel 80 H.P.		25.18	335	1,000	3,000	401.40	
	4150	105 H.P.		34.23	600	1,800	5,400	633.85	
	4200	140 H.P.		41.15	720	2,166	6,500	762.40	
	4260	200 H.P.		62.97	1,675	5,000	15,000	1,504	
	4310	300 H.P.		80.49	1,875	5,600	16,800	1,764	
	4360	410 H.P.		106.42	3,200	9,630	28,900	2,777	
	4370	500 H.P.		132.98	3,900	11,670	35,000	3,398	
	4380	700 H.P.		229.47	5,475	16,421.52	49,300	5,120	
	4400	Loader, crawler, torque conv., diesel, 1-1/2 C.Y., 80 H.P.		29.45	550	1,651	4,950	565.80	
	4450	1-1/2 to 1-3/4 C.Y., 95 H.P.		30.18	695	2,091.42	6,275	659.75	
	4510	1-3/4 to 2-1/4 C.Y., 130 H.P.		47.61	965	2,900	8,700	960.90	
	4530	2-1/2 to 3-1/4 C.Y., 190 H.P.		57.61	1,175	3,540	10,600	1,169	
	4560	3-1/2 to 5 C.Y., 275 H.P.		71.19	1,525	4,595.96	13,800	1,489	
	4610	Front end loader, 4WD, articulated frame, diesel, 1 to 1-1/4 C.Y., 70 H.P.	↓	16.58	282	846.90	2,550	302	

01 54 33 | Equipment Rental

			UNIT	HOURLY OPER. COST	RENT PER DAY	RENT PER WEEK	RENT PER MONTH	EQUIPMENT COST/DAY
20	4620	1-1/2 to 1-3/4 C.Y., 95 H.P.	Ea.	19.94	440	1,320	3,950	423.50
	4650	1-3/4 to 2 C.Y., 130 H.P.		21.00	395	1,187.72	3,575	405.55
	4710	2-1/2 to 3-1/2 C.Y., 145 H.P.		29.44	780	2,333	7,000	702.10
	4730	3 to 4-1/2 C.Y., 185 H.P.		31.99	890	2,667	8,000	789.35
	4760	5-1/4 to 5-3/4 C.Y., 270 H.P.		53.03	890	2,666	8,000	957.50
	4810	7 to 8 C.Y., 475 H.P.		[illegible]	2,550	7,667	23,000	[illegible]
	4870	9 to 11 C.Y., 620 H.P.		131.52	2,700	8,107.48	24,300	2,674
	4880	Skid-steer loader, wheeled, 10 C.F., 30 H.P. gas		9.54	169	506.07	1,525	177.55
	4890	1 C.Y., 78 H.P., diesel		18.38	420	1,265.18	3,800	400.10
	4892	Skid-steer attachment, auger		.74	145	433.50	1,300	92.65
	4893	Backhoe		.74	122	366.64	1,100	79.25
	4894	Broom		.70	140	420.25	1,250	89.70
	4895	Forks		.15	32	96	288	20.45
	4896	Grapple		.72	88.50	265.25	795	58.80
	4897	Concrete hammer		1.05	183	550	1,650	118.40
	4898	Tree spade		.60	103	309.84	930	66.75
	4899	Trencher		.65	102	305	915	66.20
	4900	Trencher, chain, boom type, gas, operator walking, 12 H.P.		4.16	206	618.25	1,850	156.95
	4910	Operator riding, 40 H.P.		16.64	450	1,343.75	4,025	401.85
	5000	Wheel type, diesel, 4′ deep, 12″ wide		68.50	965	2,891.84	8,675	1,126
	5100	6′ deep, 20″ wide		87.32	1,050	3,127.50	9,375	1,324
	5150	Chain type, diesel, 5′ deep, 8″ wide		16.25	360	1,084.44	3,250	346.90
	5200	Diesel, 8′ deep, 16″ wide		89.39	1,925	5,783.68	17,400	1,872
	5202	Rock trencher, wheel type, 6″ wide x 18″ deep		46.98	90	270	810	429.85
	5206	Chain type, 18″ wide x 7′ deep		104.38	283	850	2,550	1,005
	5210	Tree spade, self-propelled		13.65	230	690	2,075	247.20
	5250	Truck, dump, 2-axle, 12 ton, 8 C.Y. payload, 220 H.P.		23.88	395	1,185	3,550	428.05
	5300	Three axle dump, 16 ton, 12 C.Y. payload, 400 H.P.		44.50	360	1,084.44	3,250	572.90
	5310	Four axle dump, 25 ton, 18 C.Y. payload, 450 H.P.		49.85	525	1,575.02	4,725	713.80
	5350	Dump trailer only, rear dump, 16-1/2 C.Y.		5.73	151	454.43	1,375	136.70
	5400	20 C.Y.		6.18	170	511.24	1,525	151.70
	5450	Flatbed, single axle, 1-1/2 ton rating		19.00	73.50	221.02	665	196.15
	5500	3 ton rating		23.05	1,050	3,180	9,550	820.40
	5550	Off highway rear dump, 25 ton capacity		62.67	1,475	4,389.40	13,200	1,379
	5600	35 ton capacity		66.90	665	2,000	6,000	935.20
	5610	50 ton capacity		83.87	1,825	5,499.66	16,500	1,771
	5620	65 ton capacity		89.56	2,000	5,990.24	18,000	1,915
	5630	100 ton capacity		121.24	2,950	8,830.44	26,500	2,736
	6000	Vibratory plow, 25 H.P., walking	↓	6.77	300	900	2,700	234.20
40	0010	**GENERAL EQUIPMENT RENTAL** without operators R015433-10						
	0020	Aerial lift, scissor type, to 20′ high, 1200 lb. capacity, electric	Ea.	3.48	129	385.75	1,150	105
	0030	To 30′ high, 1200 lb. capacity		3.77	203	607.67	1,825	151.70
	0040	Over 30′ high, 1500 lb. capacity		5.13	243	727.50	2,175	186.60
	0070	Articulating boom, to 45′ high, 500 lb. capacity, diesel R015433-15		9.92	250	750	2,250	229.35
	0075	To 60′ high, 500 lb. capacity		13.66	300	900	2,700	289.25
	0080	To 80′ high, 500 lb. capacity		16.05	900	2,702.25	8,100	668.85
	0085	To 125′ high, 500 lb. capacity		18.34	1,525	4,603.50	13,800	1,067
	0100	Telescoping boom to 40′ high, 500 lb. capacity, diesel		11.24	315	945	2,825	278.90
	0105	To 45′ high, 500 lb. capacity		12.51	320	965	2,900	293.05
	0110	To 60′ high, 500 lb. capacity		16.36	300	900	2,700	310.90
	0115	To 80′ high, 500 lb. capacity		21.27	355	1,067	3,200	383.55
	0120	To 100′ high, 500 lb. capacity		28.71	865	2,587.75	7,775	747.25
	0125	To 120′ high, 500 lb. capacity		29.16	1,450	4,348.50	13,000	1,103
	0195	Air compressor, portable, 6.5 CFM, electric		.90	44.50	133	400	33.80
	0196	Gasoline		.65	56	167.50	505	38.70
	0200	Towed type, gas engine, 60 CFM		9.43	129	387.50	1,175	152.95
	0300	160 CFM	↓	10.47	198	595	1,775	202.80

01 54 33 | Equipment Rental

	01 54 33 | Equipment Rental	UNIT	HOURLY OPER. COST	RENT PER DAY	RENT PER WEEK	RENT PER MONTH	EQUIPMENT COST/DAY
40 0400	Diesel engine, rotary screw, 250 CFM	Ea.	12.08	175	524	1,575	201.50
0500	365 CFM		16.00	310	937	2,800	315.40
0550	450 CFM		19.95	277	832.25	2,500	326.05
0600	600 CFM		34.10	248	743.62	2,225	421.50
0700	750 CFM		34.62	435	1,306.50	3,925	538.25
0930	Air tools, breaker, pavement, 60 lb.		.57	81.50	245	735	53.50
0940	80 lb.		.56	81	242.50	730	53
0950	Drills, hand (jackhammer), 65 lb.		.67	68	203.50	610	46.05
0960	Track or wagon, swing boom, 4" drifter		54.66	1,025	3,104	9,300	1,058
0970	5" drifter		63.30	1,025	3,104	9,300	1,127
0975	Track mounted quarry drill, 6" diameter drill		101.91	1,900	5,665	17,000	1,948
0980	Dust control per drill		1.04	25	75.50	227	23.40
0990	Hammer, chipping, 12 lb.		.60	46	138	415	32.40
1000	Hose, air with couplings, 50' long, 3/4" diameter		.07	12	36	108	7.75
1100	1" diameter		.08	12.35	37	111	8
1200	1-1/2" diameter		.22	37.50	112.50	340	24.25
1300	2" diameter		.24	45	135	405	28.90
1400	2-1/2" diameter		.36	57.50	172.50	520	37.35
1410	3" diameter		.42	58.50	175	525	38.35
1450	Drill, steel, 7/8" x 2'		.08	12.90	38.73	116	8.40
1460	7/8" x 6'		.12	19.60	58.87	177	12.70
1520	Moil points		.03	7	21	63	4.40
1525	Pneumatic nailer w/accessories		.48	39.50	118	355	27.40
1530	Sheeting driver for 60 lb. breaker		.04	7.75	23.24	69.50	5
1540	For 90 lb. breaker		.13	10.50	31.50	94.50	7.35
1550	Spade, 25 lb.		.50	7.40	22.21	66.50	8.45
1560	Tamper, single, 35 lb.		.59	48.50	145.75	435	33.85
1570	Triple, 140 lb.		.89	61.50	184.87	555	44.05
1580	Wrenches, impact, air powered, up to 3/4" bolt		.43	49.50	148.25	445	33.05
1590	Up to 1-1/4" bolt		.58	79.50	238.50	715	52.30
1600	Barricades, barrels, reflectorized, 1 to 99 barrels		.03	4	12	36	2.65
1610	100 to 200 barrels		.02	4.41	13.22	39.50	2.85
1620	Barrels with flashers, 1 to 99 barrels		.03	6.40	19.16	57.50	4.10
1630	100 to 200 barrels		.03	5.10	15.34	46	3.30
1640	Barrels with steady burn type C lights		.05	8.45	25.30	76	5.45
1650	Illuminated board, trailer mounted, with generator		3.28	139	418.28	1,250	109.85
1670	Portable barricade, stock, with flashers, 1 to 6 units		.03	6.35	19.11	57.50	4.10
1680	25 to 50 units		.03	5.95	17.82	53.50	3.85
1685	Butt fusion machine, wheeled, 1.5 HP electric, 2" - 8" diameter pipe		2.63	225	675	2,025	156.05
1690	Tracked, 20 HP diesel, 4"-12" diameter pipe		11.23	560	1,685	5,050	426.85
1695	83 HP diesel, 8" - 24" diameter pipe		51.32	1,100	3,325	9,975	1,076
1700	Carts, brick, gas engine, 1000 lb. capacity		2.94	65	195	585	62.55
1800	1500 lb., 7-1/2' lift		2.92	69.50	208	625	64.95
1822	Dehumidifier, medium, 6 lb./hr., 150 CFM		1.19	76.50	229.28	690	55.35
1824	Large, 18 lb./hr., 600 CFM		2.19	585	1,750	5,250	367.55
1830	Distributor, asphalt, trailer mounted, 2000 gal., 38 H.P. diesel		10.99	355	1,058.62	3,175	299.65
1840	3000 gal., 38 H.P. diesel		12.87	380	1,136.08	3,400	330.15
1850	Drill, rotary hammer, electric		1.11	71.50	214	640	51.70
1860	Carbide bit, 1-1/2" diameter, add to electric rotary hammer		.03	41.50	125	375	25.25
1865	Rotary, crawler, 250 H.P.		135.77	2,300	6,868.12	20,600	2,460
1870	Emulsion sprayer, 65 gal., 5 H.P. gas engine		2.77	107	320.17	960	86.15
1880	200 gal., 5 H.P. engine		7.22	179	537.06	1,600	165.20
1900	Floor auto-scrubbing machine, walk-behind, 28" path		5.62	222	667	2,000	178.40
1930	Floodlight, mercury vapor, or quartz, on tripod, 1000 watt		.46	36.50	110	330	25.65
1940	2000 watt		.59	28	84.69	254	21.65
1950	Floodlights, trailer mounted with generator, 1 - 300 watt light		3.54	78.50	235.48	705	75.45
1960	2 - 1000 watt lights		4.49	87.50	262.33	785	88.35
2000	4 - 300 watt lights	↓	4.24	100	299.51	900	93.85

01 54 33 | Equipment Rental

			UNIT	HOURLY OPER. COST	RENT PER DAY	RENT PER WEEK	RENT PER MONTH	EQUIPMENT COST/DAY
40	2005	Foam spray rig, incl. box trailer, compressor, generator, proportioner	Ea.	25.46	535	1,600.84	4,800	523.85
	2015	Forklift, pneumatic tire, rough terr, straight mast, 5000 lb, 12′ lift, gas		18.59	219	655.83	1,975	279.90
	2025	8000 lb, 12′ lift		22.68	360	1,084.44	3,250	398.35
	2030	5000 lb, 12′ lift, diesel		15.41	244	733.29	2,200	269.90
	2035	8000 lb, 12′ lift, diesel		16.70	277	831.40	2,500	299.90
	2045	All terrain, telescoping boom, diesel, 5000 lb, 10′ reach, 19′ lift		17.20	233	700	2,100	277.60
	2055	6600 lb, 29′ reach, 42′ lift		21.04	233	700	2,100	308.30
	2065	10,000 lb, 31′ reach, 45′ lift		23.03	315	950	2,850	374.20
	2070	Cushion tire, smooth floor, gas, 5000 lb capacity		8.23	247	741.50	2,225	214.10
	2075	8000 lb capacity		11.33	275	826.25	2,475	255.90
	2085	Diesel, 5000 lb capacity		7.73	210	629.25	1,900	187.65
	2090	12,000 lb capacity		12.01	400	1,194.50	3,575	335
	2095	20,000 lb capacity		17.20	660	1,980	5,950	533.60
	2100	Generator, electric, gas engine, 1.5 kW to 3 kW		2.57	46.50	140	420	48.55
	2200	5 kW		3.21	91	272.50	820	80.15
	2300	10 kW		5.91	108	322.50	970	111.80
	2400	25 kW		7.38	405	1,210	3,625	301.10
	2500	Diesel engine, 20 kW		9.18	229	687.50	2,075	210.95
	2600	50 kW		15.90	370	1,110	3,325	349.20
	2700	100 kW		28.51	445	1,340.50	4,025	496.20
	2800	250 kW		54.19	750	2,249.33	6,750	883.40
	2850	Hammer, hydraulic, for mounting on boom, to 500 ft lb.		2.89	93.50	279.89	840	79.10
	2860	1000 ft lb.		4.59	139	418.28	1,250	120.40
	2900	Heaters, space, oil or electric, 50 MBH		1.46	46.50	140	420	39.65
	3000	100 MBH		2.71	46.50	140	420	49.70
	3100	300 MBH		7.90	135	405	1,225	144.20
	3150	500 MBH		13.12	200	600	1,800	224.95
	3200	Hose, water, suction with coupling, 20′ long, 2″ diameter		.02	5.65	17	51	3.55
	3210	3″ diameter		.03	14.15	42.50	128	8.75
	3220	4″ diameter		.03	28	84	252	17.05
	3230	6″ diameter		.11	40.50	121.50	365	25.20
	3240	8″ diameter		.27	53.50	160	480	34.15
	3250	Discharge hose with coupling, 50′ long, 2″ diameter		.01	6.50	19.50	58.50	4
	3260	3″ diameter		.01	7.35	22	66	4.50
	3270	4″ diameter		.02	21	62.50	188	12.65
	3280	6″ diameter		.06	29	87	261	17.90
	3290	8″ diameter		.24	37.50	112.50	340	24.40
	3295	Insulation blower		.83	117	350	1,050	76.65
	3300	Ladders, extension type, 16′ to 36′ long		.18	41.50	125	375	26.45
	3400	40′ to 60′ long		.64	120	360.50	1,075	77.20
	3405	Lance for cutting concrete		2.20	65	195	585	56.60
	3407	Lawn mower, rotary, 22″, 5 H.P.		1.05	38.50	115	345	31.40
	3408	48″ self-propelled		2.89	138	415	1,250	106.10
	3410	Level, electronic, automatic, with tripod and leveling rod		1.05	37.50	112	335	30.80
	3430	Laser type, for pipe and sewer line and grade		2.17	117	350	1,050	87.35
	3440	Rotating beam for interior control		.90	64	192.50	580	45.70
	3460	Builder's optical transit, with tripod and rod		.10	37.50	112	335	23.20
	3500	Light towers, towable, with diesel generator, 2000 watt		4.25	101	303.64	910	94.75
	3600	4000 watt		4.50	165	495	1,475	135
	3700	Mixer, powered, plaster and mortar, 6 C.F., 7 H.P.		2.05	83.50	250	750	66.40
	3800	10 C.F., 9 H.P.		2.24	124	372.50	1,125	92.40
	3850	Nailer, pneumatic		.48	33.50	100.18	300	23.85
	3900	Paint sprayers complete, 8 CFM		.85	61.50	184.87	555	43.75
	4000	17 CFM		1.60	110	330.50	990	78.85
	4020	Pavers, bituminous, rubber tires, 8′ wide, 50 H.P., diesel		31.93	570	1,704.12	5,100	596.25
	4030	10′ wide, 150 H.P.		95.62	1,950	5,835.32	17,500	1,932
	4050	Crawler, 8′ wide, 100 H.P., diesel		87.59	2,050	6,170.98	18,500	1,935
	4060	10′ wide, 150 H.P.	↓	103.97	2,350	7,048.86	21,100	2,241

01 54 33 | Equipment Rental

			UNIT	HOURLY OPER. COST	RENT PER DAY	RENT PER WEEK	RENT PER MONTH	EQUIPMENT COST/DAY	
40	4070	Concrete paver, 12′ to 24′ wide, 250 H.P.	Ea.	87.62	1,675	5,060.72	15,200	1,713	40
	4080	Placer-spreader-trimmer, 24′ wide, 300 H.P.		117.51	2,550	7,668.54	23,000	2,474	
	4100	Pump, centrifugal gas pump, 1-1/2″ diam., 65 GPM		3.92	54	162.15	485	63.80	
	4200	2″ diameter, 130 GPM		4.98	44	132.50	400	66.35	
	4300	3″ diameter, 250 GPM		5.12	56	167.50	505	74.45	
	4400	6″ diameter, 1500 GPM		22.24	91.50	275	825	232.90	
	4500	Submersible electric pump, 1-1/4″ diameter, 55 GPM		.40	36	107.50	325	24.70	
	4600	1-1/2″ diameter, 83 GPM		.44	43.50	130	390	29.55	
	4700	2″ diameter, 120 GPM		1.64	61.50	185	555	50.15	
	4800	3″ diameter, 300 GPM		3.03	109	327.50	985	89.75	
	4900	4″ diameter, 560 GPM		14.75	61.50	185	555	155	
	5000	6″ diameter, 1590 GPM		22.08	65	195	585	215.60	
	5100	Diaphragm pump, gas, single, 1-1/2″ diameter		1.13	38.50	115	345	32	
	5200	2″ diameter		3.98	91.50	275	825	86.80	
	5300	3″ diameter		4.05	95	285	855	89.40	
	5400	Double, 4″ diameter		6.03	95	285	855	105.25	
	5450	Pressure washer 5 GPM, 3000 psi		3.87	110	330	990	96.95	
	5460	7 GPM, 3000 psi		4.94	85	255	765	90.50	
	5470	High pressure water jet 10 ksi		39.55	720	2,160	6,475	748.40	
	5480	40 ksi		27.88	980	2,940	8,825	811.05	
	5500	Trash pump, self-priming, gas, 2″ diameter		3.82	108	325	975	95.55	
	5600	Diesel, 4″ diameter		6.68	162	485	1,450	150.40	
	5650	Diesel, 6″ diameter		16.85	162	485	1,450	231.80	
	5655	Grout Pump		18.70	281	841.73	2,525	317.90	
	5700	Salamanders, L.P. gas fired, 100,000 BTU		2.88	57.50	172.50	520	57.55	
	5705	50,000 BTU		1.66	23.50	71	213	27.50	
	5720	Sandblaster, portable, open top, 3 C.F. capacity		.60	132	395	1,175	83.80	
	5730	6 C.F. capacity		1.00	132	395	1,175	87	
	5740	Accessories for above		.14	24	71.26	214	15.35	
	5750	Sander, floor		.77	73.50	220	660	50.15	
	5760	Edger		.52	35	105	315	25.15	
	5800	Saw, chain, gas engine, 18″ long		1.75	63.50	190	570	52	
	5900	Hydraulic powered, 36″ long		.78	58.50	175	525	41.25	
	5950	60″ long		.78	65	195	585	45.25	
	6000	Masonry, table mounted, 14″ diameter, 5 H.P.		1.32	76.50	230	690	56.55	
	6050	Portable cut-off, 8 H.P.		1.81	77.50	232.50	700	61	
	6100	Circular, hand held, electric, 7-1/4″ diameter		.23	13.85	41.50	125	10.10	
	6200	12″ diameter		.24	41	122.50	370	26.40	
	6250	Wall saw, w/hydraulic power, 10 H.P.		3.29	98.50	296	890	85.50	
	6275	Shot blaster, walk-behind, 20″ wide		4.73	281	841.73	2,525	206.20	
	6280	Sidewalk broom, walk-behind		2.24	82.50	247.87	745	67.50	
	6300	Steam cleaner, 100 gallons per hour		3.34	82.50	247.87	745	76.30	
	6310	200 gallons per hour		4.33	100	299.51	900	94.55	
	6340	Tar Kettle/Pot, 400 gallons		16.48	127	380	1,150	207.85	
	6350	Torch, cutting, acetylene-oxygen, 150′ hose, excludes gases		.45	15.30	45.96	138	12.80	
	6360	Hourly operating cost includes tips and gas		20.92	7	20.98	63	171.55	
	6410	Toilet, portable chemical		.13	23	69.20	208	14.90	
	6420	Recycle flush type		.16	28.50	85.72	257	18.45	
	6430	Toilet, fresh water flush, garden hose,		.19	34	102.25	305	22	
	6440	Hoisted, non-flush, for high rise		.16	28	83.66	251	18	
	6465	Tractor, farm with attachment		17.37	390	1,172.50	3,525	373.40	
	6480	Trailers, platform, flush deck, 2 axle, 3 ton capacity		1.69	94.50	284	850	70.30	
	6500	25 ton capacity		6.23	143	428.61	1,275	135.55	
	6600	40 ton capacity		8.04	203	609.35	1,825	186.20	
	6700	3 axle, 50 ton capacity		8.72	225	676.48	2,025	205.05	
	6800	75 ton capacity		11.08	300	898.54	2,700	268.30	
	6810	Trailer mounted cable reel for high voltage line work		5.89	28.50	85	255	64.10	
	6820	Trailer mounted cable tensioning rig	↓	11.67	28.50	85	255	110.40	

01 54 33 | Equipment Rental

			UNIT	HOURLY OPER. COST	RENT PER DAY	RENT PER WEEK	RENT PER MONTH	EQUIPMENT COST/DAY
40	6830	Cable pulling rig	Ea.	73.77	28.50	85	255	607.15
	6850	Portable cable/wire puller, 8000 lb max pulling capacity		3.70	120	360	1,075	101.60
	6900	Water tank trailer, engine driven discharge, 5000 gallons		7.16	158	475.09	1,425	152.25
	6925	10,000 gallons		9.75	215	645.50	1,925	207.10
	6950	Water truck, off highway, 6000 gallons		71.75	835	2,504.54	7,525	1,075
	7010	Tram car for high voltage line work, powered, 2 conductor		6.88	28.50	85	255	72.05
	7020	Transit (builder's level) with tripod		.10	17.55	52.67	158	11.30
	7030	Trench box, 3000 lb., 6′ x 8′		.56	96.50	290.22	870	62.50
	7040	7200 lb., 6′ x 20′		.72	187	560	1,675	117.75
	7050	8000 lb., 8′ x 16′		1.08	186	557.71	1,675	120.15
	7060	9500 lb., 8′ x 20′		1.20	232	697.14	2,100	149.05
	7065	11,000 lb., 8′ x 24′		1.26	219	655.83	1,975	141.25
	7070	12,000 lb., 10′ x 20′		1.49	263	790.09	2,375	169.95
	7100	Truck, pickup, 3/4 ton, 2 wheel drive		9.24	61.50	184.87	555	110.85
	7200	4 wheel drive		9.48	167	500	1,500	175.85
	7250	Crew carrier, 9 passenger		12.66	108	325	975	166.25
	7290	Flat bed truck, 20,000 lb. GVW		15.26	133	397.63	1,200	201.60
	7300	Tractor, 4 x 2, 220 H.P.		22.25	215	645.50	1,925	307.10
	7410	330 H.P.		32.33	294	883.04	2,650	435.25
	7500	6 x 4, 380 H.P.		36.09	340	1,022.47	3,075	493.25
	7600	450 H.P.		44.23	415	1,239.36	3,725	601.75
	7610	Tractor, with A frame, boom and winch, 225 H.P.		24.74	293	877.88	2,625	373.50
	7620	Vacuum truck, hazardous material, 2500 gallons		12.79	310	929.52	2,800	288.25
	7625	5,000 gallons		13.02	440	1,316.82	3,950	367.55
	7650	Vacuum, HEPA, 16 gallon, wet/dry		.85	122	365	1,100	79.80
	7655	55 gallon, wet/dry		.78	25.50	76.50	230	21.50
	7660	Water tank, portable		.73	160	480.25	1,450	101.90
	7690	Sewer/catch basin vacuum, 14 C.Y., 1500 gallons		17.31	665	1,988.14	5,975	536.15
	7700	Welder, electric, 200 amp		3.81	33.50	100	300	50.50
	7800	300 amp		5.55	103	310	930	106.40
	7900	Gas engine, 200 amp		8.95	58.50	175	525	106.55
	8000	300 amp		10.13	110	330	990	147
	8100	Wheelbarrow, any size		.06	11.15	33.50	101	7.20
	8200	Wrecking ball, 4000 lb.	↓	2.50	60	180	540	56
50	0010	**HIGHWAY EQUIPMENT RENTAL** without operators R015433-10						
	0050	Asphalt batch plant, portable drum mixer, 100 ton/hr.	Ea.	88.41	1,550	4,621.78	13,900	1,632
	0060	200 ton/hr.		101.99	1,650	4,931.62	14,800	1,802
	0070	300 ton/hr.		119.86	1,925	5,783.68	17,400	2,116
	0100	Backhoe attachment, long stick, up to 185 H.P., 10.5′ long		.37	25.50	76.43	229	18.25
	0140	Up to 250 H.P., 12′ long		.41	28.50	85.72	257	20.45
	0180	Over 250 H.P., 15′ long		.56	39	116.71	350	27.85
	0200	Special dipper arm, up to 100 H.P., 32′ long		1.16	79.50	238.58	715	56.95
	0240	Over 100 H.P., 33′ long		1.44	100	299.51	900	71.45
	0280	Catch basin/sewer cleaning truck, 3 ton, 9 C.Y., 1000 gal.		35.39	420	1,265.18	3,800	536.15
	0300	Concrete batch plant, portable, electric, 200 C.Y./hr.		24.18	560	1,678.30	5,025	529.15
	0520	Grader/dozer attachment, ripper/scarifier, rear mounted, up to 135 H.P.		3.15	63.50	190.04	570	63.20
	0540	Up to 180 H.P.		4.13	95.50	287.12	860	90.50
	0580	Up to 250 H.P.		5.85	153	459.60	1,375	138.75
	0700	Pvmt. removal bucket, for hyd. excavator, up to 90 H.P.		2.16	58	174.54	525	52.20
	0740	Up to 200 H.P.		2.31	74.50	223.08	670	63.05
	0780	Over 200 H.P.		2.52	91	273.69	820	74.90
	0900	Aggregate spreader, self-propelled, 187 H.P.		50.60	740	2,220.52	6,650	848.90
	1000	Chemical spreader, 3 C.Y.		3.17	96.50	290	870	83.35
	1900	Hammermill, traveling, 250 H.P.		67.35	515	1,550	4,650	848.80
	2000	Horizontal borer, 3″ diameter, 13 H.P. gas driven		5.42	232	695	2,075	182.35
	2150	Horizontal directional drill, 20,000 lb. thrust, 78 H.P. diesel		27.58	530	1,590	4,775	538.65
	2160	30,000 lb. thrust, 115 H.P.		33.90	615	1,850	5,550	641.20
	2170	50,000 lb. thrust, 170 H.P.	↓	48.60	710	2,135	6,400	815.80

01 54 33 | Equipment Rental

			UNIT	HOURLY OPER. COST	RENT PER DAY	RENT PER WEEK	RENT PER MONTH	EQUIPMENT COST/DAY	
50	2190	Mud trailer for HDD, 1500 gallons, 175 H.P., gas	Ea.	25.50	175	525	1,575	309	50
	2200	Hydromulcher, diesel, 3000 gallon, for truck mounting		17.43	227	680	2,050	275.45	
	2300	Gas, 600 gallon		7.49	95	285	855	116.95	
	2400	Joint & crack cleaner, walk behind, 25 H.P.		3.16	45.50	136	410	52.45	
	2500	Filler, trailer mounted, 400 gallons, 20 H.P.		8.35	147	440	1,325	154.75	
	3000	Paint striper, self-propelled, 40 gallon, 22 H.P.		6.76	122	365	1,100	127.10	
	3100	120 gallon, 120 H.P.		19.23	380	1,140	3,425	381.80	
	3200	Post drivers, 6″ I-Beam frame, for truck mounting		12.41	320	960	2,875	291.30	
	3400	Road sweeper, self-propelled, 8′ wide, 90 H.P.		35.91	715	2,143.06	6,425	715.85	
	3450	Road sweeper, vacuum assisted, 4 C.Y., 220 gallons		58.28	670	2,013.96	6,050	869	
	4000	Road mixer, self-propelled, 130 H.P.		46.23	825	2,478.72	7,425	865.60	
	4100	310 H.P.		75.01	2,150	6,480.82	19,400	1,896	
	4220	Cold mix paver, incl. pug mill and bitumen tank, 165 H.P.		94.97	2,325	6,945.58	20,800	2,149	
	4240	Pavement brush, towed		3.43	100	299.51	900	87.30	
	4250	Paver, asphalt, wheel or crawler, 130 H.P., diesel		94.23	2,275	6,816.48	20,400	2,117	
	4300	Paver, road widener, gas, 1′ to 6′, 67 H.P.		46.66	975	2,917.66	8,750	956.80	
	4400	Diesel, 2′ to 14′, 88 H.P.		56.38	1,150	3,459.88	10,400	1,143	
	4600	Slipform pavers, curb and gutter, 2 track, 75 H.P.		57.83	1,250	3,769.72	11,300	1,217	
	4700	4 track, 165 H.P.		35.69	845	2,530.36	7,600	791.55	
	4800	Median barrier, 215 H.P.		58.43	1,350	4,027.92	12,100	1,273	
	4901	Trailer, low bed, 75 ton capacity		10.71	282	846.90	2,550	255.05	
	5000	Road planer, walk behind, 10″ cutting width, 10 H.P.		2.45	243	730	2,200	165.65	
	5100	Self-propelled, 12″ cutting width, 64 H.P.		8.26	190	570	1,700	180.05	
	5120	Traffic line remover, metal ball blaster, truck mounted, 115 H.P.		46.56	905	2,720	8,150	916.50	
	5140	Grinder, truck mounted, 115 H.P.		50.89	905	2,720	8,150	951.15	
	5160	Walk-behind, 11 H.P.		3.56	142	425	1,275	113.45	
	5200	Pavement profiler, 4′ to 6′ wide, 450 H.P.		216.58	1,275	3,800	11,400	2,493	
	5300	8′ to 10′ wide, 750 H.P.		331.58	1,325	3,975	11,900	3,448	
	5400	Roadway plate, steel, 1″ x 8′ x 20′		.09	61	182.50	550	37.20	
	5600	Stabilizer, self-propelled, 150 H.P.		41.14	1,025	3,100	9,300	949.10	
	5700	310 H.P.		76.18	1,300	3,900	11,700	1,389	
	5800	Striper, truck mounted, 120 gallon paint, 460 H.P.		48.74	340	1,015	3,050	592.95	
	5900	Thermal paint heating kettle, 115 gallons		7.71	61.50	185	555	98.65	
	6000	Tar kettle, 330 gallon, trailer mounted		12.27	96.50	290	870	156.20	
	7000	Tunnel locomotive, diesel, 8 to 12 ton		29.76	620	1,859.04	5,575	609.85	
	7005	Electric, 10 ton		29.25	705	2,117.24	6,350	657.40	
	7010	Muck cars, 1/2 C.Y. capacity		2.30	26.50	80.04	240	34.40	
	7020	1 C.Y. capacity		2.51	35	104.31	315	40.95	
	7030	2 C.Y. capacity		2.66	39	116.71	350	44.60	
	7040	Side dump, 2 C.Y. capacity		2.87	48	144.59	435	51.90	
	7050	3 C.Y. capacity		3.85	53	159.05	475	62.65	
	7060	5 C.Y. capacity		5.62	68.50	205.53	615	86.10	
	7100	Ventilating blower for tunnel, 7-1/2 H.P.		2.14	52.50	158.02	475	48.70	
	7110	10 H.P.		2.42	55	165.25	495	52.40	
	7120	20 H.P.		3.54	71.50	214.82	645	71.30	
	7140	40 H.P.		6.14	94.50	284.02	850	105.90	
	7160	60 H.P.		8.69	102	304.68	915	130.45	
	7175	75 H.P.		10.37	158	475.09	1,425	177.95	
	7180	200 H.P.		20.78	310	934.68	2,800	353.20	
	7800	Windrow loader, elevating	↓	53.94	1,650	4,975	14,900	1,427	
60	0010	**LIFTING AND HOISTING EQUIPMENT RENTAL** without operators R015433-10							60
	0150	Crane, flatbed mounted, 3 ton capacity	Ea.	14.41	201	604.19	1,825	236.10	
	0200	Crane, climbing, 106′ jib, 6000 lb. capacity, 410 fpm R312316-45		39.72	2,600	7,800	23,400	1,878	
	0300	101′ jib, 10,250 lb. capacity, 270 fpm		46.43	2,275	6,800	20,400	1,731	
	0500	Tower, static, 130′ high, 106′ jib, 6200 lb. capacity at 400 fpm		45.16	2,250	6,715	20,100	1,704	
	0520	Mini crawler spider crane, up to 24″ wide, 1990 lb. lifting capacity		12.50	550	1,652.48	4,950	430.50	
	0525	Up to 30″ wide, 6450 lb. lifting capacity		14.52	655	1,962.32	5,875	508.65	
	0530	Up to 52″ wide, 6680 lb. lifting capacity	↓	23.10	800	2,401.26	7,200	665.05	

01 54 33 | Equipment Rental

			UNIT	HOURLY OPER. COST	RENT PER DAY	RENT PER WEEK	RENT PER MONTH	EQUIPMENT COST/DAY
60	0535	Up to 55" wide, 8920 lb. lifting capacity	Ea.	25.79	885	2,659.46	7,975	738.25
	0540	Up to 66" wide, 13,350 lb. lifting capacity		34.92	1,375	4,131.20	12,400	1,106
	0600	Crawler mounted, lattice boom, 1/2 C.Y., 15 tons at 12' radius		36.96	830	2,483	7,450	792.30
	0700	3/4 C.Y., 20 tons at 12' radius		50.42	930	2,790	8,375	961.35
	0800	1 C.Y., 25 tons at 12' radius		67.42	985	2,950	8,850	1,129
	0900	1-1/2 C.Y., 40 tons at 12' radius		66.31	1,125	3,375	10,100	1,206
	1000	2 C.Y., 50 tons at 12' radius		88.77	1,325	4,000	12,000	1,510
	1100	3 C.Y., 75 tons at 12' radius		75.26	2,325	7,000	21,000	2,002
	1200	100 ton capacity, 60' boom		85.91	2,675	8,000	24,000	2,287
	1300	165 ton capacity, 60' boom		106.10	3,000	9,000	27,000	2,649
	1400	200 ton capacity, 70' boom		138.21	3,825	11,500	34,500	3,406
	1500	350 ton capacity, 80' boom		182.20	4,175	12,500	37,500	3,958
	1600	Truck mounted, lattice boom, 6 x 4, 20 tons at 10' radius		39.76	1,950	5,850	17,600	1,488
	1700	25 tons at 10' radius		42.73	2,325	7,000	21,000	1,742
	1800	8 x 4, 30 tons at 10' radius		45.54	2,500	7,500	22,500	1,864
	1900	40 tons at 12' radius		48.55	2,725	8,200	24,600	2,028
	2000	60 tons at 15' radius		53.69	1,650	4,950	14,900	1,419
	2050	82 tons at 15' radius		59.43	1,775	5,350	16,100	1,545
	2100	90 tons at 15' radius		66.39	1,950	5,825	17,500	1,696
	2200	115 tons at 15' radius		74.90	2,175	6,525	19,600	1,904
	2300	150 tons at 18' radius		81.09	2,700	8,100	24,300	2,269
	2350	165 tons at 18' radius		87.03	2,425	7,275	21,800	2,151
	2400	Truck mounted, hydraulic, 12 ton capacity		29.50	390	1,175	3,525	471
	2500	25 ton capacity		36.36	485	1,450	4,350	580.85
	2550	33 ton capacity		50.67	900	2,700	8,100	945.35
	2560	40 ton capacity		49.47	900	2,700	8,100	935.80
	2600	55 ton capacity		53.78	915	2,750	8,250	980.20
	2700	80 ton capacity		75.71	1,475	4,400	13,200	1,486
	2720	100 ton capacity		74.96	1,550	4,675	14,000	1,535
	2740	120 ton capacity		102.81	1,825	5,500	16,500	1,922
	2760	150 ton capacity		109.92	2,050	6,125	18,400	2,104
	2800	Self-propelled, 4 x 4, with telescoping boom, 5 ton		15.14	430	1,285	3,850	378.10
	2900	12-1/2 ton capacity		21.42	430	1,285	3,850	428.30
	3000	15 ton capacity		34.42	450	1,350	4,050	545.35
	3050	20 ton capacity		24.02	650	1,950	5,850	582.20
	3100	25 ton capacity		36.69	1,425	4,250	12,800	1,144
	3150	40 ton capacity		44.90	660	1,975	5,925	754.20
	3200	Derricks, guy, 20 ton capacity, 60' boom, 75' mast		22.74	1,425	4,250	12,800	1,032
	3300	100' boom, 115' mast		36.04	2,000	6,000	18,000	1,488
	3400	Stiffleg, 20 ton capacity, 70' boom, 37' mast		25.41	615	1,850	5,550	573.25
	3500	100' boom, 47' mast		39.32	665	2,000	6,000	714.55
	3550	Helicopter, small, lift to 1250 lb. maximum, w/pilot		99.14	2,150	6,435	19,300	2,080
	3600	Hoists, chain type, overhead, manual, 3/4 ton		.14	10.25	30.70	92	7.30
	3900	10 ton		.79	6.20	18.59	56	10
	4000	Hoist and tower, 5000 lb. cap., portable electric, 40' high		5.12	142	426	1,275	126.20
	4100	For each added 10' section, add		.12	31.50	95	285	19.95
	4200	Hoist and single tubular tower, 5000 lb. electric, 100' high		6.96	105	315	945	118.65
	4300	For each added 6'-6" section, add		.21	38.50	115	345	24.65
	4400	Hoist and double tubular tower, 5000 lb., 100' high		7.57	105	315	945	123.60
	4500	For each added 6'-6" section, add		.23	41.50	125	375	26.80
	4550	Hoist and tower, mast type, 6000 lb., 100' high		8.24	94.50	284	850	122.70
	4570	For each added 10' section, add		.13	31.50	95	285	20.05
	4600	Hoist and tower, personnel, electric, 2000 lb., 100' @ 125 fpm		17.50	25	75	225	155
	4700	3000 lb., 100' @ 200 fpm		20.02	25	75	225	175.15
	4800	3000 lb., 150' @ 300 fpm		22.22	25	75	225	192.75
	4900	4000 lb., 100' @ 300 fpm		22.98	25	75	225	198.85
	5000	6000 lb., 100' @ 275 fpm	↓	24.70	25	75	225	212.60
	5100	For added heights up to 500', add	L.F.	.01	3.33	10	30	2.10

01 54 33 | Equipment Rental

		01 54 33 Equipment Rental	UNIT	HOURLY OPER. COST	RENT PER DAY	RENT PER WEEK	RENT PER MONTH	EQUIPMENT COST/DAY	
60	5200	Jacks, hydraulic, 20 ton	Ea.	.05	19.65	59	177	12.20	60
	5500	100 ton		.40	26	78.50	236	18.90	
	6100	Jacks, hydraulic, climbing w/50′ jackrods, control console, 30 ton cap.		2.17	31	93	279	35.90	
	6150	For each added 10′ jackrod section, add		.05	5	15	45	3.40	
	6300	50 ton capacity		3.48	33.50	100	300	47.85	
	6350	For each added 10′ jackrod section, add		.06	5	15	45	3.50	
	6500	125 ton capacity		9.10	51.50	155	465	103.85	
	6550	For each added 10′ jackrod section, add		.61	5	15	45	7.90	
	6600	Cable jack, 10 ton capacity with 200′ cable		1.82	35.50	107	320	35.95	
	6650	For each added 50′ of cable, add	↓	.22	15	45	135	10.75	
70	0010	**WELLPOINT EQUIPMENT RENTAL** without operators R015433-10							70
	0020	Based on 2 months rental							
	0100	Combination jetting & wellpoint pump, 60 H.P. diesel	Ea.	15.67	298	895	2,675	304.35	
	0200	High pressure gas jet pump, 200 H.P., 300 psi	"	33.83	275	825	2,475	435.65	
	0300	Discharge pipe, 8″ diameter	L.F.	.01	1.40	4.20	12.60	.90	
	0350	12″ diameter		.01	2.07	6.20	18.60	1.35	
	0400	Header pipe, flows up to 150 GPM, 4″ diameter		.01	.73	2.20	6.60	.50	
	0500	400 GPM, 6″ diameter		.01	1.07	3.20	9.60	.70	
	0600	800 GPM, 8″ diameter		.01	1.40	4.20	12.60	.95	
	0700	1500 GPM, 10″ diameter		.01	1.73	5.20	15.60	1.15	
	0800	2500 GPM, 12″ diameter		.03	2.07	6.20	18.60	1.45	
	0900	4500 GPM, 16″ diameter		.03	2.40	7.20	21.50	1.70	
	0950	For quick coupling aluminum and plastic pipe, add	↓	.03	9.35	28	84	5.85	
	1100	Wellpoint, 25′ long, with fittings & riser pipe, 1-1/2″ or 2″ diameter	Ea.	.07	132	395	1,175	79.55	
	1200	Wellpoint pump, diesel powered, 4″ suction, 20 H.P.		7.00	150	450	1,350	146	
	1300	6″ suction, 30 H.P.		9.39	167	500	1,500	175.15	
	1400	8″ suction, 40 H.P.		12.73	250	750	2,250	251.80	
	1500	10″ suction, 75 H.P.		18.77	265	795	2,375	309.20	
	1600	12″ suction, 100 H.P.		27.24	298	895	2,675	396.90	
	1700	12″ suction, 175 H.P.	↓	38.98	315	950	2,850	501.80	
80	0010	**MARINE EQUIPMENT RENTAL** without operators R015433-10							80
	0200	Barge, 400 Ton, 30′ wide x 90′ long	Ea.	17.63	1,200	3,588.98	10,800	858.85	
	0240	800 Ton, 45′ wide x 90′ long		22.14	1,475	4,415.22	13,200	1,060	
	2000	Tugboat, diesel, 100 H.P.		29.57	238	712.63	2,150	379.10	
	2040	250 H.P.		57.41	430	1,291	3,875	717.50	
	2080	380 H.P.		124.99	1,300	3,873	11,600	1,774	
	3000	Small work boat, gas, 16-foot, 50 H.P.		11.35	48	143.56	430	119.50	
	4000	Large, diesel, 48-foot, 200 H.P.	↓	74.68	1,375	4,105.38	12,300	1,418	

Crews - Maintenance

Crew No.	Bare Costs		In-House Costs		Incl. Subs O&P		Cost Per Labor-Hour		
Crew A-3H	**Hr.**	**Daily**	**Hr.**	**Daily**	**Hr.**	**Daily**	**Bare Costs**	**In House**	**Incl. O&P**
1 Equip. Oper. (crane)	$59.20	$ 473.60	$74.30	$ 594.40	$92.65	$ 741.20	$ 59.20	$ 74.30	$ 92.65
1 Hyd. Crane, 12 Ton (Daily)		724.85		724.85		797.34	90.61	90.61	99.67
8 L.H., Daily Totals		$1198.45		$1319.25		$1538.54	$149.81	$164.91	$192.32
Crew A-3I	**Hr.**	**Daily**	**Hr.**	**Daily**	**Hr.**	**Daily**	**Bare Costs**	**In House**	**Incl. O&P**
1 Equip. Oper. (crane)	$59.20	$ 473.60	$74.30	$ 594.40	$92.65	$ 741.20	$ 59.20	$ 74.30	$ 92.65
1 Hyd. Crane, 25 Ton (Daily)		801.40		801.40		881.54	100.18	100.18	110.19
8 L.H., Daily Totals		$1275.00		$1395.80		$1622.74	$159.38	$174.47	$202.84
Crew A-3J	**Hr.**	**Daily**	**Hr.**	**Daily**	**Hr.**	**Daily**	**Bare Costs**	**In House**	**Incl. O&P**
1 Equip. Oper. (crane)	$59.20	$ 473.60	$74.30	$ 594.40	$92.65	$ 741.20	$ 59.20	$ 74.30	$ 92.65
1 Hyd. Crane, 40 Ton (Daily)		1272.00		1272.00		1399.20	159.00	159.00	174.90
8 L.H., Daily Totals		$1745.60		$1866.40		$2140.40	$218.20	$233.30	$267.55
Crew A-3K	**Hr.**	**Daily**	**Hr.**	**Daily**	**Hr.**	**Daily**	**Bare Costs**	**In House**	**Incl. O&P**
1 Equip. Oper. (crane)	$59.20	$ 473.60	$74.30	$ 594.40	$92.65	$ 741.20	$ 54.88	$ 68.88	$ 85.88
1 Equip. Oper. (oiler)	50.55	404.40	63.45	507.60	79.10	632.80			
1 Hyd. Crane, 55 Ton (Daily)		1362.00		1362.00		1498.20			
1 P/U Truck, 3/4 Ton (Daily)		142.15		142.15		156.37	94.01	94.01	103.41
16 L.H., Daily Totals		$2382.15		$2606.15		$3028.57	$148.88	$162.88	$189.29
Crew A-3L	**Hr.**	**Daily**	**Hr.**	**Daily**	**Hr.**	**Daily**	**Bare Costs**	**In House**	**Incl. O&P**
1 Equip. Oper. (crane)	$59.20	$ 473.60	$74.30	$ 594.40	$92.65	$ 741.20	$ 54.88	$ 68.88	$ 85.88
1 Equip. Oper. (oiler)	50.55	404.40	63.45	507.60	79.10	632.80			
1 Hyd. Crane, 80 Ton (Daily)		2101.00		2101.00		2311.10			
1 P/U Truck, 3/4 Ton (Daily)		142.15		142.15		156.37	140.20	140.20	154.22
16 L.H., Daily Totals		$3121.15		$3345.15		$3841.47	$195.07	$209.07	$240.09
Crew A-3M	**Hr.**	**Daily**	**Hr.**	**Daily**	**Hr.**	**Daily**	**Bare Costs**	**In House**	**Incl. O&P**
1 Equip. Oper. (crane)	$59.20	$ 473.60	$74.30	$ 594.40	$92.65	$ 741.20	$ 54.88	$ 68.88	$ 85.88
1 Equip. Oper. (oiler)	50.55	404.40	63.45	507.60	79.10	632.80			
1 Hyd. Crane, 100 Ton (Daily)		2227.00		2227.00		2449.70			
1 P/U Truck, 3/4 Ton (Daily)		142.15		142.15		156.37	148.07	148.07	162.88
16 L.H., Daily Totals		$3247.15		$3471.15		$3980.07	$202.95	$216.95	$248.75
Crew A-16	**Hr.**	**Daily**	**Hr.**	**Daily**	**Hr.**	**Daily**	**Bare Costs**	**In House**	**Incl. O&P**
1 Maintenance Laborer	$31.60	$252.80	$40.85	$326.80	$50.65	$405.20	$31.60	$40.85	$50.65
1 Lawn Mower, Riding		106.10		106.10		116.71	13.26	13.26	14.59
8 L.H., Daily Totals		$358.90		$432.90		$521.91	$44.86	$54.11	$65.24
Crew A-17	**Hr.**	**Daily**	**Hr.**	**Daily**	**Hr.**	**Daily**	**Bare Costs**	**In House**	**Incl. O&P**
1 Maintenance Laborer	$31.60	$252.80	$40.85	$326.80	$50.65	$405.20	$31.60	$40.85	$50.65
1 Flatbed Truck, Gas, 1.5 Ton		196.15		196.15		215.76	24.52	24.52	26.97
8 L.H., Daily Totals		$448.95		$522.95		$620.97	$56.12	$65.37	$77.62
Crew A-18	**Hr.**	**Daily**	**Hr.**	**Daily**	**Hr.**	**Daily**	**Bare Costs**	**In House**	**Incl. O&P**
1 Maintenance Laborer	$31.60	$252.80	$40.85	$326.80	$50.65	$405.20	$31.60	$40.85	$ 50.65
1 Farm Tractor w/ Attachment		373.40		373.40		410.74	46.67	46.67	51.34
8 L.H., Daily Totals		$626.20		$700.20		$815.94	$78.28	$87.53	$101.99
Crew B-1	**Hr.**	**Daily**	**Hr.**	**Daily**	**Hr.**	**Daily**	**Bare Costs**	**In House**	**Incl. O&P**
1 Labor Foreman (outside)	$44.10	$ 352.80	$57.00	$ 456.00	$70.65	$ 565.20	$42.77	$55.27	$68.52
2 Laborers	42.10	673.60	54.40	870.40	67.45	1079.20			
24 L.H., Daily Totals		$1026.40		$1326.40		$1644.40	$42.77	$55.27	$68.52

Crew No.	Bare Costs		In-House Costs		Incl. Subs O&P		Cost Per Labor-Hour		
Crew B-17	**Hr.**	**Daily**	**Hr.**	**Daily**	**Hr.**	**Daily**	**Bare Costs**	**In House**	**Incl. O&P**
2 Laborers	$42.10	$ 673.60	$54.40	$ 870.40	$67.45	$1079.20	$46.54	$59.60	$74.03
1 Equip. Oper. (light)	53.00	424.00	66.50	532.00	82.95	663.60			
1 Truck Driver (heavy)	48.95	391.60	63.10	504.80	78.25	626.00			
1 Backhoe Loader, 48 H.P.		213.75		213.75		235.13			
1 Dump Truck, 8 C.Y., 220 H.P.		428.05		428.05		470.86	20.06	20.06	22.06
32 L.H., Daily Totals		$2131.00		$2549.00		$3074.78	$66.59	$79.66	$96.09

Crew B-18	**Hr.**	**Daily**	**Hr.**	**Daily**	**Hr.**	**Daily**	**Bare Costs**	**In House**	**Incl. O&P**
1 Labor Foreman (outside)	$44.10	$ 352.80	$57.00	$ 456.00	$70.65	$ 565.20	$42.77	$55.27	$68.52
2 Laborers	42.10	673.60	54.40	870.40	67.45	1079.20			
1 Vibrating Plate, Gas, 21"		165.35		165.35		181.88	6.89	6.89	7.58
24 L.H., Daily Totals		$1191.75		$1491.75		$1826.29	$49.66	$62.16	$76.10

Crew B-21	**Hr.**	**Daily**	**Hr.**	**Daily**	**Hr.**	**Daily**	**Bare Costs**	**In House**	**Incl. O&P**
1 Labor Foreman (outside)	$44.10	$ 352.80	$57.00	$ 456.00	$70.65	$ 565.20	$48.76	$62.54	$77.66
1 Skilled Worker	54.85	438.80	70.35	562.80	87.40	699.20			
1 Laborer	42.10	336.80	54.40	435.20	67.45	539.60			
.5 Equip. Oper. (crane)	59.20	236.80	74.30	297.20	92.65	370.60			
.5 S.P. Crane, 4x4, 5 Ton		189.05		189.05		207.96	6.75	6.75	7.43
28 L.H., Daily Totals		$1554.25		$1940.25		$2382.55	$55.51	$69.29	$85.09

Crew B-25B	**Hr.**	**Daily**	**Hr.**	**Daily**	**Hr.**	**Daily**	**Bare Costs**	**In House**	**Incl. O&P**
1 Labor Foreman (outside)	$44.10	$ 352.80	$57.00	$ 456.00	$70.65	$ 565.20	$47.15	$60.22	$ 74.83
7 Laborers	42.10	2357.60	54.40	3046.40	67.45	3777.20			
4 Equip. Oper. (medium)	56.75	1816.00	71.20	2278.40	88.80	2841.60			
1 Asphalt Paver, 130 H.P.		2117.00		2117.00		2328.70			
2 Tandem Rollers, 10 Ton		473.50		473.50		520.85			
1 Roller, Pneum. Whl., 12 Ton		345.75		345.75		380.32	30.59	30.59	33.64
96 L.H., Daily Totals		$7462.65		$8717.05		$10413.88	$77.74	$90.80	$108.48

Crew B-30	**Hr.**	**Daily**	**Hr.**	**Daily**	**Hr.**	**Daily**	**Bare Costs**	**In House**	**Incl. O&P**
1 Equip. Oper. (medium)	$56.75	$ 454.00	$71.20	$ 569.60	$88.80	$ 710.40	$ 51.55	$ 65.80	$ 81.77
2 Truck Drivers (heavy)	48.95	783.20	63.10	1009.60	78.25	1252.00			
1 Hyd. Excavator, 1.5 C.Y.		687.55		687.55		756.30			
2 Dump Trucks, 12 C.Y., 400 H.P.		1145.80		1145.80		1260.38	76.39	76.39	84.03
24 L.H., Daily Totals		$3070.55		$3412.55		$3979.09	$127.94	$142.19	$165.80

Crew B-34P	**Hr.**	**Daily**	**Hr.**	**Daily**	**Hr.**	**Daily**	**Bare Costs**	**In House**	**Incl. O&P**
1 Pipe Fitter	$65.55	$ 524.40	$81.05	$ 648.40	$101.35	$ 810.80	$56.52	$ 71.05	$ 88.57
1 Truck Driver (light)	47.25	378.00	60.90	487.20	75.55	604.40			
1 Equip. Oper. (medium)	56.75	454.00	71.20	569.60	88.80	710.40			
1 Flatbed Truck, Gas, 3 Ton		820.40		820.40		902.44			
1 Backhoe Loader, 48 H.P.		213.75		213.75		235.13	43.09	43.09	47.40
24 L.H., Daily Totals		$2390.55		$2739.35		$3263.17	$99.61	$114.14	$135.97

Crew B-34Q	**Hr.**	**Daily**	**Hr.**	**Daily**	**Hr.**	**Daily**	**Bare Costs**	**In House**	**Incl. O&P**
1 Pipe Fitter	$65.55	$ 524.40	$81.05	$ 648.40	$101.35	$ 810.80	$ 57.33	$ 72.08	$ 89.85
1 Truck Driver (light)	47.25	378.00	60.90	487.20	75.55	604.40			
1 Equip. Oper. (crane)	59.20	473.60	74.30	594.40	92.65	741.20			
1 Flatbed Trailer, 25 Ton		135.55		135.55		149.10			
1 Dump Truck, 8 C.Y., 220 H.P.		428.05		428.05		470.86			
1 Hyd. Crane, 25 Ton		580.85		580.85		638.93	47.69	47.69	52.45
24 L.H., Daily Totals		$2520.45		$2874.45		$3415.30	$105.02	$119.77	$142.30

Crew No.	Bare Costs		In-House Costs		Incl. Subs O&P		Cost Per Labor-Hour		
Crew B-34R	**Hr.**	**Daily**	**Hr.**	**Daily**	**Hr.**	**Daily**	**Bare Costs**	**In House**	**Incl. O&P**
1 Pipe Fitter	$65.55	$ 524.40	$81.05	$ 648.40	$101.35	$ 810.80	$ 57.33	$ 72.08	$ 89.85
1 Truck Driver (light)	47.25	378.00	60.90	487.20	75.55	604.40			
1 Equip. Oper. (crane)	59.20	473.60	74.30	594.40	92.65	741.20			
1 Flatbed Trailer, 25 Ton		135.55		135.55		149.10			
1 Dump Truck, 8 C.Y., 220 H.P.		428.05		428.05		470.86			
1 Hyd. Crane, 25 Ton		580.85		580.85		638.93			
1 Hyd. Excavator, 1 C.Y.		795.30		795.30		874.83	80.82	80.82	88.91
24 L.H., Daily Totals		$3315.75		$3669.75		$4290.13	$138.16	$152.91	$178.76

Crew B-34S	**Hr.**	**Daily**	**Hr.**	**Daily**	**Hr.**	**Daily**	**Bare Costs**	**In House**	**Incl. O&P**
2 Pipe Fitters	$65.55	$1048.80	$81.05	$1296.80	$101.35	$1621.60	$ 59.81	$ 74.88	$ 93.40
1 Truck Driver (heavy)	48.95	391.60	63.10	504.80	78.25	626.00			
1 Equip. Oper. (crane)	59.20	473.60	74.30	594.40	92.65	741.20			
1 Flatbed Trailer, 40 Ton		186.20		186.20		204.82			
1 Truck Tractor, 6x4, 380 H.P.		493.25		493.25		542.58			
1 Hyd. Crane, 80 Ton		1486.00		1486.00		1634.60			
1 Hyd. Excavator, 2 C.Y.		951.30		951.30		1046.43	97.40	97.40	107.14
32 L.H., Daily Totals		$5030.75		$5512.75		$6417.23	$157.21	$172.27	$200.54

Crew B-34T	**Hr.**	**Daily**	**Hr.**	**Daily**	**Hr.**	**Daily**	**Bare Costs**	**In House**	**Incl. O&P**
2 Pipe Fitters	$65.55	$1048.80	$81.05	$1296.80	$101.35	$1621.60	$ 59.81	$ 74.88	$ 93.40
1 Truck Driver (heavy)	48.95	391.60	63.10	504.80	78.25	626.00			
1 Equip. Oper. (crane)	59.20	473.60	74.30	594.40	92.65	741.20			
1 Flatbed Trailer, 40 Ton		186.20		186.20		204.82			
1 Truck Tractor, 6x4, 380 H.P.		493.25		493.25		542.58			
1 Hyd. Crane, 80 Ton		1486.00		1486.00		1634.60	67.67	67.67	74.44
32 L.H., Daily Totals		$4079.45		$4561.45		$5370.80	$127.48	$142.55	$167.84

Crew B-35	**Hr.**	**Daily**	**Hr.**	**Daily**	**Hr.**	**Daily**	**Bare Costs**	**In House**	**Incl. O&P**
1 Labor Foreman (outside)	$44.10	$ 352.80	$57.00	$ 456.00	$70.65	$ 565.20	$52.54	$66.53	$ 82.82
1 Skilled Worker	54.85	438.80	70.35	562.80	87.40	699.20			
1 Welder (plumber)	64.45	515.60	79.70	637.60	99.65	797.20			
1 Laborer	42.10	336.80	54.40	435.20	67.45	539.60			
1 Equip. Oper. (crane)	59.20	473.60	74.30	594.40	92.65	741.20			
1 Equip. Oper. (oiler)	50.55	404.40	63.45	507.60	79.10	632.80			
1 Welder, Electric, 300 amp		106.40		106.40		117.04			
1 Hyd. Excavator, .75 C.Y.		694.25		694.25		763.67	16.68	16.68	18.35
48 L.H., Daily Totals		$3322.65		$3994.25		$4855.92	$69.22	$83.21	$101.16

Crew B-37	**Hr.**	**Daily**	**Hr.**	**Daily**	**Hr.**	**Daily**	**Bare Costs**	**In House**	**Incl. O&P**
1 Labor Foreman (outside)	$44.10	$ 352.80	$57.00	$ 456.00	$70.65	$ 565.20	$44.25	$56.85	$70.57
4 Laborers	42.10	1347.20	54.40	1740.80	67.45	2158.40			
1 Equip. Oper. (light)	53.00	424.00	66.50	532.00	82.95	663.60			
1 Tandem Roller, 5 Ton		255.65		255.65		281.21	5.33	5.33	5.86
48 L.H., Daily Totals		$2379.65		$2984.45		$3668.42	$49.58	$62.18	$76.43

Crew B-47H	**Hr.**	**Daily**	**Hr.**	**Daily**	**Hr.**	**Daily**	**Bare Costs**	**In House**	**Incl. O&P**
1 Skilled Worker Foreman (out)	$56.85	$ 454.80	$72.95	$ 583.60	$90.55	$ 724.40	$55.35	$71.00	$ 88.19
3 Skilled Workers	54.85	1316.40	70.35	1688.40	87.40	2097.60			
1 Flatbed Truck, Gas, 3 Ton		820.40		820.40		902.44	25.64	25.64	28.20
32 L.H., Daily Totals		$2591.60		$3092.40		$3724.44	$80.99	$96.64	$116.39

Crew B-55	**Hr.**	**Daily**	**Hr.**	**Daily**	**Hr.**	**Daily**	**Bare Costs**	**In House**	**Incl. O&P**
2 Laborers	$42.10	$ 673.60	$54.40	$ 870.40	$67.45	$1079.20	$43.82	$ 56.57	$ 70.15
1 Truck Driver (light)	47.25	378.00	60.90	487.20	75.55	604.40			
1 Truck-Mounted Earth Auger		390.20		390.20		429.22			
1 Flatbed Truck, Gas, 3 Ton		820.40		820.40		902.44	50.44	50.44	55.49
24 L.H., Daily Totals		$2262.20		$2568.20		$3015.26	$94.26	$107.01	$125.64

Crew No.	Bare Costs		In-House Costs		Incl. Subs O&P		Cost Per Labor-Hour		
Crew B-80	**Hr.**	**Daily**	**Hr.**	**Daily**	**Hr.**	**Daily**	**Bare Costs**	**In House**	**Incl. O&P**
1 Labor Foreman (outside)	$44.10	$ 352.80	$57.00	$ 456.00	$70.65	$ 565.20	$46.61	$59.70	$ 74.15
1 Laborer	42.10	336.80	54.40	435.20	67.45	539.60			
1 Truck Driver (light)	47.25	378.00	60.90	487.20	75.55	604.40			
1 Equip. Oper. (light)	53.00	424.00	66.50	532.00	82.95	663.60			
1 Flatbed Truck, Gas, 3 Ton		820.40		820.40		902.44			
1 Earth Auger, Truck-Mtd.		200.50		200.50		220.55	31.90	31.90	35.09
32 L.H., Daily Totals		$2512.50		$2931.30		$3495.79	$78.52	$91.60	$109.24
Crew B-80C	**Hr.**	**Daily**	**Hr.**	**Daily**	**Hr.**	**Daily**	**Bare Costs**	**In House**	**Incl. O&P**
2 Laborers	$42.10	$ 673.60	$54.40	$ 870.40	$67.45	$1079.20	$43.82	$56.57	$70.15
1 Truck Driver (light)	47.25	378.00	60.90	487.20	75.55	604.40			
1 Flatbed Truck, Gas, 1.5 Ton		196.15		196.15		215.76			
1 Manual Fence Post Auger, Gas		53.85		53.85		59.23	10.42	10.42	11.46
24 L.H., Daily Totals		$1301.60		$1607.60		$1958.60	$54.23	$66.98	$81.61
Crew D-1	**Hr.**	**Daily**	**Hr.**	**Daily**	**Hr.**	**Daily**	**Bare Costs**	**In House**	**Incl. O&P**
1 Bricklayer	$52.05	$416.40	$68.00	$544.00	$84.10	$ 672.80	$46.90	$61.27	$75.78
1 Bricklayer Helper	41.75	334.00	54.55	436.40	67.45	539.60			
16 L.H., Daily Totals		$750.40		$980.40		$1212.40	$46.90	$61.27	$75.78
Crew D-7	**Hr.**	**Daily**	**Hr.**	**Daily**	**Hr.**	**Daily**	**Bare Costs**	**In House**	**Incl. O&P**
1 Tile Layer	$49.50	$396.00	$62.25	$498.00	$77.55	$ 620.40	$44.15	$55.52	$69.17
1 Tile Layer Helper	38.80	310.40	48.80	390.40	60.80	486.40			
16 L.H., Daily Totals		$706.40		$888.40		$1106.80	$44.15	$55.52	$69.17
Crew D-8	**Hr.**	**Daily**	**Hr.**	**Daily**	**Hr.**	**Daily**	**Bare Costs**	**In House**	**Incl. O&P**
3 Bricklayers	$52.05	$1249.20	$68.00	$1632.00	$84.10	$2018.40	$47.93	$62.62	$77.44
2 Bricklayer Helpers	41.75	668.00	54.55	872.80	67.45	1079.20			
40 L.H., Daily Totals		$1917.20		$2504.80		$3097.60	$47.93	$62.62	$77.44
Crew E-11	**Hr.**	**Daily**	**Hr.**	**Daily**	**Hr.**	**Daily**	**Bare Costs**	**In House**	**Incl. O&P**
2 Painters, Struc. Steel	$45.85	$ 733.60	$64.80	$1036.80	$79.00	$1264.00	$46.70	$62.63	$77.10
1 Building Laborer	42.10	336.80	54.40	435.20	67.45	539.60			
1 Equip. Oper. (light)	53.00	424.00	66.50	532.00	82.95	663.60			
1 Air Compressor, 250 cfm		201.50		201.50		221.65			
1 Sandblaster, Portable, 3 C.F.		83.80		83.80		92.18			
1 Set Sand Blasting Accessories		15.35		15.35		16.89	9.40	9.40	10.33
32 L.H., Daily Totals		$1795.05		$2304.65		$2797.92	$56.10	$72.02	$87.43
Crew G-1	**Hr.**	**Daily**	**Hr.**	**Daily**	**Hr.**	**Daily**	**Bare Costs**	**In House**	**Incl. O&P**
1 Roofer Foreman (outside)	$48.20	$ 385.60	$69.40	$ 555.20	$84.35	$ 674.80	$43.17	$62.19	$75.55
4 Roofers Composition	46.20	1478.40	66.55	2129.60	80.85	2587.20			
2 Roofer Helpers	34.60	553.60	49.85	797.60	60.55	968.80			
1 Application Equipment		192.85		192.85		212.13			
1 Tar Kettle/Pot		207.85		207.85		228.63			
1 Crew Truck		166.25		166.25		182.88	10.12	10.12	11.14
56 L.H., Daily Totals		$2984.55		$4049.35		$4854.44	$53.30	$72.31	$86.69
Crew G-3	**Hr.**	**Daily**	**Hr.**	**Daily**	**Hr.**	**Daily**	**Bare Costs**	**In House**	**Incl. O&P**
2 Sheet Metal Workers	$62.30	$ 996.80	$78.30	$1252.80	$97.60	$1561.60	$52.20	$66.35	$82.53
2 Building Laborers	42.10	673.60	54.40	870.40	67.45	1079.20			
32 L.H., Daily Totals		$1670.40		$2123.20		$2640.80	$52.20	$66.35	$82.53
Crew G-5	**Hr.**	**Daily**	**Hr.**	**Daily**	**Hr.**	**Daily**	**Bare Costs**	**In House**	**Incl. O&P**
1 Roofer Foreman (outside)	$48.20	$ 385.60	$69.40	$ 555.20	$84.35	$ 674.80	$41.96	$60.44	$73.43
2 Roofers Composition	46.20	739.20	66.55	1064.80	80.85	1293.60			
2 Roofer Helpers	34.60	553.60	49.85	797.60	60.55	968.80			
1 Application Equipment		192.85		192.85		212.13	4.82	4.82	5.30
40 L.H., Daily Totals		$1871.25		$2610.45		$3149.34	$46.78	$65.26	$78.73

Crew No.	Bare Costs		In-House Costs		Incl. Subs O&P		Cost Per Labor-Hour		
Crew G-8	**Hr.**	**Daily**	**Hr.**	**Daily**	**Hr.**	**Daily**	**Bare Costs**	**In House**	**Incl. O&P**
1 Roofer Composition	$46.20	$369.60	$66.55	$532.40	$80.85	$ 646.80	$46.20	$ 66.55	$ 80.85
1 Telescoping Boom Lift, to 80'		383.55		383.55		421.90	47.94	47.94	52.74
8 L.H., Daily Totals		$753.15		$915.95		$1068.70	$94.14	$114.49	$133.59
Crew L-2	**Hr.**	**Daily**	**Hr.**	**Daily**	**Hr.**	**Daily**	**Bare Costs**	**In House**	**Incl. O&P**
1 Carpenter	$53.15	$425.20	$68.70	$549.60	$85.15	$ 681.20	$46.55	$60.63	$75.05
1 Carpenter Helper	39.95	319.60	52.55	420.40	64.95	519.60			
16 L.H., Daily Totals		$744.80		$970.00		$1200.80	$46.55	$60.63	$75.05
Crew L-4	**Hr.**	**Daily**	**Hr.**	**Daily**	**Hr.**	**Daily**	**Bare Costs**	**In House**	**Incl. O&P**
2 Skilled Workers	$54.85	$ 877.60	$70.35	$1125.60	$87.40	$1398.40	$49.88	$64.42	$79.92
1 Helper	39.95	319.60	52.55	420.40	64.95	519.60			
24 L.H., Daily Totals		$1197.20		$1546.00		$1918.00	$49.88	$64.42	$79.92
Crew Q-1	**Hr.**	**Daily**	**Hr.**	**Daily**	**Hr.**	**Daily**	**Bare Costs**	**In House**	**Incl. O&P**
1 Plumber	$64.45	$515.60	$79.70	$ 637.60	$99.65	$ 797.20	$58.00	$71.72	$89.67
1 Plumber Apprentice	51.55	412.40	63.75	510.00	79.70	637.60			
16 L.H., Daily Totals		$928.00		$1147.60		$1434.80	$58.00	$71.72	$89.67
Crew Q-2	**Hr.**	**Daily**	**Hr.**	**Daily**	**Hr.**	**Daily**	**Bare Costs**	**In House**	**Incl. O&P**
2 Plumbers	$64.45	$1031.20	$79.70	$1275.20	$99.65	$1594.40	$60.15	$74.38	$93.00
1 Plumber Apprentice	51.55	412.40	63.75	510.00	79.70	637.60	60.15	74.38	93.00
24 L.H., Daily Totals		$1443.60		$1785.20		$2232.00	$60.15	$74.38	$93.00
Crew Q-4	**Hr.**	**Daily**	**Hr.**	**Daily**	**Hr.**	**Daily**	**Bare Costs**	**In House**	**Incl. O&P**
1 Plumber Foreman (inside)	$64.95	$ 519.60	$80.30	$ 642.40	$100.45	$ 803.60	$61.35	$75.86	$94.86
1 Plumber	64.45	515.60	79.70	637.60	99.65	797.20			
1 Welder (plumber)	64.45	515.60	79.70	637.60	99.65	797.20			
1 Plumber Apprentice	51.55	412.40	63.75	510.00	79.70	637.60			
1 Welder, Electric, 300 amp		106.40		106.40		117.04	3.33	3.33	3.66
32 L.H., Daily Totals		$2069.60		$2534.00		$3152.64	$64.67	$79.19	$98.52
Crew Q-5	**Hr.**	**Daily**	**Hr.**	**Daily**	**Hr.**	**Daily**	**Bare Costs**	**In House**	**Incl. O&P**
1 Steamfitter	$65.55	$524.40	$81.05	$ 648.40	$101.35	$ 810.80	$59.00	$72.95	$91.22
1 Steamfitter Apprentice	52.45	419.60	64.85	518.80	81.10	648.80			
16 L.H., Daily Totals		$944.00		$1167.20		$1459.60	$59.00	$72.95	$91.22
Crew Q-6	**Hr.**	**Daily**	**Hr.**	**Daily**	**Hr.**	**Daily**	**Bare Costs**	**In House**	**Incl. O&P**
2 Steamfitters	$65.55	$1048.80	$81.05	$1296.80	$101.35	$1621.60	$61.18	$75.65	$94.60
1 Steamfitter Apprentice	52.45	419.60	64.85	518.80	81.10	648.80	61.18	75.65	94.60
24 L.H., Daily Totals		$1468.40		$1815.60		$2270.40	$61.18	$75.65	$94.60
Crew Q-7	**Hr.**	**Daily**	**Hr.**	**Daily**	**Hr.**	**Daily**	**Bare Costs**	**In House**	**Incl. O&P**
1 Steamfitter Foreman (inside)	$66.05	$ 528.40	$81.65	$ 653.20	$102.15	$ 817.20	$62.40	$77.15	$96.49
2 Steamfitters	65.55	1048.80	81.05	1296.80	101.35	1621.60			
1 Steamfitter Apprentice	52.45	419.60	64.85	518.80	81.10	648.80			
32 L.H., Daily Totals		$1996.80		$2468.80		$3087.60	$62.40	$77.15	$96.49
Crew Q-8	**Hr.**	**Daily**	**Hr.**	**Daily**	**Hr.**	**Daily**	**Bare Costs**	**In House**	**Incl. O&P**
1 Steamfitter Foreman (inside)	$66.05	$ 528.40	$81.65	$ 653.20	$102.15	$ 817.20	$62.40	$77.15	$ 96.49
1 Steamfitter	65.55	524.40	81.05	648.40	101.35	810.80			
1 Welder (steamfitter)	65.55	524.40	81.05	648.40	101.35	810.80			
1 Steamfitter Apprentice	52.45	419.60	64.85	518.80	81.10	648.80			
1 Welder, Electric, 300 amp		106.40		106.40		117.04	3.33	3.33	3.66
32 L.H., Daily Totals		$2103.20		$2575.20		$3204.64	$65.72	$80.47	$100.15

Crew No.	Bare Costs		In-House Costs		Incl. Subs O&P		Cost Per Labor-Hour		
Crew Q-9	**Hr.**	**Daily**	**Hr.**	**Daily**	**Hr.**	**Daily**	**Bare Costs**	**In House**	**Incl. O&P**
1 Sheet Metal Worker	$62.30	$498.40	$78.30	$ 626.40	$97.60	$ 780.80	$56.08	$70.47	$87.85
1 Sheet Metal Apprentice	49.85	398.80	62.65	501.20	78.10	624.80			
16 L.H., Daily Totals		$897.20		$1127.60		$1405.60	$56.08	$70.47	$87.85
Crew Q-10	**Hr.**	**Daily**	**Hr.**	**Daily**	**Hr.**	**Daily**	**Bare Costs**	**In House**	**Incl. O&P**
2 Sheet Metal Workers	$62.30	$ 996.80	$78.30	$1252.80	$97.60	$1561.60	$58.15	$73.08	$91.10
1 Sheet Metal Apprentice	49.85	398.80	62.65	501.20	78.10	624.80			
24 L.H., Daily Totals		$1395.60		$1754.00		$2186.40	$58.15	$73.08	$91.10
Crew Q-13	**Hr.**	**Daily**	**Hr.**	**Daily**	**Hr.**	**Daily**	**Bare Costs**	**In House**	**Incl. O&P**
1 Sprinkler Foreman (inside)	$63.75	$ 510.00	$79.00	$ 632.00	$98.75	$ 790.00	$60.21	$74.60	$93.26
2 Sprinkler Installers	63.25	1012.00	78.35	1253.60	97.95	1567.20			
1 Sprinkler Apprentice	50.60	404.80	62.70	501.60	78.40	627.20			
32 L.H., Daily Totals		$1926.80		$2387.20		$2984.40	$60.21	$74.60	$93.26
Crew Q-14	**Hr.**	**Daily**	**Hr.**	**Daily**	**Hr.**	**Daily**	**Bare Costs**	**In House**	**Incl. O&P**
1 Asbestos Worker	$58.65	$469.20	$74.80	$ 598.40	$93.00	$ 744.00	$52.77	$67.30	$83.67
1 Asbestos Apprentice	46.90	375.20	59.80	478.40	74.35	594.80			
16 L.H., Daily Totals		$844.40		$1076.80		$1338.80	$52.77	$67.30	$83.67
Crew Q-15	**Hr.**	**Daily**	**Hr.**	**Daily**	**Hr.**	**Daily**	**Bare Costs**	**In House**	**Incl. O&P**
1 Plumber	$64.45	$ 515.60	$79.70	$ 637.60	$99.65	$ 797.20	$58.00	$71.72	$89.67
1 Plumber Apprentice	51.55	412.40	63.75	510.00	79.70	637.60			
1 Welder, Electric, 300 amp		106.40		106.40		117.04	6.65	6.65	7.32
16 L.H., Daily Totals		$1034.40		$1254.00		$1551.84	$64.65	$78.38	$96.99
Crew Q-16	**Hr.**	**Daily**	**Hr.**	**Daily**	**Hr.**	**Daily**	**Bare Costs**	**In House**	**Incl. O&P**
2 Plumbers	$64.45	$1031.20	$79.70	$1275.20	$99.65	$1594.40	$60.15	$74.38	$93.00
1 Plumber Apprentice	51.55	412.40	63.75	510.00	79.70	637.60			
1 Welder, Electric, 300 amp		106.40		106.40		117.04	4.43	4.43	4.88
24 L.H., Daily Totals		$1550.00		$1891.60		$2349.04	$64.58	$78.82	$97.88
Crew Q-20	**Hr.**	**Daily**	**Hr.**	**Daily**	**Hr.**	**Daily**	**Bare Costs**	**In House**	**Incl. O&P**
1 Sheet Metal Worker	$62.30	$ 498.40	$78.30	$ 626.40	$97.60	$ 780.80	$57.13	$71.46	$89.17
1 Sheet Metal Apprentice	49.85	398.80	62.65	501.20	78.10	624.80			
.5 Electrician	61.35	245.40	75.40	301.60	94.45	377.80			
20 L.H., Daily Totals		$1142.60		$1429.20		$1783.40	$57.13	$71.46	$89.17
Crew R-3	**Hr.**	**Daily**	**Hr.**	**Daily**	**Hr.**	**Daily**	**Bare Costs**	**In House**	**Incl. O&P**
1 Electrician Foreman	$61.85	$ 494.80	$76.05	$ 608.40	$95.20	$ 761.60	$61.12	$75.44	$ 94.39
1 Electrician	61.35	490.80	75.40	603.20	94.45	755.60			
.5 Equip. Oper. (crane)	59.20	236.80	74.30	297.20	92.65	370.60			
.5 S.P. Crane, 4x4, 5 Ton		189.05		189.05		207.96	9.45	9.45	10.40
20 L.H., Daily Totals		$1411.45		$1697.85		$2095.76	$70.57	$84.89	$104.79
Crew R-26	**Hr.**	**Daily**	**Hr.**	**Daily**	**Hr.**	**Daily**	**Bare Costs**	**In House**	**Incl. O&P**
2 Electricians	$61.35	$ 981.60	$75.40	$1206.40	$94.45	$1511.20	$61.35	$75.40	$ 94.45
1 Articulating Boom Lift, to 45'		229.35		229.35		252.29	14.33	14.33	15.77
16 L.H., Daily Totals		$1210.95		$1435.75		$1763.48	$75.68	$89.73	$110.22

Travel Costs

The following table is used to estimate the cost for in-house staff or the contractor's personnel to travel to and from the job site. The amount incurred must be added to the in-house total labor cost or contractor's billing application to calculate the total cost with travel.

In-House Staff–Travel Times versus Costs

Travel Time Versus Costs	Crew Abbr.	Skwk	Clab	Clam	Asbe	Bric	Carp	Elec	Pord	Plum	Rofc	Shee	Stpi
	Rate with Mark-ups	$70.35	$54.40	$40.85	$74.80	$68.00	$68.70	$75.40	$56.75	$79.70	$66.55	$78.30	$81.05
Round Trip Travel Time (hours)	Crew Size												
0.50	1	35.18	27.20	20.43	37.40	34.00	34.35	37.70	28.38	39.85	33.28	39.15	40.53
	2	70.35	54.40	40.85	74.80	68.00	68.70	75.40	56.75	79.70	66.55	78.30	81.05
	3	105.53	81.60	61.28	112.20	102.00	103.05	113.10	85.13	119.55	99.83	117.45	121.58
0.75	1	52.76	40.80	30.64	56.10	51.00	51.53	56.55	42.56	59.78	49.91	58.73	60.79
	2	105.53	81.60	61.28	112.20	102.00	103.05	113.10	85.13	119.55	99.83	117.45	121.58
	3	158.29	122.40	91.91	168.30	153.00	154.58	169.65	127.69	179.33	149.74	176.18	182.36
1.00	1	70.35	54.40	40.85	74.80	68.00	68.70	75.40	56.75	79.70	66.55	78.30	81.05
	2	140.70	108.80	81.70	149.60	136.00	137.40	150.80	113.50	159.40	133.10	156.60	162.10
	3	211.05	163.20	122.55	224.40	204.00	206.10	226.20	170.25	239.10	199.65	234.90	243.15
1.50	1	105.53	81.60	61.28	112.20	102.00	103.05	113.10	85.13	119.55	99.83	117.45	121.58
	2	211.05	163.20	122.55	224.40	204.00	206.10	226.20	170.25	239.10	199.65	234.90	243.15
	3	316.58	244.80	183.83	336.60	306.00	309.15	339.30	255.38	358.65	299.48	352.35	364.73
2.00	1	140.70	108.80	81.70	149.60	136.00	137.40	150.80	113.50	159.40	133.10	156.60	162.10
	2	281.40	217.60	163.40	299.20	272.00	274.80	301.60	227.00	318.80	266.20	313.20	324.20
	3	422.10	326.40	245.10	448.80	408.00	412.20	452.40	340.50	478.20	399.30	469.80	486.30

Installing Contractors–Travel Times versus Costs

Travel Time Versus Costs	Crew Abbr.	Skwk	Clab	Clam	Asbe	Bric	Carp	Elec	Pord	Plum	Rofc	Shee	Stpi
	Rate with O & P	$87.40	$67.45	$50.65	$93.00	$84.10	$85.15	$94.45	$70.50	$99.65	$80.85	$97.60	$101.35
Round Trip Travel Time (hours)	Crew Size												
0.50	1	43.70	33.73	25.33	46.50	42.05	42.58	47.23	35.25	49.83	40.43	48.80	50.68
	2	87.40	67.45	50.65	93.00	84.10	85.15	94.45	70.50	99.65	80.85	97.60	101.35
	3	131.10	101.18	75.98	139.50	126.15	127.73	141.68	105.75	149.48	121.28	146.40	152.03
0.75	1	65.55	50.59	37.99	69.75	63.08	63.86	70.84	52.88	74.74	60.64	73.20	76.01
	2	131.10	101.18	75.98	139.50	126.15	127.73	141.68	105.75	149.48	121.28	146.40	152.03
	3	196.65	151.76	113.96	209.25	189.23	191.59	212.51	158.63	224.21	181.91	219.60	228.04
1.00	1	87.40	67.45	50.65	93.00	84.10	85.15	94.45	70.50	99.65	80.85	97.60	101.35
	2	174.80	134.90	101.30	186.00	168.20	170.30	188.90	141.00	199.30	161.70	195.20	202.70
	3	262.20	202.35	151.95	279.00	252.30	255.45	283.35	211.50	298.95	242.55	292.80	304.05
1.50	1	131.10	101.18	75.98	139.50	126.15	127.73	141.68	105.75	149.48	121.28	146.40	152.03
	2	262.20	202.35	151.95	279.00	252.30	255.45	283.35	211.50	298.95	242.55	292.80	304.05
	3	393.30	303.53	227.93	418.50	378.45	383.18	425.03	317.25	448.43	363.83	439.20	456.08
2.00	1	174.80	134.90	101.30	186.00	168.20	170.30	188.90	141.00	199.30	161.70	195.20	202.70
	2	349.60	269.80	202.60	372.00	336.40	340.60	377.80	282.00	398.60	323.40	390.40	405.40
	3	524.40	404.70	303.90	558.00	504.60	510.90	566.70	423.00	597.90	485.10	585.60	608.10

Table H1010-101 General Contractor's Overhead

There are two distinct types of overhead on a construction project: project overhead and main office overhead. Project overhead includes those costs at a construction site not directly associated with the installation of construction materials. Examples of project overhead costs include the following:

1. Superintendent
2. Construction office and storage trailers
3. Temporary sanitary facilities
4. Temporary utilities
5. Security fencing
6. Photographs
7. Clean up
8. Performance and payment bonds

The above project overhead items are also referred to as General Requirements and therefore are estimated in Division 1. Division 1 is the first division listed in the CSI MasterFormat but it is usually the last division estimated. The sum of the costs in Divisions 1 through 49 is referred to as the sum of the direct costs.

All construction projects also include indirect costs. The primary components of indirect costs are the contractor's main office overhead and profit. The amount of the main office overhead expense varies depending on the following:

1. Owner's compensation
2. Project managers and estimator's wages
3. Clerical support wages
4. Office rent and utilities
5. Corporate legal and accounting costs
6. Advertising
7. Automobile expenses
8. Association dues
9. Travel and entertainment expenses

These costs are usually calculated as a percentage of annual sales volume. This percentage can range from 35% for a small contractor doing less than $500,000 to 5% for a large contractor with sales in excess of $100 million.

Table H1010-102 Main Office Expense

A general contractor's main office expense consists of many items not detailed in the front portion of the data set. The percentage of main office expense declines with increased annual volume of the contractor. Typical main office expense ranges from 2% to 20% with the median about 7.2% of total volume. This equals about 7.7% of direct costs. The following are approximate percentages of total overhead for different items usually included in a general contractor's main office overhead. With different accounting procedures, these percentages may vary.

Item	Typical Range	Average
Managers', clerical and estimators' salaries	40 % to 55 %	48%
Profit sharing, pension and bonus plans	2 to 20	12
Insurance	5 to 8	6
Estimating and project management (not including salaries)	5 to 9	7
Legal, accounting and data processing	0.5 to 5	3
Automobile and light truck expense	2 to 8	5
Depreciation of overhead capital expenditures	2 to 6	4
Maintenance of office equipment	0.1 to 1.5	1
Office rental	3 to 5	4
Utilities including phone and light	1 to 3	2
Miscellaneous	5 to 15	8
Total		100%

Table H1010-201 Architectural Fees

Tabulated below are typical percentage fees by project size for good professional architectural service. Fees may vary from those listed depending upon degree of design difficulty and economic conditions in any particular area.

Rates can be interpolated horizontally and vertically. Various portions of the same project requiring different rates should be adjusted proportionately. For alterations, add 50% to the fee for the first $500,000 of project cost and add 25% to the fee for project cost over $500,000.

Architectural fees tabulated below include Structural, Mechanical and Electrical Engineering Fees. They do not include the fees for special consultants such as kitchen planning, security, acoustical, interior design, etc.

Building Types	Total Project Size in Thousands of Dollars						
	100	250	500	1,000	5,000	10,000	50,000
Factories, garages, warehouses, repetitive housing	9.0%	8.0%	7.0%	6.2%	5.3%	4.9%	4.5%
Apartments, banks, schools, libraries, offices, municipal buildings	12.2	12.3	9.2	8.0	7.0	6.6	6.2
Churches, hospitals, homes, laboratories, museums, research	15.0	13.6	12.7	11.9	9.5	8.8	8.0
Memorials, monumental work, decorative furnishings	—	16.0	14.5	13.1	10.0	9.0	8.3

Table H1010-202 Engineering Fees

Typical **Structural Engineering Fees** based on type of construction and total project size. These fees are included in Architectural Fees.

Type of Construction	Total Project Size (in thousands of dollars)			
	$500	$500-$1,000	$1,000-$5,000	Over $5000
Industrial buildings, factories & warehouses	Technical payroll times 2.0 to 2.5	1.60%	1.25%	1.00%
Hotels, apartments, offices, dormitories, hospitals, public buildings, food stores	↓	2.00%	1.70%	1.20%
Museums, banks, churches and cathedrals	↓	2.00%	1.75%	1.25%
Thin shells, prestressed concrete, earthquake resistive	↓	2.00%	1.75%	1.50%
Parking ramps, auditoriums, stadiums, convention halls, hangars & boiler houses	↓	2.50%	2.00%	1.75%
Special buildings, major alterations, underpinning & future expansion	↓	Add to above 0.5%	Add to above 0.5%	Add to above 0.5%

For complex reinforced concrete or unusually complicated structures, add 20% to 50%.

Table H1010-203 Mechanical and Electrical Fees

Typical **Mechanical and Electrical Engineering Fees** based on the size of the subcontract. The fee structure for both is shown below. These fees are included in Architectural Fees.

Type of Construction	Subcontract Size							
	$25,000	$50,000	$100,000	$225,000	$350,000	$500,000	$750,000	$1,000,000
Simple structures	6.4%	5.7%	4.8%	4.5%	4.4%	4.3%	4.2%	4.1%
Intermediate structures	8.0	7.3	6.5	5.6	5.1	5.0	4.9	4.8
Complex structures	10.1	9.0	9.0	8.0	7.5	7.5	7.0	7.0

For renovations, add 15% to 25% to applicable fee.

Table H1010-301 Builder's Risk Insurance

Builder's risk insurance is insurance on a building during construction. Premiums are paid by the owner or the contractor. Blasting, collapse and underground insurance would raise total insurance costs.

Table H1010-302 Performance Bond

This table shows the cost of a performance bond for a construction job scheduled to be completed in 12 months. Add 1% of the premium cost per month for jobs requiring more than 12 months to complete. The rates are "standard" rates offered to contractors that the bonding company considers financially sound and capable of doing the work. Preferred rates are offered by some bonding companies based upon financial strength of the contractor. Actual rates vary from contractor to contractor and from bonding company to bonding company. Contractors should prequalify through a bonding agency before submitting a bid on a contract that requires a bond.

Contract Amount	Building Construction Class B Projects	Highways & Bridges	
		Class A New Construction	Class A-1 Highway Resurfacing
First $ 100,000 bid	$25.00 per M	$15.00 per M	$9.40 per M
Next 400,000 bid	$ 2,500 plus $15.00 per M	$ 1,500 plus $10.00 per M	$ 940 plus $7.20 per M
Next 2,000,000 bid	8,500 plus 10.00 per M	5,500 plus 7.00 per M	3,820 plus 5.00 per M
Next 2,500,000 bid	28,500 plus 7.50 per M	19,500 plus 5.50 per M	15,820 plus 4.50 per M
Next 2,500,000 bid	47,250 plus 7.00 per M	33,250 plus 5.00 per M	28,320 plus 4.50 per M
Over 7,500,000 bid	64,750 plus 6.00 per M	45,750 plus 4.50 per M	39,570 plus 4.00 per M

Table H1010-401 Workers' Compensation Insurance Rates by Trade

The table below tabulates the national averages for workers' compensation insurance rates by trade and type of building. The average "Insurance Rate" is multiplied by the "% of Building Cost" for each trade. This produces the "Workers' Compensation" cost by % of total labor cost, to be added for each trade by building type to determine the weighted average workers' compensation rate for the building types analyzed.

Trade	Insurance Rate (% Labor Cost) Range	Insurance Rate Average	% of Building Cost Office Bldgs.	% of Building Cost Schools & Apts.	% of Building Cost Mfg.	Workers' Compensation Office Bldgs.	Workers' Compensation Schools & Apts.	Workers' Compensation Mfg.
Excavation, Grading, etc.	2.3 % to 16.9%	9.6%	4.8%	4.9%	4.5%	0.46%	0.47%	0.43%
Piles & Foundations	3.3 to 28.0	15.7	7.1	5.2	8.7	1.11	0.82	1.37
Concrete	3.1 to 25.8	14.4	5.0	14.8	3.7	0.72	2.13	0.53
Masonry	3.3 to 52.1	27.7	6.9	7.5	1.9	1.91	2.08	0.53
Structural Steel	4.4 to 31.8	18.1	10.7	3.9	17.6	1.94	0.71	3.19
Miscellaneous & Ornamental Metals	2.9 to 21.4	12.2	2.8	4.0	3.6	0.34	0.49	0.44
Carpentry & Millwork	3.4 to 29.1	16.2	3.7	4.0	0.5	0.60	0.65	0.08
Metal or Composition Siding	4.8 to 124.6	64.7	2.3	0.3	4.3	1.49	0.19	2.78
Roofing	4.8 to 110.4	57.6	2.3	2.6	3.1	1.32	1.50	1.79
Doors & Hardware	3.1 to 29.1	16.1	0.9	1.4	0.4	0.14	0.23	0.06
Sash & Glazing	3.9 to 21.6	12.8	3.5	4.0	1.0	0.45	0.51	0.13
Lath & Plaster	2.7 to 30.1	16.4	3.3	6.9	0.8	0.54	1.13	0.13
Tile, Marble & Floors	2.0 to 19.0	10.5	2.6	3.0	0.5	0.27	0.32	0.05
Acoustical Ceilings	1.7 to 29.7	15.7	2.4	0.2	0.3	0.38	0.03	0.05
Painting	3.3 to 37.4	20.3	1.5	1.6	1.6	0.30	0.32	0.32
Interior Partitions	3.4 to 29.1	16.2	3.9	4.3	4.4	0.63	0.70	0.71
Miscellaneous Items	1.9 to 97.7	10.3	5.2	3.7	9.7	0.54	0.38	1.00
Elevators	1.3 to 9.0	5.1	2.1	1.1	2.2	0.11	0.06	0.11
Sprinklers	1.8 to 14.6	8.2	0.5	—	2.0	0.04	—	0.16
Plumbing	1.4 to 13.3	7.4	4.9	7.2	5.2	0.36	0.53	0.38
Heat., Vent., Air Conditioning	2.8 to 15.8	9.3	13.5	11.0	12.9	1.26	1.02	1.20
Electrical	1.7 to 10.7	6.2	10.1	8.4	11.1	0.63	0.52	0.69
Total	1.3 % to 124.6%	—	100.0%	100.0%	100.0%	15.54%	14.79%	16.13%
		Overall Weighted Average 15.49%						

Table H1010-402 Workers' Compensation Insurance Rates by States

The table below lists the weighted average Workers' Compensation base rate for each state with a factor comparing this with the national average of 9.9%.

State	Weighted Average	Factor	State	Weighted Average	Factor	State	Weighted Average	Factor
Alabama	12.4%	126	Kentucky	10.5%	108	North Dakota	6.4%	65
Alaska	8.6	88	Louisiana	17.5	178	Ohio	5.5	56
Arizona	7.7	79	Maine	8.0	81	Oklahoma	7.4	76
Arkansas	5.1	52	Maryland	9.7	99	Oregon	7.5	76
California	19.6	200	Massachusetts	8.8	90	Pennsylvania	17.2	175
Colorado	6.2	63	Michigan	6.3	64	Rhode Island	10.0	102
Connecticut	12.4	126	Minnesota	13.4	137	South Carolina	16.2	165
Delaware	11.8	120	Mississippi	9.2	94	South Dakota	7.6	77
District of Columbia	7.7	78	Missouri	12.0	122	Tennessee	6.0	61
Florida	8.4	86	Montana	7.1	73	Texas	4.9	50
Georgia	30.0	307	Nebraska	11.7	120	Utah	5.7	58
Hawaii	8.4	86	Nevada	7.6	77	Vermont	8.9	91
Idaho	7.8	79	New Hampshire	8.8	90	Virginia	6.7	69
Illinois	17.5	179	New Jersey	14.2	145	Washington	7.3	75
Indiana	3.1	32	New Mexico	13.0	132	West Virginia	4.2	42
Iowa	9.9	101	New York	15.6	159	Wisconsin	11.2	115
Kansas	5.6	57	North Carolina	11.6	119	Wyoming	4.8	49
			Weighted Average for U.S. is	9.9% of payroll = 100%				

The weighted average skilled worker rate for 35 trades is 9.8%. For bidding purposes, apply the full value of Workers' Compensation directly to total labor costs, or if labor is 38%, materials 42% and overhead and profit 20% of total cost, carry 38/80 x 9.8% = 4.66% of cost (before overhead and profit) into overhead. Rates vary not only from state to state but also with the experience rating of the contractor.

Rates are the most current available at the time of publication.

General Conditions — RH1020-300 Overhead & Miscellaneous Data

Unemployment Taxes and Social Security Taxes

State unemployment tax rates vary not only from state to state, but also with the experience rating of the contractor. The federal unemployment tax rate is 6.0% of the first $7,000 of wages. This is reduced by a credit of up to 5.4% for timely payment to the state. The minimum federal unemployment tax is 0.6% after all credits.

Social security (FICA) for 2020 is estimated at time of publication to be 7.65% of wages up to $137,100.

General Conditions — RH1020-400 Overtime

Overtime

One way to improve the completion date of a project or eliminate negative float from a schedule is to compress activity duration times. This can be achieved by increasing the crew size or working overtime with the proposed crew.

To determine the costs of working overtime to compress activity duration times, consider the following examples. Below is an overtime efficiency and cost chart based on a five, six, or seven day week with an eight through twelve hour day. Payroll percentage increases for time and one half and double times are shown for the various working days.

Days per Week	Hours per Day	Production Efficiency					Payroll Cost Factors	
		1st Week	2nd Week	3rd Week	4th Week	Average 4 Weeks	@ 1-1/2 Times	@ 2 Times
5	8	100%	100%	100%	100%	100%	1.000	1.000
	9	100	100	95	90	96	1.056	1.111
	10	100	95	90	85	93	1.100	1.200
	11	95	90	75	65	81	1.136	1.273
	12	90	85	70	60	76	1.167	1.333
6	8	100	100	95	90	96	1.083	1.167
	9	100	95	90	85	93	1.130	1.259
	10	95	90	85	80	88	1.167	1.333
	11	95	85	70	65	79	1.197	1.394
	12	90	80	65	60	74	1.222	1.444
7	8	100	95	85	75	89	1.143	1.286
	9	95	90	80	70	84	1.183	1.365
	10	90	85	75	65	79	1.214	1.429
	11	85	80	65	60	73	1.240	1.481
	12	85	75	60	55	69	1.262	1.524

General Conditions — RH1030-100 General

Sales Tax by State

State sales tax on materials is tabulated below (5 states have no sales tax). Many states allow local jurisdictions, such as a county or city, to levy additional sales tax.

Some projects may be sales tax exempt, particularly those constructed with public funds.

State	Tax (%)	State	Tax (%)	State	Tax (%)	State	Tax (%)
Alabama	4	Illinois	6.25	Montana	0	Rhode Island	7
Alaska	0	Indiana	7	Nebraska	5.5	South Carolina	6
Arizona	5.6	Iowa	6	Nevada	6.85	South Dakota	4.5
Arkansas	6.5	Kansas	6.5	New Hampshire	0	Tennessee	7
California	7.25	Kentucky	6	New Jersey	6.625	Texas	6.25
Colorado	2.9	Louisiana	4.45	New Mexico	5.125	Utah	4.85
Connecticut	6.35	Maine	5.5	New York	4	Vermont	6
Delaware	0	Maryland	6	North Carolina	4.75	Virginia	4.3
District of Columbia	6	Massachusetts	6.25	North Dakota	5	Washington	6.5
Florida	6	Michigan	6	Ohio	5.75	West Virginia	6
Georgia	4	Minnesota	6.875	Oklahoma	4.5	Wisconsin	5
Hawaii	4	Mississippi	7	Oregon	0	Wyoming	4
Idaho	6	Missouri	4.225	Pennsylvania	6	Average	5.06 %

Table H1040-101 Steel Tubular Scaffolding

On new construction, tubular scaffolding is efficient up to 60′ high or five stories. Above this it is usually better to use hung scaffolding if construction permits. Swing scaffolding operations may interfere with tenants. In this case, the tubular is more practical at all heights.

In repairing or cleaning the front of an existing building, the cost of tubular scaffolding per S.F. of building front goes up as the height increases above the first tier. The first tier cost is relatively high due to leveling and alignment.

The minimum efficient crew for erection is three workers. For heights over 50′, a crew of four is more efficient. Use two or more on top and two at the bottom for handing up or hoisting. Four workers can erect and dismantle about nine frames per hour up to five stories. From five to eight stories, they will average six frames per hour. With 7′ horizontal spacing, this will run about 400 S.F. and 265 S.F. of wall surface, respectively. Time for placing planks must be added to the above. On heights above 50′, five planks can be placed per labor-hour.

The cost per 1,000 S.F. of building front in the table below was developed by pricing the materials required for a typical tubular scaffolding system eleven frames long and two frames high. Planks were figured five wide for standing plus two wide for materials.

Frames are 5′ wide and usually spaced 7′ O.C. horizontally. Sidewalk frames are 6′ wide. Rental rates will be lower for jobs over three months' duration.

For jobs under twenty-five frames, add 50% to rental cost. These figures do not include accessories which are listed separately below. Large quantities for long periods can reduce rental rates by 20%.

Item	Unit	Monthly Rent	Per 1,000 S.F. of Building Front	
			No. of Pieces	Rental per Month
5′ Wide Standard Frame, 6′ 4″ High	Ea.	$ 6.30	24	$151.20
Leveling Jack & Plate	↓	1.98	24	47.52
Cross Brace		0.99	44	43.56
Side Arm Bracket, 21″		1.98	12	23.76
Guardrail Post		0.99	12	11.88
Guardrail, 7′ section		0.99	22	21.78
Stairway Section		35.00	2	70.00
Walk-Through Frame Guardrail	▼	2.48	2	4.96
			Total	$374.66
			Per C.S.F., 1 Use/Mo.	$ 37.47

Scaffolding is often used as falsework over 15′ high during construction of cast-in-place concrete beams and slabs. Two-foot wide scaffolding is generally used for heavy beam construction. The span between frames depends upon the load to be carried, with a maximum span of 5′.

Heavy duty shoring frames with a capacity of 10,000#/leg can be spaced up to 10′ O. C., depending upon form support design and loading.

Scaffolding used as horizontal shoring requires less than half the material required with conventional shoring.

On new construction, erection is done by carpenters.

Rolling towers supporting horizontal shores can reduce labor and speed the job. For maintenance work, catwalks with spans up to 70′ can be supported by the rolling towers.

City Adjustments RJ1010-010

General: The following information on current city cost indexes is calculated for over 930 zip code locations in the United States and Canada. Index figures for both material and installation are based upon the 30 major city average of 100 and represent the cost relationship on July 1, 2020.

In addition to index adjustment, the user should consider:

1. productivity
2. management efficiency
3. competitive conditions
4. automation
5. restrictive union practices
6. unique local requirements
7. regional variations due to specific building codes

The weighted-average index is calculated from about 66 materials, 6 equipment types and 21 building trades. The component contribution of these in a model building is tabulated in Table J1010-011 below.

If the systems component distribution of a building is unknown, the weighted-average index can be used to adjust the cost for any city.

Table J1010-011 Labor, Material, and Equipment Cost Distribution for Weighted Average Listed by System Division

Division No.	Building System	Percentage	Division No.	Building System	Percentage
A	Substructure	6.1%	D10	Services: Conveying	4.0%
B10	Shell: Superstructure	19.2	D20-40	Mechanical	22.5
B20	Exterior Closure	12.2	D50	Electrical	11.9
B30	Roofing	2.5	E	Equipment & Furnishings	2.1
C	Interior Construction	15.8	G	Site Work	3.7
				Total weighted average (Div. A-G)	100.0%

How to Use the Component Indexes: Table J1010-012 below shows how the average costs obtained from this data set for each division should be adjusted for your particular building in your city. The example is a building adjusted for New York, NY. Indexes for other cities are tabulated in Division RJ1030. These indexes should also be used when compiling data on the form in Division K1010.

Table J1010-012 Adjustment of "Book" Costs to a Particular City

Systems Division Number	System Description	New York, NY							
		City Cost Index	Cost Factor	Book Cost	Adjusted Cost	City Cost Index	Cost Factor	Book Cost	Adjusted Cost
A	Substructure	143.8	1.438	$ 111,800	$ 160,800				
B10	Shell: Superstructure	132.4	1.324	308,800	408,900				
B20	Exterior Closure	144.1	1.441	209,300	301,600				
B30	Roofing	134.5	1.345	46,100	62,000				
C	Interior Construction	141.1	1.411	30,300	42,800				
D10	Services: Conveying	111.7	1.117	59,800	66,800				
D20-40	Mechanical	137.8	1.378	369,800	509,600				
D50	Electrical	148.0	1.480	206,900	306,200				
E	Equipment	105.9	1.059	31,200	33,000				
G	Site Work	117.2	1.172	88,000	103,100				
Total Cost, (Div. A-G)				$1,462,000	$1,994,800				

Alternate Method: Use the weighted-average index for your city rather than the component indexes. **For example:**

(Total weighted-average index A-G)/100 x (total cost) = Adjusted city cost
New York, NY 1.371 x $1,462,000 = $2,004,400

City to City: To convert known or estimated costs from one city to another, cost indexes can be used as follows:

$$\text{Unknown City Cost} = \text{Known Cost} \times \frac{\text{Unknown City Index}}{\text{Known City Index}}$$

For example: If the building cost in Boston, MA, is $2,000,000, how much would a duplicated building cost in Los Angeles, CA?

$$\text{L.A. Cost} = \$2{,}000{,}000 \times \frac{\text{(Los Angeles) } 116.3}{\text{(Boston) } 120.3} = \$1{,}933{,}500$$

The table below lists both the RSMeans City Cost Index based on January 1, 1993 = 100 as well as the computed value of an index based on January 1, 2020 costs. Since the January 1, 2020 figure is estimated, space is left to write in the actual index figures as they become available through the quarterly *RSMeans Construction Cost Indexes*. To compute the actual index based on January 1, 2020 = 100, divide the Quarterly City Cost Index for a particular year by the actual January 1, 2020 Quarterly City Cost Index. Space has been left to advance the index figures as the year progresses.

Table J1020-011 Historical Cost Indexes

Year	Historical Cost Index Jan. 1, 1993 = 100		Current Index Based on Jan. 1, 2020 = 100		Year	Historical Cost Index Jan. 1, 1993 = 100	Current Index Based on Jan. 1, 2020 = 100		Year	Historical Cost Index Jan. 1, 1993 = 100	Current Index Based on Jan. 1, 2020 = 100	
	Est.	Actual	Est.	Actual		Actual	Est.	Actual		Actual	Est.	Actual
Jan 2020*	239.1				July 2004	143.7	63.2		July 1986	84.2	37.1	
Oct 2019*					2003	132	58.1		1985	82.6	36.3	
July 2019		232.2			2002	128.7	56.6		1984	82.0	36.1	
April 2019		230.8			2001	125.1	55.0		1983	80.2	35.3	
Jan 2019		229.6	100.0	100.0	2000	120.9	53.2		1982	76.1	33.5	
July 2018		222.9	98.1		1999	117.6	51.7		1981	70.0	30.8	
2017		213.6	94.0		1998	115.1	50.6		1980	62.9	27.7	
2016		207.3	91.2		1997	112.8	49.6		1979	57.8	25.4	
2015		206.2	90.7		1996	110.2	48.5		1978	53.5	23.5	
2014		204.9	90.1		1995	107.6	47.3		1977	49.5	21.8	
2013		201.2	88.5		1994	104.4	45.9		1976	46.9	20.6	
2012		194.6	85.6		1993	101.7	44.7		1975	44.8	19.7	
2011		191.2	84.1		1992	99.4	43.7		1974	41.4	18.2	
2010		183.5	80.7		1991	96.8	42.6		1973	37.7	16.6	
2009		180.1	79.2		1990	94.3	41.5		1972	34.8	15.3	
2008		180.4	79.4		1989	92.1	40.5		1971	32.1	14.1	
2007		169.4	74.5		1988	89.9	39.5		1970	28.7	12.6	
2006		162.0	71.3		1987	87.7	38.6		1969	26.9	11.8	
2005		151.6	66.7									

To find the **current cost** from a project built previously in either the same city or a different city, the following formula is used:

$$\text{Present Cost (City X)} = \frac{\text{Current HCI} \times \text{CCI (City X)}}{\text{Previous HCI} \times \text{CCI (City Y)}} \times \text{Former Cost (City Y)}$$

For example: Find the construction cost of a building to be built in San Francisco, CA, as of January 1, 2020 when the identical building cost $500,000 in Boston, MA, on July 1, 1969.

$$\text{Jan. 1, 2020 (San Francisco)} = \frac{\text{(San Francisco) } 239.1 \times 134.3}{\text{(Boston) } 26.9 \times 120.3} \times \$500{,}000 = \$4{,}961{,}500$$

Note: The City Cost Indexes for Canada can be used to convert U.S. national averages to local costs in Canadian dollars.

To Project Future Construction Costs: Using the results of the last five years average percentage increase as a basis, an average increase of 1.6% could be used.

The historical index figures above are compiled from the Means Construction Index Service.

Example:

To estimate and compare the cost of a building in Toronto, ON in 2020 with the known cost of $600,000 (US$) in New York, NY in 2020:

INDEX Toronto = 115.6

INDEX New York = 137.1

$$\frac{\text{INDEX Toronto}}{\text{INDEX New York}} \times \text{Cost New York} = \text{Cost Toronto}$$

$$\frac{115.6}{137.1} \times \$600{,}000 = .834 \times \$600{,}000 = \$505{,}908$$

The construction cost of the building in Toronto is $505,908 (CN$).

*Historical Cost Index updates and other resources are provided on the following website:
http://info.thegordiangroup.com/RSMeans.html

How to Use the City Cost Indexes

What you should know before you begin

RSMeans City Cost Indexes (CCI) are an extremely useful tool for when you want to compare costs from city to city and region to region.

This publication contains average construction cost indexes for 317 U.S. and Canadian cities covering over 930 three-digit zip code locations, as listed directly under each city.

Keep in mind that a City Cost Index number is a percentage ratio of a specific city's cost to the national average cost of the same item at a stated time period.

In other words, these index figures represent relative construction factors (or, if you prefer, multipliers) for material and installation costs, as well as the weighted average for Total In Place costs for each Uniformat II Element. Installation costs include both labor and equipment rental costs.

The 30 City Average Index is the average of 30 major U.S. cities and serves as a national average.

Index figures for both material and installation are based on the 30 major city average of 100 and represent the cost relationship as of July 1, 2019. The index for each element is computed from representative material and labor quantities for that element. The weighted average for each city is a weighted total of the components listed above it. It does not include relative productivity between trades or cities.

As changes occur in local material prices, labor rates, and equipment rental rates (including fuel costs), the impact of these changes should be accurately measured by the change in the City Cost Index for each particular city (as compared to the 30 city average).

Therefore, if you know (or have estimated) building costs in one city today, you can easily convert those costs to expected building costs in another city.

In addition, by using the Historical Cost Index, you can easily convert national average building costs at a particular time to the approximate building costs for some other time. The City Cost Indexes can then be applied to calculate the costs for a particular city.

Quick calculations

Location Adjustment Using the City Cost Indexes:

$$\frac{\text{Index for City A}}{\text{Index for City B}} \times \text{Cost in City B} = \text{Cost in City A}$$

Time Adjustment for the National Average Using the Historical Cost Index:

$$\frac{\text{Index for Year A}}{\text{Index for Year B}} \times \text{Cost in Year B} = \text{Cost in Year A}$$

Adjustment from the National Average:

$$\frac{\text{Index for City A}}{100} \times \text{National Average Cost} = \text{Cost in City A}$$

Since each of the other RSMeans data sets contains many different items, any *one* item multiplied by the particular city index may give incorrect results. However, the larger the number of items compiled, the closer the results should be to actual costs for that particular city.

The City Cost Indexes for Canadian cities are calculated using Canadian material and equipment prices and labor rates in Canadian dollars. Therefore, indexes for Canadian cities can be used to convert U.S. national average prices to local costs in Canadian dollars.

How to use this section

1. Compare costs from city to city.

In using the RSMeans Indexes, remember that an index number is not a fixed number but a ratio: It's a percentage ratio of a building component's cost at any stated time to the national average cost of that same component at the same time period. Put in the form of an equation:

$$\frac{\text{Specific City Cost}}{\text{National Average Cost}} \times 100 = \text{City Index Number}$$

Therefore, when making cost comparisons between cities, do not subtract one city's index number from the index number of another city and read the result as a percentage difference. Instead, divide one city's index number by that of the other city. The resulting number may then be used as a multiplier to calculate cost differences from city to city.

The formula used to find cost differences between cities for the purpose of comparison is as follows:

$$\frac{\text{City A Index}}{\text{City B Index}} \times \text{City B Cost (Known)} = \text{City A Cost (Unknown)}$$

In addition, you can use RSMeans CCI to calculate and compare costs division by division between cities using the same basic formula. (Just be sure that you're comparing similar divisions.)

2. Compare a specific city's construction costs with the national average.

When you're studying construction location feasibility, it's advisable to compare a prospective project's cost index with an index of the national average cost.

For example, divide the weighted average index of construction costs of a specific city by that of the 30 City Average, which = 100.

$$\frac{\text{City Index}}{100} = \text{\% of National Average}$$

As a result, you get a ratio that indicates the relative cost of construction in that city in comparison with the national average.

3. Convert U.S. national average to actual costs in Canadian City.

$$\frac{\text{Index for Canadian City}}{100} \times \text{National Average Cost} = \text{Cost in Canadian City in \\$ CAN}$$

4. **Adjust construction cost data based on a national average.**

When you use a source of construction cost data which is based on a national average (such as RSMeans cost data), it is necessary to adjust those costs to a specific location.

$$\frac{\text{City Index}}{100} \times \frac{\text{Cost Based on}}{\text{National Average Costs}} = \frac{\text{City Cost}}{\text{(Unknown)}}$$

5. **When applying the City Cost Indexes to demolition projects, use the appropriate division installation index.** For example, for removal of existing doors and windows, use the Division 8 (Openings) index.

What you might like to know about how we developed the Indexes

The information presented in the CCI is organized according to the Uniformat II classification system.

To create a reliable index, RSMeans researched the building type most often constructed in the United States and Canada. Because it was concluded that no one type of building completely represented the building construction industry, nine different types of buildings were combined to create a composite model.

The exact material, labor, and equipment quantities are based on detailed analyses of these nine building types, and then each quantity is weighted in proportion to expected usage. These various material items, labor hours, and equipment rental rates are thus combined to form a composite building representing as closely as possible the actual usage of materials, labor, and equipment in the North American building construction industry.

The following structures were chosen to make up that composite model:

1. Factory, 1 story
2. Office, 2–4 stories
3. Store, Retail
4. Town Hall, 2–3 stories
5. High School, 2–3 stories
6. Hospital, 4–8 stories
7. Garage, Parking
8. Apartment, 1–3 stories
9. Hotel/Motel, 2–3 stories

For the purposes of ensuring the timeliness of the data, the components of the index for the composite model have been streamlined. They currently consist of:

- specific quantities of 66 commonly used construction materials;
- specific labor-hours for 21 building construction trades; and
- specific days of equipment rental for 6 types of construction equipment (normally used to install the 66 material items by the 21 trades.) Fuel costs and routine maintenance costs are included in the equipment cost.

Material and equipment price quotations are gathered quarterly from cities in the United States and Canada. These prices and the latest negotiated labor wage rates for 21 different building trades are used to compile the quarterly update of the City Cost Index.

The 30 major U.S. cities used to calculate the national average are:

Atlanta, GA	Memphis, TN
Baltimore, MD	Milwaukee, WI
Boston, MA	Minneapolis, MN
Buffalo, NY	Nashville, TN
Chicago, IL	New Orleans, LA
Cincinnati, OH	New York, NY
Cleveland, OH	Philadelphia, PA
Columbus, OH	Phoenix, AZ
Dallas, TX	Pittsburgh, PA
Denver, CO	St. Louis, MO
Detroit, MI	San Antonio, TX
Houston, TX	San Diego, CA
Indianapolis, IN	San Francisco, CA
Kansas City, MO	Seattle, WA
Los Angeles, CA	Washington, DC

What the CCI does not indicate

The weighted average for each city is a total of the divisional components weighted to reflect typical usage. It does not include the productivity variations between trades or cities.

In addition, the CCI does not take into consideration factors such as the following:

- managerial efficiency
- competitive conditions
- automation
- restrictive union practices
- unique local requirements
- regional variations due to specific building codes

City Cost Indexes

RJ1030-010 Building Systems

DIV. NO.	BUILDING SYSTEMS	ALABAMA														
		BIRMINGHAM			HUNTSVILLE			MOBILE			MONTGOMERY			TUSCALOOSA		
		MAT.	INST.	TOTAL	MAT.	INST.	TOTAL	MAT.	INST.	TOTAL	MAT.	INST.	TOTAL	MAT.	INST.	TOTAL
A	Substructure	98.8	75.7	85.5	94.0	73.4	82.1	87.1	72.5	78.7	87.0	74.5	79.8	96.6	74.0	83.6
B10	Shell: Superstructure	100.3	83.2	93.9	101.3	83.7	94.7	99.3	83.0	93.2	98.5	83.1	92.8	101.2	83.6	94.6
B20	Exterior Closure	86.4	66.8	77.6	87.0	66.1	77.6	89.8	64.6	78.5	87.8	65.9	78.0	86.6	65.9	77.4
B30	Roofing	94.1	68.4	83.8	93.1	66.5	82.4	95.6	65.9	83.7	94.2	67.6	83.5	93.2	67.0	82.7
C	Interior Construction	96.8	69.4	85.4	97.3	66.3	84.4	91.9	66.8	81.5	91.8	69.1	82.3	97.3	69.0	85.5
D10	Services: Conveying	100.0	85.4	95.6	100.0	84.8	95.4	100.0	84.3	95.3	100.0	85.0	95.5	100.0	84.7	95.4
D20 - 40	Mechanical	100.0	66.3	86.8	100.0	66.9	87.0	100.0	61.4	84.8	100.0	64.3	86.0	100.0	67.1	87.1
D50	Electrical	98.4	59.7	79.3	95.4	65.9	80.8	100.4	56.8	78.9	101.1	76.1	88.8	95.0	59.7	77.6
E	Equipment & Furnishings	100.0	69.4	98.2	100.0	65.3	98.0	100.0	69.0	98.2	100.0	70.1	98.3	100.0	69.2	98.2
G	Site Work	93.3	94.2	93.9	86.7	89.0	88.2	97.7	84.9	89.2	95.6	89.5	91.5	87.4	89.1	88.5
A-G	**WEIGHTED AVERAGE**	97.4	71.8	86.6	97.1	71.7	86.4	96.7	68.7	84.9	96.3	73.1	86.5	97.1	71.4	86.3

DIV. NO.	BUILDING SYSTEMS	ALASKA									ARIZONA					
		ANCHORAGE			FAIRBANKS			JUNEAU			FLAGSTAFF			MESA/TEMPE		
		MAT.	INST.	TOTAL	MAT.	INST.	TOTAL	MAT.	INST.	TOTAL	MAT.	INST.	TOTAL	MAT.	INST.	TOTAL
A	Substructure	115.2	118.1	116.9	117.6	117.8	117.7	121.2	118.1	119.4	94.4	73.9	82.5	91.6	75.0	82.0
B10	Shell: Superstructure	124.6	110.0	119.2	122.0	110.0	117.5	114.9	110.0	113.1	102.1	70.4	90.3	103.0	71.7	91.3
B20	Exterior Closure	134.2	118.7	127.3	129.2	117.2	123.9	133.6	118.7	126.9	118.6	61.5	93.1	99.0	60.6	81.8
B30	Roofing	177.8	117.9	153.7	187.4	116.6	158.9	189.5	117.9	160.7	101.9	70.6	89.4	99.0	69.5	87.2
C	Interior Construction	127.6	116.2	122.8	129.8	115.1	123.7	128.7	116.2	123.5	100.2	66.4	86.1	95.6	69.0	84.5
D10	Services: Conveying	100.0	114.2	104.3	100.0	113.7	104.1	100.0	114.2	104.3	100.0	83.0	94.9	100.0	83.4	95.0
D20 - 40	Mechanical	100.5	107.1	103.1	100.2	107.9	103.2	100.5	107.1	103.1	100.3	77.0	91.1	100.1	77.0	91.1
D50	Electrical	117.2	109.4	113.3	124.2	109.4	116.9	110.2	109.4	109.8	101.3	60.7	81.3	94.2	62.9	78.7
E	Equipment & Furnishings	100.0	113.8	100.8	100.0	112.8	100.7	100.0	113.8	100.8	100.0	68.1	98.2	100.0	71.9	98.4
G	Site Work	118.8	122.1	121.0	119.4	125.0	123.1	136.4	122.1	126.9	90.4	90.0	90.1	96.9	91.0	93.0
A-G	**WEIGHTED AVERAGE**	118.5	113.1	116.2	118.8	113.0	116.3	116.8	113.1	115.2	102.4	70.5	88.9	98.7	71.5	87.2

DIV. NO.	BUILDING SYSTEMS	ARIZONA									ARKANSAS					
		PHOENIX			PRESCOTT			TUCSON			FORT SMITH			JONESBORO		
		MAT.	INST.	TOTAL	MAT.	INST.	TOTAL	MAT.	INST.	TOTAL	MAT.	INST.	TOTAL	MAT.	INST.	TOTAL
A	Substructure	91.4	76.0	82.6	92.6	74.9	82.4	89.3	73.9	80.4	83.0	71.7	76.5	80.9	75.8	77.9
B10	Shell: Superstructure	104.2	72.4	92.4	101.8	71.2	90.4	103.2	70.9	91.2	94.5	71.3	85.8	88.8	79.3	85.2
B20	Exterior Closure	99.3	63.7	83.4	102.5	60.6	83.8	93.1	61.5	79.0	85.7	61.3	74.8	84.2	62.5	74.5
B30	Roofing	100.3	71.0	88.5	100.6	71.1	88.8	99.9	69.0	87.5	104.0	60.5	86.5	108.7	60.7	89.4
C	Interior Construction	101.4	68.8	87.8	99.0	68.9	86.5	92.7	66.5	81.8	91.6	62.6	79.5	91.2	63.0	79.5
D10	Services: Conveying	100.0	84.8	95.4	100.0	83.0	94.9	100.0	83.4	95.0	100.0	79.6	93.9	100.0	67.4	90.2
D20 - 40	Mechanical	100.0	78.5	91.6	100.3	76.8	91.0	100.1	74.8	90.2	100.0	48.0	79.6	100.4	51.2	81.1
D50	Electrical	99.8	60.7	80.5	101.0	62.9	82.2	96.2	60.6	78.7	93.1	57.0	75.3	99.0	62.3	81.0
E	Equipment & Furnishings	100.0	70.8	98.3	100.0	72.1	98.4	100.0	67.9	98.2	100.0	63.2	97.9	100.0	63.4	97.9
G	Site Work	97.4	94.4	95.4	78.4	90.0	86.1	92.3	91.0	91.4	80.2	85.4	83.6	98.6	102.1	100.9
A-G	**WEIGHTED AVERAGE**	100.6	72.3	88.6	99.9	71.3	87.8	97.7	70.2	86.1	94.1	62.6	80.8	93.8	66.6	82.3

DIV. NO.	BUILDING SYSTEMS	ARKANSAS									CALIFORNIA					
		LITTLE ROCK			PINE BLUFF			TEXARKANA			ANAHEIM			BAKERSFIELD		
		MAT.	INST.	TOTAL	MAT.	INST.	TOTAL	MAT.	INST.	TOTAL	MAT.	INST.	TOTAL	MAT.	INST.	TOTAL
A	Substructure	83.2	73.8	77.8	78.7	71.9	74.8	85.1	72.2	77.7	90.1	128.5	112.3	92.3	128.4	113.1
B10	Shell: Superstructure	94.5	72.5	86.3	96.8	72.1	87.6	91.8	71.8	84.3	104.2	124.0	111.6	100.4	122.9	108.8
B20	Exterior Closure	85.4	62.4	75.2	95.7	61.4	80.4	83.6	61.6	73.8	95.3	136.9	113.8	92.6	135.2	111.6
B30	Roofing	98.6	61.9	83.9	98.0	60.9	83.1	98.6	60.8	83.4	109.1	136.3	120.1	108.7	124.2	114.9
C	Interior Construction	90.3	64.6	79.6	93.4	62.8	80.6	96.7	58.3	80.7	101.0	134.3	114.9	93.9	133.3	110.3
D10	Services: Conveying	100.0	80.0	94.0	100.0	79.6	93.9	100.0	66.5	89.9	100.0	117.8	105.4	100.0	115.1	104.5
D20 - 40	Mechanical	99.9	49.1	79.9	100.0	51.4	80.9	100.0	54.5	82.1	99.9	130.9	112.1	100.1	128.8	111.4
D50	Electrical	100.1	66.7	83.7	95.9	57.5	77.0	97.6	57.5	77.9	91.5	113.3	102.2	107.6	108.1	107.8
E	Equipment & Furnishings	100.0	63.8	97.9	100.0	63.2	97.9	100.0	63.2	97.9	100.0	138.1	102.2	100.0	137.7	102.2
G	Site Work	86.8	90.2	89.1	86.1	85.4	85.7	95.6	87.6	90.3	98.8	105.6	103.3	96.3	107.1	103.5
A-G	**WEIGHTED AVERAGE**	94.5	65.4	82.2	96.0	63.6	82.3	94.8	63.3	81.5	99.3	126.6	110.9	98.9	124.6	109.8

DIV. NO.	BUILDING SYSTEMS	CALIFORNIA														
		FRESNO			LOS ANGELES			OAKLAND			OXNARD			REDDING		
		MAT.	INST.	TOTAL	MAT.	INST.	TOTAL	MAT.	INST.	TOTAL	MAT.	INST.	TOTAL	MAT.	INST.	TOTAL
A	Substructure	94.1	132.6	116.3	90.0	129.6	112.8	101.9	140.3	124.1	97.5	127.5	114.8	113.7	132.6	124.6
B10	Shell: Superstructure	101.2	128.2	111.2	93.4	124.6	105.1	104.4	133.8	115.4	98.0	123.8	107.6	105.5	128.0	113.9
B20	Exterior Closure	98.1	142.4	117.9	101.2	138.1	117.7	117.0	153.8	133.4	95.3	135.8	113.4	121.7	140.9	130.3
B30	Roofing	99.1	131.4	112.1	103.4	134.8	116.0	110.9	155.7	128.9	108.9	134.8	119.3	128.7	139.0	132.8
C	Interior Construction	93.9	146.8	116.0	103.5	134.6	116.4	99.0	164.6	126.3	91.5	133.9	109.2	112.1	148.9	127.4
D10	Services: Conveying	100.0	127.5	108.3	100.0	118.7	105.6	100.0	128.9	108.7	100.0	118.1	105.4	100.0	127.5	108.3
D20 - 40	Mechanical	100.1	130.1	111.9	99.9	131.0	112.1	100.2	169.6	127.5	100.0	131.0	112.2	100.3	130.1	112.0
D50	Electrical	97.0	107.4	102.2	96.9	131.0	113.7	99.1	161.8	130.0	101.4	117.9	109.5	101.3	120.8	110.9
E	Equipment & Furnishings	100.0	155.1	103.2	100.0	138.7	102.2	100.0	172.8	104.2	100.0	138.1	102.2	100.0	155.7	103.2
G	Site Work	99.2	103.3	101.9	93.2	109.5	104.0	118.0	106.0	110.0	99.5	101.7	100.9	122.5	102.7	109.3
A-G	**WEIGHTED AVERAGE**	98.5	129.4	111.5	98.5	129.8	111.7	103.4	152.0	123.9	98.0	126.7	110.1	107.6	131.5	117.7

DIV. NO.	BUILDING SYSTEMS	CALIFORNIA														
		RIVERSIDE			SACRAMENTO			SAN DIEGO			SAN FRANCISCO			SAN JOSE		
		MAT.	INST.	TOTAL	MAT.	INST.	TOTAL	MAT.	INST.	TOTAL	MAT.	INST.	TOTAL	MAT.	INST.	TOTAL
A	Substructure	93.9	128.1	113.6	88.3	135.6	115.6	95.2	123.3	111.4	113.0	141.3	129.3	108.7	137.9	125.6
B10	Shell: Superstructure	105.1	124.0	112.2	98.4	126.0	108.7	94.8	120.3	104.3	111.1	136.4	120.5	102.6	135.9	115.0
B20	Exterior Closure	91.8	136.6	111.8	107.5	143.7	123.7	98.4	131.9	113.3	126.2	153.3	138.3	112.9	151.2	130.0
B30	Roofing	109.3	136.3	120.1	122.4	142.3	130.4	108.3	121.2	113.5	116.7	155.7	132.3	108.2	155.8	127.3
C	Interior Construction	101.2	134.3	115.0	104.8	150.7	123.9	103.1	125.9	112.6	104.3	164.7	129.5	95.4	164.4	124.1
D10	Services: Conveying	100.0	116.8	105.0	100.0	128.1	108.5	100.0	114.3	104.3	100.0	122.5	106.8	100.0	128.2	108.5
D20 - 40	Mechanical	99.9	130.9	112.1	100.1	131.4	112.4	100.0	129.4	111.5	100.1	182.8	132.6	99.9	170.5	127.7
D50	Electrical	91.3	114.4	102.7	94.8	120.8	107.6	103.8	104.3	104.0	99.2	182.8	140.4	100.7	176.6	138.1
E	Equipment & Furnishings	100.0	138.1	102.2	100.0	158.2	103.3	100.0	125.8	101.5	100.0	173.9	104.3	100.0	173.5	104.2
G	Site Work	97.6	103.7	101.6	98.7	111.6	107.3	105.7	108.1	107.2	118.1	113.2	114.9	130.6	97.7	108.7
A-G	**WEIGHTED AVERAGE**	99.3	126.6	110.8	100.8	133.0	114.4	99.7	121.7	109.0	107.4	158.4	129.0	102.6	153.5	124.2

DIV. NO.	BUILDING SYSTEMS	CALIFORNIA									COLORADO					
		SANTA BARBARA			STOCKTON			VALLEJO			COLORADO SPRINGS			DENVER		
		MAT.	INST.	TOTAL	MAT.	INST.	TOTAL	MAT.	INST.	TOTAL	MAT.	INST.	TOTAL	MAT.	INST.	TOTAL
A	Substructure	97.3	127.8	114.9	96.1	133.3	117.5	94.2	139.9	120.6	111.5	72.3	88.9	115.6	77.1	93.4
B10	Shell: Superstructure	98.4	123.7	107.8	102.3	128.5	112.0	102.1	130.2	112.6	100.4	73.0	90.2	103.6	74.6	92.8
B20	Exterior Closure	94.4	135.0	112.5	101.0	140.9	118.8	96.0	151.7	120.8	102.1	64.5	85.3	102.9	66.2	86.5
B30	Roofing	106.1	134.2	117.4	112.3	138.9	123.0	113.2	153.8	129.5	109.6	70.0	93.7	107.2	72.0	93.1
C	Interior Construction	92.2	133.9	109.6	101.7	151.2	122.3	105.7	163.7	129.8	98.2	65.0	84.4	102.3	68.5	88.2
D10	Services: Conveying	100.0	117.0	105.1	100.0	124.1	107.2	100.0	127.9	108.4	100.0	84.6	95.4	100.0	88.8	96.6
D20 - 40	Mechanical	100.0	131.0	112.2	99.9	131.7	112.4	100.2	147.2	118.7	100.1	72.2	89.1	99.9	73.8	89.7
D50	Electrical	94.5	113.7	103.9	97.5	114.3	105.8	92.4	127.5	109.7	103.2	73.1	88.4	105.0	80.4	92.9
E	Equipment & Furnishings	100.0	137.9	102.2	100.0	157.9	103.3	100.0	171.0	104.1	100.0	63.9	97.9	100.0	67.6	98.1
G	Site Work	99.4	103.6	102.2	100.5	102.7	102.0	100.4	111.3	107.7	100.2	86.6	91.1	105.7	102.5	103.6
A-G	WEIGHTED AVERAGE	97.2	126.1	109.4	100.7	131.3	113.7	100.2	141.6	117.7	101.2	71.5	88.6	103.0	75.4	91.3

DIV. NO.	BUILDING SYSTEMS	COLORADO												CONNECTICUT		
		FORT COLLINS			GRAND JUNCTION			GREELEY			PUEBLO			BRIDGEPORT		
		MAT.	INST.	TOTAL	MAT.	INST.	TOTAL	MAT.	INST.	TOTAL	MAT.	INST.	TOTAL	MAT.	INST.	TOTAL
A	Substructure	120.2	73.9	93.5	112.2	74.6	90.5	105.3	73.9	87.2	103.4	71.8	85.2	110.0	118.7	115.0
B10	Shell: Superstructure	101.1	73.5	90.8	107.5	74.1	95.0	98.0	73.5	88.8	106.9	73.6	94.5	100.5	119.6	107.6
B20	Exterior Closure	109.4	65.3	89.7	120.4	64.2	95.3	99.9	64.9	84.3	99.5	62.8	83.1	100.2	128.9	113.0
B30	Roofing	109.2	70.6	93.7	111.1	70.1	94.7	108.4	70.6	93.2	110.6	68.3	93.6	99.7	121.5	108.5
C	Interior Construction	96.9	68.1	84.9	104.1	67.6	88.9	96.2	67.5	84.3	98.8	65.8	85.1	96.1	119.2	105.7
D10	Services: Conveying	100.0	84.0	95.2	100.0	85.3	95.6	100.0	84.0	95.2	100.0	85.4	95.6	100.0	114.0	104.2
D20 - 40	Mechanical	100.0	71.7	88.9	100.0	74.8	90.1	100.0	71.7	88.9	100.0	72.2	89.1	100.1	118.6	107.4
D50	Electrical	99.6	80.6	90.2	95.5	54.6	75.3	99.6	80.6	90.2	96.4	63.3	80.0	93.0	105.0	98.9
E	Equipment & Furnishings	100.0	67.5	98.1	100.0	66.2	98.1	100.0	67.5	98.1	100.0	63.2	97.9	100.0	113.6	100.8
G	Site Work	111.6	91.0	97.9	135.4	92.0	106.5	97.6	91.0	93.2	126.3	84.5	98.5	104.5	97.9	100.1
A-G	WEIGHTED AVERAGE	102.2	73.5	90.1	105.7	70.5	90.8	99.3	73.4	88.4	101.8	70.0	88.3	99.3	117.0	106.8

DIV. NO.	BUILDING SYSTEMS	CONNECTICUT														
		BRISTOL			HARTFORD			NEW BRITAIN			NEW HAVEN			NORWALK		
		MAT.	INST.	TOTAL	MAT.	INST.	TOTAL	MAT.	INST.	TOTAL	MAT.	INST.	TOTAL	MAT.	INST.	TOTAL
A	Substructure	105.6	118.6	113.1	106.8	120.0	114.4	106.7	118.6	113.6	107.8	118.2	113.8	108.9	118.4	114.4
B10	Shell: Superstructure	99.6	119.5	107.0	103.9	119.5	109.7	97.0	119.5	105.4	97.4	119.2	105.5	100.3	119.4	107.4
B20	Exterior Closure	96.7	128.9	111.0	95.5	128.8	110.4	97.4	128.9	111.4	111.9	128.9	119.4	96.6	128.1	110.7
B30	Roofing	99.9	118.1	107.2	104.5	118.7	110.2	99.9	118.2	107.2	100.0	118.3	107.3	99.9	121.3	108.5
C	Interior Construction	96.1	119.1	105.7	97.3	119.6	106.6	96.1	119.1	105.7	96.1	119.2	105.7	96.1	118.8	105.6
D10	Services: Conveying	100.0	114.0	104.2	100.0	114.4	104.3	100.0	114.0	104.2	100.0	114.0	104.2	100.0	114.0	104.2
D20 - 40	Mechanical	100.1	118.6	107.4	100.0	118.6	107.3	100.1	118.6	107.4	100.1	118.6	107.4	100.1	118.6	107.4
D50	Electrical	93.0	103.7	98.3	92.7	109.4	100.9	93.0	103.7	98.3	92.9	106.0	99.4	93.0	104.8	98.8
E	Equipment & Furnishings	100.0	113.5	100.8	100.0	113.9	100.8	100.0	113.5	100.8	100.0	113.6	100.8	100.0	113.6	100.8
G	Site Work	103.3	97.9	99.7	101.5	102.7	102.3	103.6	97.9	99.8	103.5	98.6	100.2	104.2	97.8	99.9
A-G	WEIGHTED AVERAGE	98.5	116.7	106.2	99.6	118.0	107.3	98.1	116.7	106.0	99.9	117.0	107.1	98.8	116.7	106.4

DIV. NO.	BUILDING SYSTEMS	CONNECTICUT						D.C.			DELAWARE			FLORIDA		
		STAMFORD			WATERBURY			WASHINGTON			WILMINGTON			DAYTONA BEACH		
		MAT.	INST.	TOTAL	MAT.	INST.	TOTAL	MAT.	INST.	TOTAL	MAT.	INST.	TOTAL	MAT.	INST.	TOTAL
A	Substructure	110.1	118.6	115.0	110.0	118.6	115.0	102.8	82.5	91.1	101.2	105.8	103.8	95.9	68.5	80.1
B10	Shell: Superstructure	100.5	119.7	107.7	100.5	119.6	107.6	103.6	87.1	97.5	102.6	113.7	106.7	100.1	77.1	91.5
B20	Exterior Closure	96.9	128.1	110.9	96.9	128.9	111.2	103.2	87.3	96.1	92.5	102.5	97.0	87.9	62.8	76.7
B30	Roofing	99.9	121.3	108.5	99.9	118.6	107.4	104.2	85.2	96.5	105.2	112.3	108.1	101.3	66.4	87.3
C	Interior Construction	96.1	119.2	105.7	96.0	119.5	105.8	99.7	73.0	88.6	90.6	101.4	95.1	91.9	60.7	78.9
D10	Services: Conveying	100.0	114.0	104.2	100.0	114.0	104.2	100.0	97.2	99.2	100.0	104.5	101.4	100.0	84.2	95.3
D20 - 40	Mechanical	100.1	118.6	107.4	100.1	118.6	107.4	100.0	88.4	95.4	100.1	119.4	107.7	99.9	75.6	90.4
D50	Electrical	93.0	150.1	121.1	92.6	107.1	99.7	97.7	100.0	98.8	95.1	109.7	102.3	96.7	58.9	78.1
E	Equipment & Furnishings	100.0	115.7	100.9	100.0	113.7	100.8	100.0	73.5	98.5	100.0	97.9	99.9	100.0	60.0	97.7
G	Site Work	104.9	97.9	100.2	104.2	97.9	100.0	103.4	97.5	99.5	103.1	107.6	106.1	116.3	84.4	95.1
A-G	WEIGHTED AVERAGE	99.0	123.2	109.2	98.9	117.2	106.6	101.1	87.5	95.4	97.9	109.7	102.9	97.2	69.5	85.5

DIV. NO.	BUILDING SYSTEMS	FLORIDA														
		FORT LAUDERDALE			JACKSONVILLE			MELBOURNE			MIAMI			ORLANDO		
		MAT.	INST.	TOTAL	MAT.	INST.	TOTAL	MAT.	INST.	TOTAL	MAT.	INST.	TOTAL	MAT.	INST.	TOTAL
A	Substructure	95.6	64.0	77.4	96.5	68.0	80.1	108.6	69.2	85.8	95.5	66.0	78.5	110.3	70.3	87.2
B10	Shell: Superstructure	96.4	75.7	88.7	99.0	76.5	90.6	109.4	78.3	97.8	96.7	75.8	88.9	99.3	77.6	91.2
B20	Exterior Closure	89.7	57.5	75.3	87.8	60.9	75.8	90.1	62.9	78.0	89.0	58.4	75.3	91.5	62.9	78.7
B30	Roofing	106.4	60.5	88.0	101.7	63.4	86.3	101.9	64.9	87.0	106.2	61.7	88.3	109.3	67.1	92.4
C	Interior Construction	92.3	59.3	78.6	91.9	59.9	78.6	91.3	62.6	79.3	93.5	59.6	79.4	95.3	60.9	81.0
D10	Services: Conveying	100.0	83.4	95.0	100.0	80.5	94.1	100.0	84.2	95.3	100.0	84.1	95.2	100.0	84.5	95.3
D20 - 40	Mechanical	100.0	66.7	86.9	99.9	63.1	85.5	99.9	74.2	89.8	100.0	63.8	85.8	100.0	56.4	82.9
D50	Electrical	94.7	67.4	81.3	96.4	61.9	79.4	97.7	62.7	80.5	98.4	77.9	88.3	98.4	63.1	81.0
E	Equipment & Furnishings	100.0	60.5	97.7	100.0	58.8	97.6	100.0	60.2	97.7	100.0	61.5	97.8	100.0	60.3	97.7
G	Site Work	94.9	72.5	80.0	116.3	84.6	95.2	124.6	84.6	98.0	96.3	78.2	84.3	115.0	89.2	97.8
A-G	WEIGHTED AVERAGE	96.1	66.4	83.6	96.9	66.6	84.1	100.1	70.3	87.5	96.7	68.0	84.6	99.0	66.6	85.3

DIV. NO.	BUILDING SYSTEMS	FLORIDA														
		PANAMA CITY			PENSACOLA			ST. PETERSBURG			TALLAHASSEE			TAMPA		
		MAT.	INST.	TOTAL	MAT.	INST.	TOTAL	MAT.	INST.	TOTAL	MAT.	INST.	TOTAL	MAT.	INST.	TOTAL
A	Substructure	101.7	71.0	84.0	116.4	69.1	89.1	103.1	70.0	84.0	96.0	70.0	81.0	101.4	71.9	84.4
B10	Shell: Superstructure	100.7	79.9	92.9	104.6	78.5	94.9	99.9	80.3	92.6	100.1	76.7	91.4	98.9	82.0	92.6
B20	Exterior Closure	96.2	62.7	81.2	102.5	62.2	84.5	108.7	61.4	87.6	87.0	62.2	76.0	91.1	63.5	78.8
B30	Roofing	102.0	65.0	87.1	101.9	63.8	86.6	106.3	63.2	89.0	96.6	64.9	83.9	106.5	64.5	89.7
C	Interior Construction	91.5	66.5	81.1	91.3	61.5	78.9	92.4	60.7	79.2	96.0	61.4	81.6	93.6	63.5	81.0
D10	Services: Conveying	100.0	82.5	94.8	100.0	82.2	94.6	100.0	82.2	94.7	100.0	84.2	95.2	100.0	82.2	94.7
D20 - 40	Mechanical	99.9	63.2	85.5	99.9	62.7	85.3	100.0	59.9	84.2	100.0	65.5	86.5	100.0	59.9	84.3
D50	Electrical	95.5	58.6	77.3	99.0	50.7	75.2	95.0	60.6	78.0	103.7	58.6	81.4	94.7	63.1	79.1
E	Equipment & Furnishings	100.0	66.1	98.0	100.0	60.5	97.7	100.0	59.4	97.7	100.0	61.0	97.8	100.0	63.5	97.9
G	Site Work	129.5	84.7	99.7	130.2	84.4	99.8	109.1	84.5	92.7	109.4	89.3	96.0	109.2	86.5	94.1
A-G	WEIGHTED AVERAGE	98.6	68.3	85.8	101.2	65.8	86.2	99.7	66.8	85.8	98.2	67.7	85.3	97.6	68.4	85.3

City Cost Indexes

RJ1030-010 Building Systems

DIV. NO.	BUILDING SYSTEMS	GEORGIA														
		ALBANY			ATLANTA			AUGUSTA			COLUMBUS			MACON		
		MAT.	INST.	TOTAL	MAT.	INST.	TOTAL	MAT.	INST.	TOTAL	MAT.	INST.	TOTAL	MAT.	INST.	TOTAL
A	Substructure	95.5	70.7	81.2	107.3	77.3	90.0	100.7	76.2	86.6	95.2	70.6	81.0	95.0	73.6	82.7
B10	Shell: Superstructure	103.6	84.1	96.3	98.4	78.4	90.9	95.9	77.9	89.2	103.2	84.2	96.1	99.1	84.2	93.5
B20	Exterior Closure	84.8	70.8	78.5	91.3	70.4	82.0	85.6	70.3	78.8	84.7	71.1	78.6	88.7	71.2	80.9
B30	Roofing	100.9	66.9	87.2	96.8	73.7	87.5	95.2	70.9	85.4	100.8	67.9	87.6	99.2	69.9	87.4
C	Interior Construction	87.6	69.4	80.0	97.4	74.4	87.8	92.6	74.5	85.1	87.5	68.8	79.7	83.1	69.4	77.4
D10	Services: Conveying	100.0	88.4	96.5	100.0	89.0	96.7	100.0	88.6	96.6	100.0	88.4	96.5	100.0	88.5	96.5
D20 - 40	Mechanical	100.0	68.9	87.8	100.0	69.4	88.0	100.1	69.0	87.9	100.1	66.8	87.0	100.1	68.2	87.6
D50	Electrical	95.5	61.9	79.0	97.6	71.5	84.7	98.3	67.3	83.1	95.7	67.7	81.9	94.2	62.7	78.6
E	Equipment & Furnishings	100.0	66.0	98.0	100.0	73.5	98.5	100.0	74.6	98.5	100.0	66.3	98.1	100.0	66.3	98.1
G	Site Work	107.8	76.0	86.6	98.8	96.0	96.9	95.1	91.9	93.0	107.6	76.1	86.6	108.4	90.5	96.5
A-G	WEIGHTED AVERAGE	96.5	71.9	86.1	98.2	75.0	88.4	96.0	73.9	86.6	96.5	72.2	86.2	95.2	73.1	85.8

DIV. NO.	BUILDING SYSTEMS	GEORGIA						HAWAII						IDAHO		
		SAVANNAH			VALDOSTA			HONOLULU			STATES & POSS., GUAM			BOISE		
		MAT.	INST.	TOTAL	MAT.	INST.	TOTAL	MAT.	INST.	TOTAL	MAT.	INST.	TOTAL	MAT.	INST.	TOTAL
A	Substructure	99.5	73.9	84.7	94.9	60.0	74.7	146.6	120.0	131.2	174.9	72.7	115.9	93.4	87.9	90.2
B10	Shell: Superstructure	101.2	84.9	95.1	102.1	74.1	91.6	128.7	114.8	123.5	148.2	69.9	119.0	106.5	83.9	98.1
B20	Exterior Closure	82.3	71.2	77.3	92.1	76.7	85.2	129.5	121.8	126.1	148.1	36.9	98.4	111.6	84.5	99.5
B30	Roofing	98.7	69.6	87.0	101.1	63.1	85.8	147.0	119.5	135.9	151.5	62.0	115.6	101.5	87.0	95.7
C	Interior Construction	93.8	73.3	85.3	84.9	50.2	70.4	126.2	127.6	126.8	152.2	47.7	108.7	94.3	78.1	87.6
D10	Services: Conveying	100.0	88.7	96.6	100.0	88.4	96.5	100.0	112.1	103.6	100.0	68.8	90.6	100.0	89.7	96.9
D20 - 40	Mechanical	100.1	64.8	86.3	100.1	68.5	87.7	100.4	112.3	105.1	102.6	33.5	75.5	100.1	74.2	89.9
D50	Electrical	99.1	63.7	81.7	94.1	57.8	76.2	109.8	124.5	117.1	158.1	35.9	97.9	97.2	68.2	82.9
E	Equipment & Furnishings	100.0	73.0	98.4	100.0	34.6	96.2	100.0	124.6	101.4	100.0	44.9	96.8	100.0	79.6	98.8
G	Site Work	107.3	81.4	90.1	116.9	76.0	89.7	151.3	106.7	121.6	181.1	97.3	125.4	88.3	91.9	90.7
A-G	WEIGHTED AVERAGE	97.3	72.7	86.9	96.7	66.2	83.8	119.1	118.5	118.9	137.1	51.4	100.9	101.0	79.9	92.1

DIV. NO.	BUILDING SYSTEMS	IDAHO									ILLINOIS					
		LEWISTON			POCATELLO			TWIN FALLS			CHICAGO			DECATUR		
		MAT.	INST.	TOTAL	MAT.	INST.	TOTAL	MAT.	INST.	TOTAL	MAT.	INST.	TOTAL	MAT.	INST.	TOTAL
A	Substructure	103.1	84.7	92.5	95.7	87.6	91.0	98.3	81.9	88.8	118.2	144.9	133.6	94.4	110.1	103.5
B10	Shell: Superstructure	102.9	86.0	96.6	113.6	83.4	102.3	114.1	80.1	101.4	100.4	151.6	119.5	102.8	117.8	108.4
B20	Exterior Closure	118.9	83.9	103.3	108.4	82.2	96.7	117.2	78.0	99.7	97.4	163.1	126.7	79.9	121.9	98.7
B30	Roofing	160.2	83.0	129.2	101.1	73.5	90.0	101.9	79.9	93.1	95.1	150.0	117.2	95.9	112.6	102.6
C	Interior Construction	131.7	79.4	109.9	95.2	76.2	87.3	96.9	73.1	87.0	98.7	163.3	125.6	95.3	118.4	104.9
D10	Services: Conveying	100.0	94.3	98.3	100.0	89.7	96.9	100.0	87.2	96.2	100.0	125.5	107.7	100.0	107.5	102.3
D20 - 40	Mechanical	100.9	85.1	94.6	100.0	73.6	89.6	100.0	72.2	89.1	99.9	137.4	114.6	99.9	98.8	99.5
D50	Electrical	86.9	78.2	82.6	94.4	66.0	80.4	89.6	70.9	80.4	97.9	136.2	116.8	95.3	103.7	99.5
E	Equipment & Furnishings	100.0	79.6	98.8	100.0	79.4	98.8	100.0	76.1	98.6	100.0	158.0	103.3	100.0	116.7	101.0
G	Site Work	95.1	87.0	89.7	89.1	91.9	90.9	96.1	91.6	93.1	104.5	104.6	104.6	96.7	96.7	96.7
A-G	WEIGHTED AVERAGE	108.2	83.5	97.8	102.0	78.4	92.0	103.2	76.8	92.1	100.1	145.6	119.3	96.6	110.1	102.3

DIV. NO.	BUILDING SYSTEMS	ILLINOIS														
		EAST ST. LOUIS			JOLIET			PEORIA			ROCKFORD			SPRINGFIELD		
		MAT.	INST.	TOTAL	MAT.	INST.	TOTAL	MAT.	INST.	TOTAL	MAT.	INST.	TOTAL	MAT.	INST.	TOTAL
A	Substructure	91.1	110.7	102.4	110.6	136.8	125.7	95.8	110.0	104.0	96.7	121.3	110.9	92.8	108.5	101.9
B10	Shell: Superstructure	99.4	124.6	108.8	97.3	141.7	113.9	97.4	119.5	105.7	97.7	135.2	111.7	100.3	115.6	106.0
B20	Exterior Closure	74.2	123.5	96.2	95.0	155.8	122.1	97.2	121.6	108.1	87.5	141.5	111.6	85.2	121.2	101.3
B30	Roofing	90.6	109.7	98.2	99.6	143.1	117.0	96.5	112.1	102.8	98.9	127.7	110.5	98.2	113.8	104.4
C	Interior Construction	89.1	114.7	99.8	95.7	158.1	121.7	93.6	118.5	104.0	93.6	129.8	108.7	98.1	116.0	105.6
D10	Services: Conveying	100.0	107.9	102.4	100.0	118.0	105.4	100.0	103.3	101.0	100.0	112.8	103.9	100.0	107.7	102.3
D20 - 40	Mechanical	99.9	99.2	99.6	100.0	130.0	111.8	100.0	100.2	100.0	100.1	116.1	106.4	99.9	103.7	101.4
D50	Electrical	92.7	99.7	96.2	96.9	133.3	114.8	97.7	91.5	94.6	98.0	125.6	111.6	98.1	88.7	93.5
E	Equipment & Furnishings	100.0	111.2	100.6	100.0	157.0	103.3	100.0	112.8	100.7	100.0	124.6	101.4	100.0	113.4	100.8
G	Site Work	105.0	97.2	99.8	100.0	97.4	98.3	98.5	95.7	96.6	98.1	96.6	97.1	100.5	100.3	100.3
A-G	WEIGHTED AVERAGE	93.8	110.4	100.9	98.3	138.8	115.4	97.6	108.8	102.3	96.6	125.5	108.8	97.5	108.3	102.1

DIV. NO.	BUILDING SYSTEMS	INDIANA														
		ANDERSON			BLOOMINGTON			EVANSVILLE			FORT WAYNE			GARY		
		MAT.	INST.	TOTAL	MAT.	INST.	TOTAL	MAT.	INST.	TOTAL	MAT.	INST.	TOTAL	MAT.	INST.	TOTAL
A	Substructure	103.5	80.1	90.0	97.4	82.1	88.6	96.0	88.4	91.6	107.4	78.2	90.6	106.6	107.8	107.3
B10	Shell: Superstructure	99.1	83.3	93.2	99.2	78.0	91.3	92.9	81.8	88.8	100.0	81.1	93.0	99.7	109.2	103.2
B20	Exterior Closure	86.7	75.0	81.5	95.8	74.7	86.4	94.1	77.4	86.6	88.1	72.2	81.0	87.3	112.1	98.4
B30	Roofing	109.6	75.4	95.9	95.9	79.1	89.1	100.1	82.9	93.2	109.3	77.5	96.5	108.0	106.7	107.5
C	Interior Construction	91.1	75.8	84.8	93.7	82.2	88.9	90.1	77.3	84.8	91.0	72.8	83.4	90.6	111.0	99.1
D10	Services: Conveying	100.0	88.6	96.6	100.0	88.9	96.7	100.0	93.0	97.9	100.0	89.0	96.7	100.0	107.7	102.3
D20 - 40	Mechanical	100.0	77.0	91.0	99.7	79.6	91.8	99.9	78.6	91.5	100.0	72.2	89.1	100.0	106.2	102.4
D50	Electrical	87.7	83.2	85.5	99.9	86.6	93.3	95.7	81.4	88.7	88.4	74.9	81.7	99.5	110.1	104.7
E	Equipment & Furnishings	100.0	77.5	98.7	100.0	83.4	99.0	100.0	77.2	98.7	100.0	73.5	98.5	100.0	109.6	100.6
G	Site Work	100.4	88.8	92.7	87.7	87.8	87.8	93.0	115.3	107.8	102.3	88.6	93.2	101.2	92.7	95.6
A-G	WEIGHTED AVERAGE	96.0	79.8	89.1	97.8	81.1	90.7	95.5	82.7	90.1	96.6	76.3	88.0	97.5	108.1	102.0

DIV. NO.	BUILDING SYSTEMS	INDIANA												IOWA		
		INDIANAPOLIS			MUNCIE			SOUTH BEND			TERRE HAUTE			CEDAR RAPIDS		
		MAT.	INST.	TOTAL	MAT.	INST.	TOTAL	MAT.	INST.	TOTAL	MAT.	INST.	TOTAL	MAT.	INST.	TOTAL
A	Substructure	101.3	86.6	92.8	101.6	79.6	88.9	104.8	82.2	91.7	94.1	86.8	89.9	104.3	85.1	93.2
B10	Shell: Superstructure	96.3	80.3	90.3	101.2	83.0	94.4	102.3	91.3	98.2	93.1	81.6	88.8	93.6	88.7	91.7
B20	Exterior Closure	94.1	78.5	87.1	91.1	75.1	84.0	89.0	76.3	83.3	102.0	74.0	89.5	94.0	80.4	87.9
B30	Roofing	98.8	82.9	92.4	98.3	76.9	89.7	103.3	79.6	93.8	100.2	80.4	92.2	105.7	80.7	95.7
C	Interior Construction	99.4	84.4	93.2	88.4	75.5	83.0	92.3	77.7	86.2	90.3	76.7	84.6	99.0	82.9	92.3
D10	Services: Conveying	100.0	94.7	98.4	100.0	87.5	96.2	100.0	90.5	97.1	100.0	90.3	97.1	100.0	94.7	98.4
D20 - 40	Mechanical	100.1	79.9	92.2	99.7	76.9	90.7	99.9	75.0	90.1	99.9	76.5	90.7	100.0	81.2	92.6
D50	Electrical	101.9	86.7	94.4	91.4	74.0	82.9	101.3	83.9	92.7	94.1	83.5	88.8	98.8	80.1	89.5
E	Equipment & Furnishings	100.0	85.9	99.2	100.0	77.3	98.7	100.0	75.3	98.6	100.0	76.4	98.6	100.0	83.1	99.0
G	Site Work	101.1	91.4	94.7	88.1	87.6	87.8	99.0	93.7	95.5	94.8	115.6	108.7	99.9	91.2	94.1
A-G	WEIGHTED AVERAGE	98.7	83.2	92.1	96.2	78.2	88.6	98.4	81.8	91.4	96.2	81.7	90.1	98.0	83.7	92.0

DIV. NO.	BUILDING SYSTEMS	IOWA														
		COUNCIL BLUFFS			DAVENPORT			DES MOINES			DUBUQUE			SIOUX CITY		
		MAT.	INST.	TOTAL	MAT.	INST.	TOTAL	MAT.	INST.	TOTAL	MAT.	INST.	TOTAL	MAT.	INST.	TOTAL
A	Substructure	106.9	79.4	91.0	101.5	95.9	98.2	97.8	94.0	95.6	100.7	83.5	90.8	105.1	86.7	94.5
B10	Shell: Superstructure	96.9	85.2	92.5	93.0	101.2	96.1	94.8	94.5	94.7	91.0	87.6	89.7	93.4	93.0	93.3
B20	Exterior Closure	95.7	77.6	87.6	92.8	93.2	93.0	86.4	89.7	87.9	93.2	71.6	83.6	91.5	72.7	83.1
B30	Roofing	105.0	74.6	92.8	105.1	92.4	100.0	98.7	88.0	94.4	105.3	78.8	94.7	105.1	74.6	92.8
C	Interior Construction	94.4	74.0	85.9	97.2	94.7	96.2	97.1	89.7	94.0	96.3	79.6	89.4	98.0	75.5	88.6
D10	Services: Conveying	100.0	93.8	98.1	100.0	97.2	99.2	100.0	96.6	99.0	100.0	93.7	98.1	100.0	93.7	98.1
D20 - 40	Mechanical	100.0	72.8	89.4	100.0	93.9	97.6	99.8	86.8	94.7	100.0	75.6	90.4	100.0	81.2	92.6
D50	Electrical	103.9	82.7	93.4	96.9	87.6	92.3	105.1	84.3	94.8	102.5	76.9	89.9	98.8	74.0	86.6
E	Equipment & Furnishings	100.0	72.5	98.4	100.0	94.6	99.7	100.0	88.1	99.3	100.0	81.4	98.9	100.0	74.0	98.5
G	Site Work	102.9	88.3	93.2	98.4	94.5	95.8	99.0	99.3	99.2	98.1	88.6	91.8	108.4	93.1	98.2
A-G	**WEIGHTED AVERAGE**	98.9	79.2	90.6	97.1	94.5	96.0	97.2	90.2	94.3	97.2	79.9	89.9	97.7	81.5	90.9

DIV. NO.	BUILDING SYSTEMS	IOWA			KANSAS											
		WATERLOO			DODGE CITY			KANSAS CITY			SALINA			TOPEKA		
		MAT.	INST.	TOTAL	MAT.	INST.	TOTAL	MAT.	INST.	TOTAL	MAT.	INST.	TOTAL	MAT.	INST.	TOTAL
A	Substructure	109.5	80.0	92.5	112.5	80.7	94.2	92.6	97.3	95.3	100.8	79.9	88.7	96.5	83.2	88.8
B10	Shell: Superstructure	94.5	84.6	90.8	97.7	87.5	93.9	101.4	103.0	102.0	95.1	87.4	92.2	99.0	89.8	95.5
B20	Exterior Closure	90.5	76.5	84.3	104.3	62.2	85.5	90.7	98.7	94.3	104.6	58.6	84.1	89.0	67.9	79.6
B30	Roofing	105.6	78.1	94.6	97.7	68.7	86.1	90.3	99.5	94.0	97.1	63.2	83.5	94.0	76.0	86.8
C	Interior Construction	92.7	70.7	83.5	90.9	61.9	78.8	88.0	98.1	92.2	90.2	62.7	78.7	99.2	66.9	85.8
D10	Services: Conveying	100.0	94.5	98.3	100.0	89.8	96.9	100.0	94.0	98.2	100.0	88.5	96.6	100.0	82.3	94.7
D20 - 40	Mechanical	99.9	80.1	92.1	100.0	71.1	88.7	99.9	99.1	99.6	100.0	70.5	88.5	100.0	73.2	89.5
D50	Electrical	95.4	61.6	78.7	92.9	67.9	80.6	106.2	98.1	102.2	92.7	73.8	83.4	105.1	71.9	88.7
E	Equipment & Furnishings	100.0	64.0	97.9	100.0	57.5	97.6	100.0	98.4	99.9	100.0	60.7	97.7	100.0	64.9	98.0
G	Site Work	106.8	92.5	97.3	110.6	91.3	97.8	96.7	90.0	92.2	101.1	91.9	94.9	99.1	94.1	95.8
A-G	**WEIGHTED AVERAGE**	96.8	77.4	88.6	98.6	73.3	87.9	97.3	98.6	97.8	97.2	73.5	87.2	98.6	76.5	89.3

DIV. NO.	BUILDING SYSTEMS	KANSAS			KENTUCKY											
		WICHITA			BOWLING GREEN			LEXINGTON			LOUISVILLE			OWENSBORO		
		MAT.	INST.	TOTAL	MAT.	INST.	TOTAL	MAT.	INST.	TOTAL	MAT.	INST.	TOTAL	MAT.	INST.	TOTAL
A	Substructure	96.6	78.4	86.1	87.3	78.4	82.2	93.4	82.4	87.0	90.4	80.8	84.8	90.2	87.5	88.7
B10	Shell: Superstructure	96.2	84.8	91.9	94.8	79.7	89.2	93.6	82.8	89.6	94.4	81.9	89.8	88.7	82.1	86.3
B20	Exterior Closure	92.6	57.6	77.0	94.8	70.6	84.0	80.5	72.4	76.9	80.9	72.5	77.1	101.1	79.1	91.3
B30	Roofing	96.2	61.9	82.4	87.5	79.7	84.3	104.3	75.3	92.6	101.0	76.3	91.1	100.2	76.4	90.6
C	Interior Construction	96.5	58.9	80.8	88.8	75.3	83.2	85.8	72.3	80.2	89.2	76.0	83.7	88.0	74.8	82.5
D10	Services: Conveying	100.0	89.1	96.7	100.0	84.9	95.5	100.0	94.9	98.5	100.0	91.9	97.6	100.0	95.3	98.6
D20 - 40	Mechanical	99.8	71.0	88.5	99.9	77.0	90.9	100.0	76.4	90.8	99.9	80.0	92.1	99.9	77.1	91.0
D50	Electrical	95.9	73.8	85.0	94.6	74.7	84.8	92.5	74.7	83.7	96.4	77.9	87.3	94.1	73.7	84.0
E	Equipment & Furnishings	100.0	55.1	97.4	100.0	80.0	98.8	100.0	71.1	98.3	100.0	80.2	98.9	100.0	75.1	98.6
G	Site Work	97.2	98.1	97.8	79.2	89.5	86.0	95.8	92.8	93.8	88.1	94.5	92.4	93.2	115.5	108.0
A-G	**WEIGHTED AVERAGE**	97.0	72.6	86.7	94.6	77.2	87.3	93.1	78.1	86.7	93.9	79.6	87.8	94.7	81.0	88.9

DIV. NO.	BUILDING SYSTEMS	LOUISIANA														
		ALEXANDRIA			BATON ROUGE			LAKE CHARLES			MONROE			NEW ORLEANS		
		MAT.	INST.	TOTAL	MAT.	INST.	TOTAL	MAT.	INST.	TOTAL	MAT.	INST.	TOTAL	MAT.	INST.	TOTAL
A	Substructure	88.4	67.5	76.3	91.5	74.5	81.7	93.3	70.1	79.9	88.2	66.7	75.8	93.0	72.9	81.4
B10	Shell: Superstructure	91.2	66.4	82.0	95.7	72.2	86.9	89.0	67.7	81.0	91.1	66.0	81.7	97.5	64.8	85.3
B20	Exterior Closure	94.1	64.7	81.0	87.5	67.5	78.6	91.6	66.2	80.3	91.9	63.2	79.1	100.3	62.7	83.5
B30	Roofing	99.1	65.6	85.7	97.7	69.8	86.5	97.1	67.4	85.2	99.1	64.7	85.3	94.9	68.4	84.3
C	Interior Construction	97.1	60.3	81.8	95.0	70.4	84.7	96.3	66.5	83.9	97.0	60.1	81.6	97.4	67.7	85.0
D10	Services: Conveying	100.0	81.7	94.5	100.0	85.7	95.7	100.0	85.4	95.6	100.0	81.6	94.5	100.0	88.6	96.6
D20 - 40	Mechanical	100.0	63.2	85.6	99.9	64.1	85.9	100.2	64.5	86.2	100.0	61.9	85.0	100.1	62.5	85.4
D50	Electrical	94.3	61.4	78.1	101.7	58.0	80.2	97.7	65.9	82.0	95.9	57.3	76.9	102.1	70.6	86.6
E	Equipment & Furnishings	100.0	58.9	97.6	100.0	72.2	98.4	100.0	67.9	98.2	100.0	58.6	97.6	100.0	70.8	98.3
G	Site Work	100.8	87.6	92.0	102.4	91.2	94.9	103.5	87.0	92.5	100.8	87.5	92.0	103.8	94.1	97.3
A-G	**WEIGHTED AVERAGE**	95.9	65.6	83.1	96.6	69.3	85.1	95.6	68.2	84.1	95.8	64.4	82.5	99.0	68.5	86.1

DIV. NO.	BUILDING SYSTEMS	LOUISIANA			MAINE											
		SHREVEPORT			AUGUSTA			BANGOR			LEWISTON			PORTLAND		
		MAT.	INST.	TOTAL	MAT.	INST.	TOTAL	MAT.	INST.	TOTAL	MAT.	INST.	TOTAL	MAT.	INST.	TOTAL
A	Substructure	92.2	68.3	78.4	92.5	93.5	93.1	78.6	91.9	86.3	84.0	92.0	88.6	90.4	93.4	92.2
B10	Shell: Superstructure	95.7	66.0	84.6	102.7	91.1	98.4	90.3	90.8	90.5	94.3	90.8	93.0	100.6	90.7	96.9
B20	Exterior Closure	88.7	63.1	77.3	99.9	92.4	96.6	108.6	92.0	101.2	97.9	92.0	95.2	98.1	92.0	95.4
B30	Roofing	97.5	65.7	84.7	111.2	100.2	106.8	109.1	99.6	105.3	108.9	99.6	105.1	113.8	100.2	108.4
C	Interior Construction	96.8	60.7	81.7	99.9	83.6	93.1	93.3	81.8	88.5	96.2	81.8	90.2	96.6	82.0	90.5
D10	Services: Conveying	100.0	85.6	95.7	100.0	101.1	100.3	100.0	103.2	101.0	100.0	103.2	101.0	100.0	103.7	101.1
D20 - 40	Mechanical	99.9	62.8	85.3	100.0	74.3	89.9	100.3	74.1	90.0	100.3	74.1	90.0	100.0	74.1	89.9
D50	Electrical	101.8	65.5	83.9	101.6	76.8	89.3	100.0	69.4	84.9	101.9	76.7	89.5	105.0	76.7	91.0
E	Equipment & Furnishings	100.0	59.7	97.7	100.0	74.7	98.5	100.0	72.2	98.4	100.0	72.5	98.4	100.0	72.6	98.4
G	Site Work	102.5	92.2	95.7	88.8	98.2	95.1	90.8	95.7	94.1	87.7	95.7	93.0	88.3	100.8	96.6
A-G	**WEIGHTED AVERAGE**	97.1	66.3	84.1	100.4	85.6	94.2	97.1	83.9	91.5	97.5	84.9	92.2	99.6	85.4	93.6

DIV. NO.	BUILDING SYSTEMS	MARYLAND						MASSACHUSETTS								
		BALTIMORE			HAGERSTOWN			BOSTON			BROCKTON			FALL RIVER		
		MAT.	INST.	TOTAL	MAT.	INST.	TOTAL	MAT.	INST.	TOTAL	MAT.	INST.	TOTAL	MAT.	INST.	TOTAL
A	Substructure	108.5	83.0	93.8	92.3	84.1	87.5	102.9	133.8	120.7	93.7	122.7	110.5	91.9	120.2	108.2
B10	Shell: Superstructure	105.0	88.3	98.7	99.0	93.7	97.0	103.0	139.1	116.5	97.8	126.9	108.7	97.4	121.1	106.3
B20	Exterior Closure	103.6	75.5	91.1	92.8	86.4	90.0	108.5	145.7	125.1	101.4	133.7	115.8	101.6	131.3	114.8
B30	Roofing	102.7	81.5	94.2	101.3	82.5	93.7	107.4	134.8	118.4	103.4	126.1	112.5	103.2	122.0	110.8
C	Interior Construction	99.1	76.6	89.7	94.4	80.4	88.6	99.2	147.1	119.2	93.7	130.4	109.0	93.5	129.5	108.5
D10	Services: Conveying	100.0	88.1	96.4	100.0	92.6	97.8	100.0	116.4	104.9	100.0	108.5	102.6	100.0	109.1	102.7
D20 - 40	Mechanical	100.1	82.0	93.0	99.9	83.1	93.3	100.1	127.4	110.8	100.4	103.9	101.7	100.4	103.5	101.6
D50	Electrical	99.6	86.9	93.3	98.0	80.8	89.5	102.6	129.5	115.9	101.3	97.7	99.5	101.2	97.7	99.4
E	Equipment & Furnishings	100.0	77.0	98.7	100.0	77.1	98.7	100.0	139.7	102.3	100.0	118.3	101.1	100.0	118.2	101.0
G	Site Work	101.1	97.4	98.7	92.0	89.8	90.6	91.8	102.1	98.7	90.4	98.1	95.5	89.3	98.2	95.2
A-G	**WEIGHTED AVERAGE**	101.8	83.2	93.9	97.4	85.3	92.3	101.9	134.0	115.5	98.5	116.7	106.2	98.3	114.9	105.3

City Cost Indexes RJ1030-010 Building Systems

DIV. NO.	BUILDING SYSTEMS	MASSACHUSETTS														
		HYANNIS			LAWRENCE			LOWELL			NEW BEDFORD			PITTSFIELD		
		MAT.	INST.	TOTAL	MAT.	INST.	TOTAL	MAT.	INST.	TOTAL	MAT.	INST.	TOTAL	MAT.	INST.	TOTAL
A	Substructure	83.2	120.0	104.5	95.1	123.7	111.6	89.8	124.4	109.8	85.2	120.2	105.4	91.3	105.7	99.6
B10	Shell: Superstructure	92.5	120.8	103.1	97.8	128.9	109.4	96.5	127.4	108.0	96.0	121.2	105.4	96.7	107.9	100.8
B20	Exterior Closure	98.5	131.5	113.2	102.9	134.6	117.0	93.0	133.4	111.0	100.9	133.3	115.4	93.2	110.9	101.1
B30	Roofing	102.7	123.0	110.9	104.2	126.2	113.0	104.0	126.4	113.0	103.2	122.1	110.7	104.0	101.8	103.1
C	Interior Construction	89.8	129.3	106.2	92.9	130.7	108.6	97.5	131.3	111.6	93.4	129.5	108.4	97.5	108.4	102.1
D10	Services: Conveying	100.0	108.5	102.6	100.0	108.5	102.6	100.0	109.7	102.9	100.0	109.1	102.7	100.0	99.2	99.8
D20 - 40	Mechanical	100.4	104.2	101.9	100.1	119.7	107.8	100.1	121.2	108.4	100.4	103.4	101.5	100.1	94.2	97.8
D50	Electrical	98.9	97.7	98.3	99.7	122.9	111.1	100.1	119.3	109.6	102.1	97.7	99.9	100.1	96.1	98.2
E	Equipment & Furnishings	100.0	118.3	101.1	100.0	119.5	101.1	100.0	119.5	101.1	100.0	118.2	101.0	100.0	101.7	100.1
G	Site Work	87.9	98.3	94.8	91.3	98.1	95.8	89.9	98.3	95.5	87.6	98.2	94.7	91.1	96.9	95.0
A-G	WEIGHTED AVERAGE	95.7	115.0	103.8	98.4	124.2	109.3	97.5	123.8	108.6	97.7	115.2	105.1	97.7	102.6	99.7

DIV. NO.	BUILDING SYSTEMS	MASSACHUSETTS						MICHIGAN								
		SPRINGFIELD			WORCESTER			ANN ARBOR			DEARBORN			DETROIT		
		MAT.	INST.	TOTAL	MAT.	INST.	TOTAL	MAT.	INST.	TOTAL	MAT.	INST.	TOTAL	MAT.	INST.	TOTAL
A	Substructure	92.0	115.2	105.4	91.7	123.2	109.9	90.5	99.5	95.7	89.2	100.0	95.4	103.5	103.0	103.2
B10	Shell: Superstructure	99.1	116.8	105.7	99.1	128.0	109.8	100.3	109.6	103.8	100.1	109.9	103.7	104.0	98.8	102.1
B20	Exterior Closure	93.1	118.1	104.2	92.9	131.2	110.0	93.4	97.6	95.2	93.3	99.2	95.9	101.6	99.1	100.5
B30	Roofing	104.0	107.6	105.4	104.0	119.6	110.3	105.6	98.5	102.7	103.9	101.2	102.8	102.8	103.5	103.1
C	Interior Construction	97.4	121.5	107.4	97.5	128.0	110.2	92.4	104.2	97.3	92.3	103.6	97.0	97.2	105.2	100.5
D10	Services: Conveying	100.0	100.9	100.3	100.0	102.7	100.8	100.0	92.1	97.6	100.0	92.6	97.8	100.0	101.9	100.6
D20 - 40	Mechanical	100.1	98.5	99.5	100.1	104.7	101.9	100.0	91.3	96.6	100.0	100.0	100.0	99.9	102.7	101.0
D50	Electrical	100.1	93.0	96.6	100.1	104.1	102.1	97.3	103.0	100.1	97.3	96.1	96.7	99.9	103.5	101.6
E	Equipment & Furnishings	100.0	117.7	101.0	100.0	115.7	100.9	100.0	105.7	100.3	100.0	105.4	100.3	100.0	105.9	100.3
G	Site Work	90.5	97.2	95.0	90.4	98.1	95.5	81.5	91.6	88.2	81.2	91.7	88.2	97.3	101.4	100.0
A-G	WEIGHTED AVERAGE	98.2	108.5	102.5	98.1	117.0	106.1	97.1	99.8	98.2	96.9	100.9	98.6	100.7	102.0	101.3

DIV. NO.	BUILDING SYSTEMS	MICHIGAN														
		FLINT			GRAND RAPIDS			KALAMAZOO			LANSING			MUSKEGON		
		MAT.	INST.	TOTAL	MAT.	INST.	TOTAL	MAT.	INST.	TOTAL	MAT.	INST.	TOTAL	MAT.	INST.	TOTAL
A	Substructure	89.5	88.3	88.8	93.0	82.3	86.8	90.6	81.9	85.6	98.2	86.5	91.5	89.4	83.1	85.8
B10	Shell: Superstructure	100.6	101.6	101.0	99.1	81.8	92.7	100.8	81.5	93.6	100.8	98.5	99.9	98.8	83.0	92.9
B20	Exterior Closure	93.4	87.0	90.5	89.5	74.2	82.7	90.9	77.7	85.0	89.3	85.3	87.5	86.0	76.2	81.6
B30	Roofing	103.3	83.8	95.5	98.0	71.5	87.4	96.3	78.9	89.3	102.2	83.6	94.7	95.1	73.5	86.4
C	Interior Construction	92.1	82.2	88.0	97.2	75.0	88.0	85.6	74.9	81.2	96.5	75.2	87.6	83.3	77.0	80.7
D10	Services: Conveying	100.0	89.9	97.0	100.0	95.9	98.8	100.0	96.5	98.9	100.0	91.1	97.3	100.0	97.1	99.1
D20 - 40	Mechanical	100.0	83.3	93.4	100.0	79.6	92.0	100.1	78.2	91.5	99.9	83.7	93.5	99.9	83.4	93.4
D50	Electrical	97.3	89.1	93.2	102.2	82.9	92.7	96.4	74.1	85.4	99.1	87.6	93.4	96.8	73.0	85.1
E	Equipment & Furnishings	100.0	80.5	98.9	100.0	71.2	98.3	100.0	72.1	98.4	100.0	71.8	98.4	100.0	74.4	98.5
G	Site Work	71.9	91.3	84.8	93.6	86.0	88.6	94.6	81.8	86.0	90.6	96.0	94.2	91.9	81.9	85.2
A-G	WEIGHTED AVERAGE	96.8	88.6	93.3	97.9	79.9	90.3	95.8	78.6	88.6	98.0	86.8	93.3	94.3	80.0	88.3

DIV. NO.	BUILDING SYSTEMS	MICHIGAN			MINNESOTA											
		SAGINAW			DULUTH			MINNEAPOLIS			ROCHESTER			SAINT PAUL		
		MAT.	INST.	TOTAL	MAT.	INST.	TOTAL	MAT.	INST.	TOTAL	MAT.	INST.	TOTAL	MAT.	INST.	TOTAL
A	Substructure	88.7	86.7	87.6	96.1	98.1	97.3	96.0	112.6	105.6	95.8	97.5	96.8	100.1	112.7	107.4
B10	Shell: Superstructure	100.2	100.3	100.3	99.3	108.7	102.8	98.1	119.4	106.0	98.6	111.1	103.3	100.1	120.7	107.8
B20	Exterior Closure	93.6	80.1	87.6	96.3	106.5	100.8	106.3	119.9	112.3	94.6	106.7	100.0	96.4	127.0	110.1
B30	Roofing	104.2	81.0	94.9	98.1	101.2	99.3	101.1	116.6	107.3	105.6	87.5	98.3	102.1	119.8	109.2
C	Interior Construction	91.2	79.5	86.4	96.2	100.9	98.2	98.9	116.0	106.0	94.0	95.7	94.7	94.4	119.9	105.0
D10	Services: Conveying	100.0	88.9	96.7	100.0	94.8	98.5	100.0	105.0	101.5	100.0	103.0	100.9	100.0	107.9	102.4
D20 - 40	Mechanical	100.0	78.3	91.5	99.8	94.5	97.7	100.0	110.6	104.2	100.0	91.6	96.7	99.9	116.7	106.5
D50	Electrical	95.4	84.5	90.0	100.3	101.7	101.0	107.2	110.3	108.7	103.0	93.1	98.1	102.9	115.1	108.9
E	Equipment & Furnishings	100.0	79.2	98.8	100.0	92.9	99.6	100.0	111.4	100.7	100.0	93.3	99.6	100.0	115.1	100.9
G	Site Work	74.9	91.2	85.7	99.0	101.5	100.7	94.9	108.8	104.2	96.4	97.4	97.0	97.2	102.4	100.7
A-G	WEIGHTED AVERAGE	96.5	85.1	91.7	98.6	101.3	99.7	100.7	114.1	106.3	98.3	98.7	98.5	99.0	117.6	106.9

DIV. NO.	BUILDING SYSTEMS	MINNESOTA			MISSISSIPPI											
		ST. CLOUD			BILOXI			GREENVILLE			JACKSON			MERIDIAN		
		MAT.	INST.	TOTAL	MAT.	INST.	TOTAL	MAT.	INST.	TOTAL	MAT.	INST.	TOTAL	MAT.	INST.	TOTAL
A	Substructure	94.9	110.6	104.0	107.3	68.3	84.8	104.2	69.8	84.4	99.6	69.9	82.4	102.8	68.1	82.7
B10	Shell: Superstructure	93.1	118.5	102.6	93.8	74.7	86.7	95.0	79.0	89.1	97.3	75.1	89.0	91.0	75.1	85.1
B20	Exterior Closure	90.2	122.1	104.4	83.7	62.6	74.3	108.4	62.1	87.7	87.1	61.5	75.7	81.8	62.1	73.0
B30	Roofing	105.0	111.1	107.4	100.2	63.3	85.4	98.2	62.2	83.8	98.4	63.2	84.3	99.8	62.6	84.9
C	Interior Construction	85.7	116.7	98.6	92.6	62.1	79.9	92.6	62.5	80.1	94.4	63.5	81.5	90.0	61.1	78.0
D10	Services: Conveying	100.0	102.1	100.6	100.0	72.9	91.9	100.0	72.4	91.7	100.0	72.8	91.8	100.0	72.7	91.8
D20 - 40	Mechanical	99.5	110.5	103.8	100.0	55.3	82.4	100.0	57.4	83.3	100.0	59.1	83.9	100.0	57.6	83.3
D50	Electrical	102.7	115.1	108.8	101.0	53.7	77.7	96.8	55.9	76.6	102.7	55.9	79.6	100.1	55.6	78.2
E	Equipment & Furnishings	100.0	111.5	100.7	100.0	63.9	97.9	100.0	62.4	97.8	100.0	64.8	98.0	100.0	62.4	97.8
G	Site Work	93.3	100.7	98.2	106.3	85.3	92.3	108.2	85.6	93.1	101.7	90.1	94.0	102.3	85.6	91.2
A-G	WEIGHTED AVERAGE	95.1	114.2	103.2	96.2	63.9	82.6	98.7	65.5	84.7	97.3	65.6	83.9	94.6	64.5	81.9

DIV. NO.	BUILDING SYSTEMS	MISSOURI														
		CAPE GIRARDEAU			COLUMBIA			JOPLIN			KANSAS CITY			SPRINGFIELD		
		MAT.	INST.	TOTAL	MAT.	INST.	TOTAL	MAT.	INST.	TOTAL	MAT.	INST.	TOTAL	MAT.	INST.	TOTAL
A	Substructure	90.3	85.3	87.4	87.5	87.0	87.2	100.0	80.4	88.7	96.7	103.2	100.5	93.4	81.0	86.2
B10	Shell: Superstructure	92.2	97.4	94.1	94.1	101.0	96.6	90.6	88.4	89.8	95.6	109.0	100.6	99.2	90.6	96.0
B20	Exterior Closure	98.4	80.9	90.6	101.8	87.4	95.4	87.6	80.8	84.6	93.5	104.0	98.2	90.7	82.6	87.1
B30	Roofing	96.0	86.8	92.3	90.6	86.0	88.7	91.9	80.9	87.5	91.8	105.4	97.2	94.7	74.3	86.5
C	Interior Construction	97.8	79.9	90.4	89.0	83.5	86.7	95.7	74.9	87.0	97.7	104.2	100.4	95.1	78.2	88.0
D10	Services: Conveying	100.0	96.1	98.8	100.0	95.6	98.7	100.0	80.4	94.1	100.0	97.7	99.3	100.0	92.6	97.8
D20 - 40	Mechanical	100.0	98.7	99.5	100.0	96.9	98.8	100.1	71.1	88.7	100.0	103.4	101.4	100.0	69.4	88.0
D50	Electrical	99.2	98.5	98.9	97.8	81.4	89.7	92.9	66.1	79.7	103.3	101.2	102.2	101.3	68.8	85.3
E	Equipment & Furnishings	100.0	81.1	98.9	100.0	78.6	98.8	100.0	72.8	98.4	100.0	103.8	100.2	100.0	74.3	98.5
G	Site Work	89.7	89.2	89.4	95.6	93.1	94.0	102.3	92.2	95.6	95.8	98.3	97.5	95.1	91.1	92.4
A-G	WEIGHTED AVERAGE	97.0	91.2	94.5	96.1	90.7	93.8	95.1	77.7	87.7	97.9	103.8	100.4	97.6	79.0	89.7

DIV. NO.	BUILDING SYSTEMS	MISSOURI						MONTANA								
		ST. JOSEPH			ST. LOUIS			BILLINGS			BUTTE			GREAT FALLS		
		MAT.	INST.	TOTAL	MAT.	INST.	TOTAL	MAT.	INST.	TOTAL	MAT.	INST.	TOTAL	MAT.	INST.	TOTAL
A	Substructure	93.3	93.4	93.4	97.8	102.1	100.3	110.6	77.1	91.3	119.4	77.0	94.9	124.3	77.0	97.0
B10	Shell: Superstructure	92.1	104.8	96.8	97.6	111.7	102.8	108.1	81.2	98.0	104.7	81.0	95.9	108.8	81.1	98.4
B20	Exterior Closure	90.0	91.4	90.6	94.1	108.4	100.5	99.1	82.0	91.5	94.4	81.0	88.4	99.0	82.0	91.4
B30	Roofing	92.2	88.6	90.7	94.4	105.5	98.8	108.5	71.6	93.7	108.1	71.5	93.4	108.8	71.5	93.8
C	Interior Construction	95.4	93.3	94.5	100.6	102.1	101.2	95.2	71.3	85.2	94.6	70.9	84.7	97.9	71.3	86.8
D10	Services: Conveying	100.0	94.9	98.5	100.0	103.0	100.9	100.0	95.1	98.5	100.0	95.1	98.5	100.0	95.1	98.5
D20 - 40	Mechanical	100.1	87.5	95.2	100.0	105.5	102.2	100.0	76.4	90.7	100.0	71.2	88.7	100.0	71.2	88.7
D50	Electrical	101.5	75.8	88.9	101.8	98.5	100.2	102.3	75.2	88.9	109.2	71.3	90.5	101.7	70.8	86.5
E	Equipment & Furnishings	100.0	87.5	99.3	100.0	101.3	100.1	100.0	65.2	98.0	100.0	65.0	98.0	100.0	65.0	98.0
G	Site Work	97.8	87.1	90.7	94.6	97.4	96.4	92.5	93.3	93.0	97.5	93.1	94.6	101.0	93.1	95.8
A-G	WEIGHTED AVERAGE	96.1	90.9	93.9	98.8	104.5	101.2	101.6	78.4	91.8	101.5	76.5	90.9	102.9	76.6	91.8

DIV. NO.	BUILDING SYSTEMS	MONTANA						NEBRASKA								
		HELENA			MISSOULA			GRAND ISLAND			LINCOLN			NORTH PLATTE		
		MAT.	INST.	TOTAL	MAT.	INST.	TOTAL	MAT.	INST.	TOTAL	MAT.	INST.	TOTAL	MAT.	INST.	TOTAL
A	Substructure	103.0	78.0	88.6	96.3	77.6	85.5	108.0	77.5	90.4	90.0	82.6	85.7	109.8	75.0	89.7
B10	Shell: Superstructure	103.4	80.7	94.9	97.5	82.2	91.8	96.2	84.0	91.7	93.9	86.4	91.1	96.6	83.3	91.6
B20	Exterior Closure	96.6	81.0	89.6	97.5	81.2	90.2	98.8	74.0	87.7	91.3	76.9	84.9	90.1	76.7	84.2
B30	Roofing	103.3	72.1	90.8	107.2	74.2	93.9	102.2	79.1	92.9	100.2	81.2	92.6	96.0	78.0	88.8
C	Interior Construction	97.7	70.9	86.6	94.3	73.2	85.5	88.2	69.1	80.3	101.2	77.3	91.2	89.8	75.0	83.7
D10	Services: Conveying	100.0	95.4	98.6	100.0	96.0	98.8	100.0	92.8	97.8	100.0	93.2	97.9	100.0	59.2	87.7
D20 - 40	Mechanical	100.0	71.2	88.7	100.0	71.2	88.7	100.1	79.5	92.0	100.0	79.6	92.0	100.0	73.7	89.6
D50	Electrical	109.0	70.8	90.2	107.0	70.2	88.9	95.9	65.7	81.0	109.2	65.7	87.8	93.9	65.7	80.0
E	Equipment & Furnishings	100.0	65.1	98.0	100.0	65.0	98.0	100.0	66.1	98.1	100.0	74.2	98.5	100.0	75.9	98.6
G	Site Work	90.9	96.7	94.8	80.8	93.1	89.0	108.9	90.4	96.5	95.7	95.4	95.5	104.7	90.0	94.9
A-G	WEIGHTED AVERAGE	100.9	76.7	90.7	98.6	77.1	89.5	97.4	76.8	88.7	98.3	79.7	90.5	96.3	75.6	87.6

DIV. NO.	BUILDING SYSTEMS	NEBRASKA			NEVADA									NEW HAMPSHIRE		
		OMAHA			CARSON CITY			LAS VEGAS			RENO			MANCHESTER		
		MAT.	INST.	TOTAL	MAT.	INST.	TOTAL	MAT.	INST.	TOTAL	MAT.	INST.	TOTAL	MAT.	INST.	TOTAL
A	Substructure	92.8	81.2	86.1	100.4	87.8	93.1	99.0	106.5	103.4	104.7	87.4	94.7	99.6	102.2	101.1
B10	Shell: Superstructure	94.5	80.9	89.5	109.3	89.1	101.8	119.3	106.0	114.4	114.7	91.3	106.0	101.5	96.2	99.5
B20	Exterior Closure	91.3	80.2	86.3	107.3	70.1	90.7	109.3	107.3	108.4	111.7	70.3	93.3	92.9	100.3	96.2
B30	Roofing	96.6	80.4	90.1	118.5	81.9	103.8	129.8	102.7	119.0	114.5	81.0	101.0	112.8	110.0	111.7
C	Interior Construction	99.6	76.7	90.1	98.4	75.4	88.8	95.2	106.0	99.7	94.6	75.6	86.7	95.4	99.0	96.9
D10	Services: Conveying	100.0	91.9	97.6	100.0	103.4	101.0	100.0	105.3	101.6	100.0	103.1	100.9	100.0	111.9	103.6
D20 - 40	Mechanical	100.0	80.3	92.3	100.0	77.0	91.0	100.1	103.0	101.3	100.0	77.1	91.0	100.1	87.3	95.1
D50	Electrical	103.9	83.7	94.0	102.2	88.3	95.3	103.5	107.8	105.6	99.4	88.3	93.9	93.9	79.3	86.7
E	Equipment & Furnishings	100.0	76.9	98.7	100.0	75.3	98.6	100.0	103.3	100.2	100.0	75.1	98.6	100.0	92.6	99.6
G	Site Work	89.6	93.9	92.4	83.9	97.3	92.8	78.5	95.5	89.8	75.5	92.5	86.8	88.0	101.5	97.0
A-G	WEIGHTED AVERAGE	97.6	81.5	90.8	102.9	82.5	94.3	105.0	105.1	105.0	103.6	82.6	94.7	98.2	94.5	96.6

DIV. NO.	BUILDING SYSTEMS	NEW HAMPSHIRE						NEW JERSEY								
		NASHUA			PORTSMOUTH			CAMDEN			ELIZABETH			JERSEY CITY		
		MAT.	INST.	TOTAL	MAT.	INST.	TOTAL	MAT.	INST.	TOTAL	MAT.	INST.	TOTAL	MAT.	INST.	TOTAL
A	Substructure	89.3	100.7	95.9	80.6	100.5	92.1	86.0	126.3	109.2	78.3	136.2	111.7	76.5	133.6	109.5
B10	Shell: Superstructure	98.0	96.3	97.4	92.7	97.0	94.3	101.0	121.9	108.8	93.9	135.1	109.3	96.7	132.1	109.9
B20	Exterior Closure	94.5	99.6	96.8	90.6	95.4	92.8	95.1	133.0	112.0	101.3	143.6	120.2	91.6	143.4	114.7
B30	Roofing	108.6	109.4	108.9	108.1	108.4	108.2	102.2	132.7	114.5	104.3	144.4	120.4	103.5	142.8	119.3
C	Interior Construction	98.3	97.4	97.9	95.8	97.0	96.3	96.2	141.7	115.2	97.8	154.7	121.5	95.3	155.0	120.1
D10	Services: Conveying	100.0	111.6	103.5	100.0	110.8	103.3	100.0	111.2	103.4	100.0	129.5	108.9	100.0	129.5	108.9
D20 - 40	Mechanical	100.2	87.3	95.1	100.2	86.3	94.7	100.0	126.3	110.3	100.0	137.1	114.6	100.0	135.7	114.0
D50	Electrical	93.2	79.3	86.4	91.7	75.3	83.6	98.4	132.1	115.0	96.6	138.5	117.2	100.5	142.3	121.1
E	Equipment & Furnishings	100.0	92.5	99.6	100.0	92.2	99.6	100.0	137.2	102.1	100.0	148.0	102.8	100.0	148.1	102.8
G	Site Work	88.0	96.4	93.6	82.8	97.1	92.3	87.6	99.7	95.6	107.3	101.7	103.6	93.2	101.7	98.9
A-G	WEIGHTED AVERAGE	97.5	93.7	95.9	94.9	92.4	93.8	98.0	127.8	110.6	97.5	138.3	114.7	96.5	137.8	114.0

DIV. NO.	BUILDING SYSTEMS	NEW JERSEY									NEW MEXICO					
		NEWARK			PATERSON			TRENTON			ALBUQUERQUE			FARMINGTON		
		MAT.	INST.	TOTAL	MAT.	INST.	TOTAL	MAT.	INST.	TOTAL	MAT.	INST.	TOTAL	MAT.	INST.	TOTAL
A	Substructure	94.1	137.5	119.1	88.4	136.1	115.9	97.9	126.8	114.6	92.9	75.4	82.8	95.2	75.4	83.8
B10	Shell: Superstructure	101.9	134.9	114.2	95.3	134.9	110.0	103.3	120.4	109.6	105.2	80.6	96.0	103.9	80.6	95.2
B20	Exterior Closure	95.7	143.6	117.1	94.1	143.6	116.2	95.6	132.9	112.3	99.6	62.3	82.9	105.4	62.3	86.1
B30	Roofing	105.4	145.0	121.3	104.1	135.7	116.8	103.0	135.3	116.0	101.1	74.6	90.4	101.4	74.6	90.6
C	Interior Construction	98.9	155.1	122.3	99.6	154.7	122.5	96.8	144.4	116.6	94.9	65.9	82.8	95.2	65.9	83.0
D10	Services: Conveying	100.0	129.6	108.9	100.0	129.5	108.9	100.0	111.5	103.5	100.0	87.4	96.2	100.0	87.4	96.2
D20 - 40	Mechanical	100.0	139.2	115.4	100.0	135.5	114.0	100.1	130.8	112.1	100.3	69.0	88.0	100.2	69.0	88.0
D50	Electrical	105.0	142.3	123.4	100.5	138.5	119.2	102.6	128.7	115.4	87.9	69.5	78.8	85.9	69.5	77.8
E	Equipment & Furnishings	100.0	148.2	102.8	100.0	148.0	102.8	100.0	138.8	102.2	100.0	66.9	98.1	100.0	66.9	98.1
G	Site Work	109.0	106.5	107.4	105.0	101.7	102.8	88.6	105.7	100.0	91.5	97.1	95.2	98.7	97.1	97.7
A-G	WEIGHTED AVERAGE	100.3	139.7	117.0	98.0	137.7	114.8	99.7	128.9	112.0	98.6	72.5	87.6	99.0	72.5	87.8

DIV. NO.	BUILDING SYSTEMS	NEW MEXICO									NEW YORK					
		LAS CRUCES			ROSWELL			SANTA FE			ALBANY			BINGHAMTON		
		MAT.	INST.	TOTAL	MAT.	INST.	TOTAL	MAT.	INST.	TOTAL	MAT.	INST.	TOTAL	MAT.	INST.	TOTAL
A	Substructure	94.6	68.4	79.5	98.7	75.4	85.3	95.7	76.8	84.8	85.6	110.6	100.0	102.6	98.7	100.4
B10	Shell: Superstructure	102.7	75.2	92.4	105.8	80.6	96.4	100.5	80.1	92.9	99.0	118.5	106.3	96.8	117.9	104.7
B20	Exterior Closure	85.6	61.5	74.9	113.6	62.3	90.7	93.4	62.3	79.5	90.2	116.3	101.8	90.9	105.6	97.5
B30	Roofing	92.4	69.2	83.1	105.4	74.6	93.0	103.6	75.7	92.4	105.6	109.7	107.2	109.6	93.1	103.0
C	Interior Construction	96.0	65.1	83.2	94.2	65.9	82.4	98.1	65.9	84.7	93.8	106.9	99.3	92.2	95.5	93.6
D10	Services: Conveying	100.0	84.7	95.4	100.0	87.4	96.2	100.0	87.6	96.3	100.0	103.8	101.1	100.0	97.4	99.2
D20 - 40	Mechanical	100.4	68.7	88.0	100.0	69.0	87.8	100.3	69.0	88.0	100.1	110.3	104.1	100.6	97.0	99.2
D50	Electrical	90.1	83.8	87.0	89.2	69.5	79.5	100.3	69.5	85.2	100.0	108.4	104.1	98.7	101.0	99.8
E	Equipment & Furnishings	100.0	66.9	98.1	100.0	66.9	98.1	100.0	67.0	98.1	100.0	104.4	100.3	100.0	90.7	99.5
G	Site Work	98.5	77.3	84.4	100.6	97.1	98.3	99.9	102.4	101.6	80.7	104.8	96.8	94.7	88.1	90.3
A-G	WEIGHTED AVERAGE	96.9	71.4	86.1	100.8	72.5	88.9	99.0	72.9	88.0	96.8	111.2	102.8	97.3	101.5	99.1

City Cost Indexes — RJ1030-010 Building Systems

DIV. NO.	BUILDING SYSTEMS	NEW YORK														
		BUFFALO			HICKSVILLE			NEW YORK			RIVERHEAD			ROCHESTER		
		MAT.	INST.	TOTAL	MAT.	INST.	TOTAL	MAT.	INST.	TOTAL	MAT.	INST.	TOTAL	MAT.	INST.	TOTAL
A	Substructure	105.5	115.4	111.2	98.0	151.4	128.8	100.3	165.1	137.7	99.7	155.5	131.9	97.7	103.0	100.8
B10	Shell: Superstructure	98.9	112.6	104.0	102.8	167.0	126.7	98.2	176.1	127.2	103.8	162.5	125.7	103.8	111.4	106.6
B20	Exterior Closure	105.5	119.2	111.6	102.5	174.9	134.8	102.4	187.2	140.3	103.8	172.9	134.6	97.1	103.8	100.1
B30	Roofing	102.5	110.8	105.8	108.8	154.3	127.1	105.1	168.5	130.6	109.9	150.8	126.3	116.6	99.2	109.6
C	Interior Construction	101.1	117.2	107.8	92.6	164.3	122.4	99.0	189.1	136.5	92.8	160.6	121.1	98.6	101.3	99.7
D10	Services: Conveying	100.0	105.3	101.6	100.0	129.1	108.7	100.0	135.2	110.6	100.0	115.6	104.7	100.0	98.8	99.6
D20 - 40	Mechanical	100.0	102.4	100.9	99.8	161.0	123.8	100.2	178.3	130.9	100.0	155.9	121.9	99.9	88.1	95.3
D50	Electrical	99.5	105.3	102.4	98.0	141.0	119.2	96.8	184.5	140.0	99.5	134.9	116.9	99.5	89.9	94.8
E	Equipment & Furnishings	100.0	115.4	100.9	100.0	154.6	103.1	100.0	189.7	105.2	100.0	154.2	103.1	100.0	99.4	100.0
G	Site Work	97.5	102.3	100.7	113.5	119.5	117.5	106.8	112.1	110.3	114.9	119.0	117.7	90.2	104.8	99.9
A-G	**WEIGHTED AVERAGE**	100.8	110.3	104.8	99.9	157.1	124.1	99.7	175.2	131.6	100.6	153.5	122.9	100.3	99.2	99.8

DIV. NO.	BUILDING SYSTEMS	NEW YORK														
		SCHENECTADY			SYRACUSE			UTICA			WATERTOWN			WHITE PLAINS		
		MAT.	INST.	TOTAL	MAT.	INST.	TOTAL	MAT.	INST.	TOTAL	MAT.	INST.	TOTAL	MAT.	INST.	TOTAL
A	Substructure	93.1	109.1	102.3	97.7	97.4	97.5	90.8	95.2	93.3	100.0	99.5	99.7	87.8	143.2	119.8
B10	Shell: Superstructure	98.6	118.9	106.2	98.6	108.6	102.4	96.0	106.1	99.8	97.2	109.9	102.0	86.9	162.4	115.0
B20	Exterior Closure	92.4	116.3	103.1	96.2	100.8	98.2	93.7	99.1	96.1	101.8	104.7	103.1	91.3	159.7	121.8
B30	Roofing	100.4	109.0	103.9	104.2	92.8	99.6	92.9	92.9	92.9	93.2	96.1	94.4	103.0	147.4	120.9
C	Interior Construction	93.2	106.8	98.9	92.2	90.0	91.3	93.9	87.6	91.3	92.6	92.6	92.6	94.7	159.4	121.6
D10	Services: Conveying	100.0	103.4	101.0	100.0	96.2	98.8	100.0	90.6	97.2	100.0	97.3	99.2	100.0	124.8	107.4
D20 - 40	Mechanical	100.2	106.2	102.6	100.3	94.4	98.0	100.3	92.5	97.2	100.3	87.8	95.4	100.4	145.9	118.3
D50	Electrical	98.7	108.4	103.5	98.6	101.0	99.8	96.6	101.0	98.8	98.6	90.2	94.4	88.8	170.6	129.1
E	Equipment & Furnishings	100.0	104.3	100.2	100.0	86.7	99.2	100.0	83.5	99.1	100.0	90.0	99.4	100.0	148.5	102.8
G	Site Work	82.4	99.7	93.9	93.3	98.1	96.5	71.6	97.7	88.9	79.2	98.1	91.8	99.0	105.5	103.3
A-G	**WEIGHTED AVERAGE**	97.0	109.9	102.4	97.8	98.3	98.0	96.0	96.6	96.2	97.7	96.8	97.4	93.8	152.8	118.8

DIV. NO.	BUILDING SYSTEMS	NEW YORK			NORTH CAROLINA											
		YONKERS			ASHEVILLE			CHARLOTTE			DURHAM			FAYETTEVILLE		
		MAT.	INST.	TOTAL	MAT.	INST.	TOTAL	MAT.	INST.	TOTAL	MAT.	INST.	TOTAL	MAT.	INST.	TOTAL
A	Substructure	96.8	143.2	123.5	101.8	69.1	83.0	104.7	70.2	84.8	104.4	71.0	85.1	105.6	69.9	85.0
B10	Shell: Superstructure	94.6	162.4	119.9	101.1	78.6	92.7	102.6	78.1	93.5	114.9	78.7	101.4	118.4	78.2	103.4
B20	Exterior Closure	97.5	159.8	125.3	81.1	65.7	74.2	81.9	65.7	74.7	84.7	65.7	76.2	81.6	63.0	73.3
B30	Roofing	103.4	148.1	121.3	99.0	64.4	85.1	93.6	65.1	82.2	102.7	64.4	87.3	98.5	62.5	84.0
C	Interior Construction	98.3	159.6	123.8	88.4	61.1	77.0	93.3	61.1	79.9	92.4	61.1	79.4	88.9	60.4	77.0
D10	Services: Conveying	100.0	128.4	108.5	100.0	85.6	95.7	100.0	85.6	95.7	100.0	85.6	95.7	100.0	84.3	95.3
D20 - 40	Mechanical	100.4	146.0	118.3	100.4	61.3	85.0	99.9	63.1	85.5	100.5	61.3	85.1	100.2	59.5	84.2
D50	Electrical	94.9	170.6	132.2	100.8	57.8	79.6	100.0	60.3	80.4	96.5	57.4	77.2	100.6	57.4	79.3
E	Equipment & Furnishings	100.0	148.1	102.8	100.0	58.6	97.6	100.0	58.9	97.6	100.0	58.6	97.6	100.0	58.4	97.6
G	Site Work	104.2	105.5	105.0	96.5	77.0	83.5	98.5	81.2	87.0	98.4	85.7	89.9	96.0	85.5	89.0
A-G	**WEIGHTED AVERAGE**	97.9	153.0	121.2	96.4	66.6	83.8	97.4	67.6	84.8	100.1	67.2	86.2	100.2	66.1	85.8

DIV. NO.	BUILDING SYSTEMS	NORTH CAROLINA												NORTH DAKOTA		
		GREENSBORO			RALEIGH			WILMINGTON			WINSTON-SALEM			BISMARCK		
		MAT.	INST.	TOTAL	MAT.	INST.	TOTAL	MAT.	INST.	TOTAL	MAT.	INST.	TOTAL	MAT.	INST.	TOTAL
A	Substructure	103.7	71.0	84.8	106.6	71.8	86.5	101.8	68.1	82.3	105.6	71.0	85.6	103.6	87.0	94.0
B10	Shell: Superstructure	108.5	78.7	97.4	103.0	78.1	93.7	100.8	78.2	92.4	106.6	78.7	96.2	97.7	88.7	94.3
B20	Exterior Closure	83.7	65.7	75.7	80.1	64.6	73.2	77.1	63.0	70.8	83.8	65.7	75.7	92.1	85.0	88.9
B30	Roofing	102.5	64.4	87.2	97.1	64.3	83.9	99.0	62.5	84.3	102.5	64.4	87.2	109.5	87.1	100.5
C	Interior Construction	92.5	61.1	79.4	93.4	60.9	79.9	88.8	60.4	77.0	92.5	61.1	79.4	100.2	71.9	88.5
D10	Services: Conveying	100.0	82.6	94.8	100.0	85.4	95.6	100.0	84.3	95.3	100.0	85.6	95.7	100.0	97.6	99.3
D20 - 40	Mechanical	100.4	61.3	85.0	100.0	60.6	84.5	100.4	59.5	84.3	100.4	61.3	85.0	99.9	77.0	90.9
D50	Electrical	95.7	57.8	77.0	99.1	56.3	78.0	101.1	55.8	78.8	95.7	57.8	77.0	98.7	73.4	86.3
E	Equipment & Furnishings	100.0	58.6	97.6	100.0	58.6	97.6	100.0	58.3	97.6	100.0	58.6	97.6	100.0	72.9	98.4
G	Site Work	98.2	85.8	89.9	98.8	90.3	93.1	97.7	76.8	83.8	98.6	85.8	90.1	99.8	97.7	98.4
A-G	**WEIGHTED AVERAGE**	98.5	67.2	85.3	97.4	67.0	84.5	96.0	65.2	83.0	98.2	67.3	85.2	98.9	81.6	91.6

DIV. NO.	BUILDING SYSTEMS	NORTH DAKOTA									OHIO					
		FARGO			GRAND FORKS			MINOT			AKRON			CANTON		
		MAT.	INST.	TOTAL	MAT.	INST.	TOTAL	MAT.	INST.	TOTAL	MAT.	INST.	TOTAL	MAT.	INST.	TOTAL
A	Substructure	101.0	85.1	91.8	109.1	82.8	93.9	109.0	83.1	94.0	100.1	88.0	93.1	100.7	82.4	90.1
B10	Shell: Superstructure	100.3	87.7	95.6	97.5	86.5	93.4	97.5	87.1	93.7	98.4	82.6	92.5	98.5	75.6	90.0
B20	Exterior Closure	97.6	90.6	94.5	99.1	81.4	91.2	96.6	82.8	90.4	95.7	88.3	92.4	95.4	78.8	88.0
B30	Roofing	104.3	86.5	97.1	107.6	84.2	98.2	107.3	84.9	98.3	105.0	91.1	99.4	106.0	88.2	98.9
C	Interior Construction	99.3	68.8	86.6	96.3	68.6	84.7	95.8	67.8	84.2	102.2	83.2	94.3	98.7	72.9	88.0
D10	Services: Conveying	100.0	97.0	99.1	100.0	89.8	96.9	100.0	96.6	99.0	100.0	91.5	97.5	100.0	90.6	97.2
D20 - 40	Mechanical	100.0	74.1	89.8	100.1	72.3	89.2	100.1	72.1	89.1	100.0	87.8	95.2	100.0	78.7	91.6
D50	Electrical	98.5	69.4	84.2	97.3	68.3	83.0	100.1	72.6	86.6	98.6	81.8	90.4	97.9	85.0	91.5
E	Equipment & Furnishings	100.0	67.9	98.2	100.0	67.7	98.1	100.0	68.0	98.2	100.0	80.9	98.9	100.0	72.2	98.4
G	Site Work	101.2	97.6	98.8	111.2	93.5	99.4	108.7	94.0	98.9	97.4	94.4	95.4	97.6	94.1	95.3
A-G	**WEIGHTED AVERAGE**	99.7	80.3	91.5	99.4	77.6	90.2	99.3	78.6	90.5	99.4	86.0	93.8	98.9	79.9	90.9

DIV. NO.	BUILDING SYSTEMS	OHIO														
		CINCINNATI			CLEVELAND			COLUMBUS			DAYTON			LORAIN		
		MAT.	INST.	TOTAL	MAT.	INST.	TOTAL	MAT.	INST.	TOTAL	MAT.	INST.	TOTAL	MAT.	INST.	TOTAL
A	Substructure	94.4	83.2	87.9	98.1	93.6	95.5	100.9	82.3	90.1	87.8	82.2	84.5	96.9	85.7	90.4
B10	Shell: Superstructure	92.5	80.5	88.0	99.1	87.3	94.7	98.7	79.1	91.4	90.7	77.5	85.8	98.2	81.4	91.9
B20	Exterior Closure	90.2	79.2	85.3	99.4	95.6	97.7	94.0	83.3	89.2	85.2	76.0	81.1	93.8	91.2	92.6
B30	Roofing	102.8	81.4	94.2	102.0	96.7	99.9	93.4	82.8	89.2	108.9	80.0	97.3	105.9	89.8	99.4
C	Interior Construction	97.2	78.9	89.6	98.0	89.5	94.5	96.5	77.2	88.5	96.2	75.5	87.5	98.7	77.2	89.7
D10	Services: Conveying	100.0	89.5	96.9	100.0	97.5	99.2	100.0	89.6	96.9	100.0	86.1	95.8	100.0	93.6	98.1
D20 - 40	Mechanical	99.9	76.9	90.9	100.0	89.9	96.0	100.0	84.1	93.7	100.7	81.2	93.0	100.0	88.4	95.4
D50	Electrical	96.3	74.2	85.4	98.3	92.7	95.5	99.9	80.2	90.2	94.9	75.7	85.4	98.0	78.2	88.2
E	Equipment & Furnishings	100.0	79.8	98.8	100.0	88.7	99.3	100.0	77.7	98.7	100.0	77.0	98.7	100.0	70.4	98.3
G	Site Work	91.2	98.3	95.9	95.8	96.4	96.2	102.8	92.3	95.8	91.2	94.4	93.3	96.6	94.5	95.2
A-G	**WEIGHTED AVERAGE**	96.1	80.0	89.3	99.1	91.6	95.9	98.4	82.0	91.4	94.8	79.2	88.2	98.4	84.7	92.6

DIV. NO.	BUILDING SYSTEMS	OHIO									OKLAHOMA					
		SPRINGFIELD			TOLEDO			YOUNGSTOWN			ENID			LAWTON		
		MAT.	INST.	TOTAL	MAT.	INST.	TOTAL	MAT.	INST.	TOTAL	MAT.	INST.	TOTAL	MAT.	INST.	TOTAL
A	Substructure	89.3	82.0	85.1	97.7	87.5	91.8	99.4	84.5	90.8	90.5	69.2	78.2	88.7	69.5	77.6
B10	Shell: Superstructure	91.0	77.5	85.9	97.9	85.7	93.3	98.3	79.2	91.1	96.8	62.0	83.9	99.5	62.0	85.5
B20	Exterior Closure	84.9	75.6	80.7	92.8	90.3	91.7	95.2	84.8	90.6	94.5	57.0	77.8	90.8	57.0	75.7
B30	Roofing	108.8	79.7	97.1	90.7	90.4	90.6	106.2	87.7	98.8	101.1	65.2	86.7	100.8	65.2	86.5
C	Interior Construction	95.4	75.4	87.0	95.1	85.6	91.1	98.7	78.2	90.1	94.0	56.7	78.4	95.6	56.7	79.4
D10	Services: Conveying	100.0	85.9	95.8	100.0	92.2	97.7	100.0	90.4	97.1	100.0	82.6	94.8	100.0	82.6	94.8
D20 - 40	Mechanical	100.7	80.9	92.9	100.0	93.6	97.5	100.0	83.9	93.7	100.0	67.5	87.2	100.0	67.5	87.2
D50	Electrical	94.9	80.2	87.7	100.0	103.4	101.7	98.0	74.3	86.3	95.9	71.0	83.7	97.6	69.2	83.6
E	Equipment & Furnishings	100.0	77.2	98.7	100.0	84.7	99.1	100.0	74.3	98.5	100.0	55.1	97.4	100.0	55.0	97.4
G	Site Work	91.6	94.3	93.4	100.4	90.8	94.0	97.3	94.2	95.2	101.3	91.6	94.9	97.1	93.1	94.4
A-G	WEIGHTED AVERAGE	94.8	79.7	88.4	97.6	91.2	94.9	98.7	81.9	91.6	97.0	66.0	83.9	97.3	65.8	84.0

DIV. NO.	BUILDING SYSTEMS	OKLAHOMA									OREGON					
		MUSKOGEE			OKLAHOMA CITY			TULSA			EUGENE			MEDFORD		
		MAT.	INST.	TOTAL	MAT.	INST.	TOTAL	MAT.	INST.	TOTAL	MAT.	INST.	TOTAL	MAT.	INST.	TOTAL
A	Substructure	84.5	68.1	75.0	90.2	74.5	81.1	90.0	69.5	78.2	107.8	100.1	103.4	110.2	100.0	104.3
B10	Shell: Superstructure	92.4	68.0	83.3	94.5	65.4	83.6	95.8	70.6	86.4	108.6	97.7	104.6	108.4	97.5	104.3
B20	Exterior Closure	93.6	50.9	74.6	92.4	57.7	76.9	89.3	58.0	75.4	95.1	101.0	97.7	100.2	101.0	100.5
B30	Roofing	96.8	63.0	83.3	92.1	67.7	82.3	96.9	64.7	84.0	117.8	102.5	111.6	118.5	94.7	109.0
C	Interior Construction	92.4	50.1	74.8	92.7	63.3	80.5	93.1	56.1	77.7	97.8	95.3	96.8	99.9	94.9	97.8
D10	Services: Conveying	100.0	79.1	93.7	100.0	83.0	94.9	100.0	79.1	93.7	100.0	102.8	100.8	100.0	102.8	100.8
D20 - 40	Mechanical	100.1	61.3	84.9	99.9	67.8	87.3	100.1	63.7	85.8	100.1	98.6	99.5	100.1	105.2	102.1
D50	Electrical	95.4	67.2	81.5	103.4	71.1	87.5	97.3	67.0	82.4	100.7	99.8	100.2	104.1	82.6	93.5
E	Equipment & Furnishings	100.0	54.9	97.4	100.0	65.0	98.0	100.0	55.7	97.5	100.0	96.1	99.8	100.0	95.3	99.7
G	Site Work	90.1	87.7	88.5	94.6	98.4	97.2	95.7	87.1	90.0	97.3	98.5	98.1	104.5	98.5	100.5
A-G	WEIGHTED AVERAGE	95.0	62.9	81.4	96.4	68.7	84.7	95.9	65.8	83.2	101.7	98.7	100.5	103.3	97.4	100.8

DIV. NO.	BUILDING SYSTEMS	OREGON						PENNSYLVANIA								
		PORTLAND			SALEM			ALLENTOWN			ALTOONA			ERIE		
		MAT.	INST.	TOTAL	MAT.	INST.	TOTAL	MAT.	INST.	TOTAL	MAT.	INST.	TOTAL	MAT.	INST.	TOTAL
A	Substructure	110.1	100.2	104.4	104.5	101.3	102.6	91.6	104.3	98.9	97.4	90.3	93.3	97.5	92.3	94.5
B10	Shell: Superstructure	110.3	97.9	105.6	114.7	97.6	108.3	97.6	114.4	103.8	93.2	102.2	96.6	94.1	103.8	97.7
B20	Exterior Closure	95.4	101.0	97.9	97.1	101.0	98.8	93.1	96.0	94.4	85.1	85.6	85.3	79.3	88.7	83.5
B30	Roofing	117.7	99.5	110.4	114.4	100.3	108.7	104.2	110.3	106.6	102.9	90.4	97.8	103.4	88.8	97.5
C	Interior Construction	96.7	96.0	96.4	101.3	95.9	99.1	91.3	106.6	97.6	86.4	87.2	86.7	88.1	88.4	88.3
D10	Services: Conveying	100.0	102.8	100.8	100.0	103.2	101.0	100.0	98.2	99.5	100.0	96.1	98.8	100.0	97.5	99.2
D20 - 40	Mechanical	100.1	111.5	104.6	100.1	106.9	102.8	100.3	114.5	105.9	99.8	83.6	93.4	99.8	93.6	97.4
D50	Electrical	100.9	108.7	104.7	109.3	99.8	104.6	98.0	95.8	96.9	88.7	107.7	98.0	90.3	93.1	91.7
E	Equipment & Furnishings	100.0	96.5	99.8	100.0	96.2	99.8	100.0	110.0	100.6	100.0	82.6	99.0	100.0	84.6	99.1
G	Site Work	100.2	98.5	99.1	94.1	102.2	99.5	91.8	96.7	95.1	94.0	96.5	95.7	91.0	96.9	94.9
A-G	WEIGHTED AVERAGE	102.1	102.8	102.4	104.4	100.8	102.9	96.7	105.8	100.6	93.3	92.8	93.1	93.2	93.9	93.5

DIV. NO.	BUILDING SYSTEMS	PENNSYLVANIA														
		HARRISBURG			PHILADELPHIA			PITTSBURGH			READING			SCRANTON		
		MAT.	INST.	TOTAL	MAT.	INST.	TOTAL	MAT.	INST.	TOTAL	MAT.	INST.	TOTAL	MAT.	INST.	TOTAL
A	Substructure	93.5	95.6	94.7	93.8	131.0	115.3	104.6	101.4	102.8	85.1	102.0	94.9	94.2	94.9	94.6
B10	Shell: Superstructure	103.1	106.2	104.3	101.5	132.2	113.0	103.8	104.9	104.2	97.2	116.9	104.6	99.7	107.7	102.6
B20	Exterior Closure	89.9	85.5	88.0	104.0	139.5	119.9	100.6	102.1	101.3	96.0	95.1	95.6	93.3	94.9	94.0
B30	Roofing	98.9	102.6	100.4	103.5	139.6	118.0	98.6	98.3	98.5	109.4	105.1	107.7	104.1	91.2	98.9
C	Interior Construction	98.8	86.6	93.7	98.0	146.1	118.0	99.6	100.5	99.9	88.0	90.1	88.9	92.2	89.4	91.0
D10	Services: Conveying	100.0	96.7	99.0	100.0	116.0	104.8	100.0	103.8	101.1	100.0	97.5	99.2	100.0	96.1	98.8
D20 - 40	Mechanical	100.1	91.7	96.8	100.1	141.2	116.3	100.0	99.8	99.9	100.3	106.8	102.8	100.3	96.0	98.6
D50	Electrical	99.7	85.8	92.9	98.8	159.0	128.4	97.1	111.3	104.1	97.8	91.3	94.6	98.0	95.2	96.6
E	Equipment & Furnishings	100.0	84.6	99.1	100.0	144.3	102.6	100.0	97.1	99.8	100.0	82.6	99.0	100.0	83.7	99.1
G	Site Work	87.8	100.0	95.9	99.1	101.1	100.5	103.5	96.0	98.5	101.2	107.5	105.4	92.4	96.7	95.3
A-G	WEIGHTED AVERAGE	98.7	93.0	96.3	100.2	138.7	116.5	100.7	102.7	101.5	96.5	101.5	98.6	97.4	96.5	97.1

DIV. NO.	BUILDING SYSTEMS	PENNSYLVANIA			PUERTO RICO			RHODE ISLAND			SOUTH CAROLINA					
		YORK			SAN JUAN			PROVIDENCE			CHARLESTON			COLUMBIA		
		MAT.	INST.	TOTAL	MAT.	INST.	TOTAL	MAT.	INST.	TOTAL	MAT.	INST.	TOTAL	MAT.	INST.	TOTAL
A	Substructure	86.0	94.7	91.0	83.2	39.3	57.9	91.6	118.1	106.9	104.6	70.9	85.2	106.0	72.0	86.4
B10	Shell: Superstructure	97.0	106.9	100.7	110.6	33.3	81.8	100.8	119.2	107.7	104.1	80.5	95.3	102.3	79.9	94.0
B20	Exterior Closure	93.9	88.0	91.2	104.9	24.6	69.1	98.1	122.0	108.8	89.7	69.3	80.6	83.9	69.2	77.3
B30	Roofing	95.8	103.0	98.7	135.1	28.3	92.2	108.5	124.6	115.0	96.8	64.3	83.8	93.2	64.5	81.6
C	Interior Construction	89.0	87.6	88.4	102.1	22.5	69.0	96.4	125.5	108.5	92.7	68.1	82.5	93.1	67.9	82.6
D10	Services: Conveying	100.0	97.1	99.1	100.0	27.6	78.2	100.0	108.7	102.6	100.0	70.9	91.3	100.0	71.1	91.3
D20 - 40	Mechanical	100.2	92.7	97.3	88.5	18.8	61.1	100.2	114.8	105.9	100.4	58.4	83.9	100.0	58.3	83.6
D50	Electrical	93.7	82.9	88.4	84.4	25.0	55.2	102.3	95.5	99.0	98.1	58.9	78.8	98.5	62.8	80.9
E	Equipment & Furnishings	100.0	84.5	99.1	100.0	19.6	95.4	100.0	122.3	101.3	100.0	65.5	98.0	100.0	65.7	98.0
G	Site Work	83.5	95.1	91.2	80.4	88.3	85.7	89.6	102.5	98.2	108.4	84.6	92.5	106.6	89.0	94.9
A-G	WEIGHTED AVERAGE	95.2	93.0	94.3	98.5	29.8	69.4	99.3	115.0	105.9	98.7	68.2	85.8	97.6	68.9	85.5

DIV. NO.	BUILDING SYSTEMS	SOUTH CAROLINA									SOUTH DAKOTA					
		FLORENCE			GREENVILLE			SPARTANBURG			ABERDEEN			PIERRE		
		MAT.	INST.	TOTAL	MAT.	INST.	TOTAL	MAT.	INST.	TOTAL	MAT.	INST.	TOTAL	MAT.	INST.	TOTAL
A	Substructure	94.5	70.8	80.8	94.8	70.6	80.8	95.0	71.0	81.2	105.7	80.2	91.0	101.2	73.1	85.0
B10	Shell: Superstructure	100.3	80.0	92.7	100.9	80.0	93.1	101.1	80.4	93.4	96.6	79.3	90.2	97.5	76.6	89.7
B20	Exterior Closure	89.4	69.2	80.4	86.5	69.2	78.8	87.3	69.3	79.3	96.6	70.7	85.1	92.1	73.9	84.0
B30	Roofing	97.2	64.3	84.0	97.1	64.3	83.9	97.1	64.3	83.9	103.2	79.9	93.8	104.1	73.9	92.0
C	Interior Construction	89.4	67.9	80.5	90.3	67.9	81.0	90.7	67.9	81.2	94.9	66.8	83.2	99.6	45.7	77.1
D10	Services: Conveying	100.0	70.9	91.3	100.0	70.9	91.3	100.0	70.9	91.3	100.0	86.2	95.8	100.0	88.5	96.5
D20 - 40	Mechanical	100.4	58.2	83.9	100.4	58.2	83.9	100.4	58.2	83.9	100.0	54.2	82.0	100.0	76.8	90.9
D50	Electrical	96.1	62.8	79.7	98.2	59.8	79.3	98.2	59.8	79.3	101.1	65.3	83.4	103.9	47.7	76.2
E	Equipment & Furnishings	100.0	65.6	98.0	100.0	65.5	98.0	100.0	65.5	98.0	100.0	73.9	98.5	100.0	35.5	96.3
G	Site Work	116.7	84.3	95.1	112.0	84.8	93.9	111.7	84.8	93.8	100.1	93.3	95.5	97.4	96.2	96.6
A-G	WEIGHTED AVERAGE	96.9	68.6	84.9	97.0	68.2	84.8	97.2	68.3	85.0	98.6	70.1	86.5	99.0	68.5	86.1

DIV. NO.	BUILDING SYSTEMS	SOUTH DAKOTA						TENNESSEE								
		RAPID CITY			SIOUX FALLS			CHATTANOOGA			JACKSON			JOHNSON CITY		
		MAT.	INST.	TOTAL	MAT.	INST.	TOTAL	MAT.	INST.	TOTAL	MAT.	INST.	TOTAL	MAT.	INST.	TOTAL
A	Substructure	102.8	76.3	87.5	93.4	85.8	89.0	99.1	68.9	81.7	98.8	63.8	78.6	86.3	64.0	73.5
B10	Shell: Superstructure	97.4	80.5	91.1	95.2	87.4	92.3	94.1	75.6	87.2	96.2	72.6	87.4	88.6	74.1	83.2
B20	Exterior Closure	95.3	75.5	86.5	88.4	77.7	83.6	94.3	58.8	78.5	97.9	47.3	75.3	119.1	47.9	87.3
B30	Roofing	103.7	80.3	94.3	103.3	85.6	96.2	99.3	62.7	84.6	98.1	55.8	81.1	95.4	56.3	79.7
C	Interior Construction	96.7	61.1	81.9	101.0	79.1	91.9	98.0	56.6	80.8	91.5	43.7	71.6	97.9	56.3	80.6
D10	Services: Conveying	100.0	88.1	96.4	100.0	88.5	96.5	100.0	69.2	90.7	100.0	64.8	89.4	100.0	77.7	93.3
D20 - 40	Mechanical	100.0	76.8	90.9	99.9	70.2	88.2	100.1	59.3	84.1	100.1	61.4	84.9	100.0	55.5	82.5
D50	Electrical	97.6	47.7	73.0	102.6	63.4	83.3	100.8	82.9	92.0	100.5	65.7	83.3	91.5	40.8	66.5
E	Equipment & Furnishings	100.0	48.5	97.0	100.0	75.8	98.6	100.0	57.5	97.6	100.0	40.6	96.6	100.0	60.4	97.7
G	Site Work	98.0	92.6	94.4	92.2	98.6	96.5	106.2	93.3	97.6	101.8	93.5	96.3	113.1	80.5	91.4
A-G	**WEIGHTED AVERAGE**	98.3	72.0	87.2	97.7	78.4	89.5	98.0	68.1	85.3	97.6	61.4	82.3	98.1	58.6	81.4

DIV. NO.	BUILDING SYSTEMS	TENNESSEE									TEXAS					
		KNOXVILLE			MEMPHIS			NASHVILLE			ABILENE			AMARILLO		
		MAT.	INST.	TOTAL	MAT.	INST.	TOTAL	MAT.	INST.	TOTAL	MAT.	INST.	TOTAL	MAT.	INST.	TOTAL
A	Substructure	93.3	67.8	78.6	98.6	75.4	85.2	93.2	75.3	82.9	86.4	66.4	74.9	87.1	64.5	74.1
B10	Shell: Superstructure	93.6	75.7	86.9	89.7	76.3	84.7	99.8	77.6	91.5	102.8	65.8	89.0	98.8	62.8	85.4
B20	Exterior Closure	84.2	52.3	70.0	94.1	58.1	78.0	93.2	59.2	78.0	93.0	61.0	78.7	94.8	60.2	79.4
B30	Roofing	92.8	63.2	80.9	93.0	68.2	83.1	96.9	66.5	84.7	98.4	62.6	84.0	100.2	61.5	84.7
C	Interior Construction	93.1	59.2	79.0	97.5	62.2	82.8	99.1	66.3	85.4	94.2	59.5	79.8	98.2	52.8	79.3
D10	Services: Conveying	100.0	81.6	94.5	100.0	82.6	94.8	100.0	82.7	94.8	100.0	81.3	94.4	100.0	75.5	92.6
D20 - 40	Mechanical	100.0	60.5	84.5	100.1	70.3	88.4	99.9	75.1	90.2	100.1	52.1	81.3	99.9	51.1	80.7
D50	Electrical	97.0	54.9	76.2	102.9	63.6	83.5	93.5	63.2	78.6	97.9	54.0	76.3	100.9	58.9	80.2
E	Equipment & Furnishings	100.0	62.0	97.8	100.0	64.7	98.0	100.0	69.1	98.2	100.0	60.0	97.7	100.0	51.0	97.2
G	Site Work	93.5	83.2	86.7	90.9	95.1	93.7	103.5	97.9	99.8	92.5	87.3	89.1	91.3	91.3	91.3
A-G	**WEIGHTED AVERAGE**	94.8	63.7	81.6	96.6	69.8	85.3	98.0	71.8	86.9	97.8	61.4	82.4	98.2	60.0	82.1

DIV. NO.	BUILDING SYSTEMS	TEXAS														
		AUSTIN			BEAUMONT			CORPUS CHRISTI			DALLAS			EL PASO		
		MAT.	INST.	TOTAL	MAT.	INST.	TOTAL	MAT.	INST.	TOTAL	MAT.	INST.	TOTAL	MAT.	INST.	TOTAL
A	Substructure	92.9	64.5	76.5	90.0	67.1	76.8	111.6	63.8	84.0	94.4	72.0	81.4	80.7	67.5	73.1
B10	Shell: Superstructure	98.9	62.3	85.3	98.1	68.4	87.0	99.1	71.8	88.9	99.9	74.2	90.3	97.7	64.9	85.5
B20	Exterior Closure	90.1	60.0	76.7	98.5	62.1	82.2	84.6	61.4	74.2	97.5	61.8	81.6	91.6	62.0	78.4
B30	Roofing	95.1	63.1	82.2	92.8	63.4	80.9	97.6	61.2	83.0	87.2	67.5	79.2	91.8	63.2	80.3
C	Interior Construction	96.3	52.9	78.2	95.5	57.2	79.6	101.4	53.1	81.3	96.9	62.1	82.4	93.7	57.4	78.6
D10	Services: Conveying	100.0	79.6	93.9	100.0	80.8	94.2	100.0	81.1	94.3	100.0	85.0	95.5	100.0	79.3	93.8
D20 - 40	Mechanical	99.9	58.7	83.8	100.0	63.0	85.5	100.0	48.7	79.9	99.9	62.3	85.1	99.8	65.3	86.3
D50	Electrical	97.5	57.2	77.6	100.1	63.4	82.0	94.0	63.4	78.9	95.2	63.2	79.4	97.5	50.5	74.3
E	Equipment & Furnishings	100.0	52.6	97.3	100.0	54.3	97.4	100.0	51.9	97.2	100.0	63.1	97.9	100.0	56.5	97.5
G	Site Work	95.5	89.9	91.8	92.3	91.7	91.9	139.4	82.0	101.2	104.7	97.3	99.8	89.8	92.6	91.7
A-G	**WEIGHTED AVERAGE**	97.2	61.4	82.1	97.9	65.5	84.2	98.9	61.4	83.1	98.2	68.0	85.4	95.9	63.6	82.3

DIV. NO.	BUILDING SYSTEMS	TEXAS														
		FORT WORTH			HOUSTON			LAREDO			LUBBOCK			ODESSA		
		MAT.	INST.	TOTAL	MAT.	INST.	TOTAL	MAT.	INST.	TOTAL	MAT.	INST.	TOTAL	MAT.	INST.	TOTAL
A	Substructure	90.1	67.9	77.3	91.9	69.5	79.0	88.4	63.6	74.1	90.2	65.3	75.9	86.5	66.6	75.0
B10	Shell: Superstructure	101.4	65.5	88.0	101.1	70.0	89.5	97.1	63.2	84.5	105.9	73.0	93.7	102.2	65.8	88.6
B20	Exterior Closure	89.2	59.8	76.1	98.3	63.7	82.8	90.7	60.2	77.1	91.9	62.3	78.7	93.0	60.6	78.5
B30	Roofing	89.4	64.1	79.3	86.1	68.1	78.8	93.8	62.2	81.1	88.0	62.6	77.8	98.4	61.9	83.7
C	Interior Construction	95.4	59.1	80.3	104.7	60.0	86.1	94.7	52.4	77.1	98.9	54.6	80.4	94.2	58.9	79.5
D10	Services: Conveying	100.0	81.7	94.5	100.0	86.2	95.9	100.0	78.9	93.7	100.0	81.7	94.5	100.0	75.2	92.5
D20 - 40	Mechanical	99.9	55.4	82.4	100.0	65.3	86.3	100.0	58.9	83.8	99.7	53.1	81.4	100.1	52.5	81.4
D50	Electrical	100.2	58.6	79.7	101.8	67.8	85.1	96.0	57.1	76.8	96.7	60.1	78.6	98.0	58.9	78.7
E	Equipment & Furnishings	100.0	60.4	97.7	100.0	59.2	97.7	100.0	51.5	97.2	100.0	52.3	97.3	100.0	60.4	97.7
G	Site Work	98.2	92.6	94.5	106.4	95.3	99.0	99.8	86.0	90.6	112.4	86.1	94.9	92.5	88.0	89.5
A-G	**WEIGHTED AVERAGE**	97.6	62.9	82.9	100.4	68.2	86.8	96.4	61.1	81.5	99.2	62.9	83.9	97.7	61.8	82.6

DIV. NO.	BUILDING SYSTEMS	TEXAS									UTAH					
		SAN ANTONIO			WACO			WICHITA FALLS			LOGAN			OGDEN		
		MAT.	INST.	TOTAL	MAT.	INST.	TOTAL	MAT.	INST.	TOTAL	MAT.	INST.	TOTAL	MAT.	INST.	TOTAL
A	Substructure	89.2	65.9	75.8	83.4	66.1	73.4	87.0	66.6	75.3	93.0	76.6	83.5	92.3	76.6	83.3
B10	Shell: Superstructure	99.1	61.5	85.1	100.6	64.4	87.1	101.3	65.8	88.1	105.9	77.6	95.3	106.4	77.6	95.7
B20	Exterior Closure	91.8	60.0	77.6	87.1	60.8	75.3	87.2	60.6	75.3	114.3	61.5	90.8	99.5	61.5	82.6
B30	Roofing	86.6	65.1	78.0	92.2	63.0	80.5	92.2	62.2	80.1	103.1	70.7	90.1	101.9	70.7	89.3
C	Interior Construction	103.4	53.1	82.4	78.1	58.7	70.0	78.2	60.2	70.7	92.0	65.8	81.1	91.1	65.8	80.6
D10	Services: Conveying	100.0	81.4	94.4	100.0	81.3	94.4	100.0	75.2	92.5	100.0	86.7	96.0	100.0	86.7	96.0
D20 - 40	Mechanical	100.0	59.8	84.2	100.0	58.7	83.8	100.0	52.1	81.2	100.0	69.2	87.9	100.0	69.2	87.9
D50	Electrical	97.5	59.8	78.9	103.0	54.1	78.9	105.0	56.0	80.8	95.2	71.3	83.4	95.5	71.3	83.6
E	Equipment & Furnishings	100.0	51.9	97.2	100.0	60.1	97.7	100.0	60.2	97.7	100.0	67.1	98.1	100.0	67.1	98.1
G	Site Work	100.1	93.7	95.9	96.2	87.9	90.7	97.1	88.0	91.0	99.6	89.3	92.8	87.9	89.3	88.9
A-G	**WEIGHTED AVERAGE**	98.3	62.3	83.1	94.4	62.4	80.9	95.0	61.6	80.9	100.9	71.7	88.5	98.8	71.7	87.4

DIV. NO.	BUILDING SYSTEMS	UTAH						VERMONT						VIRGINIA		
		PROVO			SALT LAKE CITY			BURLINGTON			RUTLAND			ALEXANDRIA		
		MAT.	INST.	TOTAL	MAT.	INST.	TOTAL	MAT.	INST.	TOTAL	MAT.	INST.	TOTAL	MAT.	INST.	TOTAL
A	Substructure	94.2	76.3	83.8	98.1	76.6	85.7	106.1	95.8	100.2	92.1	94.4	93.4	102.2	79.5	89.1
B10	Shell: Superstructure	104.7	77.6	94.6	110.7	77.6	98.4	103.7	90.7	98.8	99.3	90.8	96.1	105.0	89.3	99.2
B20	Exterior Closure	115.7	61.5	91.5	122.7	61.5	95.4	97.6	85.6	92.3	90.9	85.7	88.6	93.7	75.8	85.7
B30	Roofing	105.0	70.7	91.2	110.2	70.7	94.3	107.5	85.8	98.8	101.3	85.2	94.8	102.1	82.1	94.1
C	Interior Construction	94.2	65.8	82.4	92.8	65.8	81.6	101.4	82.9	93.7	98.3	82.7	91.8	94.1	72.6	85.2
D10	Services: Conveying	100.0	86.7	96.0	100.0	86.7	96.0	100.0	90.6	97.2	100.0	90.2	97.0	100.0	88.7	96.6
D20 - 40	Mechanical	100.0	69.2	87.9	100.1	69.2	88.0	100.0	68.3	87.6	100.2	68.2	87.7	100.4	86.2	94.8
D50	Electrical	95.7	68.6	82.3	98.1	71.3	84.9	107.6	53.6	81.0	108.4	53.6	81.4	96.6	100.2	98.4
E	Equipment & Furnishings	100.0	67.0	98.1	100.0	67.1	98.1	100.0	76.7	98.7	100.0	76.5	98.6	100.0	71.1	98.3
G	Site Work	96.3	87.8	90.6	87.8	89.3	88.8	94.0	101.2	98.8	90.4	96.2	94.3	115.9	88.5	97.7
A-G	**WEIGHTED AVERAGE**	101.2	71.2	88.5	103.5	71.7	90.1	101.9	79.8	92.5	98.9	79.4	90.7	99.6	84.8	93.3

DIV. NO.	BUILDING SYSTEMS	VIRGINIA														
		ARLINGTON			NEWPORT NEWS			NORFOLK			PORTSMOUTH			RICHMOND		
		MAT.	INST.	TOTAL	MAT.	INST.	TOTAL	MAT.	INST.	TOTAL	MAT.	INST.	TOTAL	MAT.	INST.	TOTAL
A	Substructure	103.5	78.8	89.2	101.5	69.3	82.9	107.8	73.8	88.2	99.9	69.7	82.5	97.0	74.9	84.2
B10	Shell: Superstructure	103.8	89.7	98.5	103.1	79.4	94.3	105.9	80.3	96.4	101.5	79.6	93.3	103.9	82.8	96.0
B20	Exterior Closure	103.4	75.0	90.7	96.3	57.8	79.1	94.6	57.7	78.1	98.6	56.9	80.0	93.9	67.4	82.1
B30	Roofing	104.3	81.4	95.1	105.1	70.5	91.2	104.4	73.2	91.9	105.1	70.8	91.3	101.1	74.2	90.3
C	Interior Construction	93.3	71.8	84.4	92.5	58.7	78.5	92.5	58.6	78.4	91.2	58.9	77.8	96.7	62.4	82.5
D10	Services: Conveying	100.0	86.2	95.9	100.0	86.6	96.0	100.0	86.4	95.9	100.0	86.8	96.0	100.0	86.1	95.8
D20 - 40	Mechanical	100.4	85.6	94.6	100.4	66.8	87.2	100.1	66.8	87.0	100.4	67.1	87.4	100.0	69.1	87.9
D50	Electrical	94.4	100.2	97.2	93.6	64.2	79.1	98.0	60.7	79.6	92.1	60.7	76.7	96.8	70.4	83.8
E	Equipment & Furnishings	100.0	71.0	98.3	100.0	58.8	97.6	100.0	58.8	97.6	100.0	58.7	97.6	100.0	59.3	97.7
G	Site Work	125.3	86.5	99.5	107.9	87.8	94.5	107.8	93.4	98.2	106.2	87.7	93.9	103.8	92.3	96.2
A-G	**WEIGHTED AVERAGE**	100.4	84.2	93.6	98.8	68.3	85.9	99.9	68.7	86.7	98.3	67.8	85.4	99.2	72.8	88.1

DIV. NO.	BUILDING SYSTEMS	VIRGINIA			WASHINGTON											
		ROANOKE			EVERETT			RICHLAND			SEATTLE			SPOKANE		
		MAT.	INST.	TOTAL	MAT.	INST.	TOTAL	MAT.	INST.	TOTAL	MAT.	INST.	TOTAL	MAT.	INST.	TOTAL
A	Substructure	111.6	81.1	94.0	103.6	107.3	105.7	88.4	84.7	86.3	106.3	111.4	109.3	91.1	84.5	87.3
B10	Shell: Superstructure	106.9	90.0	100.6	114.5	102.4	110.0	92.3	85.3	89.7	114.0	106.3	111.1	94.9	84.9	91.2
B20	Exterior Closure	92.9	67.3	81.5	101.0	103.1	101.9	95.8	81.9	89.6	109.3	108.0	108.7	96.2	81.9	89.8
B30	Roofing	103.6	77.7	93.2	112.6	106.7	110.2	162.0	84.4	130.8	111.7	111.9	111.8	158.3	84.7	128.7
C	Interior Construction	93.0	60.5	83.0	105.5	101.9	104.0	112.6	78.2	98.3	107.3	106.5	107.0	111.8	78.4	97.9
D10	Services: Conveying	100.0	80.3	94.1	100.0	103.5	101.1	100.0	94.1	98.2	100.0	104.6	101.4	100.0	94.1	98.2
D20 - 40	Mechanical	100.4	67.9	87.6	100.0	105.2	102.0	100.3	108.6	103.6	99.9	123.0	109.0	100.3	83.7	93.8
D50	Electrical	96.2	60.8	78.8	107.0	104.1	105.6	87.2	92.8	90.0	106.2	118.9	112.5	85.4	77.1	81.3
E	Equipment & Furnishings	100.0	70.4	98.3	100.0	104.8	100.3	100.0	78.2	98.7	100.0	109.5	100.5	100.0	77.4	98.7
G	Site Work	106.9	86.2	93.1	92.2	105.7	101.2	103.4	86.1	91.9	97.1	109.1	105.1	102.8	86.1	91.7
A-G	**WEIGHTED AVERAGE**	100.1	73.6	88.9	105.1	104.0	104.6	99.8	90.0	95.6	106.3	112.5	108.9	100.1	82.4	92.6

DIV. NO.	BUILDING SYSTEMS	WASHINGTON									WEST VIRGINIA					
		TACOMA			VANCOUVER			YAKIMA			CHARLESTON			HUNTINGTON		
		MAT.	INST.	TOTAL	MAT.	INST.	TOTAL	MAT.	INST.	TOTAL	MAT.	INST.	TOTAL	MAT.	INST.	TOTAL
A	Substructure	104.8	106.8	106.0	114.2	98.8	105.3	109.6	95.2	101.3	98.4	92.1	94.8	105.5	91.3	97.3
B10	Shell: Superstructure	115.8	101.2	110.4	115.5	99.3	109.4	115.3	90.6	106.1	97.0	97.3	97.1	101.3	99.2	100.5
B20	Exterior Closure	101.1	100.8	101.0	106.9	106.1	106.5	99.5	84.2	92.6	86.1	90.3	88.0	88.5	92.9	90.4
B30	Roofing	112.3	102.6	108.4	112.2	99.8	107.2	112.5	86.4	102.0	101.4	87.7	95.9	105.3	87.4	98.1
C	Interior Construction	105.0	101.7	103.7	100.6	95.7	98.6	104.6	94.3	100.3	93.3	90.0	91.9	91.3	91.1	91.2
D10	Services: Conveying	100.0	98.9	99.7	100.0	102.3	100.7	100.0	95.1	98.5	100.0	93.1	97.9	100.0	92.8	97.8
D20 - 40	Mechanical	100.0	100.6	100.2	100.0	110.7	104.2	100.0	107.4	102.9	100.1	81.4	92.8	100.6	90.6	96.7
D50	Electrical	106.8	97.7	102.3	113.1	107.3	110.2	109.9	92.9	101.5	98.0	87.1	92.6	96.6	89.6	93.1
E	Equipment & Furnishings	100.0	104.4	100.3	100.0	96.1	99.8	100.0	103.6	100.2	100.0	87.7	99.3	100.0	88.3	99.3
G	Site Work	95.0	105.3	101.9	104.4	92.8	96.7	97.9	103.9	101.9	99.9	92.6	95.0	105.0	87.9	93.6
A-G	**WEIGHTED AVERAGE**	105.4	101.3	103.7	106.6	102.8	105.0	105.6	95.4	101.3	96.5	89.4	93.5	97.8	92.2	95.4

DIV. NO.	BUILDING SYSTEMS	WEST VIRGINIA						WISCONSIN								
		PARKERSBURG			WHEELING			EAU CLAIRE			GREEN BAY			KENOSHA		
		MAT.	INST.	TOTAL	MAT.	INST.	TOTAL	MAT.	INST.	TOTAL	MAT.	INST.	TOTAL	MAT.	INST.	TOTAL
A	Substructure	101.8	87.3	93.4	101.8	88.9	94.4	98.0	98.2	98.1	101.5	99.2	100.2	109.5	103.9	106.3
B10	Shell: Superstructure	103.3	94.6	100.1	103.5	97.8	101.4	94.9	102.0	97.5	98.1	102.2	99.6	102.4	104.3	103.1
B20	Exterior Closure	91.0	85.4	88.5	100.6	86.4	94.3	89.0	97.5	92.8	103.7	99.2	101.7	93.0	107.6	99.5
B30	Roofing	103.9	87.1	97.1	104.1	86.6	97.1	104.2	94.2	100.2	106.3	97.6	102.8	105.6	103.6	104.8
C	Interior Construction	92.8	86.0	90.0	93.3	85.6	90.1	96.4	95.3	96.0	94.9	99.0	96.6	93.2	108.6	99.6
D10	Services: Conveying	100.0	92.1	97.6	100.0	84.6	95.4	100.0	94.2	98.3	100.0	95.6	98.7	100.0	96.2	98.9
D20 - 40	Mechanical	100.3	88.3	95.6	100.4	89.8	96.2	100.0	88.2	95.4	100.2	83.4	93.6	100.0	95.9	98.4
D50	Electrical	96.7	88.3	92.5	94.1	89.3	91.7	104.3	84.4	94.5	99.0	81.2	90.2	100.9	97.1	99.0
E	Equipment & Furnishings	100.0	85.3	99.2	100.0	82.9	99.0	100.0	93.7	99.6	100.0	96.5	99.8	100.0	105.6	100.3
G	Site Work	109.5	87.9	95.1	110.3	87.6	95.2	96.9	98.8	98.2	99.4	95.2	96.6	103.9	98.5	100.3
A-G	**WEIGHTED AVERAGE**	98.6	88.6	94.4	99.6	89.6	95.4	97.5	94.1	96.1	99.4	93.4	96.9	99.4	102.0	100.5

DIV. NO.	BUILDING SYSTEMS	WISCONSIN												WYOMING		
		LA CROSSE			MADISON			MILWAUKEE			RACINE			CASPER		
		MAT.	INST.	TOTAL	MAT.	INST.	TOTAL	MAT.	INST.	TOTAL	MAT.	INST.	TOTAL	MAT.	INST.	TOTAL
A	Substructure	89.7	97.7	94.3	99.6	100.5	100.1	95.8	109.3	103.6	100.8	104.7	103.0	104.2	73.6	86.5
B10	Shell: Superstructure	92.6	100.1	95.4	102.5	99.5	101.4	97.3	104.6	100.0	100.6	104.3	102.0	103.2	72.9	91.9
B20	Exterior Closure	85.4	95.4	89.8	95.2	99.9	97.3	97.8	115.3	105.6	93.9	107.6	100.0	102.1	65.9	85.9
B30	Roofing	103.6	93.0	99.3	106.5	100.6	104.1	104.8	112.7	108.0	105.6	102.1	104.2	113.0	66.5	94.3
C	Interior Construction	94.9	96.3	95.5	97.1	98.6	97.7	101.8	115.1	107.3	94.8	108.3	100.4	104.5	55.9	84.3
D10	Services: Conveying	100.0	92.6	97.8	100.0	97.1	99.1	100.0	104.3	101.3	100.0	96.2	98.9	100.0	98.3	99.5
D20 - 40	Mechanical	100.0	88.1	95.3	99.8	95.3	98.0	99.8	107.0	102.6	99.8	96.0	98.3	100.0	70.5	88.4
D50	Electrical	104.6	84.4	94.7	101.5	94.0	97.8	100.2	102.4	101.3	100.4	98.0	99.2	94.1	60.3	77.5
E	Equipment & Furnishings	100.0	93.7	99.6	100.0	94.4	99.7	100.0	112.1	100.7	100.0	105.6	100.3	100.0	48.3	97.0
G	Site Work	90.7	98.8	96.1	96.3	105.4	102.3	93.4	96.2	95.2	98.0	102.1	100.8	98.1	94.7	95.8
A-G	**WEIGHTED AVERAGE**	95.9	93.5	94.9	99.7	98.1	99.0	99.2	107.9	102.9	98.7	102.3	100.2	101.5	68.9	87.7

DIV. NO.	BUILDING SYSTEMS	WYOMING						CANADA								
		CHEYENNE			ROCK SPRINGS			CALGARY, ALBERTA			EDMONTON, ALBERTA			HALIFAX, NOVA SCOTIA		
		MAT.	INST.	TOTAL	MAT.	INST.	TOTAL	MAT.	INST.	TOTAL	MAT.	INST.	TOTAL	MAT.	INST.	TOTAL
A	Substructure	99.1	76.5	86.1	99.7	75.5	85.7	135.4	101.6	115.9	141.6	102.3	118.9	111.3	89.7	98.8
B10	Shell: Superstructure	104.2	76.3	93.8	102.3	75.1	92.2	131.3	99.5	119.4	135.4	99.5	122.0	131.2	93.1	117.0
B20	Exterior Closure	104.4	66.9	87.7	128.3	63.5	99.4	147.7	88.5	121.3	148.1	88.5	121.5	137.2	85.4	114.1
B30	Roofing	107.4	68.8	91.9	109.2	71.7	94.1	132.1	100.5	119.4	143.8	100.5	126.4	137.5	89.7	118.3
C	Interior Construction	101.3	65.1	86.2	102.9	63.1	86.3	101.6	93.9	98.4	100.5	94.0	97.8	102.6	86.9	96.1
D10	Services: Conveying	100.0	90.6	97.2	100.0	90.1	97.0	131.1	92.5	119.5	131.1	93.6	119.9	131.1	66.5	111.7
D20 - 40	Mechanical	100.1	73.3	89.6	100.0	73.9	89.8	105.1	90.2	99.2	104.9	90.4	99.2	104.9	82.3	96.0
D50	Electrical	95.2	69.0	82.3	92.5	66.0	79.4	107.1	93.9	100.6	115.0	93.9	104.6	113.3	90.6	102.1
E	Equipment & Furnishings	100.0	62.8	97.9	100.0	62.6	97.8	131.1	94.1	129.0	131.1	94.1	129.0	131.1	87.6	128.6
G	Site Work	92.0	89.9	90.6	89.9	89.4	89.6	123.7	118.1	120.0	123.4	121.6	122.2	104.2	103.2	103.5
A-G	**WEIGHTED AVERAGE**	101.1	72.7	89.1	103.4	71.4	89.9	119.7	95.6	109.5	121.8	95.9	110.9	117.9	88.0	105.2

City Cost Indexes

RJ1030-010 Building Systems

DIV. NO.	BUILDING SYSTEMS	CANADA														
		HAMILTON, ONTARIO			KITCHENER, ONTARIO			LAVAL, QUEBEC			LONDON, ONTARIO			MONTREAL, QUEBEC		
		MAT.	INST.	TOTAL	MAT.	INST.	TOTAL	MAT.	INST.	TOTAL	MAT.	INST.	TOTAL	MAT.	INST.	TOTAL
A	Substructure	114.1	100.9	106.5	107.3	95.2	100.3	115.2	85.7	98.2	120.1	98.3	107.5	123.6	96.6	108.1
B10	Shell: Superstructure	130.9	102.6	120.3	121.1	95.6	111.6	112.9	84.5	102.3	132.2	101.1	120.6	140.2	98.1	124.5
B20	Exterior Closure	127.7	100.2	115.4	103.3	97.3	100.7	122.0	76.3	101.6	133.1	98.9	117.8	131.2	87.4	111.6
B30	Roofing	134.2	101.2	120.9	116.4	97.7	108.9	115.3	87.6	104.2	130.0	98.4	117.3	125.8	98.6	114.9
C	Interior Construction	102.4	95.1	99.4	91.4	89.0	90.4	98.0	82.2	91.4	99.7	90.3	95.8	103.4	91.9	98.6
D10	Services: Conveying	131.1	92.3	119.5	131.1	90.0	118.8	131.1	77.1	114.9	131.1	92.2	119.4	131.1	81.8	116.3
D20 - 40	Mechanical	105.2	92.1	100.0	104.2	91.3	99.1	104.3	86.5	97.3	105.1	89.2	98.8	105.1	81.2	95.7
D50	Electrical	108.5	103.6	106.1	109.6	99.8	104.8	110.9	67.0	89.3	103.5	102.0	102.8	107.9	84.1	96.2
E	Equipment & Furnishings	131.1	94.0	129.0	131.1	86.8	128.6	131.1	80.3	128.2	131.1	87.2	128.6	131.1	89.8	128.8
G	Site Work	106.9	106.7	106.8	95.5	99.7	98.3	98.7	93.6	95.3	104.8	106.4	105.9	114.6	105.1	108.3
A-G	**WEIGHTED AVERAGE**	116.3	98.8	108.9	108.6	94.6	102.7	110.7	81.6	98.4	116.3	96.5	108.0	119.1	90.1	106.9

DIV. NO.	BUILDING SYSTEMS	CANADA														
		OSHAWA, ONTARIO			OTTAWA, ONTARIO			QUEBEC CITY, QUEBEC			REGINA, SASKATCHEWAN			SASKATOON, SASKATCHEWAN		
		MAT.	INST.	TOTAL	MAT.	INST.	TOTAL	MAT.	INST.	TOTAL	MAT.	INST.	TOTAL	MAT.	INST.	TOTAL
A	Substructure	129.8	99.1	112.0	123.3	98.7	109.1	127.1	96.4	109.4	142.7	99.5	117.8	112.9	91.8	100.7
B10	Shell: Superstructure	118.3	98.0	110.7	133.1	102.3	121.6	136.2	98.7	122.2	139.9	97.7	124.1	107.4	89.8	100.9
B20	Exterior Closure	111.8	100.6	106.8	128.7	98.8	115.4	128.5	89.9	111.2	143.6	84.9	117.4	119.3	84.0	103.5
B30	Roofing	117.4	103.8	111.9	137.5	99.4	122.2	124.9	98.6	114.3	144.1	87.4	121.3	116.9	85.2	104.2
C	Interior Construction	94.3	93.0	93.8	105.8	88.6	98.6	102.4	91.8	98.0	108.3	92.1	101.6	98.3	91.5	95.5
D10	Services: Conveying	131.1	90.4	118.9	131.1	90.0	118.8	131.1	81.5	116.2	131.1	68.3	112.2	131.1	66.5	111.7
D20 - 40	Mechanical	104.2	91.9	99.3	105.1	90.7	99.4	105.1	81.2	95.7	104.8	86.9	97.8	104.4	86.7	97.5
D50	Electrical	110.6	103.8	107.3	104.4	101.9	103.2	114.7	84.1	99.6	110.8	91.5	101.3	114.6	91.4	103.2
E	Equipment & Furnishings	131.1	90.3	128.8	131.1	86.1	128.6	131.1	89.8	128.8	131.1	91.6	128.9	131.1	91.3	128.9
G	Site Work	106.6	99.5	101.9	108.0	106.8	107.2	116.6	104.3	108.4	129.1	119.4	122.7	113.2	92.5	99.5
A-G	**WEIGHTED AVERAGE**	110.8	97.2	105.0	117.5	96.8	108.7	118.7	90.4	106.7	123.1	92.4	110.2	109.9	88.5	100.9

DIV. NO.	BUILDING SYSTEMS	CANADA														
		ST. CATHARINES, ONTARIO			ST. JOHN'S, NEWFOUNDLAND			THUNDER BAY, ONTARIO			TORONTO, ONTARIO			VANCOUVER, BRITISH COLUMBIA		
		MAT.	INST.	TOTAL	MAT.	INST.	TOTAL	MAT.	INST.	TOTAL	MAT.	INST.	TOTAL	MAT.	INST.	TOTAL
A	Substructure	103.7	97.0	99.8	139.3	94.9	113.7	109.8	96.2	101.9	114.7	107.2	110.4	125.4	99.1	110.2
B10	Shell: Superstructure	112.5	97.3	106.8	138.5	94.4	122.0	114.1	96.2	107.4	131.4	107.8	122.6	131.7	101.7	120.5
B20	Exterior Closure	103.1	99.5	101.5	145.0	87.4	119.3	106.5	99.0	103.1	127.8	107.9	118.9	128.5	87.4	110.1
B30	Roofing	116.4	101.7	110.5	140.4	96.4	122.7	116.7	99.0	109.6	135.9	109.9	125.5	142.1	90.5	121.3
C	Interior Construction	90.2	94.2	91.9	105.8	80.4	95.2	91.9	93.4	92.5	100.1	101.9	100.9	104.9	91.4	99.3
D10	Services: Conveying	131.1	66.6	111.7	131.1	66.8	111.8	131.1	66.9	111.8	131.1	94.8	120.2	131.1	90.9	119.0
D20 - 40	Mechanical	104.2	91.1	99.0	104.8	83.0	96.2	104.2	91.2	99.1	105.1	99.6	103.0	105.2	87.4	98.2
D50	Electrical	111.3	101.8	106.6	115.6	81.4	98.7	109.6	101.2	105.4	106.8	103.8	105.3	108.7	78.6	93.9
E	Equipment & Furnishings	131.1	93.7	129.0	131.1	82.8	128.4	131.1	91.9	128.9	131.1	100.5	129.4	131.1	89.7	128.8
G	Site Work	96.3	96.0	96.1	117.9	111.5	113.7	100.9	95.9	97.6	109.5	107.3	108.1	116.6	122.5	120.5
A-G	**WEIGHTED AVERAGE**	106.6	95.6	101.9	122.6	87.4	107.7	107.8	95.0	102.4	116.0	104.2	111.0	117.9	92.4	107.2

DIV. NO.	BUILDING SYSTEMS	CANADA														
		WINDSOR, ONTARIO			WINNIPEG, MANITOBA											
		MAT.	INST.	TOTAL	MAT.	INST.	TOTAL	MAT.	INST.	TOTAL	MAT.	INST.	TOTAL	MAT.	INST.	TOTAL
A	Substructure	105.6	95.9	100.0	136.7	78.3	103.0									
B10	Shell: Superstructure	113.3	96.1	106.9	140.5	78.5	117.4									
B20	Exterior Closure	102.9	98.7	101.0	139.1	65.7	106.3									
B30	Roofing	116.4	98.6	109.3	136.3	70.8	110.0									
C	Interior Construction	90.5	91.6	91.0	106.2	64.6	88.9									
D10	Services: Conveying	131.1	66.6	111.8	131.1	63.7	110.9									
D20 - 40	Mechanical	104.2	90.7	98.9	105.2	64.6	89.2									
D50	Electrical	114.5	100.4	107.6	113.1	63.4	88.6									
E	Equipment & Furnishings	131.1	89.7	128.8	131.1	64.7	127.3									
G	Site Work	92.3	95.9	94.7	114.3	115.1	114.8									
A-G	**WEIGHTED AVERAGE**	107.1	94.4	101.8	122.0	71.2	100.5									

Costs shown in RSMeans cost data publications are based on national averages for materials and installation. To adjust these costs to a specific location, simply multiply the base cost by the factor and divide by 100 for that city. The data is arranged alphabetically by state and postal zip code numbers. For a city not listed, use the factor for a nearby city with similar economic characteristics.

STATE/ZIP	CITY	MAT.	INST.	TOTAL
ALABAMA				
350-352	Birmingham	97.4	71.8	86.6
354	Tuscaloosa	97.1	71.4	86.3
355	Jasper	97.8	69.8	86.0
356	Decatur	97.1	69.2	85.3
357-358	Huntsville	97.1	71.7	86.4
359	Gadsden	97.3	70.2	85.9
360-361	Montgomery	96.3	73.1	86.5
362	Anniston	95.9	67.7	84.0
363	Dothan	96.2	72.7	86.3
364	Evergreen	95.8	68.3	84.2
365-366	Mobile	96.7	68.7	84.9
367	Selma	95.9	72.5	86.0
368	Phenix City	96.8	70.8	85.8
369	Butler	96.1	70.6	85.3
ALASKA				
995-996	Anchorage	118.5	113.1	116.2
997	Fairbanks	118.8	113.0	116.3
998	Juneau	116.8	113.1	115.2
999	Ketchikan	128.4	113.4	122.1
ARIZONA				
850,853	Phoenix	100.6	72.3	88.6
851,852	Mesa/Tempe	98.7	71.5	87.2
855	Globe	99.5	71.4	87.6
856-857	Tucson	97.7	70.2	86.1
859	Show Low	99.7	71.5	87.8
860	Flagstaff	102.4	70.5	88.9
863	Prescott	99.9	71.3	87.8
864	Kingman	98.3	70.4	86.5
865	Chambers	98.3	71.8	87.1
ARKANSAS				
716	Pine Bluff	96.0	63.6	82.3
717	Camden	94.0	64.7	81.6
718	Texarkana	94.8	63.3	81.5
719	Hot Springs	93.5	64.4	81.2
720-722	Little Rock	94.5	65.4	82.2
723	West Memphis	93.3	69.8	83.3
724	Jonesboro	93.8	66.6	82.3
725	Batesville	91.6	63.6	79.8
726	Harrison	92.9	62.8	80.2
727	Fayetteville	90.5	64.1	79.3
728	Russellville	91.7	63.1	79.6
729	Fort Smith	94.1	62.6	80.8
CALIFORNIA				
900-902	Los Angeles	98.5	129.8	111.7
903-905	Inglewood	93.8	128.8	108.6
906-908	Long Beach	95.4	128.8	109.5
910-912	Pasadena	94.8	128.6	109.1
913-916	Van Nuys	97.8	128.6	110.8
917-918	Alhambra	96.7	128.6	110.2
919-921	San Diego	99.7	121.7	109.0
922	Palm Springs	97.4	125.3	109.2
923-924	San Bernardino	95.0	126.3	108.2
925	Riverside	99.3	126.6	110.8
926-927	Santa Ana	97.0	126.3	109.4
928	Anaheim	99.3	126.6	110.9
930	Oxnard	98.0	126.7	110.1
931	Santa Barbara	97.2	126.1	109.4
932-933	Bakersfield	98.9	124.6	109.8
934	San Luis Obispo	98.2	125.8	109.9
935	Mojave	95.4	124.4	107.7
936-938	Fresno	98.5	129.4	111.5
939	Salinas	99.1	137.2	115.2
940-941	San Francisco	107.4	158.4	129.0
942,956-958	Sacramento	100.8	133.0	114.4
943	Palo Alto	99.2	153.8	122.3
944	San Mateo	101.5	152.0	122.9
945	Vallejo	100.2	141.6	117.7
946	Oakland	103.4	152.0	123.9
947	Berkeley	103.0	151.9	123.7
948	Richmond	102.4	146.2	120.9
949	San Rafael	104.5	149.1	123.4
950	Santa Cruz	104.6	137.5	118.5

STATE/ZIP	CITY	MAT.	INST.	TOTAL
CALIFORNIA (CONT'D)				
951	San Jose	102.6	153.5	124.2
952	Stockton	100.7	131.3	113.7
953	Modesto	100.6	130.6	113.3
954	Santa Rosa	101.2	148.4	121.2
955	Eureka	102.5	134.1	115.9
959	Marysville	101.7	131.3	114.2
960	Redding	107.6	131.5	117.7
961	Susanville	107.3	131.6	117.5
COLORADO				
800-802	Denver	103.0	75.4	91.3
803	Boulder	98.7	77.0	89.6
804	Golden	100.9	74.7	89.8
805	Fort Collins	102.2	73.5	90.1
806	Greeley	99.3	73.4	88.4
807	Fort Morgan	99.3	73.5	88.4
808-809	Colorado Springs	101.2	71.5	88.6
810	Pueblo	101.8	70.0	88.3
811	Alamosa	103.6	70.3	89.5
812	Salida	103.1	70.9	89.5
813	Durango	103.9	65.6	87.8
814	Montrose	102.6	65.8	87.1
815	Grand Junction	105.7	70.5	90.8
816	Glenwood Springs	103.9	66.8	88.2
CONNECTICUT				
060	New Britain	98.1	116.7	106.0
061	Hartford	99.6	118.0	107.3
062	Willimantic	98.7	117.1	106.5
063	New London	95.2	117.2	104.5
064	Meriden	97.2	117.0	105.6
065	New Haven	99.9	117.0	107.1
066	Bridgeport	99.3	117.0	106.8
067	Waterbury	98.9	117.2	106.6
068	Norwalk	98.8	116.7	106.4
069	Stamford	99.0	123.2	109.2
D.C.				
200-205	Washington	101.1	87.5	95.4
DELAWARE				
197	Newark	98.3	109.7	103.1
198	Wilmington	97.9	109.7	102.9
199	Dover	99.0	109.6	103.5
FLORIDA				
320,322	Jacksonville	96.9	66.6	84.1
321	Daytona Beach	97.2	69.5	85.5
323	Tallahassee	98.2	67.7	85.3
324	Panama City	98.6	68.3	85.8
325	Pensacola	101.2	65.8	86.2
326,344	Gainesville	98.8	65.8	84.8
327-328,347	Orlando	99.0	66.6	85.3
329	Melbourne	100.1	70.3	87.5
330-332,340	Miami	96.7	68.0	84.6
333	Fort Lauderdale	96.1	66.4	83.6
334,349	West Palm Beach	95.1	64.1	82.0
335-336,346	Tampa	97.6	68.4	85.3
337	St. Petersburg	99.7	66.8	85.8
338	Lakeland	97.1	67.6	84.6
339,341	Fort Myers	96.5	67.7	84.3
342	Sarasota	98.8	67.0	85.4
GEORGIA				
300-303,399	Atlanta	98.2	75.0	88.4
304	Statesboro	97.8	69.7	85.9
305	Gainesville	96.2	65.6	83.2
306	Athens	95.6	66.9	83.5
307	Dalton	97.5	71.4	86.5
308-309	Augusta	96.0	73.9	86.6
310-312	Macon	95.2	73.1	85.8
313-314	Savannah	97.3	72.7	86.9
315	Waycross	96.8	70.0	85.5
316	Valdosta	96.7	66.2	83.8
317,398	Albany	96.5	71.9	86.1
318-319	Columbus	96.5	72.2	86.2

STATE/ZIP	CITY	MAT.	INST.	TOTAL
HAWAII				
967	Hilo	114.8	118.4	116.3
968	Honolulu	119.1	118.5	118.9
STATES & POSS.				
969	Guam	137.1	51.4	100.9
IDAHO				
832	Pocatello	102.0	78.4	92.0
833	Twin Falls	103.2	76.8	92.1
834	Idaho Falls	100.5	79.9	91.8
835	Lewiston	108.2	83.5	97.8
836-837	Boise	101.0	79.9	92.1
838	Coeur d'Alene	108.1	84.4	98.1
ILLINOIS				
600-603	North Suburban	98.3	142.9	117.1
604	Joliet	98.3	138.8	115.4
605	South Suburban	98.3	142.8	117.1
606-608	Chicago	100.1	145.6	119.3
609	Kankakee	95.0	134.9	111.9
610-611	Rockford	96.6	125.5	108.8
612	Rock Island	94.6	99.5	96.7
613	La Salle	95.9	127.0	109.0
614	Galesburg	95.7	108.5	101.1
615-616	Peoria	97.6	108.8	102.3
617	Bloomington	95.0	109.5	101.1
618-619	Champaign	98.5	110.5	103.6
620-622	East St. Louis	93.8	110.4	100.9
623	Quincy	95.7	104.2	99.3
624	Effingham	95.1	109.1	101.0
625	Decatur	96.6	110.1	102.3
626-627	Springfield	97.5	108.3	102.1
628	Centralia	92.8	111.2	100.5
629	Carbondale	92.5	109.6	99.7
INDIANA				
460	Anderson	96.0	79.8	89.1
461-462	Indianapolis	98.7	83.2	92.1
463-464	Gary	97.5	108.1	102.0
465-466	South Bend	98.4	81.8	91.4
467-468	Fort Wayne	96.6	76.3	88.0
469	Kokomo	94.2	81.1	88.6
470	Lawrenceburg	92.5	79.6	87.0
471	New Albany	93.9	77.3	86.9
472	Columbus	96.1	80.5	89.5
473	Muncie	96.2	78.2	88.6
474	Bloomington	97.8	81.1	90.7
475	Washington	94.6	84.8	90.4
476-477	Evansville	95.5	82.7	90.1
478	Terre Haute	96.2	81.7	90.1
479	Lafayette	95.8	78.1	88.3
IOWA				
500-503,509	Des Moines	97.2	90.2	94.3
504	Mason City	95.3	74.0	86.3
505	Fort Dodge	95.5	73.7	86.3
506-507	Waterloo	96.8	77.4	88.6
508	Creston	95.7	84.2	90.8
510-511	Sioux City	97.7	81.5	90.9
512	Sibley	96.8	62.4	82.3
513	Spencer	98.2	62.3	83.0
514	Carroll	95.5	82.4	89.9
515	Council Bluffs	98.9	79.2	90.6
516	Shenandoah	96.1	81.9	90.1
520	Dubuque	97.2	79.9	89.9
521	Decorah	96.4	73.7	86.8
522-524	Cedar Rapids	98.0	83.7	92.0
525	Ottumwa	96.3	78.7	88.9
526	Burlington	95.7	84.3	90.9
527-528	Davenport	97.1	94.5	96.0
KANSAS				
660-662	Kansas City	97.3	98.6	97.8
664-666	Topeka	98.6	76.5	89.3
667	Fort Scott	96.0	77.1	88.0
668	Emporia	96.1	76.5	87.8
669	Belleville	97.9	70.7	86.4
670-672	Wichita	97.0	72.6	86.7
673	Independence	97.2	76.0	88.2
674	Salina	97.2	73.5	87.2
675	Hutchinson	92.5	71.9	83.8
676	Hays	96.5	72.0	86.1
677	Colby	97.3	74.0	87.4

STATE/ZIP	CITY	MAT.	INST.	TOTAL
KANSAS (CONT'D)				
678	Dodge City	98.6	73.3	87.9
679	Liberal	96.5	71.7	86.0
KENTUCKY				
400-402	Louisville	93.9	79.6	87.8
403-405	Lexington	93.1	78.1	86.7
406	Frankfort	95.4	80.3	89.0
407-409	Corbin	90.8	78.0	85.4
410	Covington	94.3	77.8	87.3
411-412	Ashland	93.2	89.7	91.7
413-414	Campton	94.4	78.0	87.5
415-416	Pikeville	95.8	85.4	91.4
417-418	Hazard	93.8	78.7	87.4
420	Paducah	92.4	83.2	88.6
421-422	Bowling Green	94.6	77.2	87.3
423	Owensboro	94.7	81.0	88.9
424	Henderson	92.2	81.9	87.8
425-426	Somerset	91.6	79.5	86.5
427	Elizabethtown	91.3	75.6	84.7
LOUISIANA				
700-701	New Orleans	99.0	68.5	86.1
703	Thibodaux	96.0	67.0	83.8
704	Hammond	93.7	64.4	81.3
705	Lafayette	95.4	67.9	83.8
706	Lake Charles	95.6	68.2	84.1
707-708	Baton Rouge	96.6	69.3	85.1
710-711	Shreveport	97.1	66.3	84.1
712	Monroe	95.8	64.4	82.5
713-714	Alexandria	95.9	65.6	83.1
MAINE				
039	Kittery	92.6	87.0	90.2
040-041	Portland	99.6	85.4	93.6
042	Lewiston	97.5	84.9	92.2
043	Augusta	100.4	85.6	94.2
044	Bangor	97.1	83.9	91.5
045	Bath	95.9	85.4	91.4
046	Machias	95.4	85.3	91.1
047	Houlton	95.6	85.2	91.2
048	Rockland	94.6	85.4	90.7
049	Waterville	96.0	85.2	91.4
MARYLAND				
206	Waldorf	97.3	86.6	92.8
207-208	College Park	97.3	87.7	93.2
209	Silver Spring	96.5	86.7	92.4
210-212	Baltimore	101.8	83.2	93.9
214	Annapolis	100.4	86.3	94.4
215	Cumberland	96.9	83.8	91.4
216	Easton	98.6	74.8	88.5
217	Hagerstown	97.4	85.3	92.3
218	Salisbury	99.0	66.9	85.4
219	Elkton	95.9	86.2	91.8
MASSACHUSETTS				
010-011	Springfield	98.2	108.5	102.5
012	Pittsfield	97.7	102.6	99.7
013	Greenfield	95.9	109.0	101.4
014	Fitchburg	94.6	116.2	103.7
015-016	Worcester	98.1	117.0	106.1
017	Framingham	94.0	124.2	106.8
018	Lowell	97.5	123.8	108.6
019	Lawrence	98.4	124.2	109.3
020-022, 024	Boston	101.9	134.0	115.5
023	Brockton	98.5	116.7	106.2
025	Buzzards Bay	93.2	115.2	102.5
026	Hyannis	95.7	115.0	103.8
027	New Bedford	97.7	115.2	105.1
MICHIGAN				
480,483	Royal Oak	95.1	100.3	97.3
481	Ann Arbor	97.1	99.8	98.2
482	Detroit	100.7	102.0	101.3
484-485	Flint	96.8	88.6	93.3
486	Saginaw	96.5	85.1	91.7
487	Bay City	96.6	85.3	91.8
488-489	Lansing	98.0	86.8	93.3
490	Battle Creek	95.5	80.4	89.1
491	Kalamazoo	95.8	78.6	88.6
492	Jackson	94.0	90.7	92.6
493,495	Grand Rapids	97.9	79.9	90.3
494	Muskegon	94.3	80.0	88.3

STATE/ZIP	CITY	MAT.	INST.	TOTAL
MICHIGAN (CONT'D)				
496	Traverse City	93.4	77.7	86.8
497	Gaylord	94.5	80.6	88.6
498-499	Iron Mountain	96.5	81.7	90.2
MINNESOTA				
550-551	Saint Paul	99.0	117.6	106.9
553-555	Minneapolis	100.7	114.1	106.3
556-558	Duluth	98.6	101.3	99.7
559	Rochester	98.3	98.7	98.5
560	Mankato	95.7	98.5	96.9
561	Windom	94.3	91.2	93.0
562	Willmar	94.0	98.0	95.7
563	St. Cloud	95.1	114.2	103.2
564	Brainerd	95.6	98.6	96.8
565	Detroit Lakes	97.4	92.4	95.3
566	Bemidji	96.7	95.4	96.1
567	Thief River Falls	96.3	89.3	93.3
MISSISSIPPI				
386	Clarksdale	95.3	55.2	78.4
387	Greenville	98.7	65.5	84.7
388	Tupelo	96.6	59.0	80.7
389	Greenwood	96.6	55.5	79.2
390-392	Jackson	97.3	65.6	83.9
393	Meridian	94.6	64.5	81.9
394	Laurel	96.1	57.8	79.9
395	Biloxi	96.2	63.9	82.6
396	McComb	94.4	55.8	78.0
397	Columbus	95.9	59.1	80.4
MISSOURI				
630-631	St. Louis	98.8	104.5	101.2
633	Bowling Green	96.4	95.5	96.0
634	Hannibal	95.3	91.6	93.7
635	Kirksville	98.9	86.5	93.7
636	Flat River	97.3	94.2	96.0
637	Cape Girardeau	97.0	91.2	94.5
638	Sikeston	95.7	87.8	92.3
639	Poplar Bluff	95.1	87.7	92.0
640-641	Kansas City	97.9	103.8	100.4
644-645	St. Joseph	96.1	90.9	93.9
646	Chillicothe	93.7	93.8	93.7
647	Harrisonville	93.2	100.8	96.4
648	Joplin	95.1	77.7	87.7
650-651	Jefferson City	96.4	89.9	93.7
652	Columbia	96.1	90.7	93.8
653	Sedalia	96.4	91.5	94.3
654-655	Rolla	94.1	94.3	94.2
656-658	Springfield	97.6	79.0	89.7
MONTANA				
590-591	Billings	101.6	78.4	91.8
592	Wolf Point	101.4	76.9	91.1
593	Miles City	99.2	77.0	89.8
594	Great Falls	102.9	76.6	91.8
595	Havre	100.4	74.8	89.6
596	Helena	100.9	76.7	90.7
597	Butte	101.5	76.5	90.9
598	Missoula	98.6	77.1	89.5
599	Kalispell	98.1	76.2	88.9
NEBRASKA				
680-681	Omaha	97.6	81.5	90.8
683-685	Lincoln	98.3	79.7	90.5
686	Columbus	96.0	80.7	89.6
687	Norfolk	97.4	77.5	89.0
688	Grand Island	97.4	76.8	88.7
689	Hastings	97.1	78.2	89.1
690	McCook	96.3	73.1	86.5
691	North Platte	96.3	75.6	87.6
692	Valentine	98.6	69.2	86.2
693	Alliance	98.7	73.3	87.9
NEVADA				
889-891	Las Vegas	105.0	105.1	105.0
893	Ely	103.7	93.2	99.2
894-895	Reno	103.6	82.6	94.7
897	Carson City	102.9	82.5	94.3
898	Elko	102.3	85.2	95.0
NEW HAMPSHIRE				
030	Nashua	97.5	93.7	95.9
031	Manchester	98.2	94.5	96.6

STATE/ZIP	CITY	MAT.	INST.	TOTAL
NEW HAMPSHIRE (CONT'D)				
032-033	Concord	98.0	93.2	96.0
034	Keene	94.5	89.0	92.2
035	Littleton	94.4	80.2	88.4
036	Charleston	94.1	88.4	91.7
037	Claremont	93.1	88.4	91.1
038	Portsmouth	94.9	92.4	93.8
NEW JERSEY				
070-071	Newark	100.3	139.7	117.0
072	Elizabeth	97.5	138.3	114.7
073	Jersey City	96.5	137.8	114.0
074-075	Paterson	98.0	137.7	114.8
076	Hackensack	96.1	138.3	114.0
077	Long Branch	95.8	130.9	110.7
078	Dover	96.4	137.5	113.7
079	Summit	96.4	138.3	114.1
080,083	Vineland	96.3	129.8	110.5
081	Camden	98.0	127.8	110.6
082,084	Atlantic City	96.8	133.2	112.2
085-086	Trenton	99.7	128.9	112.0
087	Point Pleasant	98.1	130.8	111.9
088-089	New Brunswick	98.7	135.6	114.3
NEW MEXICO				
870-872	Albuquerque	98.6	72.5	87.6
873	Gallup	99.0	72.5	87.8
874	Farmington	99.0	72.5	87.8
875	Santa Fe	99.0	72.9	88.0
877	Las Vegas	97.2	72.5	86.8
878	Socorro	96.8	72.5	86.6
879	Truth/Consequences	96.6	69.4	85.1
880	Las Cruces	96.9	71.4	86.1
881	Clovis	99.3	72.4	87.9
882	Roswell	100.8	72.5	88.9
883	Carrizozo	101.6	72.5	89.3
884	Tucumcari	99.9	72.4	88.3
NEW YORK				
100-102	New York	99.7	175.2	131.6
103	Staten Island	95.6	176.1	129.6
104	Bronx	93.8	175.3	128.3
105	Mount Vernon	94.1	150.6	118.0
106	White Plains	93.8	152.8	118.8
107	Yonkers	97.9	153.0	121.2
108	New Rochelle	94.4	144.8	115.7
109	Suffern	94.1	130.2	109.4
110	Queens	100.0	177.3	132.6
111	Long Island City	101.6	177.3	133.6
112	Brooklyn	101.9	177.3	133.8
113	Flushing	102.1	177.3	133.9
114	Jamaica	100.3	177.3	132.8
115,117,118	Hicksville	99.9	157.1	124.1
116	Far Rockaway	102.2	177.3	133.9
119	Riverhead	100.6	153.5	122.9
120-122	Albany	96.8	111.2	102.8
123	Schenectady	97.0	109.9	102.4
124	Kingston	100.6	133.8	114.6
125-126	Poughkeepsie	99.8	136.8	115.4
127	Monticello	99.1	134.6	114.1
128	Glens Falls	92.0	107.3	98.5
129	Plattsburgh	97.0	96.6	96.8
130-132	Syracuse	97.8	98.3	98.0
133-135	Utica	96.0	96.6	96.2
136	Watertown	97.7	96.8	97.4
137-139	Binghamton	97.3	101.5	99.1
140-142	Buffalo	100.8	110.3	104.8
143	Niagara Falls	97.4	107.5	101.6
144-146	Rochester	100.3	99.2	99.8
147	Jamestown	96.4	96.0	96.3
148-149	Elmira	96.3	102.5	98.9
NORTH CAROLINA				
270,272-274	Greensboro	98.5	67.2	85.3
271	Winston-Salem	98.2	67.3	85.2
275-276	Raleigh	97.4	67.0	84.5
277	Durham	100.1	67.2	86.2
278	Rocky Mount	96.0	66.6	83.6
279	Elizabeth City	96.9	68.3	84.8
280	Gastonia	97.0	66.8	84.2
281-282	Charlotte	97.4	67.6	84.8
283	Fayetteville	100.2	66.1	85.8
284	Wilmington	96.0	65.2	83.0
285	Kinston	94.4	65.5	82.2

STATE/ZIP	CITY	MAT.	INST.	TOTAL
NORTH CAROLINA (CONT'D)				
286	Hickory	94.7	67.3	83.1
287-288	Asheville	96.4	66.6	83.8
289	Murphy	95.5	64.7	82.5
NORTH DAKOTA				
580-581	Fargo	99.7	80.3	91.5
582	Grand Forks	99.4	77.6	90.2
583	Devils Lake	99.2	78.8	90.6
584	Jamestown	99.2	78.8	90.6
585	Bismarck	98.9	81.6	91.6
586	Dickinson	99.9	77.9	90.6
587	Minot	99.3	78.6	90.5
588	Williston	98.4	77.7	89.6
OHIO				
430-432	Columbus	98.4	82.0	91.4
433	Marion	94.3	86.7	91.1
434-436	Toledo	97.6	91.2	94.9
437-438	Zanesville	94.9	85.7	91.0
439	Steubenville	96.1	90.0	93.5
440	Lorain	98.4	84.7	92.6
441	Cleveland	99.1	91.6	95.9
442-443	Akron	99.4	86.0	93.8
444-445	Youngstown	98.7	81.9	91.6
446-447	Canton	98.9	79.9	90.9
448-449	Mansfield	96.4	84.2	91.3
450	Hamilton	94.8	79.5	88.3
451-452	Cincinnati	96.1	80.0	89.3
453-454	Dayton	94.8	79.2	88.2
455	Springfield	94.8	79.7	88.4
456	Chillicothe	94.3	88.5	91.8
457	Athens	97.1	84.8	91.9
458	Lima	97.2	81.7	90.7
OKLAHOMA				
730-731	Oklahoma City	96.4	68.7	84.7
734	Ardmore	95.2	65.6	82.7
735	Lawton	97.3	65.8	84.0
736	Clinton	96.4	65.5	83.3
737	Enid	97.0	66.0	83.9
738	Woodward	95.3	62.9	81.6
739	Guymon	96.3	63.3	82.4
740-741	Tulsa	95.9	65.8	83.2
743	Miami	92.7	65.1	81.1
744	Muskogee	95.0	62.9	81.4
745	McAlester	92.4	60.5	78.9
746	Ponca City	93.1	64.0	80.8
747	Durant	93.1	65.1	81.3
748	Shawnee	94.5	65.2	82.2
749	Poteau	92.3	64.9	80.7
OREGON				
970-972	Portland	102.1	102.8	102.4
973	Salem	104.4	100.8	102.9
974	Eugene	101.7	98.7	100.5
975	Medford	103.3	97.4	100.8
976	Klamath Falls	103.5	97.3	100.9
977	Bend	102.6	100.2	101.6
978	Pendleton	98.6	100.3	99.3
979	Vale	96.2	87.0	92.3
PENNSYLVANIA				
150-152	Pittsburgh	100.7	102.7	101.5
153	Washington	97.5	100.2	98.6
154	Uniontown	97.8	99.8	98.6
155	Bedford	98.8	92.1	96.0
156	Greensburg	98.7	96.3	97.7
157	Indiana	97.6	97.7	97.7
158	Dubois	99.3	96.0	97.9
159	Johnstown	98.9	91.9	95.9
160	Butler	91.3	100.6	95.2
161	New Castle	91.4	97.8	94.1
162	Kittanning	91.8	97.9	94.4
163	Oil City	91.3	97.8	94.0
164-165	Erie	93.2	93.9	93.5
166	Altoona	93.3	92.8	93.1
167	Bradford	94.9	97.1	95.8
168	State College	94.5	94.6	94.5
169	Wellsboro	95.5	92.9	94.4
170-171	Harrisburg	98.7	93.0	96.3
172	Chambersburg	95.1	89.4	92.7
173-174	York	95.2	93.0	94.3
175-176	Lancaster	93.8	94.5	94.1

STATE/ZIP	CITY	MAT.	INST.	TOTAL
PENNSYLVANIA (CONT'D)				
177	Williamsport	92.5	92.3	92.4
178	Sunbury	94.6	92.7	93.8
179	Pottsville	93.7	96.1	94.7
180	Lehigh Valley	94.9	111.6	102.0
181	Allentown	96.7	105.8	100.6
182	Hazleton	94.4	95.9	95.0
183	Stroudsburg	94.3	104.7	98.7
184-185	Scranton	97.4	96.5	97.1
186-187	Wilkes-Barre	94.0	95.5	94.7
188	Montrose	93.8	96.8	95.1
189	Doylestown	94.0	127.1	108.0
190-191	Philadelphia	100.2	138.7	116.5
193	Westchester	95.8	127.9	109.3
194	Norristown	94.7	128.0	108.8
195-196	Reading	96.5	101.5	98.6
PUERTO RICO				
009	San Juan	98.5	29.8	69.4
RHODE ISLAND				
028	Newport	97.0	114.7	104.5
029	Providence	99.3	115.0	105.9
SOUTH CAROLINA				
290-292	Columbia	97.6	68.9	85.5
293	Spartanburg	97.2	68.3	85.0
294	Charleston	98.7	68.2	85.8
295	Florence	96.9	68.6	84.9
296	Greenville	97.0	68.2	84.8
297	Rock Hill	96.7	66.4	83.9
298	Aiken	97.6	67.1	84.7
299	Beaufort	98.3	57.5	81.1
SOUTH DAKOTA				
570-571	Sioux Falls	97.7	78.4	89.5
572	Watertown	97.3	69.0	85.4
573	Mitchell	96.3	60.9	81.3
574	Aberdeen	98.6	70.1	86.5
575	Pierre	99.0	68.5	86.1
576	Mobridge	96.9	62.9	82.5
577	Rapid City	98.3	72.0	87.2
TENNESSEE				
370-372	Nashville	98.0	71.8	86.9
373-374	Chattanooga	98.0	68.1	85.3
375,380-381	Memphis	96.6	69.8	85.3
376	Johnson City	98.1	58.6	81.4
377-379	Knoxville	94.8	63.7	81.6
382	McKenzie	96.3	55.1	78.9
383	Jackson	97.6	61.4	82.3
384	Columbia	94.8	66.5	82.8
385	Cookeville	96.2	56.9	79.6
TEXAS				
750	McKinney	97.2	63.5	82.9
751	Waxahachie	97.2	63.5	83.0
752-753	Dallas	98.2	68.0	85.4
754	Greenville	97.4	62.9	82.8
755	Texarkana	96.7	62.0	82.1
756	Longview	97.5	61.0	82.1
757	Tyler	97.8	61.6	82.5
758	Palestine	94.2	60.4	79.9
759	Lufkin	94.7	62.6	81.1
760-761	Fort Worth	97.6	62.9	82.9
762	Denton	97.3	62.9	82.7
763	Wichita Falls	95.0	61.6	80.9
764	Eastland	94.2	60.3	79.8
765	Temple	92.5	58.5	78.2
766-767	Waco	94.4	62.4	80.9
768	Brownwood	97.4	58.7	81.1
769	San Angelo	97.1	59.4	81.2
770-772	Houston	100.4	68.2	86.8
773	Huntsville	98.7	62.8	83.6
774	Wharton	99.8	64.3	84.8
775	Galveston	97.7	66.5	84.5
776-777	Beaumont	97.9	65.5	84.2
778	Bryan	95.2	64.9	82.4
779	Victoria	99.7	62.9	84.2
780	Laredo	96.4	61.1	81.5
781-782	San Antonio	98.3	62.3	83.1
783-784	Corpus Christi	98.9	61.4	83.1
785	McAllen	99.7	56.5	81.4
786-787	Austin	97.2	61.4	82.1

STATE/ZIP	CITY	MAT.	INST.	TOTAL
TEXAS (CONT'D)				
788	Del Rio	99.7	60.4	83.1
789	Giddings	96.1	60.4	81.0
790-791	Amarillo	98.2	60.0	82.1
792	Childress	97.6	60.3	81.8
793-794	Lubbock	99.2	62.9	83.9
795-796	Abilene	97.8	61.4	82.4
797	Midland	99.7	62.2	83.8
798-799,885	El Paso	95.9	63.6	82.3
UTAH				
840-841	Salt Lake City	103.5	71.7	90.1
842,844	Ogden	98.8	71.7	87.4
843	Logan	100.9	71.7	88.5
845	Price	101.4	70.0	88.1
846-847	Provo	101.2	71.2	88.5
VERMONT				
050	White River Jct.	98.2	79.8	90.4
051	Bellows Falls	96.7	92.1	94.8
052	Bennington	97.0	88.6	93.4
053	Brattleboro	97.4	92.1	95.2
054	Burlington	101.9	79.8	92.5
056	Montpelier	100.2	84.4	93.5
057	Rutland	98.9	79.4	90.7
058	St. Johnsbury	98.1	79.3	90.2
059	Guildhall	96.8	79.0	89.3
VIRGINIA				
220-221	Fairfax	99.5	83.6	92.8
222	Arlington	100.4	84.2	93.6
223	Alexandria	99.6	84.8	93.3
224-225	Fredericksburg	98.3	78.7	90.0
226	Winchester	98.9	79.5	90.7
227	Culpeper	98.7	83.0	92.1
228	Harrisonburg	99.0	68.8	86.3
229	Charlottesville	99.4	71.0	87.4
230-232	Richmond	99.2	72.8	88.1
233-235	Norfolk	99.9	68.7	86.7
236	Newport News	98.8	68.3	85.9
237	Portsmouth	98.3	67.8	85.4
238	Petersburg	98.7	72.6	87.7
239	Farmville	97.9	65.3	84.1
240-241	Roanoke	100.1	73.6	88.9
242	Bristol	98.2	59.3	81.8
243	Pulaski	97.8	67.1	84.8
244	Staunton	98.6	67.9	85.6
245	Lynchburg	98.7	73.8	88.2
246	Grundy	98.1	58.8	81.5
WASHINGTON				
980-981,987	Seattle	106.3	112.5	108.9
982	Everett	105.1	104.0	104.6
983-984	Tacoma	105.4	101.3	103.7
985	Olympia	104.0	101.8	103.1
986	Vancouver	106.6	102.8	105.0
988	Wenatchee	105.6	88.7	98.5
989	Yakima	105.6	95.4	101.3
990-992	Spokane	100.1	82.4	92.6
993	Richland	99.8	90.0	95.6
994	Clarkston	99.1	81.9	91.8
WEST VIRGINIA				
247-248	Bluefield	97.0	88.1	93.2
249	Lewisburg	98.7	86.1	93.4
250-253	Charleston	96.5	89.4	93.5
254	Martinsburg	96.7	84.1	91.4
255-257	Huntington	97.8	92.2	95.4
258-259	Beckley	95.2	87.7	92.0
260	Wheeling	99.6	89.6	95.4
261	Parkersburg	98.6	88.6	94.4
262	Buckhannon	98.3	88.8	94.3
263-264	Clarksburg	98.9	89.5	94.9
265	Morgantown	98.8	89.6	94.9
266	Gassaway	98.3	89.0	94.3
267	Romney	98.3	87.2	93.6
268	Petersburg	98.0	85.9	92.9
WISCONSIN				
530,532	Milwaukee	99.2	107.9	102.9
531	Kenosha	99.4	102.0	100.5
534	Racine	98.7	102.3	100.2
535	Beloit	98.6	98.9	98.7
537	Madison	99.7	98.1	99.0

STATE/ZIP	CITY	MAT.	INST.	TOTAL
WISCONSIN (CONT'D)				
538	Lancaster	96.7	94.3	95.7
539	Portage	95.2	98.0	96.4
540	New Richmond	95.4	94.6	95.0
541-543	Green Bay	99.4	93.4	96.9
544	Wausau	94.7	93.4	94.2
545	Rhinelander	97.9	92.2	95.5
546	La Crosse	95.9	93.5	94.9
547	Eau Claire	97.5	94.1	96.1
548	Superior	95.1	96.5	95.7
549	Oshkosh	95.4	91.9	93.9
WYOMING				
820	Cheyenne	101.1	72.7	89.1
821	Yellowstone Nat'l Park	99.0	71.9	87.6
822	Wheatland	100.3	67.1	86.2
823	Rawlins	102.0	71.5	89.1
824	Worland	99.7	70.7	87.5
825	Riverton	101.0	68.6	87.3
826	Casper	101.5	68.9	87.7
827	Newcastle	99.6	71.1	87.5
828	Sheridan	102.5	70.2	88.8
829-831	Rock Springs	103.4	71.4	89.9
CANADIAN FACTORS (reflect Canadian currency)				
ALBERTA				
	Calgary	119.7	95.6	109.5
	Edmonton	121.8	95.9	110.9
	Fort McMurray	121.9	88.4	107.7
	Lethbridge	117.0	87.9	104.7
	Lloydminster	111.5	84.7	100.2
	Medicine Hat	111.7	84.0	100.0
	Red Deer	112.1	84.0	100.2
BRITISH COLUMBIA				
	Kamloops	113.0	83.5	100.5
	Prince George	114.0	82.8	100.8
	Vancouver	117.9	92.4	107.2
	Victoria	114.1	87.6	102.9
MANITOBA				
	Brandon	120.6	70.5	99.4
	Portage la Prairie	111.6	69.1	93.6
	Winnipeg	122.0	71.2	100.5
NEW BRUNSWICK				
	Bathurst	110.7	62.3	90.2
	Dalhousie	110.2	62.5	90.0
	Fredericton	119.3	69.0	98.0
	Moncton	110.9	76.0	96.2
	Newcastle	110.7	62.9	90.5
	Saint John	113.2	75.7	97.3
NEWFOUNDLAND				
	Corner Brook	124.7	68.4	100.9
	St. John's	122.6	87.4	107.7
NORTHWEST TERRITORIES				
	Yellowknife	131.4	83.2	111.0
NOVA SCOTIA				
	Bridgewater	111.3	70.3	94.0
	Dartmouth	121.5	70.2	99.8
	Halifax	117.9	88.0	105.2
	New Glasgow	119.6	70.2	98.7
	Sydney	117.8	70.2	97.7
	Truro	110.8	70.3	93.7
	Yarmouth	119.5	70.2	98.7
ONTARIO				
	Barrie	116.5	87.1	104.1
	Brantford	113.1	90.7	103.6
	Cornwall	113.1	87.3	102.2
	Hamilton	116.3	98.8	108.9
	Kingston	114.1	87.4	102.8
	Kitchener	108.6	94.6	102.7
	London	116.3	96.5	108.0
	North Bay	122.5	85.2	106.8
	Oshawa	110.8	97.2	105.0
	Ottawa	117.5	96.8	108.7
	Owen Sound	116.7	85.4	103.4
	Peterborough	113.1	87.1	102.1
	Sarnia	113.2	91.3	103.9

STATE/ZIP	CITY	MAT.	INST.	TOTAL
ONTARIO (CONT'D)				
	Sault Ste. Marie	108.6	88.7	100.2
	St. Catharines	106.6	95.6	101.9
	Sudbury	106.1	93.6	100.8
	Thunder Bay	107.8	95.0	102.4
	Timmins	113.3	85.3	101.4
	Toronto	116.0	104.2	111.0
	Windsor	107.1	94.4	101.8
PRINCE EDWARD ISLAND				
	Charlottetown	121.3	60.2	95.5
	Summerside	121.9	57.9	94.9
QUEBEC				
	Cap-de-la-Madeleine	110.6	80.0	97.7
	Charlesbourg	110.6	80.0	97.7
	Chicoutimi	110.4	87.6	100.7
	Gatineau	110.3	79.8	97.4
	Granby	110.5	79.7	97.5
	Hull	110.5	79.8	97.5
	Joliette	110.8	80.0	97.8
	Laval	110.7	81.6	98.4
	Montreal	119.1	90.1	106.9
	Quebec City	118.7	90.4	106.7
	Rimouski	110.2	87.6	100.6
	Rouyn-Noranda	110.3	79.8	97.4
	Saint-Hyacinthe	109.8	79.8	97.1
	Sherbrooke	110.6	79.8	97.6
	Sorel	110.8	80.0	97.8
	Saint-Jerome	110.4	79.8	97.4
	Trois-Rivieres	120.2	79.9	103.2
SASKATCHEWAN				
	Moose Jaw	109.0	63.9	89.9
	Prince Albert	108.2	62.2	88.8
	Regina	123.1	92.4	110.2
	Saskatoon	109.9	88.5	100.9
YUKON				
	Whitehorse	132.2	69.2	105.6

L1010-101 Minimum Design Live Loads in Pounds per S.F. for Various Building Codes

Occupancy or Use	Uniform (psf)	Concentrated (lbs)
1. Access floor systems		
Office use	50	2000
Computer use	100	2000
2. Armories and drill rooms	100	-
3. Assembly areas		
Fixed seats (fastened to floor)	60	
Follow spot, projections and control rooms	50	
Lobbies	100	
Movable seats	100	-
Stage floors	150	
Platforms (assembly)	100	
Other assembly areas	100	
4. Balconies and decks	Same as occupancy served	-
5. Catwalks	40	300
6. Cornices	60	-
7. Corridors		
First floor	100	-
Other floors	Same as occupancy served except as indicated	
8. Dining rooms and restaurants	100	-
9. Elevator machine room grating (on an area of 2 in by 2 in)	-	300
10. Finish light floorplate construction (on area of 1 in by 1 in)	-	200
11. Fire escapes	100	
On single-family dwellings only		-
12. Garages (passenger vehicles only)	40	Note a
13. Hospitals		
Corridors above first floor	80	1000
Operating rooms, laboratories	60	1000
Patient rooms	40	1000
14. Libraries		
Corridors above first floor	80	1000
Reading rooms	60	1000
Stacks	150, Note b	1000
15. Manufacturing		
Heavy	250	3000
Light	125	2000
16. Marquees	75	-
17. Office buildings		
Corridors above first floor	80	2000
Lobbies and first-floor corridors	100	2000
Offices	50	2000
18. Penal institutions		
Cell blocks	40	
Corridors	100	
19. Recreational uses:		
Bowling alleys, poolrooms, and similar uses	75	
Dance halls and ballrooms	100	
Gymnasiums	100	
Reviewing stands, grandstands and bleachers	100	
Stadiums and arenas with fixed seats (fastened to floor)	60	
20. Residential		
One- and two- family dwellings		
Uninhabitable attics without storage, Note c	10	
Uninhabitable attics with storage, Note c, d, e	20	
Habitable attics and sleeping areas, Note e	30	
All other areas	0	
Hotels and multifamily dwellings		
Private rooms and corridors serving them	40	
Public rooms and corridors serving them	100	

L1010-101 Minimum Design Live Loads in Pounds per S.F. for Various Building Codes (cont.)

Occupancy or Use	Uniform (psf)	Concentrated (lbs)
21. Roofs		
All roof surfaces subject to maintenance workers		300
Awnings and canopies:		
Fabric construction supported by a skeleton structure	5	-
All other construction	20	
Ordinary flat, pitched and curved roofs (not occupiable)	20	
Where primary roof members are exposed to a work floor, at single panel point of lower cord of roof trusses or any point along primary structural members supporting roofs:		
Over manufacturing, storage warehouses, and repair garages		2000
All other primary roof members		300
Occupiable roofs:		
Roof gardens	100	
Assembly areas	100	
All other similar areas	Note f	Note f
22. Schools		
Classrooms	40	1000
Corridors	80	1000
First-floor corridors	100	1000
23. Scuttles, skylight ribs and accessible ceilings	-	200
24. Sidewalks, vehicular driveways and yards, subject to trucking	250	8000, Note g
25. Stairs and exits		
One- and two- family dwellings	40	300, Note h
All others	100	3000, Note h
26. Storage warehouses		
Heavy	250	-
Light	125	
27. Stores		
Retail		
First floor	100	1000
Upper floors	75	1000
Wholesale, all floors	125	1000
28. Walkways and elevated platforms (other than exitways)	60	-
29. Yards and terraces, pedestrians	100	-

Notes:

a. Floors in garages or portions of buildings used for the storage of motor vehicles shall be designed for the uniformly distributed live loads of Table 1607.1 or the following concentrated loads: (1) for garages restricted to passenger vehicles accommodating not more than nine passengers, 3,000 pounds acting on an area of 4.5 in. by 4.5 in.; (2) for mechanical parking structures without slab or deck that are used for storing passenger vehicles only, 2,250 pounds per wheel.

b. The loading applies to stack room floors that support nonmobile, double-faced library books stacks, subject to the following limitations:
1. The nominal bookstack unit height shall not exceed 90 inches;
2. The nominal shelf depth shall not exceed 12 inches for each face; and
3. Parallel rows of double-faced book stacks shall be separated by aisles not less than 36 inches wide.

c. Uninhabitable attics without storage are those where the maximum clear height between the joists and rafters is less than 42 inches, or where there are not two or more adjacent trusses with web configurations capable of accommodating an assumed rectangle 42 inches in height by 24 inches in width, or greater, within the plane of the trusses. This live load need not be assumed to act concurrently with any other live load requirements.

d. Uninhabitable attics with storage are those where the maximum clear height between the joists and rafters is 42 inches or greater, or where there are two or more adjacent trusses with web configurations capable of accommodating an assumed rectangle 42 inches in height by 24 inches in width, or greater, within the plane of the trusses.

The live load need only be applied to those portions of the joists or truss bottom chords where both of the following conditions are met:
i. The attic area is accessible from an opening not less than 20 inches in width by 30 inches in length that is located where the clear height in the attic is a minimum of 30 inches; and
ii. The slopes of the joists or truss bottom chords are no greater than two units vertical in 12 units horizontal.

The remaining portions of the joists or truss bottom chords shall be designed for a uniformly distributed concurrent live load of not less than 10 lb/ft^2.

e. Attic spaces served by stairways other than the pull-down type shall be designed to support the minimum live load specified for habitable attics and sleeping rooms.

f. Areas of occupiable roofs, other than roof gardens and assembly areas, shall be designed for approporate loads as approved by the building official.

g. The concentrated wheel load shall be applied on an area of 4.5 inches by 4.5 inches.

h. The minimum concentrated load on stair treads shall be applied on an area of 2 in by 2 in. This load need not be assumed to act concurrently with the uniform load.

Table L1010-201 Design Weight Per S.F. for Walls and Partitions

Type	Wall Thickness	Description	Weight Per S.F.
Brick	4″	Clay brick, high absorption	34 lb.
		Clay brick, medium absorption	39
		Clay brick, low absorption	46
		Sand-lime brick	38
		Concrete brick, heavy aggregate	46
		Concrete brick, light aggregate	33
	8″	Clay brick, high absorption	69
		Clay brick, medium absorption	79
		Clay brick, low absorption	89
		Sand-lime brick	74
		Concrete brick, heavy aggregate	89
		Concrete brick, light aggregate	68
	12″	Common brick	120
		Pressed brick	130
		Sand-lime brick	105
		Concrete brick, heavy aggregate	130
		Concrete brick, light aggregate	98
	16″	Clay brick, high absorption	134
		Clay brick, medium absorption	155
		Clay brick, low absorption	173
		Sand-lime brick	138
		Concrete brick, heavy aggregate	174
		Concrete brick, light aggregate	130
Concrete block	4″	Solid conc. block, stone aggregate	45
		Solid conc. block, lightweight	34
		Hollow conc. block, stone aggregate	30
		Hollow conc. block, lightweight	20
	6″	Solid conc. block, stone aggregate	50
		Solid conc. block, lightweight	37
		Hollow conc. block, stone aggregate	42
		Hollow conc. block, lightweight	30
	8″	Solid conc. block, stone aggregate	67
		Solid conc. block, lightweight	48
		Hollow conc. block, stone aggregate	55
		Hollow conc. block, lightweight	38
	10″	Solid conc. block, stone aggregate	84
		Solid conc. block, lightweight	62
		Hollow conc. block, stone aggregate	55
		Hollow conc. block, lightweight	38
	12″	Solid conc. block, stone aggregate	108
		Solid conc. block, lightweight	72
		Hollow conc. block, stone aggregate	85
		Hollow conc. block, lightweight	55
Drywall	6″	Drywall on wood studs	10
Clay tile	2″	Split terra cotta furring	10 lb.
		Non load bearing clay tile	11
	3″	Split terra cotta furring	12
		Non load bearing clay tile	18
	4″	Non load bearing clay tile	20
		Load bearing clay tile	24
	6″	Non load bearing clay tile	30
		Load bearing clay tile	36
	8″	Non load bearing clay tile	36
		Load bearing clay tile	42
	12″	Non load bearing clay tile	46
		Load bearing clay tile	58
Gypsum block	2″	Hollow gypsum block	9.5
		Solid gypsum block	12
	3″	Hollow gypsum block	10
		Solid gypsum block	18
	4″	Hollow gypsum block	15
		Solid gypsum block	24
	5″	Hollow gypsum block	18
	6″	Hollow gypsum block	24
Structural facing tile	2″	Facing tile	15
	4″	Facing tile	25
	6″	Facing tile	38
Glass	4″	Glass block	18
	1″	Structural glass	15
Plaster	1″	Gypsum plaster (1 side)	5
		Cement plaster (1 side)	10
		Gypsum plaster on lath	8
		Cement plaster on lath	13
Plaster partition (2 finished faces)	2″	Solid gypsum on metal lath	18
		Solid cement on metal lath	25
		Solid gypsum on gypsum lath	18
		Gypsum on lath & metal studs	18
	3″	Gypsum on lath & metal studs	19
	4″	Gypsum on lath & metal studs	20
	6″	Gypsum on lath & wood studs	18
Concrete	6″	Reinf concrete, stone aggregate	75
		Reinf. concrete, lightweight	36-60
	8″	Reinf. concrete, stone aggregate	100
		Reinf. concrete, lightweight	48-80
	10″	Reinf. concrete, stone aggregate	125
		Reinf. concrete, lightweight	60-100
	12″	Reinf concrete stone aggregate	150
		Reinf concrete, lightweight	72-120

Table L1010-202 Design Weight per S.F. for Roof Coverings

Type		Description	Weight lb. Per S.F.
Sheathing	Gypsum	1″ thick	4
	Wood	¾″ thick	3
Insulation	per 1″	Loose	.5
		Poured in place	2
		Rigid	1.5
Built-up	Tar & gravel	3 ply felt	5.5
		5 ply felt	6.5

Type	Wall Thickness	Description	Weight lb. Per S.F.
Metal	Aluminum	Corr. & ribbed, .024″ to .040″	.4-.8
	Copper	or tin	1.5-2.5
	Steel	Corrugated, 29 ga. to 12 ga.	.6-5.0
Shingles	Asphalt	Strip shingles	1.7-2.8
	Clay	Tile	8-16
	Slate	¼″ thick	9.5
	Wood		2-3

Table L1010-203 Design Weight per Square Foot for Floor Fills and Finishes

Type		Description	Weight lb. per S.F.
Floor fill	per 1"	Cinder fill	5
		Cinder concrete	9
		Lightweight concrete	3-9
		Stone concrete	12
		Sand	8
		Gypsum	6
Terrazzo	1"	Terrazzo, 2" stone concrete	25
Marble	and mortar	on stone concrete fill	33
Resilient	1/16"-1/4"	Linoleum, asphalt, vinyl tile	2
Tile	3/4"	Ceramic or quarry	10

Type		Description	Weight lb. per S.F.
Wood	Single 7/8"	On sleepers, light concrete fill	16
		On sleepers, stone concrete fill	25
	Double 7/8"	On sleepers, light concrete fill	19
		On sleepers, stone concrete fill	28
	3"	Wood block on mastic, no fill	15
		Wood block on ½" mortar	16
	per 1"	Hardwood flooring (25/32")	4
		Underlayment (Plywood per 1")	3
Asphalt	1-1/2"	Mastic flooring	18
	2"	Block on ½" mortar	30

Table L1010-204 Design Weight Per Cubic Foot for Miscellaneous Materials

Type		Description	Weight lb. per C.F.
Bituminous	Coal, piled	Anthracite	47-58
		Bituminous	40-54
		Peat, turf, dry	47
		Coke	75
	Petroleum	Unrefined	54
		Refined	50
		Gasoline	42
	Pitch		69
	Tar	Bituminous	75
Concrete	Plain	Stone aggregate	144
		Slag aggregate	132
		Expanded slag aggregate	100
		Haydite (burned clay agg.)	90
		Vermiculite & perlite, load bearing	70-105
		Vermiculite & perlite, non load bear	35-50
	Reinforced	Stone aggregate	150
		Slag aggregate	138
		Lightweight aggregates	30-120
Earth	Clay	Dry	63
		Damp, plastic	110
		and gravel, dry	100
	Dry	Loose	76
		Packed	95
	Moist	Loose	78
		Packed	96
	Mud	Flowing	108
		Packed	115
	Riprap	Limestone	80-85
		Sandstone	90
		Shale	105
	Sand & gravel	Dry, loose	90-105
		Dry, packed	100-120
		Wet	118-120
Gases	Air	0C., 760 mm.	.0807
	Gas	Natural	.0385
Liquids	Alcohol	100%	49
	Water	4°C., maximum density	62.5
		Ice	56
		Snow, fresh fallen	8
		Sea water	64

Type		Description	Weight lb. per C.F.
Masonry	Ashlar	Granite	165
		Limestone, crystalline	165
		Limestone, oolitic	135
		Marble	173
		Sandstone	144
	Rubble, in mortar	Granite	155
		Limestone, crystalline	147
		Limestone, oolitic	138
		Marble	156
		Sandstone & Bluestone	130
	Brick	Pressed	140
		Common	120
		Soft	100
	Cement	Portland, loose	90
		Portland set	183
	Lime	Gypsum, loose	53-64
	Mortar	Set	103
Metals	Aluminum	Cast, hammered	165
	Brass	Cast, rolled	534
	Bronze	7.9 to 14% Sn	509
	Copper	Cast, rolled	556
	Iron	Cast, pig	450
		Wrought	480
	Lead		710
	Monel		556
	Steel	Rolled	490
	Tin	Cast, hammered	459
	Zinc	Cast rolled	440
Timber	Cedar	White or red	24.2
	Fir	Douglas	23.7
		Eastern	25
	Maple	Hard	44.5
		White	33
	Oak	Red or Black	47.3
		White	47.3
	Pine	White	26
		Yellow, long leaf	44
		Yellow, short leaf	38
	Redwood	California	26
	Spruce	White or black	27

Table L1010-225 Design Weight Per S.F. for Structural Floor and Roof Systems

Type Slab		Description	Weight in Pounds per S.F. Slab Depth in Inches											
Concrete Slab	Reinforced	Stone aggregate	1″	12.5	2″	25	3″	37.5	4″	50	5″	62.5	6″	75
		Lightweight sand aggregate	↓	9.5	↓	19	↓	28.5	↓	38	↓	47.5	↓	57
		All lightweight aggregate	↓	9.0	↓	18	↓	27.0	↓	36	↓	45.0	↓	54
	Plain, nonreinforced	Stone aggregate	↓	12.0	↓	24	↓	36.0	↓	48	↓	60.0	↓	72
		Lightweight sand aggregate	↓	9.0	↓	18	↓	27.0	↓	36	↓	45.0	↓	54
		All lightweight aggregate	↓	8.5	↓	17	↓	25.5	↓	34	↓	42.5	↓	51
Concrete Waffle	19″ x 19″	5″ wide ribs @ 24″ O.C.	6+3	77	8+3	92	10+3	100	12+3	118				
			6+4 ½	96	8+4½	110	10+4½	119	12+4½	136				
	30″ x 30″	6″ wide ribs @ 36″ O.C.	8+3	83	10+3	95	12+3	109	14+3	118	16+3	130	20+3	155
			8+4½	101	10+4½	113	12+4½	126	14+4½	137	16+4½	149	20+4½	173
Concrete Joist	20″ wide form	5″ wide rib	8+3	60	10+3	67	12+3	74	14+3	81				
		6″ wide rib	↓	63	↓	70	↓	78	↓	86	16+3	94	20+3	111
		7″ wide rib									↓	99	↓	118
		5″ wide rib	8+4½	79	10+4½	85	12+4½	92	14+4½	99				
		6″ wide rib	↓	82	↓	89	↓	97	↓	104	16+4½	113	20+4½	130
		7″ wide rib									↓	118	↓	136
	30″ wide form	5″ wide rib	8+3	54	10+3	58	12+3	63	14+3	68				
		6″ wide rib	↓	56	↓	61	↓	67	↓	72	16+3	78	20+3	91
		7″ wide rib									↓	83	↓	96
		5″ wide rib	8+4½	72	10+4½	77	12+4½	82	14+4½	87				
		6″ wide rib	↓	75	↓	80	↓	85	↓	91	16+4½	97	20+4½	109
		7″ wide rib									↓	101	↓	115
Wood Joists	Incl. Subfloor	12″ O.C.	2x6	6	2x8	6	2x10	7	2x12	8	3x8	8	3x12	11
		16″ O.C.	↓	5	↓	6	↓	6	↓	7	↓	7	↓	9

Table L1010-226 Superimposed Dead Load Ranges

Component	Load Range (PSF)
Ceiling	5-10
Partitions	20-30
Mechanical	4-8

Table L1010-301 Design Loads for Structures for Wind Load

Wind Loads: Structures are designed to resist the wind force from any direction. Usually 2/3 is assumed to act on the windward side, 1/3 on the leeward side.

For more than 1/3 openings, add 10 psf for internal wind pressure or 5 psf for suction, whichever is critical.

For buildings and structures, use psf values from Table L1010-301.

For glass over 4 S.F. use values in Table L1010-302 after determining 30′ wind velocity from Table L1010-303.

Type Structure	Height Above Grade	Horizontal Load in Lb. per S.F.
Buildings	Up to 50 ft. 50 to 100 ft. Over 100 ft.	15 20 20 + .025 per ft.
Ground signs & towers	Up to 50 ft. Over 50 ft.	15 20
Roof structures		30
Glass	See Table below	

Table L1010-302 Design Wind Load in PSF for Glass at Various Elevations

Height	Velocity in Miles per Hour and Design Load in Pounds per S.F.																					
From Grade	Vel.	PSF	Vel.	PSF	Vel.	PSF	Vel.	PSF	Vel.	PSF	Vel.	PSF	Vel.	PSF	Vel.	PSF	Vel.	PSF	Vel.	PSF	Vel.	PSF
To 10 ft.	42	6	46	7	49	8	52	9	55	10	59	11	62	12	66	14	69	15	76	19	83	22
10-20	52	9	58	11	61	11	65	14	70	16	74	18	79	20	83	22	87	24	96	30	105	35
20-30*	60	12	67	14	70	16	75	18	80	20	85	23	90	26	95	29	100	32	110	39	120	46
30-60	66	14	74	18	77	19	83	22	88	25	94	28	99	31	104	35	110	39	121	47	132	56
60-120	73	17	82	21	85	12	92	27	98	31	104	35	110	39	116	43	122	48	134	57	146	68
120-140	81	21	91	26	95	29	101	33	108	37	115	42	122	48	128	52	135	48	149	71	162	84
240-480	90	26	100	32	104	35	112	40	119	45	127	51	134	57	142	65	149	71	164	86	179	102
480-960	98	31	110	39	115	42	123	49	131	55	139	62	148	70	156	78	164	86	180	104	197	124
Over 960	98	31	110	39	115	42	123	49	131	55	139	62	148	70	156	78	164	86	180	104	197	124

*Determine appropriate wind at 30′ elevation Fig. L1010-303 below.

Table L1010-303 Design Wind Velocity at 30 Ft. Above Ground

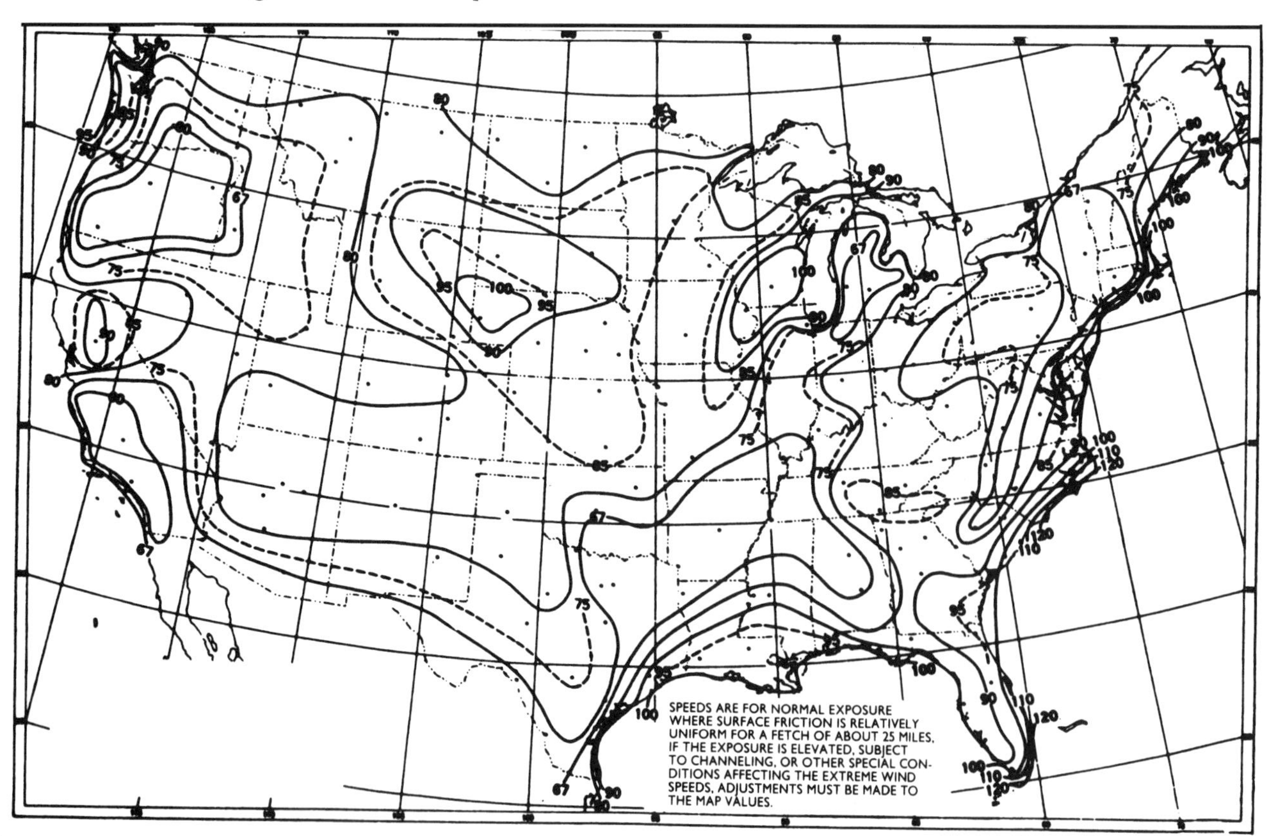

Table L1010-401 Snow Load in Pounds Per Square Foot on the Ground

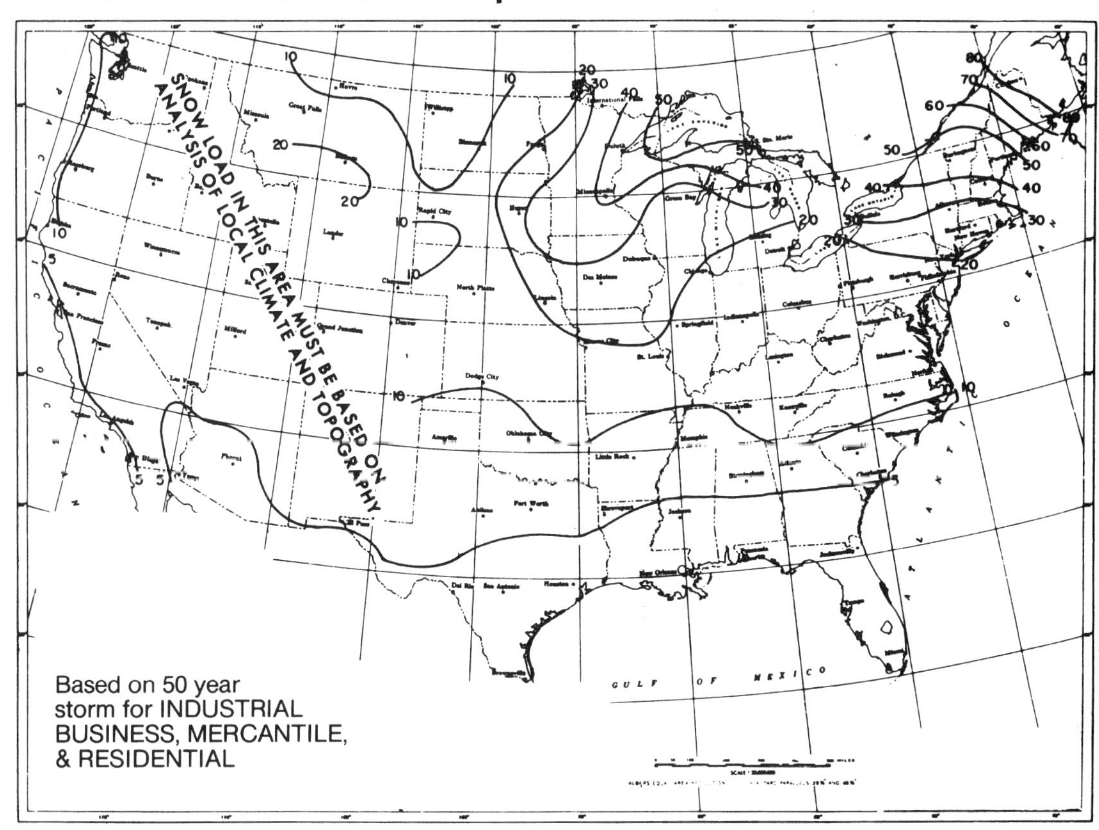

Table L1010-402 Snow Load in Pounds Per Square Foot on the Ground

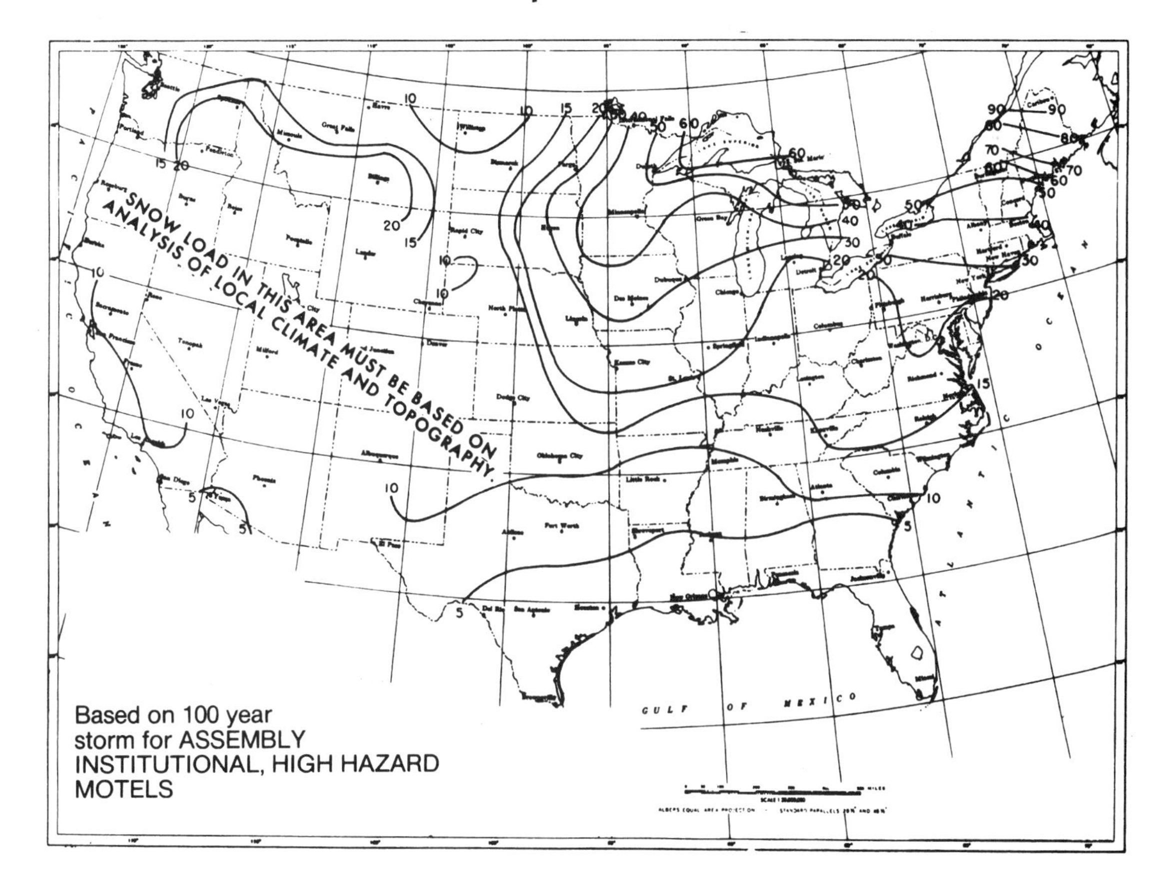

Table L1010-403 Snow Loads

To convert the ground snow loads on the previous page to roof snow loads, the ground snow loads should be multiplied by the following factors depending upon the roof characteristics.

Note, in all cases $\frac{\alpha - 30}{50}$ is valid only for > 30 degrees.

Description	Sketch	Formula	Angle	Conversion Factor = C_f	
				Sheltered	Exposed
Simple flat and shed roofs	Load Diagram	For $\alpha > 30°$ $C_f = 0.8 - \frac{\alpha - 30}{50}$	0° to 30° 40° 50° 60° 70° to 90°	0.8 0.6 0.4 0.2 0	0.6 0.45 0.3 0.15 0
				Case I	Case II
Simple gable and hip roofs	Case I Load Diagram (C_f) Case II Load Diagram (C_f)	$C_f = 0.8 - \frac{\alpha - 30}{50}$ $C_f = 1.25\left(0.8 - \left(\frac{\alpha - 30}{50}\right)\right)$	10° 20° 30° 40° 50° 60°	0.8 0.8 0.8 0.6 0.4 0.2	— — 1.0 0.75 0.5 0.25
Valley areas of Two span roofs	α_1, α_2, L_1, L_2 Case I (C_f; L_1, L_2) Case II (0.5, 1.0; $\frac{L_1}{2}$, $\frac{L_2}{2}$) (0.5, 1.5; $\frac{L_1}{4}$, $\frac{L_2}{4}$)	$\beta = \frac{\alpha_1 + \alpha_2}{2}$ $C_f = 0.8 - \frac{\alpha - 30}{50}$	$\beta \le 10°$ use Case I only $\beta > 10°$ $\beta < 20°$ use Case I & II $\beta \ge 20°$ use Case I, II & III		
Lower level of multi level roofs (or on an adjacent building not more than 15 ft. away)	h C_f, 0.8 W = 2h	$C_f = 15\frac{h}{g}$ h = difference in roof height in feet g = ground snow load in psf w = width of drift For h < 5, w = 10 h >15, w = 30	When $15\frac{h}{g} < .8$, use 0.8 When $15\frac{h}{g} > 3.0$, use 3.0		

Summary of above:

1. For flat roofs or roofs up to 30°, use 0.8 x ground snow load for the roof snow load.
2. For roof pitches in excess of 30°, conversion factor becomes lower than 0.8.
3. For exposed roofs there is a further 25% reduction of conversion factor.
4. For steep roofs a more highly loaded half span must be considered.
5. For shallow roof valleys conversion factor is 0.8.
6. For moderate roof valleys, conversion factor is 1.0 for half the span.
7. For steep roof valleys, conversion factor is 1.5 for one quarter of the span.
8. For roofs adjoining vertical surfaces the conversion factor is up to 3.0 for part of the span.
9. If snow load is less than 30 psf, use water load on roof for clogged drain condition.

Table L1020-101 Floor Area Ratios

The table below lists commonly used gross to net area and net to gross area ratios expressed in % for various building types.

Building Type	Gross to Net Ratio	Net to Gross Ratio	Building Type	Gross to Net Ratio	Net to Gross Ratio
Apartment	156	64	School Buildings (campus type)		
Bank	140	72	Administrative	150	67
Church	142	70	Auditorium	142	70
Courthouse	162	61	Biology	161	62
Department Store	123	81	Chemistry	170	59
Garage	118	85	Classroom	152	66
Hospital	183	55	Dining Hall	138	72
Hotel	158	63	Dormitory	154	65
Laboratory	171	58	Engineering	164	61
Library	132	76	Fraternity	160	63
Office	135	75	Gymnasium	142	70
Restaurant	141	70	Science	167	60
Warehouse	108	93	Service	120	83
			Student Union	172	59

The gross area of a building is the total floor area based on outside dimensions.

The net area of a building is the usable floor area for the function intended and excludes such items as stairways, corridors, and mechanical rooms.

In the case of a commercial building, it might be considered the "leasable area."

Table L1020-201 Partition/Door Density

Building Type		Stories	Partition/Density	Doors	Description of Partition
Apartments		1 story	9 SF/LF	90 SF/door	Plaster, wood doors & trim
		2 story	8 SF/LF	80 SF/door	Drywall, wood studs, wood doors & trim
		3 story	9 SF/LF	90 SF/door	Plaster, wood studs, wood doors & trim
		5 story	9 SF/LF	90 SF/door	Plaster, metal studs, wood doors & trim
		6-15 story	8 SF/LF	80 SF/door	Drywall, metal studs, wood doors & trim
Bakery		1 story	50 SF/LF	500 SF/door	Conc. block, paint, door & drywall, wood studs
		2 story	50 SF/LF	500 SF/door	Conc. block, paint, door & drywall, wood studs
Bank		1 story	20 SF/LF	200 SF/door	Plaster, wood studs, wood doors & trim
		2-4 story	15 SF/LF	150 SF/door	Plaster, metal studs, wood doors & trim
Bottling Plant		1 story	50 SF/LF	500 SF/door	Conc. block, drywall, metal studs, wood trim
Bowling Alley		1 story	50 SF/LF	500 SF/door	Conc. block, wood & metal doors, wood trim
Bus Terminal		1 story	15 SF/LF	150 SF/door	Conc. block, ceramic tile, wood trim
Cannery		1 story	100 SF/LF	1000 SF/door	Drywall on metal studs
Car Wash		1 story	18 SF/LF	180 SF/door	Concrete block, painted & hollow metal door
Dairy Plant		1 story	30 SF/LF	300 SF/door	Concrete block, glazed tile, insulated cooler doors
Department Store		1 story	60 SF/LF	600 SF/door	Drywall, metal studs, wood doors & trim
		2-5 story	60 SF/LF	600 SF/door	30% concrete block, 70% drywall, wood studs
Dormitory		2 story	9 SF/LF	90 SF/door	Plaster, concrete block, wood doors & trim
		3-5 story	9 SF/LF	90 SF/door	Plaster, concrete block, wood doors & trim
		6-15 story	9 SF/LF	90 SF/door	Plaster, concrete block, wood doors & trim
Funeral Home		1 story	15 SF/LF	150 SF/door	Plaster on concrete block & wood studs, paneling
		2 story	14 SF/LF	140 SF/door	Plaster, wood studs, paneling & wood doors
Garage Sales & Service		1 story	30 SF/LF	300 SF/door	50% conc. block, 50% drywall, wood studs
Hotel		3-8 story	9 SF/LF	90 SF/door	Plaster, conc. block, wood doors & trim
		9-15 story	9 SF/LF	90 SF/door	Plaster, conc. block, wood doors & trim
Laundromat		1 story	25 SF/LF	250 SF/door	Drywall, wood studs, wood doors & trim
Medical Clinic		1 story	6 SF/LF	60 SF/door	Drywall, wood studs, wood doors & trim
		2-4 story	6 SF/LF	60 SF/door	Drywall, metal studs, wood doors & trim
Motel		1 story	7 SF/LF	70 SF/door	Drywall, wood studs, wood doors & trim
		2-3 story	7 SF/LF	70 SF/door	Concrete block, drywall on metal studs, wood paneling
Movie Theater	200-600 seats	1 story	18 SF/LF	180 SF/door	Concrete block, wood, metal, vinyl trim
	601-1400 seats		20 SF/LF	200 SF/door	Concrete block, wood, metal, vinyl trim
	1401-22000 seats		25 SF/LF	250 SF/door	Concrete block, wood, metal, vinyl trim
Nursing Home		1 story	8 SF/LF	80 SF/door	Drywall, metal studs, wood doors & trim
		2-4 story	8 SF/LF	80 SF/door	Drywall, metal studs, wood doors & trim
Office		1 story	20 SF/LF	200-500 SF/door	30% concrete block, 70% drywall on wood studs
		2 story	20 SF/LF	200-500 SF/door	30% concrete block, 70% drywall on metal studs
		3-5 story	20 SF/LF	200-500 SF/door	30% concrete block, 70% movable partitions
		6-10 story	20 SF/LF	200-500 SF/door	30% concrete block, 70% movable partitions
		11-20 story	20 SF/LF	200-500 SF/door	30% concrete block, 70% movable partitions
Parking Ramp (Open)		2-8 story	60 SF/LF	600 SF/door	Stair and elevator enclosures only
Parking Garage		2-8 story	60 SF/LF	600 SF/door	Stair and elevator enclosures only
Pre-engineered	Steel	1 story	0		
	Store	1 story	60 SF/LF	600 SF/door	Drywall on metal studs, wood doors & trim
	Office	1 story	15 SF/LF	150 SF/door	Concrete block, movable wood partitions
	Shop	1 story	15 SF/LF	150 SF/door	Movable wood partitions
	Warehouse	1 story	0		
Radio & TV Broadcasting		1 story	25 SF/LF	250 SF/door	Concrete block, metal and wood doors
& TV Transmitter		1 story	40 SF/LF	400 SF/door	Concrete block, metal and wood doors
Self Service Restaurant		1 story	15 SF/LF	150 SF/door	Concrete block, wood and aluminum trim
Cafe & Drive-in Restaurant		1 story	18 SF/LF	180 SF/door	Drywall, metal studs, ceramic & plastic trim
Restaurant with seating		1 story	25 SF/LF	250 SF/door	Concrete block, paneling, wood studs & trim
Supper Club		1 story	25 SF/LF	250 SF/door	Concrete block, paneling, wood studs & trim
Bar or Lounge		1 story	24 SF/LF	240 SF/door	Plaster or gypsum lath, wooded studs
Retail Store or Shop		1 story	60 SF/LF	600 SF/door	Drywall metal studs, wood doors & trim
Service Station	Masonry	1 story	15 SF/LF	150 SF/door	Concrete block, paint, door & drywall, wood studs
	Metal panel	1 story	15 SF/LF	150 SF/door	Concrete block, paint, door & drywall, wood studs
	Frame	1 story	15 SF/LF	150 SF/door	Drywall, wood studs, wood doors & trim
Shopping Center	(strip)	1 story	30 SF/LF	300 SF/door	Drywall, metal studs, wood doors & trim
	(group)	1 story	40 SF/LF	400 SF/door	50% concrete block, 50% drywall, wood studs
		2 story	40 SF/LF	400 SF/door	50% concrete block, 50% drywall, wood studs
Small Food Store		1 story	30 SF/LF	300 SF/door	Concrete block drywall, wood studs, wood trim
Store/Apt. above	Masonry	2 story	10 SF/LF	100 SF/door	Plaster, metal studs, wood doors & trim
	Frame	2 story	10 SF/LF	100 SF/door	Plaster, metal studs, wood doors & trim
	Frame	3 story	10 SF/LF	100 SF/door	Plaster, metal studs, wood doors & trim
Supermarkets		1 story	40 SF/LF	400 SF/door	Concrete block, paint, drywall & porcelain panel
Truck Terminal		1 story	0		
Warehouse		1 story	0		

Table L1020-301 Occupancy Determinations

Function of Space	SF/Person Required
Accessory storage areas, mechanical equipment rooms	300
Agriculture Building	300
Aircraft Hangars	500
Airport Terminal	
Baggage claim	20
Baggage handling	300
Concourse	100
Waiting areas	15
Assembly	
Gaming floors (Keno, slots, etc.)	11
Exhibit gallery and museum	30
Assembly w/ fixed seats	load determined by seat number
Assembly w/o fixed seats	
Concentrated (chairs only-not fixed)	7
Standing space	5
Unconcentrated (tables and chairs)	15
Bowling centers, allow 5 persons for each lane including 15 feet of runway, and for additional areas	7
Business areas	100
Courtrooms-other than fixed seating areas	40
Day care	35
Dormitories	50
Educational	
Classroom areas	20
Shops and other vocational room areas	50
Exercise rooms	50
Fabrication and manufacturing areas where hazardous materials are used	200
Industrial areas	100
Institutional areas	
Inpatient treatment areas	240
Outpatient areas	100
Sleeping areas	120
Kitchens commercial	200
Library	
Reading rooms	50
Stack area	100
Mercantile	
Areas on other floors	60
Basement and grade floor areas	30
Storage, stock, shipping areas	300
Parking garages	200
Residential	200
Skating rinks, swimming pools	
Rink and pool	50
Decks	15
Stages and platforms	15
Warehouses	500

Table L1020-302 Length of Exitway Access Travel (ft.)

Occupancy Type	Without Sprinkler System (feet)	With Sprinkler System (feet)
A, E, F-1, M, R, S-1	200	250
I-1	Not Permitted	250
B	200	300
F-2, S-2, U	300	400
H-1	Not Permitted	75
H-2	Not Permitted	100
H-3	Not Permitted	150
H-4	Not Permitted	175
H-5	Not Permitted	200
I-2, I-3, I-4	Not Permitted	200

Note:

Refer to the 2012 *International Building Code* Section 1016 Exit Access Travel Distance for any exceptions or additions to the information above.

Table L1020-303 Capacity Per Unit Egress Width*

Use Group	Without Fire Suppression System (Inches per Person)*		With Fire Suppression System (Inches per Person)*	
	Stairways	Doors, Ramps and Corridors	Stairways	Doors, Ramps and Corridors
Assembly, Business, Educational, Factory Industrial, Mercantile, Residential, Storage	0.3	0.2	0.2	0.15
Institutional—1	0.3	0.2	0.2	0.15
Institutional—2	—	0.7	0.3	0.2
Institutional—3	0.3	0.2	0.2	0.15
High Hazard	0.7	0.4	0.3	0.2

* 1″=25.4 mm

Table L1030-401 Resistances ("R") of Building and Insulating Materials

Material	Wt./Lbs. per C.F.	R per Inch	R Listed Size
Air Spaces and Surfaces			
Enclosed non-reflective spaces, E=0.82			
50° F mean temp., 30°/10° F diff.			
.5"			.90/.91
.75"			.94/1.01
1.50"			.90/1.02
3.50"			.91/1.01
Inside vert. surface (still air)			0.68
Outside vert. surface (15 mph wind)			0.17
Building Boards			
Asbestos cement, 0.25" thick	120		0.06
Gypsum or plaster, 0.5" thick	50		0.45
Hardboard regular	50	1.37	
Tempered	63	1.00	
Laminated paper	30	2.00	
Particle board	37	1.85	
	50	1.06	
	63	0.85	
Plywood (Douglas Fir), 0.5" thick	34		0.62
Shingle backer, .375" thick	18		0.94
Sound deadening board, 0.5" thick	15		1.35
Tile and lay-in panels, plain or acoustical, 0.5" thick	18		1.25
Vegetable fiber, 0.5" thick	18		1.32
	25		1.14
Wood, hardwoods	48	0.91	
Softwoods	32	1.25	
Flooring Carpet with fibrous pad			2.08
With rubber pad			1.23
Cork tile, 1/8" thick			0.28
Terrazzo			0.08
Tile, resilient			0.05
Wood, hardwood, 0.75" thick			0.68
Subfloor, 0.75" thick			0.94
Glass			
Insulation, 0.50" air space			2.04
Single glass			0.91
Insulation Blanket or Batt, mineral, glass or rock fiber, approximate thickness			
3.0" to 3.5" thick			11
3.5" to 4.0" thick			13
6.0" to 6.5" thick			19
6.5" to 7.0" thick			22
8.5" to 9.0" thick			30
Boards			
Cellular glass	8.5	2.63	
Fiberboard, wet felted			
Acoustical tile	21	2.70	
Roof insulation	17	2.94	
Fiberboard, wet molded			
Acoustical tile	23	2.38	
Mineral fiber with resin binder	15	3.45	
Polystyrene, extruded,			
cut cell surface	1.8	4.00	
smooth skin surface	2.2	5.00	
	3.5	5.26	
Bead boards	1.0	3.57	
Polyurethane	1.5	6.25	
Wood or cane fiberboard, 0.5" thick			1.25

Material	Wt./Lbs. per C.F.	R per Inch	R Listed Size
Insulation Loose Fill			
Cellulose	2.3	3.13	
	3.2	3.70	
Mineral fiber, 3.75" to 5" thick	2-5		11
6.5" to 8.75" thick			19
7.5" to 10" thick			22
10.25" to 13.75" thick			30
Perlite	5-8	2.70	
Vermiculite	4-6	2.27	
Wood fiber	2-3.5	3.33	
Masonry Brick, Common	120	0.20	
Face	130	0.11	
Cement mortar	116	0.20	
Clay tile, hollow			
1 cell wide, 3" width			0.80
4" width			1.11
2 cells wide, 6" width			1.52
8" width			1.85
10" width			2.22
3 cells wide, 12" width			2.50
Concrete, gypsum fiber	51	0.60	
Lightweight	120	0.19	
	80	0.40	
	40	0.86	
Perlite	40	1.08	
Sand and gravel or stone	140	0.08	
Concrete block, lightweight			
3 cell units, 4"-15 lbs. ea.			1.68
6"-23 lbs. ea.			1.83
8"-28 lbs. ea.			2.12
12"-40 lbs. ea.			2.62
Sand and gravel aggregates,			
4"-20 lbs. ea.			1.17
6"-33 lbs. ea.			1.29
8"-38 lbs. ea.			1.46
12"-56 lbs. ea.			1.81
Plastering Cement Plaster,			
Sand aggregate	116	0.20	
Gypsum plaster, Perlite aggregate	45	0.67	
Sand aggregate	105	0.18	
Vermiculite aggregate	45	0.59	
Roofing			
Asphalt, felt, 15 lb.			0.06
Rolled roofing	70		0.15
Shingles	70		0.44
Built-up roofing .375" thick	70		0.33
Cement shingles	120		0.21
Vapor-permeable felt			0.06
Vapor seal, 2 layers of mopped 15 lb. felt			0.12
Wood, shingles 16"-7.5" exposure			0.87
Siding			
Aluminum or steel (hollow backed) oversheathing			0.61
With .375" insulating backer board			1.82
Foil backed			2.96
Wood siding, beveled, ½" x 8"			0.81

Table L1030-501 Weather Data and Design Conditions

City	Latitude (1)		Winter Temperatures (1)			Winter Degree Days (2)	Summer (Design Dry Bulb) Temperatures and Relative Humidity		
	°	1′	Med. of Annual Extremes	99%	97½%		1%	2½%	5%
UNITED STATES									
Albuquerque, NM	35	0	5.1	12	16	4,400	96/61	94/61	92/61
Atlanta, GA	33	4	11.9	17	22	3,000	94/74	92/74	90/73
Baltimore, MD	39	2	7	14	17	4,600	94/75	91/75	89/74
Birmingham, AL	33	3	13	17	21	2,600	96/74	94/75	92/74
Bismarck, ND	46	5	-32	-23	-19	8,800	95/68	91/68	88/67
Boise, ID	43	3	1	3	10	5,800	96/65	94/64	91/64
Boston, MA	42	2	-1	6	9	5,600	91/73	88/71	85/70
Burlington, VT	44	3	-17	-12	-7	8,200	88/72	85/70	82/69
Charleston, WV	38	2	3	7	11	4,400	92/74	90/73	87/72
Charlotte, NC	35	1	13	18	22	3,200	95/74	93/74	91/74
Casper, WY	42	5	-21	-11	-5	7,400	92/58	90/57	87/57
Chicago, IL	41	5	-8	-3	2	6,600	94/75	91/74	88/73
Cincinnati, OH	39	1	0	1	6	4,400	92/73	90/72	88/72
Cleveland, OH	41	2	-3	1	5	6,400	91/73	88/72	86/71
Columbia, SC	34	0	16	20	24	2,400	97/76	95/75	93/75
Dallas, TX	32	5	14	18	22	2,400	102/75	100/75	97/75
Denver, CO	39	5	-10	-5	1	6,200	93/59	91/59	89/59
Des Moines, IA	41	3	-14	-10	-5	6,600	94/75	91/74	88/73
Detroit, MI	42	2	-3	3	6	6,200	91/73	88/72	86/71
Great Falls, MT	47	3	-25	-21	-15	7,800	91/60	88/60	85/59
Hartford, CT	41	5	-4	3	7	6,200	91/74	88/73	85/72
Houston, TX	29	5	24	28	33	1,400	97/77	95/77	93/77
Indianapolis, IN	39	4	-7	-2	2	5,600	92/74	90/74	87/73
Jackson, MS	32	2	16	21	25	2,200	97/76	95/76	93/76
Kansas City, MO	39	1	-4	2	6	4,800	99/75	96/74	93/74
Las Vegas, NV	36	1	18	25	28	2,800	108/66	106/65	104/65
Lexington, KY	38	0	-1	3	8	4,600	93/73	91/73	88/72
Little Rock, AR	34	4	11	15	20	3,200	99/76	96/77	94/77
Los Angeles, CA	34	0	36	41	43	2,000	93/70	89/70	86/69
Memphis, TN	35	0	10	13	18	3,200	98/77	95/76	93/76
Miami, FL	25	5	39	44	47	200	91/77	90/77	89/77
Milwaukee, WI	43	0	-11	-8	-4	7,600	90/74	87/73	84/71
Minneapolis, MN	44	5	-22	-16	-12	8,400	92/75	89/73	86/71
New Orleans, LA	30	0	28	29	33	1,400	93/78	92/77	90/77
New York, NY	40	5	6	11	15	5,000	92/74	89/73	87/72
Norfolk, VA	36	5	15	20	22	3,400	93/77	91/76	89/76
Oklahoma City, OK	35	2	4	9	13	3,200	100/74	97/74	95/73
Omaha, NE	41	2	-13	-8	-3	6,600	94/76	91/75	88/74
Philadelphia, PA	39	5	6	10	14	4,400	93/75	90/74	87/72
Phoenix, AZ	33	3	27	31	34	1,800	109/71	107/71	105/71
Pittsburgh, PA	40	3	-1	3	7	6,000	91/72	88/71	86/70
Portland, ME	43	4	-10	-6	-1	7,600	87/72	84/71	81/69
Portland, OR	45	4	18	17	23	4,600	89/68	85/67	81/65
Portsmouth, NH	43	1	-8	-2	2	7,200	89/73	85/71	83/70
Providence, RI	41	4	-1	5	9	6,000	89/73	86/72	83/70
Rochester, NY	43	1	-5	1	5	6,800	91/73	88/71	85/70
Salt Lake City, UT	40	5	0	3	8	6,000	97/62	95/62	92/61
San Francisco, CA	37	5	36	38	40	3,000	74/63	71/62	69/61
Seattle, WA	47	4	22	22	27	5,200	85/68	82/66	78/65
Sioux Falls, SD	43	4	-21	-15	-11	7,800	94/73	91/72	88/71
St. Louis, MO	38	4	-3	3	8	5,000	98/75	94/75	91/75
Tampa, FL	28	0	32	36	40	680	92/77	91/77	90/76
Trenton, NJ	40	1	4	11	14	5,000	91/75	88/74	85/73
Washington, DC	38	5	7	14	17	4,200	93/75	91/74	89/74
Wichita, KS	37	4	-3	3	7	4,600	101/72	98/73	96/73
Wilmington, DE	39	4	5	10	14	5,000	92/74	89/74	87/73
ALASKA									
Anchorage	61	1	-29	-23	-18	10,800	71/59	68/58	66/56
Fairbanks	64	5	-59	-51	-47	14,280	82/62	78/60	75/59
CANADA									
Edmonton, Alta.	53	3	-30	-29	-25	11,000	85/66	82/65	79/63
Halifax, N.S.	44	4	-4	1	5	8,000	79/66	76/65	74/64
Montreal, Que.	45	3	-20	-16	-10	9,000	88/73	85/72	83/71
Saskatoon, Sask.	52	1	-35	-35	-31	11,000	89/68	86/66	83/65
St. John, Nwf.	47	4	1	3	7	8,600	77/66	75/65	73/64
Saint John, N.B.	45	2	-15	-12	-8	8,200	80/67	77/65	75/64
Toronto, Ont.	43	4	-10	-5	-1	7,000	90/73	87/72	85/71
Vancouver, B.C.	49	1	13	15	19	6,000	79/67	77/66	74/65
Winnipeg, Man.	49	5	-31	-30	-27	10,800	89/73	86/71	84/70

(1) Handbook of Fundamentals, ASHRAE, Inc., NY 1989
(2) Local Climatological Annual Survey, USDC Env. Science Services Administration, Asheville, NC

Table L1030-502 Maximum Depth of Frost Penetration in Inches

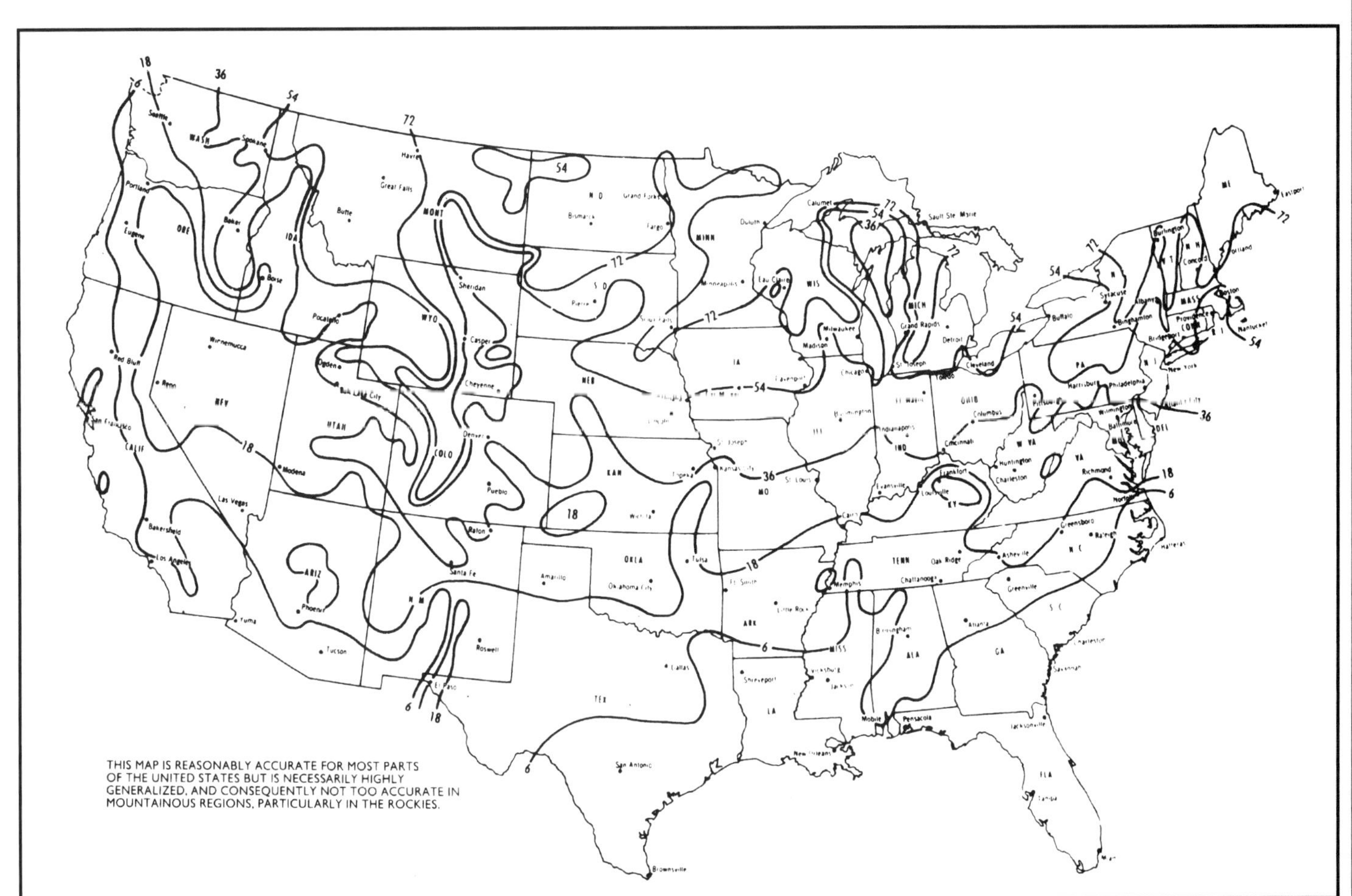

Table L1040-101 Fire-Resisting Ratings of Structural Elements (in hours)

Description of the Structural Element	Type of Construction									
	No. 1 Fireproof		No. 2 Non Combustible			No. 3 Exterior Masonry Wall			No. 4 Frame	
			Protected		Unprotected	Heavy Timber	Ordinary		Pro-tected	Unpro-tected
							Pro-tected	Unpro-tected		
	1A	1B	2A	2B	2C	3A	3B	3C	4A	4B
Exterior, Bearing Walls	4	3	2	1½	1	2	2	2	1	1
Nonbearing Walls	2	2	1½	1	1	2	2	2	1	1
Interior Bearing Walls and Partitions	4	3	2	1	0	2	1	0	1	0
Fire Walls and Party Walls	4	3	2	2	2	2	2	2	2	2
Fire Enclosure of Exitways, Exit Hallways and Stairways	2	2	2	2	2	2	2	2	1	1
Shafts other than Exitways, Hallways and Stairways	2	2	2	2	2	2	2	2	1	1
Exitway access corridors and Vertical separation of tenant space	1	1	1	1	0	1	1	0	1	0
Columns, girders, trusses (other than roof trusses) and framing:										
Supporting more than one floor	4	3	2	1	0	—	1	0	1	0
Supporting one floor only	3	2	1½	1	0	—	1	0	1	0
Structural members supporting wall	3	2	1½	1	0	1	1	0	1	0
Floor construction including beams	3	2	1½	1	0	—	1	0	1	0
Roof construction including beams, trusses and framing arches and roof deck 15′ or less in height to lowest member	2	1½	1	1	0	—	1	0	1	0

Note: a. Codes include special requirements and exceptions that are not included in the table above.
b. Each type of construction has been divided into sub-types which vary according to the degree of fire resistance required. Sub-types (A) requirements are more severe than those for sub-types (B).
c. Protected construction means all structural members are chemically treated, covered or protected so that the unit has the required fire resistance.

Type No. 1, Fireproof Construction — Buildings and structures of fireproof construction are those in which the walls, partitions, structural elements, floors, ceilings, roofs, and the exitways are protected with approved noncombustible materials to afford the fire-resistance rating specified in Table L1040-101; except as otherwise specifically regulated. Fire-resistant treated wood may be used as specified.

Type No. 2, Noncombustible Construction — Buildings and structures of noncombustible construction are those in which the walls, partitions, structural elements, floors, ceilings, roofs and the exitways are approved noncombustible materials meeting the fire-resistance rating requirements specified in Table L1040-101; except as modified by the fire limit restrictions. Fire-retardant treated wood may be used as specified.

Type No. 3, Exterior Masonry Wall Construction — Buildings and structures of exterior masonry wall construction are those in which the exterior, fire and party walls are masonry or other approved noncombustible materials of the required fire-resistance rating and structural properties. The floors, roofs, and interior framing are wholly or partly wood or metal or other approved construction. The fire and party walls are ground-supported; except that girders and their supports, carrying walls of masonry shall be protected to afford the same degree of fire-resistance rating of the supported walls. All structural elements have the required fire-resistance rating specified in Table L1040-101.

Type No. 4, Frame Construction — Buildings and structures of frame construction are those in which the exterior walls, bearing walls, partitions, floor and roof construction are wholly or partly of wood stud and joist assemblies with a minimum nominal dimension of two inches or of other approved combustible materials. Fire stops are required at all vertical and horizontal draft openings in which the structural elements have required fire-resistance ratings specified in Table L1040-101.

Table L1040-201 Fire Resistance Ratings

Fire Hazard for Fire Walls

The degree of fire hazard of buildings relating to their intended use is defined by "Fire Rating" the occupancy type. Such a rating system is listed in Table L1040-201 below. This type of rating determines the requirements for fire walls and fire separation walls (exterior fire exposure). For mixed use occupancy, use the higher Fire Rating requirement of the components.

Group	Fire-Resistance Rating (hours)
A, B, E, H-4[1], I, R-1, R-2, U	3
F-1[2], H-3[1], H-5[1], M, S-1	3
H-1[1], H-2[1]	4
F-2[3], S-2, R-3, R-4	2

Note: *The difference in "Fire Hazards" is determined by their occupancy and use.

1. High Hazard: Industrial and storage buildings in which the combustible contents might cause fires to be unusually intense or where explosives, combustible gases or flammable liquids are manufactured or stored.
2. Moderate Hazard: Mercantile buildings, industrial and storage buildings in which combustible contents might cause fires of moderate intensity.
3. Low Hazard: Business buildings that ordinarily do not burn rapidly.

Note:
In Type II or V construction, walls shall be permitted to have a 2-hour fire-resistance rating.

Table 706.4
Excerpted from the 2012 *International Building Code*, Copyright 2011. Washington, D.C.: International Code Council. Reproduced with permission.

Table L1040-202 Interior Finish Classification

Flame Spread for Interior Finishes

The flame spreadability of a material is the burning characteristic of the material relative to the fuel contributed by its combustion and the density of smoke developed. The flame spread classification of a material is based on a ten minute test on a scale of 0 to 100. Cement asbestos board is assigned a rating of 0 and select red oak flooring a rating of 100.

The three classes are listed in Table L1040-202.

The flame spread ratings for interior finish walls and ceilings shall not be greater than the Class listed in Table L1040-203.

Finish Class	Flame Spread Index	Smoke Developed Index
A	0-25	0-450
B	26-75	0-450
C	76-200	0-450

Section 803.1.1 Interior wall and ceiling finish materials

Excerpted from the 2012 *International Building Code*, Copyright 2011. Washington, D.C.: International Code Council. Reproduced with permission.

Table L1040-203 Interior Finish Requirements by Class

Group	Sprinklered			Nonsprinklered		
	Interior exit stairways, interior exit ramps and exit passageways Note a, b	Corridors and enclosure for exit access stairways and exit access ramps	Rooms and enclosed spaces Note c	Interior exit stairways, interior exit ramps and exit passageways Note a, b	Corridors and enclosure for exit access stairways and exit access ramps	Rooms and enclosed spaces Note c
A-1 & A-2	B	B	C	A	A (d)	B (e)
A-3 (f), A-4, A-5	B	B	C	A	A (d)	C
B, E, M, R-1	B	C	C	A	B	C
R-4	B	C	C	A	B	B
F	C	C	C	B	C	C
H	B	B	C	A	A	B
I-1	B	C	C	A	B	B
I-2	B	B	B (h, i)	A	A	B
I-3	A	A (j)	C	A	A	B
I-4	B	B	B (h, i)	A	A	B
R-2	C	C	C	B	B	C
R-3	C	C	C	C	C	C
S	C	C	C	B	B	C
U	No Restrictions			No Restrictions		

Notes:

a. Class C interior finish materials shall be permitted for wainscotting or paneling of not more than 1,000 square feet of applied surface area in the grade lobby where applied directly to a noncombustible base or over furring strips applied to a noncombustible base and fireblocked as required by IBC Section 803.11.1.

b. In other Group I-2 occupancies in buildings less than three stories above grade plane of other than Group I-3, Class B interior finish for nonsprinklered buildings and Class C interior finish for sprinklered buildings shall be permitted in interior exit stairways and ramps.

c. Requirements for rooms and enclosed spaces shall be based upon spaces enclosed by partitions. Where a fire-resistance rating is required for structural elements, the enclosing partitions shall extend from the floor to the ceiling. Partitions that do not comply with this shall be considered enclosed spaces and the rooms or spaces on both sides shall be considered one. In determining the applicable requirements for rooms and enclosed spaces, the specific occupancy thereof shall be the governing factor regardless of the group classification of the building or structure.

d. Lobby areas in Group A-1, A-2, and A-3 occupancies shall not be less than Class B materials.

e. Class C interior finish materials shall be permitted in places of assembly with an occupant load of 300 persons or less.

f. For places of religious worship, wood used for ornamental purposes, trusses, paneling or chancel furnishing shall be permitted.

g. Class B material is required where the building exceeds two stories.

h. Class C interior finish materials shall be permitted in administrative spaces.

i. Class C interior finish materials shall be permitted in rooms with a capacity of four persons or less.

j. Class B materials shall be permitted as wainscotting extending not more than 48 inches above the finished floor in corridors and exit access stairways and ramps.

Section 803.9

RL1090-100 Conversion Factors

Description: This table is primarily for converting customary U.S. units in the left hand column to SI metric units in the right hand column. In addition, conversion factors for some commonly encountered Canadian and non-SI metric units are included.

Table L1090-101 Metric Conversion Factors

	If You Know		Multiply By		To Find
Length	Inches	x	25.4[a]	=	Millimeters
	Feet	x	0.3048[a]	=	Meters
	Yards	x	0.9144[a]	=	Meters
	Miles (statute)	x	1.609	=	Kilometers
Area	Square inches	x	645.2	=	Square millimeters
	Square feet	x	0.0929	=	Square meters
	Square yards	x	0.8361	=	Square meters
Volume (Capacity)	Cubic inches	x	16,387	=	Cubic millimeters
	Cubic feet	x	0.02832	=	Cubic meters
	Cubic yards	x	0.7646	=	Cubic meters
	Gallons (U.S. liquids)[b]	x	0.003785	=	Cubic meters[c]
	Gallons (Canadian liquid)[b]	x	0.004546	=	Cubic meters[c]
	Ounces (U.S. liquid)[b]	x	29.57	=	Milliliters[c, d]
	Quarts (U.S. liquid)[b]	x	0.9464	=	Liters[c, d]
	Gallons (U.S. liquid)[b]	x	3.785	=	Liters[c, d]
Force	Kilograms force[d]	x	9.807	=	Newtons
	Pounds force	x	4.448	=	Newtons
	Pounds force	x	0.4536	=	Kilograms force[d]
	Kips	x	4448	=	Newtons
	Kips	x	453.6	=	Kilograms force[d]
Pressure, Stress, Strength (Force per unit area)	Kilograms force per square centimeter[d]	x	0.09807	=	Megapascals
	Pounds force per square inch (psi)	x	0.006895	=	Megapascals
	Kips per square inch	x	6.895	=	Megapascals
	Pounds force per square inch (psi)	x	0.07031	=	Kilograms force per square centimeter[d]
	Pounds force per square foot	x	47.88	=	Pascals
	Pounds force per square foot	x	4.882	=	Kilograms force per square meter[d]
Flow	Cubic feet per minute	x	0.4719	=	Liters per second
	Gallons per minute	x	0.0631	=	Liters per second
	Gallons per hour	x	1.05	=	Milliliters per second
Bending Moment Or Torque	Inch-pounds force	x	0.01152	=	Meter-kilograms force[d]
	Inch-pounds force	x	0.1130	=	Newton-meters
	Foot-pounds force	x	0.1383	=	Meter-kilograms force[d]
	Foot-pounds force	x	1.356	=	Newton-meters
	Meter-kilograms force[d]	x	9.807	=	Newton-meters
Mass	Ounces (avoirdupois)	x	28.35	=	Grams
	Pounds (avoirdupois)	x	0.4536	=	Kilograms
	Tons (metric)	x	1000	=	Kilograms
	Tons, short (2000 pounds)	x	907.2	=	Kilograms
	Tons, short (2000 pounds)	x	0.9072	=	Megagrams[e]
Mass per Unit Volume	Pounds mass per cubic foot	x	16.02	=	Kilograms per cubic meter
	Pounds mass per cubic yard	x	0.5933	=	Kilograms per cubic meter
	Pounds mass per gallon (U.S. liquid)[b]	x	119.8	=	Kilograms per cubic meter
	Pounds mass per gallon (Canadian liquid)[b]	x	99.78	=	Kilograms per cubic meter
Temperature	Degrees Fahrenheit	(F-32)/1.8		=	Degrees Celsius
	Degrees Fahrenheit	(F+459.67)/1.8		=	Degrees Kelvin
	Degrees Celsius	C+273.15		=	Degrees Kelvin

[a]The factor given is exact
[b]One U.S. gallon = 0.8327 Canadian gallon
[c]1 liter = 1000 milliliters = 1000 cubic centimeters
1 cubic decimeter = 0.001 cubic meter
[d]Metric but not SI unit
[e]Called "tonne" in England and "metric ton" in other metric countries

Life Cycle Costing (LCC)
for Evaluating Economic Performance of Building Investments

Introduction

A complete analysis of maintenance costs requires consideration of life cycle costs, or LCC, as it is often called. Under current conditions, facilities managers have to wisely spend their limited operating and maintenance budgets. When deciding how to allocate limited budgets, maintenance professionals are constantly faced with the economic decision of when to stop maintaining and start replacing. Once the decision to replace has been made, LCC analysis can enhance the choice of alternatives as well.

The American Society of Testing and Materials (ASTM) Subcommittee E06.81 on Building Economics has developed a series of standards for evaluating the economic performance of building investments. These standards, which have been endorsed by the American Association of Cost Engineers (AACE), are intended to provide uniformity in both the terminology and calculation methods utilized for Life Cycle Costing. Without these standards as a basis for calculations, life cycle cost comparisons could be misleading or incomplete. This chapter is based on the current ASTM standards which are listed in the references at the end of this section.

Definition of Life Cycle Costing

(ASTM Standard Terminology of Building Economics E833-89)

A technique of economic evaluation that sums over a given study period the cost of initial investment less resale value, replacements, operations, energy use, and maintenance and repair of an investment decision. The costs are expressed in either lump sum present value terms or in equivalent uniform annual values.

This technique for life cycle costing is important to the facilities manager, as it is a valuable tool for evaluating investment alternatives and supports his or her decision-making.

To understand Life Cycle Costing, one should be familiar with the notions of compounding, discounting, present value, and equivalent uniform annual value.

Compounding

Definition of Compounding *(Webster's New Collegiate Dictionary)*

The process of computing the value of an original principal sum based on interest calculated on the sum of the original principal and accrued interest.

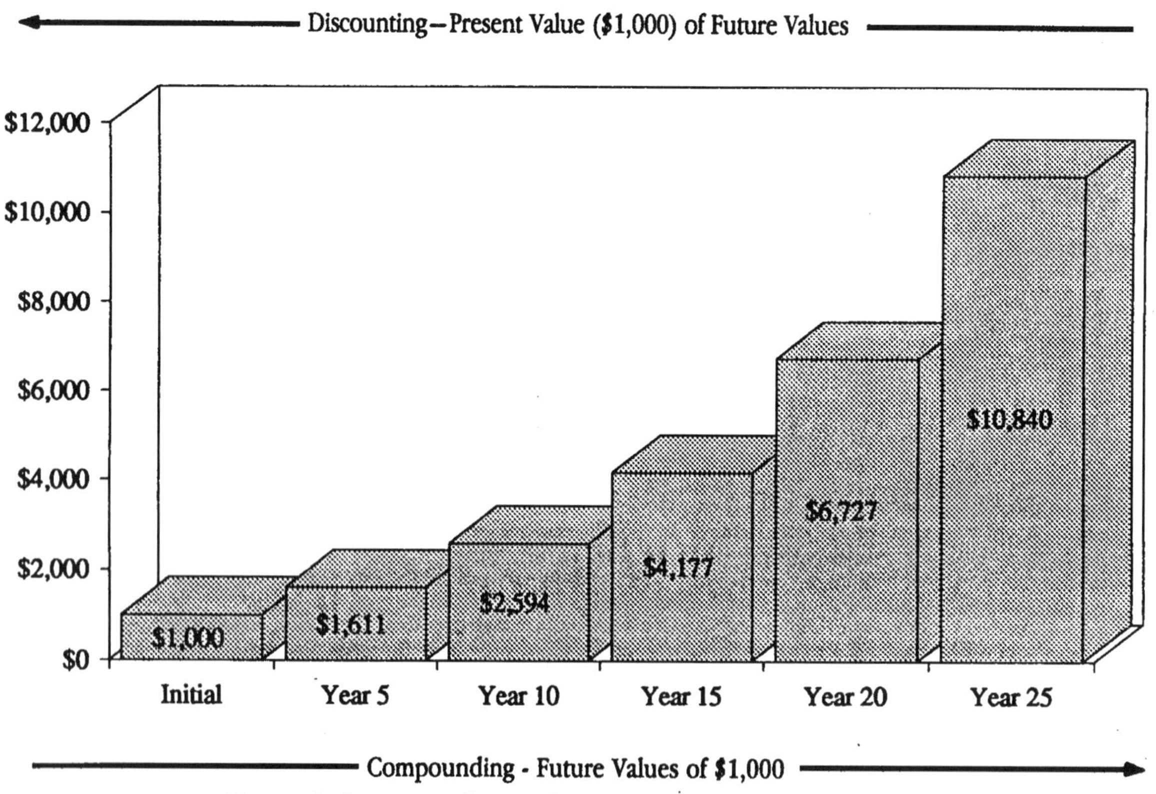

Figure 1: Compounding and Discounting at 10% Per Annum

Figure 1 illustrates compounding. As shown, a principal sum of $1,000 at a compound interest rate of 10% per annum increases to $1,611 in 5 years, $2,594 in 10 years, $4,177 in 15 years, $6,727 in 20 years, and $10,840 in 25 years; this process (compounding) is usually well understood in that it represents the amount accumulated in a bank account for a specific amount at a given interest rate and calculated on the sum of the original principal and accrued interest.

Discounting

ASTM Definition of Discounting

A technique for converting cash flows that occur over time to equivalent amounts at a common time (base time).

It is also evident from Figure 1 that if one had to pay out or meet an obligation of either $1,611 in 5 years, $2,594 in 10 years, $4,177 in 15 years, $6,727 in 20 years, or $10,840 in 25 years, the amount to be invested initially at 10% interest would be $1,000. This amount is referred to as the present value (PV) of a future amount and it is calculated by "discounting" a future value (F) in a specific year at a given rate of interest.

Present Value

ASTM Definition of Present Value

The value of a benefit or cost found by discounting future cash flows to the base time. (Syn. present worth)

The basic mathematics for calculating present values are relatively simple:

Given:

P = Present sum of money
F = Future sum of money
i = Interest or discount rate (expressed as a decimal and not a percentage)
n = Number of years

The future value (F) of a present sum of money (P) may be computed as shown in Table 1, a relatively simple process.

Year	Future Value (F)
1	$P (1 + i)$
2	$P (1 + i) (1 + i) = P (1 + i)^2$
3	$P (1 + i)^2 (1 + i) = P (1 + i)^3$
n	$P (1 + i)^n$

Table 1: The Future Value of a Present Sum of Money

Since $F = P (1 + i)^n$

Therefore $$P = \frac{F}{(1 + 1)^n} = F \times \frac{1}{(1 + i)^n}$$

Example

What is the present value (P) of an anticipated maintenance expense of $300 in year 3 (F) if the interest rate is 10%?

$$P = F \times \frac{1}{(1 + i)^n} = 300 \times \frac{1}{(1 + 0.10)^n} = 300 \times \frac{1}{1.33}$$

$$= 300 \times 0.75$$

$$= \$225$$

From an interest standpoint, this tells us that $225 is the amount that would have to be deposited today into an account paying 10% interest per annum in order to provide $300 at the end of year 3 to meet the anticipated expense. From the economist's point of view, discounting using the investor's minimum acceptable rate of return (10% in this case) makes the economic outlook the same, whether it is the future amount of $300 or a present value of $225.

Note that $(1 + i)^n$ is commonly referred to as the *Single Compound Amount (SCA)* factor, and $1/(1 + i)^n$ as the *Single Present Value Factor (SPV)*. Both factors may be readily found in standard discount factor tables for a wide range of interest rates and time periods, thus simplifying calculations. The use of these factors in calculating life cycle costs by the formula method is discussed later in this section.

A life cycle cost analysis can be performed in *constant dollars* (dollars of constant purchasing power, net of general inflation) or *current dollars* (actual prices inclusive of general inflation). If you use constant dollars, you must use a "real" discount rate (i.e., a rate which reflects the time value of money, net of general inflation) and differential escalation rates (also net of general inflation) for future cost items. Constant dollars are generally tied to a specific year. If you use current dollars, the discount rates and escalation rate must include general inflation (i.e., they must be "nominal" or "market" rates). In this chapter, all costs are expressed in current dollars and rates in nominal terms. When taxes are included in the analysis, the current dollar approach is generally preferred.

The Cash Flow Method for Calculating Life Cycle Costs

There are basically two approaches to calculating life cycle costs—the *cash flow* method and the *formula* method which, when applicable, is somewhat more simple.

To illustrate the cash flow method, a relatively simple example follows.

Example

A facility manager is considering the purchase of maintenance equipment required for a four-year period for which initial costs, energy costs, and maintenance costs of alternative proposals vary.

The financial criteria on which the economic evaluation will be based are the following:

Interest/Discount Rate	–	10%
Energy Escalation Rate	–	8% per annum
Labor and Material Escalation Rate	–	4% per annum
Period of Study	–	4 years

The following data is provided in one of the alternative proposals to be analyzed:

- Initial Capital Cost – $1,000
- Maintenance Costs – A fixed annual cost of $100 per year quoted by the supplier (no escalation to be considered)
- Annual Energy Costs – Initially $100 per year and subject to 8% annual escalation – i.e., $108 for year one.
- Salvage Value – At the end of the four-year period, the equipment has no further useful life and the supplier agrees to purchase it as scrap for $50.

What is the present value of this alternative proposal?

The steps to resolve this problem are the following:

Step 1 – Prepare a Cash Flow Diagram.

Step 2 – Establish a Time Schedule of Costs.

Step 3 – Calculate Annual Net Cash Flows.

Step 4 – Calculate Present Value Factors.

Step 5 – Calculate Present Values of Annual Cash Flows and Life Cycle Costs.

Step 1: Prepare a Cash Flow Diagram

To better understand the problem, cash flow diagrams are often prepared. Revenues are noted as vertical lines above a horizontal time axis, and disbursements are noted as vertical lines below it.

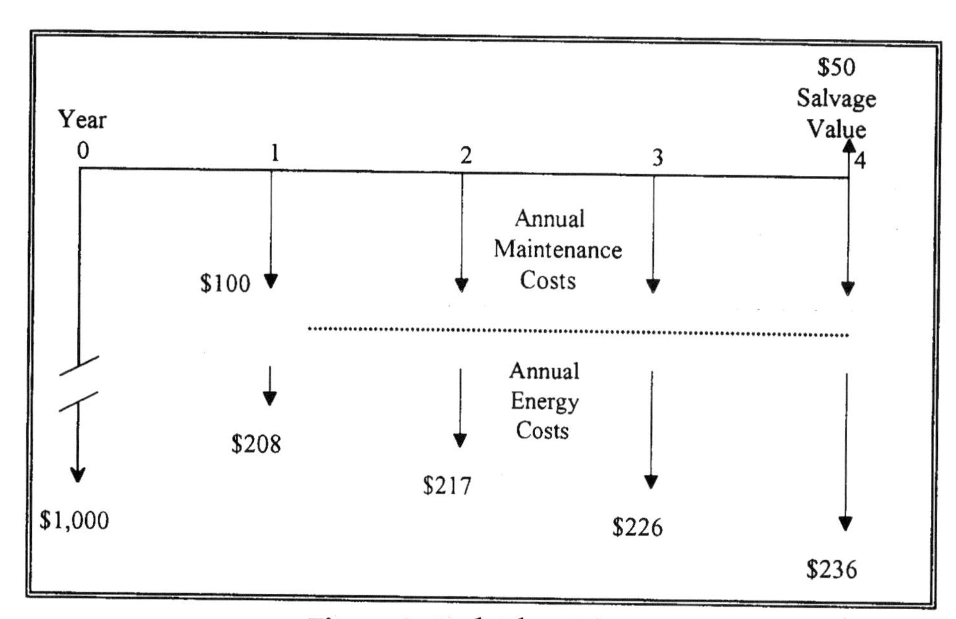

Figure 2: Cash Flow Diagram

Step 2: Establish a Time Schedule of Costs

Table 2 below presents revenues and disbursements for each year of the study:

A Year	B Capital Costs	C Maintenance Costs	D Energy Cost (8% escalation)
0	1,000		-
1	-	100	108
2	-	100	117
3	-	100	126
4	(50)	100	136

Table 2

Note that the disbursements in this example are considered positive, and revenues for the salvage value are negative.

Step 3: Calculate Annual Net Cash Flows

Table 3 below provides the net cash flow (NCF) for each year, which could be a disbursement or a revenue.

A Year	B Capital Costs	C Maintenance Costs	D Energy Cost (8% escalation)	E NCF (B + C + D)
0	1,000	-	-	1,000
1	-	100	108	208
2	-	100	117	217
3	-	100	126	226
4	(50)	100	136	186

Table 3

Step 4: Calculate Present Value Factors

As in a previous example, present value factors based on 10% interest will be calculated for each year to convert the annual net cash flows to present values.

Year n	PV Factor $\frac{1}{(1+i)^n}$	
0	1	
1	$\frac{1}{(1+0.1)^1}$	= 0.91
2	$\frac{1}{(1+0.1)^2}$	= 0.83
3	$\frac{1}{(1+0.1)^3}$	= 0.75
4	$\frac{1}{(1+0.1)^4}$	= 0.68

Table 4

Step 5: Calculate Present Values of Annual Cash Flows and Life Cycle Costs

Annual net cash flows are converted to present values of the base year. Their sum (Total Present Value) will represent the life cycle costs of the alternative.

Year n	NCF	PV Factor	PV $
0	1,000	1	1,000
1	208	0.91	189
2	217	0.83	180
3	226	0.75	170
4	186	0.68	126
PV Life Cycle Cost	-	-	1,665

Table 5

The total life cycle cost of this alternative in terms of today's dollar (if this year is the base year) is $1,665. This means that if a present sum of $1,665 were deposited today at an interest rate of 10%, all expenses could be paid over a four-year period, at which time the bank balance would be zero.

Once the costs of the other proposals have been calculated in present value dollars, the facilities manager will be in a position to quantify the differences in life cycle costs and decide which proposal represents the lowest cost of ownership in the long run.

The approach outlined is a practical one when considering the purchase of equipment such as elevators, controls, etc.; quotations should not only include capital costs, but also the costs of long-term maintenance contracts which could make a higher initial cost proposal lower over the life cycle of the equipment.

The Formula Method for Calculating Life Cycle Costs

It is also possible to solve the previous problem utilizing the formula method if the problem is adaptable to this technique. Business calculators or discount factor tables such as those published by ASTM can be employed to generate the present value factors, and results may be obtained more quickly.

In some cases, the complexity of the calculations (such as when taxes on net revenues are considered) will necessitate the "cash flow" method, with calculations done manually or using a spreadsheet program.

With the formula method approach, the calculations for the life cycle cost of the maintenance equipment alternative analyzed would be as follows:

1.	PV of Initial Investment	$1,000
2.	PV of Maintenance Costs Annual Cost (A) x UPV factor 100 x 3.17	$ 317
3.	PV of Energy Costs Base Year Cost (A) x UPV* 100 x 3.82	$ 382
4.	Salvage (Scrap) Value Future Value (F) x SPV (50) x 0.68	(34)
5.	Life Cycle Cost / Total Present Value	$1,665

Note: The above factors can be found in Discount Factor Tables.

UPV–Uniform Present Value
SPV–Single Present Value
UPV*–Uniform Present Value Modified (*)

The calculated life cycle cost of $1,665 is identical to that obtained with the cash flow method, but is arrived at with less effort.

Equivalent Uniform Annual Value

ASTM Definition of Equivalent Uniform Annual Value

A uniform annual amount equivalent to the project costs or benefits, taking into account the time value of money throughout the study period. (Syn. annual value)

As noted previously in the definition of Life Cycle Costs, these may be expressed as a lump sum present value with a base time reference, or in equivalent uniform annual values (EUAV) over the study period; decision-makers may prefer one or the other, or require both for the analysis of investment alternatives.

To express life cycle costs in annual values, it is necessary to first determine the lump sum present value of an alternative and multiply it by the Uniform Capital Recovery (UCR) factor given the interest rate and time period. This process is identical to calculating annual payments for a mortgage.

Example

The present value life cycle cost for one of the maintenance equipment alternatives analyzed amounted to $1,665; what is the equivalent uniform annual value "A"?

Given: P	(a principal sum or present value life cycle cost)	= $1,665
	i	= 10%
	n	= 4 years
	UCR Factor	= 0.32 (from standard factor tables for i)
		= 10%, n = 4 years
	A	= equivalent uniform annual value – to be determined
Solution:	A	= P x UCR
		= $1,665 x 0.32
		= $530

The life cycle cost of the alternative may be expressed as a lump sum, $1,665, or an annual value of $530 based on a four-year time period.

Other Methods of Evaluating Economic Performance

While calculating life cycle costs is a helpful tool for the facilities manager evaluating alternatives, there are other approaches for evaluating economic performance of building investments. Among these other methods are *payback, net benefits,* and *internal rate of return;* these are described in detail in an ASTM compilation of *Building Economics Standards* which are endorsed by the American Association of Cost Engineers (AACE). These standards, as well as adjuncts available, such as discount factor tables and a life cycle cost analysis computer program, are included in the References at the end of this section.

ASTM has also developed a Standard Classification of Building Elements and Related Site Work entitled UNIFORMAT II (Designation E-1557-93). One of the many applications for this classification for facilities managers is its use to structure operating and maintenance cost data by building elements or systems, thus facilitating life cycle costing studies. This classification can also be the basis of life cycle cost plans for new facilities that should be developed during the design period with the assistance of the facility manager.

References (Annual Book of ASTM Standards Vol. 04.07)

ASTM Standards

E 833 Terminology of Building Economics

E 917 Practice for Measuring Life Cycle Costs of Buildings and Building Systems

E 964 Practice for Measuring Benefit-to-Cost and Savings-to-Investment Ratios for Buildings and Building Systems

E 1057 Practice for Measuring Internal Rates of Return and Adjusted Internal Rates of Return for Investments in Buildings and Building Systems

E 1074 Practice for Measuring Net Benefits for Investments in Buildings and Building Systems

E 1121 Practice for Measuring Payback for Investments in Buildings and Building Systems

E 1185 Guide for Selecting Economic Methods for Evaluating Investments in Buildings and Building Systems

E 1369 Guide for Selecting Techniques for Treating Uncertainty and Risk in the Economic Evaluation of Buildings and Building Systems

E 1557 Standard Classification for Building Elements and Related Site Work–UNIFORMAT II

ASTM Adjuncts

Discount Factor Tables, Adjunct to Practices E 917, E 964,E 1057, and E 1074.

Computer Program and User's Guide to Building Maintenance, Repair, and Replacement Database for Life Cycle Cost Analysis, Adjunct to Practices E 917, E 964, E 1057, and E 1121.

Note: The Standards noted above, other than E 1557, are available in one publication entitled *ASTM Standards on Building Economics, Second Edition.*

Abbreviations

Abbreviation	Meaning
A	Area Square Feet; Ampere
AAFES	Army and Air Force Exchange Service
ABS	Acrylonitrile Butadiene Stryrene; Asbestos Bonded Steel
A.C., AC	Alternating Current; Air-Conditioning; Asbestos Cement; Plywood Grade A & C
ACI	American Concrete Institute
ACR	Air Conditioning Refrigeration
ADA	Americans with Disabilities Act
AD	Plywood, Grade A & D
Addit.	Additional
Adh.	Adhesive
Adj.	Adjustable
af	Audio-frequency
AFFF	Aqueous Film Forming Foam
AFUE	Annual Fuel Utilization Efficiency
AGA	American Gas Association
Agg.	Aggregate
A.H., Ah	Ampere Hours
A hr.	Ampere-hour
A.H.U., AHU	Air Handling Unit
A.I.A.	American Institute of Architects
AIC	Ampere Interrupting Capacity
Allow.	Allowance
alt., alt	Alternate
Alum.	Aluminum
a.m.	Ante Meridiem
Amp.	Ampere
Anod.	Anodized
ANSI	American National Standards Institute
APA	American Plywood Association
Approx.	Approximate
Apt.	Apartment
Asb.	Asbestos
A.S.B.C.	American Standard Building Code
Asbe.	Asbestos Worker
ASCE	American Society of Civil Engineers
A.S.H.R.A.E.	American Society of Heating, Refrig. & AC Engineers
ASME	American Society of Mechanical Engineers
ASTM	American Society for Testing and Materials
Attchmt.	Attachment
Avg., Ave.	Average
AWG	American Wire Gauge
AWWA	American Water Works Assoc.
Bbl.	Barrel
B&B, BB	Grade B and Better; Balled & Burlapped
B&S	Bell and Spigot
B.&W.	Black and White
b.c.c.	Body-centered Cubic
B.C.Y.	Bank Cubic Yards
BE	Bevel End
B.F.	Board Feet
Bg. cem.	Bag of Cement
BHP	Boiler Horsepower; Brake Horsepower
B.I.	Black Iron
bidir.	bidirectional
Bit., Bitum.	Bituminous
Bit., Conc.	Bituminous Concrete
Bk.	Backed
Bkrs.	Breakers
Bldg., bldg	Building
Blk.	Block
Bm.	Beam
Boil.	Boilermaker
bpm	Blows per Minute
BR	Bedroom
Brg., brng.	Bearing
Brhe.	Bricklayer Helper
Bric.	Bricklayer
Brk., brk	Brick
brkt	Bracket
Brs.	Brass
Brz.	Bronze
Bsn.	Basin
Btr.	Better
BTU	British Thermal Unit
BTUH	BTU per Hour
Bu.	Bushels
BUR	Built-up Roofing
BX	Interlocked Armored Cable
°C	Degree Centigrade
c	Conductivity, Copper Sweat
C	Hundred; Centigrade
C/C	Center to Center, Cedar on Cedar
C-C	Center to Center
Cab	Cabinet
Cair.	Air Tool Laborer
Cal.	Caliper
Calc	Calculated
Cap.	Capacity
Carp.	Carpenter
C.B.	Circuit Breaker
C.C.A.	Chromate Copper Arsenate
C.C.F.	Hundred Cubic Feet
cd	Candela
cd/sf	Candela per Square Foot
CD	Grade of Plywood Face & Back
CDX	Plywood, Grade C & D, exterior glue
Cefi.	Cement Finisher
Cem.	Cement
CF	Hundred Feet
C.F.	Cubic Feet
CFM	Cubic Feet per Minute
CFRP	Carbon Fiber Reinforced Plastic
c.g.	Center of Gravity
CHW	Chilled Water; Commercial Hot Water
C.I., CI	Cast Iron
C.I.P., CIP	Cast in Place
Circ.	Circuit
C.L.	Carload Lot
CL	Chain Link
Clab.	Common Laborer
Clam	Common Maintenance Laborer
C.L.F.	Hundred Linear Feet
CLF	Current Limiting Fuse
CLP	Cross Linked Polyethylene
cm	Centimeter
CMP	Corr. Metal Pipe
CMU	Concrete Masonry Unit
CN	Change Notice
Col.	Column
CO_2	Carbon Dioxide
Comb.	Combination
comm.	Commercial, Communication
Compr.	Compressor
Conc.	Concrete
Cont., cont	Continuous; Continued, Container
Corkbd.	Cork Board
Corr.	Corrugated
Cos	Cosine
Cot	Cotangent
Cov.	Cover
C/P	Cedar on Paneling
CPA	Control Point Adjustment
Cplg.	Coupling
CPM	Critical Path Method
CPVC	Chlorinated Polyvinyl Chloride
C.Pr.	Hundred Pair
CRC	Cold Rolled Channel
Creos.	Creosote
Crpt.	Carpet & Linoleum Layer
CRT	Cathode-ray Tube
CS	Carbon Steel, Constant Shear Bar Joist
Csc	Cosecant
C.S.F.	Hundred Square Feet
CSI	Construction Specifications Institute
CT	Current Transformer
CTS	Copper Tube Size
Cu	Copper, Cubic
Cu. Ft.	Cubic Foot
cw	Continuous Wave
C.W.	Cool White; Cold Water
Cwt.	100 Pounds
C.W.X.	Cool White Deluxe
C.Y.	Cubic Yard (27 cubic feet)
C.Y./Hr.	Cubic Yard per Hour
Cyl.	Cylinder
d	Penny (nail size)
D	Deep; Depth; Discharge
Dis., Disch.	Discharge
Db	Decibel
Dbl.	Double
DC	Direct Current
DDC	Direct Digital Control
Demob.	Demobilization
d.f.t.	Dry Film Thickness
d.f.u.	Drainage Fixture Units
D.H.	Double Hung
DHW	Domestic Hot Water
DI	Ductile Iron
Diag.	Diagonal
Diam., Dia	Diameter
Distrib.	Distribution
Div.	Division
Dk.	Deck
D.L.	Dead Load; Diesel
DLH	Deep Long Span Bar Joist
dlx	Deluxe
Do.	Ditto
DOP	Dioctyl Phthalate Penetration Test (Air Filters)
Dp., dp	Depth
D.P.S.T.	Double Pole, Single Throw
Dr.	Drive
DR	Dimension Ratio
Drink.	Drinking
D.S.	Double Strength
D.S.A.	Double Strength A Grade
D.S.B.	Double Strength B Grade
Dty.	Duty
DWV	Drain Waste Vent
DX	Deluxe White, Direct Expansion
dyn	Dyne
e	Eccentricity
E	Equipment Only; East; Emissivity
Ea.	Each
EB	Encased Burial
Econ.	Economy
E.C.Y	Embankment Cubic Yards
EDP	Electronic Data Processing
EIFS	Exterior Insulation Finish System
E.D.R.	Equiv. Direct Radiation
Eq.	Equation
EL	Elevation
Elec.	Electrician; Electrical
Elev.	Elevator; Elevating
EMT	Electrical Metallic Conduit; Thin Wall Conduit
Eng.	Engine, Engineered
EPDM	Ethylene Propylene Diene Monomer
EPS	Expanded Polystyrene
Eqhv.	Equip. Oper., Heavy
Eqlt.	Equip. Oper., Light
Eqmd.	Equip. Oper., Medium
Eqmm.	Equip. Oper., Master Mechanic
Eqol.	Equip. Oper., Oilers
Equip.	Equipment
ERW	Electric Resistance Welded

Abbreviations

Abbreviation	Meaning
E.S.	Energy Saver
Est.	Estimated
esu	Electrostatic Units
E.W.	Each Way
EWT	Entering Water Temperature
Excav.	Excavation
excl	Excluding
Exp., exp	Expansion, Exposure
Ext., ext	Exterior; Extension
Extru.	Extrusion
f.	Fiber Stress
F	Fahrenheit; Female; Fill
Fab., fab	Fabricated; Fabric
FBGS	Fiberglass
F.C.	Footcandles
f.c.c.	Face-centered Cubic
f'c.	Compressive Stress in Concrete; Extreme Compressive Stress
F.E.	Front End
FEP	Fluorinated Ethylene Propylene (Teflon)
F.G.	Flat Grain
F.H.A.	Federal Housing Administration
Fig.	Figure
Fin.	Finished
FIPS	Female Iron Pipe Size
Fixt.	Fixture
FJP	Finger jointed and primed
Fl. Oz.	Fluid Ounces
Flr.	Floor
Flrs.	Floors
FM	Frequency Modulation; Factory Mutual
Fmg.	Framing
FM/UL	Factory Mutual/Underwriters Labs
Fdn.	Foundation
FNPT	Female National Pipe Thread
Fori.	Foreman, Inside
Foro.	Foreman, Outside
Fount.	Fountain
fpm	Feet per Minute
FPT	Female Pipe Thread
Fr	Frame
F.R.	Fire Rating
FRK	Foil Reinforced Kraft
FSK	Foil/Scrim/Kraft
FRP	Fiberglass Reinforced Plastic
FS	Forged Steel
FSC	Cast Body; Cast Switch Box
Ft., ft	Foot; Feet
Ftng.	Fitting
Ftg.	Footing
Ft lb.	Foot Pound
Furn.	Furniture
FVNR	Full Voltage Non-Reversing
FVR	Full Voltage Reversing
FXM	Female by Male
Fy.	Minimum Yield Stress of Steel
g	Gram
G	Gauss
Ga.	Gauge
Gal., gal.	Gallon
Galv., galv	Galvanized
GC/MS	Gas Chromatograph/Mass Spectrometer
Gen.	General
GFI	Ground Fault Interrupter
GFRC	Glass Fiber Reinforced Concrete
Glaz.	Glazier
GPD	Gallons per Day
gpf	Gallon per Flush
GPH	Gallons per Hour
gpm, GPM	Gallons per Minute
GR	Grade
Gran.	Granular
Grnd.	Ground
GVW	Gross Vehicle Weight
GWB	Gypsum Wall Board
H	High Henry
HC	High Capacity
H.D., HD	Heavy Duty; High Density
H.D.O.	High Density Overlaid
HDPE	High Density Polyethylene Plastic
Hdr.	Header
Hdwe.	Hardware
H.I.D., HID	High Intensity Discharge
Help.	Helper Average
HEPA	High Efficiency Particulate Air Filter
Hg	Mercury
HIC	High Interrupting Capacity
HM	Hollow Metal
HMWPE	High Molecular Weight Polyethylene
HO	High Output
Horiz.	Horizontal
H.P., HP	Horsepower; High Pressure
H.P.F.	High Power Factor
Hr.	Hour
Hrs./Day	Hours per Day
HSC	High Short Circuit
Ht.	Height
Htg.	Heating
Htrs.	Heaters
HVAC	Heating, Ventilation & Air-Conditioning
Hvy.	Heavy
HW	Hot Water
Hyd.; Hydr.	Hydraulic
Hz	Hertz (cycles)
I.	Moment of Inertia
IBC	International Building Code
I.C.	Interrupting Capacity
ID	Inside Diameter
I.D.	Inside Dimension; Identification
I.F.	Inside Frosted
I.M.C.	Intermediate Metal Conduit
In.	Inch
Incan.	Incandescent
Incl.	Included; Including
Int.	Interior
Inst.	Installation
Insul., insul	Insulation/Insulated
I.P.	Iron Pipe
I.P.S., IPS	Iron Pipe Size
IPT	Iron Pipe Threaded
I.W.	Indirect Waste
J	Joule
J.I.C.	Joint Industrial Council
K	Thousand; Thousand Pounds; Heavy Wall Copper Tubing, Kelvin
K.A.H.	Thousand Amp. Hours
kcmil	Thousand Circular Mils
KD	Knock Down
K.D.A.T.	Kiln Dried After Treatment
kg	Kilogram
kG	Kilogauss
kgf	Kilogram Force
kHz	Kilohertz
Kip	1000 Pounds
KJ	Kilojoule
K.L.	Effective Length Factor
K.L.F.	Kips per Linear Foot
Km	Kilometer
KO	Knock Out
K.S.F.	Kips per Square Foot
K.S.I.	Kips per Square Inch
kV	Kilovolt
kVA	Kilovolt Ampere
kVAR	Kilovar (Reactance)
KW	Kilowatt
KWh	Kilowatt-hour
L	Labor Only; Length; Long; Medium Wall Copper Tubing
Lab.	Labor
lat	Latitude
Lath.	Lather
Lav.	Lavatory
lb.; #	Pound
L.B., LB	Load Bearing; L Conduit Body
L. & E.	Labor & Equipment
lb./hr.	Pounds per Hour
lb./L.F.	Pounds per Linear Foot
lbf/sq.in.	Pound-force per Square Inch
L.C.L.	Less than Carload Lot
L.C.Y.	Loose Cubic Yard
Ld.	Load
LE	Lead Equivalent
LED	Light Emitting Diode
L.F.	Linear Foot
L.F. Hdr	Linear Feet of Header
L.F. Nose	Linear Foot of Stair Nosing
L.F. Rsr	Linear Foot of Stair Riser
Lg.	Long; Length; Large
L & H	Light and Heat
LH	Long Span Bar Joist
L.H.	Labor Hours
L.L., LL	Live Load
L.L.D.	Lamp Lumen Depreciation
lm	Lumen
lm/sf	Lumen per Square Foot
lm/W	Lumen per Watt
LOA	Length Over All
log	Logarithm
L-O-L	Lateralolet
long.	Longitude
L.P., LP	Liquefied Petroleum; Low Pressure
L.P.F.	Low Power Factor
LR	Long Radius
L.S.	Lump Sum
Lt.	Light
Lt. Ga.	Light Gauge
L.T.L.	Less than Truckload Lot
Lt. Wt.	Lightweight
L.V.	Low Voltage
M	Thousand; Material; Male; Light Wall Copper Tubing
M^2CA	Meters Squared Contact Area
m/hr.; M.H.	Man-hour
mA	Milliampere
Mach.	Machine
Mag. Str.	Magnetic Starter
Maint.	Maintenance
Marb.	Marble Setter
Mat; Mat'l.	Material
Max.	Maximum
MBF	Thousand Board Feet
MBH	Thousand BTU's per hr.
MC	Metal Clad Cable
MCC	Motor Control Center
M.C.F.	Thousand Cubic Feet
MCFM	Thousand Cubic Feet per Minute
M.C.M.	Thousand Circular Mils
MCP	Motor Circuit Protector
MD	Medium Duty
MDF	Medium-density fibreboard
M.D.O.	Medium Density Overlaid
Med.	Medium
MF	Thousand Feet
M.F.B.M.	Thousand Feet Board Measure
Mfg.	Manufacturing
Mfrs.	Manufacturers
mg	Milligram
MGD	Million Gallons per Day
MGPH	Million Gallons per Hour
MH, M.H.	Manhole; Metal Halide; Man-Hour
MHz	Megahertz
Mi.	Mile
MI	Malleable Iron; Mineral Insulated
MIPS	Male Iron Pipe Size
mj	Mechanical Joint
m	Meter
mm	Millimeter
Mill.	Millwright
Min., min.	Minimum, Minute

Abbreviations

Abbreviation	Meaning
Misc.	Miscellaneous
ml	Milliliter, Mainline
M.L.F.	Thousand Linear Feet
Mo.	Month
Mobil.	Mobilization
Mog.	Mogul Base
MPH	Miles per Hour
MPT	Male Pipe Thread
MRGWB	Moisture Resistant Gypsum Wallboard
MRT	Mile Round Trip
ms	Millisecond
M.S.F.	Thousand Square Feet
Mstz.	Mosaic & Terrazzo Worker
M.S.Y.	Thousand Square Yards
Mtd., mtd., mtd	Mounted
Mthe.	Mosaic & Terrazzo Helper
Mtng.	Mounting
Mult.	Multi; Multiply
MUTCD	Manual on Uniform Traffic Control Devices
M.V.A.	Million Volt Amperes
M.V.A.R.	Million Volt Amperes Reactance
MV	Megavolt
MW	Megawatt
MXM	Male by Male
MYD	Thousand Yards
N	Natural; North
nA	Nanoampere
NA	Not Available; Not Applicable
N.B.C.	National Building Code
NC	Normally Closed
NEMA	National Electrical Manufacturers Assoc.
NEHB	Bolted Circuit Breaker to 600V.
NFPA	National Fire Protection Association
NLB	Non-Load-Bearing
NM	Non-Metallic Cable
nm	Nanometer
No.	Number
NO	Normally Open
N.O.C.	Not Otherwise Classified
Nose.	Nosing
NPT	National Pipe Thread
NQOD	Combination Plug-on/Bolt on Circuit Breaker to 240V.
N.R.C., NRC	Noise Reduction Coefficient/ Nuclear Regulator Commission
N.R.S.	Non Rising Stem
ns	Nanosecond
NTP	Notice to Proceed
nW	Nanowatt
OB	Opposing Blade
OC	On Center
OD	Outside Diameter
O.D.	Outside Dimension
ODS	Overhead Distribution System
O.G.	Ogee
O.H.	Overhead
O&P	Overhead and Profit
Oper.	Operator
Opng.	Opening
Orna.	Ornamental
OSB	Oriented Strand Board
OS&Y	Outside Screw and Yoke
OSHA	Occupational Safety and Health Act
Ovhd.	Overhead
OWG	Oil, Water or Gas
Oz.	Ounce
P.	Pole; Applied Load; Projection
p.	Page
Pape.	Paperhanger
P.A.P.R.	Powered Air Purifying Respirator
PAR	Parabolic Reflector
P.B., PB	Push Button
Pc., Pcs.	Piece, Pieces
P.C.	Portland Cement; Power Connector
P.C.F.	Pounds per Cubic Foot
PCM	Phase Contrast Microscopy
PDCA	Painting and Decorating Contractors of America
P.E., PE	Professional Engineer; Porcelain Enamel; Polyethylene; Plain End
P.E.C.I.	Porcelain Enamel on Cast Iron
Perf.	Perforated
PEX	Cross Linked Polyethylene
Ph.	Phase
P.I.	Pressure Injected
Pile.	Pile Driver
Pkg.	Package
Pl.	Plate
Plah.	Plasterer Helper
Plas.	Plasterer
plf	Pounds Per Linear Foot
Pluh.	Plumber Helper
Plum.	Plumber
Ply.	Plywood
p.m.	Post Meridiem
Pntd.	Painted
Pord.	Painter, Ordinary
pp	Pages
PP, PPL	Polypropylene
P.P.M.	Parts per Million
Pr.	Pair
P.E.S.B.	Pre-engineered Steel Building
Prefab.	Prefabricated
Prefin.	Prefinished
Prop.	Propelled
PSF, psf	Pounds per Square Foot
PSI, psi	Pounds per Square Inch
PSIG	Pounds per Square Inch Gauge
PSP	Plastic Sewer Pipe
Pspr.	Painter, Spray
Psst.	Painter, Structural Steel
P.T.	Potential Transformer
P. & T.	Pressure & Temperature
Ptd.	Painted
Ptns.	Partitions
Pu	Ultimate Load
PVC	Polyvinyl Chloride
Pvmt.	Pavement
PRV	Pressure Relief Valve
Pwr.	Power
Q	Quantity Heat Flow
Qt.	Quart
Quan., Qty.	Quantity
Q.C.	Quick Coupling
r	Radius of Gyration
R	Resistance
R.C.P.	Reinforced Concrete Pipe
Rect.	Rectangle
recpt.	Receptacle
Reg.	Regular
Reinf.	Reinforced
Req'd.	Required
Res.	Resistant
Resi.	Residential
RF	Radio Frequency
RFID	Radio-frequency Identification
Rgh.	Rough
RGS	Rigid Galvanized Steel
RHW	Rubber, Heat & Water Resistant; Residential Hot Water
rms	Root Mean Square
Rnd.	Round
Rodm.	Rodman
Rofc.	Roofer, Composition
Rofp.	Roofer, Precast
Rohe.	Roofer Helpers (Composition)
Rots.	Roofer, Tile & Slate
R.O.W.	Right of Way
RPM	Revolutions per Minute
R.S.	Rapid Start
Rsr	Riser
RT	Round Trip
S.	Suction; Single Entrance; South
SBS	Styrene Butadiere Styrene
SC	Screw Cover
SCFM	Standard Cubic Feet per Minute
Scaf.	Scaffold
Sch., Sched.	Schedule
S.C.R.	Modular Brick
S.D.	Sound Deadening
SDR	Standard Dimension Ratio
S.E.	Surfaced Edge
Sel.	Select
SER, SEU	Service Entrance Cable
S.F.	Square Foot
S.F.C.A.	Square Foot Contact Area
S.F. Flr.	Square Foot of Floor
S.F.G.	Square Foot of Ground
S.F. Hor.	Square Foot Horizontal
SFR	Square Feet of Radiation
S.F. Shlf.	Square Foot of Shelf
S4S	Surface 4 Sides
Shee.	Sheet Metal Worker
Sin.	Sine
Skwk.	Skilled Worker
SL	Saran Lined
S.L.	Slimline
Sldr.	Solder
SLH	Super Long Span Bar Joist
S.N.	Solid Neutral
SO	Stranded with oil resistant inside insulation
S-O-L	Socketolet
sp	Standpipe
S.P.	Static Pressure; Single Pole; Self-Propelled
Spri.	Sprinkler Installer
spwg	Static Pressure Water Gauge
S.P.D.T.	Single Pole, Double Throw
SPF	Spruce Pine Fir; Sprayed Polyurethane Foam
S.P.S.T.	Single Pole, Single Throw
SPT	Standard Pipe Thread
Sq.	Square; 100 Square Feet
Sq. Hd.	Square Head
Sq. In.	Square Inch
S.S.	Single Strength; Stainless Steel
S.S.B.	Single Strength B Grade
sst, ss	Stainless Steel
Sswk.	Structural Steel Worker
Sswl.	Structural Steel Welder
St.; Stl.	Steel
STC	Sound Transmission Coefficient
Std.	Standard
Stg.	Staging
STK	Select Tight Knot
STP	Standard Temperature & Pressure
Stpi.	Steamfitter, Pipefitter
Str.	Strength; Starter; Straight
Strd.	Stranded
Struct.	Structural
Sty.	Story
Subj.	Subject
Subs.	Subcontractors
Surf.	Surface
Sw.	Switch
Swbd.	Switchboard
S.Y.	Square Yard
Syn.	Synthetic
S.Y.P.	Southern Yellow Pine
Sys.	System
t.	Thickness
T	Temperature; Ton
Tan	Tangent
T.C.	Terra Cotta
T & C	Threaded and Coupled
T.D.	Temperature Difference
TDD	Telecommunications Device for the Deaf
T.E.M.	Transmission Electron Microscopy
temp	Temperature, Tempered, Temporary
TFFN	Nylon Jacketed Wire

Abbreviations

Abbreviation	Meaning
TFE	Tetrafluoroethylene (Teflon)
T. & G.	Tongue & Groove; Tar & Gravel
Th., Thk.	Thick
Thn.	Thin
Thrded	Threaded
Tilf.	Tile Layer, Floor
Tilh.	Tile Layer, Helper
THHN	Nylon Jacketed Wire
THW.	Insulated Strand Wire
THWN	Nylon Jacketed Wire
T.L., TL	Truckload
T.M.	Track Mounted
Tot.	Total
T-O-L	Threadolet
tmpd	Tempered
TPO	Thermoplastic Polyolefin
T.S.	Trigger Start
Tr.	Trade
Transf.	Transformer
Trhv.	Truck Driver, Heavy
Trlr	Trailer
Trlt.	Truck Driver, Light
TTY	Teletypewriter
TV	Television
T.W.	Thermoplastic Water Resistant Wire
UCI	Uniform Construction Index
UF	Underground Feeder
UGND	Underground Feeder
UHF	Ultra High Frequency
U.I.	United Inch
U.L., UL	Underwriters Laboratory
Uld.	Unloading
Unfin.	Unfinished
UPS	Uninterruptible Power Supply
URD	Underground Residential Distribution
US	United States
USGBC	U.S. Green Building Council
USP	United States Primed
UTMCD	Uniform Traffic Manual For Control Devices
UTP	Unshielded Twisted Pair
V	Volt
VA	Volt Amperes
VAT	Vinyl Asbestos Tile
V.C.T.	Vinyl Composition Tile
VAV	Variable Air Volume
VC	Veneer Core
VDC	Volts Direct Current
Vent.	Ventilation
Vert.	Vertical
V.F.	Vinyl Faced
V.G.	Vertical Grain
VHF	Very High Frequency
VHO	Very High Output
Vib.	Vibrating
VLF	Vertical Linear Foot
VOC	Volatile Organic Compound
Vol.	Volume
VRP	Vinyl Reinforced Polyester
W	Wire; Watt; Wide; West
w/	With
W.C., WC	Water Column; Water Closet
W.F.	Wide Flange
W.G.	Water Gauge
Wldg.	Welding
W. Mile	Wire Mile
W-O-L	Weldolet
W.R.	Water Resistant
Wrck.	Wrecker
WSFU	Water Supply Fixture Unit
W.S.P.	Water, Steam, Petroleum
WT., Wt.	Weight
WWF	Welded Wire Fabric
XFER	Transfer
XFMR	Transformer
XHD	Extra Heavy Duty
XHHW	Cross-Linked Polyethylene Wire Insulation
XLPE	
XLP	Cross-linked Polyethylene
Xport	Transport
Y	Wye
yd	Yard
yr	Year
Δ	Delta
%	Percent
~	Approximately
Ø	Phase; diameter
@	At
#	Pound; Number
<	Less Than
>	Greater Than
Z	Zone

Other Data & Services

A tradition of excellence in construction cost information and services since 1942

For more information visit our website at RSMeans.com

Unit prices according to the latest MasterFormat®

Cost Data Selection Guide

The following table provides definitive information on the content of each cost data publication. The number of lines of data provided in each unit price or assemblies division, as well as the number of crews, is listed for each data set. The presence of other elements such as reference tables, square foot models, equipment rental costs, historical cost indexes, and city cost indexes, is also indicated. You can use the table to help select the RSMeans data set that has the quantity and type of information you most need in your work.

Unit Cost Divisions	Building Construction	Mechanical	Electrical	Commercial Renovation	Square Foot	Site Work Landsc.	Green Building	Interior	Concrete Masonry	Open Shop	Heavy Construction	Light Commercial	Facilities Construction	Plumbing	Residential
1	609	444	465	564	0	533	198	365	495	608	550	310	1078	450	217
2	754	278	87	710	0	970	181	397	219	753	737	479	1197	285	274
3	1745	341	232	1265	0	1537	1043	355	2274	1745	1930	538	2028	317	445
4	960	22	0	920	0	724	180	613	1158	928	614	532	1175	0	446
5	1890	158	155	1094	0	853	1788	1107	729	1890	1026	980	1907	204	746
6	2462	18	18	2121	0	110	589	1544	281	2458	123	2151	2135	22	2671
7	1593	215	128	1633	0	580	761	532	523	1590	26	1326	1693	227	1046
8	2140	80	3	2733	0	255	1138	1813	105	2142	0	2328	2966	0	1552
9	2125	86	45	1943	0	313	464	2216	424	2062	15	1779	2379	54	1544
10	1088	17	10	684	0	232	32	898	136	1088	34	588	1179	237	224
11	1096	199	166	540	0	135	56	924	29	1063	0	230	1116	162	108
12	539	0	2	297	0	219	147	1546	14	506	0	272	1565	23	216
13	740	149	157	252	0	365	124	250	77	716	266	109	756	115	103
14	273	36	0	223	0	0	0	257	0	273	0	12	293	16	6
21	127	0	41	37	0	0	0	293	0	127	0	121	665	685	259
22	1165	7543	160	1226	0	2010	1061	849	20	1154	2109	875	7505	9400	719
23	1170	6906	546	940	0	157	865	775	38	1153	98	887	5143	1919	486
25	0	0	14	14	0	0	0	0	0	0	0	0	0	0	0
26	1513	491	10465	1293	0	860	646	1159	55	1439	649	1361	10246	399	636
27	95	0	448	102	0	0	0	71	0	95	39	67	389	0	56
28	143	79	223	124	0	0	28	97	0	127	0	70	209	57	41
31	1511	733	610	807	0	3263	286	7	1216	1456	3280	607	1568	660	616
32	896	49	8	937	0	4523	408	417	361	867	1941	486	1800	140	533
33	1255	1088	565	260	0	3078	33	0	241	532	3213	135	1726	2101	161
34	107	0	47	4	0	190	0	0	31	62	221	0	136	0	0
35	18	0	0	0	0	327	0	0	0	18	442	0	84	0	0
41	63	0	0	34	0	8	0	22	0	62	31	0	69	14	0
44	75	79	0	0	0	0	0	0	0	0	0	0	75	75	0
46	23	16	0	0	0	274	261	0	0	23	264	0	33	33	0
48	8	0	36	2	0	0	21	0	0	8	15	8	21	0	8
Totals	26183	19027	14631	20759	0	21516	10310	16507	8426	24945	17623	16251	51136	17595	13113

Assem Div	Building Construction	Mechanical	Elec-trical	Commercial Renovation	Square Foot	Site Work Landscape	Assemblies	Green Building	Interior	Concrete Masonry	Heavy Construction	Light Commercial	Facilities Construction	Plumbing	Asm Div	Residential
A		15	0	188	164	577	598	0	0	536	571	154	24	0	1	378
B		0	0	848	2554	0	5661	56	329	1976	368	2094	174	0	2	211
C		0	0	647	954	0	1334	0	1641	146	0	844	251	0	3	591
D		1057	941	712	1858	72	2538	330	824	0	0	1345	1104	1088	4	851
E		0	0	85	261	0	301	0	5	0	0	258	5	0	5	391
F		0	0	0	114	0	143	0	0	0	0	114	0	0	6	357
G		527	447	318	312	3378	792	0	0	535	1349	205	293	677	7	307
															8	760
															9	80
															10	0
															11	0
															12	0
Totals		1599	1388	2798	6217	4027	11367	386	2799	3193	2288	5014	1851	1765		3926

Reference Section	Building Construction Costs	Mechanical	Electrical	Commercial Renovation	Square Foot	Site Work Landscape	Assem.	Green Building	Interior	Concrete Masonry	Open Shop	Heavy Construction	Light Commercial	Facilities Construction	Plumbing	Resi.
Reference Tables	yes	yes	yes	yes	no	yes	yes	yes	yes	yes	yes	yes	yes	yes	yes	yes
Models					111			25					50			28
Crews	582	582	582	561		582		582	582	582	560	582	560	561	582	560
Equipment Rental Costs	yes	yes	yes	yes		yes		yes	yes	yes	yes	yes	yes	yes	yes	yes
Historical Cost Indexes	yes	yes	yes	yes	yes	yes	yes	yes	yes	yes	yes	yes	yes	yes	yes	no
City Cost Indexes	yes	yes	yes	yes	yes	yes	yes	yes	yes	yes	yes	yes	yes	yes	yes	yes

2020 Seminar Schedule

877.620.6245

Note: call for exact dates, locations, and details as some cities are subject to change.

Location	Dates
Seattle, WA	January and August
Dallas/Ft. Worth, TX	January
Austin, TX	February
Jacksonville, FL	February
Anchorage, AK	March and September
Las Vegas, NV	March
Washington, D.C.	April and September
Charleston, SC	April
Toronto	May
Denver, CO	May
San Francisco, CA	June
Bethesda, MD	June
Dallas, TX	September
Raleigh, NC	October
Baltimore, MD	November
Orlando, FL	November
San Diego, CA	December
San Antonio, TX	December

Gordian also offers a suite of online RSMeans data self-paced offerings.
Check our website at RSMeans.com/products/training.aspx for more information.

Facilities Construction Estimating

In this two-day course, professionals working in facilities management can get help with their daily challenges to establish budgets for all phases of a project.

Some of what you'll learn:

- Determining the full scope of a project
- Understanding of Means data and what is included in prices
- Identifying appropriate factors to be included in your estimate
- Creative solutions to estimating issues
- Organizing estimates for presentation and discussion
- Special estimating techniques for repair/remodel and maintenance projects
- Appropriate use of contingency, city cost indexes, and reference notes
- Techniques to get to the correct estimate quickly

Who should attend: facility managers, engineers, contractors, facility tradespeople, planners, and project managers.

Mechanical & Electrical Estimating

This two-day course teaches attendees how to prepare more accurate and complete mechanical/electrical estimates, avoid the pitfalls of omission and double-counting, and understand the composition and rationale within the RSMeans mechanical/electrical database.

Some of what you'll learn:

- The unique way mechanical and electrical systems are interrelated
- M&E estimates—conceptual, planning, budgeting, and bidding stages
- Order of magnitude, square foot, assemblies, and unit price estimating
- Comparative cost analysis of equipment and design alternatives

Who should attend: architects, engineers, facilities managers, mechanical and electrical contractors, and others who need a highly reliable method for developing, understanding, and evaluating mechanical and electrical contracts.

Construction Cost Estimating: Concepts and Practice

This one or two day introductory course to improve estimating skills and effectiveness starts with the details of interpreting bid documents and ends with the summary of the estimate and bid submission.

Some of what you'll learn:

- Using the plans and specifications to create estimates
- The takeoff process—deriving all tasks with correct quantities
- Developing pricing using various sources; how subcontractor pricing fits in
- Summarizing the estimate to arrive at the final number
- Formulas for area and cubic measure, adding waste and adjusting productivity to specific projects
- Evaluating subcontractors' proposals and prices
- Adding insurance and bonds
- Understanding how labor costs are calculated
- Submitting bids and proposals

Who should attend: project managers, architects, engineers, owners' representatives, contractors, and anyone who's responsible for budgeting or estimating construction projects.

sessing Scope of Work for Facilities nstruction Estimating

two-day practical training program addresses the vital rtance of understanding the scope of projects in order oduce accurate cost estimates for facilities repair and deling.

e of what you'll learn:

- cussions of site visits, plans/specs, record drawings of ilities, and site-specific lists
- view of CSI divisions, including means, methods, terials, and the challenges of scoping each topic
- rcises in scope identification and scope writing for urate estimating of projects
- ids-on exercises that require scope, take-off, d pricing

should attend: corporate and government estimators, ners, facility managers, and others who need to duce accurate project estimates.

Maintenance & Repair Estimating for Facilities

This two-day course teaches attendees how to plan, budget, and estimate the cost of ongoing and preventive maintenance and repair for existing buildings and grounds.

Some of what you'll learn:

- The most financially favorable maintenance, repair, and replacement scheduling and estimating
- Auditing and value engineering facilities
- Preventive planning and facilities upgrading
- Determining both in-house and contract-out service costs
- Annual, asset-protecting M&R plan

Who should attend: facility managers, maintenance supervisors, buildings and grounds superintendents, plant managers, planners, estimators, and others involved in facilities planning and budgeting.

ractical Project Management for onstruction Professionals

this two-day course, acquire the essential knowledge nd develop the skills to effectively and efficiently xecute the day-to-day responsibilities of the construction project manager.

Some of what you'll learn:

- General conditions of the construction contract
- Contract modifications: change orders and construction change directives
- Negotiations with subcontractors and vendors
- Effective writing: notification and communications
- Dispute resolution: claims and liens

Who should attend: architects, engineers, owners' representatives, and project managers.

Life Cycle Cost Estimating for Facilities Asset Managers

Life Cycle Cost Estimating will take the attendee through choosing the correct RSMeans database to use and then correctly applying RSMeans data to their specific life cycle application. Conceptual estimating through RSMeans' new building models, conceptual estimating of major existing building projects through RSMeans' renovation models, pricing specific renovation elements, estimating repair, replacement and preventive maintenance costs today and forward up to 30 years will be covered.

Some of what you'll learn:

- Cost implications of managing assets
- Planning projects and initial & life cycle costs
- How to use RSMeans data online

Who should attend: facilities owners and managers and anyone involved in the financial side of the decision making process in the planning, design, procurement, and operation of facilities real assets.

Please bring a laptop with ability to access the internet.

Building Systems and the Construction Process

This one-day course was written to assist novices and those outside the industry in obtaining a solid understanding of the construction process - from both a building systems and construction administration approach.

Some of what you'll learn:

- Various systems used and how components come together to create a building
- Start with foundation and end with the physical systems of the structure such as HVAC and Electrical
- Focus on the process from start of design through project closeout

This training session requires you to bring a laptop computer to class.

Who should attend: building professionals or novices to help make the crossover to the construction industry; suited for anyone responsible for providing high level oversight on construction projects.

Training for our Online Estimating Solution

Construction estimating is vital to the decision-making process at each state of every project. Our online solution works the way you do. It's systematic, flexible and intuitive. In this one-day class you will see how you can estimate any phase of any project faster and better.

Some of what you'll learn:

- Customizing our online estimating solution
- Making the most of RSMeans "Circle Reference" numbers
- How to integrate your cost data
- Generating reports, exporting estimates to MS Excel, sharing, collaborating and more

Also offered as a self-paced or on-site training program!

Training for our CD Estimating Solution

This one-day course helps users become more familiar with the functionality of the CD. Each menu, icon, screen, and function found in the program is explain in depth. Time is devoted to hands-on estimating exercises.

Some of what you'll learn:

- Searching the database using all navigation methods
- Exporting RSMeans data to your preferred spreadsheet format
- Viewing crews, assembly components, and much more
- Automatically regionalizing the database

This training session requires you to bring a laptop computer to class.

When you register for this course you will receive an outline for your lapto requirements.

Also offered as a self-paced or on-site training program!

Site Work Estimating with RSMeans data

This one-day program focuses directly on site work costs. Accurately scoping, quantifying, and pricing site preparation, underground utility work, and improvements to exterior site elements are often the most difficult estimating tasks on any project. Some of what you'll learn:

- Evaluation of site work and understanding site scope including: site clearing, grading, excavation, disposal and trucking of materials, backfill and compaction, underground utilities, paving, sidewalks, and seeding & planting.
- Unit price site work estimates—Correct use of RSMeans site work cost data to develop a cost estimate.
- Using and modifying assemblies—Save valuable time when estimating site work activities using custom assemblies.

Who should attend: Engineers, contractors, estimators, project managers, owner's representatives, and others who are concerned with the proper preparation and/or evaluation of site work estimates.

Please bring a laptop with ability to access the internet.

Facilities Estimating Using the CD

This two-day class combines hands-on skill-building with best estimating practices and real-life problems. You will learn key concepts, tips, pointer and guidelines to save time and avoid cost oversights and errors.

Some of what you'll learn:

- Estimating process concepts
- Customizing and adapting RSMeans cost data
- Establishing scope of work to account for all known variables
- Budget estimating: when, why, and how
- Site visits: what to look for and what you can't afford to overlook
- How to estimate repair and remodeling variables

This training session requires you to bring a laptop computer to class.

Who should attend: facility managers, architects, engineers, contractors, facility tradespeople, planners, project managers, and anyone involved with JOC, SABRE or IDIQ.

Registration Information

Register early to save up to $100!!!

Register 45+ days before the date of a class and save $50 off each class. This savings cannot be combined with any other promotion or discounting of the regular price of classes!

How to register

By Phone

Register by phone at 877.620.6245

Online

Register online at RSMeans.com/products/seminars.aspx

Note: Purchase Orders or Credits Cards are required to register.

Two-day seminar registration fee - $1,200*.

One-Day Construction Cost Estimating or Building Systems and the Construction Process - $765*.

Government pricing

All federal government employees save off the regular seminar price. Other promotional discounts cannot be combined with the government discount. Call 781.422.5115 for government pricing.

CANCELLATION POLICY:

If you are unable to attend a seminar, substitutions may be made at any time before the session starts by notifying the seminar registrar at 781.422.5115 or your sales representative.

If you cancel twenty-one (21) days or more prior to the seminar, there will be no penalty and your registration fees will be refunded. These cancellations must be received by the seminar registrar or your sales representative and will be confirmed to be eligible for cancellation.

If you cancel fewer than twenty-one (21) days prior to the seminar, you will forfeit the registration fee.

In the unfortunate event of an RSMeans cancellation, RSMeans will work with you to reschedule your attendance in the same seminar at a later date or will fully refund your registration fee. RSMeans cannot be responsible for any non-refundable travel expenses incurred by you or another as a result of your registration, attendance at, or cancellation of an RSMeans seminar.

Any on-demand training modules are not eligible for cancellation, substitution, transfer, return or refund.

AACE approved courses

Many seminars described and offered here have been approved for 14 hours (1.4 recertification credits) of credit by the AACE International Certification Board toward meeting the continuing education requirements for recertification as a Certified Cost Engineer/Certified Cost Consultant.

AIA Continuing Education

We are registered with the AIA Continuing Education System (AIA/CES) and are committed to developing quality learning activities in accordance with the CES criteria. Many seminars meet the AIA/CES criteria for Quality Level 2. AIA members may receive 14 learning units (LUs) for each two-day RSMeans course.

Daily course schedule

The first day of each seminar session begins at 8:30 a.m. and ends at 4:30 p.m. The second day begins at 8:00 a.m. and ends at 4:00 p.m. Participants are urged to bring a hand-held calculator since many actual problems will be worked out in each session.

Continental breakfast

Your registration includes the cost of a continental breakfast and a morning and afternoon refreshment break. These informal segments allow you to discuss topics of mutual interest with other seminar attendees. (You are free to make your own lunch and dinner arrangements.)

Hotel/transportation arrangements

We arrange to hold a block of rooms at most host hotels. To take advantage of special group rates when making your reservation, be sure to mention that you are attending the RSMeans Institute data seminar. You are, of course, free to stay at the lodging place of your choice. (Hotel reservations and transportation arrangements should be made directly by seminar attendees.)

Important

Class sizes are limited, so please register as soon as possible.

***Note: Pricing subject to change.**